8th International Conference on Thermal Equipment, Renewable Energy and Rural Development (TE-RE-RD 2019)

E3S Web of Conferences Volume 112 (2019)

Targoviste, Romania
6-8 June 2019

Editors:

Gabriel-Paul Negreanu
Nicolae-Valentin Vlăduţ

ISBN: 978-1-5108-9212-5

Printed from e-media with permission by:

Curran Associates, Inc.
57 Morehouse Lane
Red Hook, NY 12571

Some format issues inherent in the e-media version may also appear in this print version.

No changes have been made to the content of these proceedings. There may be changes to pagination and minor adjustments for aesthetics.

Printed with permission by Curran Associates, Inc. (2019)

For additional information, please contact EDP Sciences – Web of Conferences at the address below.

EDP Sciences – Web of Conferences
17, Avenue du Hoggar
Parc d'Activité de Courtabœuf
BP 112
F-91944 Les Ulis Cedex A
France

Phone: +33 (0) 1 69 18 75 75
Fax: +33 (0) 1 69 28 84 91

contact-edps@webofconferences.org

Additional copies of this publication are available from:

Curran Associates, Inc.
57 Morehouse Lane
Red Hook, NY 12571 USA
Phone: 845-758-0400
Fax: 845-758-2633
Email: curran@proceedings.com
Web: www.proceedings.com

TABLE OF CONTENTS

PREFACE

The 8-th International Conference on Thermal Equipment, Renewable Energy and Rural Development (TE-RE-RD 2019 - http://www.tererd.pub.ro/) was organized in the period 06-06 June 2019 at the International Conference Center of the University Valahia of Târgoviște (Romania). The social Programme and participants accommodation were hosted by Hotel Nova of Târgoviște.

TE-RE-RD is an International pluridisciplinary Conference aimed to be a forum for information exchange between scientists from academic, research and development communities around the world on topics related to thermal machines and equipments and technologies, equipments, technologies and policies for rural development, renewable energy sources conversion and other technical domains.

Another target of the Conference is to offer to the young people involved in research activities (PhD, Master, and regular students) the opportunity to disseminate their work results and learn from experimented scientists.

We also welcome people and entities from industry, agriculture, administration, trade, politics and media who want to cooperate with academic and R&D organizations in the fields of the Conference.

The main Organizers, University POLITEHNICA of Bucharest (Faculty of Mechanical Engineering and Mechatronics and Faculty of Biotechnical Systems Engineering), National Institute of Research – Development for Machines and Installations Designed to Agriculture and Food Industry and the Romanian Society of Agricultural Mechanical Engineers produce every year since 2012 an itinerant Conference in academic cities or touristic resorts. This year, the chosen partner was University Valahia of Târgoviște who offered his Conference Center and a part of his staff for the event success.

The main topics of the Conference were:

- Processes in thermal equipment
- Steam and hot water generators
- Steam and gas turbines
- Internal combustion engines
- Environmental protection
- Solar, wind, geothermal, biomass
- Building efficiency
- Equipment for rural development
- Politics of rural development
- Agricultural engineering
- Energy of rural space

This year the Organizers used the Conference Management Toolkit provided from Microsoft in order to manage the paper submissions and reviewing process. We received 96 papers distributed to 85 reviewers. 79 papers were accepted to be presented and 16 papers were rejected or retired by the authors after revision process.

The Conference was developed on 4 tracks:

- 19 papers in the section Thermal Equipments
- 14 papers in the section Renewable Energy
- 34 papers in the section Rural Development
- 10 papers in the section Miscellaneous

Besides the representatives of the Organizers, the Conference also welcomed researchers from University Angel Kanchev of Ruse (Bulgaria), Technical University of Varna (Bulgaria), Technical University of Sofia (Bulgaria), University of Food Technologies of Plovdiv (Bulgaria), SEC Biomass Center of Kiev (Ukraine), Institute of Engineering Thermophysics of Kiev (Ukraine), "Ferdinand I" Technical Military Academy of Bucharest (Romania), University of Craiova (Romania), University "Dunărea de Jos" of Galați (Romania), INCD Comoti (Romania), SCDCPN Dăbuleni (Romania), INCD ICEMENERG (Romania), INOE2000-IHP Bucharest (Romania), Encon Services (USA), Bioterpena System (Romania).

We hope that next edition the number and the value of papers will increase, the current participants will come back and new authors will attend.

See you again in 2020,

Prof. Gabriel Negreanu

Dr. Valentin Vlăduț

Editors of the Conference Proceedings TE-RE-RD 2019

OPENING ADDRESS

Dear Vice-Rector of University Valahia of Târgovişte,

Dear Dean of Technical Faculty - University of Food Technologies of Plovdiv,

Dear Dean of Faculty of Mechanical Engineering and Mechatronics - University Politehnica of Bucharest,

Dear researchers, ladies and gentleman

On behalf of the Organizing Committee, please allow me to welcome you at the 8th International Conference on Thermal Equipment, Renewable Energy and Rural Development (TE-RE-RD 2019), organized by a consortia of Universities, Research Institute and NGO in the old and beautiful city Târgovişte.

The first ten editions (2001-2011) of the Conference were organized at National level only in University Politehnica. Since 2012 the Conference became International and was organized in several locations in Romania, excepting 2017 when was hosted by the Golden Sands resort in Bulgaria.

Dear participants and guests,

The first reason why the city of Târgovişte, the capital of Dâmboviţa County, was selected as a Conference host is related to the industry and agriculture development of the area, close to the Conference topics. Here I can remember the Thermal Power Plant of Doiceşti (the first TPP powered with pulverized coal), the metallurgic industry in Târgovişte, the cement factory in Fieni and the agriculture, with multiple examples of "energy from waste" projects.

The second reason is historical one: more than 200 years, Târgovişte was the Capital of Valahia. Here are a lot of beautiful monuments and buildings that wait to be visited by you.

Ladies and gentlemen,

The novelty of this edition of our Conference is that the accepted and presented papers will be published under the auspices of the prestigious Publishing House - E3S Web of Conferences and will be indexed in Web of Sciences, Scopus and other important international scientific databases.

Finally, I wish to thank to all participants to the Conference for their presence here, and especially to the members of the Scientific and Organizing Committees, who worked hard to make it possible.

Prof. Lucian Mihăescu

Chairman of TE-RE-RD 2019

Statement of Peer review

In submitting conference proceedings to *Web of Conferences*, the editors of the proceedings certify to the Publisher that

1. They adhere to its **Policy on Publishing Integrity** in order to safeguard good scientific practice in publishing.
2. All articles have been subjected to peer review administered by the proceedings editors.
3. Reviews have been conducted by expert referees, who have been requested to provide unbiased and constructive comments aimed, whenever possible, at improving the work.
4. Proceedings editors have taken all reasonable steps to ensure the quality of the materials they publish and their decision to accept or reject a paper for publication has been based only on the merits of the work and the relevance to the journal.

Title, date and place of the conference

8-th International Conference on Thermal Equipment, Renewable Energy and Rural Development
TE-RE-RD 2019
06-08 June 2019, Târgoviște, Romania

Proceedings editor(s):

Prof. Dr. Ing. Gabriel-Paul NEGREANU
Dr. Ing. Nicolae-Valentin VLĂDUȚ

Date and editor's signature

18.07.2019

https://doi.org/10.1051/e3sconf/201911201001

Waste Heat Recovery for a Diesel Tractor Engine: A Comparison Study

Bogdan Radu*, *Alexandru* Racovitza

UPB, Dept.of Thermo., Eng., Therm. and Ref. Mach., Spl. Independenței, nr 313, s.6,Romania

Abstract. One of the major concerns of nowadays is related to the Waste Heat Recovery (WHR) from the thermal engines, especially regarding the high rating power engines. For Diesel tractor engines, long time operated at high loads, the amount of the saved energy is linked to a significant fuel economy. The proposed study is based on the AMESIM simulation code and performs a comparison between two methods of heat recovery: the one using a supplementary turbine (turbo-compound method) and that using an Organic Rankine Cycle (ORC). The tested engine is an IVECO Cursor 10 Diesel engine type, equipping CASE tractors designed for agricultural activities. The energy savings provided by the engine exhaust gas heat recovery could be further used in supplementing the engine traction, either to be delivered to the auxiliary systems, leading in both cases to a decrease of the fuel consumption.

Key words: Diesel tractor engine, Waste Heat Recovery, Turbo-compound, Organic Rankine Cycle, fuel consumption decrease.

1 Introduction

By nowadays the society development has reached unprecedented limits comparing to the last decades, but its actual features could not be replaced by the efforts to reduce the consumption of the resources. Therefore human kind is concerned to increase the efficiency of all the processes, to reduce fuel consumption, to recycle materials etc. On this issue, the energy recovery from the internal combustion engines, especially from the heavy ones, with high costs is a dominant preoccupation in the present.

The waste heat recovery (WHR) from the vehicles engines could be classified in two categories: engine-power train-applied or engine-bottoming technologies, depending if they are directly applied or retrofitted to the engine-power train system, or if they recover wasted engine energy. These methods will allow the transfer of the recovered energy to the propulsion system or to the auxiliary systems of the engine. For heavy-duty Diesel engines turbocharging is a necessity to increase the engine power. Thus, any modification of the exhaust system would affect the operation of this group, as of the entire engine assembly.

Speaking about the multiplicity of the WHR methods, two of them play a significant role: the insert of a supplementary power gas turbine along with the one motioning the turbocharging group (the turbocompund method) and the use of a new energy circuit to transfer this energy to another work agent applying a Rankine, or an Organic Rankine Cycle (RC or ORC).

Heavy engines represent a key issue in the research studies regarding the energy recovery. The accent in [1] is put on the HDDE (Heavy-Duty Diesel Engines), on both automotive and stationary applications. Analyzing studies developed by the greatest companies, like Cummins, BMW or Honda reveal a multitude of results without a conclusion suggesting a certain beneficiary system. There are work papers proposing even more sophisticated energy recovery methods, such as the use of the Kalina Cycle [2], which uses a blend of two miscible fluids, ammonia and water for instance, participating in various fractions in order to ensure a better overtake of the waste energy.

Once the technical aspects are simpler in case of using the turbocompund method (designing a new turbine with the natural result of influencing the engine operation), more problems are to occur by applying the Rankine Cycle release. Authors of paper [3] identify no less than 6 different ways to implement such an energy recovery system. Additionally, they are analyzing 8 different types of cycle working fluids. Paper [4] expresses more than 30 sorts of fluid agents. Other authors also investigate several types of agents and how to get to the best choosing mode of them [5]. Their results are inconclusive referring to all the involving criteria and similar comments are given also by [6]. Authors of paper [7] went even far, by searching among 9 cycle agents from the thermodynamic results and the presumed costs point of view. As expected, after using a multi-objective optimization process, the conclusions showed that a comprehensive result remains highly difficult to be obtained.

The energy recovery for the engines with low temperature combustion (LTC) is presented in paper [8]. In case of an engine fueled with lean mixtures of natural gas and by split-injection of Diesel fuel, using also various fractions of exhaust gas recirculation (EGR) and various values of injection timings, an increase of 18% of the rating power was obtained, together with the drop of the emissions.

Another category of studies refers to the control of an ORC system [9,10,11]. These papers highlight numerical models regarding the possibilities to manage these control options in order to increase the engine performances. Paper [12] reveals a theoretical and a comparative study between using the two above-mentioned energy recovery methods, trying to identify also the optimal adjustments (corresponding to the injection timing, to the combustion period, to the designing elements of the turbine) for the improvement of the engine operation. The main conclusions were related to the fact that is no possible to increase the engine efficiency for all the analyzed regimes (the break specific fuel consumption – BSFC has opposite variation from one to another of the regimes), meanwhile highlighting a slight benefit when using the turbocompund method.

2. Iveco Cursor 10 Diesel engine simulation using AMESIM numerical code

Iveco engines are worldwide spread into the mechanical agriculture industry, being appreciated for their technical qualities and reliability. The group of 6-cylinders versions is formed by the following types: Cursor 9, 8.71 l swept volume, rating between 200 and 305 kW, Cursor 10, 10.3 l swept volume, rating between 265 and 315 kW, Cursor 13, 12.9 swept volume, rating between 325 and 427 kW and Cursor 16, 15.9 swept volume, rating between 480 and 570 kW. This engines family easily covers a wide range of agricultural machinery size.

The proposed engine to be analyzed is Cursor 10 -400 with characteristics presented in Table 1. and different versions of this engine are used also for commercial trucks.

Table 1 Engine Cursor 10 -400 specifications

Maximum power [kW/hp]	294/400
Rated speed [rpm]	2100
Number of cylinder	6
Compression ratio	16.5:1
Bore [mm]	125
Stroke [mm]	140
Cubic capacity [l]	10.38
Fuel system	Direct injection mechanical
Admission	turbocharged

One of the basic requests of these machines is their operation at high load of the engine. Due to the fact that the total weight of the vehicle is not an important issue and for certain reasons being brought at a minimum value, the design of a supplementary system for the energy recovery should not become a difficult task. In paper [1] is mentioning that an agricultural tractor engine should operate for 80% of its period at 80 to 100% load and covering a speed within 80 and 100% of its nominal value. With the elements listed in Table 1, the AMESIM model is presented in Fig. 1.

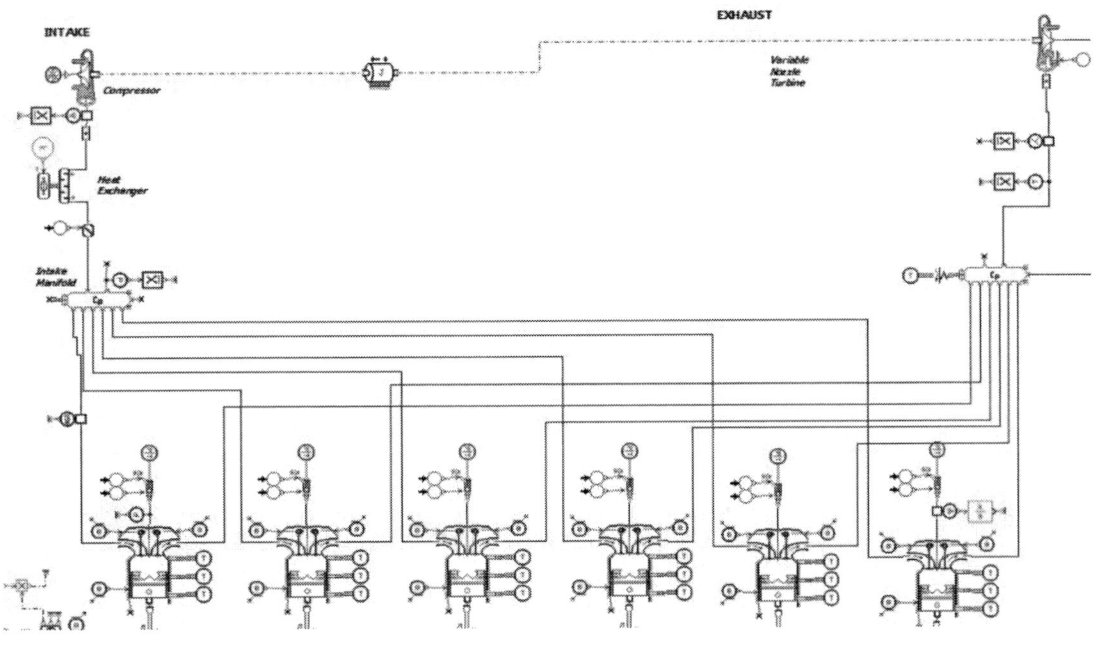

Fig. 1. AMESIM model for Cursor 10 Diesel type engine

The model is formed by several elements describing the different parts of the engine. The cylinder, as an important component of the engine is detailed in Fig. 2.

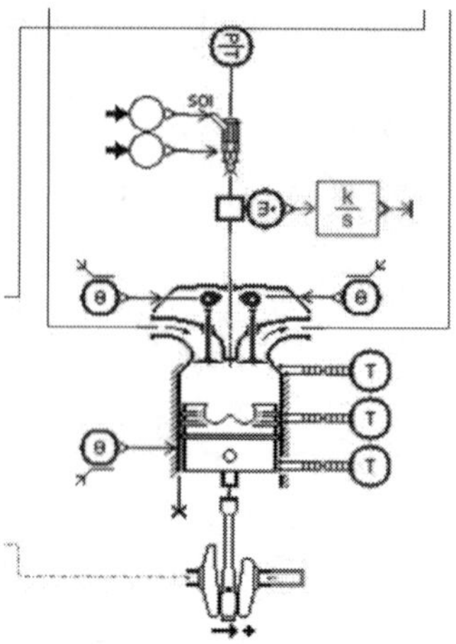

Fig. 2. Piston-cylinder system

The processes developing inside the cylinder are defining the specificity of Diesel engine. Therefore, the gas exchange part of the procedures, given by the cams and the valves motion, the fueling system formed by an injector delivering the fuel dose upon a characteristic well-established by the user and the crankshaft curve with the afferent connecting-rod, transforming the translation motion into rotation, all these elements have to be added into the model. Inside the cylinder the gas exchange processes are defined by the finite volumes hypothesis and the air-fuel mixture formation, the autoignition delay and the combustion are those proposed by Chmela [14] and implemented using the AMESIM numerical code.

Related to the turbocharging processes the models proposed by AMESIM are used, by applying its operating maps. Initially, an engine model has been performed so that the chosen values for this paper could be obtained. At 2100 rpm the engine rated a power of 294 kW, corresponding to an injected fuel cycle dose of 151 mg. This is further related to a fuel consumption of 56.5 kg/h and consequently to a BSFC of 192 g/kWh, a value confirmed by [13]. In-cylinder pressure and temperature variations are highlighted in Fig. 3, with the maximum corresponding values of 18 MPa and 2500 K.

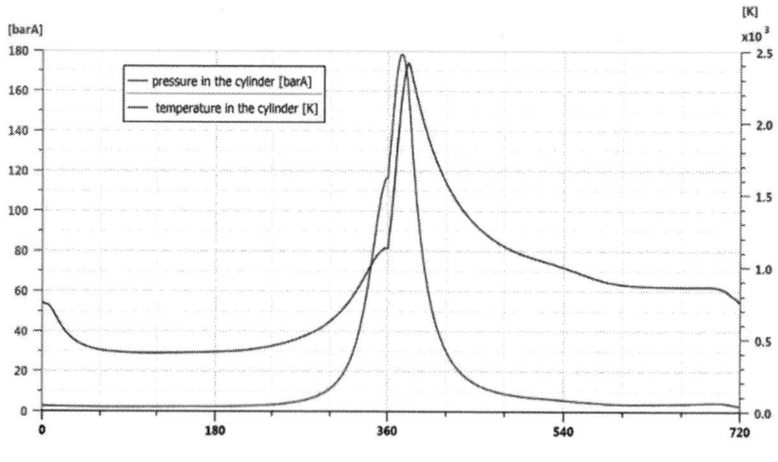

Fig. 3. In-cylinder pressure and temperature variations

Among the other characteristic variables corresponding to the top rating power, the turbocharging pressure of 0.25 MPa, the turbocharging group speed of 86400 rpm, the exhaust gas flow of 384 g/s and the exhaust gas temperature of 905 K, obtained by the charge combustion under a relative air-fuel ratio of 1.54, all these are also mentioned in the model.

The engine operation at full load for an excessive long period is less probably, therefore an 80% load coefficient regime has been adopted, for a lower than 1800 rpm speed. At this speed the rated power is 240 kW, thus a 200kW power will be used for the chosen engine operating regime. This power is obtained using a cycle fuel dose of 118 mg, respectively a fuel consumption of 38 kg/h and a BSFC value of 190 g/kWh. The turbocharging group speed is 79000 rpm, the exhaust gas reaching a temperature of 809 K, while the gas flow measures 308 g/s and the relative air-fuel ratio is positioned at 1.86. In this scenario the top firing pressure becomes 17.5 MPa and the in-cylinder temperature is reaching its top value of 2580 K (see Fig. 4.).

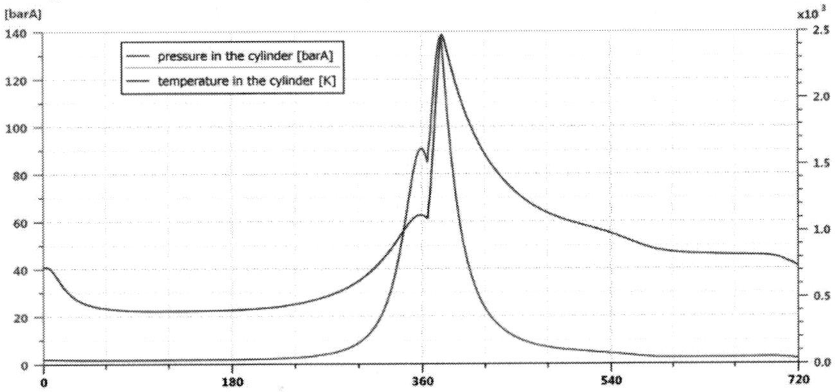

Fig. 4. In-cylinder pressure and temperature variations for the operating regime

3. WHR using the power turbine (the turbocompound method)

The system modeling in this case is performed by adding supplementary elements to the initial version of the model, corresponding to the power turbine issues (see Fig. 5.)

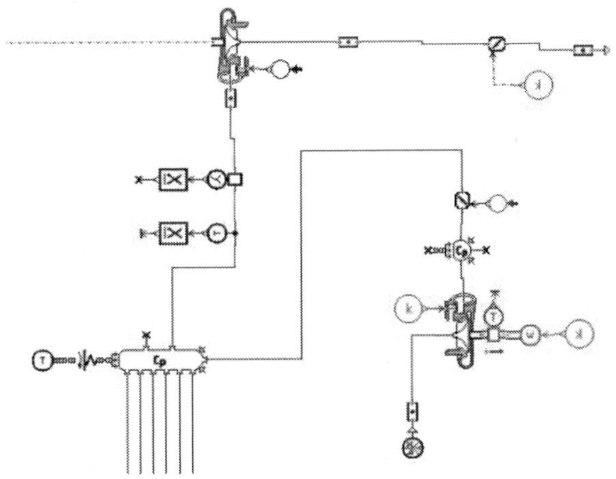

Fig. 5. The power turbine model

The supplementary power turbine is fueled by the exhaust gas flow in parallel with the turbocharging group turbine through a flap system adjusting the gas flow value. Downstream the main turbine a control volume is positioned and the power turbine is connected to a constant speed generator. The analysis of the system operation will be followed as a result of varying the flap flowing section and the power turbine speed.

 https://doi.org/10.1051/e3sconf/201911201001

The obtained results

For the given load regime, the simulation for the energy recovery used a parametric computation of the throttle opening degree toward the supplementary turbine (Table 2).

Analyzing the obtained data it appears that the supplementary turbine did not significantly influence the engine performances, because the relative air-fuel ratio is very high (1.8) related to the original engine design. By reducing the relative air-fuel ratio down to 1.36, the power delivered by the turbine did not decrease. The throttle opening over 60% did not lead to a gain of power for the supplementary turbine, moreover, a working limit of the gas flow corresponding to this speed level being reached. The obtained results concerning other speeds did not conduct to significant differences comparing to the references either.

Table 2 Turbocompound results at 80000 rpm speed for the power turbine

Throttle opening [%]	Turbocompressor speed [rpm]	Engine power [kW]	Relative air-fuel ratio [-]	Exhaust gas temperature [K]	Turbine power [kW]
20	71771	202	1.6	861	0.48
40	62914	200	1.36	920	3.32
60	62341	199.7	1.34	926	3.69
80	62244	199.7	1.34	927	3.72
100	62244	199.7	1.34	927	3.73

4. WHR using a system based on ORC

This energy recovery method uses the residual heat contained by the exhaust gas downstream the turbocharging group turbine, transferring it to the Rankine installation fluid so that it could be heated and vaporized. The entire system is presented in Fig. 6.

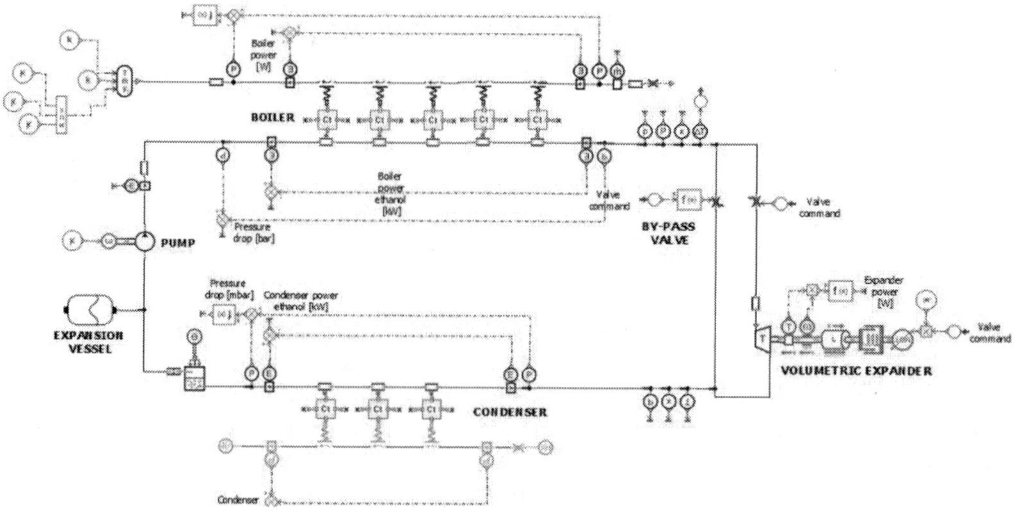

Fig. 6. The model for the ORC system

The fluid agent is vaporized, overheated and than is expended into a turbine. The system needs also a condenser to bring back the agent into the liquid phase. In what concerns the heating and the vaporizing part of the installation, the following elements were used (see Fig. 7).

In order to simulate the exhaust gas flow, a boiler has been used, to transform three signals in specific parameters for the exhaust gas, the mass gas flow, its temperature and composition, as given in the previous model. The exhaust gas will pass through a heat exchange system formed by 5 consecutive elements, which in fact simulate a single exchange system, but due to the need of numerical computation convergence it appears preferable to be divided. Concerning the organic fluid agent, a pump for the liquid phase is insert, to push the agent in the other side of the heat exchange system, on a reversed sense comparing to the exhaust gas. The organic fluid will be heated and vaporized while flowing through the heat exchange pipes.

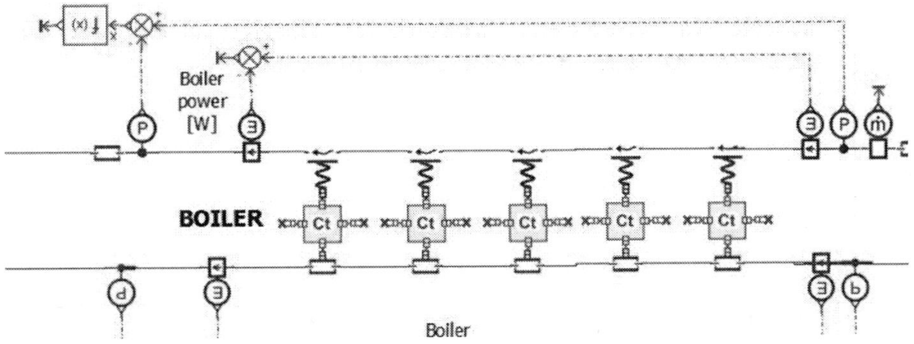

Fig. 7. The boiler model

The gain of the mechanical work was possible using a turbine-type element, connected to an electric generator (as seen in Fig. 8.).

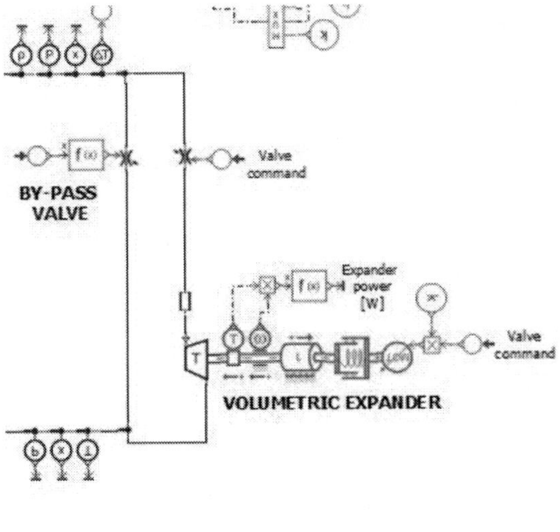

Fig. 8. The model for the power turbine

The connection of the boiler together with the turbine is made using two throttles, working simultaneously oppositely, meaning that when the overheating temperature increases, the first one opens the circuit towards the turbine and the second one is closing the flowing circuit from the boiler. For the condenser a model has been proposed, very similar to the one used for the boiler (see Fig. 9). The condensation part is completed by a liquid-vapors separator and an expansion vessel.

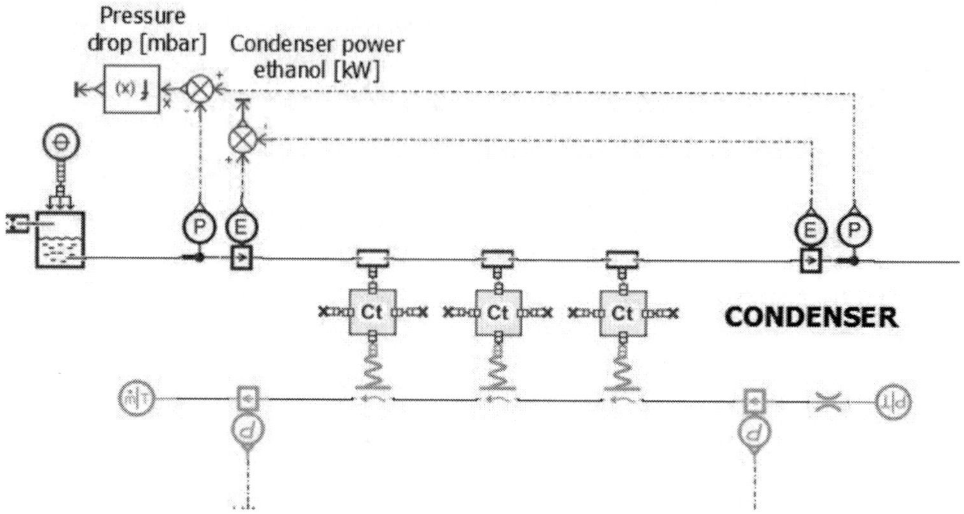

Fig. 9. The model for the condenser

The obtained results

The heating and vaporizing processes are very unsteady from modeling point of view, because of the density and the heat transfer coefficient rapid variations which occur during this period. Thus, in the vaporizing initial period (see Fig. 10), the temperature is very much varying and very fast in correlation with the pressure until they stabilize inside the system.

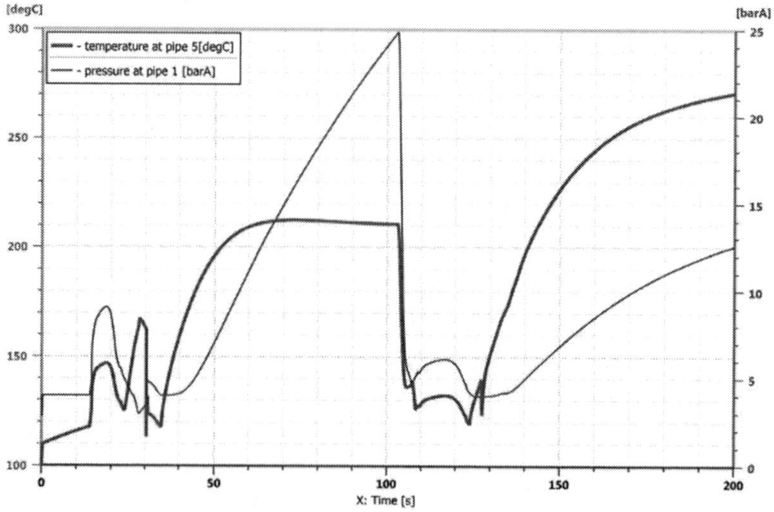

Fig. 10. Agent temperature and pressure variations at pipe no.5 (detail)

 https://doi.org/10.1051/e3sconf/201911201001

After a great number of integration iterations the system is stabilized, the pressure reaching about 1.4 MPa, corresponding to a temperature of 268°C, inside the domain of the overheated vapors for ethanol (Fig. 11).

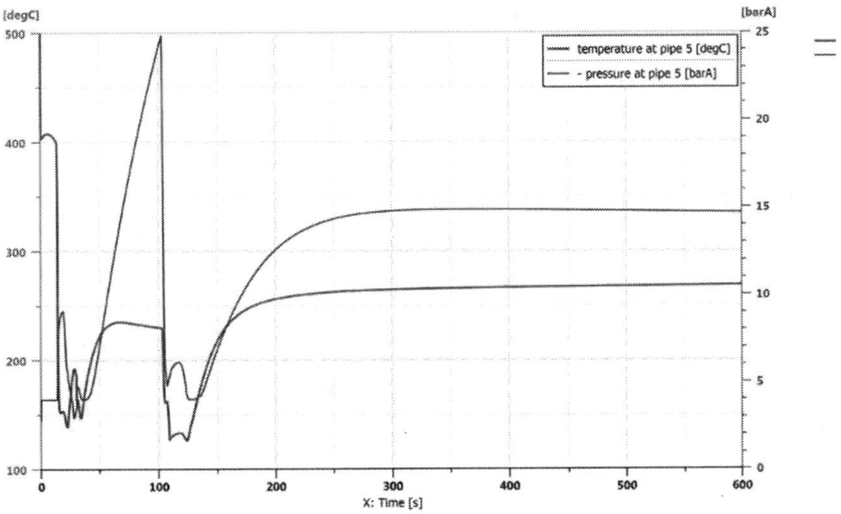

Fig. 11. Agent temperature and pressure variations at pipe no.5

For the turbine producing the mechanical work it has been usefully considered the use of a variable speed (during the computation period), as a function of the flap opening the receiving gas flow circuit, the final value of the speed reaching 6000 rpm. The power extracted by the organic agent from the exhaust gas is reflected by the exhaust gas temperature at the level of the boiler's 5 pipes (see Fig. 11). The heat transfer toward the organic fluid is relatively reduced, with a gain of power of about 5 kW. The power benefit of the turbine turns to an almost neglected value of 60 W, therefore the system efficiency is barely reduced.

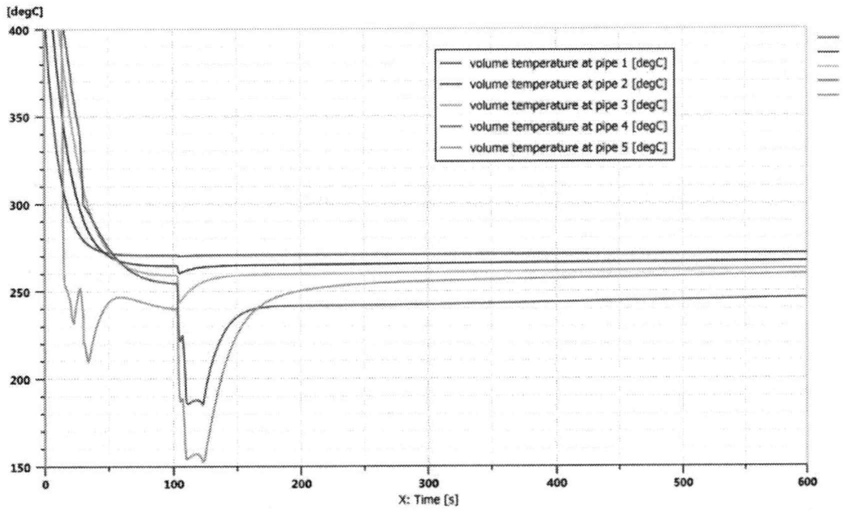

Fig. 12. Exhaust gas temperature variation

Conclusions

To add a supplementary turbine in order to produce mechanical work to supply the vehicle traction does not represent a difficult technical task for the heavy-duty engines manufacturers. The problems occur when this turbine has to be connected to the propulsion system of the engine and when optimal operating regimes must be identified. Using the proposed model, a power extra gain of about 3.7 kW should reveal itself as a supplement for the engine 200 kW maximum rating power, without affecting its complementary performances.

The energy recovery applying a system based on the Organic Rankine Cycle consists in a highly difficult design project. The exhaust gas temperature could become a major constrain especially when choosing the working fluid agent. Correlations between the exhaust gas flow, its enthalpy and the geometrical dimensions of the turbocharging turbine, the functional characteristics of the agent pump and those of the organic agent expansion turbine must be established.

The proposed model has not been adjusted yet from all these points of view, therefore until now the results are not sufficiently satisfactory.

References

1. S. Lion, C. Michos, I. Vlaskos, C. Rouaud, R. Taccani Renewable and Sustainable Energy Reviews **79** (2017)
2. D. Ziviani, A. Beyene, M. Venturini Applied Energy **121** (2014)
3. V. Apostol, H. Pop, A. Dobrovicescu, T. Prisecaru, Ana Alexandru, Mălina Prisecaru Procedia Engineering **100** (2015)
4. A Hattiangadi *Working Fluid Design for Organic Rankine Cycle* Thesis TU Delft
5. H.Tian, G. Shu, H. Wei, X. Liang, L. Liu Energy **47,** (2012)
6. D. Jung , S. Park ., K. Min . Applied Thermal Engineering **81,** (2015)
7. A. Ganjehkaviri ,M. Jaafar, Entropy **16**, (2014) .
8. K. Srinivasan, P.Mago, S. Krishnan, Energy **35** (2010)
9. S. Trabucchi,C. De Servi, F. Cassela, P. Colonna Energy Procedia **129** (2017)
10. J. Galindo,V. Dolz, L. Royo-Pascual,A. Brizard Energy Procedia **129** (2017)
11. X. Liu, A. Yebi, P. Anschel, J. Shutty, B. Xu, M. Hoffman, S. Onori Energy Procedia **129** (2017)
12. A.M. Andwari, A. Pesiridis, V. Esfahanian, A.Salavati-Zadeh,A. Karvountzis ,Vi. Muralidharan, Energies **10**, 1087 (2017)
13. G. Biaggini, W. Knecht, FISITA World Automotive Congress **F2000A066**
14. F. Chmela, C. Orthaber, *SAE paper 1999-01-0186*, **(1999)**

E3S Web of Conferences **112**, 01002 (2019) https://doi.org/10.1051/e3sconf/201911201002
TE-RE-RD 2019

Influences of Using B20 Reformulated Type Fuel on a Heavy-Duty Diesel Engine Performances

Adnan Kadhim Rashid[1], *Bogdan* Radu[1,*], *Alexandru* Racovitza[1], *Radu* Chiriac[1,2]

[1]UPB, Dept.of Thermo., Eng., Therm. and Ref. Mach., Spl. Independenței, nr 313, s.6,Romania
[2] CMGPCE-EA7341, CNAM, Paris, France

Abstract. Starting from the need to replace up to 20% of the energetic fuel content as European Regulations have already stipulated for the years to come, B20 mixture fuel proves to receive an increasing recognition nowadays as an appropriate and alternative fuel for Diesel engines. Studies have provided that B20 increases engine efficiency under specific operating conditions together with a significant drop of the main emissions' levels. This paper is proposing a numerical analysis of the operating behavior of an IVECO Cursor 10 heavy-duty Diesel engine fueled with Diesel-Biodiesel B20 fuel, featuring the projections of the AMESIM simulation code.

Key words: Heavy-Duty Diesel Engine, Biodiesel-Diesel Fuel Mixture, High Efficiency, Emissions Decrease

1 Introduction

The benefits of using the biofuels of the fourth generation production by now in Diesel engines are obvious when comparing to the use of Diesel classic petroleum based fuel. Among the other promising fuels, Biodiesel has especially become an appropriate source to replace partially or even totally Diesel fuel. The advantages of Biodiesel compared to Diesel fuel are: a higher combustion efficiency, a higher autoignition capacity (a higher cetane number, within 42 and 67 units), a better biodegradability, together with lower carbon monoxide (CO) emissions [1,2,3]. The disadvantages of using Biodiesel are also to be mentioned, including slightly higher NOx emissions, cold start problems, lower heating value (LHV) and injection difficulties due to its higher viscosity.
Actually, Biodiesel is more expensive to be produced than Diesel fuel, animal and vegetable organic sources needing to be collected from more and more extended areas. Thus, current worldwide production of vegetable oil and animal fat is not enough to replace liquid fossil fuel. Therefore, of more increasing importance are the blends of other fuels prepared with Diesel fuel, like the bio-alcohols [4,5].
In its content Biodiesel has no aromatics, no sulfur, and provides about 10% to 11% weight oxygen. These characteristics lead to an important exhaust gas emission reduction in carbon monoxide, unburned Hydrocarbon (UHC) and particulate matter (PM). Regarding NOx emission, due to the more restricting standards, using Biodiesel with Diesel fuel and even with light fractions of gasoline, from multiple literature studies it might appears necessary to apply certain fueling methods combined with different several modes of air-fuel mixture

formation and ignition, such as HCCI (Homogeneous Charge Compression Ignition), PCCI (Premixed Charge Compression Ignition), PPCI (Partially Premixed Compression Ignition) or RCCI (Reactive Controlled Compression Ignition) modes [6,7].

The critical problem of Biodiesel higher viscosity affects the characteristics of this fuel injection, because the higher the viscosity is, the poorest the fuel atomization becomes. High viscosity usually causes larger droplet sizes, lower vaporization rate, narrower injection spray angle, and longer in-cylinder penetration of the fuel spray. These aspects normally lead to overall poorer combustion, higher emissions, and increased oil dilution [8].

Combustion and emission characteristics of Diesel engine operated using Diesel fuel blended with Biodiesel were largely investigated. It has been proved that the higher the oil ratio in the blend is, the higher the break specific fuel consumption (BSFC) is and the lower the UHC and CO emissions are. The main exception for the emissions levels trend is the increase of the NOx emission, with no modifications brought to the engine and with no changing the air-fuel ratio. As it has been already mentioned, to improve fuel consumption and NOx emissions, there are some common techniques to be applied, such as modifying the fuel properties, by using metal based and oxygenated additives, antioxidants, using cold flow improving elements, or improving the exhaust gas treatment [9,10].

Although Biodiesel is perfectly miscible in Diesel fuel at any ratios, the most common commercialized and tested blends between FAME (Fatty Acid Methyl Ester) and pure Diesel fuel are within the range B5-B30 (%vol. Biodiesel-Diesel fuel), one of them being especially revealed as highly appropriate to be on large scale used, namely the B20 fueling mixture [11].

The main goal of this paper is to numerically analyze the engine system efficiency, performances and emissions, based on the use of a heavy-duty IVECO Cursor 10 Diesel type engine, when fueled with Diesel fuel and with Biodiesel-Diesel B20 fueling mixture.

2. Differences between Diesel fuel and Biodiesel

Biodiesel appeared as an alternative fuel to Diesel classic fuel, because its properties are comparable to those of the fossil fuel. Paper [1] extensively analyses the properties of both fuels, taking into account the fact that it has been developed a multitude of manufacturing sources for Biodiesel. Thus, it is constantly mentioned that viscosity, which the injection characteristics significantly depend on, has higher value for Biodiesel; Biodiesel has also a higher cetane number and a larger density comparing to Diesel fuel. Its pour point is high enough to generate difficulties when used at low temperatures cold starting. Biodiesel qualities as fuel are standardized in DIN EN 14214, its characteristics being expressed for the maximum and the minimum temperature scale values.

Cetane number values for different types of Biodiesel are varying in-between 37 and 62 units [12]. This fact could lead to an important variation of the autoignition delay when using Biodiesel as a single fuel. Same study [12] statues that Biodiesel low heating value almost constant of 37 MJ/kg for every tested type of.

The exposed elements show a close similarity between the physical properties of both analyzed fuel types (Biodiesel and Diesel fuel), making difficult to adopt a simulation model to highlight different occurring aspects during the in-cylinder combustion process, especially when using Diesel fuel – Biodiesel blends.

For a Diesel engine receiving liquid phase injected fuel, this one will penetrate the combustion chamber under a spray of droplets, which would vaporize, would mix with the air and than would burn. If using a mechanical injection type, describing the IVECO CURSOR 10 engine, Table 1, the viscosity of the fluid should generate a specific injection characteristic for each fuel. After the spray penetrates the combustion chamber volume, the vaporization rate of the fuel droplets becomes more important. Paper [8] describes an analysis of the vaporization duration of a droplet in case of a single fuel or a fuel mixture. The results reveal that the vaporization duration of a 100% Biodiesel droplet is within 7 to 20% greater than for a similar droplet of Diesel fuel and within 1.2 to 4% when speaking of a B20 (vol.20% Biodiesel + vol.80% Diesel fuel) fuel droplet, comparing to the same reference.

Table 1. Engine Cursor 10 -400 specifications [13]

Maximum power [kW/hp]	294/400
Rated speed [rpm]	2100
Number of cylinder	6
Compression ratio	16.5:1
Bore [mm]	125
Stroke [mm]	140
Cubic capacity [l]	10.38
Fuel system	Direct injection mechanical
Admission	turbocharged

In this context, in order to highlight the different combustion features separating pure Diesel fuel from a Biodiesel-Diesel fuel mixture it is necessary to adapt a computer model so that it could emphasize the injection modes, the vaporization and the unequal heating value of the fuel types.

2. Characteristics of the fuel injection process

The selected engine is an IVECO CURSOR 10-400 heavy-duty Diesel engine, with a maximal rating power of 294 kW, corresponding to 2100 rpm. The engine uses a mechanical injection system, with a maximum injection pressure of 180 MPa. Therefore, the following model has been applied (see Fig. 1.).

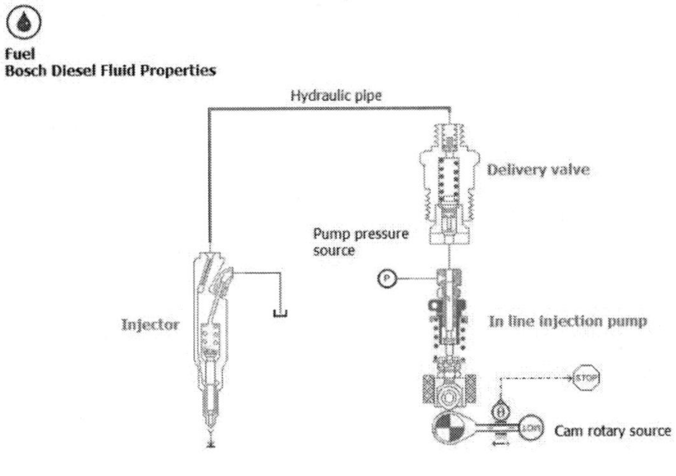

Fig. 1. Injection system model

The injection system consists in a pumping module containing the came acting on the tappet and consequently further on the piston of the pumping element (see Fig. 2).

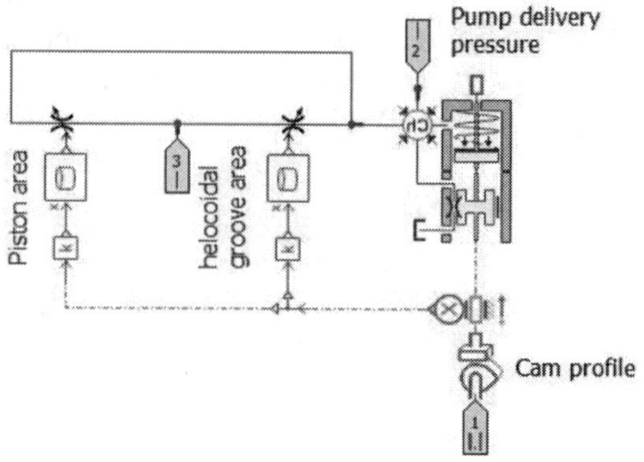

Fig. 2. Pumping element

The fuel flow is controlled by the piston compressing surface and by the section offered by the piston helical control edge. The engine has a pump-injector assembly and the high-pressure pipe is very short. In case of using pure Diesel fuel, the injection pressure together with the needle lift are plotted in Fig. 3. The maximum of the injection pressure is 1800 bar.

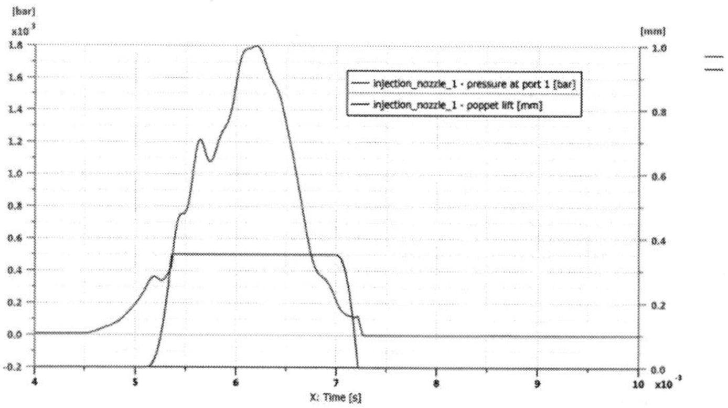

Fig. 3. Injection pressure and needle lift at Diesel

In case of using B20 fuel, the injection characteristic is highlighted in Fig. 4. The injection maximum pressure and the needle lift have both increased values than in the previous case.

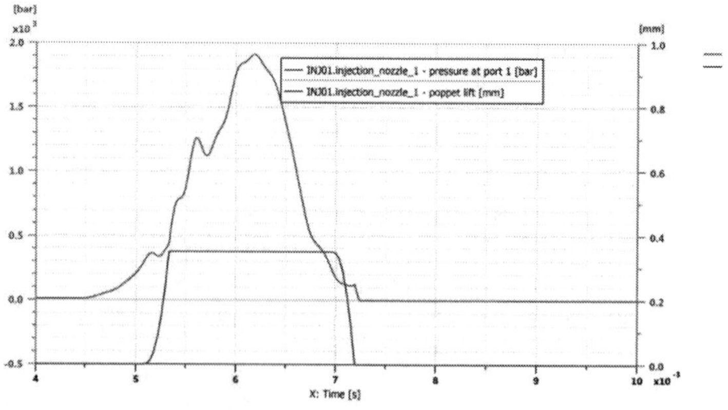

Fig. 4. Injection pressure and needle lift at B20

The system operating characteristics are comparatively presented in Fig.5. The maximum injection pressure is with 100 bar higher when using Biodiesel B20 fuel than in the classic situation for Diesel fuel due to its high viscosity.

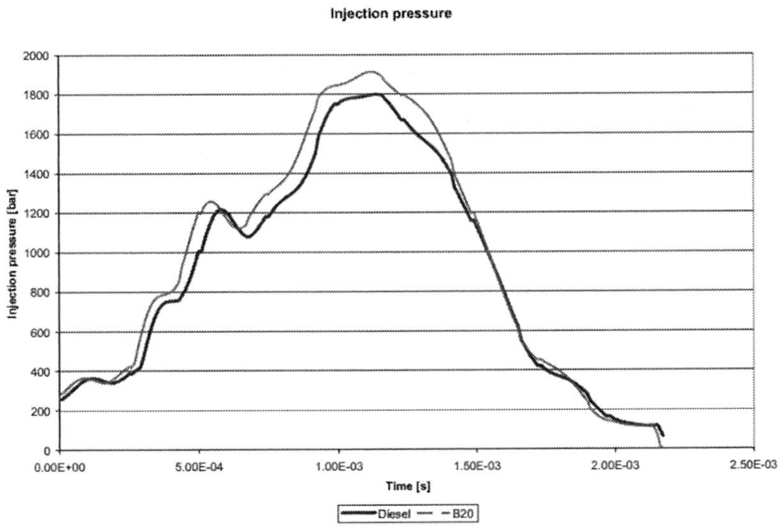

Fig. 5. Injection pressure variation vs. time

Referring to the injection fuel flows, small differences occur between the two operating situations (see Fig. 6). The image contains also the polynomial interpolation functions, necessary to generate a function for the injector model in engine model.

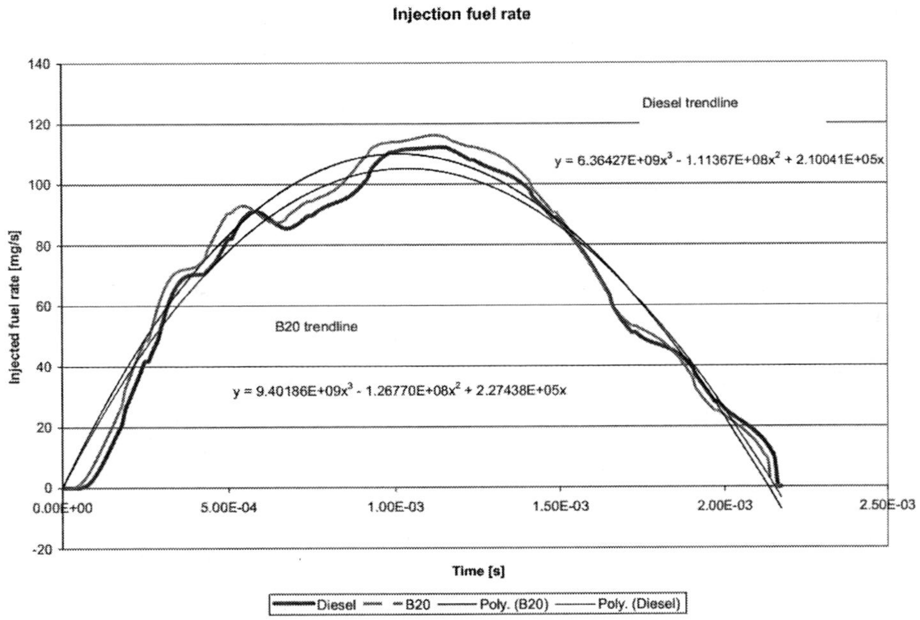

Fig. 6. Injection fuel rate and trendline

From the total injected mass point of view, the differences are rather small, from 146 mg Diesel injected fuel to about 152 mg of B20 injected fuel. Following the simulations, one common scenario has been used, the one in which both fuel types on-cycle injected quantities are equally to 146 mg.

3. Modelling in-cylinder processes

The fuel is injected inside the cylinder through the pumping element, which applies as a main parameter an interpolation function of the injected fuel flow. This one is taken from the injection system model for each fuel apart. In order to describe the development of the in-cylinder processes, the AMESIM numerical code uses Chmela [14] proposed model, related to the vaporization and the mixing rates of the fuel droplets sprayed into the cylinder. The heat release rate for the applied model is based on the dependence on four parameters: the vaporization specific time, the mixing duration, the constant of the injection turbulence dissipation and one constant depending on the effect of the residual burned gas on the combustion process. In the analysis of the air-fuel mixture formation it is assumed that the only difference between the two cases (Diesel fuel and Biodiesel B20 mixture) represents the vaporization duration of the droplets, for the use of Biodiesel this duration is reduced as it is reported by reference [8]. The other values depending on the autoignition process have been kept constant when comparatively analyzing the two fuels.

3.1. Reference results obtained for Diesel fuel

When fuelling with Diesel fuel the obtained rated power was 294 kW at 2100 rpm, corresponding to fuel dose amount of 148 mg per cycle and cylinder, meaning a BSFC value of 190 g/kWh. The cylinder peak fire pressure was 23 MPa, the relative air-fuel ratio was 1.5 and the supercharging pressure was 2.4 bar.

3.2 Results obtained for B20

The different properties of B20 compared to Diesel fuel were: a lower heat value of 41.4 MJ/kg for Biodiesel vs. 44 MJ/kg for Diesel fuel, the vaporization duration for Biodiesel decreased by 3% [8] comparing to Diesel fuel and the injection characteristic for B20 shaped as in Fig. 6. When using B20 fuel, a maximum rating power is 270 kW. The supercharging pressure was 2.31 bar, the peak fire pressure was 21 MPa and the relative air-fuel ratio was 1.46.

3.3 Comparison between Diesel fuel and B20 fuel

The heat release characteristic (see Fig. 7) has the same shape for both types of the used fuels, only the released energy being different due to the higher heating value of Diesel fuel.

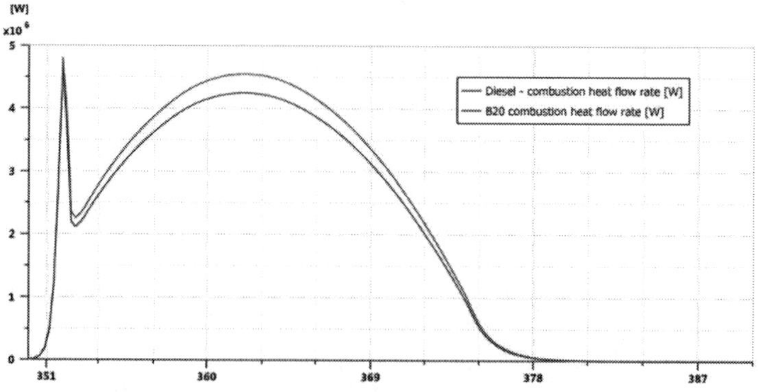

Fig. 7. Comparison of the heat release rates for Diesel fuel and B20

Corresponding to the first premixed air-fuel mixtures combustion stage, the variation of the released energy is practically the same in both cases (see Fig. 8). The unit used for this graph are watts and, in correlation with lower heat value for B20, indicates the influence of the faster B20 vaporization rate versus Diesel fuel because a larger amount of fuel is vaporised in this period of time.

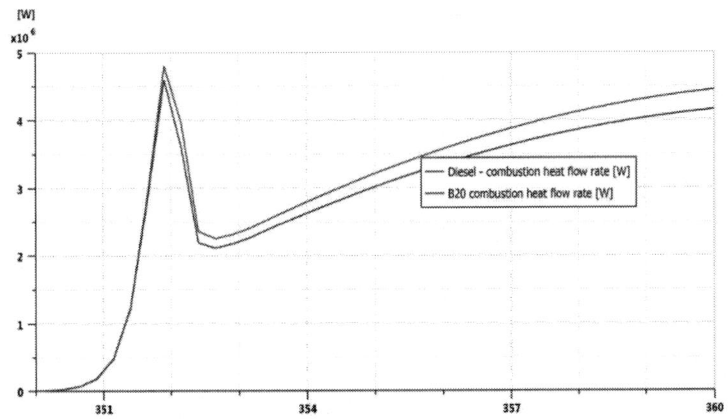

Fig. 8. Combustion heat release rate for Diesel fuel and B20 (detail)

The in-cylinder processes are affected by the injection characteristic, by the vaporization rate and by the lower heating value of each fuels. Thus, the peak fire pressure (showed in Fig.9) is 22 MPa in case of Diesel fuel vs. 21 MPa in case of B20 fuel.

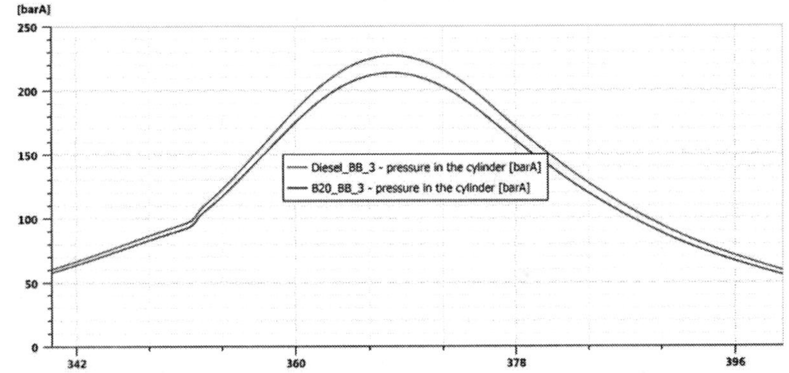

Fig. 9. The indicated pressure vs. crankshaft position

The in-cylinder temperatures have similar variation as the pressure traces following in Fig. 10. This is explainable by the way in which the turbocharging system responses when replacing the original Diesel fuel, respectfully to the fact that the turbocharging group speed decreases from 83800 rpm, when using Diesel fuel to 81700 rpm, correspondingly when using B20. In these conditions the relative air-fuel ratio drops from 1.5 to 1.46 and the temperature variation during the combustion process is very similar between the two fuels.

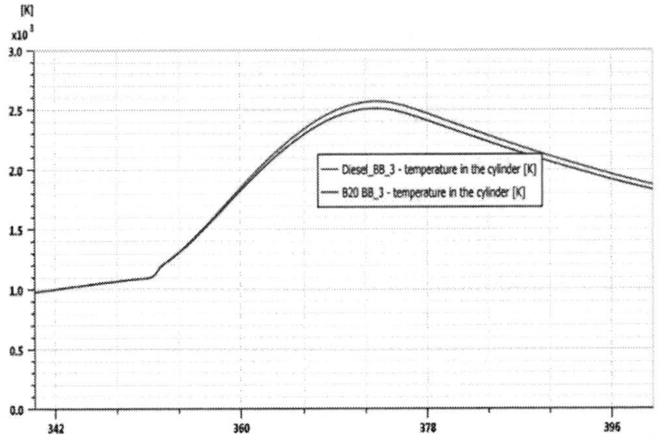

Fig. 10. In-cylinder temperature variation

The obtained results concerning the in-cylinder processes allow a further evaluation of the way in which the pollutant emissions are generated.

Conclusions

The next conclusions can be summarized as follow:

1. Despite the different of some physical properties, existing between the two fuels considered, no major changes have been identified by simulation on the injection characteristics; an increase of 5.5% in maximum injection pressure form 1800 bar for Diesel fuel to 1900 bar for Biodiesel B20 which does not change by more than 4% the fuel amount per cycle and cylinder;
2. This slight alteration of the fuel amount when using B20 alters the relative air-fuel ratio which in turns associated with the reduction of the lower heating value modifies the heat release characteristics and the in-cylinder pressure and temperature; the corresponding peak fire pressure and the peak fire temperature are reduced by 4.5% and respectively by 4%.
3. As a general conclusion, taking into account these model results and the experimental knowledge of the authors, the use of a fuel that includes 20% biodiesel in a conventional engine without constructive adaptations, will not generate essential differences in the engine running.

References

1. P. Saxenaa, S. Jawaleb, M. Joshipurac ELSEVIER Proc. Eng. **51** (2013)
2. K. A. Sorate, P. Bhale ELSEVIER Ren. and Sust. Energy Reviews **41** (2015)
3. P. Tamilselvana, N. Nallusamyb, S. Rajkumarb ELSEVIER Ren. and Sust. Energy Reviews **79** (2017)
4. I. Yahuza, H.Dandakonta, M.E.Ibrahim J. of Bio-process. & Biotech., **vol.6**/2016
5. Ramona Stefanescu Mihaila ELSEVIER Ethanol Sc. and Eng., **2019**, pp.523-535
6. A. Racovitza, B. Radu, M. Aldhaidhawi, R. Chiriac *EAEC-ESFA*, Springerlinks **2016**
7. M. Gharbanpour , R. Rasekhi ELSEVIER Fuel **106** (2013)
8. Qubeissi, M. ELSEVIER Applied Thermal Engineering **(2018)**
9. Li Han, Y. Wenming, Z. Dezhi, Yu Wenbin ELSEVIER, App. Therm Eng. **137** (2018)
10. P.S. Varun, K.T. Samaresh, S. Rituparn, K.Naresh ELSEVIER, Renewable and Sustainable Energy Reviews **79** (2017)
11. M. Aldhaidhawi Study on Performance, *Emissions and Combustion Charac. of a Diesel Engine Fueled with Biodiesel B20*, PhD Thesis, Politehnica Univ. Buc., **(2017)**
12. I.Barabás and I.A. Todoruţ Biodiesel http://cdn.intechweb.org/pdfs/23666.pdf
13. http://www.powertechengines.com/IvecoData/CURSOR-10.pdf
14. F. Chmela, C. Orthaber, *SAE paper 1999-01-0186*, **(1999)**

Reduce deposits on heat exchange surfaces when burning inferior coal

Marin Bică[1] , *Andrei* Stoian[2] , *Dragoş* Tutunea[1] , *Mădălina* Călbureanu [1*]

[1]University Of Craiova, Faculty of Mechanics, Romania
[2]Turceni Thermal Plant, Romania

Abstract. The phenomenon of soiling of heat exchange surfaces occurs in all types of combustion plants. The deposit of solids (combustion gases particles) worsens heat transfer by reducing combustion efficiency, increasing fuel consumption and pollution. Deposits are influenced by combustion temperature, gas flow rate, dimensions of solids drift by combustion gases as well as the composition of fuel mixtures used (eg inferior coal with flame retardant gas or tar fuel). In the present paper are presented techniques and methods adopted for the reduction of soiling of the heat exchange surfaces and the results obtained on high energy power plants (steam boilers). Key words: deposits, heat exchange, combustion temperature, inferior coal.

1 General considerations

The conditions in the combustion plants vary greatly - from relatively cold temperatures to temperatures above 1500°C in the radiant part of the boiler and above 1800°C inside the flame. The dimensions of the solid particles trained by the combustion gases can also vary from less than one micron to a few millimeters in diameter.

The shape and consistency of the particles vary - depending on the equipment and the fuels - from spherical shapes to very irregular shapes and from solids to fluids. The deposits formed on the heat exchange surfaces have the effect of reducing the heat transferred to the working fluid.

The dirty area of the combustion heat exchanger parts is recorded in all areas of the industry. Wherever heat is transferred from the burnt gas stream containing corrosive or reactive particles, the accumulation of deposits on the exposed surfaces, not only on the surfaces of the heat exchangers, can cause problems.

As a result, unwanted solids build-up can occur on pipes or fittings in equipment such as dryers or reactors, as well as in combustion chambers and chimneys.

Expansion of harmful effects depends on a multitude of factors. The operating temperature, fluid velocities, particle sizes and the concentration of any solid substances entrained in the

* Corresponding author: Mădălina Călbureanu, madalina.calbureanu@gmail.com

gas stream, as well as the chemical composition of components in the fuel mixture, are liable to cause the soiling of the heat exchange surfaces.

The presence of deposits on the surfaces of the heat exchangers will prevent an efficient heat transfer that will affect the efficiency of the whole plant. In addition, unwanted accumulation of deposits on structures may cause mechanical damage that may result in unscheduled shutdowns (damages). They lead to decreases in production and increased maintenance costs.

2 Formation of deposits mechanisms

In an attempt to understand the effects of different deposit variables in a combustion system, it is useful to have a general understanding of the mechanisms involved, so that the needs and requirements for a good design and operation of the equipment can be appreciated.

The mechanisms involved will vary depending on the type of deposit particles, vapor diffusion or chemical reaction.

It is convenient to analyze these mechanisms separately, although they will rarely appear isolated, and some interactions, to a greater or lesser extent, occur in industrial conditions.

In this research a model for the growth of the deposition layer has not been achieved. It will be a concern in future research.

2.1 Molecular diffusion

It is the result of the individual and disordered displacement of molecules, determined by the movement of thermal agitation and elastic collisions with the other molecules in a mixture of several molecular species. Molecular diffusion is therefore a mechanism that occurs at the molecular level and consists in transporting a molecular species within a mixture when there is a potential difference in the mixture, which produces an imbalance within the mixture.

The potential difference can be determined by several causes:

- **the existence of a concentration gradient in the mixture** – which causes a *Fick diffusion*;
- **the existence of a temperature gradient in the mixture** – which causes a *thermal diffusion*;
- **the existence of a mixture pressure gradient** – which causes a *pressure diffusion.*

2.2 Vapor diffusion

The transport of particles to heat exchanging surfaces in a combustion system may result from the vaporization of the inorganic components of the ash and their subsequent diffusion to the surfaces of the equipment. Flame-breaking of inorganic components is the usual origin of these vapors. Generally, the diffusion principles (except for inertial collision) will apply to the molecules of these vapors, but some significant consequences may result from the temperature gradients existing in the equipment. Vapor diffusion from warmer areas to cooler areas can be changed in stages, from vapor to liquid droplet and ultimately to solid particles. The conditions are illustrated in Figure 1.

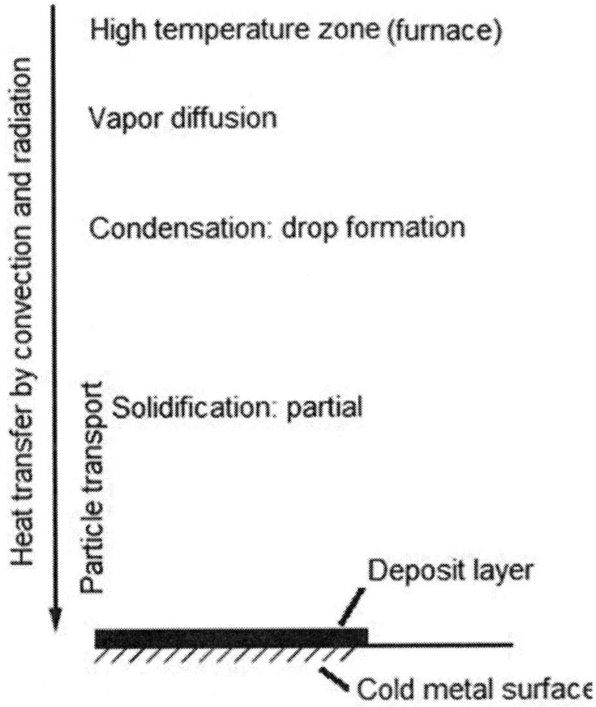

Figure 1. Changing the phases along the temperature gradient in the combustion chamber

The location of the various areas, as they are presented very generally in the above figure, will undergo a permanent change. Because of the change in temperature, the distribution increases as the thickness of the deposit layer. Due to these temperature changes, the vapor and drip areas will move to the cold area, which, after the accumulation process has begun, will be the outer layer of the deposit. If deposition was due only to vapor diffusion, the deposition thickness becomes asymptotic over time since the outer layer is above the dew point.

Phase change can lead to amplification of other effects such as:

- droplet fusion
- agglomeration of particles during transport by cementation due to their sticky or melt state
- condensation of vapors either on the chiller's surfacees of the equipment or in the chiller's area.

2.3 Chemical reactions

Given the complexity of most fuels, such as fossil fuels and coal, or to a lesser extent household waste or biomass, the possibilities of chemical reactions are very high. The type of chemical reactions is influenced by the conditions in the combustion plants.

3 Cleaning of heat exchange surfaces

In most combustion systems, the phenomena of dirt and grease are likely to occur and steps must be taken to restore the efficiency of the heat exchange surfaces to or near the project values.

There are a number of techniques and technologies used to clean these surfaces. They can be divided into two large groups:

- hot cleaning tehniques and tehnologies (during the operating installation)
- cold cleaning techniques and tehnologies (wth the equipment off)

To facilitate cleaning, additives are used. In coal-fired combustion plants the use of an additive can lead to the transformation of deposits so that they are suitable for a hot-stripping method, or if the accumulations are high that it is necessary to stop the plant, the use of additives during operation may reduce the time necessary for cold cleaning operations.

3.1 Cleaning during plant operation

The purpose of this cleaning method is to keep the heat exchange surfaces in a reasonable state of cleanness to ensure efficiency and operation in the plant's parameters. Hot soot removal processes use soot blowing or chemical cleaning processes.

3.2 Soot blowing

Most soot blowers are rotating and retractable in length. Rotary blowers use multidirectional elements able to rotate so that the depositing layer can be removed with a jet spring.
Steam blowers are mainly used in the area of superheaters.
Air blowers. Air use requires a separate system with compressor and *water jets*. It achieves a considerably higher amount of deposition than the use of air or steam. In addition, there is a hardening effect on hot deposition that leads to fracture of the deposition and breakage of its bond with the heat exchange surface. These effects make it easier to remove deposits. Unwanted effects also occur: large pieces of slag that can damage the heat exchange surfaces may occur.

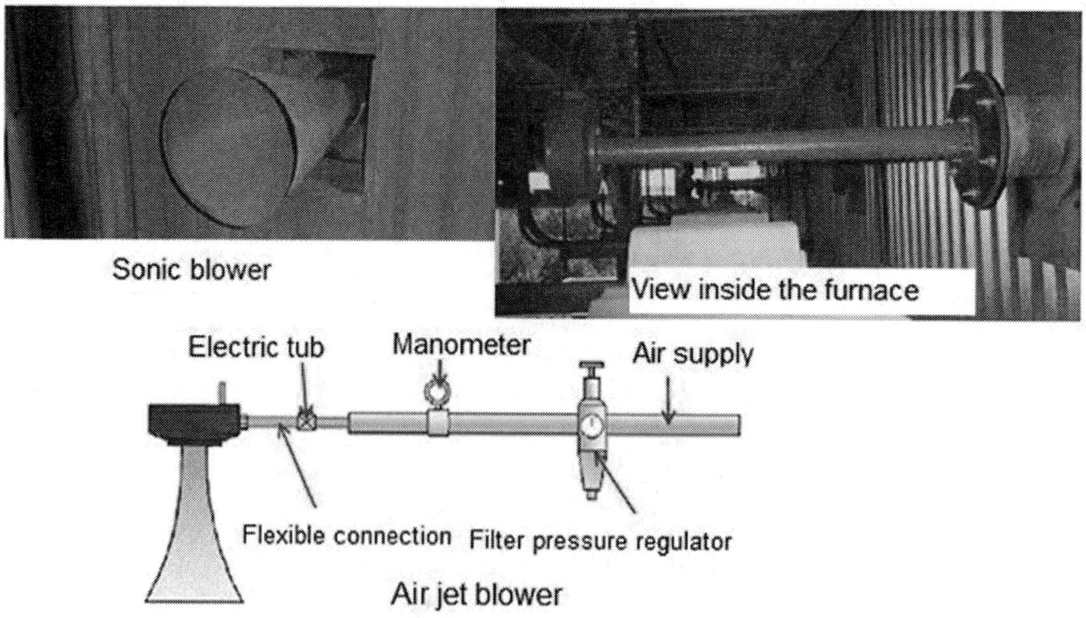

Figure 2 Types of blowers

Multijet blowers are also used to clean the cold parts of the plant, convective parts, extended surface economisers and regenerative preheaters.Where high temperatures are recorded, it may be necessary to use the air to cool the various components of the blower. High or low frequency sounds are used to clean the dirty surfaces of heat exchangers - special attention should be paid to low-frequency sounds.

Figure 3 Acoustic soot blowers

Sonic blowers are not very effective for hard deposits, but are useful for loose, soft deposits. Unlike steam or air blowers, sonic cleaning devices are being used more and more frequently lately. The operating costs of the audio devices are usually lower than those of traditional devices.

3.3 Shot cleaning of the deposits

Shot cleansing uses the kinetic energy of the shots to dislodge deposits. Technology involves "bombarding" surfaces that need to be cleaned with metallic shots or made of other material. The basic principle of using this technology is not the cleaning of deposits and especially the prevention of their formation.
Cleaned surfaces should be inspected periodically to see if damage has occurred due to shots. In case of such damage, it is necessary to reposition the distribution system or to apply protective shields.

3.4 Chemical cleaning

Chemical cleaning during operation or after stopping is routinely used. Hot cleaning is done by spraying on the deposit's surface chemicals that remove them from the chemical reactions that take place. There may be problems with equipment caused by thermal shocks. Cleaning after stopping is more efficient and presents a lower risk of damage to the structure. Gas ammonia and steam have been successfully used.

3.5 Cold cleaning

The cleaning methods described below are those that can be applied when the combustion equipment is switched off either for routine maintenance or especially when cleaning is required as a result of reducing the plant's operating load.
It provides a good opportunity to clean the plant and allows a visual inspection to verify the results of cleaning techniques applied to determine if cleaning causes degradation of heat exchange surfaces.

3.6 Dry cleaning

Completely dry cleaning by manual or mechanical means should be used if cleaning of the heat exchange surfaces with water or steam is not possible or is considered inappropriate. Manual dry cleaning is an activity involving a lot of dust and many hazards, so special precautions must be taken.
An alternative method to manual dry cleaning is dry cleaning by blasting or sanding. It is necessary to have enough space under the equipment to collect both the shots and the dislocated deposits.
Manual or mechanical dry cleaning can be supplemented by using compressed air lances or by suction equipment.

3.7 Washing

Water washing can be an effective way of eliminating unwanted deposits. The cleaning technique applies to the entire system with large amounts of water. Washing can be done by spraying or by pressure water jets. Adding wetting agents or detergents can improve the washing process.

3.8 Steam cleaning

This technology involves the intake of steam into the combustion gas space in contact with the surfaces that have accumulated deposits. To protect the system, it will be necessary to fill the heat exchanger pipes with cold water. Frequent site inspections are also required. Problems caused by corrosion of metal parts can be encountered. For the intake of steam it is necessary to build special installations, but the existing blowers are usually used.

Conclusions

The boilers of the Turceni Thermal Power Plant have cleaned the heat-exchange surfaces by several processes. It has been found that on the heat exchange surfaces of the flue gas path of deposits with variable thicknesses which can not be completely removed by mechanical cleaning or washing, respectively:

- in the furnace, in the upper part – hard deposits on boiling pipes very difficult to remove by mechanical cleaning;
- on the S3 pipes – heavy deposits difficult to remove by mechanical cleaning;
- on the SI 2 pipes – dust-like deposits, relatively easy to remove by mechanical cleaning;
- to rotary air preheaters (PAR) – hard deposits containing sulfur compounds, difficult to remove by washing with water jet under pressure.

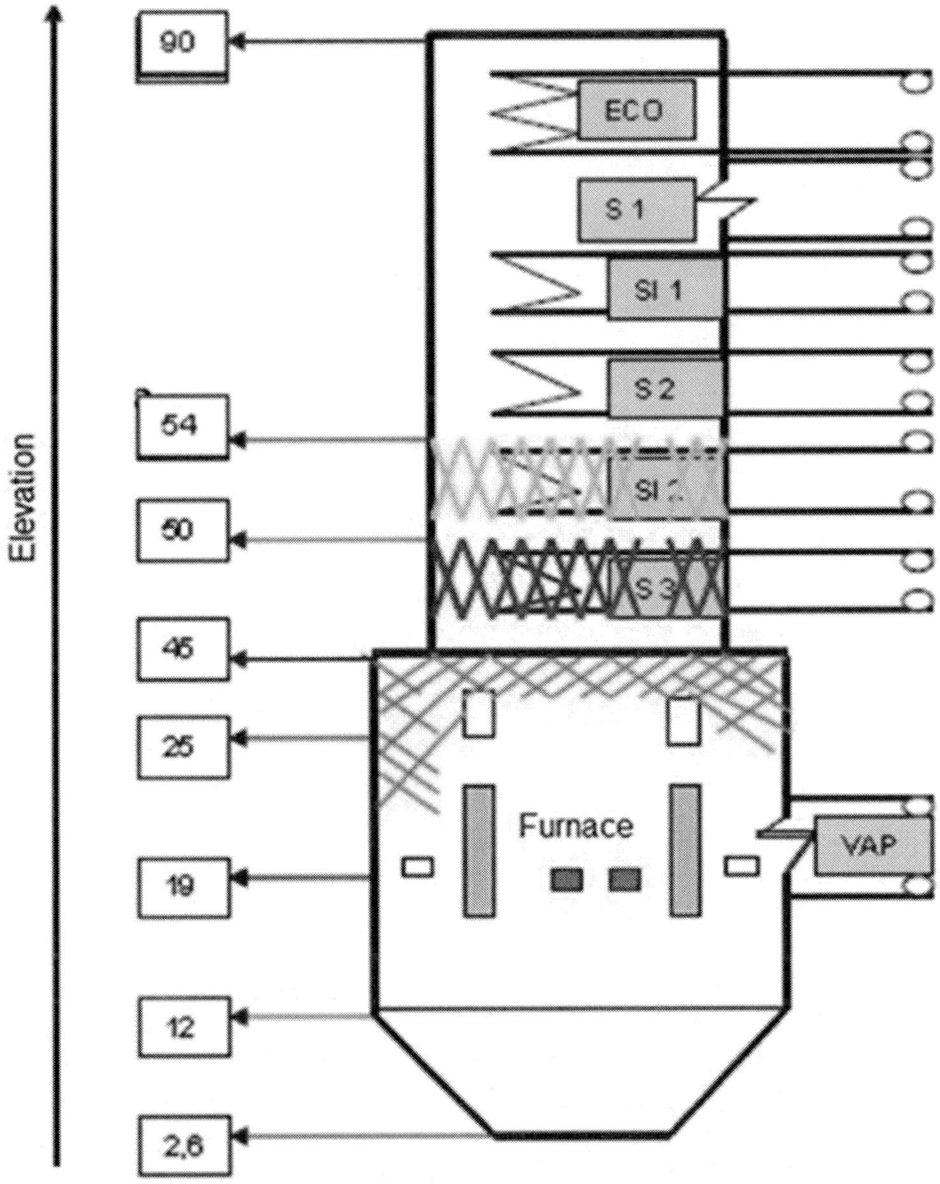

Figure 4. Benson type tower with forced circulation and flow rate of 1035 tones/h

In the experimental researches, the THERMA-CHEM (UK license) technology was used to clean and reduce the pollutant emissions, which ensured the following goals during the boiler operation:

- Reduction / elimination of deposits on the heat exchange surfaces located on the flue gas path and implicitly improvement of the heat exchange;
- Reduce fuel consumption, increase efficiency and load availability of the boiler,
- Improving the boiler and combustion process by eliminating the unburnt material,
- Reduction of low temperature corrosion (acidic corrosion),
- Reduction of emissions into the environment of: NOX, SO2, SO3, CO and CO2.

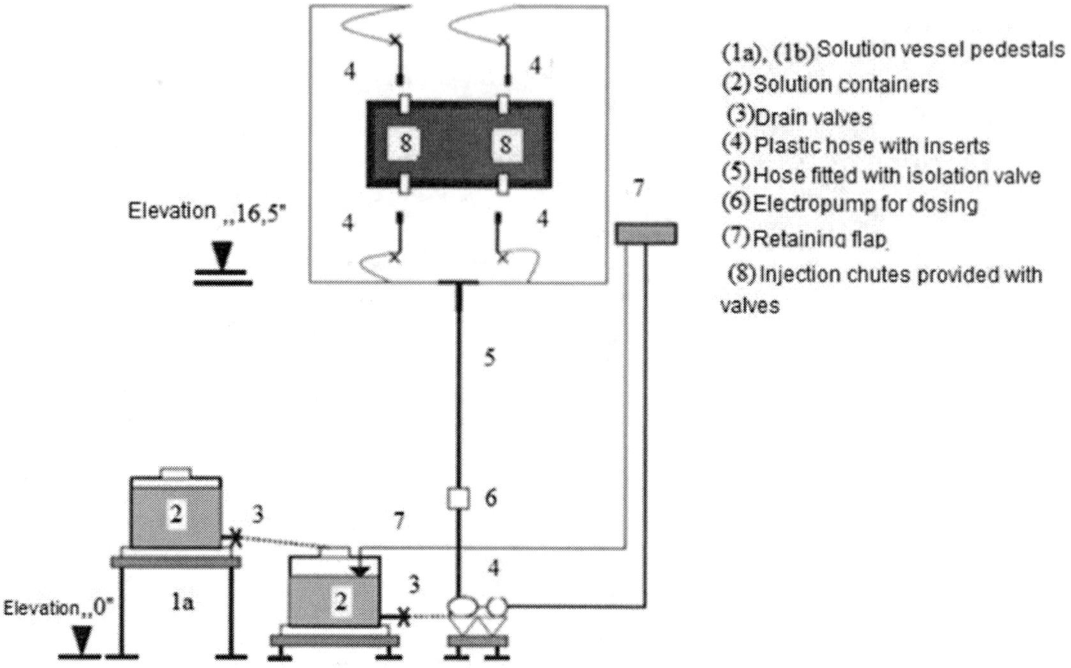

Figure 5 Schematic diagram of the cleaning installation

The benefits of cleaning the heat exchange surfaces are obvious: exemplified by the graphs for recording the temperature of the combustion gases before and after cleaning.

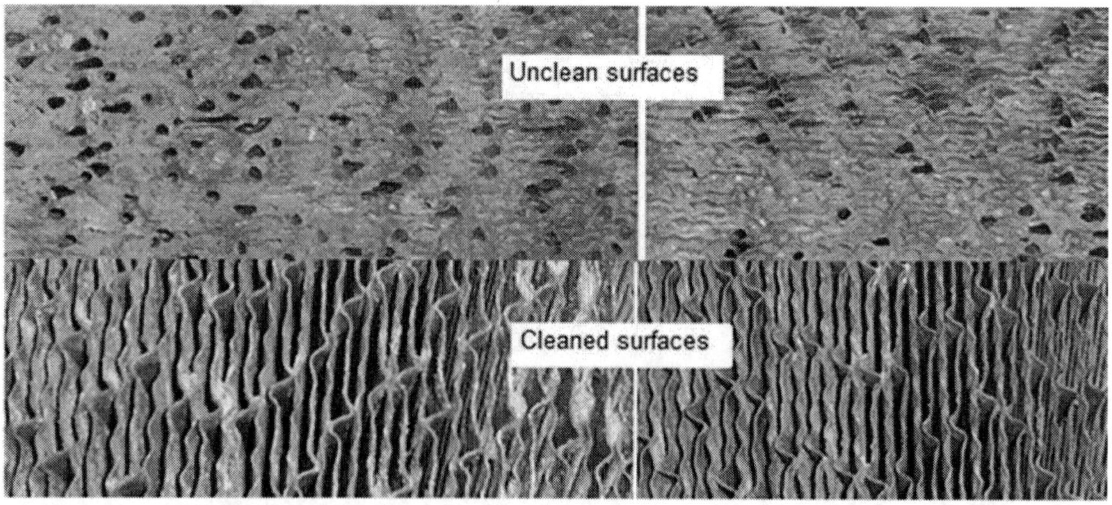

Figure 6. Surface condition before and after cleaning

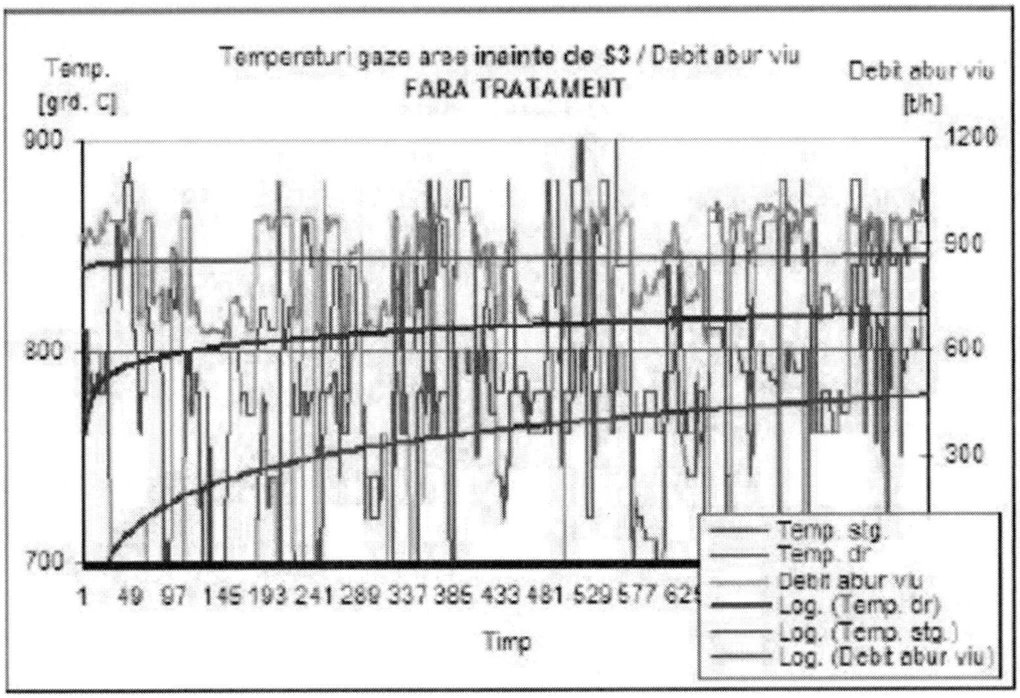

Figure 7. During the period prior to the cleaning in the combustion chamber, at a relatively constant load, the temperature of the combustion gas has a ASCENDING evolution, indicating a worsening of the heat exchange over time as a result of soiling of the heat exchange surfaces through the appearance of the slag deposits.

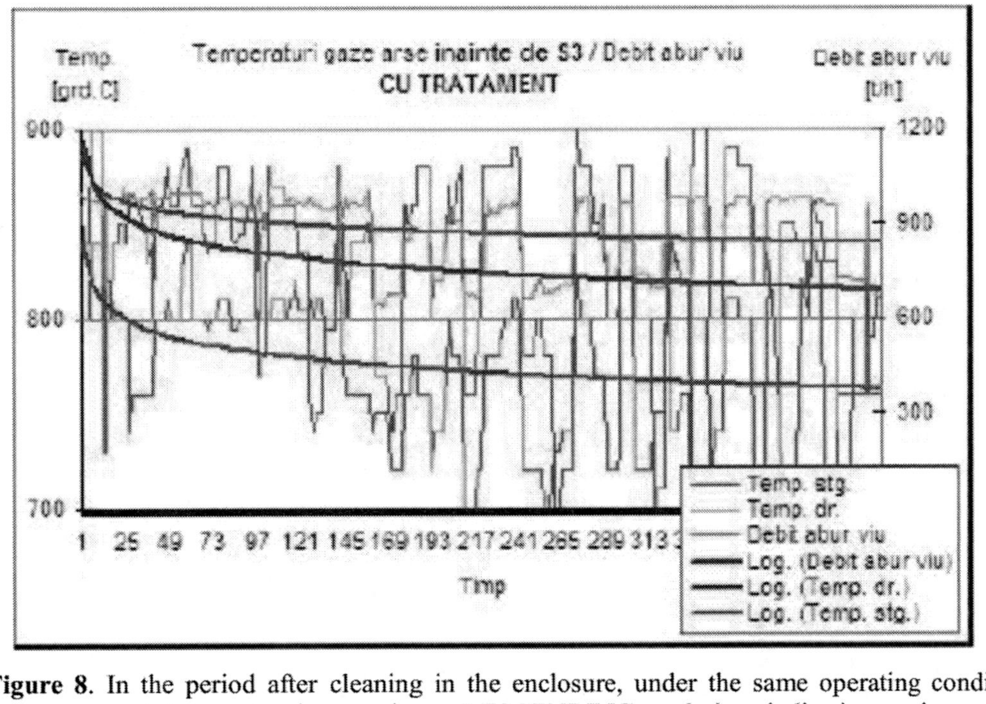

Figure 8. In the period after cleaning in the enclosure, under the same operating conditions, the temperature of the combustion gas has a DESCENDING evolution, indicating an improvement in heat exchange over time due to the removal of deposit deposits (by eliminating in particular unbound carbon, which constitutes generally the element with the largest insulating properties) and their partial elimination.

A simple calculation made at the Turceni Thermal Power Plant:
- considering an increase in efficiency of only 1% (corresponding to the reduction of the temperature to the chimney), it results a saving of 1620 USD / day:
1% x 400 tonnes / h lignite x 24h / day x 15 USD / lignite tonne = 1440 USD / day.
1% x 5000 Nm3 / h x 24 h / day x 150 USD / 1000 Nm 3 = 180 USD / day
- reduction of injections, mean with approx. 20 - 40 kg of water / 1t of live steam in overheaters for steam temperature control has the effect of improving the boiler efficiency by 0.5-1%, resulting savings of approx. 800-1600 USD / day.

References

1.D. Ştefănescu, A. Leca, *Transfer de caldura si masa*, Editura didactica si pedagogica, Bucuresti, (1983)

2.M. Bică, *Bazele termotehnicii si masini termice*, Universitatea din Craiova, (1996)

3.A. Leca, s.,a., *Transfer de caldura si masa*, Editura tehnica, Bucuresti, (1988)

4.M. Bică, M. Naghi, *Transfer de caldura si masa*, Editura Universitaria, Craiova (1999)

5.*Manualul inginerului. Termotehnica*, Editura tehnica, Bucuresti, (1986)

6.R. Vilcu, *Termodinamica chimica*, Editura tehnica, Bucuresti, (1985)

7.M. Bică, *Termotehnică şi maşini termice*, Editura Universitaria Craiova, (2000)

8. M. Motocu, V. Bănică, L. Danciu, V. Osiac: *Optimizarea activitatii termocentralei Turceni în concordanţă cu standardele de mediu*, Revista Energetica, iulie (2005), ISSN 14353-2365.

9.K. Schroder, *Centrale electrice de mare putere*, Editura tehnica, Bucuresti, (1974)

10.A.I. Reynolds, *Curgeri turbulente in tehnica*, Editura tehnica, Bucuresti, (1982)

11.ESDU *Fouling in cooling systems using seawater*. ESDU 03004, ESDU

International plc, London, UK, (2003).

12. ESDU *Fouling in cooling systems using freshwater*. ESDU 06017, ESDU International plc, London, UK, (2006).

E3S Web of Conferences **112**, 01004 (2019) https://doi.org/10.1051/e3sconf/201911201004
TE-RE-RD 2019

Considerations on the combustion of biofuels in compression ignition engines

Um Min Allah Fazal[1], *Marin* Bică[*1], *Alexandru* Gruia[1], *Dragoş* Tutunea[1], *Mădălina* Călbureanu[1]

[1]University of Craiova, Faculty of Mechanics, Romania.

Abstract. Biofuels are considered to be attractive alternative fuels due to rising oil prices and the depletion of fossil sources. It is produced by a process of transesterification of vegetable oils and animal fats with an alcohol in the presence of a catalyst. At present, biodiesel and bioethanol are considered solutions, and almost all EU countries have prepared a policy on the production and use of biodiesel in the transport and heating sectors. Biodiesel has certain advantages, such as a higher number of cetan, biodegradability, low sulfur content and aromatic content, and lower carbon monoxide emissions, unburned hydrocarbons and particulate matter. One of the main issues related to biodiesel and biodiesel blends is the high viscosity of the fuel. Among the biodiesel properties, density, viscosity and heating value affect engine performance and emission characteristics. The presented paper highlights the differences that occur in comparison to the classical fuels and the performances obtained when using biodiesel.

1 General consideration

The development of the transports required the improvement of Diesel engines, especially since the steam engines that used the railway transport locomotives did not respond to the needs (tasks and speeds) The emergence of oil, the rapid development of the oil industry, led to the neglect of the use of biodiesel, due to the undeniable qualities the refined fuels and the low price at that time. The increase in oil consumption during World War II has led to the depletion of their production. The idea of using biofuels has become a solution. For engines evolved at that time, the high viscosity of biodiesel did not allow it to be used. In 1937, G. Chavanne, a Belgian, invented transesterification of vegetable oil, a process that enabled the efficient use of these fuels, which proved to be environmentally friendly. In the 1980s, environmental pollution increased, global temperature increased - global warming. Since then, the European Union has become the largest biofuel producer. The first motors were built to work with vegetable oils without being subjected to further processing. Biodiesel has several advantages such as: since is made from animal fats or vegetable oil reduce the dependence of petroleum products, can be used in existing engines with little or

* Corresponding author: Marin Bică, marinbica52@gmail.com

no modifications, reduce the greenhouse gases by 78% according to the experts, can be produces in local refineries reducing transportation cost, is safe to handle (flash point over 150ºC comparative with 52 ºC for diesel fuel) and positive economic impact since biodiesel is produced locally. The disadvantages include the variation in biodiesel quality, problems in cold weather, the rising price for raw materials used for biodiesel production, the increase of using fertilizers for biodiesel crops and clogging the fuel filters in the engine. There is an increase in nitrogen oxides and degradation of tropical rainforests replaced by palm plantations [1].

2 Physical-chemical properties

The structure of the injection jet from combustion chamber determines the increases of engine performance and reduces pollutant emissions. The geometry of the injector, size and nozzle is selected function of the fuel properties, the combustion chamber architecture and the engine's energy solution. The larger angle of the injection cone decreases the penetration and interference between the jets. Hiroyasu and Arai proposed the relationship for determining the angle of the injection jet cone:

$$\theta = 83.5\left(\frac{1}{d}\right)^{-0.22}\left(\frac{d}{D_O}\right)^{0.15}\left(\frac{\rho_g}{\rho_l}\right)^{0.26}, \text{[degrees]} \tag{1}$$

d- diameter of the spray jet, [m]
D_0 - diameter of the cylinder, [m]
ρ_g - gas density in the combustion chamber, [kg / m^3]
ρ_l - density of spray liquid, [kg / m^3]

The spraying angle is based on:
- the geometry of the spray jet hole;
- kinematic viscosity of the fuel - affects only the limits of the jet;
- injection pressure - increases the angle of the jet;
- fuel density - decreases the angle of the jet;
- temperature - decreases the angle of the injection jet.

The characteristics of the spraying of biodiesel-diesel mixtures were determined by using an atmospheric pressure spray gun using two types of injectors, figures 1, 2 and 3. The density and viscosity of the fuel are important in determining the spray characteristics. The kinematic density and viscosity of the fuel mixture has an increase as the biodiesel concentration of the mixture increases in the blend.

Figure 1. Bosch common rail piezoelectric injector, BMK type engine

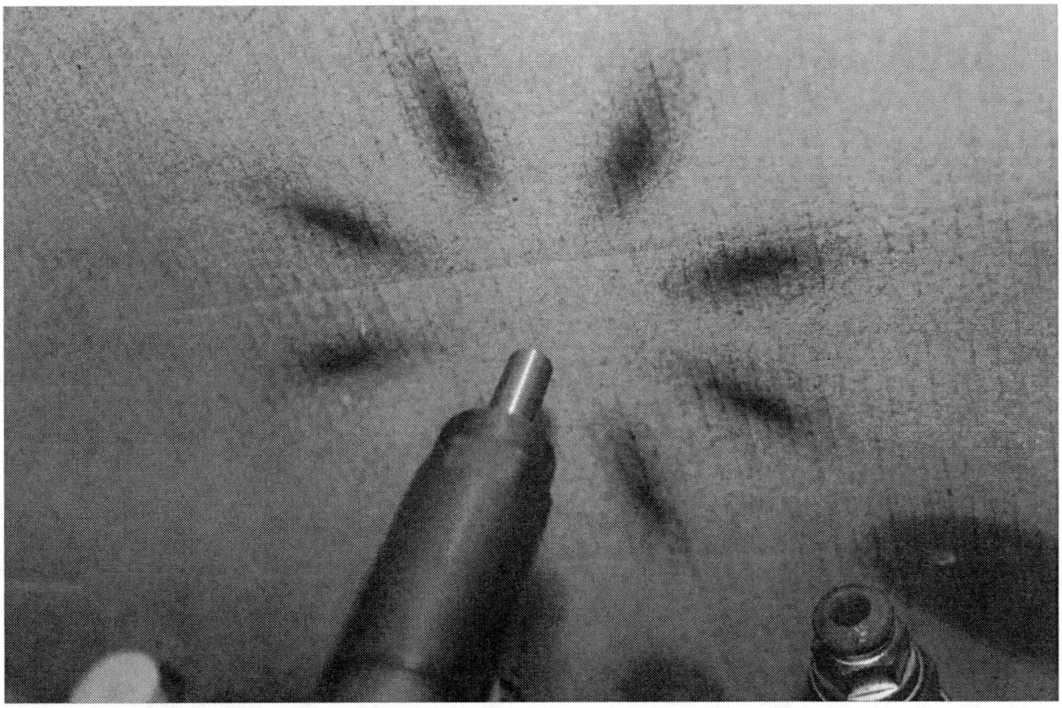

Figure 2. Fuel spraying

Figure 3. Injector Delphi.Tip VM Motori engine 2,5 l. Spray in the atmosphere

Table 1. Biofuel properties

The determined size	Value	Testing method	EN 14214
Density at 15°C , Kg/m^3	875	EN ISO 3675	860-900
Refractive index	1,4562	ASTM D 1298	≥1,4900
Viscosity at 40°C, mm^2/s	4,719	EN ISO 3104	3,5-5
Acidity, mg KOH/g	0,370	≤0,5	PeEN 14104
Iodine index, gI_2/100g	128	PrEN14111	<120
Saponificat index, gKOH/g	187,4	AOAC (1999)	185-190
Flash point, °C	178	ASTM D 6751	Min 101
Freezing point, °C	-7	EN 2315	-
Cloud point, °C	-2	-	-
Stability at oxidation at 110°C, h	8	EN 14112	Min 6
Distillation, °C	307	ASTM D 6751	Max 360

3 Influence of parameters determined on biodiesel quality

3.1 Density

In general density is a key parameter that affects the engine operation parameters and pollutant emissions. Researchers define the density as the mass per unit volume of a fluid. In diesel engines the system of injection delivers the biodiesel by volume not by mass, this property influence the quantity of the fuel injected by cycle. Also the air-fuel mixture in the combustion chamber is influenced by the fuel density due to the variation in the energy content. The density of biodiesel is higher than those of petroleum fuel. For all types of biodiesel, the tendency to change the density is the same, over time it increases with storage time. This phenomenon happens due to the change in biodiesel structure in the oxidation process. Samples containing hydrocarbons with shorter chains and saturated fatty acids tend to crystallize more quickly. This has as a result the decrease in volume and thus the increase in density. The mass of the sample also increases as a result of insoluble sediment formation, which also leads to increased density.

3.2 Kinematic viscosity

It is a parameter that expresses the friction between the interstices of a fluid being a major parameter in assessing the quality of liquid fuels. Viscosity influence the atomization of the fuel when is injected in the combustion chamber due to the formation of large droplets resulting in formation of deposits in the engine. The transesterification process reduce the viscosity of vegetable oil / animal fats which is approximately 10 times higher than petrodiesel fuel to a value closer to that of diesel fuel [2,3]. High viscosity and poor cold properties is the main reason why the use of vegetable oil has been abandoned. The majority of biodiesel and diesel standards include biodiesel as a key parameter to be used to monitor biodiesel composition. Researchers reported that viscosity is dependent of the chain length (number of atoms of carbon) and the degree of saturation. Viscosity increases with chain length (number of carbon atoms) and degree of saturation. This also applies to the alcoholic portion because the viscosity of the ethyl ester is slightly higher than that of the methyl ester [2]. Viscosity of fuels is taken into account in engine design specifications, is directly responsible for flow and lubrication properties; the increase in viscosity leads to a tempering of the injector nozzles. Biodiesel with high viscosity leads to the formation of too large droplets that will penetrate the wall opposite the injector. If the surface of the cylinder is cold, the combustion reaction will be interrupted and black smoke will be formed (the intermediate product of combustion consisting of aldehydes and acidic caustic-odorous acids). Incomplete combustion results in lower engine power.

3.3 Flash point

An increased flammability of biodiesel indicates that it has a low probability of accidental ignition. Changing the flash point can be influenced by the residual methanol in the biodiesel production process. The methanol residue may cause degradation of engine elastomers and seals, as well as the corrosion of the aluminum and zinc present in the injectors. The flash point gives indications of the presence of unreacted methyl alcohol in the biodiesel process. It is also important in relation to the legal requirements and safety measures involved in fuel handling and storage and are usually specified in the regulations on fire safety measures [4].

3.4 Cloud point

It is the temperature at which the first crystals appear in the mass of biodiesel. They notice visually and under these conditions biodiesel can only be used at a higher temperature. The cloud point variations leads to problems related to the decrease of the flow through the injection pump (decrease of the engine power), the clogging of the filters and the supply pipes. The parameter is an indicator of the temperature at which saturates can flocculate (coagulated as vials), which leads to increased viscosity; at the same time, sets the temperature at which a fuel can no longer be filtered within a specified time limit; shows the injection system blocking potential; shows the destination for use in cold weather or cold season; allows analysis of the need to add additives; can be controlled by mixing raw materials; the parameter also characterizes the determination of the fog point (precipitation of crystals). The cloud point is relative to the degree of saturation and the number of carbon atoms in the fatty acid chain.

Saturated fat compounds have increased melting points than unsaturated fatty compounds and crystallize in a mixture at superior temperatures than unsaturated compounds. Thus, biodiesel fuels produced by fats or vegetable oils with significant amounts of saturated fatty compounds will have greater cloud points. The increase in the flash point can be improved by the addition of additives [5].

3.5 Freezing point

It is the temperature at which biodiesel gels and does not flow. The mixture can no longer be pumped to the engine and it is necessary to heat it. A high value can limit its use in cold climates. An important role in modifying the freezing point is paraffinic substances which, upon cooling, form a network of crystals, thus preventing the movement of the oil. Apart from the presence of paraffin, the freezing point also depends on a number of factors including: the content of resinous and asphaltic substances, the water content of the oils, the preheating temperature, and the cooling temperature of the oils [6].

3.6 Acidity index

Indicates the amount of free acids in biodiesel. They come from two sources: from acids used in the production of biodiesel and which have not been completely eliminated and from oxidative degradation. Increased acidity can cause corrosion, deposits in the feed system or clogging of filters due to increased viscosity at the same time may be a symptom of the presence of water in the product.

3.7 Iodine index

The iodine value is an indicator of stability and a measure of total unsaturation; increased iodine value is closely linked to polymerization to oligomers and leads to damage to the injector; low-value iodine oils are less sensitive to oxidation processes.

3.8 Distillation temperature

In the case of unusually low distillation temperatures, these were effectively combined with a low boiling point and flash point indicating high residual amounts of methanol and / or glycerin content. Lower the 10% distillate temperature value is, easier will be the engine to start in cold climate. The temperature at which it distills 50% of the volume of biodiesel exerts an influence on the good running of the engine, that is, these average fractions influence the quality of the fuel mixture in the case of a heated engine leading to a stable engine operation and a smooth passage from the running regime empty at the medium speed. The wear of the piston rings, of the bearings, of the cylinders liners is influenced by the heavy fraction characterized by the temperature at which it distills 90% of the fuel, since the heavy fractions in a high percentage remain unvaporized and dissolve the oil thus creating a possibility of dry friction between the engine organs. Also, heavy fractions in liquid form are not in contact with air and do not burns completely in the combustion chamber, creating calamines that produce engine wear [7].

3.9 Saponification index

This property indicates the quantity of saponifiable units (for example acyl groups) per unit of biodiesel weight. A high level shows a higher quantity of low molecular weight fatty acids in the vegetable oil or vice versa. The saponification index is used to measure the average molecular weight of biodiesel and is indicated in milligrams of potassium hydroxide (mg KOH / biodiesel).The saponification process is a chemical reaction in which is produced metal salt or soap. In this reaction the hydroxide ion, -OH reacts with the methyl esters, free fatty acid or triglycerides. If the water is present in biodiesel this reaction cannot be removed. There is always a little water and there is always a soap that is formed when biodiesel is produced [8,10]. However is a possibility when are used solid (heterogenous) catalysts which do not give the free metal ions necessary to form soap. This solution which involves the biodiesel production and glycerin free can be used at large scale eliminating the quality problems of biodiesel. In fact, many of these heterogeneous catalysts present themselves problems by the fact that eliminate metal ions in the liquid and thus requiring the cleaning of the reaction products. The soap residue should be removed from the biodiesel after the reaction, and this can be done either by washing with water using a liquid-mixed solid adsorbent or by passing the liquid through an ion exchange resin bed.

3.10 Refractive index

It changes when the temperature of the biodiesel is near the point of disruption being a significant parameter for assessing its state. An increased refractive index indicates the presence of unsaturated substances.

3.11 Oxidation stability

It is expressed as the ability of a fuel to resist chemically to the changes occurring over time over the storage period. It is a major factor in the commercialization of large scale of biodiesel. One of the major problems which restrict the biodiesel usage in internal combustion engines is its poor oxidative stability. Biodiesel can be affected during long period of storage by the oxidation process in contact with air and in contact with water by hydrolytic degradation. Biodiesel may be affected by oxidation during storage (in contact with air) and hydrolytic degradation (in contact with water). These two processes can be referred as the biodiesel stability at oxidation and hydrolytic. The stability of biodiesel can be characterized by two major aspects: long period of storage (stability) or stability in condition of pressure and temperature when biodiesel is used in the engine fueling system. For the renewable fuels the oxidation stability is very important and biodiesel producers have to find methods to improve the biodiesel quality by using additives or extend the guarantee by more than six mounts. By analyzing biodiesel composition which is composed by fatty acids of different carbon chain especially polyunsaturated the tendency to oxidize is very high. After the oxidation process begin the hydro peroxides composed by hydrogen atom and two oxygen atoms are added to the chain of fatty acids. Also the oxidation process can be started by the materials used for recipients used to storage biodiesel or even light. The chemical reaction produces hydroperoxides which can react to produce aldehydes ketones and short chain fatty acids. Hydroperoxides in turn can polymerize to form large molecules. Thus, the process of oxidation increases the viscosity of biodiesel. In addition, oxidation increases the acid value of the fuel, the color can change from yellow to brown with the appearance of a paint smell and solid deposits appears in the engine fuel system (pipes and filters) [11,12]. For example, calculations were made for burning diesel and biodiesel. Results are shown in tables and graphics.

Table 2. Elementary composition of diesel fuel, colza biodiesel and FAME (Biodiesel)

Elements	Colza biodiesel	Diesel	Biodiesel
Carbon (%)	76	86.8	76.2
Hydrogen (%)	14.5	12.6	12.6
Oxygen (%)	6.3	0	11.2
Azote (%)	3.2	0	0

*Laboratory tests were performed with colza biodiesel

Table 3. Values V_{air}, V_i and V_{O2} for full burning of colza biodiesel

λ	**V_{air}[kmolair/kg.comb]**	**V_i [kmol/kg.comb]**	**V_{O2} [kmol/kg.comb]**
1.2	12.489	12.495	0.437
1.4	14.518	14.524	0.874
1.6	16.664	16.67	1.31
1.8	18.74	18.7468	1.75
2	20.83	20.8368	2.18

Table 4. Values for $V_{CO_2}, V_{H_2O}, V_{N_2}$ şi V_g at the full burning of colza biodiesel

λ	V_{CO_2}	V_{H_2O}	V_{N_2}	V_g
1.2	1.42	1.62	9.987	15.11
1.4	1.42	1.62	11.63	18.836
1.6	1.42	1.62	13.16	22.449
1.8	1.42	1.62	14.924	26.296
2	1.42	1.62	16.569	30.024

Table 5. Values V_{air}, V_i şi $V_{O2}$for complete burning of Diesel

λ	**V_{air}[kmolair/kg.comb]**	**V_i [kmol/kg.comb]**	**V_{O2} [kmol/kg.comb]**
1.2	13.296	13.30	0.465
1.4	15.512	15.517	0.93
1.6	17.728	17.733	1.396
1.8	19.944	19.95	1.86
2	22.16	22.165	2.326

Table 6. Values $V_{CO_2}, V_{H_2O}, V_{N_2}$ şi $V_g$for complete burning of Diesel

λ	V_{CO2}	V_{H2O}	V_{N2}	V_g
1.2	1.62	1.41	10.5	15.746
1.4	1.62	1.41	12.25	19.702
1.6	1.62	1.41	14.005	23.673
1.8	1.62	1.41	15.75	27.634
2	1.62	1.41	17.506	31.606

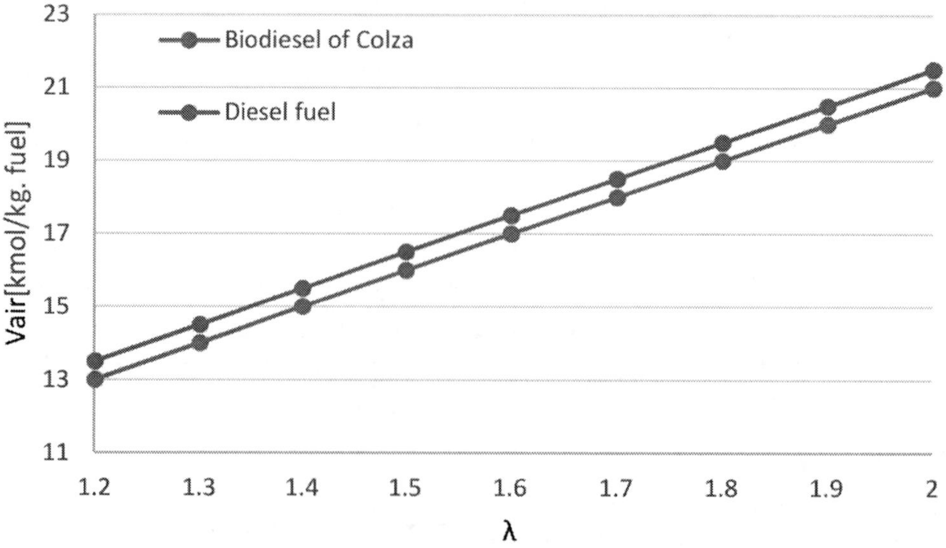

Figure 4. The air required for burning ($\frac{m_N^3}{kg.comb}$)

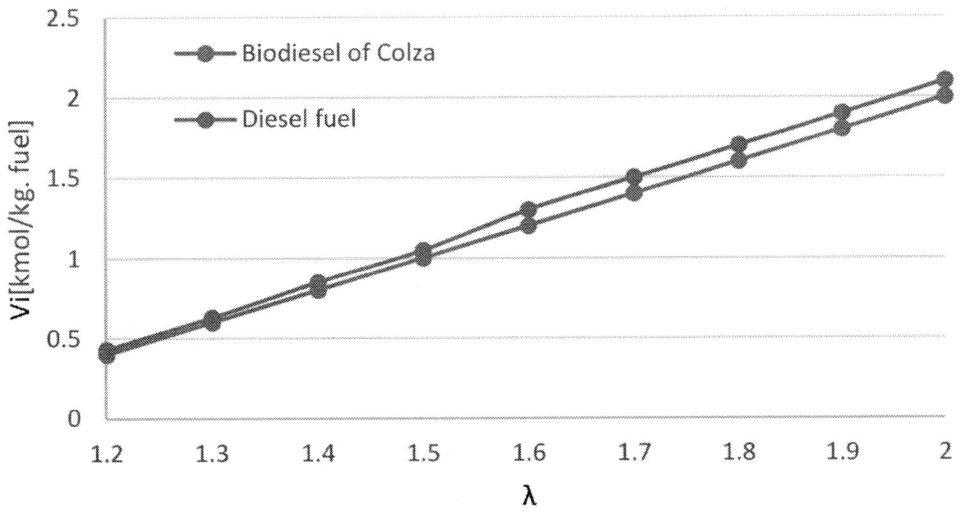

Figure 6. Comparative chart between the excess air λ and the number of kilograms of the substance in the reaction

4 Conclusions

Based on the experimental results the biodiesel spray has a narrower angle and the penetration is larger than diesel fuel. The biodiesel properties that affect the injection of fuel are fuel vaporization, higher injection pressure and worse atomization. The atomization of biodiesel is the consequence of higher values of surface tension and viscosity comparative with diesel fuel. The results are in accordance with similar rapports from various authors [6,9,12,14].Calculation of combustion indicates that less air is required for biodiesel due to higher content of oxygen.

References

[1] Aronsson, U. (2011): *Processes in optical diesel engines*, Doctoral Thesis, Lund University. Sweden.

[2]Information on http://www.combustiontechnology.co.za/training/incomplete_combustion.htm Accessed on 31/01/2016

[3] Bennett, S. (2014): *Modern diesel technology: Diesel engines*, Cengage Learning, Second edition, Stanford, USA.

[4] Baumgarten, C. (2006): *Mixture formation in internal combustion engines*, Springer. Hannover, Germany.

[5] Weber, J. (2008): *Optimization methods for the mixture formation and combustion process in diesel engines*, Cuvillier Verlag. Gottingen, Germany

[6] Hiroyasu, H. Arai, M. (1990): *Structure of fuel sprays in diesel engines*, SAE Technical Paper. ID 900475

[7] Neerinckx, M. Polfliet, T. (2015): *Experimental comparison of evaporating diesel sprays from different nozzle geometries*, Mater Thesis. Ghent University, Belgium.

[8] Martinez-Martinez, S. Sanchez, F. Bermudez, V. Riesco-Avila, J. M. (2010): *Liquid sprays characteristics in diesel engines*, Fuel injection, InTech, Daniela Siano (Ed.), China.

[9] Agarwal, A. K.Som, S. Shukla, P. C. Goyal, H. Longman, D. (2015): *In-nozzle flow and spray characteristics for mineral diesel*, Karanja and Jatropha biodiesels, Applied energy. Vol. **156**, pp. 138-148.

[10] Som, S.Longman, D. E. Ramirez, A. I. A, Aggarwal, S. K. (2010): *A comparison of injector flow and spray characteristics of biodiesel with petrodiesel*, Fuel. **89**, pp. 4014-4024.

[11] Flynn, P. F. Durrett, R. P. Hunter, G. L. Loye, A. O. Akinyemi, O. C. Dec, J. E. Westbrook, C. K. (1999): *Diesel Combustion: An integrated view combining laser diagnostics, chemical kinetics and empirical validation*, SAE Technical Papers, UCRL-JC-133512.

[12] Agarwal, A. K. Agarwal, A. Singh, A. P. (2015): *Time resolved in-situ biodiesel combustion visualization using engine endoscopy*, Measurement. **69**, 236-249

[13] Tutunea, D. (2013): *Thermal investigation of biodiesel blends derived from rapeseed oil, Journal of Thermal Analysis and Calorimetry*. **111**, pp. 869-875.

[14] Hoekman, S. K.Broch, A. Robbins, C. Ceniceros, E. Natarajan, M. (2012): *Review of biodiesel composition properties.*

https://doi.org/10.1051/e3sconf/201911201005

Determining the energy performance of a rotating machine for fluid flow

Albertino Giovani Roza[1], *Ammar* Almaslamani[1], *Elena Beatrice* Tănase[1], *Nicolae* Băran[1] and *Mihaela* Constantin [1,*]

[1] University Politehnica of Bucharest, Faculty of Mechanical Engineering and Mechatronics, Splaiul Independenței, no. 313, Bucharest, Romania

Abstract. The rotating machine, to which the paper relates, refers to the following:
- A rotating machine construction solution that can be used as a power machine (steam engine, pneumatic engine)
- The same construction solution can be used as a working machine (pump, fan, and compressor). The constructive solution and the operating principle of the rotating machine are presented, the flow rate and the machine power computation model is indicated. At the end of the paper, the machine effective efficiency is experimentally determined.

1 Introduction

The machines are aggregates for converting the energies from one to another by using a mobile body (piston, profiled rotor, and palette).

The machine subdivided into more two broad categories:

a- Force machine (motor machine)

b- Working a machine which converts mechanical energy into another form of energy (compressors, fans, pumps).

Regarding the flow parameters variation, working and force machines that are crossed by fluids, are classified as follows:

a. Hydraulic machines, which are driven or circulate liquids, where the thermal phenomena are neglected.

b. Thermal machine, which circulates gases or vapors (or are driven by these) to which thermal the phenomena that appear cannot be neglected.

The most common are thermal machines, after the principle of operation, are classified as follows [1]:

a. Machines with a reciprocating motion piston

b. Rotating machines (profiled rotors or palettes)

c. Machines (motors) with reaction

Regarding the nature of the motor fluid, thermal machines are classified as follows:

a Steam machines (piston or rotor)

b. Gas machine (internal combustion engines, gas turbines, jet engines).

 https://doi.org/10.1051/e3sconf/201911201005

2 Presentation of the rotating machine

The machine can function as follows:

- pump, blower, and fan, (working machine); the machine receives a motor torque at the shaft (9) and the discharge pressure (p_r) will be higher than the suction fluid pressure (p_a);
- As a power machine (hydraulic motor, pneumatic engine, steam or gas engine) in this case, the machine delivers a motor torque outward through the shaft (9).

The present paper presents the operation of the machine as a rotating volumetric pump with profiled rotors [1, 2].

The machine (Figure 1) consists of has two identical rotors (2, 5) with a special shape which rotates within a case (1, 4) with the same speed. The rotation of the two rotors is synchronous and is provided by a cylindrical gear consisting of two inclined gear wheels located in or out the machine.

The gear wheels provide a rotational movement so that the rotating pistons (6) of the upper rotor enters the cavities (10) of the lower rotor. The gear wheels have the same division diameter and are mounted on shafts 7 and 9.

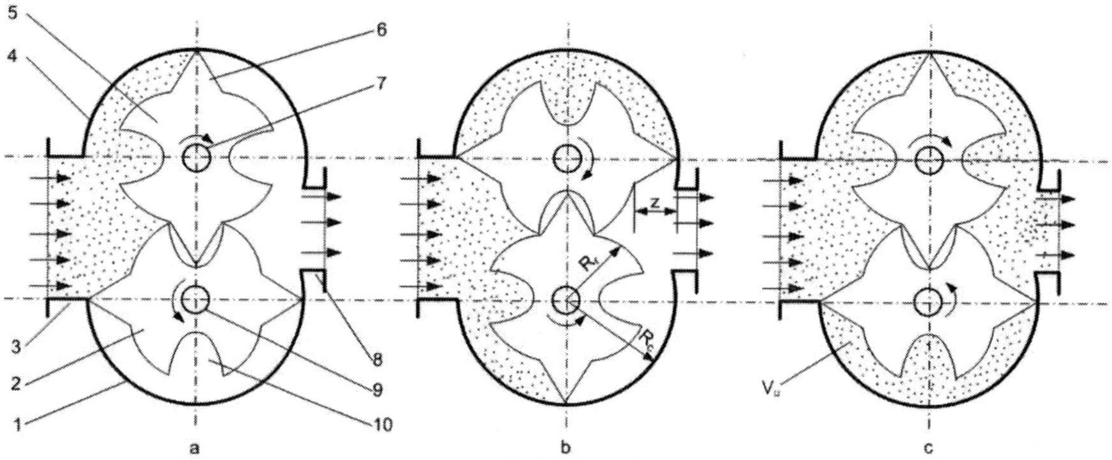

Fig. 1. The functioning principle of the rotating volumetric machine
1- bottom case; 2- bottom rotor; 3-suction chamber; 4- top case; 5- top rotor; 6- piston; 7-driven shaft; 8- drain chamber; 9-driving shaft; 10-cavity in which the top rotor piston enters
a, b, c, the rotor's position after a 90° rotation

The sucked fluid (Figure 1.a) is transported to the discharge and after a 90 ° rotation of both rotors - the position in Figure 1b and then the position in Figure 1c is reached.

After a 180 ° rotation of the fluid in the useful volume V_u (figure 1.c), i.e. in the space between the pistons, the bottom case (1) and the bottom rotor (3) will be forced to leave the case and penetrate into the drain pipe.

3 Calculation of the volumetric flow rate and the driving power of the machine

The fluid entering the suction chamber is transported to the discharge chamber regardless of its nature or composition.

This type of rotating machine belongs to the category of driving machines (not actuating) and can transport: clean liquids or suspensions, polyphase fluids (water + ash, water + fuel oil, etc.), wastewater, rheological fluids, gases, vapors etc.

The construction solution is simple, does not require special execution technologies and has a higher degree of reliability. This type of machine can be used in the following fields:

energy industry, petrochemical industry, wastewater treatment plants, mining, and agriculture.

When operating as a pump or compressor, at a full rotation of the shaft (9), two volumes [3, 4] will be transported from aspiration to discharge:

$$V_u = \left(\pi R_c^2 - \pi R_r^2\right) l \left[m^3 / rot\right] \tag{1}$$

The case radius (R_c) is the sum of the rotor radius (R_r) and the piston height (z),:

$$R_c = R_r + z \ \left[m\right] \tag{2}$$

it results:

$$V_u = l \cdot z\left(z + 2R_r\right) \ \left[m^3 / rot\right] \tag{3}$$

The volumetric flow rate of a single rotor:

$$\dot{V}_u = l \cdot z\left(z + 2R_r\right) \cdot \frac{n_r}{60} \ \left[m^3 / s\right] \tag{4}$$

The flow rate for the whole machine with two identical rotors is:

$$\dot{V}_{um} = 2\dot{V}_u = l \cdot z\left(z + 2R_r\right) \cdot \frac{n_r}{60} \ \left[m^3 / s\right] \tag{5}$$

From the above relations, it is observed that the volumetric flow rate increases linearly with the rotor radius and length (1, R_r) and with the speed (n).

The driving power is obtained (theoretical) with the relation [4]:

$$P_m = V_m \cdot \Delta p\left[W\right] \tag{6}$$

where the flow rate from the previous relations is replaced and the following is obtained:

$$P_m = lz\Delta p\left(2R_r + z\right) \cdot \frac{n_r}{30} \left[W\right] \tag{7}$$

4 The experimental installation scheme

The scheme of the experimental installation is presented in figure 2.

The working fluid (air, water, oil) is taken from the tank (2) by means of the profiled rotors (5). The machine can carry both gases (air) and liquids (water, oil).

The speed of the electric motor (6) acting on the pump is controlled by a frequency converter (7).

At the pump outlet, the fluid passes through an electromagnetic flow meter (10) and a flow control valve (11). Then the liquid is pumped to h = 2 m and finally reaches the tank (2).

The liquid layer in the tank has a height of 0.5 m, height that ensures that the pump will always be "drowned" (figure 2).

This pump transmits any fluid, but the electromagnetic flow meter determines the state: the electrical conductivity of the fluid to be measured is greater than 200 μS / m;

consequently, for the reported flows subject to the measurements, reference is made to water only.

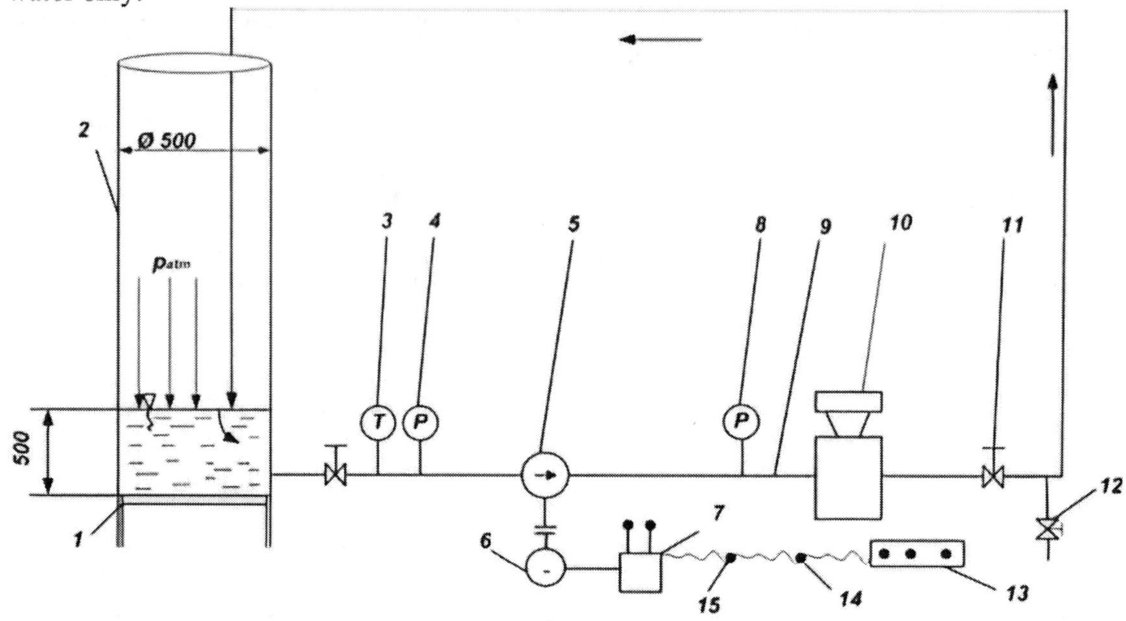

Fig. 2. The experimental installation scheme
1 - tank support; 2- water tank; 3 - thermometer; 4 - manometer; 5 - pump; 6 - electric motor; 7 - speed regulating device; 8 - manometer; 9 - duct Ø 50 x 3 mm; 10 - electromagnetic flow meter; 11 - flow control valve; 12 - fluid circuit drain valve; 13 - 380 VAC source; 14 – multimeter; 15 - ammeter.

The pipeline path is built of 50 x 3 mm transparent Plexiglas, which allows a better flow visualization. On the fluid circulation pipe, there are manometers, thermometers, an electromagnetic flow meter; the pump speed can be changed by means of a frequency inverter.

5 The purpose of experimental researches

The effective efficiency of the rotating machine is to be determined; for this purpose, the fluid flow velocity through the pipes is calculated to determine the linear and local pressure losses.

5.1 Determining the fluid flow velocity through the pipes

In the flow calculation equation, the dimensions of the built model are replaced:
- $l = 0.05$ m; $z = 0.03$ m; $R_f = 0.05$ m;
-the machine speed was chosen: 100, 150, 200 and 250 [rpm];
- the working fluid is water with a density of 1000 [kg / m^3].

The values of the water flow rate flowing through the pump are shown in Table 1; the water circuit in the installation is through a Ø 50 x 3 mm pipe so that the inside diameter is 44 mm. It is possible to calculate the water velocity in the circuit [5, 6].

$$w = \frac{\dot{V}}{\pi \frac{d^2}{4}} = 4\frac{\dot{V}}{\pi d^2}[m/s] \tag{8}$$

The computation results are presented in Table 1.

Table 1. Water flow rate and velocity values.

n_r [rot/min] / Fluid (water)	100	150	200	250
$\dot{V}\left[m^3/s\right]$	0.00204	0.00306	0.00408	0.00510
w [m/s]	1.34138	2.01207	2.68276	3.35345

5.2 Determining the power absorbed at the rotating machine coupler

After measuring the voltage (U) and the intensity of three-phase electric current (I), the electrical power absorbed by the electric motor is determined [7, 8]:

$$P_{m,e} = \sqrt{3}UI\cos\varphi[W] \tag{9}$$

The electric motor has the following characteristics:
- Electricity efficiency: $\eta_{m,e}$=0.747;
- The factor of the power: cosφ =0.71.

The power consumed by the electric motor ($P_{m,e}$) is also determined by calculation, taking into account the efficiency of the electric motor, determining the power at the coupling of the machine (P_{cm}).

$$P_{cm} = P_{m,e} \cdot \eta_{m,e}\left[W\right] \tag{10}$$

Subsequently, from P_{cm}, the power to overcome the hydrostatic load of the pump and of the linear and local hydraulic resistances on the installation circuit is removed; the remaining value must be equal to the power consumption of viscous friction (P_{tf}). The efficiency of the volumetric pump will be:

$$\eta_e = \frac{P_{\text{H}} + P_{\Delta p}}{P_{cm}} \tag{11}$$

With relations (1) and (2), based on experimental data (U, I), the results are shown in Table 2, the working fluid being water.

Table 2. $P_{m,e}$ and P_{cm} values

No	n_r [rot/min]	I [A]	$P_{m,e}$ [W]	P_{cm} [W]	U [V]
1	100	0.56	264.61	197.66	384.70
2	150	0.60	283.14	211.50	384.20
3	200	0.75	354.67	264.93	385.00
4	250	1.02	481.60	359.75	384.40

5.3 Determining the power to overcome the hydrostatic load and the hydraulic resistances inside the circuit

5.3.1 Determining the power to overcome the hydrostatic load

For the correct calculation, n is selected 200 [rot/min].

$$P_H = \dot{V} \cdot \Delta p_H \; [W] \quad (12)$$

where:
$\dot{V}$ - volumetric flow rate [m^3 / s]
Δp_H - pressure increase [N / m^2]
$\Delta p_H = \rho g H = 10^3 \cdot 9.81 \cdot 1.5 = 14715$ [N / m^2]

$$P_H = 4.08 \cdot 10^{-3} \cdot 14715 = 60 [W] \quad (13)$$

5.3.2 Determining the power consumed to overcome local hydraulic resistances inside the circuit [9-11]

Determining of linear pressure losses is made when n_r=200 [rot/min].
The following are known:
- fluid flow rate inside a Ø 50 x 3 mm pipe: $\dot{V} = 4.08 \cdot 10^{-3} [m^3 / s]$
- water velocity inside the pipe: $w = 2.68 [m/s]$;
- length of the path: l = 6 m.

The power to overcome linear hydraulic resistances:

$$P_{lin} = \dot{V} \cdot \Delta p = 4.08 \cdot 10^{-3} \cdot 10283.89 = 42 [W] \quad (14)$$

The power consumed to overcome local hydraulic resistances:

$$P_{loc} = \dot{V} \cdot \Delta p = 4.08 \cdot 10^{-3} \cdot 23701.92 = 96.7 [W] \quad (15)$$

So:

$$P_{\Delta p} = P_{lin} + P_{loc} = 42 + 96.7 = 138.7 [W] \quad (16)$$

6 Determining the effective efficiency of the rotating working machine

Relation (11) is resumed:

$$\eta_e = \frac{P_H + P_{\Delta p}}{P_{cm}} = \frac{60 + 138.7}{264.93} = 0.75 \quad (17)$$

0.75 is a good value compared to other literature results [11, 12].
The energy balance of the installation is as follows:

$$P_{cm} = 264.93W \begin{cases} \rightarrow P_H = 60W \\ \rightarrow P_{\Delta p} = 138.7W \\ \rightarrow P_f = 66.23W \end{cases} \quad (18)$$

P_f is the power absorbed by the viscous friction between the case and the rotors, calculated as a difference:

$$P_f = P_{cm} - \left(P_H + P_{\Delta p}\right)[W] \quad (19)$$

Similarly, the calculations are made for n_r: 150, 200, 250 [rot/min] and presented in Table 3 results.

Table 3. Values of η_e for different machine speeds

n_r [rot/min]	P_{cm} [W]	$\Delta P_{\Delta pt}$ [W]	P_H [W]	P_t [W]	η_e
100	197.66	17.57	30.01	47.58	0.24
150	211.50	57.11	45.02	102.53	0.48
200	264.93	138.7	60.03	198.74	0.75
250	359.75	255.2	75.04	258	0.719

Considering the data in Table 3, the function $\eta_e = f(n_r)$ was plotted.

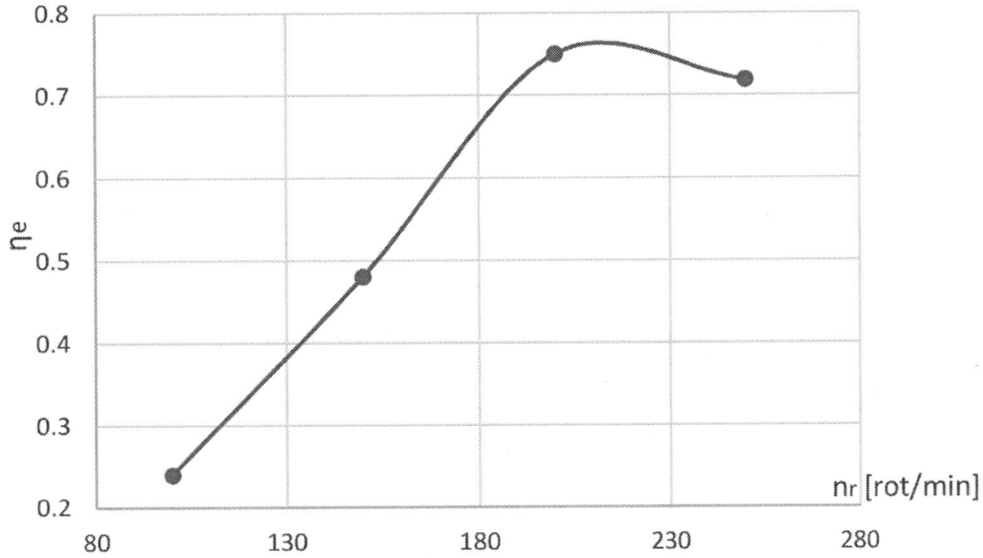

Fig. 3. The graph of the function $\eta_e = f(n_r)$

From Figure 3 it is confirmed that with the speed increase, which is the ω increase, the machine efficiency will increase to a certain value and then begin to decrease; this is because increasing the speed increases the flow rate, thus increasing pressure loss on the hydraulic circuit.

The values obtained for η_e are satisfactory with the data from the literature [12-15].

7 Conclusions

- Rotating working machines have the advantage of transforming the moment received to the engine shaft into potential pressure with minimal energy loss;
- The presented solution can convey clean or suspended liquids as well as rheological fluids or gases;
- The effective machine efficiency is influenced by the fluid nature, machine speed, and fluid viscosity.

 https://doi.org/10.1051/e3sconf/201911201005

References

1. N. Băran, I. Căluşaru, A. Detzortzis, JMSE A 2, *Research Regarding the Testing of a New Type of Rotating Machine with Profiled Rotors*, **3**, 372-376, (2012).
2. A. Motorga, Ph.D. Thesis, *Influenţa parametrilor constructivi şi funcţionali asupra performanţelor maşinilor rotative cu rotoare profilate*, University Politehnica of Bucharest, (2011).
3. N. Băran, *Maşini termice rotative -Maşini de forţă, Maşini de lucru*, Editura MATRIX ROM, Bucureşti, (2001).
4. N. Băran, *Maşini termice rotative de lucru*, Editura Matrix Rom, Bucureşti, 2003.
5. Al. Dobrovicescu, N. Băran, ş.a., *Elemente de termodinamică tehnică*, Editura POLITEHNICA-PRESS, Bucureşti, (2009).
6. C. Isbăşoiu, *Tratat de mecanica fluidelor*, Editura AGIR, Bucureşti, (2011).
7. M. Preda, P. Cristea, *Bazele Electrotehnicii-Circuite electrice*, **2**, Editura Didactică şi Pedagogică, Bucureşti, (1980).
8. C-tin Ghiţă, *Maşini şi acţionări electrice*, Editura I.C.P.E., Bucureşti, (1997).
9. G. Bar-Meir, *Basics of fluid mechanics*, Chicago, (2011).
10. P.G. Kiselev, *Îndrumar pentru calcule hidraulice*, Editura Tehnică, Bucureşti, (1998).
11. I.E. Idelcik, *Îndrumar pentru calculul rezistenţelor hidraulice*, Editura Tehnică, Bucureşti, (1994).
12. M. Exarhu, *Maşini şi instalaţii hidraulice şi pneumatice*, Editura ANDOR TIP S.R.L., Bucureşti, (2006).
13. A. Beazit, *Maşini hidropneumatice navale*, Editura Academiei Navale Mircea cel Bătrân, Constanţa, (2010).
14. Y. H. Kang, H.H. Vu, Journal of Mech. Sci. and Tech., *A newly developed rotor profile for lobe pumps: Generation and numerical performance assessment*, Taiwan, Marc, **28**, Issue 3, 915-926, (2014).
15. V. Burchiu, L. Gheorghiu, A. Dudau, *Ghidul utilizatorului de pompe*, Editura AtlasPress, Bucureşti (2006).

E3S Web of Conferences **112**, 01006 (2019) https://doi.org/10.1051/e3sconf/201911201006
TE-RE-RD 2019

Experimental research on a hot air generator working on the TLUD principle

Ioan Pavel[1,*], *Radu-Iulian* Rădoi[1], *Gabriela* Matache[1], *Ana-Maria Carla* Popescu[1], and *Ioan* Caba[2]

[1]Hydraulics and Pneumatics Research Institute INOE 2000-IHP, 040558, Cutitul de Argint 14, Bucharest, Romania
[2]National Institute of Research - Development for Machines and Installations Designed to Agriculture and Food Industry - INMA, 71592, 6 Ion Ionescu de la Brad, Blv, Bucharest, Romania

Abstract. The current paper presents the processes of data acquisition and air temperature monitoring in the key points of a hot air generator working on the TLUD principle, which uses temperature sensors, a data acquisition board and an application developed in LabView that numerically and graphically displays temperature variation during burning. The results of the monitoring were obtained on an experimental model, developed by research scientists at INOE 2000-IHP. The data acquired is then used to establish the algorithm for opening the air supply valves for gasification and combustion air, in order to optimize combustion by maintaining the optimum temperature at the funnel.

1 Introduction

The use of renewable energies is a characteristic of contemporary society, which is facing the decrease of classical resources, the rise in fuel prices and the pollution generated by their burning. However, the use of renewable resources to provide energy needs is still far from its full potential; this is also because of the lack of efficient conversion tools able to exploit various resources in mono-source or combined energy systems [1].

In the process of gasification, the inputs are biomass and air, while the outputs are fuel gas, biochar and ash, with a negative balance of carbon emissions due to carbon sequestration by biochar. The TLUD gasification principle (Figure 1) occurs when the biomass layer is introduced into a reactor and rests on a grid through which the air flow for gasification circulates from bottom to top.

Priming gasification process is done by igniting the top layer of the biomass in the reactor. The combustion front descends continuously by consuming the biomass in the reactor. Due to the heat radiated from the oxidation front the biomass is heated, dried, and then it enters a fast pyrolysis process that releases volatiles and there remains unconverted carbon.

* Corresponding author: pavel.ihp@fluidas.ro

 https://doi.org/10.1051/e3sconf/201911201006

By the time the combustion front reaches the grill all the volatiles in the biomass were gasified and some of the fixed carbon was reduced; about 10 - 20% of the initial mass remain on the grill, in the form of sterile "green charcoal", called *biochar*.
The proportion of biochar, in which most of the ash from biomass is incorporated, depends both on the carbon stored in the biomass and on the temperature maintained in the carbon reduction reaction; a high temperature ensures the reduction of more carbon. If the supply with gasification air continues, we switch to an updraft gasification process in which a layer of incandescent charcoal is kept on the grill; of this, there results mainly CO and little CO2, which passing through the hot charcoal layer enters the reduction reaction $C+CO_2 \Rightarrow 2CO$. This second phase is called charcoal gasification.

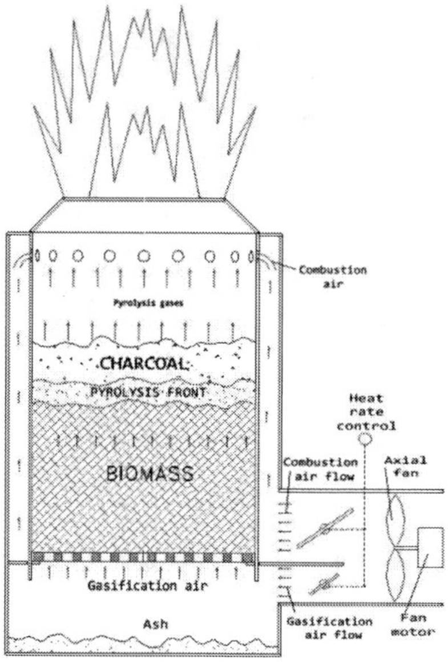

Fig. 1. The TLUD gasification principle [2].

Compared to direct combustion or gasification processes of wood and pellets, the TLUD gasification process is characterized by very low values of the superficial gas velocity through the oxidation front, resulting in a very low content of atmospheric particulate matter (PM), PM < 5 mg/MJ, well below the required standard in the EU since 2015 for biomass combustion processes, which is 25 mg/MJ.
The calculation algorithm of thermal power is [1]:
Reactor diameter:

$$D_r = 0.35\ m \tag{1}$$

Biomass layer height (batch height):

$$H_{rbm} = 0.6\ m \tag{2}$$

Reactor section:

$$S_r = \frac{\pi \cdot D_r^2}{4} = \frac{\pi \cdot 0.35^2}{4} = \frac{0.3847}{4} = 0.0961\ m^2 \tag{3}$$

Biomass volume in the reactor:

$$V_{rbm} = H_{rbm} \cdot S_r = 0.6 \cdot 0.0961 = 0.0577\ m^3 \tag{4}$$

Biomass layer density: 600 kg/m^3 (pellets).
Initial mass in the reactor:

$$M_{bm0} = 600 \cdot 0.05766 = 34.6\ kg \tag{5}$$

The specific hourly consumption of gasified biomass is 85 kg/m^2 · h; therefore, for the surface of the reactor we will have a specific consumption of:

$$C_{hbmg} = 85 \cdot 0.0961 = 8.17\ kg/h \tag{6}$$

The running time will be:

$$t_g = \frac{34.6}{8.17} = 4.23\ h \tag{7}$$

Energy from gasified biomass:

$$E_{gbm} = M_{bm0} \cdot P_{Cbm} = 34.6 \cdot 17 = 588.2\ MJ \tag{8}$$

Thermal power of hot gases:

$$P_g = \frac{E_{gbm}}{t_g \cdot 3.6} \cdot \eta_{gTLUD} = \frac{588.2 \cdot 0.93}{4.23 \cdot 3.6} = 35.922\ kWth \tag{9}$$

and burner thermal power, which takes into account the efficiency of combustion (*comb*) of the combustible gas ($\eta_{comb} = 0.95$) and the insulation (*insul*) efficiency (losses outward – $\eta_{insul} = 0.96$), will be:

$$P_{comb} = P_g \cdot \eta_{comb} \cdot \eta_{insul} = 35.922 \cdot 0.95 \cdot 0.96 = 32.76\ kWth \tag{10}$$

The gasification process is done with a reduced intensity with a specific hourly consumption of biomass of 80 – 150 kg/m^{2}h, which leads to reduced specific powers of 250 – 350 kW/m^2 of the reactor. The slow process maintains the superficial speed of the generator gas produced at very low values, $v_{sup} \leq 0.06$ m/s, which ensures reduced traction of free ash, and also at concentrations of $PM_{2.5}$ of maximum 5 mg/MJ_{bm} when leaving the burner; this value is at least five times lower than current standards required for solid fuel heat generators. [3, 4, 5, 6, 7]

The stages of the gasification process take place simultaneously in different areas of the reactor. These stages are: drying, pyrolysis, oxidation and reduction.

Drying is necessary because the moisture content of the biomass is variable, ranging from 5 to 55%. At temperatures above 100° C, the water is removed and turned into steam. During the drying process, the biomass does not suffer any decomposition.

Pyrolysis takes place in the temperature range of 150 - 700° C, and it consists of thermal decomposition of biomass in the absence of oxygen.

Oxidation takes place by the aid of the air introduced into the oxidation zone. The air contains, together with oxygen, water vapor, inert gases, nitrogen and argon that do not react with the biomass components. Oxidation takes place at 700-2000° C.

Reduction occurs in the reduction zone of the reactor. Here several chemical reactions take place at a temperature of 800 - 1000° C and in the absence of oxygen.

The generator gas is a mixture of combustible and non-combustible gases. The combustible gases are: carbon monoxide (15 - 30%), hydrogen (10 - 20%), methane (2 - 4%). The non-combustible gases are: nitrogen (45 - 60%), water vapors (6 - 8%), carbon dioxide (5 - 15%) [8].

2 The objective of testing

Until now, the TLUD gasification principle has been applied mostly to small and very small food preparation equipment. The prototype of hot air generator on the TLUD principle under research is a 32-kW generator, so it is necessary to test the principle for such a size, in different operating modes.

Testing the prototype of hot air generator on the TLUD principle aims to acquire temperature data from key points, in two variants, that is with and without getting biochar.

Temperature monitoring is done for the following areas (Figure 2):

- Environment
- Combustion air inlet and gasification air inlet
- Hot air outlet
- Funnel.

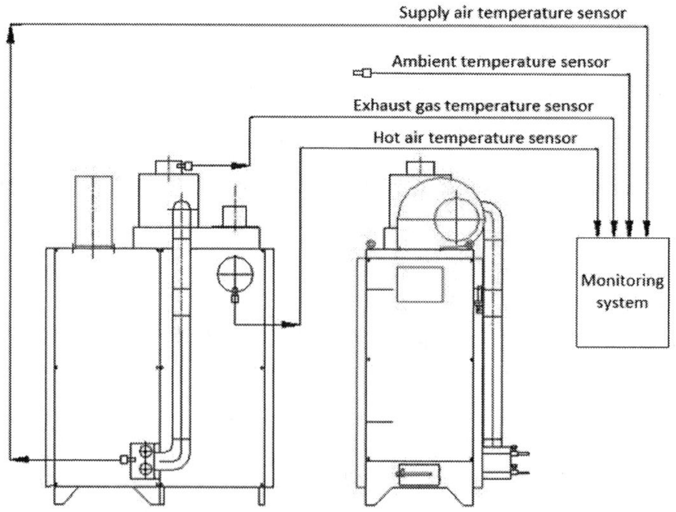

Fig. 2. Positioning of temperature sensors.

The acquired data is then used to establish the opening algorithm for gasification air supply valves and combustion air supply valves, in order to optimize combustion by keeping the optimum temperature at the funnel.

The generator consists of a reactor (7) in which the gasification of fuel from biomass (in this case pellets) occurs, a burner (6) and a heat exchanger (1). It is equipped with a fan for forced draft along the funnel (3) and another one for the circulation of hot air through the heat exchanger (5). The gasification air is controlled at the inlet (9), and the combustion air - at the inlet (8). The hot air leaves the equipment through the outlet (2), and the smoke - through the funnel (4).

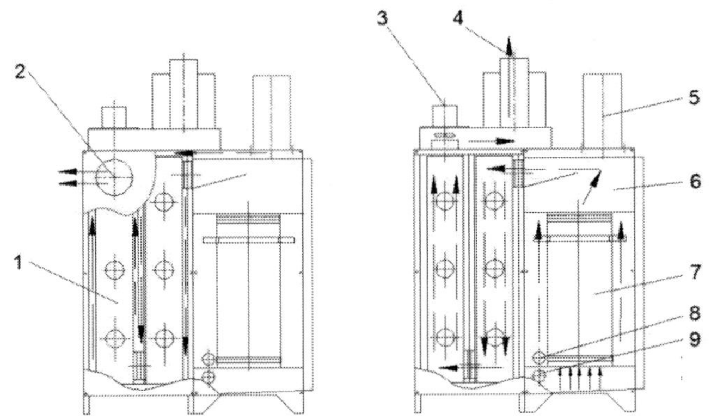

Fig. 3. Hot air generator working on the TLUD principle.

The correct operation of the hot air generator implies keeping a temperature of approximately 200° C at the funnel to avoid condensation and tar deposition. It is important to adjust the air at the outlets 8 and 9 in order to meet this requirement. If the temperature is lower, tar is deposited, and if it is higher, unnecessary energy is lost to the atmosphere.

Opening of the valves for combustion air and gasification air intake has been made by measuring the opening slit in such a way that the ratio of air set to the valves is Qa / Qg = 2.62, where Qa is the combustion air flow rate, and Qg is the gasification air flow rate. Specialized literature [9], [10] indicates that the flow rate ratio must be in the range 2.44 ... 3.3 to get the best results. From the same sources it is known that the superficial velocities of the synthesis gas are from 0.03 m/s to 0.04 m/s (PM < 5 mg/MJ), far below the critical level of 0.1 m/s where the specific $PM_{2.5}$ concentration increases above 2.4 mg/MJ. Upon leaving the reactor, the flue gas has a flow rate of 0.007 m^3/s and an average temperature of 695° C [11].

Testing consists of data acquisition and monitoring of air temperatures at key points for the two operating modes: minimum and maximum power. Power control is done by controlling the opening of the combustion air and gasification air intake valves.

3 Data acquisition

Recording the temperature evolution at various points of the hot air generator is done by means of Pt1000 temperature probes. They are connected to a data acquisition board via signal amplifiers of 4 ... 20 mA output, and voltage conversion for the analog input of the data acquisition board is achieved by means of resistors (Figure 4).

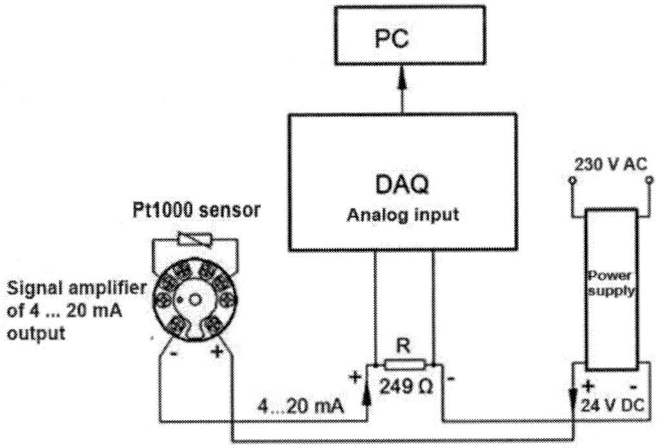

Fig. 4. Connection diagram.

A 24 V DC power supply was used to power the acquisition board and the amplifier [12]. The technical features of the signal amplifier code 2000.35.010 transducer RTD-4...20 mA are according to the technical sheet [13].

The signal from the temperature sensor is amplified in a 4 ... 20 mA output signal. The data acquisition board (DAQ M Series NI USB-621*x* [14]) allows inputs of 0-10 V, and the amplified signal of 4-20 mA was converted into voltage by means of a resistor of 249 Ω, as calculated:

$$20mA = 0.02A \tag{11}$$

$$0.02A \times 249\Omega = 5V \tag{12}$$

(the middle of the permissible voltage range of the acquisition board, which is 0...10 V)

An application in LabView (Figure 5) was created to display and record data. The application displays the temperature variation during the burning process numerically and graphically. When the recording stops, the application allows the user to store the temperature versus time values in a database. These data may be further processed.

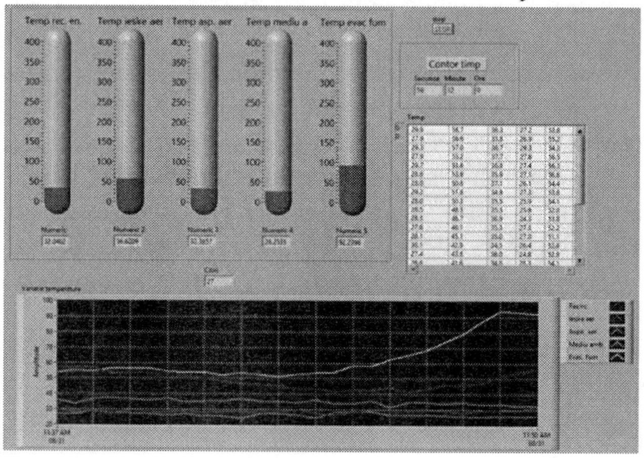

Fig. 5. Graphical user interface of the data acquisition software application.

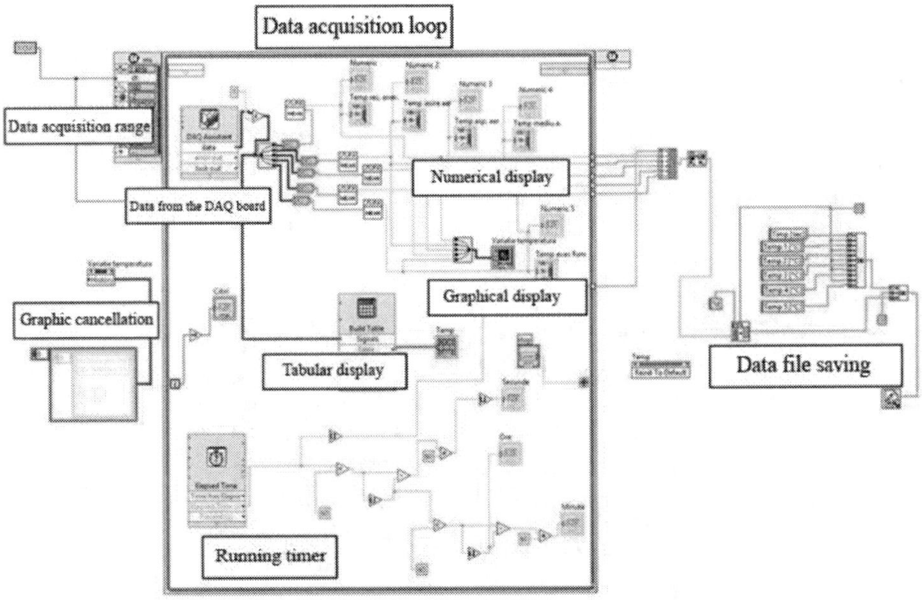

Fig. 6. Functional diagram of the application developed in LabView.

The application developed in LabView comprises the following functional blocks (Figure 6): a data acquisition loop with the functionality to set the range of data acquisition, an input block for data from the acquisition board in which there are also performed *temperature - output signal* scaling, numerical display, graphical display, running timer (seconds, minutes, hours), data recording in text file, cancellation of the graphic (when starting a new test), and a block for data display in tabular format.

4 Actual testing with data acquisition

The contents of a 15 kg bag of pellets were introduced into the reactor. Initiation of the combustion process was done at the top of the material in the reactor, by using firelighters, and from ignition to steady gasification operation mode (Figure 7) it took about 12 minutes according to the data acquisition.

Fig. 7. The steady gasification operation mode

4.1 Test no. 1- Data acquisition in the variant of gasification with biochar production under minimal operating conditions

 https://doi.org/10.1051/e3sconf/201911201006

After entering the gasification operation mode, the flame has moved from the combustion material in the reactor to the burner (Figure 8.a). The gas produced in the reactor mixed with the combustion air, and a flame similar to a stove one was generated. Successive adjustment of the valves for gasification air and combustion air intake has been done (Figure 8.b) keeping the ratio Qa / Qg=2.62, and it was found that the power of the hot air generator was adjustable. When opening the valves one can see on the graphical interface, through the acquisition system (Figure 8.c), that the flame increases almost instantaneously. The correct operation of the TLUD (Top-Lit UpDraft) gasification principle has thus been demonstrated.

(a) (b) (c)

Fig. 8. (a) Adjustment of air intake valves, (b) Steady gasification operation mode, (c) Data acquisition system.

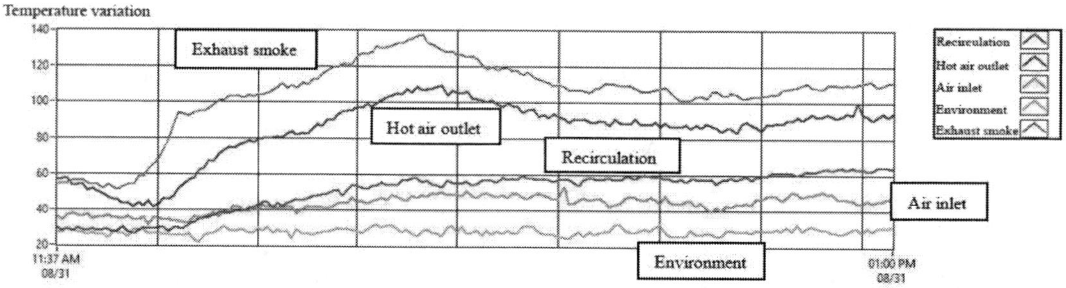

The data acquisition resulted in the following graph and related database.

Fig. 9. The data acquisition graph for the test with obtaining of biochar.

A good adjustment of the power of the hot air generator has been obtained by setting at minimum the combustion air and the gasification air. The temperature of the heating air at the output of the generator, for the case of minimum operating conditions, registered values of approx. 110° C, and on the funnel, a temperature of approx. 140° C was obtained. The air intake valves were set to approximately 1-2 mm opening.

The combustion process (Figure 9) took 83 minutes for a 15 kg bag of pellets under minimum power operating conditions (the intake valves - almost closed), and it was interrupted when the flame became blue (a sign that gasification of the material was over and burning of the biochar was about to start). Thus, about 25% of the volume of the material initially introduced for combustion was obtained as biochar.

The biochar obtained can be seen in Figure 10.

Fig. 10. Biochar obtained following gasification.

4.2 Test no. 2 – Variant of gasification with burning of the biochar

The contents of a 15 kg bag of pellets were introduced into the reactor. Initiation of the combustion process was done by using firelighters, and from ignition to steady gasification operation mode) it took about 12 minutes according to the data acquisition.

Under maximum power test conditions, that is when combustion air and gasification air intake valves open to more than 10 mm, a heating air temperature of approx. 200° C was obtained, and a temperature of approx. 250° C was obtained on the funnel, according to the data acquisition graph (Figure 11).

In Test no. 2 the combustion process has not been interrupted any more to obtain biochar (when the flame became blue), and thus an operating time approximately equal to that of Test no. 1 was obtained, this time at maximum power, and the amount of energy obtained was higher by about 25% compared to the variant when the reactor was shutdown in order to produce biochar. The final result of the total combustion (including the biochar) was a very small amount of ash of about 50 g resulting from the combustion of a 15 kg bag of pellets.

When the pyrolytic front was close to the bottom of the reactor and the thickness of the biochar layer decreased, while maintaining the gasification air flow and the minimum resistance created by the material, the combustion process accelerated for a short time (until all the material including the biochar was consumed by combustion), and after that, although the combustion process became less intense, the generator still provided hot air for about 20 minutes due to thermal inertia.

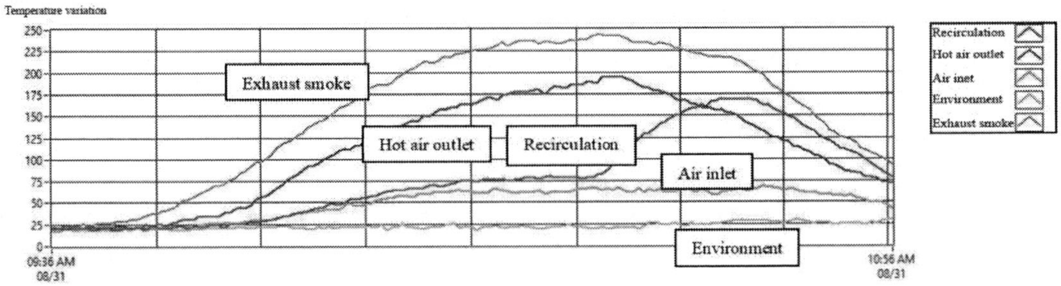

Fig. 11. The data acquisition graph for the test with burning of the biochar.

During the tests, emission of pollutants was measured and no emissions of smoke or unburnt gases were found beyond legal limit.

5 Conclusions

The experimental model of 32 kW- hot air generator has been tested in the variant of minimum operating power with biochar production and in the variant of maximum operating power with gasification of the biochar, demonstrating the correct functioning of the TLUD principle and adjustability of the power of the generator.

The acquired data can be used in determining an algorithm for automatic opening of combustion air and gasification air intake valves, thus achieving system automation and correct operation of the TLUD principle. Another version can be with fixed adjustment of the air intake valves, keeping the Qa / Qg=2.62 ratio, for an average value of the power of the generator, provided that the fan flow rate for the forced draft on the funnel and the flow rate for the circulation of hot air through the heat exchanger can be varied. In this case, electronic air flow control is required in order to keep the temperature at the funnel around 200° C.

Based on this research and the acquired data a simulation model can be developed to optimize the automation of combustion air and gasification air supply, so that the advantages of the TLUD principle (carbon emission negative balance, or obtaining of biochar) get optimal.

This paper has been developed in INOE 2000-IHP, as part of a project co-financed by the European Union through the European Regional Development Fund, under Competitiveness Operational Programme 2014-2020, Priority Axis 1: Research, technological development and innovation (RD&I) to support economic competitiveness and business development, Action 1.2.3 – Partnerships for knowledge transfer, project title: *Eco-innovative technologies for recovery of biomass wastes*, project acronym: *ECOVALDES*, SMIS code: 105693, Financial agreement no. 129/23.09.2016.

References

1. L. Dumitrescu, C. Cristescu, V. Barbu, M. Mateescu, *Proc. of International Symposium ISB-INMA'TEH*, pp. 987-992 (2018)
2. G. Matache, I. Pavel, A. Pantiru, M. Ciciu, Hidraulica, 2, pp. 70-75 (2016) http://www.hidraulica.fluidas.ro/2016/nr2/70-75.pdf
3. A. Belonio, *Dual- reactor rice husk gasifier for 6-ton capacity recirculating-type paddy dryer* (Iloilo City, Philippines : Central Philippine University) https://stoves.bioenergylists.org/stovesdoc/Belonio/paddygas/Dual_Reactor_RH_Gasifier.pdf
4. H. Mukunda, S. Dasappa, P. J. Paul, N. K. S. Rajan, M. Yagnaraman, D. R. Kumar, M. Deogaonkar, Current Science **98**, 5, pp. 627-638 (2010)
5. E. Murad, A. Culamet, G. Zamfiroiu, *Proc. of International Symposium HERVEX 2011*, pp. 37-43 (2011)
6. J. Porteiro, D. Patino, J. Collazo, E. Granada, J. Moran, J.L. Miguez, Fuel **89**, 1, pp. 26-35 (2010)
7. S. Varunkunar, *Packed bed gasification-combustion in biomass domestic stove and combustion systems* (PhD Thesis, Bangalore, India: Indian Institute of Science, Department of Aerospace Engineering, 2012)

8. G. Matache, I. Pavel, A. Pantiru, M. Ciciu, *Proc. of 5th International Conference of Thermal Equipment, Renewable Energies and Rural Development TE-RE-RD*, pp. 277-282 (2016)
9. F. Sundblad, *An improved cooking stove for the urban and peri-urban areas in Zambia* (Master of Science Thesis in Industrial Design Engineering, Gothenburg, Sweden: Chalmers University of Technology, 2014)
10. D. Supramono, F. Inayati, *Proceeding of the 13th International Conference on QIR (Quality in Research)*, pp. 393-398 (2013)
11. E. Maican, I.C. Duţu, G. Matache, C. Dumitrescu, I. Pavel, INMATEH, **53**, 3, pp. 5-12 (2017) http://www.inmateh.eu/INMATEH_3_2017/53-01%20Maican%20E.pdf
12. http://www.farnell.com/datasheets/43587.pdf?_ga=2.205171536.1703443844.1508838846-1396001712.1508838846
13. http://www.adelaida.ro/convertor-rtd-to-4-20ma-2300.10.176-pixsys.html
14. https://www.ni.com/pdf/manuals/371931f.pdf

Experimental study of heat pump type air-water for heating system performance

Penka Zlateva[1,*], *Krastin* Yordanov[1]

[1] Department of Thermal Engineering, Shipbuilding Faculty, Technical University of Varna, 1 Studentska Str., 9010 Varna, Bulgaria

Abstract. The operation of an air-to-water heat pump system providing space heating production for an administrative building in Varna has been explored. For this purpose, discontinuous operational mode of the heat pump system during wintertime is considered. Study has been performed by use of real values for external ambient temperature and operating characteristics of the heat pump system. The seasonal efficiency of the heat pump operation has been calculated. Also, the minimal external ambient temperature until the heat pump may work in heating regime and operation duration at outdoor air temperatures higher than the design air temperature can be determined. The economic analysis of the heat pump system exploitation has been made finally.

1 Introduction

Demand for environmental solutions for building heating is growing up. It is caused by both global warming and uncertainty about electrical energy prices. Heat pump systems use electricity for their work, and some of this electricity comes from renewable sources [1-2]. This makes them both effective for well-known environmental issues related to direct and indirect emissions of carbon dioxide in the atmosphere, and efficient for heating and cooling. Moreover, they satisfy both consumers' expectations of both quality and cost effectiveness [3-4]. The ever-increasing energy consumption of buildings has caused a great interest in improving the seasonal characteristics of heat pumps [5-6]. Also, the temperature of the outside air, which has a great impact on the heating characteristics of the heat pump, is also essential. Northeastern Bulgaria with its Black Sea climate, gives the possibility of maximum efficiency of the heat pump systems, in particular air-water. As known, the use of outside air is to heat the water through the heat pump to which the heaters are powered. An important point in the investigation of energy efficiency buildings is the precise operation of the heat source and the determination of its operating efficiency for the entire heating season. In the present work calculations are made, which aim to determine the efficiency of the air-to-water heat pump heating system.

* Corresponding author: pzlateva@tu-varna.bg

2 Presentation

This paper presents the results of a methodology [7], which determines the operational seasonal efficiency of an air-water heat pump system used for heating an administrative building located in Varna. During the winter heating season, the building is heated with the Wesper (Systemair) model VLH 804 LN with discontinuous operation. The operating schedule of the air-water heating system for heating is from Monday to Friday from 6 a.m. to 6 p.m. An important point for the accuracy of the calculations is to obtain accurate information about the change in external ambient temperature over time. For the external ambient temperature in the survey used data measured in 5 minutes and obtained from a meteorological station of the National Institute of Meteorology and Hydrology-Varna for the heating season from 29.10.2017 to 19.04.2018, fig.1 [8].

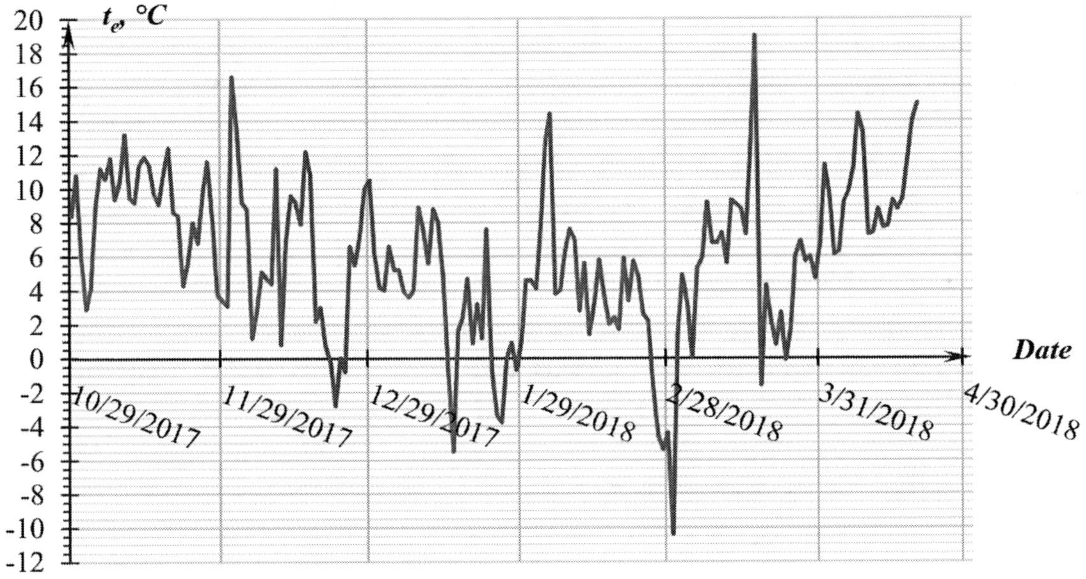

Fig. 1. Graph of external ambient temperature for the period from 29.10.2017 to 19.04.2018

Also, from fig. 1 for the heating season (from 29.10.2017 to 19.04.2018) the number of hours during the heating season in which the external ambient temperature is in the i-th interval [7] shown in Table 1 by months is calculated.

Table 1. Calculated counts of hours by months during the heating season and their total sum for the i-th interval

t_e °C	τ, *h*							τ_i *h*
	29.10.2017	*11.2017*	*12.2017*	*01.2018*	*02.2018*	*03.2018*	*19.04.2018*	
-5	0,0	0,00	0,00	0,83	13,67	0,50	0,00	**15,00**
…	…	…	…	…	…	…	…	…
0	0,0	0,00	3,17	17,50	2,00	29,67	0,00	**52,33**
1	0,0	0,0	10,8	14,8	2,7	7,7	0,0	**36,00**
…	…	…	…	…	…	…	…	…
15	0,0	13,0	8,7	1,0	1,8	7,3	15,5	**47,33**
…	…	…	…	…	…	…	…	…
19	0,0	1,2	0,0	0,0	0,0	3,0	3,5	**7,67**
							$\Sigma\tau_i$=	**1475,33**

In fig. 2 is the graphical dependence of the external ambient temperature (t_e) on the number of hours during the heating season (τ_i) and the number of hours during which the heat pump operates ($\tau_{w,\ i\text{-}HP}$).

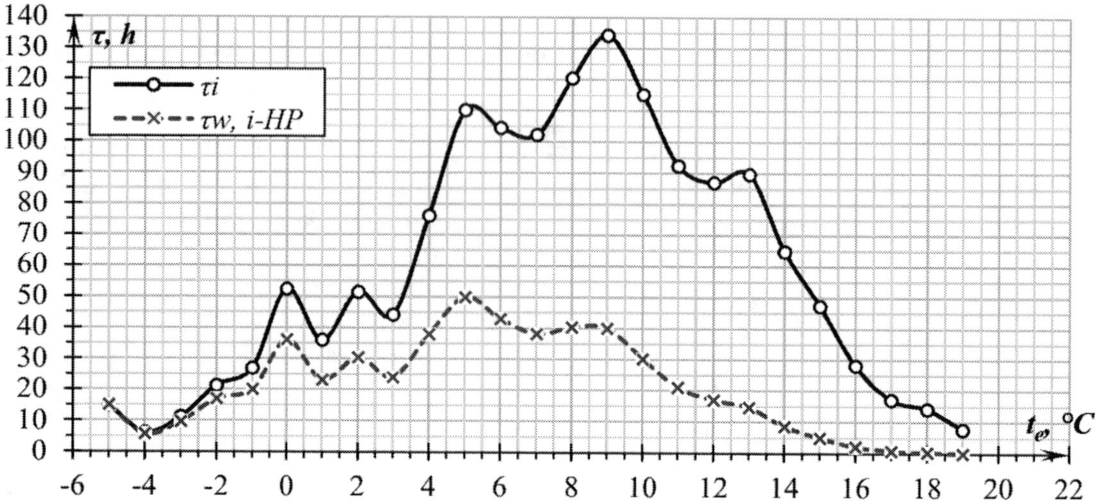

Fig. 2. Graphical dependence of external ambient temperature (t_e) on the number of hours during the heating season (τ_i) and the number of hours the heat pump operates ($\tau_{w,\ i\text{-}HP}$)

3 Methodology for calculation of electrical energy consumed by the heat pump system and seasonal work efficiency

The methodology for calculating the electrical energy consumption of the heat pump system and the seasonal efficiency of the operation includes the following steps:

1) Price per 1 kWh of electrical energy consumed $E_p = 0{,}117\ €/kWh$ [9-10];
2) Calculation of heat quantity $Q_{B,i}$ produced for the i-th temperature range, *kWh*

$$Q_{B,i} = \Phi_{B,i} \cdot \tau_i, kWh \tag{1}$$

where: $\Phi_{B,i}$- the required heat capacity of the building for the i-th temperature range, kW;

3) Calculation of $\tau_{w,i-HP}$ - the number of hours the heat pump works to produce the required heat for the building, *h* [1];

$$\tau_{w,i-HP} = \frac{Q_{B,i}}{Q_{HP}}, h \tag{2}$$

4) Calculation of $E_{C,i-HP}$ - consumed electrical energy from compressors of the heat pump system for the i-th temperature range, *kWh*;

$$E_{C,i-HP} = W_c \cdot \tau_{w,i-HP}, kWh \tag{3}$$

5) Calculation of $E_{F,i-HP}$ - consumed electrical energy from fans of the heat pump for the i-th temperature range, *kWh*;

$$E_{F,i-HP} = W_F \cdot \tau_{w,i-HP}, kWh \tag{4}$$

6) Calculation of $E_{B,i-HP}$- electrical energy consumption of the building by compressors and fans of the thermo-pump system for the i-th temperature range, *kWh*;

$$E_{B,i-HP} = E_{C,i-HP} + E_{F,i-HP}, kWh \tag{5}$$

7) The seasonal coefficient of performance is presented in two temperature ranges:
   - from -5°C to +15°C

$$COP_{-5°C+15°C} = \frac{\sum_{i=-5°C}^{15°C} Q_{B,i}}{\sum_{i=-5°C}^{15°C} E_{B,i-HP}} \tag{6}$$

   - from -5°C to +19°C

$$COP_{-5°C+19°C} = \frac{\sum_{i=-5°C}^{19°C} Q_{B,i}}{\sum_{i=-5°C}^{19°C} E_{B,i-HP}} \tag{7}$$

8) The calculated value for the consumed electrical energy for the i-th temperature interval is presented in the equation:

$$PriceEE_{HP} = E_{B,i-HP} \cdot E_p, € \tag{8}$$

From the catalog of Wesper (Sistemair) model VLH 804 LN [11] has taken the data of the heating capacity (Q_{HP}) and the electrical power of the compressors (P_C) as a function of the external ambient temperature (t_e) and the temperature of supply water to the building (t_{SW}).

At external ambient temperature of 0°C and above, the heating water produced by the heat pump is at a temperature of +50°C. When the external ambient temperature is less than 0°C and higher than -5°C, the heat pump produces water for heating at +45°C.

The electrical power of the compressors (P_C) and the fans (P_F) is equal to the work performed W_c and W_F or:

$P_F = W_F = 3ul\ x\ 1{,}25kW = 3{,}75kW$
$P_C = W_c$ - Table 2

Table 2 shows the flow heat capacity (Q_{HP}) and the operation of compressors (W_C) at external ambient temperature (t_e) ranging from -5°C to 19°C on the Wesper VLH 804 LN heat pump and the supply water temperature to building (t_{SW}) - + 45°C and + 50°C. Yellow values are given after heat capacity interpolation (Q_{HP}) and the operation of compressors (W_C) depending on the external ambient temperature (t_e) - Fig. 2 and Fig. 3.

Table 2. Flow heating capacity (Q_{HP}) and the operation of the compressors (W_C) when the external ambient temperature (t_e) in the range from -5°C to 19°C in the heat pump Wesper VLH 804 LN and temperature of supply water to the building (t_{SW}) - + 45°C and + 50°C

t_e °C	t_{SW} °C	Q_{HP} kW	W_C kW	t_e °C	t_{SW} °C	Q_{HP} kW	W_C kW
-5	45	**159,1**	**65,20**	**8**	50	**218,6**	**71,33**
-4		164,2	63,38	9		223,2	71,47
-3		**169,2**	**62,20**	**10**		**227,9**	**71,60**
-2		174,0	62,00	11		233,0	71,76
-1		178,7	62,22	12		238,1	71,92
0		**183,5**	**62,40**	**13**		**243,1**	**72,08**
1	50	186,3	70,68	14		248,2	72,24
2		191,0	70,76	15		**253,3**	**72,40**
3		195,8	70,84	16		259,1	72,57
4		200,5	70,92	17		264,9	72,75
5		**205,3**	**71,00**	18		271,0	72,93
6		209,6	71,10	19		277,3	73,12
7		**213,9**	**71,20**	-	-	-	-

The data from Table 2 shows the graphical dependence of the supplied heat capacity (Q_{HP}) on the external ambient temperature (t_e) from -5°C to + 19°C on the Wesper VLH 804 LN Heat Pump (t_{SW}) - +45°C and +50°C, shown in Figure 3.

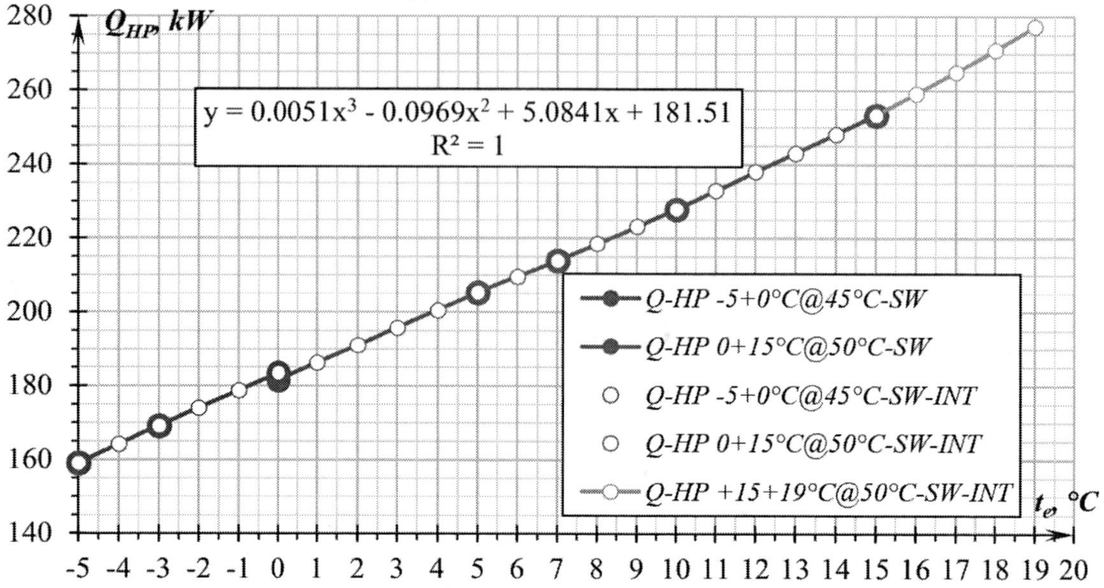

Fig. 3. Heat pump capacity (Q_{HP}) and the external ambient temperature (t_e) temperature range from -5°C to +19°C on heat pump Wesper VLH 804 LN at the temperature of the water supplied to the building (t_{SW}) - +45°C and +50°C

Table 2 shows the graphical dependence of the compressed air heat capacity (W_C) from the external ambient temperature (t_e) at -5°C to +19°C on the Wesper VLH 804 LN heat pump when delivering water to the building (t_{SW}) - +45°C and +50°C, shown in Figure 4.

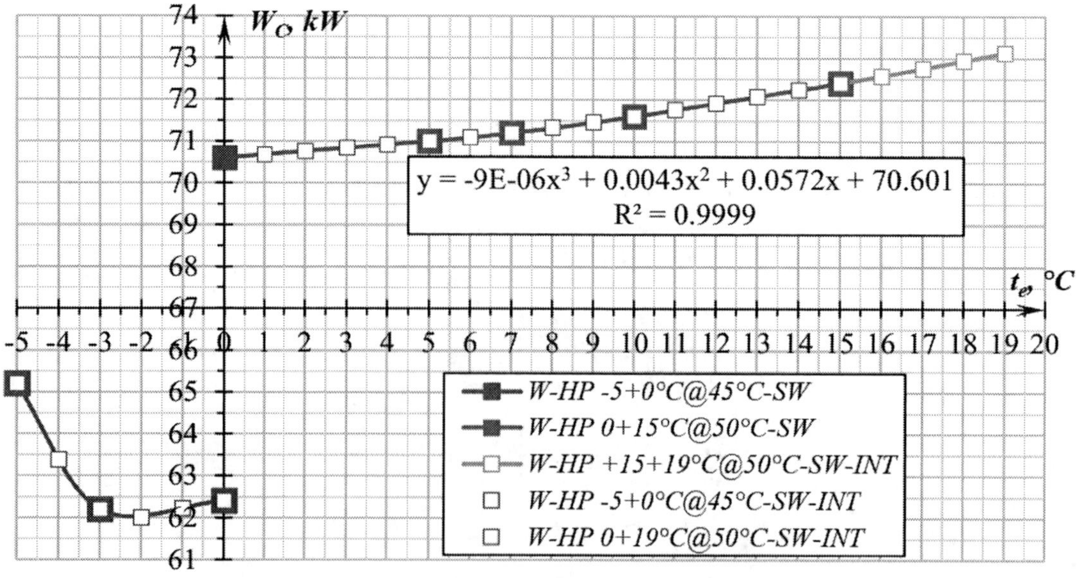

Fig.4. Graph of the heat capacity of the compressors (W_C) and the external ambient temperature (t_e) in the range of -5°C to +19°C on the Wesper VLH 804 LN at the temperature of the water supplied to the building (t_{SW}) - +45°C and +50°C

4 Results and analysis

The results of the method calculations are given in Table 3 and shown graphically in Figure 5.

Table 3. Calculation of the thermal energy for the building ($Q_{B,\,i}$), the consumption of the electrical energy consumption ($E_{B,\,i\text{-}HP}$) and its value ($Price_{E\text{-}HP}$) for the heat pump system in dependence from the external ambient temperature (t_e) from -5°C to +19°C

t_e °C	τ h	$Q_{B,\,i}$ kWh	$\tau_{w,\,i\text{-}HP}$ h	$E_{C,\,i}$ kWh	$E_{F,\,i}$ kWh	$E_{B,\,i\text{-}HP}$ kWh	COP_i -	$Price_{E\text{-}HP}$ €
-5	15,0	2399,70	15,08	983,41	56,56	1039,97	2,31	121,55
...	...	...	...	...	...	...	...	...
0	52,3	6626,97	36,11	2253,53	135,43	2388,96	2,77	279,21
1	36,0	4318,56	23,18	1638,41	86,93	1725,34	2,50	201,65
...	...	...	...	...	...	...	...	...
15	47,3	1258,12	4,97	359,60	18,63	378,23	3,33	44,21
16	28,2	560,80	2	157,05	8,12	165,16	3,40	19,30
...	...	...	...	...	...	...	...	...
19	7,7	0,00	0	0,00	0,00	0,00	0,00	0,00
Σ -5 +15°C	-	**108741,4**	-	-	-	**38719,4**	-	**4525,34**
Σ -5 +19°C	-	**109622,5**	-	-	-	**38976,5**	-	**4555,39**

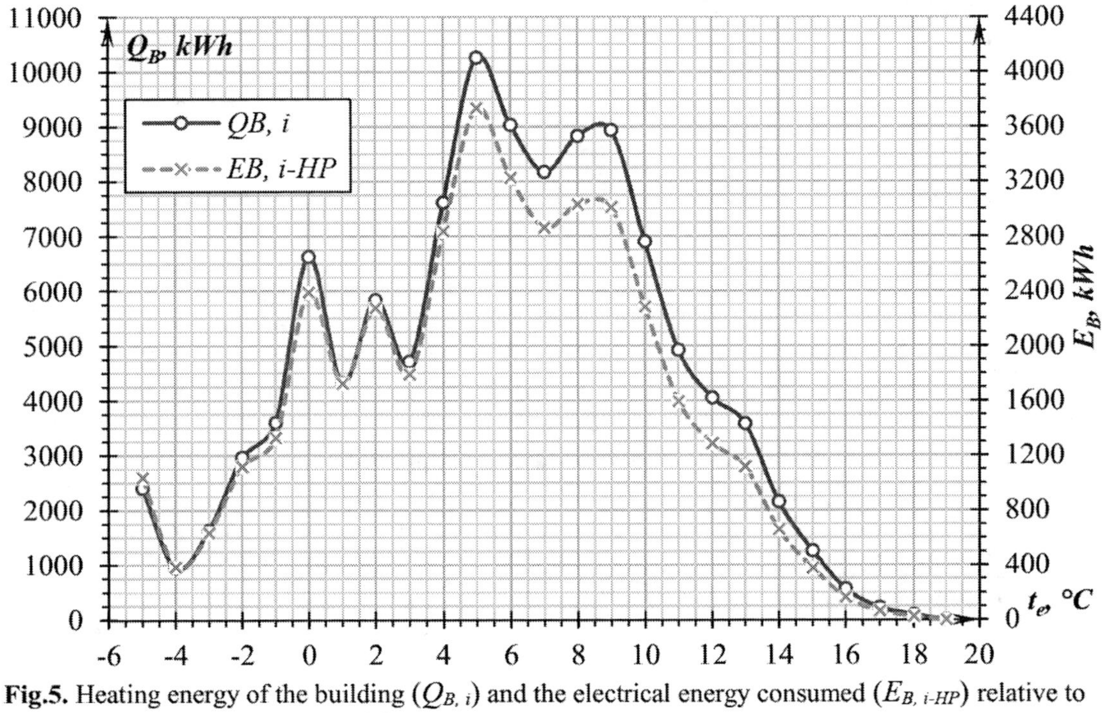

Fig.5. Heating energy of the building ($Q_{B,\,i}$) and the electrical energy consumed ($E_{B,\,i\text{-}HP}$) relative to an external ambient temperature (t_e) in the range from -5°C to +19°C

The seasonal efficiency of the heat pump operation in winter mode is determined by equation (6) for the coefficient of transformation in the temperature range of -5°C to +15°C, resulting in $COP_{-5°C+15°C} = 2{,}808$, and at a temperature range of -5°C to +19°C of equation (7), there was obtained $COP_{-5°C+19°C} = 2{,}813$. The results obtained justify the claim that the air-to-water heat pump system operates efficiently in heating mode at precisely measured external ambient temperature characteristic of the Black Sea climate zone.

Figure 6 gives a relationship between the thermal energy of the building ($Q_{B,\,i}$) and the external ambient temperature (t_e) in the range of -5°C to +19°C with the value of the electrical energy ($Price_{E\text{-}HP}$).

The value of the consumed electrical energy from the heat pump system for the winter heating season in the temperature range of -5°C to +15° C is 4525,34 € and in the temperature range of -5°C to + 19°C is 4555,39 € and it can be said that the system works economically.

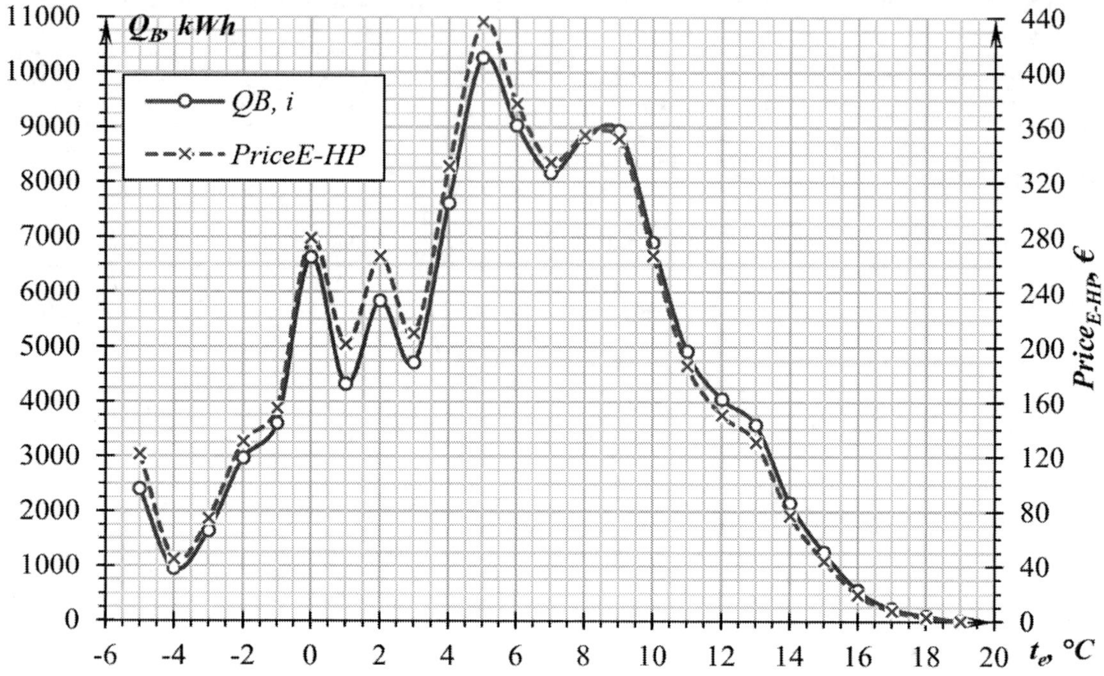

Fig. 6. Building thermal energy ($Q_{B,\,i}$) and value of the electrical energy ($Price_{E\text{-}HP}$) relative to an external ambient temperature (t_e) in the range from -5°C to + 19°C

5 Conclusion

The experimental results for the seasonal efficiency of the heating air-water heating system in the two temperature ranges from -5°C to +15°C and from -5°C to +19°C are $COP_{-5°C+15°C} = 2{,}808$ and $COP_{-5°C+19°C} = 2{,}813$ and warranted the claim that the air-to-water heat pump system operates efficiently in heating mode at the precisely measured external ambient temperature typical of the Black Sea climate zone. The economic analysis shows that the costs for the consumed electrical energy from the heat pump system for the winter heating season at the considered temperature intervals are extremely effective. The results of the study can be used both in the design of air-to-water heat pump systems and in the search for energy efficiency buildings.

References

1. G. Valtchev, N. Kalojanov, V. Rasheva, M. Minchev, S. Tasheva. Analysis of results after implementation of energy saving measures in public buildings, *Bulgarian Chemical Communications*, **48**, 283 – 289 (2016)
2. V. Rasheva, V. Kamburova, M. Velikanov. Results from an Energy Audit of a Joint-stock Company "Medica AD", *Proceedings at 7th International Conference on TE-RE-RD*, 129-134 (2018)

3. N. Penkova, N. Harryzanov, Analysis and optimization of the temperature stratification at thermal energy storage tank, *Proceedings at 5th International Conference on Energy and Sustainability*, 469-477 (2014)
4. Z. D. Kolev, S. Y. Kadirova, T. R. Nenov, Research of reversible heat pump installation for greenhouse heating, *INMATEH - Agricultural Engineering*, **2**, 77-84 (2017)
5. X. Wang, C. Zhang, Z. Zhang, B. Sun, Experimental research on performance of air-source heat pumps, *HV&AC*, **44**, 119-123 (2014)
6. Z. Zeng, J. Wu, X. Wei, Experimental Study and Thermal Performance Analysis of Air - source Heat Pump Direct Floor Radiant Heating System, *Acta Energiae Solaris Sinica*, **8**, 1151-1157 (2011)
7. D. Rusev, A. Kirov, P. Zlateva, N. Trizlova, D. Genov, Analysis of the operation of air-water heat pump unit in winter mode, *Proceedings of the annual symposium with international participation SISOM & ACOUSTICS*, 203-208 (2012)
8. National Institute of Meteorology and Hydrology-Varna, http://www.stringmeteo.com (2019)
9. Energo pro, https://www.energo-pro.bg (2019)
10. Bulgarian National Bank, http://www.bnb.bg (2019)
11. Catalog of Wesper, http://www.wesper.co.uk (2019)

E3S Web of Conferences **112**, 01008 (2019) https://doi.org/10.1051/e3sconf/201911201008
TE-RE-RD 2019

CFD simulation of forced heat transfer of gas in pipe

Zhivko Kolev[1] and *Seher* Kadirova[2]

[1]University of Ruse, Department of Thermotechnics, Hydraulics and Engineering Ecology, 7017 Ruse, Bulgaria
[2]University of Ruse, Department of Electronics, 7017 Ruse, Bulgaria

Abstract. This paper presents results from CFD simulation of heat transfer processes in ABAQUS. The investigations are realized at forced convection of air in steel pipe. Verification of the computing mesh and validation of the model, have been done. The average heat convection coefficients have been determined by methodology based on criteria equations, and on simulation methodology. Heat transfer processes between air flow in a steel pipe and the environment, have been experimentally accomplished. In order to analyze the processes of heat convection between the fluid and the internal surface of the pipe, numerical modelling is applied. A geometric model of the fluid flowing in the pipe is built. The computing mesh has been verified by increasing the number of cells and nodes. The numerical model has been validated based on experimentally measured temperature values and the simulation data. The heat convection coefficients have been investigated by analogy of the above. The results demonstrate that the numerical model is adequate and can be used to study similar heat transfer processes.

1 Introduction

Processes of heat transfer at forced convection are realized in heat exchange apparatuses [1-3]. Many experimental and numerical methods have been proposed for obtaining the parameters of heat transfer processes and characteristics of the fluid flow [1-7].
The correct defining of boundary conditions is obligatory for solving of numerical heat transfer problems [8].
Generally, in numerical simulation the used equations are discretized by cells from the division of computational domain using finite element method [3, 7, 10]. Three dimensional discrete model are being created in ABAQUS FEA engineering software in order to analyze the phenomena occurring the heat transfer process [11-13]. One of the turbulence model, which can be used for simulation of fluid flow is „Spalart-Almaras".
More accurate estimation of the heat transfer coefficient is required for the high performance heat exchanger equipment [4]. Determination of heat convection coefficient can be done in different ways [2]. The aim of the paper is determination of the heat convection coefficient between the air flow and the pipe's internal surface, by methodology based on criteria equations, as well as simulation, based on ABAQUS.

2 Methodology

2.1 Schematic diagram of the simulation model

Schematic diagram of the simulation model is presented in Figure 1. The object of investigation presented in the figure is the fluid flowing in the pipe. The input parameters for CFD simulation are: bulk fluid temperatures in the pipe's inlet and outlet ($T_{f,1}$ and $T_{f,2}$, °C); average fluid velocity in the pipe's inlet (U_{input}, m/s); manometric fluid pressure in the pipe's outlet ($p_{m,output}$, Pa); heat flux between the fluid and the pipe's internal surface (q, W/m^2); pipe's internal diameter (d, m); pipe's length (l, m). The values of the above parameters have been experimentally determined.

The selection of the boundary conditions for implementation of CFD simulations is determined by the necessity to obtain the temperature distribution in the fluid on the base of the average heat flux q. The heat flux has been calculated through the following parameters: mass flow rate; average mass specific heat capacity at constant pressure; temperature difference "$T_{f,1}$ - $T_{f,2}$"; internal tube diameter d and pipe length l. Therefore, this requires setting of the heat flux q, the inlet bulk temperature $T_{f,1}$, and the outlet bulk temperature $T_{f,2}$ as boundary conditions.

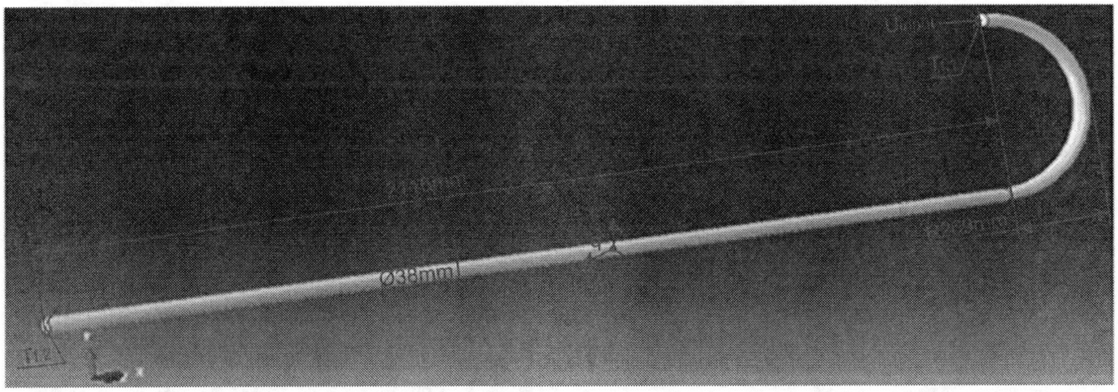

Fig. 1. Schematic diagram of the simulation model.

2.2 Verification of computing mesh and validation of CFD model

In Figure 2 is presented the scheme for verification of computing mesh and validation of CFD model. They are implemented based on pipe's internal temperatures, taken in the corresponding cross-sections.

Fig. 2. Scheme of the cross-sections for verification and validation.

The parameter for verification of the computing mesh is the average simulation value of the temperature T_{w-3}. Based on verification results, the computing mesh for simulation of the studied heat transfer processes is being chosen.
The validation of CFD model is performed by comparing simulation and experimental values of T_{w-1}, T_{w-2} and T_{w-3}, using the selected computing mesh.

2.3 Calculation of the heat convection coefficient

The average heat convection coefficient (h_c, W/m^2) between the fluid and the pipe's internal surface is calculated in two methods – based on criteria equations, and based on simulation data. According to the first method the following criteria equation has been used:

$$Nu_{fd} = 0.018.Re_{fd}^{0.8}.\varepsilon_l \qquad (1)$$

The number of Reynolds Re_{fd} is calculated on the basis of experimentally determined fluid temperatures ($T_{f,1}$ and $T_{f,2}$), and fluid velocity in the inlet pipe's cross-section (U_{input}). The coefficient ε_l is determined on the basis of the ratio l/d.
The calculation of the heat convection coefficient with simulation data is by the equation [2]:

$$h_c = q/\Delta T \qquad (2)$$

where ΔT is the cross-section temperature difference, calculated as an average value of the temperature differences in six fluid cross-sections, K (Figure 3).

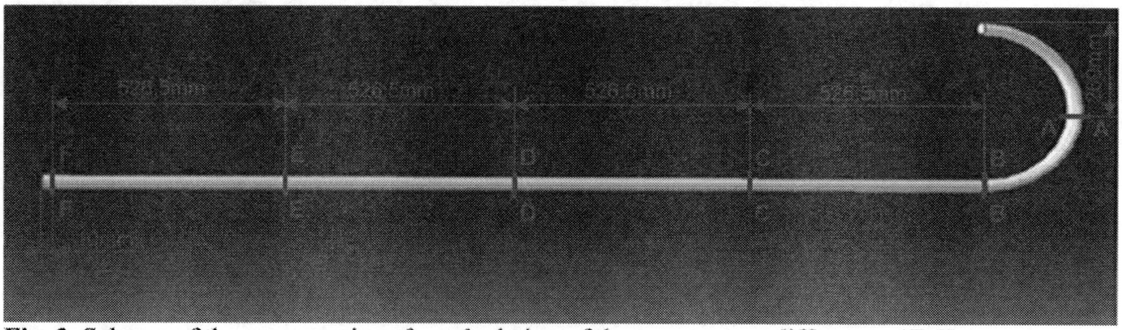

Fig. 3. Scheme of the cross-sections for calculation of the temperature difference ΔT, K.

3 Results and discussion

3.1 Input parameters of the simulated heat transfer processes

For applying of the numerical model for different cases, and for calculating the heat convection coefficient according to the above mentioned methods, three heat transfer processes have been realized and investigated via fluid flow rate regulation.
Table 1 presents the values of the input parameters for CFD simulation of the investigated heat transfer processes.

Table 1. Input parameters for simulation of the heat transfer processes.

Heat transfer processes	**$T_{f,1}$, °C**	**$T_{f,2}$, °C**	**U_{input} , m/s**	**$p_{m,output}$, Pa**	**q, W/m^2**
Process I	131	101	44.63	0	3873.52
Process II	137	102	41.88	0	4176.30
Process III	141	106	36.16	0	3734.36

The pipe's internal diameter is $d = 0.038$ m, and length is $l = 2.901$ m. They are calculated on the base of the schematic diagram of the simulation model, presented in Figure 1.

3.2 Results from verification of the computing mesh, and validation of CFD model

Verification and validation have been realized for Process I (Table 1).

➢ Results, obtained from verification of the computing mesh

Table 2 presents the parameters of the investigated variants of computing mesh and the determined simulation value of the temperature T_{w-3}, used for verification.
Figure 4 presents the fluid cross-sections for determination of T_{w-3}. This temperature is calculated as the average temperature value for computing nodes, located on the external cylindrical surface of the fluid, which in fact is the internal heat exchange surface of the pipe. The figure presents information about the typical shape and the number of the computing cells as well as the number of the computing nodes for the investigated variants of mesh. The set shape of the computing cells is "Tet" (a 4-node linear fluid tetrahedron).

Table 2. Parameters of the computing meshes and the calculated simulation value of T_{w-3}.

Variants of computing mesh	**Shape of the computing cells**	**Approximate size of the computing cells, m**	**Number of the computing cells**	**T_{w-3} (simulated), °C**
Mesh I	Tet	0.0100	21733	68.28
Mesh II	Tet	0.0050	151141	73.14
Mesh III	Tet	0.0045	209028	73.87
Mesh IV	Tet	0.0040	276137	74.19

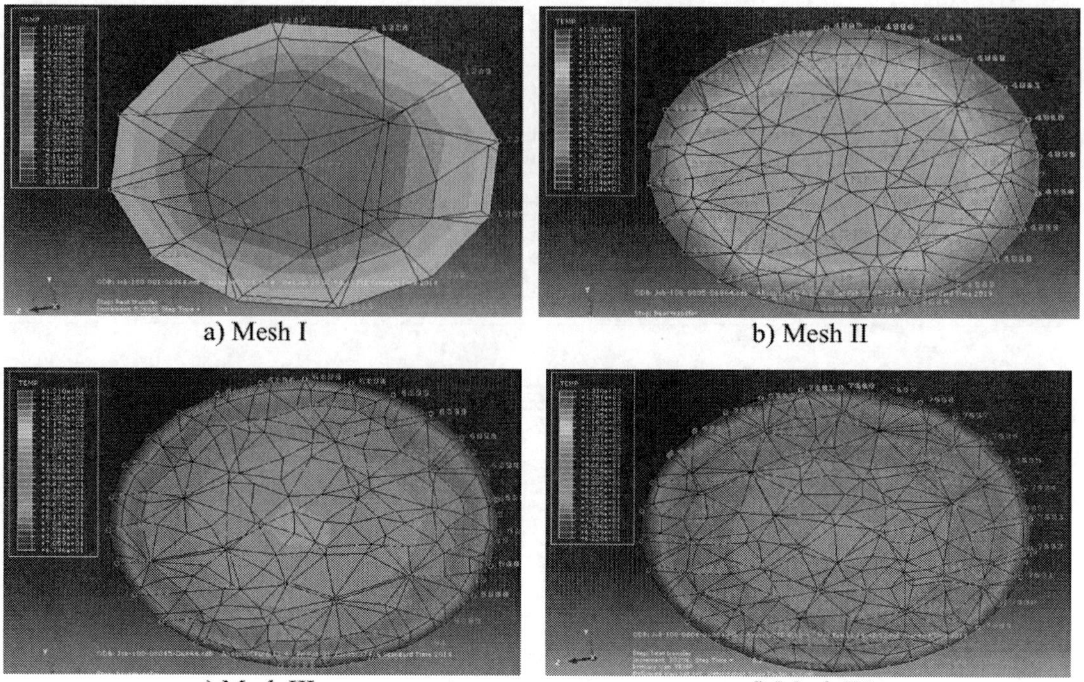

a) Mesh I b) Mesh II

c) Mesh III d) Mesh IV

Fig. 4. Fluid cross-section for determination of the simulation value of T_{w-3} at different computing mesh variants.

It was found that the difference in the values of T_{w-3} between the contiguous variants III and IV is relatively small (0.32 °C). It can be assumed that from the point of view of the model's accuracy, the further increase the mesh density is not required. Therefore, the variant Mesh IV is selected.

➢ Results from validation of CDF model

Table 3 presents results from validation of the numerical model, accomplished by comparison of experimental and simulation values of the temperatures T_{w-1}, T_{w-2} and T_{w-3}. The CFD model's error has been calculated (Table 3).

Table 3. Results from model's validation, and errors.

Temperature	**Experimental value, °C**	**Simulation value, °C**	**Model's error, ±%**
T_{w-1}	86.00	87.57	+1.79
T_{w-2}	79.00	76.52	-3.14
T_{w-3}	74.00	74.19	+0.26

From the results presented in Table 3, can be concluded that the highest calculated model's error of -3.14 % is less than the generally accepted maximum permissible engineering error of ±5 %.

3.3 Results for calculated heat convection coefficient and simulation obtained temperature field

Table 4. Results for heat convection coefficient

Investigated heat transfer process	**h_c (by criteria equations), W/(m^2.K)**	**h_c (by simulation method), W/(m^2.K)**	**Difference, %**
Process I	111.66	115.56	-3.49
Process II	110.53	110.08	+0.41
Process III	94.71	98.04	-3.52

Table 4 presents results for heat convection coefficient h_c, respectively calculated values on the base on criteria equations and values observed by simulation method. The percentage difference has also been calculated and analyzed.

Figure 5 presents the simulated temperature field in the longitudinal cross-section of the fluid, which gives information about the temperature values, used for calculation of ΔT, K.

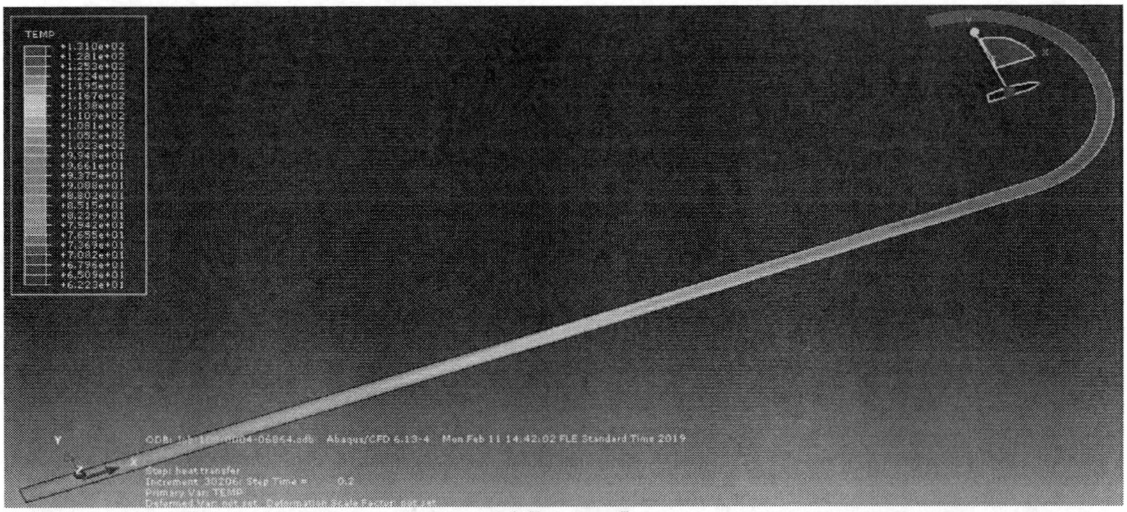

a) Process I

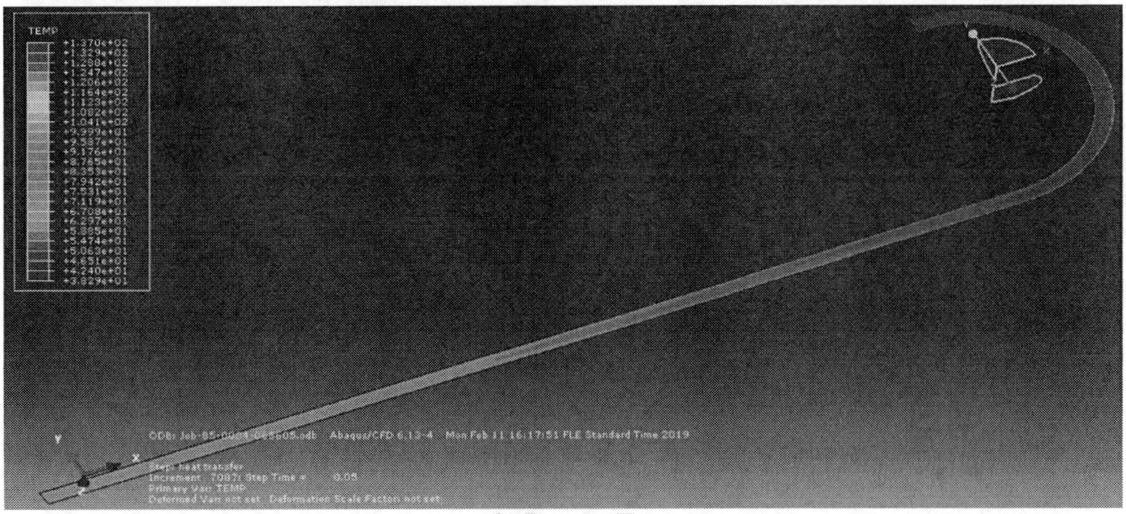

b) Process II

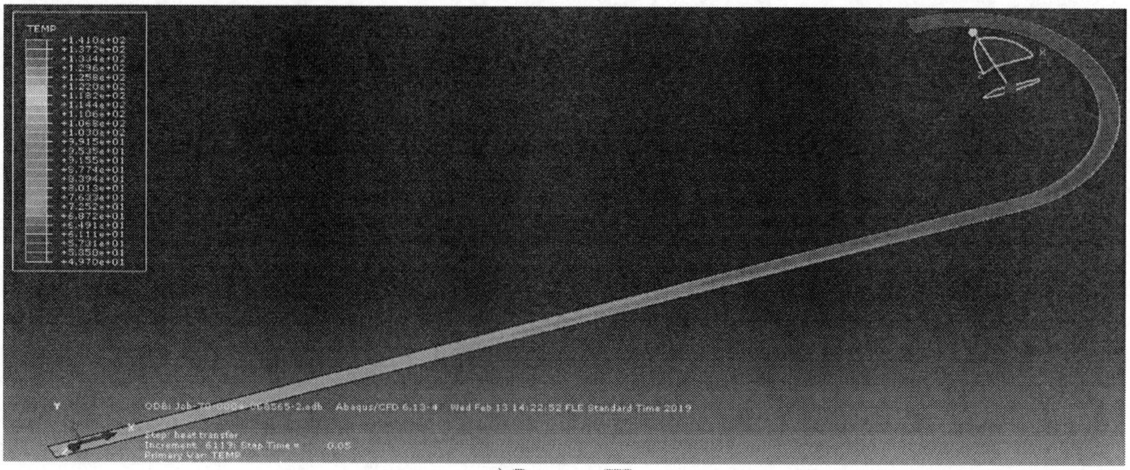

c) Process III

Fig. 5. Simulation results for temperature field in the longitudinal cross-section of the fluid.

Figure 6 presents information about the temperature distribution and the fluid velocity distribution in cross-section "D-D".

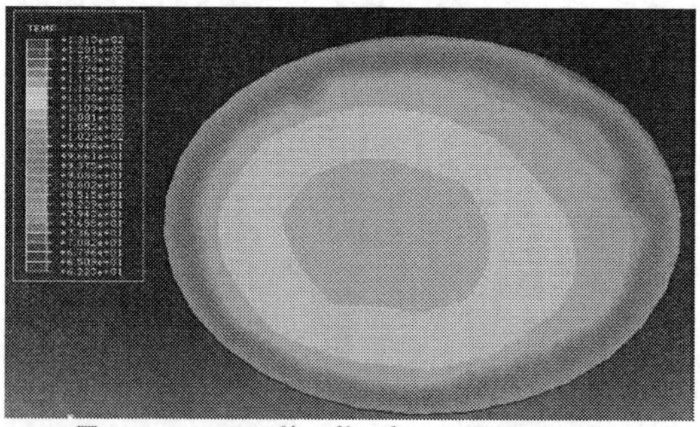

Temperature distribution – Process I

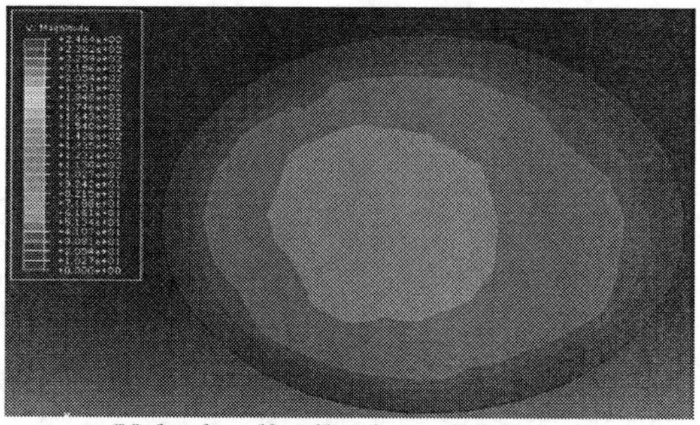

Velocity distribution – Process I

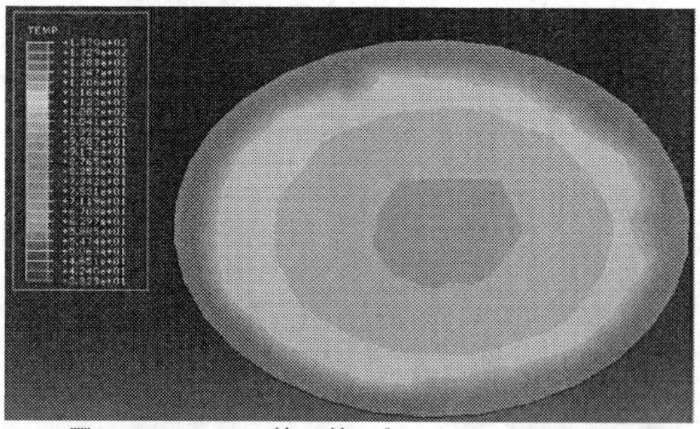

Temperature distribution – Process II

Velocity distribution – Process II

Temperature distribution – Process III

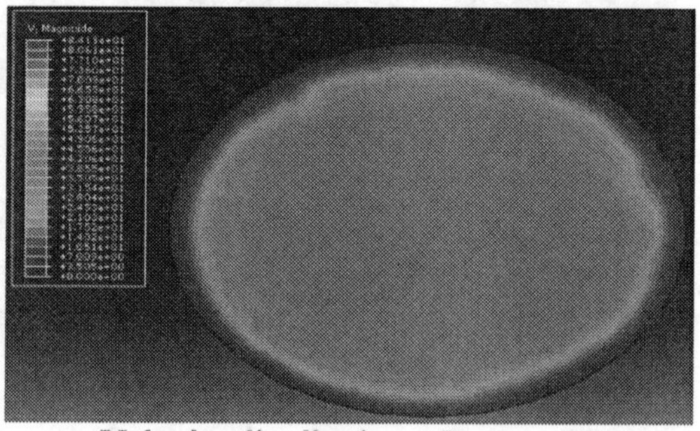

Velocity distribution – Process III

Fig. 6. Simulation results for the temperature distribution and the fluid velocity distribution in cross-section "D-D".

The results observed from numerical simulation covers the results calculated by criteria equations. The highest calculated percentage difference of -3.49 % is less than ±5 %. Therefore, it can be concluded that the proposed scheme for determination of ΔT is appropriate for simulation studies.

4 Conclusion

A methodology for investigation of heat transfer processes at forced convection of air in steel pipe by numerical simulation in ABAQUS has been proposed.
A scheme of fluid cross-sections for verification of the computing mesh and validation of CFD model has been developed.
A method for calculation of the average heat convection coefficient, via determination of simulation temperatures in the fluid cross-sections has been developed and proved.
From the results of the computing mesh verification can be concluded that with increase of the mesh density, the difference between the temperature values (used as a basis for verification) has been reduced.
From the results obtained for the validation of CFD model is seen that the maximum value of the model's error is less than the maximum permissible technical error. Therefore, the CFD model is adequate and can be used for simulation of similar heat transfer processes.
The results obtained for the heat convection coefficient indicate that the proposed method provides a sufficiently high accuracy.

References

1. A.H. Dhumal, G.M. Kerkal , K.T. Pawale, Heat transfer enhancement for tube in tube heat exchanger using twisted tape inserts, IJAERS **4**, 89 (2017)
2. K. Majewski, S. Grądziel, CFD Simulations of Heat Transfer in Internally Helically Ribbed Tubes, Chem. Process Eng. J. **37 (2)**, 251 (2016)
3. C. Oon, H. Togun, S. Kazi, A. Badarudin, M. Zubir, E. Sadeghinezhad, Numerical simulation of heat transfer to separation air flow in an annular pipe, Int. Commun. Heat Mass. J. **39**, 1176 (2012)
4. Han-Taw Chen, Yu-Jie Chiu, Chein-Shan Liu, Jiang-Ren Chang, Numerical and experimental study of natural convection heat transfer characteristics for vertical annular finned tube heat exchanger, Int. J. Heat Mass. Tran. **109**, 378 (2017)
5. E. Dimofte, F. Popescu, I. Ion, Numerical modelling of mixing fluids at different temperatures, *Proceedings of TE-RE-RD,* 35 (2016)
6. W. Osley, P. Droegemueller, P. Ellerby, CFD Investigation of Heat Transfer and Flow Patterns in Tube Side Laminar Flow and the Potential for Enhancement, Chem. Eng. Trans. J. **35**, 997 (2013)
7. K. Yordanov, T. Mechkarova, A. Stoyanova, P. Zlateva, Determination of the temperature of cathode unit of indirect plasma burner through a computer simulation model, *Proceedings of* IITI'17, 403 (2018)
8. K. Yordanov, P. Zlateva, I. Hadzhidimov, A. Stoyanova, Testing and clearing the high temperature module error from 0 to 1250 °C for measurement with 16 K-type thermocouples, *Proceedings of* SIELA, 480 (2018)
9. S. Elena-Madalina, Gh. Voicu, G. Ipate, A.G. Constantin, Modelling and numerical analysis by the finite element method of wheat milling process in roller mills, *Proceedings of TE-RE-RD*, 249 (2013)

10. W. Igo, L. Mihaescu, Turbulent forced convection air flow in a Venturi channel, *Proceedings of TE-RE-RD*, 51 (2013)
11. Z. Saternus, W. Piekarska, Numerical analysis of thermomechanical phenomena in laser welded pipe-to-flat, Procedia Eng. J. **177**, 196 (2017)
12. R. Steinheimer, B. Engel, Thermal influences during rotary draw bending of tubes from stainless steel, Procedia Eng. J. **81**, 2165 (2014)
13. K. Abdelkader, K. Youcef, T. Sabiha, Numerical simulation of free vibrations of multilayer composite beams, *Proceedings of ICAME*, 105 (2016)

Performances of a Research CFR Octane Rating Unit Engine and Dacia Single Cylinder SI Engine Ignited by a LASER System

Bogdan George Done [1], *Ion* Copae[1]

[1]Department of Military Vehicles and Transportations, Faculty of Aircraft and Military Vehicles, " Ferdinand I " Technical Military Academy of Bucharest.

Abstract. At this time, the severe legislation regarding the level limits of the waste and exhaust gases released by thermal engines and also the necessity of engines efficiency improvement boost the engine research domain to bring in front the use of new technologies that can be used to control the in-cylinder combustion process. Now, the new technologies is represented by LASER spark plug systems which can be successfully used at petrol engines. LASER spark plug technology can have many advantages for engine operation control, an ignition system that could provide improved combustion is the one using plasma generation and a Q-switched LASER that results in pulses with high MW power. The LASER spark plug device used in the current research was a LASER medium Nd:YAG/Cr4+:YAG ceramic structure made up of a 8.0-mm long, 1.0-at.% Nd:YAG ceramic, optically-bonded to a Cr4+:YAG c. It was developed and constructed similar to classical spark plug and could be assembled on a CFR Octane Rating Unit Engine as well as on a Dacia Single Cylinder SI Engine which led to several results among which: influences on in-cylinder pressure, combustion and pollutant emissions.

Keywords: LASER spark plug, combustion, pollutant emissions, new technologies, pressures.

1 Introduction

Internal combustion engines are extremely important in transportation and energy production therefore any improvement will lead to a substantial decrease in pollutants and consequently greenhouse gases. Ignition is a complex phenomenon which greatly impacts combustion [1], especially the initial stages which result in pollutant formation, flame propagation as well as quenching. The ignition source has undergone few changes over the past hundred years. The classical spark plug is made up of two electrodes having a space between them, where an electrical arc is produced due to high voltage discharge. For several years researchers have been intent on finding a LASER based ignition source [2], which can replace a classical spark plug with a pulse focused LASER beam and they also attempted to control ignition by a LASER source [3]. The development of flame kernel size simultaneous with NOx production are highly important [4] and in this situation a LASER

*Corresponding author: bogdan.done@yahoo.fr

spark plug source can improve engine combustion compared to classical spark plugs. Laser spark plug systems are intended to protect resources and the decrease CO2 emissions thus limiting the greenhouse effect. It could be obtained through lower fuel consumption achieved by the spark ignition (SI) engine system, owing to high thermodynamic capacity resulting from direct injection. One of the main drawbacks is that with classical spark ignition the place of ignition cannot be specifically chosen. LASER induced ignition could eliminate some of these difficulties. Several other ignition systems apart from LASER spark plug are reviewed [4] such as microwave ignition and high frequency ignition.

2 LASER Spark Plug

LASER spark plug is the chemical-kinetic mechanism of starting combustion by the stimulus of a LASER source. Scientific literature generally classifies energetic interactions of a LASER with a gas into four schemes [5], characterised by the nanosecond domain of the LASER pulse and the duration of the entire combustion could be several hundreds of milliseconds. It takes only a few nanoseconds for the LASER energy to be deposited followed by shock wave generation. Combustion may take from 100 ms to several seconds according to the air-fuel dosage, initial pressure, pulse energy, plasma size, plasma and initial temperature in the combustion chamber.

The following main advantages of LASER spark plug are as follows [6, 7, 8, 9 10]:

- a choice of arbitrary positioning of the ignition plasma in the combustion cylinder
- absence of quenching effects by the spark plug electrodes
- ignition of leaner mixtures than with the spark plug => lower combustion temperatures => less NOx emissions no erosion effects as in the case of the spark plugs => lifetime of a LASER spark plug system expected to be significantly longer than that of a spark plug
- high load/ignition pressures possible => increase in efficiency
- precise ignition timing possible
- exact regulation of the ignition energy deposited in the ignition plasma
- easier possibility of multipoint ignition
- shorter ignition delay time and shorter combustion time
- fuel-lean ignition possible

The disadvantages of LASER spark plug are:

- high system costs
- concept proven, but no commercial system available yet.

3 Experimental investigation

Experimental investigations were carried out in a research laboratory of the Faculty of Mechanical Engineering and Mechatronics, Department of Thermotechnics, Engines, Thermal Equipments and Refrigeration Installations, University POLITEHNICA of Bucharest.

This section describes the experimental research which was developed on an experimental single cylinder SI engine, equipped with LASER spark plug. The Dacia Single Cylinder SI Engine figure 1, indicates a speed of 2800 rev/min, a 90 % load and the CFR Octane Rating Unit Engine figure 2, shows a speed of 900 rev/min. Both experimental engines were single-cylinder and were mounted on test beds adequately instrumented. A photo of the LASER spark provided by INFLPR, Laboratory of Solid-State Quantum Electronics, Magurele, Romania is shown in figure 3, with the technical details for the LASER to be found in articles [11, 12, 13].

Fig. 1. Dacia Single Cylinder SI Engine equipped with LASER spark plug

Fig. 2. CFR Octane Rating Unit Engine equipped with LASER spark plug

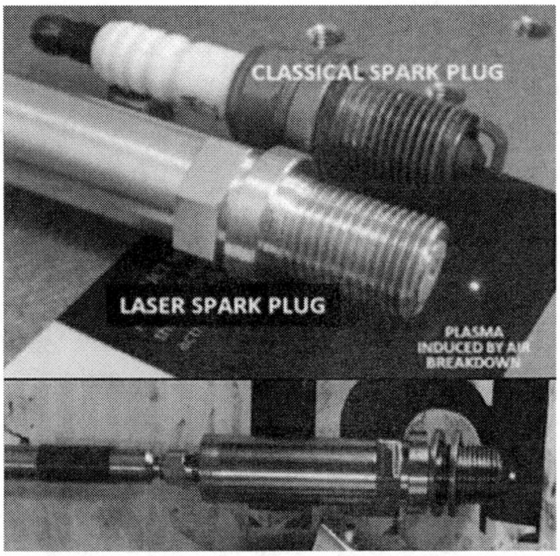

Fig. 3. A photo of a LASER spark plug is shown in comparison with a classical spark plug. The plasma induced in air by optical breakdown is visible. (Courtesy of INFLPR, Laboratory of Solid-State Quantum Electronics, Magurele, Romania)

4 Results and discussion

Figure 4 shows the pressure curves indicated for the two types of spark plugs, for the 491 cycles of the CFR engine with an advance ignition of β=28.5 degrees and for the two values of the excess air coefficient λ =1.1 and λ =1.276.

The four graphs also present the external outlines of the curves for the indicated pressure as well as the maximum values related to the upper and lower external outlines, respectively. The graphs in figures 4a and 4b highlight the existence of several maximum values which are slightly higher in the case of the LASER spark plug (22.4-26.9 bar) compared to those of the classical spark plug (22.0-26.3 bar) for a stoichiometric mix (λ =1.0). Conversely, figure 4c and figure 4d show slighly lower maximum values for the LASER spark plug (13.4-22.0 bar) compared to those of the classical spark plug (22.3-13.4 bar) for a leaner mix (λ =1.276).

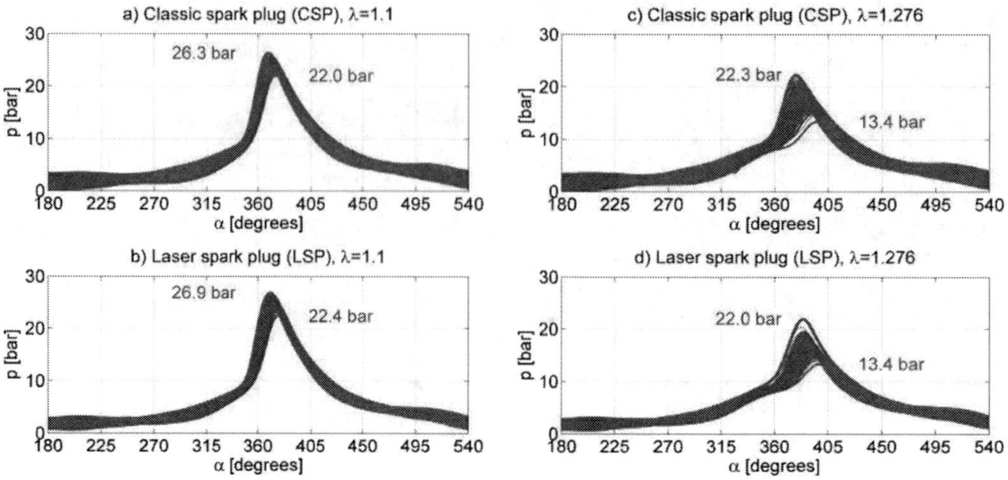

Fig. 4. Pressure curves indicated for the two types of spark plugs

The graphs in figure 4 suggest that in a leaner mix the cyclic dispersion is more intense than in the stoichiometric mix, the curves on the right having a wider spread; moreover, the maximum values variation range of the external outlines of the curves is wider at λ=1.276 (8.9 bar compared to 4.3 bar for the classical spark plug, and 8.6 bar compared to 4.5 bar for the LASER spark plug, respectively).
However, if cyclic dispersion is estimated based on the COV variation coefficient, this particular conclusion is refuted, as seen in figure 5, for four values presented in the graphs. This proves the inconsistency of utilizing the variation coefficient as a criterion for cyclic dispersion because it represents a ratio of two values (standard deviation and average value) and as a result both of them influence the COV value.

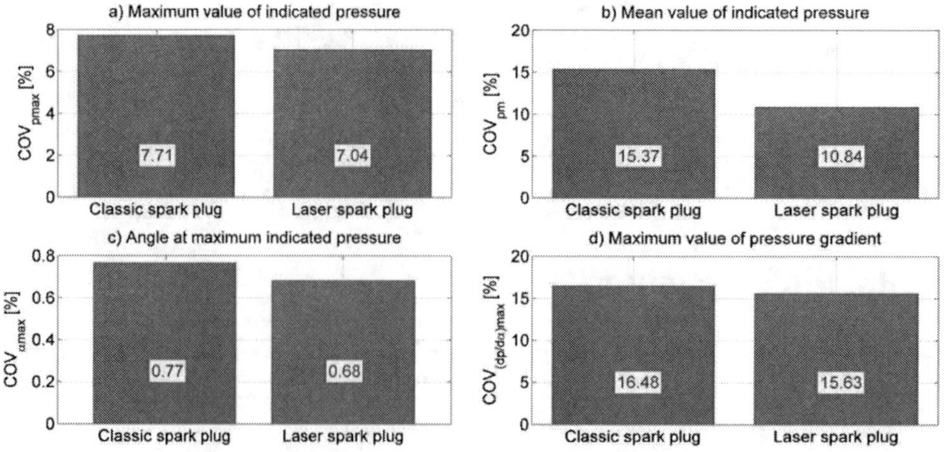

Fig. 5. Estimated cyclic dispersion

Figure 6 presents the diagrams indicated for the two types of spark plugs (classical and LASER) for 50 cycles of the Dacia mono-cylinder engine, a revolution of 2800 RPM and a 90% charge, for three values of the excess air coefficient: λ=0.9, λ=1.0 şi λ=1.1.

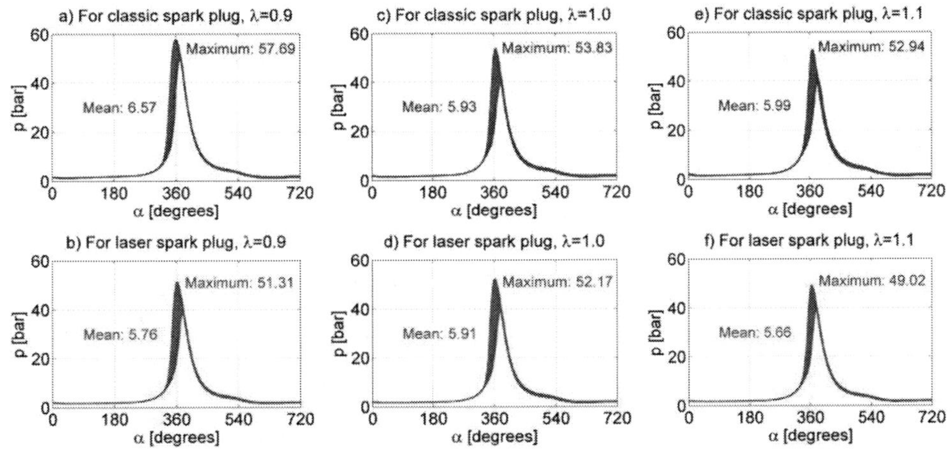

Fig.6 Maximum and average indicated pressure values of 50 functional cycles.

The graphs in figure 6 show the maximum and average indicated pressure values of the 50 functional cycles. As seen, the values are lower for the LASER spark plug compared to those of the classical ignition, for all the three combined values of the air-fuel dosage which suggests lower levels of power performance in the case of the LASER spark plug.

This fact is confirmed in figure 7, which presents the indicated pressure-volume (*p-V*) diagram for λ=0.9, similarly showing the same information for the other two values. Indeed, as shown in the graph, the power of the mono-cylinder engine P_e, is 7.9% lower in the case of the LASER spark plug (6.48 HP compared to 7.03 HP) and becomes obvious at 2800 RPM and a 90% charge.

The decrease is justified by the smaller area of the indicated diagram *p-V* (area difference A_1-A_2, detail A), an area related to mechanical work 277.03 Nm of the classical spark plug and 255.18 Nm of the LASER spark plug. The graph also presents the estimated maximum power values of the poly-cylindrical engine with 4 engines P_{max}, thus at a rotation of 5200 RPM and a 100% charge: 47.33 HP for the classical spark plug and 43.6 HP for the LASER spark plug. It is well known that the technical specification of the engine indicates a maximum power of 54 HP, the difference being caused by the wear and tear of the engine and the estimation error. Figure 7 also shows the values of the actual specific fule consumption c_e. As indicated in the graph, the specific consumption for the LASER spark plug is 6.9% lower than that of the classical spark plug.

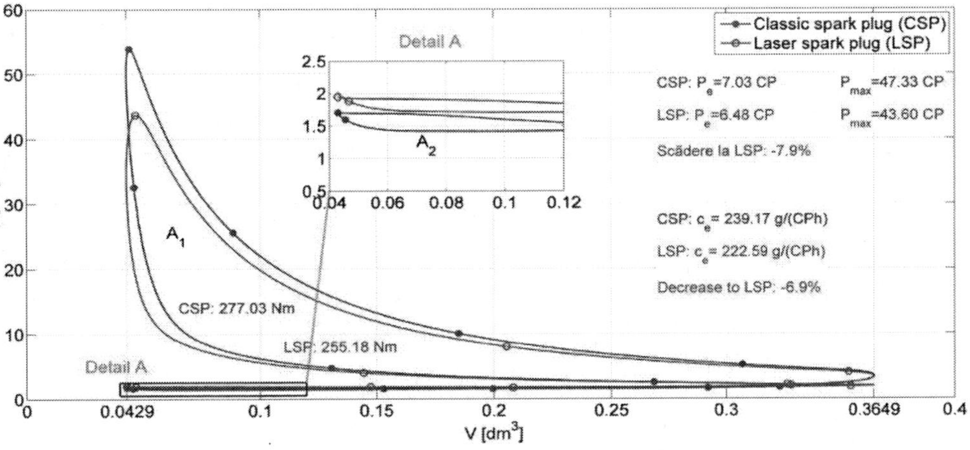

Fig. 7. Indicated pressure-volume (*p-V*) diagram

As already stated, the LASER spark plug has been chosen also because it ensures a decrease in exhaust gases, in particular nitrogen oxides. In this sense, figure 8 and figure 9 present exhaust gases values measured on the CFR engine, while figure 10 shows the values of polluting substances mentioned in the graphs in the case of the Dacia mono-cylinder engine.

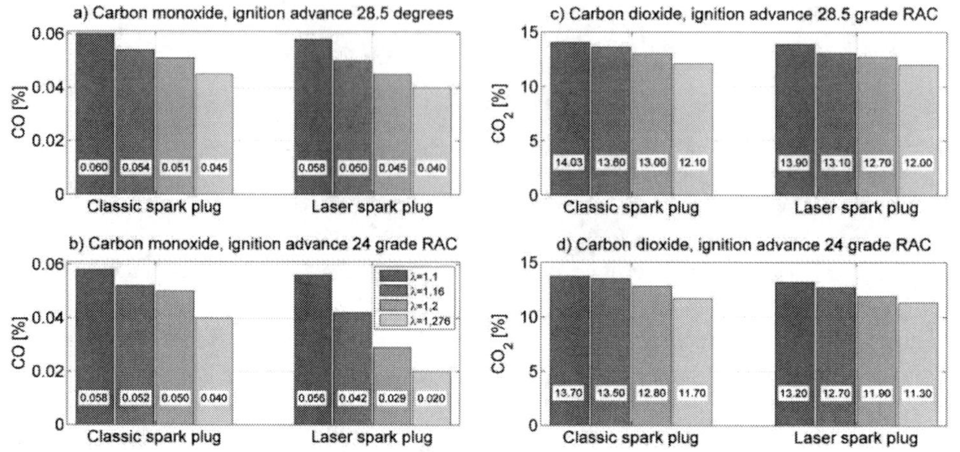

Fig. 8. Pollutant emission monoxide.

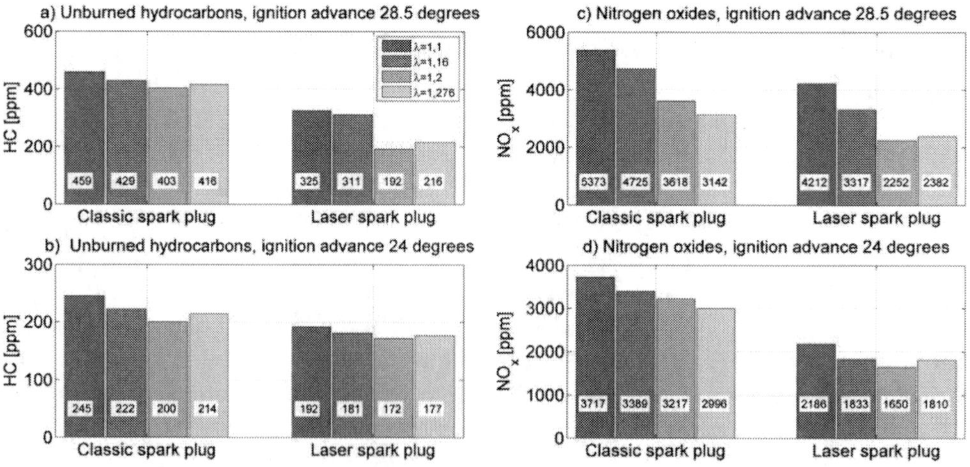

Fig. 9. Pollutant emission hydrocarbons.

The graphs in figure 8, figure 9 and figure 10 highlight the reduction of all exhaust gases for the LASER Spark plug compared to the classical spark plug, with a noticeable decrease of nitrogen oxides.

The graphs in figure 8, figure 9 and figure 10 also confirm the reduction of all exhaust gases as the air-fuel mix becomes leaner (increasing the λ coefficient) up to a value of λ=1.2, following which exhaust gases will increase (here up to λ=1.276). In addition, the graphs in figure 8 and figure 9 confirm the increase of exhaust gases values as the ignition advance rises, here from β=24 degrees to β=28.5 degrees.

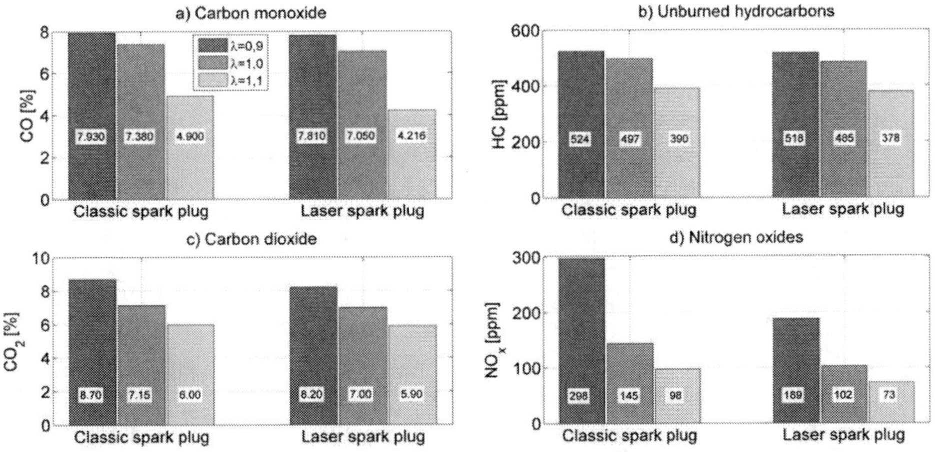

Fig.10. Pollutant emission carbon monoxide

5 Conclusions

Several important conclusions have resulted from the aforementioned experiments:
- the use of the LASER spark plug ensures a decrease of the specific fuel consumption for any composition of the air-fuel mix and for any ignition advance value;
- the study based on all the experimental data (the present paper shows only a part of the study) indicates that the use of the LASER spark plug ensures the reduction of exhaust gases down to a leaner air-fuel mix equivalent to an excess air coefficient of λ =1.2
- the use of the variation coefficient to estimate cyclic dispersion has the disadvantage of being a ratio of two values. Consequently, it cannot lead to valid conclusions;
- the study confirms the well-known fact that cyclic dispersion, performance and exhaust gases are influenced by the ignition advance and the quality of the air-fuel mix (by means of the excess air coefficient);
- the use of the LASER spark plug leads to an engine power reduction the values depend on the quality of the air-fuel mix and the ignition advance.

Acknowledgments

In the course of my extended research and comprehensive testing I was kindly provided with guidance and assistance by several highly-reputed scientists who deserve my deepest gratitude: Prof. Constantin Pana, Prof. Niculae Negurescu, Lecturer Alexandru Cernat, Lecturer Dinu Fuiorescu, Lecturer Cristian Nutu, Faculty of Mechanical Engineering and Mechatronics, Department of Thermotechnics, Engines, Thermal Equipments and Refrigeration Installations, University POLITEHNICA of Bucharest.

The experiments were developed with a LASER spark device designed at National Institute for LASER, Plasma and Radiation Physics, Laboratory of Solid-State Quantum Electronics, Magurele, Ilfov, 077125, Romania. The authors are deeply indebted Mr. Pavel Nicolaie, Mr. Dinca Mihai and Mrs. Croitoru Gabriela for their advice and assistance during the experiments.
Also, the authors wish to thank AVL GmbH for providing the necessary equipment.

References

1. M. Lackner, Winter, F., What is ignition? Combustion File 256, IFRF Online CombustionHandbook, ISSN 1607-9116, International Flame Research Foundation, Ijmuiden, The Netherlands, (2004).
2. H. Kopecek, M. Lackner, F. Winter, E. Wintner, Laser ignition of methane air mixtures at pressures up to 4 MPa, Journal of Laser Physics **13** (11), 1365 (2003).
3. H. Kopecek, Lackner, M., Wintner, E., Winter, F., Laser-Stimulated Ignition in a Homogeneous Charge Compression Ignition Engine, SAE 2004 World Congress, paper No 2004-01-0937, Detroit, MI, USA (2004).
4. J. D. Dale, M. D. Checkel, P. R. Smy, Application of High Energy Ignition Systems To Engines, Prog. Energy Combust. Sci. **23**, 379-398 (1997).
5. P.D. Ronney, Laser versus conventional ignition of flames, Optical Engineering **33** (2), 510 (1994).
6. H. Kopecek , Charareh S., Lackner M., Forsich C., Winter F., Klausner J., Herdin G., Wintner E., Laser Ignition of Methane-Air Mixtures at High Pressures and Diagnostics, Salzburg, Austria: Proceedings of ICES03, Spring Technical Conference of ASME Internal Combustion Engine Division (2003).
7. J.B. Heywood, Internal combustion engine fundamentals, McGraw-Hill international editions (1988).
8. T.X. Phuoc, Single-point versus multi-point laser ignition: Experimental measurements of combustion times and pressures, Combustion and Flame 122, 508-510 (2000).
9. M.H. Morsy, Chung S.H., Laser induced multi-point ignition with a single-shot laser using two conical cavities for hydrogen/air mixtures, Experimental Thermal and Fluid Science **27**, 491-497 (2003).
10. M. Weinrotter, Kopecek H., Wintner E., Lackner M., Winter F., Laser ignition of hydrogen-air mixtures at high pressures, Orleans, France: Proceedings of the European Combustion Meeting, European Combustion Institute (2003).
11. B. Done, Alexandru Cernat, Fuiorescu Dinu, Cristian Nutu "Aspects of the cycle variability study of a si engine with LASER plug Ignition system." U.P.B Sci. Bull. Series D, Vol. 79, Iss. **2**, 2017. ISSN 1454-2358.
12. B. Done "Experimental Investigations of LASER Ignition Use at Spark Ignition Engine" 11th International Conference Interdisciplinarity in Engineering, INTER-ENG 2017, Procedia Manufacturing **22** (2018) 659–666. 6
13. N. Pavel, T. Dascalu, G. Salamu, M. Dinca, N. Boicea, and A. Birtas, "Ignition of an automobile engine by high-peak power Nd:YAG/Cr4+:YAG laser-spark devices," Opt. Express **23**(26), 33028-33037 (2015).

E3S Web of Conferences **112**, 01010 (2019) https://doi.org/10.1051/e3sconf/201911201010
TE-RE-RD 2019

Numerical simulation of a phase change material melting process

Dorin Stanciu[1], *Camelia* Stanciu[1,*], *Valentin* Apostol[1] and *Horatiu* Pop[1]

[1] University POLITEHNICA of Bucharest, Department of Engineering Thermodynamics, Heat Engines, Thermal and Refrigeration Equipment, Bucharest, Romania

Abstract. Storage processes are usually integrated in solar energy systems applications due to daily variation of this energy source availability. Among different thermal storage solutions, phase change materials (PCM) lately became more extensively used covering a wide range of operating temperatures. In this regard, a numerical simulation of a PCM melting process is performed under ANSYS CFD environment. A particular configuration is considered consisting in a 2m length annular tube having a 5.48 cm external diameter. The tube is filled with paraffin chosen as PCM. A concentric interior tube of 2.54 cm diameter is used for transporting the heat transfer fluid (HTF) from the solar collector. Heat is transferred through the 1mm thick pipe wall to the PCM placed all around the HTF tube. The numerical results reveal the melting process of the PCM at different instances and tube sections. The time variation of the PCM liquid fraction is emphasized. The results describe the dynamic behavior of a PCM melting process and might be further integrated in any solar power plant storage charging process simulation.

1 Introduction

Energy storage represents an essential component of solar energy systems, that lately became more and more used to transform thermal energy of Sun into mechanical, electric and/or thermal one to the user. The solar energy source has the important advantage of being free and renewable. Nevertheless, it has a certain variability, not only from day to day, but even along the same day. More than that, cloudy days could affect the system predicted operation due to lack of input energy over some periods. Thus, a backup in terms of thermal energy source is required in order to maintain the system continuous operation at desired parameters. The main role of a thermal energy storage component (TES) is to ensure continuously the right levels of temperature and heat flux input to the converting system. Secondly, this component improves the overall system operation, maintaining constant parameters in operation, extending the operation interval and improving the overall system efficiency and reliability.

Thermal energy may be stored either by storing the sensible heat, as water tanks or rock bed, or the latent one, as phase change materials (PCM) use. The latest is a viable and

* Corresponding author: camelia.stanciu@upb.ro

efficient solution since it offers a high energy density and a constant temperature level corresponding to the phase change process.

The current literature offers an important number of publications reporting PCM research and results. An extensive review of 100 papers on TES is reported in [1]. More than one third is dedicated to PCM researches on modelling and experimental results. Sharma et.al. [2] presents a review containing 155 papers only on PCM systems and applications. Stages in PCM systems development, properties and criteria for PCM selection and their classification from different points of view are presented. As grouped in organic, inorganic and eutectic, PCM are employed over a wide range of temperatures. Among the most important criteria for PCM choice, one may consider the melting point, latent heat and volumetric latent heat storage capacity, as these are constraining the system operation parameters and size. Organic PCM are paraffin wax and non-paraffin. They are not corrosive, safe in use and have a very long freeze–melt cycle, so they are a reliable choice for the system. Their melting point varies from 5.5°C to 75.9°C for paraffin and reaches 127°C for non-paraffin [2] so that they cover a quite large temperature range of applications. The volumetric latent heat storage capacity for the organic compounds lies on 128–200 kg/dm^3 [2]. A double value is characteristic to inorganic PCM. These are salt hydrates and metallics. Their melting point varies between 14°C to 117°C for salt hydrates and even higher, 1300°C for carbonate salts [3], and 30°C to 125°C for metallics. The biggest problems encountered with inorganic PCM could be a more reduced charge-discharge cycle due to salt solubility in water decrease in time and necessity to discharge at a lower temperature than the fusion one, respectively the weight problem for metallics. Eutectic PCM represents a mixture of two or more compounds melting and freezing congruently. Their melting point varies between 13°C to 767°C for LiF/CaF2 (80.5/19.5 mol%) [3]. Alva et.al. [3] presents storage materials used in solar energy applications, for different temperature levels, and also taking into account the unitary costs, which lies between 0.1$/kg and 100$/kg.

Different PCM encapsulations can be encountered, namely bulk storage in tank heat exchangers, requiring extensive heat exchange, macro-encapsulation and micro-encapsulation [1]. The macro-encapsulation consists in including the PCM inside a tube, sphere, cylinder, while the micro- encapsulation consists in including the PCM in a micro-sphere incorporated at its turn in a compatible material. The macro-encapsulation is the most widespread method [1]. More data on commercially produced PCM and geometries are reported by [4].

Regarding mathematical modeling of PCM systems, also known as latent heat thermal energy storage system (LHTES), a review is reported by Verma et.al. [5]. The authors presented a synthesis of proposed methods by grouping them in two categories, namely methods based on First Law of Thermodynamics, and respectively methods based on Second Law. It was emphasized that the methods based on energy conservation have been validated, most of them, by experimental results, while the other category based on exergy analysis takes into account also the effect of time duration through which heat is supplied and the temperature of heat in storage, but they were not validated with experimental data.

Other numerical modelling of PCM behavior are based on software simulations, such as TRNSYS or ANSYS CFD. Al-Abidi et.al. [6] presents a review regarding CFD applications for LHTES in electronic cooling technology, building thermal storage, heating, ventilation and air conditioning, and other similar areas. Spherical, rectangular and cylindrical capsule geometries are discussed. One dimensional (1D) and two dimensional (2D) models are overseen, and only one 3D model is presented. Among these, for paraffin PCM, there are two 2D models that used power law for discretization and SIMPLE algorithm adopted for pressure–velocity coupling for a cylindrical PCM, and one 3D model that used the backward Euler scheme for the temperature discretization in time for a PCM bags in duct geometry. Results as melt fraction over time are presented.

Nayak et.al. [7] presents a comparison between three PCM materials (paraffin wax, sodium acetate tri-hydrate and phenolphthalein) used for an automobile application, for cold start of the engine. In their study, the PCM material is filled inside the PCM container placed in the middle of the engine cooling water heat exchanger. The simulation is done with Gambit and Fluent software tools and validated experimentally. The charge and discharge processes are presented. For example, paraffin charging period is presented for 1300s, starting from 27°C and reaching 48.1°C, while coolant water temperature varies from 70°C to 59°C.

Another application of paraffin wax is reported by Dakhil et.al. [8] for a solar heat pump. The simulation was performed in Fluent, as a 2D model using SIMPLE algorithm for discretization and k-ε model of turbulence. The PCM properties were supposed independent on temperature (melting temperature of paraffin 51°C and latent heat of 140 kJ/kg), initially being in solid state close to the melting temperature. A comparison in terms of temperature plot versus time is presented for CFD and experimental results and authors reported good agreement between them.

Li et.al. [9] studied the melting process of a paraffin blend RT27 inside a horizontal annulus considering water inside the inner tube by applying a CFD simulation. Sensitivity studies were developed to determine the optimal geometric configuration.

Among PCM, paraffins are widely used in applications involving solar energy systems.

The choice of the TES component is very important. In our previous studies, we have modelled an absorption cooling system (ACS) fueled by solar energy using water tanks for thermal energy storage. As the water tank temperature was very sensitive to ACS load, instability in system daily operation was emphasized. Thus, the importance of ensuring a constant temperature for the ACS desorber was found as being crucial not only for the system performances, but also for the system operation [10-12]. PCM storage could be used instead to maintain a constant input temperature to the ACS desorber.

Consequently, the target of the paper is to present the results for a numerical simulation of a PCM melting process performed under ANSYS CFD environment. A 3D model is presented for a tube filled in with paraffin and surrounded by an exterior one filled with a HTF (water) from solar collector. An inside understanding of the melting process is also performed, analyzing the advancement of the melting process in the PCM volume all along the PCM tube. Solid, mushy and liquid zones are emphasized at different sections along the PCM tube as time passes. Delays in melting time between two consecutive radial sections are put into evidence. The correct computation of PCM melting time in each section is essential for system evaluation accuracy, as it involves heat transfer phenomena.

2 System description

One of the most used geometries for a LHTES is shown in fig.1a and consists of many tubes, arranged in a regular manner under a pitch p, surrounded by a cylindrical shell. Inside de tubes the HTF is flowing. The space among the tubes and the cylindrical shell is filled in by the PCM. The number of tubes and length of the storage tank depend on the PCM quantity, which have to be used for energy storage. In the case of massive storage tank, the geometry of the system is too complicated for simulations. Frequently, one chooses a simplified geometry, consisting of only one tube and an annular enclosing piece of PCM, whose external diameter is $D_{pcm} = p$.

The simplified geometry used in this simulation is presented in fig.1b. The inner diameter of the tube through which the HTF is flowing is Din = 0.0254 m, while the external diameter of the annular enclosure filled in with PCM has a value of D_{pcm} = 0.0508

m. The length of the simplified storage system is L=2 m. Additionally, another five radial test sections were considered in order to analyze the melting process advancement. The first and the last ones are placed at 0.01 m for the inflow and outflow sections, respectively (z=0.01m and z=1.99m). The other three sections are positioned at z = 0.5, 1.0 and 1.5 m measured from the inflow section. Water has used as HTF, and paraffin-wax as PCM.

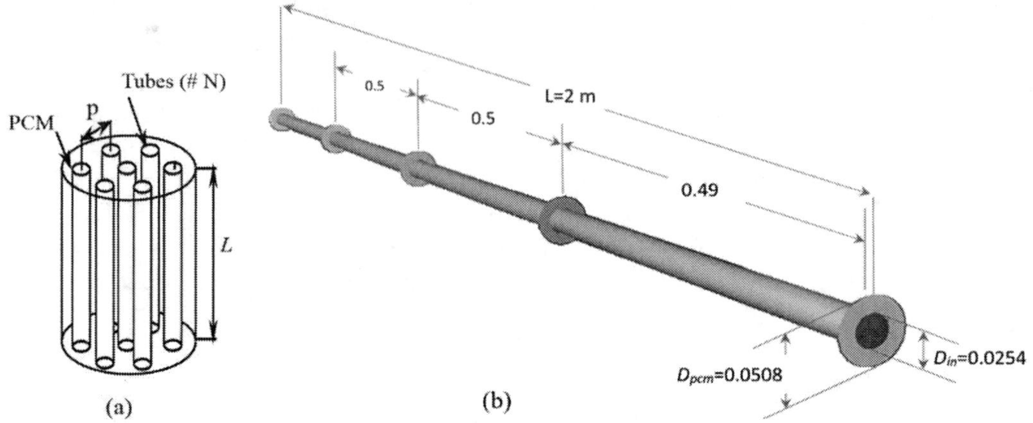

Fig. 1. The view of a thermal latent storage tank (a) and of a simplified geometry used for computation (b).

The water inlet temperature and water mass flow rate are set to $T_{htf,in}$ = 60°C and $\dot{m}_{htf}$=0.005 kg/s, respectively. The water flow is laminar and experiences the hydrodynamic and thermal developing phenomena on a certain length from the inlet section.

The melting latent heat and melting temperature of paraffin-wax are Δh_{sl} =164 kJ/kg and T_{melt} = 50°C respectively. The thermophysical properties of solid and liquid phases are shown in table 1.

Table 1. Thermophysical properties of paraffin-wax

Paraffin-wax	Specific heat at constant pressure, c_p [J/kg-K]	Thermal conductivity λ [W/m-K]	Density, ρ [kg/m^3]
Solidus	2130	0.28	818
Liquidus	2620	0.15	760

However, from numerical reasons, one has considered that the density of both solid and liquid phases have the same value, ρ = 818 kg/m^3 and the solid phase starts to melt at T_{ms} = 49.5°C and finishes melting at T_{ml} = 50.5°C. Between these temperature values, the liquid and solid phases coexist.

The heat transfer process between the HTF and PCM is obviously unsteady. Since the dynamics of the melting process is investigated, the initial temperature values used for starting the computations are set at 47°C for both the HTF and PCM domains.

3 Mathematical model and numerical procedures

The 3D mathematical model relies on the continuity, momentum and energy equations for the HTF domain, written for laminar flow and on energy equation for the PCM domain. These equations were solved in the ANSYS FLUENT v14.0 environment where the solidification & melting model was selected. This model neglects the natural convection occurring in the PCM liquid phase but considers a mushy zone as transition between the liquid and solid PCM phases. The enthalpy-porosity technique treats the mushy region (partially solidified region) as a porous medium. The first order implicit time discretization scheme and second order upwind spatial discretization algorithm were employed in calculations. For pressure-velocity coupling, the SIMPLE algorithm was set.

Fig. 2 presents some details of the final grid used in computations.

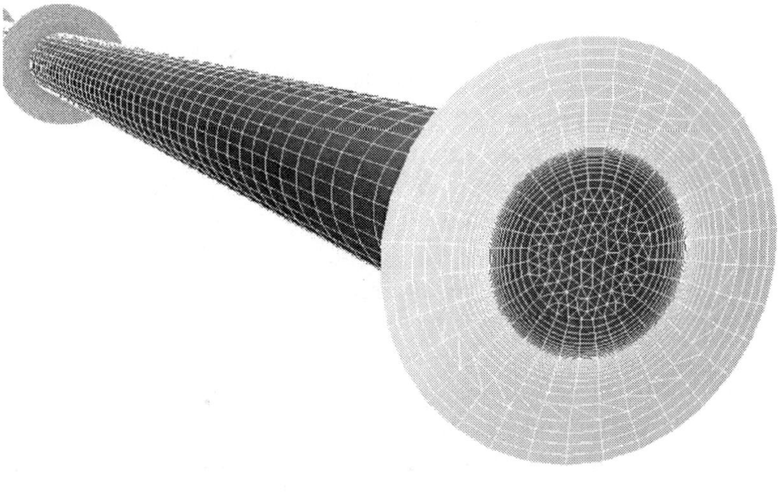

Fig. 2. Details of final grid used in numerical simulations.

This grid was obtained after establishing the independency of the numerical solution with respect to spatial discretization and contains about 627000 faces and 230000 nodes. The time step used for advancing in time the numerical solution was $\Delta\tau = 20$ seconds.

4 Results and discussions

Figs.3-6 present the temperature profiles and the distributions of liquid mass fractions in the considered sections of the system at certain time moments. Let analyze first the temperature distributions in HTF domain, and after that the dynamic of both, temperature change and melting process evolution occurring in PCM region.

In the entrance region of the duct, the HTF flow experiences hydrodynamic and thermal developing phenomena. Due to the higher *Pr* value of water, the thermal developing length is more important than the hydrodynamic one, which may represent an advantage concerning the PCM melting occurring in this first TES length. Along the time, tube walls temperatures slightly increase due to rise of the PCM temperature near the outer wall of the tube. As a result, the heat flux transferred from the HTF to PCM decreases as the time increases. Whatever the time is, the HTF temperature profiles at z = 0.01 and z = 0.5 correspond to thermal developing flow regions, which ends somewhere above this last section. On the upper radial sections, the temperature profiles reveal that the convection heat transfer takes place in the thermal developed regime, meaning that the convection heat transfer coefficient is constant.

In what follows, the time variation of temperature change inside the PCM domain and the time advancement of melting process is explained. After 600 seconds (see fig.3), the PCM temperature already increased in all testing sections over the value of the initial temperature (i.e. 320 K), but differently. As expected, the highest temperature gap is happening at z = 0.01 m, where the temperature lies between 332.5 K on the outer wall of the tube and 322 K on the outside diameter of PCM domain, approximately. As a result, in this section, the melting process is already developing. The PCM liquid phase extends until a radius of about r = 0.016m, where the temperature is 323.5 K. At lower radius values, the liquid phase of PCM is obviously superheated. At higher radius, the mushy zone extends until the PCM temperature reaches a value of 322.5 K, and above this radius, the solid region still exists. However, in all solid PCM region, the temperature increased above the initial temperature value.

One may observe that at z = 1.0 m, the melting process is in its early stage and the extent of mushy zone is lower than in the previous analyzed section, while at z = 1.99 m only the preheat of solid PCM is present.

As the time increases, heat spreads on the larger PCM area and the melting process advances, but its velocity decreases along the axis. After 1200 seconds, the temperature was further increased in all PCM volume and the liquid PCM phase lies until a radius of about 0.02 m at z = 0.01 m, and 0.0170 at z = 1.99m, respectively.

As shown in fig. 4, at t=8100 seconds, the melting is quite finished on z = 0.01 surface and has substantially advanced in all other considered sections. One also has to note that at z = 1.0 m only the liquid phase and the mushy zone exist, meaning that the melting process has already taken place in the most faraway area of PCM, whereas on z = 1.99 m a thin layer of solid PCM still exists.

As fig. 5 present, after a time of t = 10800 seconds, the melting was ended on about the first half (till z= 1 m) of the PCM volume. On the other half, excepting a short volume near the end of the tube, the mushy zone already reaches the PCM external surface. However, starting from this moment, one needs another 2100 seconds to fully melt the remaining part of solid and mushy PCM. As revealed by fig. 6, at this time the liquid PCM is entirely superheated, but the superheated degree decreases along the HTF flow direction.

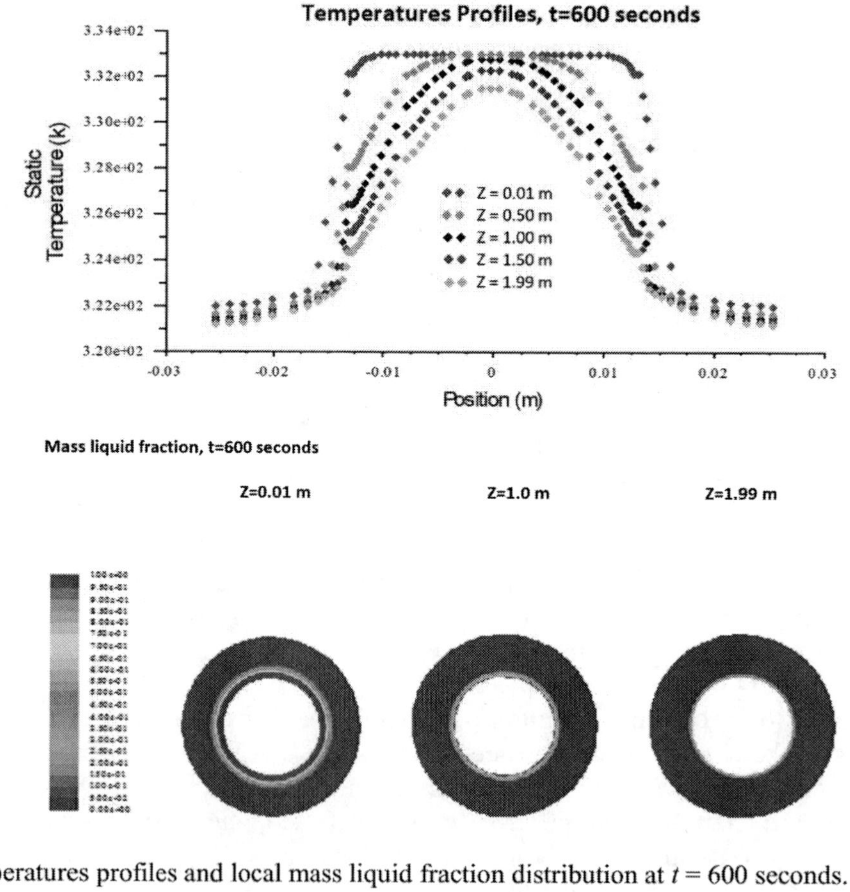

Fig. 3. Temperatures profiles and local mass liquid fraction distribution at t = 600 seconds.

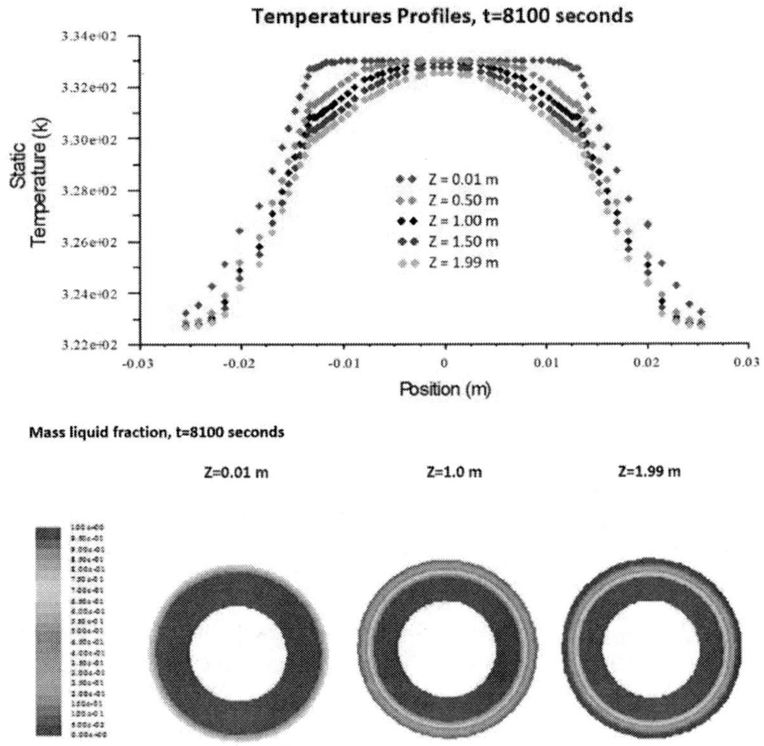

Fig. 4. Temperatures profiles and local mass liquid fraction distribution at $t = 8100$ seconds.

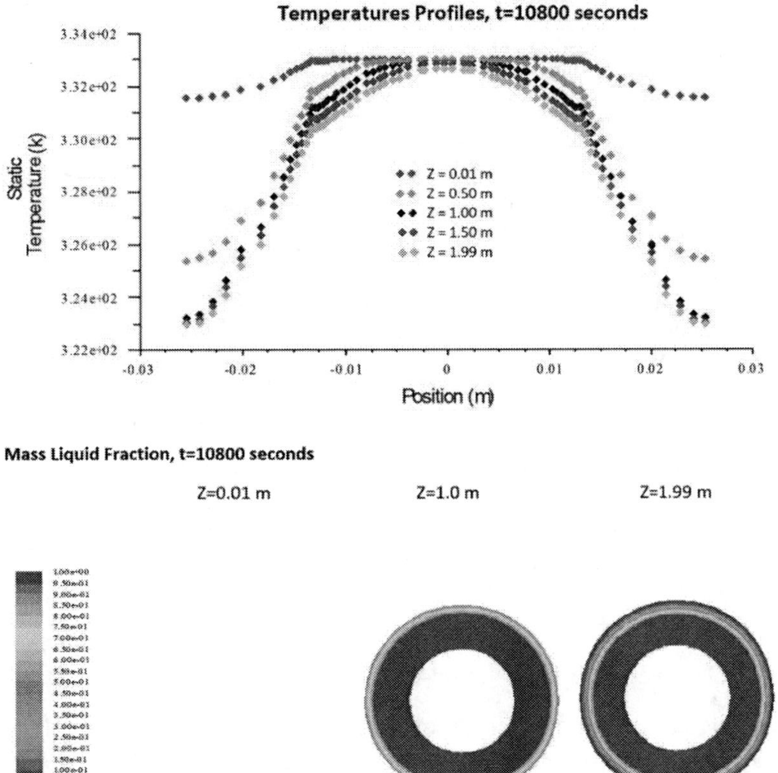

Fig. 5. Temperatures profiles and local mass liquid fraction distribution at $t = 10800$ seconds.

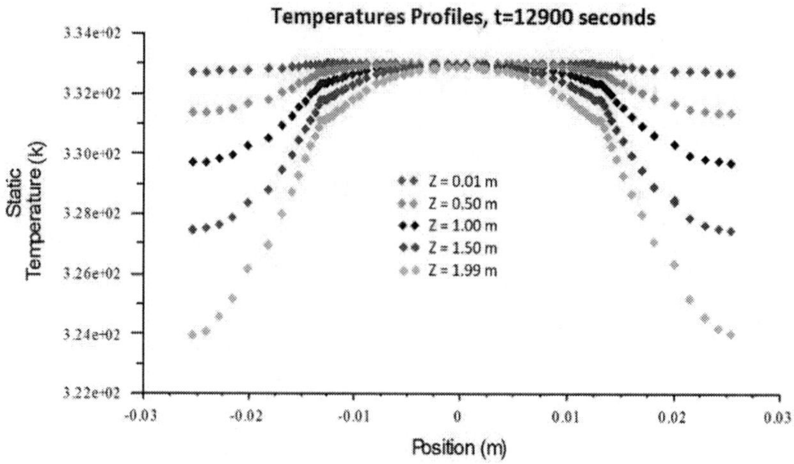

Fig. 6. Temperatures profiles at $t = 12900$ seconds. At this time, the liquefaction of the entire quantity of PCM is ended.

The time dynamics of PCM melting process is shown in fig. 7. There, the area-weighted averaged of mass liquid fraction in considered test sections is plotted vs. time. The radial variation shapes are sharper on sections placed at lower axial coordinate than those existing at higher ones. The most important shape difference appears between the radial sections of $z = 0.01$ m and $z = 0.5$ m. It is due to the fact that, on this length, both the convection heat transfer coefficient and the bulk temperature of HTF decrease with the increasing length. Once the fully thermal developed regime of HTF flow is reached, only the decrease of HTF bulk temperature along the flow direction is responsible for heat transfer decrease, so that the time shapes of area-weighted averaged of mass liquid fraction in the following sections become closer to each other.

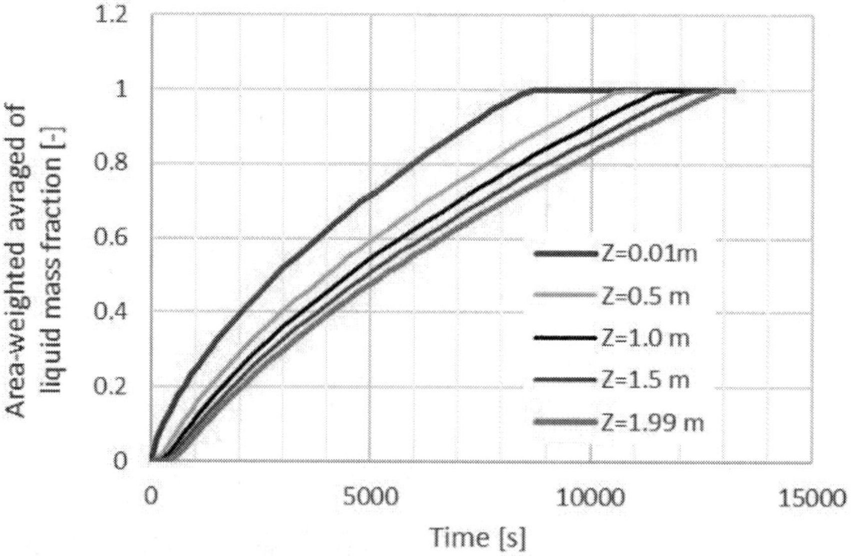

Fig. 7. Area-weighted average of mass liquid fraction vs. time.

The time needed to completely melt the PCM is considerably lower in the radial sections placed near the HTF entrance than that occurring at higher distances. For example, at $z = 0.01$ m the melting process ends after about 8580 seconds, whereas at $z = 0.05$ m it finishes after 10450 seconds. The delay in melting time between the two consecutive first radial sections is about 1780 seconds. For comparison, the melting time obtained between the following two successive sections is reduced to 1056 seconds and further it decreases to 724 seconds and 650 seconds, respectively, as presented in Table 2. This phenomenon occurs due to the fact that the HTF convection heat transfer and HTF bulk temperature decrease along the flow direction.

Table 2. Melting times in considered radial sections and melting time delays between two consecutive sections.

z [m]	0.01	0.5	1.00	1.5	1.99
Melting time [s]	8580	10450	11506	12230	12880
Delay melting time between two consecutive sections [s]	-	1870	1056	724	650

5 Conclusions

The study performed in the present paper emphasizes the melting process of a paraffin at different radial sections and time moments. The simulation was performed in CFD environment, considering a 3D model consisting in two concentric tubes, the inner one containing the water as HTF, while the outer one contains paraffin as PCM. The results obtained put into evidence the advancement of the melting process over time, analyzing the dimensions of the solid, mushy and liquid regions at different radial sections along the tube and at different time moments from the start process. As expected, the melting process is not uniform all along the tube, as the HTF convection heat transfer and HTF bulk temperature decrease along the flow direction. The delay melting time between the inlet and outlet sections is considerably high, reaching a value of 4300s, which is more than one hour. Over this period, the melt PCM in the inlet section is superheated.

The work can be further integrated in a solar energy conversion system, considering time dependent HTF temperature existing a solar collector and consequently simulating the stored energy and temperature level in the PCM component over time.

This work was supported by a grant of the Romanian Ministery of Research and Innovation, CCCDI-UEFISCDI, project number PN-III-P1-1.2-PCCDI-2017-0406/ NR. 81PCCDI / 2018 „ Innovative technologies for energy production from natural sources integrated in complex installations", within PNCDI III.

References

1. I. Sarbu, C. Sebarchievici, A Comprehensive Review of Thermal Energy Storage, Sustainability **10**, 191 (2018)
2. A. Sharma, V.V. Tyagi, C.R. Chen, D. Buddhi, Review on thermal energy storage with phase change materials and applications, Renew. Sustain. Energy Rev. **13**, 318–345 (2009)
3. G. Alva, L. Liu, X. Huang, G. Fang, Thermal energy storage materials and systems for solar energy applications, Renew. Sustain. Energy Rev. **68**, 693–706 (2017)
4. M. Kenisarin, K. Mahkamov, Solar energy storage using phase change materials, Renew. Sustain. Energy Rev. **11**, 1913–1965 (2007)

5. P. Verma, Varun, S.K. Singal, Review of mathematical modeling on latent heat thermal energy storage systems using phase-change material, Renew. Sustain. Energy Rev. **12**, 999–1031 (2008)
6. A.A. Al-Abidi, S. Bin Mat, K. Sopian, M.Y Sulaiman, A.T. Mohammed, CFD applications for latent heat thermal energy storage: A review, Renew. Sustain. Energy Rev. **20**, 353–363 (2013)
7. A.O. Nayak, G. Ramkumar, T. Manoj, R.Vinod, Comparative Study between Experimental Analysis and CFD Software Analysis of PCM material in Thermal Energy Storage System, Int. Journal of Chemical Engineering and Applications, **2**, 6 (2011)
8. S.F. Dakhil, K. Oudah, Experimental and CFD Analysis for the Solar Heat Pump by Using Phase Change Material, J. Appl. Res. Ind. Eng. Vol. **5**, No. 2 (2018) 169–180
9. S. Li, Y. Chen, Z. Sun, Numerical Simulation and Optimization of the Melting Process of Phase Change Material inside Horizontal Annulus, Energies *10*, 1249 (2017)
10. C. Stanciu, D. Stanciu, A. Gheorghian, Thermal analysis of a solar powered absorption cooling system with fully mixed thermal storage at startup, Energies **10**, 72 (2017)
11. C. Stanciu, D. Stanciu, A. Gheorghian, B. Tanase, C. Dobre, M. Spiroiu, Maximum exergetic efficiency operation of a solar powered H2O-LiBr absorption cooling system, Entropy **19**, 676 (2017)
12. C. Stanciu, D. Stanciu, B. Tanase, A. Gheorghian, C. Dobre, Storage tank mass control for optimum solar-powered absorption cooling system operation, Proc. of 8th Int. Conf. on Energy and Environment: Energy Saved Today is Asset for Future, CIEM 2017 - IEEE Power & Energy Society IEEE Xplore IEEE Catalog Number: CFP17L60-ART, p. 363-367 (2017), doi: 10.1109/CIEM.2017.8120860

A research on fuelling a truck diesel engine with hydrogen

Nikolaos Cristian Nutu[1,*], *Constantin* Pana[2], *Niculae* Negurescu[2], *Alexandru* Cernat[2], *Dinu* Fuiorescu[2], and *Ionel* Mirica[2]

[1]University Politehnica of Bucharest, Automotive Engineering Department, Bucharest, Romania
[2]University Politehnica of Bucharest, Termothechnics, Engines, Thermal Equipments, and Refrigeration Instalations Department, Bucharest, Romania

Abstract. The paper presents experimental and theoretical investigations results of a truck diesel engine fuelled with diesel fuel and hydrogen by diesel-gas method. The purpose of the paper is to reduce the level of the nitrogen oxides emission level and to improve the overall performances of the engine. The use of hydrogen represents a suitable method to reduce the classic fuels consumption, maintaining the energetic performances of the engine. The test bed situated in the Thermotechnics, Engines, Thermal Equipments and Refrigeration Instalations Department of University Politechnica of Bucharest was adapted for hydrogen fuelling of the truck diesel engine. The engine used for investigations is a turbocharged 10.2 l truck diesel engine which run at 1450 rpm and 40% engine load. By hydrogen fuelling if the investigated diesel engine a decrease has been achieved in the nitrogen oxides emission level and in the smoke emission level. Also the break energetic specific consumption decreased compared to the standard diesel engine. The theoretical investigations presented in tha paper are based on a semi-empirical corellation, used to simulate the nitrogen monoxide emission formation. The results of the experimental investigations confirm de viability of hydrogen as an alternative fuel for the diesel engine.

1 Introduction

The element hydrogen is greatly widespread in the universe, forming over 90% of the universe [1]. On the earth, hydrogen is found very little in the upper layers of the atmosphere and in very large quantities combined with oxygen and thus forming water. One km^3 of water contains 113 million tons of hydrogen [1]. Some of important properties of hydrogen are presented in the table 1, along with those of diesel fuel.

Table 1. The hydrogen properties [1].

Property [ ]	Diesel fuel	Hydrogen
Molar mass [kg/kmol]	226.44	2.016
LHV [kJ/kg]	41855	119600
Autoignition temperature [°]	200...220	565.56...582.22
Octane number [-]	-	>130
Density [kg/m^3]	800-840	0.089
Ignition energy [mJ]	-	0.02

It is complicated to provide hydrogen as unique fuel of compression ignition engines because the compression ratio being high the fuel-mixture has an uncontrolled tendency to ignite due to the very narrow range of usable dosages. This combined with low hydrogen ignition energy can result in elevated pressure rise rates and increased cycle to cycle dispersion. A method of controlling the combustion process that produces very good results is that hydrogen is used as an additive fuel, introduced in the form of vapors in the engine intake manifold. This fuelling procedure is also called the diesel gas method, whereby the air-hydrogen mixture formed in the intake manifold is admitted into the cylinders and ignited by the flame appeared in the previously injected diesel fuel sprays. It is therefore necessary to equip the engine with a separate gas fuel supply system capable of being permanently correlated with the conventional diesel fuel system.

Varde et al. in [2], studied the effects of hydrogen fuelling by introducing hydrogen into the intake manifold of a diesel engine. The main objective of this research was to reduce the particulate emission level in exhaust gases using hydrogen as addition fuel. Hydrogen substitution was equal to 10% (as a percentage of energy) and reduced the level of smoke emission at partial load. However, at the nominal load the reduction of the smoke emission level was more modest, most likely due to the reduced amount of air available in the cylinder. It has also been found that reduced hydrogen rates have had side effects on engine efficiency, but increasing the percentage of hydrogen rate notable improvements in efficiency has been obtained.

Haroun A.K. Shahad and Nabeel Abdul-Hadi conducted an experiment on a diesel engine with hydrogen injection in the intake, observing an increase in thermal efficiency of ~ 40% at engine load of 60-80%. At higher loads the thermal efficiency decreased dramatically due to incomplete combustion, the excess air ratio being low [3]. However, even the brake thermal efficiency increased, the NOx emissions increased with the increasing degree of substitution of diesel fuel with hydrogen at high engine loads, similar results were found in [4].

The paper [5] presents the effects of hydrogen addition on the energy and pollution performances of a diesel engine fueled with diesel fuel as a pilot and hydrogen-methane as a dual fuel. The effect of gaseous fuel on the maximum pressure inside the cylinder and on the heat release rate is relatively low at low and medium loads [5]. At high loads, however, the addition of gaseous fuel contributes to increasing the efficiency of the combustion process and increasing the maximum pressure inside the cylinder. Also the increase in the amount of hydrogen has led to an increase in the maximum pressure. At low and medium engine loads the authors determined that the maximum in cylinder pressure drops with the increase in the amount of gaseous fuel injected into the intake manifold (hydrogen +

methane). At high engine loads, the cylinder pressure increased due to increased temperature and intensified burning process, with a relatively high flame speed in the air-hydrogen-methane mixture [5]. Also, the burning process has become unstable and difficult to control [5]. This is confirmed by Gatts et. al. [6], which suggests that the efficiency of the combustion is related with the engine load and that hydrogen should be added at higher loads in order to improve the engines's brake thermal efficiency [6]. The nitrogen oxides emission level presented in the paper [5] increased in the case of fuelling with hydrogen compared to the classic diesel fuel supply, especially for medium and large engine loads. One reason that could explain the increased NOx level is the increase in the global temperature of the combustion process [5]. In the case of running at low engine loads, the hydrogen fueled engine produces lower levels of nitrogen oxides, but with increasing hydrogen quantity, the emission level begins to rise [5]. The authors also found that in the case of fuelling with hydrogen, the nitrogen dioxide emission is present in large quantities in the exhaust gases of the engine, which is very dangerous for humans and the environment, but it is beneficial for reducing the level of smoke emission, the NO_2 emission playing an essential role of a very good soot oxidant [5]. This is confirmed in the paper [7]. The level of unburned hydrocarbon emissions increased for almost all cases where the engine was fueled with gaseous fuel, except for hydrogen. A cause of the increase in the emission level could be the loss of a gaseous fuel quantity during the overlapping period of the valves [5], [8]. The paper [9] explains the increase in the level of unburned hydrocarbon emissions by lowering the local excess air ratio when the dual fuel is used.

In [10] by supplying hydrogen with diesel fuel, CO, HC and smoke emissions decreased due to improved combustion, but a slight increase in CO_2 emissions is observed [10].

This paper presents experimental and theoretical investigations results of a truck diesel engine fuelled with diesel fuel and hydrogen by diesel-gas method. The purpose of the paper is to reduce the level of the nitrogen oxides emission level and to improve the overall performances of the engine. The engine used for investigations is a turbocharged 10.2 l truck diesel engine and the working regimen was 1450 rpm and 40% engine load.

2 Material and method

2.1 The test bed description

For this paper the theoretical and experimental study were carried out having as a material support an experimental test bed equipped with a 10.2 l truck compression ignition engine. The engine main specifications are shown in the table 2.

Table 2. The main specifications of the D2156MTN8 engine.

Number of cylinders	6
Bore [mm]	121
Stroke [mm]	150
Displacement [l]	10.34
Compression ratio	17
Rated power [kW]	188
Maximum torque [Nm]	900
Admision type	turbocharged

The laboratory equipments are: a Hoffman eddy current dynamometer, an AVL data aquisition system, a Kistler piezoelectric pressure transducer, an Allicat Stiintific mass

 https://doi.org/10.1051/e3sconf/201911201011

flowmeter, a Meriam volume flowmeter, an AVL Dicom gas analyser and opacimeter, multiple thermocouples and thermoresistances for temperature measuring and a gravimetric fuel consumption measuring system to determine the diesel fuel consumption.

2.2 The nitrogen monoxide emission theoretical model

As presented in the abstract, the nitrogen monoxide emission level was computed with an semi-emprical mathematical equation and compared with measured results. The mathematical model which describes the NO emission formation is given by the formula 1 [11].

$$|NO| = C_1 * (\lambda - 0.89)^{2.11} * n^{-0.39} * Q_z^{2.49} * m_a^{-5.01} * \left(\int_0^{Q_z} T^{11} p^9 dQ_z\right)^{0.28} \text{[ppm]} \quad (1)$$

λ – exces air ratio;
n [rot/s] – the engine speed;
Qz [kJ/cycle] – the heat released;
m_a [g] – fresh charge mass;
T [K] – the temperature;
p [bar] – the pressure;
C_1 – calibration constant.

The calibration constant used in this model has the value of 1. Because the combustion process period can be divided in small intervals for the simulation purpose, the integral presented in the equation (1) can be replaced with a sum of pressure, temperature and the average rate of heat released [1].

$$\int_0^{Q_z} T^{11} p^9 dQ_z = \sum {T_m}^{11} * p^9 * \Delta Q_z \quad (2)$$

2.3 The working regimen and the method description

At the begining, when the experimental investigation started the engine was fuelled with diesel fuel. The purpose was to determine the standard regimen in order to maintain it as a reference. So the first investigated point was at 1450 rpm and 40% engine load, having diesel fuel as a fuel. After the reference was created the diesel fuel was partially substituted with hydrogen by introducing it in its gaseous aggregation state in the intake manifold. The main goal was to maintain the engine performances at the same level like in the reference case and to decrease the pollutant emissions and the specific fuel consumption. This was accomplished by decreasing the diesel fuel quantity and adding hydrogen. The energetic diesel fuel substitution ratio with hydrogen is given by :

$$x_c = \frac{m_{H2} H_{i_{H2}}}{m_{H2} H_{i_{H2}} + m_{dieselfuel} H_{i_{dieselfuel}}} \quad (3)$$

m_{H2}- the hydrogen quantity;
$m_{dieselfuel}$-the diesel fuel quantity;
H_i- the lower heating value.

The energetic diesel fuel substitution ratio with hydrogen was situated between [1.14-4.81].

3 Results and discussions

3.1 The maximum in cylinder pressure and the maximum rate of pressure rise

Because the diesel fuel substitution rate with hydrogen was relatively low, a good control over the combustion process was achieved so the maximum in cylinder pressure and the maximum rate of pressure rise were maintained at aproximatelly the same level like in the case of fuelling only with diesel fuel. The maximum pressure has a slightly increasing trend with increasing degree of substitution but the maximum deviation is only 1 bar. The figures 1 and 2 presents the maximum in cylinder pressure and the maximum rate of pressure rise, having on X axis the degree of substitution x_c.

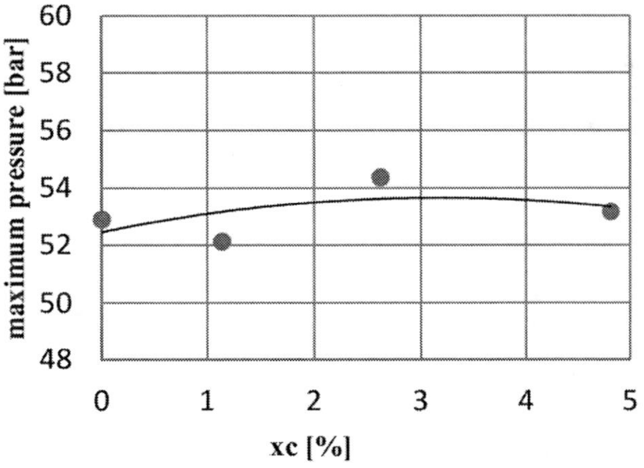

Fig . 1. The maximum in cylinder pressure.

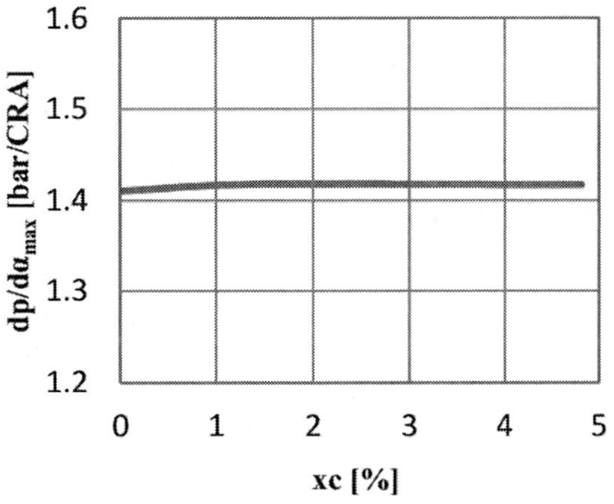

Fig. 2. The maximum rate of pressure rise.

3.2 The nitrogen oxides emission level

The measured nitrogen oxides emission level decreased for all the degrees of substitution of diesel fuel with hydrogen. An explanation for this phenomenon could be the richer local mixture in the case of fuelling with hidrogen. Also the mixture increasing degree of homogeneity tends to decrease the global temperature. The figures 3, 4, 5, 6, 7 presents the NO_x emission level for all the investigated cases. In the figure 3 is presented comparatively the measured and computed level. It can be seen that the semi-empirical equation predicts a lower level in the case of fuelling with hydrogen. The reason may be the fact that the equation predicts only the formation of NO and according to [5] when hydrogen is used as an additive fuel inside the cylinder is forming large quantities of NO_2, which represents a good catalyst for the soot emission [5]. The figures 4, 5, 6, 7 presents the forming laws of NO over the in cylinder temperature. The temperature is obtained by processing the pressure diagrams. It can be also observed that the maximum level of the NO emission is obtained at a short time after the moment when the maximum temperature is reached and the NO emission start its forming process in the first stages of combustion, in the preformed mixtures combustion period.

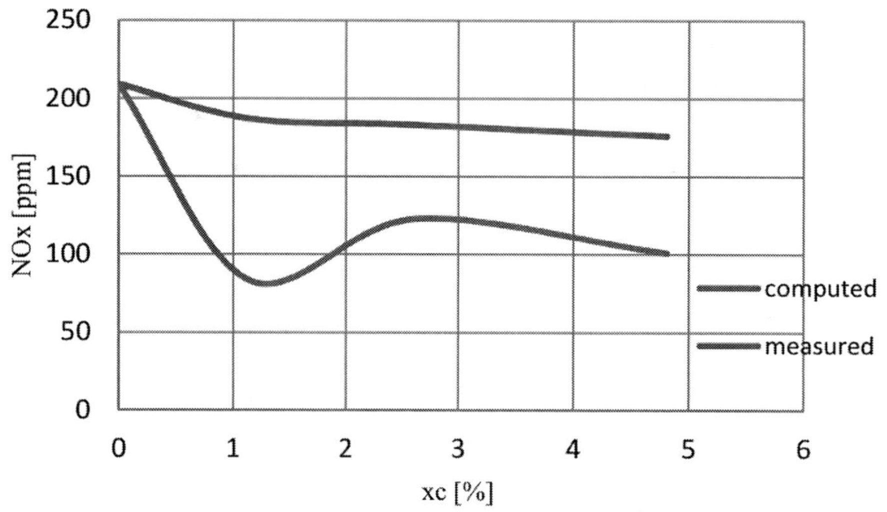

Fig. 3. The measured and computed nitrogen oxides emission level versus the degree of substitution.

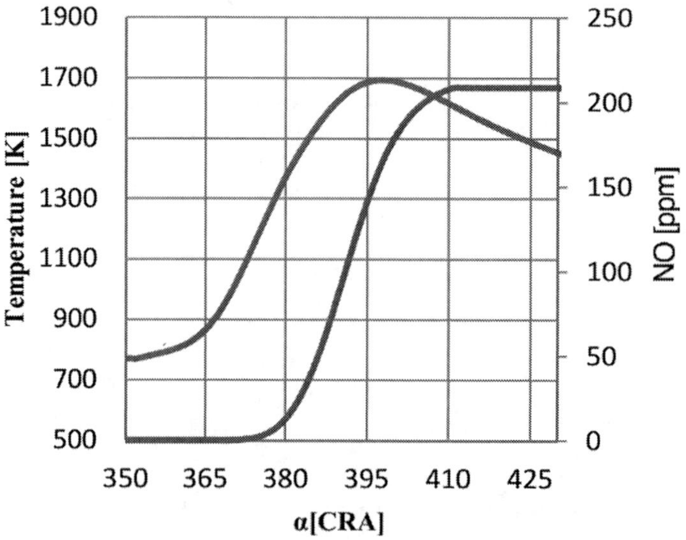

Fig. 4. The computed temperature and NO forming law for the case of x_c=0.

https://doi.org/10.1051/e3sconf/201911201011

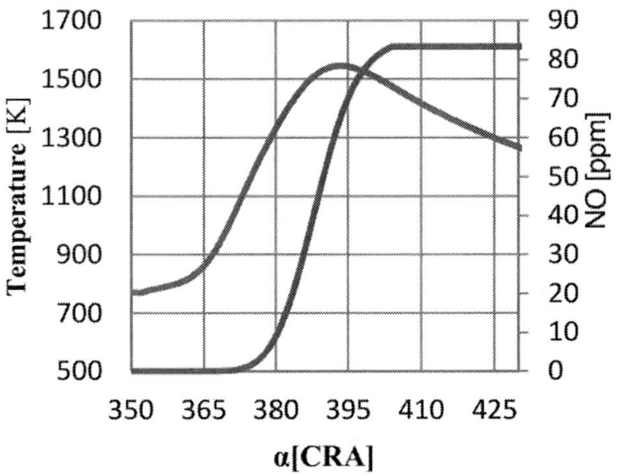

Fig. 5. The computed temperature and NO forming law for the case of x_c=1.14.

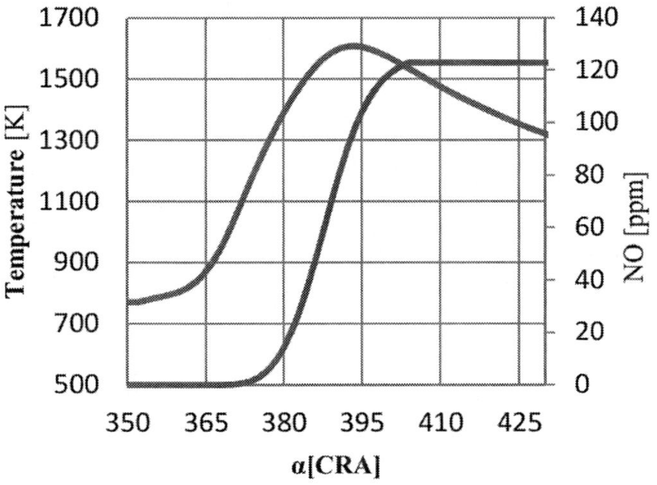

Fig. 6. The computed temperature and NO forming law for the case of x_c=2.62.

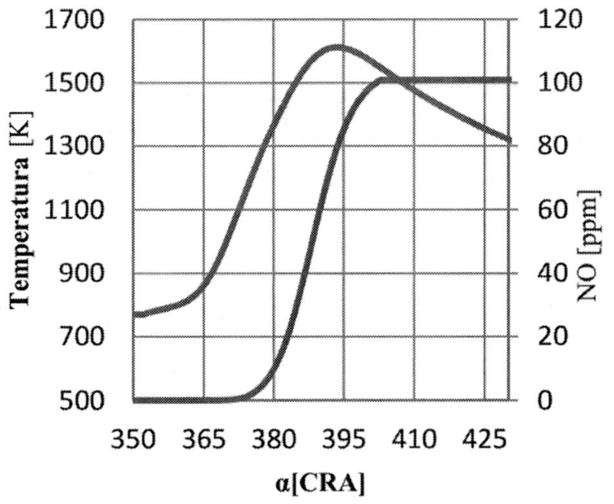

Fig. 7. The computed temperature and NO forming law for the case of x_c=4.81.

3.3. The smoke emission level

The smoke emission level decreased for all the cases when the engine was fueled with hydrogen because when the gaseous fuel is present inside the combustion chamber the number of carbon atoms is lower, the burning rate of diffusive mixtures decreases and according to [5] the nitrogen dioxides emission represents a catalyst for the soot emission. The smoke emission is presented in the figure 8, evaluated by the exhaust gases opacity.

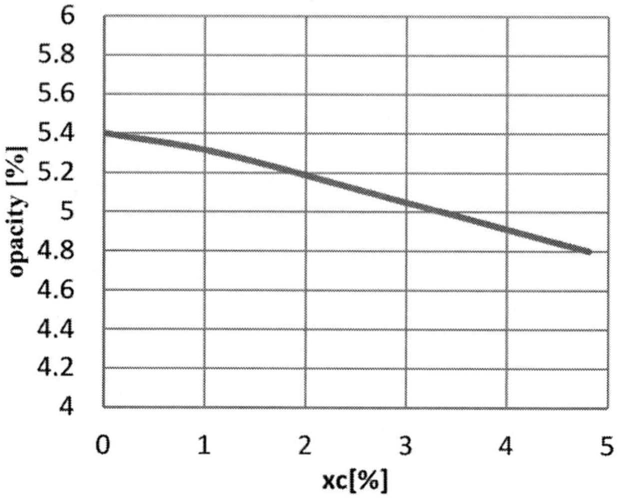

Fig. 8. The smoke emission level.

3.4. The energetic specific fuel consumption

The energetic specific fuel consumption decreased for all the cases of fuelling with hydrogen, with its lowest level at the degree of substitution x_c=1.14, because in the presence hydrogen in the combustion chamber the combustion process intensifies and become more efficient. The figure 9 presents the energetic specific fuel consumption.

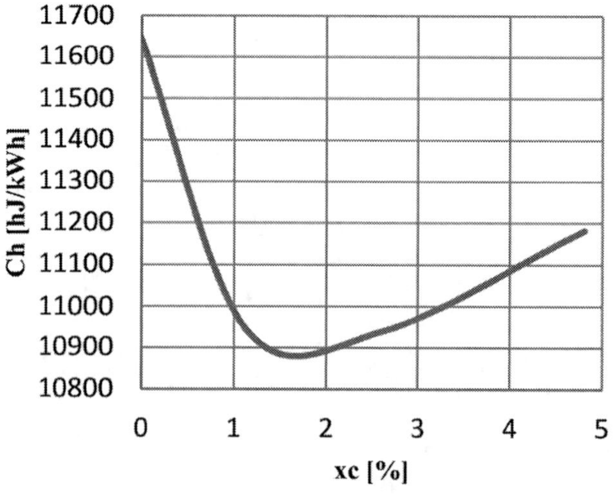

Fig. 9. The energetic specific fuel consumption.

4 Conclusions

- The maximum in cylinder pressure and the maximum rate of pressure rise were maintained at aproximatelly the same level like in the case of fuelling only with diesel fuel. The maximum pressure has a slightly increasing trend with increasing degree of substitution but the maximum deviation is only 1 bar.
- In the case of hydrogen fuelling the nitrogen oxides emission level decreased.
- The smoke emission level decreased for all the cases when the engine was fueled with hydrogen with a maximum of 11% compared with the diesel fuel case.
- The energetic specific fuel consumption decreased for all the cases of fuelling with hydrogen, with its lowest level at the degree of substitution x_c=1.14.

The authors would like to express special thanks to AVL List GmbH Graz, Austria, for offering the opportunity to use the research equipments. This work has been funded by the Romanian Ministery of Research and Innovation, CCCDI – UEFISCDI, grant PN-III-P1-1.2-PCCDI-2017-0404 / 31PCCDI/2018.

References

1. M.G. Popa, N. Negurescu, C. Pana, *Motoare Diesel. Procese,* MatrixRom, Bucharest, (2003).
2. K.S. Varde, G.A. Frame (1983), *Hydrogen aspiration* in a *direct injection type diesel engine-its effects* on *smoke* and *other engine performance parameters,* Int J Hydr Energ, 159–179, (1983).
3. A. K. Shahad Haroun, A.H. Nabeel, *Experimental Investigation of the Effect of Hydrogen Manifold Injection on the Performance of Compression Ignition Engines*, World Academy of Science, Engineering and Technology **76**, (2011).
4. W. B. Santosoa, R. A. Bakara , A. Nurb, *Combustion characteristics of diesel-hydrogen dual fuel engine at low load*, Energy Procedia, **32**, (2013).
5. J. H. Zhou, C. S. Cheung, C. W. Leung, *Combustion and Emission of a Compression Ignition Engine Fueled with Diesel and Hydrogen-Methane Mixture*, World Academy of Science, Engineering and Technology International Journal of Mechanical, Industrial Science and Engineering **7**, (2013)
6. T. Gatts, H. Li, C Liew, S. Liu, T. Spencer, T. Wayne, et al., *An experimental investigation of H2 emissions of a 2004 heavy-duty diesel engine supplemented with H2*, Int J Hydrogen Energ, **35,** (2010).
7. S.Y. Liu, H.L. Li, T. Gatts, C. Liew, S . Wayne, G. Thompson, et al., *An Investigation Of No2 Emissions From a Heavy-Duty Diesel Engine Fumigated with H2 and Natural Gas, Combust Sci Technol.*, **184**, (2012).
8. T. Gatts, S. Liu, C . Liew, B. Ralston, C. Bell, H. Li, *An experimental investigation of incomplete combustion of gaseous fuels of a heavy-duty diesel engine supplemented with hydrogen and natural gas,* Int JHydrog Energy, (2012).
9. B. Sahoo, N. Sahoo, U. Saha, *Effect of engine parameters and type of gaseous fuel on the performance of dual-fuel gas diesel engines—A critical review,* Renewable and sustainable Energy Reviews, **13, (**2009).
10. M. Talibi, P. Hellier, R. Balachandran, N. Nicos Ladommatos, *Effect of hydrogen-diesel fuel co-combustion on exhaust emissions with verification using an in-cylinder gas sampling technique*, International journal of hydrogen energy, **39,** (2014).
11. G. Pădure, M. Pădure, *Mecanisme ale formarii oxizilor de azot in motorul cu ardere interna*, Zilele Academice Timisene editia, **8**, (2001).

E3S Web of Conferences **112**, 01012 (2019) https://doi.org/10.1051/e3sconf/201911201012
TE-RE-RD 2019

Waste-Energy (Heat) Recovery System from the Gases Compressed by an Oil-Free Screw Compressor

Iulian Vlăducă[1,*], *Ramona* Stanciuc[1], *Răzvan* Bimbaşa[1] and *Sorin Gabriel* Tomescu[1]

[1]INCDT COMOTI, Bucharest, Romania

Abstract. During compression, a gas heats up, almost in all cases this heat being wasted, either by cooling the gas because it is too hot for the application, or by storing the gas and letting the compressed gas cool naturally in the storage tank. This paper presents a waste-energy (heat) recovery system from the gases compressed by an oil-free screw compressor. The gases compressed by this compressor have a very high temperature compared to an oil injected screw compressor, due to the fact that the oil used to lubricate the rotors also acts as a heat sink, the oil free variant which is used when you want a very high purity of the gas, has higher tolerances and more friction between the rotors which result in a higher gas temperature. The recovery system uses a heat exchanger to extract the waste energy from the gas and at the same time it will cool it for immediate use. Depending on the requirements, the energy recovered may be used immediately to produce useful work or stored for a later use. It may be used for heating a building, to produce steam for a turbine driving electrical generator, or in other forms.

1 Introduction

A method for rational use of energy in air compression is heat recovery, which is a secondary product in the compressed air production. Thus, the primary energy is directly transformed into compressed air, and indirectly into heat, which can be used in central heating systems. There is a great potential of energy that can be saved or recovered from compressed air systems, also having great potential for improving both the energy savings and CO_2 emissions [1].

Companies have developed their own heat recovery systems in self-contained complete systems, rotary screw compressors, boosters and blowers that are particularly well suited for heat recovery. The compressor's exhaust heat can be used as hot air for spaces or process heating purposes. For example, by the direct use of the recyclable heat via an exhaust air ducting system, up to 96% can be used, or by using a warm cooling medium that can heat water up to 70°C via a heat exchanger, up to 76% of heat energy can be recovered [2].

In a compressed air energy storage (CAES) system, proposed, the compressed air is stored in a pressurised tank, and the air is than released in an expander turbine to generate electricity.

* Corresponding author: iulian.vladuca@comoti.ro

The disadvantage is that when the compressed air is released, it cools, reducing the amount of available energy.

The key element for improving the efficiency is to develop a high temperature air storage system. The energy conversion process can be adiabatic or diabatic (nearly isothermal), and also in other thermodynamic cycle configurations. In a diabatic system, additional heating of the expanded air is required to generate additional output energy. In a adiabatic system, the compression heat is stored at a higher temperature in the storage tank and hence a generation of a larger quantity of energy output can be expected [3].

2 Heat recovery system design

The fundamental research of new gas compression solutions has led to innovative solutions in the development of oil-free helical compressors. The TURBO 2020 Core Program, carried out with the support of MCI, conducts a fundamental research in line with the general objectives of the INCDT COMOTI on the development of oil-free rotary machines based on the knowledge developed on oil-injected screw compressors. For this purpose, was purchased as a study material, an oil-free compression unit type CD14D1 [4, 5] from GHH RAND, Germany (Figure 1).

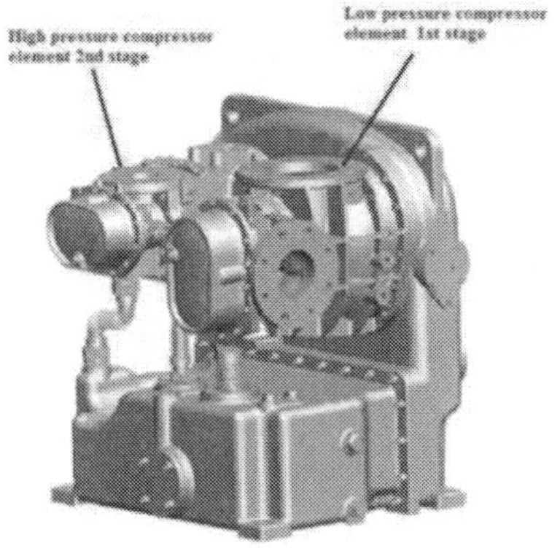

a.

b.

Fig. 1. a. Oil-free compression unit type CD14D1 [1, 2]; b. the purchased unit from GHH-Rand

The following table shows the different compressor stages and their application in the respective compressor blocks [4]:

Table 1. Assignment of compressor stages to compressor blocks.

Type		Single-stage blocks	Double-stage blocks	Booster
1st stage (low pressure)	CDA80	-	CD8D	-
	CDA82	CD9S	CD14D1	-
	CDA96	CD14S	**CD14D1**	-
	CDA128	CD23S CD26S	CD26D	-
	CDA160	CD42S	CD42D1	-
	CDA208	CD72S1	-	-
2nd stage (high pressure)	CDB54	-	CD8D	CD8HD
	CDB68	-	**CD14D1**	CD14HD1
	CDB90	-	CD26D	CD26HD
	CDB114	-	CD42D1	-

The cells with bolded frame in the left in Table 1 are the units composing the double stage CD14D1 compressor. In Table 2 are the nominal parameters of the compressor stages. In the paper the values from the columns with normal frame were used.

Table 2. Nominal limitations of using CD-D without CD8D (according to clearance classes).

Parameter	1st stage		2nd stage	
	close	open	close	open
Maximum suction temperature [°C]	46	50	51	60
Maximum discharge temperature [°C]	227	250	227	270
Maximum discharge pressure [bar absolute]	3.96	4.17	8.63	12.7

From the technological diagram shown in the figure 2, we have:

First stage of compression with CDA96 compressor unit, which gives compressed air at 4.17 bar (abs) and 250°C, and second stage of compression with CDB68 compressor unit, that give compressed air at 12.7 bar (abs) and 270°C (see tables 1 and 2 from above). The flow of the compressor at maximum load is M_{air} = 1330 m^3/h = 0.436 kg/s, at 20°C and at an air humidity of 60%, see also figure 3, the output data done with T1996WIN, Vers. 2.2.22, GHH Rand software.

In the technological diagram, figure 2, are represented the coolers for that two stages of compression in the situation of a beneficiary request for 12 bar (abs) compressed air at 60°C.

The amount of the lost heat rejected and the consumption of energy is significant. The quantity of the rejected energy in the first and second stages is given by the equation:

$$Q_{air,comp} = M_{air} \cdot c_{p,air} \cdot (T_2 - T_1) \tag{1}$$

We have the following amount of heat lost for the two compression stages:

$$Q_{air,comp,1st} = M_{air} \cdot c_{p,air} \cdot \left(T_{2,1st} - T_{1,1st}\right) \cong 88 \text{ kW} \tag{2}$$

$$Q_{air,comp,2nd} = M_{air} \cdot c_{p,air} \cdot \left(T_{2,2nd} - T_{1,2nd}\right) \cong 92.4 \text{ kW} \tag{3}$$

$$Q_{air,comp,1st+2nd} = Q_{air,comp,1st} + Q_{air,comp,2nd} \cong 180.4 \text{ kW} \tag{4}$$

Converting in kcal/h and Btu/h, we have:

$$Q_{air,comp,1st+2nd} = 155116 \text{ kcal/h, or } 615550 \text{ Btu/h} \tag{5}$$

For an air to air cooler driven by an electrical motor at 1500 rpm, with a dimension of the cooler = 1200 mm (tube length) x1500 mm x 140 mm, the total electrical power lost for the cooling the air of two compressor stages is E_{lost} = 11 kW.

If we compare with the energy consumed by the main electrical motor driving the compressor unit that is about is $E_{consumption}$ = 173 kW (calculations done with T1996WIN, Vers. 2.2.22, GHH Rand software, see figure 3, the output data), the lost energy used for cooling the air represent about:

$$\eta_{lost} = \frac{E_{lost}}{E_{consumption}} \cdot 100 = 6.35\% \tag{6}$$

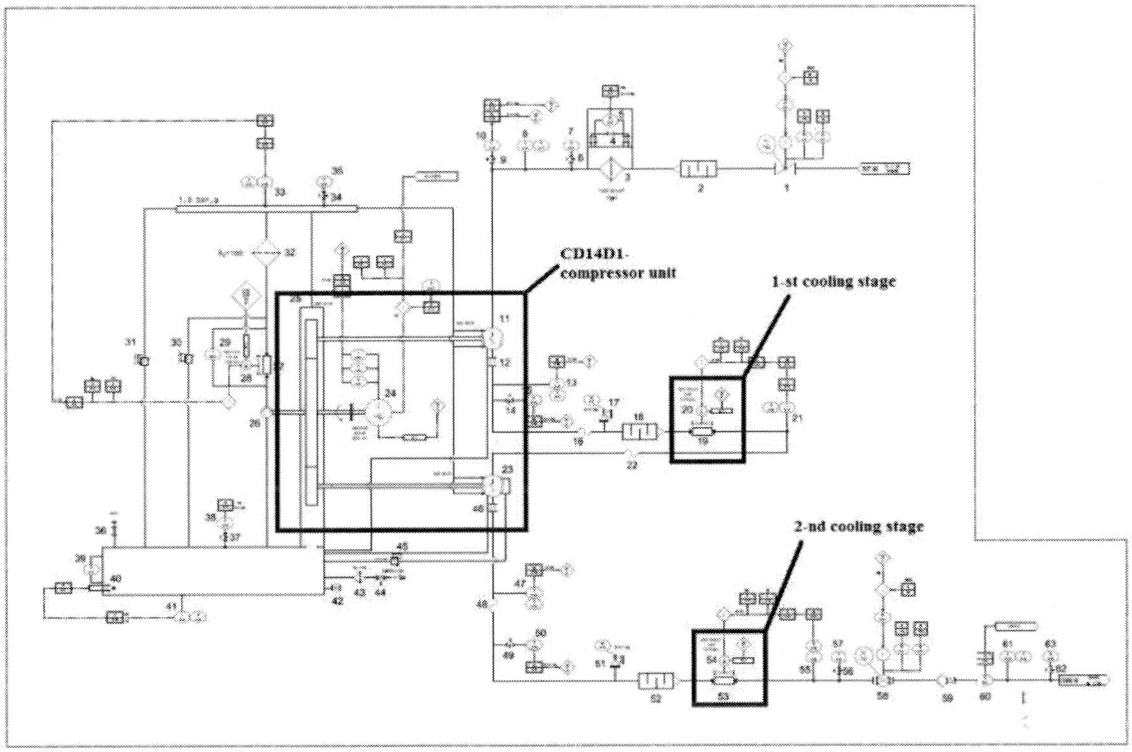

Fig. 2. The technological diagram for double oil-free compressor with CD14D1 unit

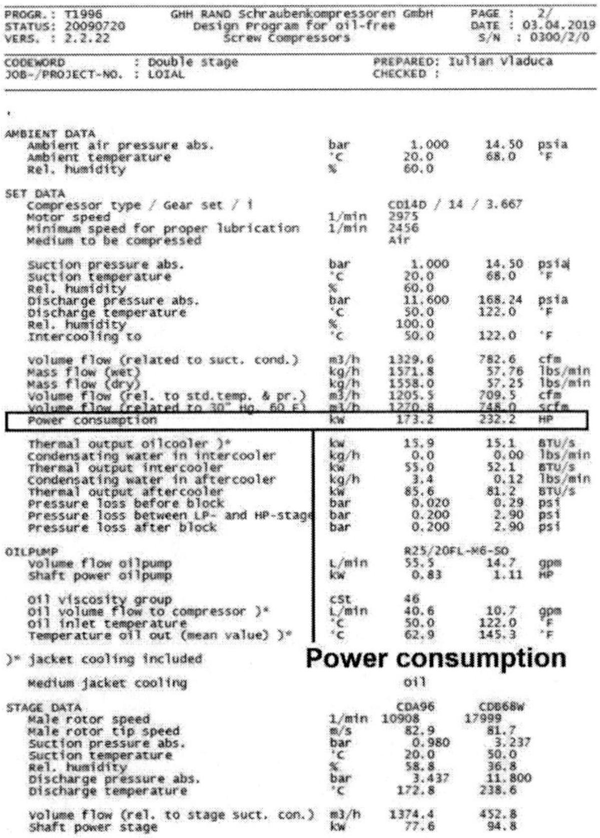

```
PROGR.: T1996        GHH RAND Schraubenkompressoren GmbH     PAGE :   2/
STATUS: 20090720        Design Program for oil-free          DATE : 03.04.2019
VERS. : 2.2.22            Screw Compressors                   S/N  : 0300/2/0

CODEWORD             : Double stage               PREPARED: Iulian Vladuca
JOB-/PROJECT-NO.     : LOIAL                      CHECKED :

AMBIENT DATA
   Ambient air pressure abs.                  bar       1.000     14.50  psia
   Ambient temperature                        °C       20.0       68.0   °F
   Rel. humidity                              %        60.0

SET DATA
   Compressor type / Gear set / i                      CD14D / 14 / 3.667
   Motor speed                                1/min    2975
   Minimum speed for proper lubrication       1/min    2456
   Medium to be compressed                             Air

   Suction pressure abs.                      bar       1.000     14.50  psia
   Suction temperature                        °C       20.0       68.0   °F
   Rel. humidity                              %        60.0
   Discharge pressure abs.                    bar      11.600    168.24  psia
   Discharge temperature                      °C       50.0      122.0   °F
   Rel. humidity                              %       100.0
   Intercooling to                            °C       50.0      122.0   °F

   Volume flow (related to suct. cond.)       m3/h   1329.6      782.6   cfm
   Mass flow (wet)                            kg/h   1571.8       57.76  lbs/min
   Mass flow (dry)                            kg/h   1558.0       57.25  lbs/min
   Volume flow (rel. to std.temp. & pr.)      m3/h   1205.5      709.5   cfm
   Volume flow (related to 30" Hg. 60 F)      m3/h   1270.8      748.0   scfm
   Power consumption                          kW      173.2      232.2   HP

   Thermal output oilcooler )*                kW       15.9       15.1   BTU/s
   Condensating water in intercooler          kg/h      0.0        0.00  lbs/min
   Thermal output intercooler                 kW       55.0       52.1   BTU/s
   Condensating water in aftercooler          kg/h      3.4        0.12  lbs/min
   Thermal output aftercooler                 kW       85.6       81.2   BTU/s
   Pressure loss before block                 bar       0.020      0.29  psi
   Pressure loss between LP- and HP-stage     bar       0.200      2.90  psi
   Pressure loss after block                  bar       0.200      2.90  psi

OILPUMP                                                R25/20FL-M6-SO
   Volume flow oilpump                        L/min    55.5       14.7   gpm
   Shaft power oilpump                        kW        0.83       1.11  HP

   Oil viscosity group                        cSt      46
   Oil volume flow to compressor )*           L/min    40.6       10.7   gpm
   Oil inlet temperature                      °C       50.0      122.0   °F
   Temperature oil out (mean value) )*        °C       62.9      145.3   °F

)* jacket cooling included

   Medium jacket cooling                               oil

STAGE DATA                                             CDA96      CDB68W
   Male rotor speed                           1/min  10908      17999
   Male rotor tip speed                       m/s      82.9       81.7
   Suction pressure abs.                      bar       0.980      3.237
   Suction temperature                        °C       20.0       50.0
   Rel. humidity                              %        58.8       36.8
   Discharge pressure abs.                    bar       3.437     11.800
   Discharge temperature                      °C      172.8      238.6

   Volume flow (rel. to stage suct. con.)     m3/h   1374.4      452.8
   Shaft power stage                          kW       77.6       94.8
```


Fig. 3. Power consumption for double oil-free compressor with CD14D1 unit, output data

In the proposed work the energy lost in cooling the air for the two stages compressor, is used in a CAES system [6, 7]. CAES system uses energy at the tip of electricity production to compress the air in pressure vessels or in depleted gas fields or in depleted or preserved salt mines. The compressor and heat exchangers are closed positioned. The compressor is driven by an electric motor and the expander drives an electric generator. After compression, the air is introduced into pressure vessels or deposits, depending on the amount of energy needed to be injected later into the grid, see figure 4. It is supposing that the compressor will work about 4 hours in a day, and the expander will work about 1 hour in the maximum grid load. That mean the storage air will be available about 20 hours/day. At a gradient of losing temperature about 1°C/h, it is supposed that the tank temperature after 20 hours to be less with 20°C, in the best insulation.

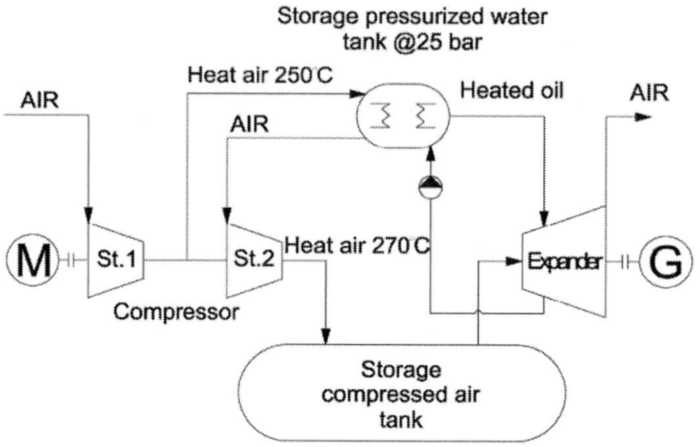

Fig. 4. The proposed recovery of heat with a CAES system

INCDT COMOTI has developed a piece of equipment for converting the energy of compressed gas into green electricity. This unique group operates using oil injection screw compressor. The thermodynamic cycle of an injection screw compressor that is a volumetric machine with positive displacement is also studied in the present [8, 10] and is used in refrigeration and air conditioning, but also in power generation. COMOTI's Expander Helical-Generator is an expander in oil-injected turbine configuration (advantageous solution for relatively small flows) that has the advantages: high efficiency due to near-isothermal compression, low maintenance costs due to its very good reliability, few pieces in motion [8, 9]. In this case, COMOTI developed the 37kW screw oil injected expander from a GHH screw compressor in collaboration with GHH -Rand company, see figure 5 [10].

Fig. 5. The 37kW screw oil injected expander [11]

3 Mathematical computation of expansion power

The process of producing mechanical energy in the turbine through irreversible adiabatic expansion of natural gas is accompanied by a decrease in temperature. Since is imposed a minimum gas temperature at the exit pipe of the expander, about 5°C, it results that the gas must be preheated before entering in the helical turbine. In our case, because the screw oil injection turbine, the oil will be heated by the first stage of the oil-free compressor, and that will cause the heating of the inlet compressed air in the turbine, in the case of lower temperatures of the compressed air, see figure 4 from above.

The expander is working with 10 bar (abs) suction pressure, because the maximum pressure of the tank is 12 bar (abs), at an air flow:

$$M_{air,exp} = 500\ m^3/h = 0.179\ kg/s \quad \text{(at 20 °C and 60\% relative humidity)} \tag{7}$$

In the equations will be used the mass flow rate in [kg/s].

In [7], the inlet pressure is the maximum allowed, but is not the working pressure. If we keep the same expansion ratio, with p_{in} = 21 bar (abs) and P_{out} = 4.5 bar (abs), we have:

$$\pi_{exp} = \frac{p_{in(abs)}}{p_{out(abs)}} = 4.67 \tag{8}$$

The output pressure p_{out}, for an input pressure p_{in} of 10 bar (abs), will be:

$$p_{out} = \frac{p_{in}}{\pi_{exp}} = 2.14\ bar\ (abs) \tag{9}$$

Working with a normal oil injected compressor, it is supposed that the exit temperature from the compressor to be around 90°C. Because the storage compressed air tank is losing the temperature with about 1°C/h, T_{in} in the expander after 20 hours will be about 70°C. The total work of the expander is given by the equation:

$$L_{exp} = c_p \cdot T_{asp} \cdot \left[\left(\pi_{exp} \right)^{\frac{k-1}{k}} - 1 \right] = 191616.7\ J/kg \tag{10}$$

Thus, the isentropic power of the expander will be:

$$P_{exp,is} = L_{exp} \cdot \frac{M_{air,exp}}{1000} = 34.3\ kW \tag{11}$$

Working with an oil-free compressor, it is supposed that the exit temperature from the compressor to be around 270°C, and with a loss of 20°C, this will be 250°C, and thus we will have:

$$L_{exp} = c_p \cdot T_{asp} \cdot \left[\left(\pi_{exp} \right)^{\frac{k-1}{k}} - 1 \right] = 29212.7\ J/kg \tag{12}$$

$$P_{exp,is} = L_{exp} \cdot \frac{M_{air,exp}}{1000} = 52.3\ kW \tag{13}$$

It is observed that in the case of using a free oil compressor, the power of the installation is increasing with about 52.5%.

If we compare the increase in the electrical energy with about 18 kW, we see that the energy lost with cooling the air is about 11 kW from the equation (6), and we have an increase with 7 kW. We did not take into consideration the efficiency of power generator and other losses in the circuit.

In real cases, the temperature at the inlet of the expander is not above 135°C. In this case, the work and power calculated with the equations above are:

$$L_{exp} = 227913.0\, J/kg \text{ and } P_{exp,is} = 40.8\, kW \quad (14)$$

The power of the installation will increase just about 19%, but the benefits of using the oil-free compressors is given by the possibility to have a bigger gradient of temperature loss in the air compressed tank, and in a longer storage time. The time increase in the case of 1°C gradient from 20 hours to about 270-70=200 hours (≅8 days), for the same power given at 70°C, and make more attractive the idea of storage compressed air at high temperature, instead the reheating of it.

4 Conclusions

The major parameters in the analysis were storage pressure, air temperature and air mass flow, the mechanical work and available power of a small scale of maximum 40 kW CAES system with an oil free compressor. Few large CAES worldwide systems are known. Studies made on CAES systems, analyse different models at small scales, up to 1 kW discharge energy and at 12 bar storage pressures [12].

INCDT COMOTI has developed a piece of equipment for converting the energy of compressed air into green electricity, using an oil injected screw compressor and an oil injected screw expander. Studies have also been made regarding a free-oil screw compressor with the characteristics described in the present work. Future research will consider replacing the oil injected screw compressor with free-oil screw compressor to develop the technology for adiabatic-CAES systems with storage of the excess heat created during compression, and reuse of it for reheating.

It is estimated that this technology will become available at large scale and also at decentralised small scale in the near-future [13], where the use of free oil compressors at high pressure is an opportunity for further research and developing studies. Future research will thus focus on improving the free oil compressor systems and also the adiabatic storage tanks, for increasing the power and efficiency of the CAES systems.

The research work presented herein was funded by the Romanian National Research Program. This work was done through the TURBOPROP 2019-2020 Program, run with the support of Ministry of Research and Innovation.

References

1. https://www.researchgate.net/publication/329378770_The_study_on_the_heat_recovery_from_air_compressors
2. https://www.kaeser.com/int-en/products/rotary-screw-compressors/heat-recovery/
3. https://www.researchgate.net/publication/326129759_Harnessing_Free_Energy_From_Nature_For_Efficient_Operation_of_Compressed_Air_Energy_Storage_System_and_Unlocking_the_Potential_of_Renewable_Power_Generation
4. GHH Rand Installation manual for CD stages and CD blocks 23152176A - March, 2009
5. Compressed Air Energy Storage (CAES) | Energy Storage Association
6. https://www.researchgate.net/publication/300495902_Compressed_Air_Energy_Storage
7. I. Mălăel, M. Sima, TURBO **2**, 24 (2017)

8. V. Petrescu, N. Toma, C. Slujitoru, TURBO **2**, 31 (2017)
9. http://www.comoti.ro/docs/jurnal/Journal%20TURBO_Vol%20IV_No%202_2017.pdf
10. COMOTI - Romanian Research and Development Institute for Gas Turbines
11. http://www.anre.ro/download.php?f=fqt%2BhQ%3D%3D&t=wOutwdHbn8%2BcmLPfvrrV5ps%3D
12. https://www.nature.com/articles/s41598-018-28025-5
13. http://ease-storage.eu/wp-content/uploads/2016/07/EASE_TD_Mechanical_DCAES.pdf

E3S Web of Conferences **112**, 01013 (2019) https://doi.org/10.1051/e3sconf/201911201013
TE-RE-RD 2019

Determination of coefficient of performance of mechanical vapour recompression heat pump

Slav Valchev[1*], *Nenko* Nenov[1] and *Vasil* Georgiev[1]

[1]University of Food Technologies, Plovdiv, Bulgaria, Department of Heat Engineering, 26 Maritza Blvd. BG-4002 Plovdiv, Bulgaria

Abstract. Mechanical vapour recompression heat pump systems are widely used in the industry - in evaporator and distillation installations, in seawater desalination and industrial wastewater treatment plants. The estimation of the energy efficiency level of this type of system is based on values of two basic parameters: specific energy consumption for production of 1 kg clean water (condensate) and actual coefficient of performance of heat pump system. The object of study is experimental determination of value of actual coefficient of performance of mechanical vapour recompression heat pump system for wastewater treatment. A mathematical regression equation between the actual coefficient of performance μ and two significant factors – temperature of secondary water vapour t_{sv} and compression ratio of water vapour in mechanical compressor of heat pump system σ is received. The analysis show that actual coefficient of performance μ highly depends of value of compression ratio σ and less depends of values of temperature of secondary vapour t_{sv}. It conclude that MVR heat pump system, in order to operate with high values of the actual coefficient of performance should be working to high values of temperature of secondary vapour and low values of compression ratio of water vapour in mechanical compressor.

1 Introduction

In process of wastewater treatment, mechanical vapour recompression (MVR) heat pump system separates wastewater flow into two flows – clean water (condensate) and concentrate sludge with high values of organic dry matters. A condensate with high temperature can be reused in various industrial processes, where a heat or clean water is needed. MVR heat pump system increases the pressure of secondary vapour, produced by the evaporator and use compressed vapour with high pressure as a primary vapour flow. Receiving of clean hot water by this method is made by low energy cost, because when the process of evaporation is started, there is no need to supply additional energy to maintenance the process.

Energy efficiency of MVR heat pump system is based on the values of two basic parameters: specific energy consumption for production of 1 kg condensate (Wh/kg) and actual coefficient of performance (-) [1, 2]. In wastewater MVR heat pump systems the

* Corresponding author: slav_vul4ev@abv.bg

difference between evaporation temperature of wastewater in separator and temperature of the compressed vapour is very small, which determines a high values of coefficient of performance of MVR heat pump systems (from 10 to 30) [3, 4, 5].
In this study a relation between actual coefficient of performance and two significant factors – temperature of secondary vapour and compression ratio of water vapour in mechanical compressor of MVR heat pump system is received.

2 Pilot MVR heat pump system

For the purpose of the study a pilot MVR heat pump system for wastewater treatment is built. The installation is located in Plovdiv, Bulgaria, at laboratory of the University Of Food Technology – Plovdiv. Schematic diagram of MVR heat pump system is shown on Figure 1.

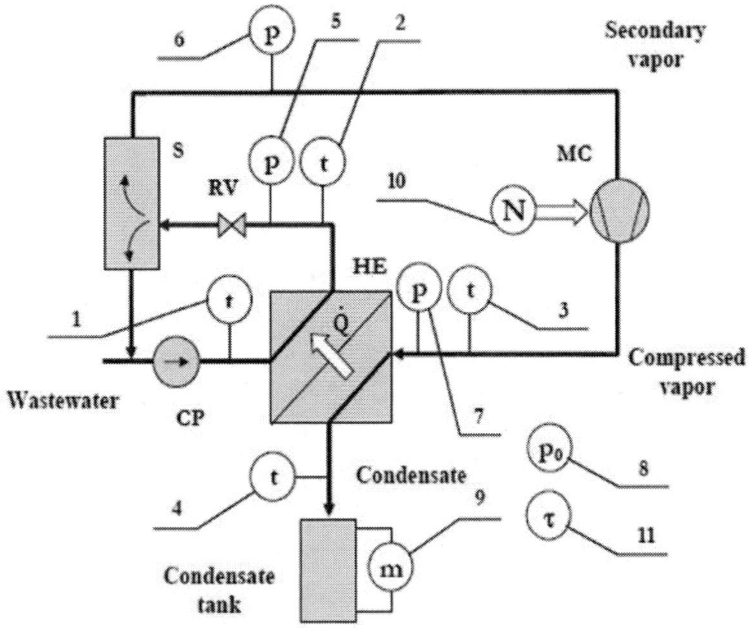

Fig. 1. Schematic diagram of mechanical vapour recompression heat pump system [6]

The wastewater goes to a plate heat exchanger HE, where it is heated by secondary vapour to a higher temperature. The temperature of wastewater at inlet of heat exchanger HE is measured with thermometer 1. Since the circulation pump CP creates a certain pressure in the heat exchanger HE, wastewater is in the state of non-boiling liquid. The temperature of wastewater at outlet of heat exchanger HE is measured with thermometer 2. The gauge pressure of wastewater at outlet of heat exchanger is measured with manometer 5. After leaving the heat exchanger HE , the wastewater passes through a reducing valve RV, which reduced wastewater pressure and when wastewater enters to the separator S, it starts to evaporate. The secondary vapour enters in suction side of mechanical compressor MC, which compresses them to a high pressure. The vacuum pressure of secondary vapour at outlet of separator S is measured with vacuum meter 6. The compressed vapour at discharge side of mechanical compressor MC enters to the plate heat exchange HE, where delivers heat to wastewater. In heat exchanger HE compressed vapour turned to condensate is discharged to condensate tank 9. The temperature of compressed vapour at outlet of mechanical compressor MC is measured with thermometer 3. The vacuum pressure of compressed vapour at outlet of mechanical compressor MC is measured with vacuum meter 7. The mass of condensate is measured by condensate tank 9 and electronic scale.

The operation time of MVR heat pump system (time to collect condensate, produced on the system) is measured by chronometer 11. The atmospheric pressure is measured by barometer 8. The power consumption of electrical energy spent by mechanical compressor MC is measured by wattmeter 10.

Measuring devices and measured parameters of MVR heat pump system are shown in Table 1.

Table 1. Measuring devices and measured parameters of pilot MVR heat pump system

Measuring device	Measured parameter	Designation	Precision measuring	Measuring range
Thermometer 1	Temperature of circulating water at heat exchanger HE inlet	t_{sv}, °C	1.0 °C	0-120 °C
Thermometer 2	Temperature of the circulating water at heat exchanger HE outlet	t_{out}, °C	1.0 °C	0-120 °C
Manometer 5	Water pressure in heat exchanger HE	p_{HE}, kPa	0.1 bar	-1 – 3 bar
Vacuum meter 6	Vacuum pressure of secondary vapour in separator S /suction side of mechanical compressor MC	$p_{v,evap}$, kPa	0.5 kPa	0-100 kPa
Vacuum meter 7	Vacuum pressure of compressed vapour/discharge side of mechanical compressor MC	$p_{v,cond}$, kPa	0.5 kPa	0-100 kPa
Barometer 8	Atmospheric pressure	p_0, kPa	1.0 mmHg	740-780 mm Hg
Water tank 9	Mass of condensate	m, kg	10.0 g	0-10 kg
Chronometer 11	Operation time of MVR heat pump system	τ, s	1 s	0-60 s
Wattmeter 10	Power consumption of electrical energy spent by mechanical compressor MC	N, W	10 W	0-999999.9 kWh

A pilot MVR heat pump system performance is between 5.0 and 16.0 kg/h evaporated water from intake wastewater flow. Absolute evaporating pressure in separator S has values from 15.5 to 25.0kPa. Absolute pressure of compressed vapour at discharge side of mechanical compressor MC has values from 20 to 43kPa. This difference in pressures provides difference between temperature of secondary vapour and temperature of compressed vapour from 5 to 13 K. These pressures (at lack of physical-chemical temperature depression) correspond to saturation evaporation temperature between 55 and 67 °C and saturation condensation temperatures between 60 and 78°C. There is also a temperature difference between 5 and 14 K of temperatures of wastewater in inlet and outlet of plate heat exchanger HE[13]. General appearance of MVR heat pump system is shown on Figure 2.

Fig. 2. General appearance of mechanical vapour recompression heat pump system [7]

3 Analysis and discussion

3.1 Basic equation

The mass flow rate of water circulating through the heat exchanger $\dot{m}_W$ is determined by the factory performance characteristic ($\dot{V}$ -Δp characteristic) of the circulator pump CP, according to the pressure Δp_{CP} :

$$\Delta p_{CP} = p_{HE} - p_{evap}, \tag{1}$$

where: - p_{HE} – absolute pressure of wastewater in heat exchanger HE, kPa;
p_{evap} – absolute pressure of secondary vapour in separator S, kPa.

Heat flux exchanged in the heat exchanger $\dot{Q}$ is determines by the equation of energy balance of wastewater flow rate through heat exchanger HE:

$$\dot{Q} = \dot{m}_W \, c(t_{out} - t_{evap}), \tag{2}$$

where: - $\dot{m}_W$ is mass flow rate of wastewater circulating through the heat exchanger, kg/s;
c – specific heat capacity of the wastewater, specified for wastewater average temperature, kJ/(kg.K);
t_{out} - temperature of wastewater of outlet of heat exchanger HE, °C;
t_{evap} - temperature of wastewater of inlet of heat exchanger HE, equal to the evaporation temperature of wastewater in separator S, °C.

A mass flow rate of vapour through the mechanical compressor $\dot{m}_K$ can be calculated from the equation of the material balance. It is equal to the mass flow rate of the collected condensate:

$$\dot{m}_K = \frac{m}{\tau} \quad (3)$$

where: - m - total quantity of condensate collected in the condensate tank, kg;
τ - operation time for collecting the condensate, s.
The process of compression of water vapour in mechanical compressor MC of MVR heat pump system is considered. It is a not isentropic process of compression of real gas - water vapour. The process of compression of water vapour occurs at low absolute pressure (between 15.5 kPa and 43.0 kPa). The compressibility factor of water vapour is calculated by the equation:

$$z = \frac{pV}{RT} \quad (4)$$

where:- z - compressibility factor, -;
p –absolute pressure of water vapour, MPa;
V –volume of water vapour, m^3;
R –individual gas constant of water vapour, J/(kg.K);
T –absolute temperature of water vapour, K;
The value of compressibility factor in this case is between 0.993 and 0.996, which is very close to 1. In this case vapour can be considered as an ideal gas. Proximate theoretical process to the real process in MVR heat pump system is isentropic compression of ideal gas. In this process submitted specific flow work l_t will match with the defined target function actual coefficient of performance of MVR heat pump system. This work for isentropic process of an ideal gas is calculated by the equation [8]:

$$l_t = \frac{k}{k-1} RT_{evap} \left[1 - \left(\frac{p_{cond}}{p_{evap}} \right)^{\frac{k-1}{k}} \right] \quad (5)$$

where: - l_t – specific flow work, J/kg;
k – specific heat ratio,-;
R –individual gas constant of water vapour, J/(kg.K);
T_{evap} – absolute temperature of evaporation of wastewater, K;
p_{cond} - absolute pressure of compressed vapour, Pa.
p_{evap} - absolute pressure on secondary vapour in the separator S, Pa .
At lack of physical-chemical temperature depression the value of evaporation temperature of wastewater is equal to temperature of the secondary vapour ($t_{evap} = t_{sv}$).
This relationship is accepted as a basis for mathematical model experiment by converting in form:

$$l = (A_1 t_{sv} + A_2)\sigma^{A_3} \quad (6)$$

where: - l – actual coefficient of performance, -;
t_{sv} - temperature of secondary vapour, °C;
σ -compression ratio of water vapour in the mechanical compressor MC,-;
A_1, A_2, A_3 – coefficients,-;
From the analysis of this relationship, the following important factors of the actual coefficient of performance l are defined: temperature of secondary water vapour t_{sv} and compression ratio of water vapour in the mechanical compressor MC σ.

Compression ratio of mechanical compressor is calculated by the equation:

$$\sigma = \frac{p_{cond}}{p_{evap}} \tag{7}$$

where: - p_{cond} is absolute pressure of compressed vapour, Pa;
p_{evap} - absolute pressure on secondary vapour in the separator S, Pa.
Guidelines for selecting the relevant factors in the objective function actual coefficient of performance μ gives dependence (8) for the theoretical one μ_{th} [9, 10]:

$$\mu_{th} = \frac{T_{cond}}{T_{cond} - T_{sv}} \tag{8}$$

where: - T_{cond} - thermodynamic temperature of compressed vapour, K;
T_{sv} - thermodynamic temperature of secondary vapour, K.
In this dependence temperatures of compressed and secondary vapour are identified as a saturation temperatures for pressures of compressed and secondary vapour by tables for water and water vapour.
A mathematical model that examined objective function is selected function analogous to (6):

$$\mu = (A_1 t_{sv} + A_2)\sigma^{A_3} \tag{9}$$

Actual coefficient of performance μ is calculated by the equation [4],[5]:

$$\mu = \frac{\dot{Q}}{|N|}, \tag{10}$$

where:- $\dot{Q}$ - heat flux exchanged in heat exchanger, W;
N – power consumption of electrical energy by mechanical compressor MC of MVR heat pump system, W.

3.2. Results

A different sample of industrial wastewater were researched: water from steam boiler blow down, water from CIP (cleaning in place) system of milk factory, water from refrigerant condenser blow down, water from washing vehicles and water from washing a catering equipment. This type of wastewater has low values of organic dry matter. In researched temperature interval no physical-chemical temperature depression was found. Therefore, it is assumed that the evaporation temperature of wastewater is equal to the temperature of the secondary vapour ($t_{evap} = t_{sv}$). With a particular type of waste water, the pilot MVR heat pump system is started and introduced into an established operating mode, including stable maintenance of operating parameters and condensate generation. For a certain period of time the volume of generated condensate is collected and measured.
A classic experiment by two significant factors is conducted. The small number of factors (n = 2) allows the study to be carried out by classical multi-term experiment. A classic full plan of experiment including all recurring combinations of levels of factors is selected. The factors whose influence has been studied are defined: temperature of secondary water vapour in separator S t_{sv} and compression ratio of water vapour in the mechanical compressor MC σ.
The scheme of the experiment is unbalanced with three levels of variation of the temperature of secondary water vapour in separator t_{sv}- 55 °C, 60 °C, 65 °C and five levels of varying of the compression ratio of water vapour in the mechanical compressor MC

σ - 1.30, 1.40, 1.50, 1.60 and 1.70. Levels of variation of two significant factors are also determined by the specific design of pilot MVR heat pump system.
The results of experimental studies are shown in Table 2, Table 3 and Table 4.

Table 2. Experimental data – Study 1

№	t_{sv}	σ	Δp_{CP}	$\dot{m}_W$	t_{out}	$\dot{Q}$	N	μ
	°C	-	kPa	kg/s	°C	W	W	-
1	54.7	1.290	165.0	0.175	64.0	6846	680	10.07
2	56.0	1.394	165.0	0.175	63.0	5148	700	7.35
3	55.3	1.500	164.0	0.185	61.0	4390	730	6.01
4	54.7	1.613	165.0	0.175	59.5	3540	750	4.72
5	55.3	1.719	166.5	0.160	60.2	3254	840	3.87
6	60.1	1.300	164.0	0.185	70.0	7685	670	11.47
7	60.6	1.390	164.5	0.180	68.6	6017	700	8.60
8	59.0	1.526	165.0	0.175	66.0	5145	800	6.43
9	59.0	1.632	165.0	0.175	65.0	4410	820	5.38
10	60.1	1.750	164.0	0.185	64.9	3725	880	4.23
11	65.9	1.288	164.5	0.180	80.0	10681	720	14.83
12	65.0	1.400	165.0	0.175	77.0	8823	760	11.61
13	66.7	1.481	165.0	0.175	77.5	7926	800	9.91
14	65.9	1.615	165.0	0.175	75.0	6711	900	7.46
15	65.0	1.712	165.0	0.175	73.5	6252	1100	5.68

Table 3. Experimental data – Study 2

№	t_{sv}	σ	Δp_{CP}	$\dot{m}_W$	t_{out}	$\dot{Q}$	N	μ
	°C	-	kPa	kg/s	°C	W	W	-
1	54.7	1.290	165.0	0.175	63.7	6626	680	9.74
2	56.0	1.394	164.0	0.185	62.5	5052	700	7.22
3	55.3	1.500	164.0	0.185	61.0	4390	730	6.01
4	54.7	1.613	165.0	0.175	60.0	3907	820	4.77
5	55.3	1.719	166.5	0.160	60.0	3120	840	3.71
6	60.1	1.300	164.0	0.185	70.2	7840	660	11.88
7	60.6	1.390	164.5	0.180	69.0	6320	700	9.03
8	59.0	1.526	165.0	0.175	66.4	5439	800	6.80
9	59.0	1.632	164.0	0.185	65.4	4971	830	5.99
10	60.1	1.750	164.0	0.185	64.8	3648	850	4.29
11	65.9	1.288	165.5	0.170	81.3	11013	710	15.51
12	65.0	1.400	165.0	0.175	78.0	9558	750	12.74
13	66.7	1.481	164.0	0.185	77.5	8377	800	10.47
14	65.9	1.615	165.0	0.175	75.5	7079	910	7.78
15	65.0	1.712	165.0	0.175	74.0	6619	1080	6.13

Table 4. Experimental data – Study 3

№	t_{sv}	σ	Δp_{CP}	$\dot{m}_W$	t_{out}	$\dot{Q}$	N	μ
	°C	-	kPa	kg/s	°C	W	W	-
1	54.7	1.290	165.0	0.175	64.0	6846	690	9.92
2	56.0	1.394	164.0	0.185	63.0	5440	710	7.66
3	55.3	1.500	165.0	0.175	61.0	4154	730	5.69
4	54.7	1.613	165.0	0.175	59.8	3760	810	4.64
5	55.3	1.719	166.5	0.160	60.3	3322	830	4.00
6	60.1	1.300	164.0	0.185	69.6	7374	670	11.01
7	60.6	1.390	164.5	0.180	68.5	5942	660	9.00
8	59.0	1.526	165.0	0.175	66.0	5145	760	6.77
9	59.0	1.632	165.0	0.175	65.1	4484	830	5.40
10	60.1	1.750	165.0	0.175	65.0	3598	860	4.18
11	65.9	1.288	165.5	0.170	81.7	11298	710	15.91
12	65.0	1.400	165.0	0.175	78.0	9558	750	12.74
13	66.7	1.481	164.0	0.185	78.0	8765	820	10.69
14	65.9	1.615	165.0	0.175	75.0	6711	880	7.63
15	65.0	1.712	165.0	0.175	74.0	6619	1060	6.24

Statistical processing of experimental data is made. This includes: checking for homogeneity of variances, check significance of odds betting in the regression equations and check the adequacy of the resulting equations was carried out [11]. A summary experimental data from studies 1, 2 and 3 are shown in Table 5. X_1 and X_2 are significant factors - temperature of secondary water vapour in separator S t_{sv} and compression ratio of water vapour in the mechanical compressor MC σ. $Y_{2,1}$, $Y_{2,2}$, $Y_{2,3}$ are the results of objective function actual coefficient of performance of MVR heat pump, measured three times and Y_2 is respective arithmetical mean value of actual coefficient of performance μ of MVR heat pump.

Table 5. Summary experimental data – Study 1,2 and 3

№	X_1	X_2	$Y_{2,1}$	$Y_{2,2}$	$Y_{2,3}$	Y_2
	°C	-	-	-	-	-
1	54.7	1.290	10.07	9.74	9.92	9.91
2	56.0	1.394	7.35	7.22	7.66	7.41
3	55.3	1.500	6.01	6.01	5.69	5.90
4	54.7	1.613	4.72	4.77	4.64	4.71
5	55.3	1.719	3.87	3.71	4.00	3.86
6	60.1	1.300	11.47	11.88	11.01	11.45
7	60.6	1.390	8.60	9.03	9.00	8.88
8	59.0	1.526	6.43	6.80	6.77	6.67
9	59.0	1.632	5.38	5.99	5.40	5.59
10	60.1	1.750	4.23	4.29	4.18	4.23
11	65.9	1.288	14.83	15.51	15.91	15.42
12	65.0	1.400	11.61	12.74	12.74	12.36
13	66.7	1.481	9.91	10.47	10.69	10.36
14	65.9	1.615	7.46	7.78	7.63	7.62
15	65.0	1.712	5.68	6.13	6.24	6.02

Objective function actual coefficient of performance obtained equation is as follows:

$$Y_2 = (1.411.X_1 - 56.47)X_2^{-3.27} \quad (11)$$

Regression equation for actual coefficient of performance is obtained by replacing parameters $Y_2 = \mu$, $X_1 = t_{sv}$ and $X_2 = \sigma$:

$$\mu = (1.411t_{sv} - 56.47)\sigma^{-3.27} \quad (12)$$

In Figure 3 a correlation between objective function actual coefficient of performance and two significant factors: compression ratio and temperature of secondary vapour in separator as a parameter is shown.

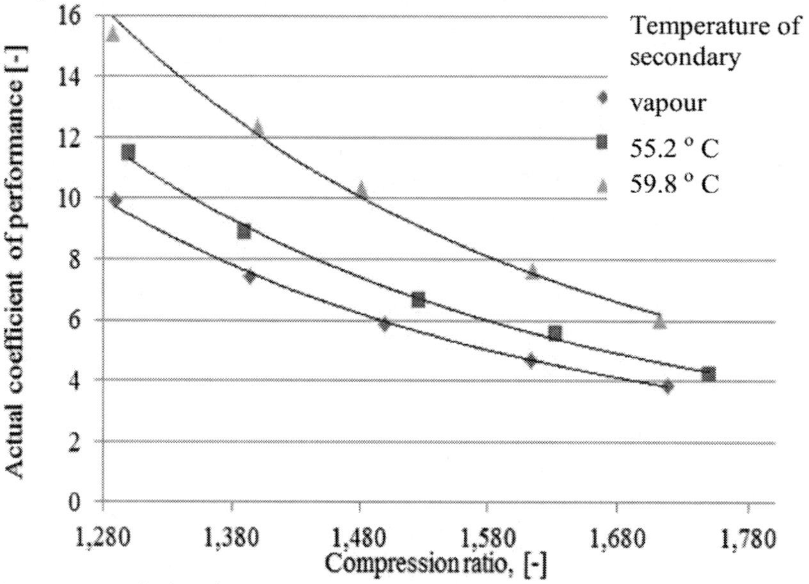

Fig. 3. Correlation between actual coefficient of performance of MVR heat pump system and two independent factors: compression ratio and temperature secondary vapour

In Table 6 a comparison of the values of theoretical μ_{th} and actual coefficient of performance of MVR heat pump system μ are given. The values of theoretical coefficient of performance of MVR heat pump system μ_{th} are calculated by equation (8).

Table 6. Comparison of the values of the theoretical and actual coefficient of performance of MVR heat pump system

t_{sv}	σ	μ_{th}	μ	μ_{th}/μ
˚C	-	-	-	-
54.7	1.290	61.5	9.91	6.21
56.0	1.394	47.0	7.41	6.34
55.3	1.500	38.6	5.90	6.54
54.7	1.613	32.8	4.71	6.96
55.3	1.719	28.9	3.86	7.49
60.1	1.300	58.8	11.45	5.14
60.6	1.390	46.8	8.88	5.27
59.0	1.526	36.6	6.67	5.49
59.0	1.632	31.6	5.59	5.65
60.1	1.750	27.5	4.23	6.50
65.9	1.288	59.5	15.42	3.86
65.0	1.400	44.9	12.36	3.63
66.7	1.481	38.0	10.36	3.67
65.9	1.615	31.2	7.62	4.09
65.0	1.712	27.7	6.02	4.60

4 Conclusions

The analysis of graphic dependencies (Fig.3) and equation (12) show that actual coefficient of performance in pilot MVR heat pump system μ highly depends of the compression ratio of water vapour in mechanical compressor MC σ. The relation between these two variables in the study area is inverse speed dependence with negative exponent value $A_3 = - 3.27$.
Less dependence on actual coefficient of performance μ proves temperature of secondary vapour t_{sv}. It is linear proportional correlation between actual coefficient of performance μ and temperature of secondary vapour t_{sv} because of positive value of coefficient $A_1 = 1.411$.
In researched interval values of actual coefficient of performance of MVR heat pump system μ are between 3.86 and 15.42, and the theoretical one μ_{th} - from 3.63 to 7.49 times greater. As a result, it can conclude that pilot MVR heat pump system, in order to operate with high values of the actual coefficient of performance μ should be working to high values of temperatures of secondary vapour t_{sv} and low values of compression ratio of water vapour in mechanical compressor σ.

References

1. H. Aybar, Desalination 142 **(2)**, 181-186 (2002)
2. H. Ettouney, Desalination 190, 1–15 (2006)
3. D. De Waal, *Feasibility Study of Heat Pumps for Waste Heat Recovery in Industry* (Master of Science Thesis, 2012)
4. W. Pang, L. Wenju, P. Qilin, L. Wenye, D. Qunte, Y. Luwei, Z. Zhentao. Presented at AIP Conf. Proc. 1573, 1122-1128 (2014)
5. C.Tuan., Y. Cheng, Y. Yeh, L. Hsu, T. Chen. Sustain. Environ.Res., 23(**2**), 129-139 (2013).
6. S. Valchev, *Application of a heat pump system for industrial waste water purification.* (Master of Science Thesis, University of Food Technology 2015).
7. S. Valchev, N. Nenov, Heat Engineering, 2, 72-75 (2014).
8. G. Kimenov, *Thermal Engineering*, (In Bulgarian - State Publishing House "Technique", Sofia, 1977).
9. S. Dichev, *Refrigeration machines.* (In Bulgarian - Academic Publishing of VIHVP - Plovdiv, 2002)
10. V. Eder, F. Moser. *Thermal pumps*, (In Bulgarian - State Publishing House "Technique", Sofia, 1984)
11. G. Raichkov, *Measurements and appliances in heat engineering*, (In Bulgarian - Academic Publishing of University of food technology - Plovdiv, 2001)

E3S Web of Conferences **112**, 01014 (2019) https://doi.org/10.1051/e3sconf/201911201014
TE-RE-RD 2019

Some aspects of cycle variability at the Diesel engine fuelled with animal fats

Adrian Nicolici[1], *Constantin Pană*[1], *Niculae* Negurescu[1], *Alexandru* Cernat[1]* and *Cristian* Nuțu[1]

[1] University Politehnica of Bucharest, Termothechnics, Engines, Thermal Equipments, and Refrigeration Instalations Department, Bucharest, Romania

Abstract. The progressive diminution of the oil reserves all over the world highlights the necessity of using alternative fuels derived from durable renewable resource. The use of the alternative fuels represents a viable solution to reduce the pollutant emissions and to replace fossil fuels. Thus, a viable solution is the use of the animal fats in mixture with the diesel fuel at the diesel engines. A D2156 MTN8 diesel engine was firstly fuelled with diesel fuel and then with different blends of diesel fuel-animal fats (5% and 10% animal fats content). In the paper are presented some results of the experimental investigations of engine fuelled with preheated animal fats. The raw animal fats effects on the combustion process and on the pollutant emissions at different engine loads and 1450 rev/min engine speed are showed. The engine cycle variability increases at the animal fats content increase. The cycle variability for maximum pressure, maximum pressure angle and indicated mean effective pressure is analysed. The cycle variability coefficients values don't exceed the recommended values of the standard diesel engine.

1 Introduction

Starting with the end of year 2014, the European Union commissioned and funded a research study named Ecofys were the use of animal fats as alternative fuel for power stations or for diesel engines is presented with great perspectives for emissions reduction of pollutant substances and greenhouse gases (GHG). Also, the advantage of a lower price comparative to vegetable oils is presented and analysed in the report [1], [2].

Due to low operating costs and high efficiency in use, durability and reliability are the most important demands for diesel engines that equip heavy duty vehicles which make them to be in common currency all over the world. Regarding the worldwide use of diesel engines due to its advantages the pollution impact on environment must be taking into consideration [3], [4], [5]. Worldwide, in recent years, many drastic measures for reduction of pollutant emissions with negative effects on health and the environment have been imposed. Many research programs have been developed to study the possibilities for

[1] Corresponding author: Alexandru Cernat : cernatt@gmail.com

reducing of the pollutant emissions from combustion gases and of new control technologies of pollutant emissions, [3], [4], [5], [6], [7], [8], [9], [10].

These important aspects and issues represent an important topic of discussion at the 2015 United Nations Climate Change Conference (COP 21 or CMP 11), conference which was held in Paris. In this context, according to some researchers, important efforts are being made in order to limit the GHG emissions to zero in 2030-2050, [3].

On the other hand, the reduction of oil reserves around the world highlights the necessity of alternative energy use, an energy obtained from sustainable and renewable resources, and implementation of new technologies for energy use, [4], [6], [7], [8], [9], [10], [11], [12], [13].

Over the past decades, the number of automotives has been greatly increased the pollution emissions level being also increased especially in the urban areas. To combat this issue, the city of Oslo, the capital of Norway, has been set to become the first big clean city, free of pollutant emissions. Since 1960, in certain cities of Denmark, the circulation of pollutant vehicles fuelled by classic fuels is forbidden, but since then various alternative solutions have being adopted for the automotives traffic in these cities, [5].

The use of animal fats in blend with diesel fuel requires preheating of animal fats at a temperature higher than 40 °C. If the temperature drops below this limit, the unsaturated fatty acids in the animal fats begin the solidification process in the fuel circuit, the risk of components clogging being present. At this temperature, the blend becomes homogeneous and it is also necessary to shake it to avoid shearing because of the density differences, [6], [12], [14], [15].

In the table 1 some of the physic-chemical properties of diesel fuel and animal fats are presented:

Table 1: The physic-chemical properties of diesel fuel and animal fats, [4], [6], [12]

Specific properties of the fuels	**Diesel Fuel**	**Animal Fat**
Density [kg/m^3]	849	920
Viscosity at 40 °C [mm^2/s]	2.96	4.8
Thick point [°C]	-12	6
Ignition point [°C]	74	170
Boiling point [°C]	191	344
Sulphur [% m]	0.036	0
Cetane number, CN [-]	49.2	56
Low heating value, Hi [MJ/kg]	42.9	39.77
Carbon [%]	85.7	76.43
Hydrogen [%]	13.3	12.59
Oxygen [%]	1.0	10.98

In the paper are presented some cyclic variability aspects of in-cylinder combustion process in a D2156-MTN8 diesel engine fuelled with animal fats-diesel fuel blends.

2 Methodology

The experimental investigations were carried on a truck diesel engine, type D2156 MTN8, 6 cylinders turbocharged, with 10.35 dm^3 displacement and compression ratio of 17.5:1, which develops a power of 188 kW and a maximum torque of 890 Nm. In order to setup the baseline of the experimental research, the engine was fuelled with only diesel fuel, all the experimental data's being registered. After baseline setup the diesel engine was fuelled with different animal fats-diesel fuel blends, the experimental procedure of data's acquisition

 https://doi.org/10.1051/e3sconf/201911201014

being repeated in terms of maintaining the engine power at the same value recorded at classic fuelling operation. Preheated animal fats are blended with diesel fuel in proportions of $5(\%)_v$ and $10(\%)_v$, the final fuel blends being heated to 40 °C. The experimental investigations were carried out at the regimes of 40%, 55% and 70% engine loads and speed of 1450 rev/min, the injection timing being maintained constant at the reference value of 22 °CA.

The test bed is equipped with specific experimental equipments used for measuring and control processes, the main components being: pressure transducer, incremental speed transducer, intake air flow meters for inlet air and fuels, data acquisition system, exhaust gas analyser, thermo-resistances and thermocouples for engine fluids temperature monitoring. The figure 1 illustrates the design of the experimental test bed.

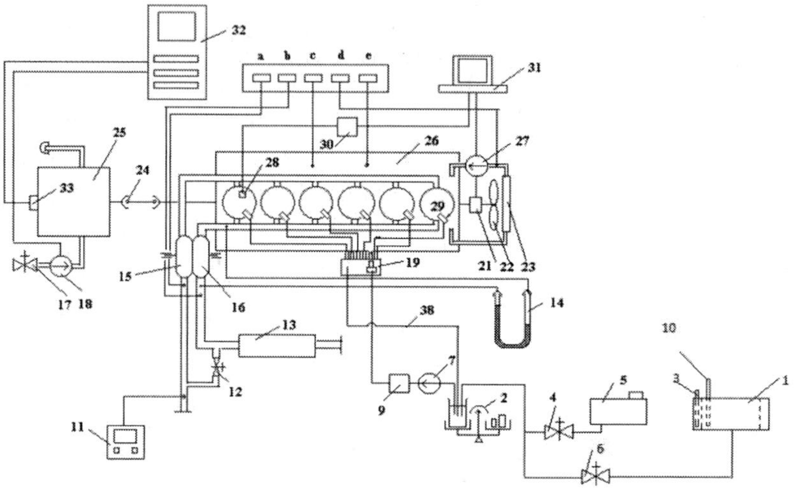

Fig. 1. The experimental test bed.

The main components of the experimental test bed are:
1. diesel fuel-animal fats blends tank; 2. fuel flow meter; 3. preheating electrical resistance; 4. diesel fuel valve; 5. diesel fuel tank; 6. animal fats-diesel fuel blends valve; 7. fuel pump; 9. fuel filter; 10. animal fats-diesel fuel blends thermometer; 11. AVL DiCom 4000 gas analyzer; 12. EGR valve; 13. Meriam air flow meter; 14. differential gauge for intake pressure measuring; 15-16. Holset Group turbo-supercharger; 17. dyno cooling water tap; 18. dyno cooling water pump; 19. fuel injection pump; 21. Kubler's incremental speed transducer; 22. fan; 23. engine cooling radiator; 24.coupling shaft; 25. Hoffman electric dyno; 26. diesel engine type D2156 MTN8; 27. engine cooling system pump; 28. in-cylinder Kistler piezoelectric pressure transducer; 29. diesel injector; 30. Kistler electric charge amplifier; 31. AVL Indicom data acquisition system; 32. dyno control cabinet; 33. dyno power cell; 38. fuel return pipeline; Shimaden SR indicators for temperatures of: a) exhaust gases, b) inlet air, c) engine oil, d) engine coolant and e) electronic indicator of engine oil pressure.

As research methodology, for each fuel blend and operating regime, a number of 100 consecutive combustion cycles were recorded by AVL data acquisition system. These cycles were further used to evaluate the indicated mean effective pressure (IMEP), the maximum pressure (p_{max}) and the angle of maximum pressure (α_{pmax}) for classic fuelling and for animal fats-diesel fuel fuelling. In order to determine the influence of the substitute ratio of diesel fuel by animal fats on the combustion cyclic variability the coefficients of variation for the indicated mean effective pressure, the maximum pressure and angle of maximum pressure were calculated.

The form of the cyclic variability coefficient, COV, reflects the ratio of the standard deviation, σ_a, to the mean, a_{mean}, for a set of 100 measured cycles, as the following relation shows [7]:

$$COV_{\ a} = \frac{\sigma a}{a_{mean}} \cdot 100 \tag{1}$$

where ''a'' represents the studied parameter (IMEP, p_{max} and α_{pmax}) [7], [8], [9].

The coefficient of cyclic variability for maximum pressure COV_{pmax} is useful to study cyclic variability during the first phase of combustion. The cyclic variability coefficient of the indicated mean effective pressure, COV_{IMEP}, reflects the engine operation stability and shows the global engine response to the variability of the combustion process. As a result of the current experimental research, a stable operation of the diesel engine is ensured when the values of the cyclic variability coefficients are below 10%, [7]. This study will show, comparatively to classic fuelling, if the fuelling with animal fats-diesel fuel blends can assure a normal engine operation so that normal drive-ability of the truck being achieved also for animal fats fuel blends.

3 Results

The maximum value of the substitution ratio, x_c, can be established in accordance with the values of COV_{IMEP} and COV_{pmax}.

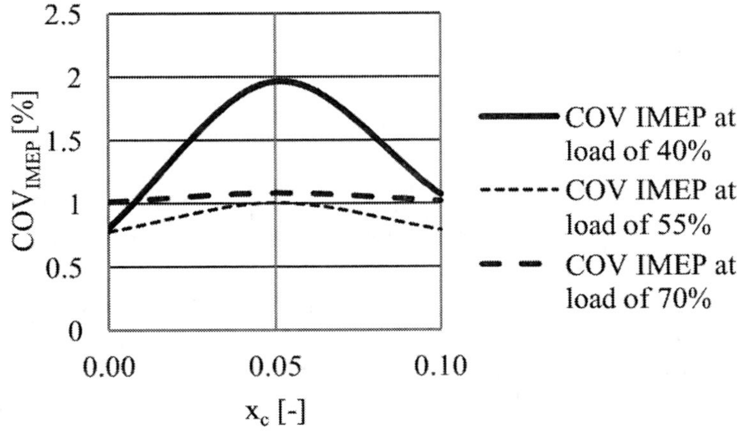

Fig. 2. The COV_{IMEP} versus substitute ratio values.

For all engine loads regimes, 40%, 55% and 70%, the COV_{IMEP} value starts to rise once with the increase of the substitute ratio comparative to classic fuelling as figure 2 shows. For the low load regime of 40% the COV increase is more significant comparative to others loads regimes, the maximum value of ~2% being registered for 5% animal fats-diesel fuel blend, but don't exceed the 10% limit value. For 10% animal fats-diesel fuel blends the COV_{IMEP} decreases and remains above values recorded for diesel fuelling. Once with the rise of engine load the influence of animal fats percent in blend with diesel fuel is not so significant, the combustion process being more stable versus low loads operation. At high load, 70%, the combustion variability is not significantly affected by animal fats use, the values of COV_{IMEP} having a very low variation versus diesel fuelling.

A similar variation is registered also for the values of COV_{pmax} calculated for all engine loads regimes, figures 3. At 40% engine load the maximum value does not exceed 2.33% and it is not necessary to limits the substitution ratio because the limit of 10% is not exceed it. At 55% and 70% engine loads the increases are reduced the maximum values remaining in the area of values registered for diesel fuelling, x_c=0.

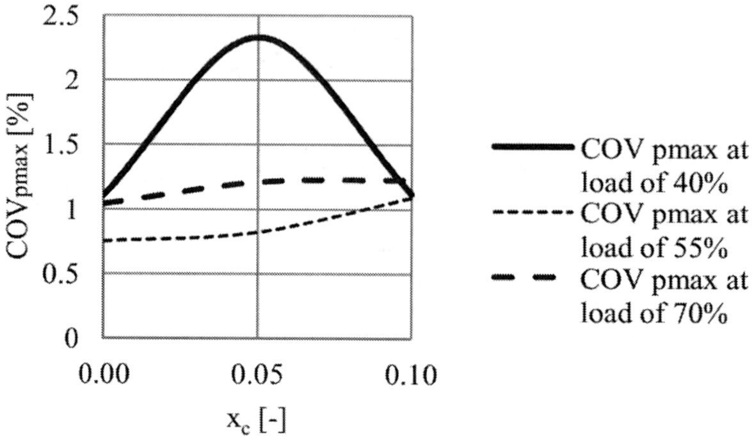

Fig. 3. The COV_{pmax} versus substitute ratio values.

For x_c=5% the increasing tendency in variation of the COV for IMEP and maximum pressure are in correlation and can be taken into consideration when $5(\%)_v$ animal fats is used in blend with diesel fuel. Also, the increase tendency for COV_{pmax} for $10(\%)_v$ animal fats-diesel fuel is present.

At the increase of the substitution ratio the angle of maximum pressure, α_{pmax}, occurs later on the combustion cycle because the increase of animal fats content leads to the aggravation of fuel atomization and autoignition delay increase. The angles of maximum pressure are: 13.4°CA at x_c = 0; 11. 8°CA at x_c = 5; 12.8°CA at x_c = 10 for 40% engine load; 13°CA at x_c = 0; 11.9°CA at x_c = 5; 12.1°CA at x_c = 10 for 55% engine load and finally, 14°CA at x_c = 0; 13°CA at x_c = 5; 12.9°CA for 70% engine load, figure 4.

The use of animal fats in small percents leads to the reduction of the dispersion between angles of maximum pressure values. The increased trend of COV values for α_{pmax} for substitution ratios bigger than x_c = 5% is in correlation with the slightly increased values for COV_{IMEP} and COV_{pmax} at 55% and 70% engine loads. Thus, at animal fats use in blend with diesel fuel the combustion variability is increased for low loads regime, the higher loads easily allowing the use of higher animal fat quantities without affecting the engine operating stability.

Thus, the slightly increase of cycle dispersion between values of IMEP and maximum pressure, but it reduce the dispersion between the cycle moment when the maximum pressure occurs per cycle. Practically, at animal fats use the combustion variability during the rapid phase of combustion is slightly increased, but dispersion between its durations from cycle to cycle tends to be reduced.

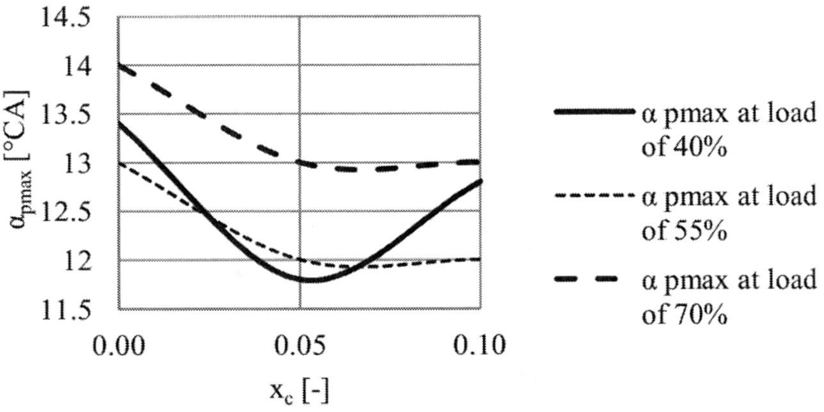

Fig.4. The maximum pressure angle versus substitute ratio values.

The NO_x formation is associated with high temperature zones corresponding to preformed mixtures. Figure 5 shows that at the 40% engine load, the NO_x emissions level is decreased with 20.4% for $5(\%)_v$ animal fat in blend with diesel and decreases with 20.8% for $10(\%)_v$ animal fats percent. The decrease of NO_x emissions level is explained by the reduction of the quantity of preformed mixture resulting from the aggravation of fuel atomization at animal fats use, which leads to lower combustion temperature compared to diesel fuel. The NO_x formation is influenced by temperature, time and mixture dosage, the increase of combustion gas temperature and combustion time leads to raised temperatures which favour the NO_x formation reactions [10]. A lower caloric power of animal fats versus diesel fuel leads also to the reduction of the combustion temperature per cycle at animal fats use and to the reduction of NO_x emission level, at high engine loads. At low engine loads, the reduction of diesel fuel dose at rise of animal fats content in blends leads to the reduction of the gases combustion temperature and autoignition delay increase.

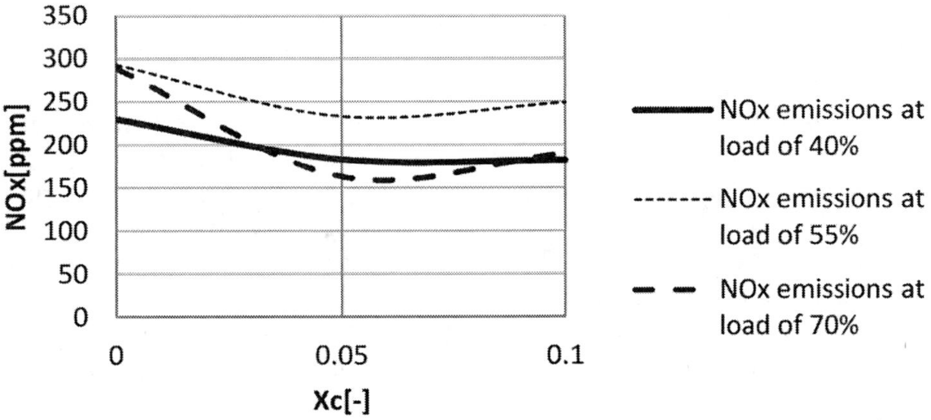

Fig 5. The NO_x emissions level versus different substitute ratio values.

The increasing tendency of NO_x emissions level at same engine loads can be limited by injection timing tune. The NO_x emission decreases are correlated with the increased stability of premixed phase of combustion, with lower values for $COV_{\alpha pmax}$, figure 4.

Versus diesel fuel fuelling the smoke emission level starts to decrease once with the rise of animal fat content in blend with diesel fuel, for all engine loads regimes, figure 6.

The reduction of smoke emission level can be explained by lower carbon content and higher oxygen content in the final fuels blend at the increase of animal fats content, due to the fact that animal fats contain more oxygen and less carbon comparative to diesel fuel.

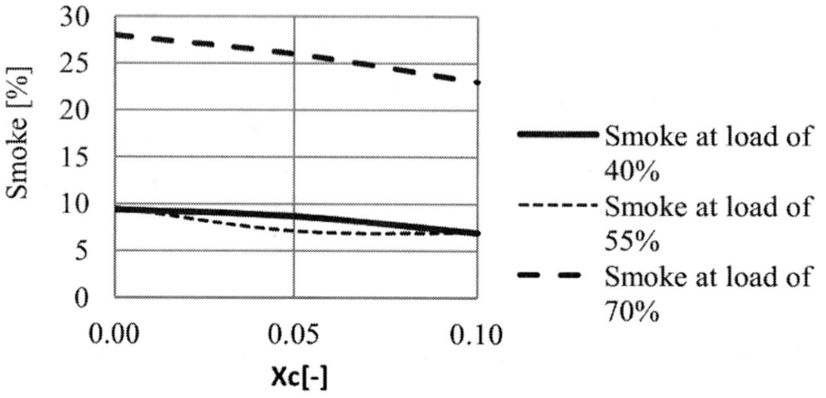

Fig. 6. The smoke emission level versus different substitute ratio values.

Thus, at 70% engine load the use of $10(\%)_v$ animal fats-diesel fuel blends leads to the reduction of smoke opacity with 14% and to the decrease of smoke number by 35%, figure 6. At 40% engine load the opacity is decreased by 25% and the smoke number is reduced by 34%, for $10(\%)_v$ animal fats-diesel fuel blend. At 55% engine load the reduction are similar.

4 Conclusions

The study of combustion cycle variability at animal fats-diesel fuel blends use at diesel engine leads to the main following conclusions:

- the values for COV_{pmax} and COV_{IMEP} slightly rise comparative to diesel fuelling, but the calculated values don't exceed 3%, being under the 10% limit, the normal diesel engine operation being assured at animal fats use and the limitation of the substitute ratio is not necessary from this point of view
- the values of $COV\alpha_{pmax}$ are reduced at animal fats use, the cyclic dispersion between angles of maximum pressure being lower comparative to diesel fuel use
- regarding the variability of the combustion process, the higher engine loads easily accept $10(\%)_v$ animal fats in final fuel blend comparative to low load regime, the COV values for maximum pressure and IMEP being smaller for 55%-70% engine loads versus values registered for 40% engine load
- the NO_x emissions level decreases with almost 20% for 40% and 70% engine loads due to reduction of the cycle combustion temperature in correlation with a smaller caloric power of animal fats versus diesel fuel
- the smoke emission levels are reduced at animal fats use, with 14% at high engine loads and with 25% respectively at low engine loads due to the rise of oxygen content and decrease of carbon content in the fuel blends once with the rise of animal fats content
- the use of animal fats-diesel fuel blends in order to improve the diesel engine pollutant performance can be assured in terms of a normal combustion variability which assures the normal drive ability of the automotive, the fuelling method being easier to apply due to the fact that it doesn't require important engine design modification.

The authors address special thanks to AVL GmbH Graz Austria for providing the necessary equipments. This work was supported by a grant of the Romanian Ministery of Research and Innovation, CCCDI – UEFISCDI, project number PN-III-P1-1.2-PCCDI-2017-0404 / 31PCCDI/2018, within PNCDI III

References

[1] C. Chudziak and S. Haye (E4tech), Final report Task 4a of ENER/C1/2013-412, *Indirect emissions from rendered animal fats used for biodiesel*, date: 21 June (2016)
[2] Al. Cernat, C. Pana, N. Negurescu, Cr. Nutu, A. Nicolici, Proceedings of the 4th International Congress of Automotive and Transport Engineering (AMMA), *Butanol effects on the fuelled diesel engine operation with preheated diesel animal fuel blends,* ISBN 978-3-319-94408-1, (2018)
[3] İ.A. Reşitoğlu, K. Altinişik, A. Keskin, Clean Techn Environ Policy, *The pollutant emissions from diesel-engine vehicles and exhaust after treatment systems*, 17: 15. https://doi.org/10.1007/s10098-014-0793-9, (2015)
[4] M. G. Popa, C. Pana, N. Negurescu, Matrixrom, Bucharest, *Diesel engines. Processes,* (2003)
[5] V. Emil Lucian, *Biocombustibili–Ghid de obtinere a energiei din deseuri* , Ed. Universitara Bucuresti, ISBN 978-606-28-0439-8, (2016)
[6] S. Alonso J. Gobernado Arribas S. Miñambr, International J. Energ. and Environ Eng, *Study of combustion in residential oil burning equipment of animal by-products and derived products not intended for human consumption*, Vol. **4**, pp 1-13, (2013)
[7] J. B. Heywood, editor: McGraw-Hill Book Company New York, *Internal Combustion Engines Fundamentals*, (1988)
[8] A. C. Polk, C. D. Carpenter, K. K. Srinivasan, et al., Fuel, *An investigation of diesel–ignited propane dual fuel combustion in a heavy-duty diesel engine*, **132,** pp. 135–148, (2014)
[9] B. Ashok, S. Denis, C. Ramesh, C., Kumar, Alex Eng J, *LPG diesel dual fuel engine – A critical review*, **54**, pp. 105–126, (2015)
[10] RK. Srinivasa, A. Ramakrishna, PV.Rao, Int Jl of Mech Eng and Techn, *Performance and emission characteristics of DI-CI diesel engine with preheated chicken fat biodiesel*, volume **4**, issue 3, (2013)
[11] M. S. Kumar, A. Kerihuel, J. Bellettre and M.Tazerout, J of Eng from Gas Turb and Pow, *A Comparative Study of Different Methods of Using Animal Fat as a Fuel in a Compression Ignition Engine*, **128**, (2006)
[12] T. Mata, A. Mendes, N. Caetano A. Martins, Ch Eng Trans, *Properties and sustainability of biodiesel from animal fats and fish oil*, volume **38**:175-80, (2014)
[13] A. Kumar, J. Kerihuel, M. Bellettre, M. Tazerout, M. Senthyl, Ren En, *The use of preheated animal fat as fuel in a compression ignition engine,* voulme **30** (issue 9), pp.1443-56, (2005)
[14] M. Czechlowski, W. Golimowsski, T. Sek J. Szymanowicz, Mainte and reli Eksploatacja I Niezawodnosc, *Exhaust opacity in a diesel engine powered with animal fats,–* **17,** Pp. 49-53, (2015)
[15] J. Alonso, A. Minambres,G. Arriba, J of Comb, *Study of Thermal Energy Production by Combustion of Lard and Diesel Mixtures*, Volume **2012**, ID 912581, pp. 1-9, (2012)

Study of the fire dynamics in a burning car and analysis of the possibilities for transfer of fire to a nearby vehicle

Angel Terziev[1,*]

[1]Department "Power Engineering and Power Machines", Technical University of Sofia, 8 Kliment Ohridski blvd., 1000 Sofia, Bulgaria

Abstract. A full scale experiment was carried out on the process of vehicle combustion and the ignition of an adjacent one. For this purpose, specialized measuring equipment (thermocouples and infrared camera) is used to determine the temperature field of the burning car and nearby vehicle. The burning car has 16 thermocouples mounted on it and one thermocouple mounted on the adjacent car in order to examine the heat transfer process. The temperature field between cars at different points of time is detected by an infrared camera, with a clear change in the temperature gradient in the presence of wind. It is one of the main reasons for the increased momentum of heat transfer. The research shows that with the available combustible load of the car (2,1 MW or around the average for the currently produced cars), the ignition of the neighbouring vehicle is expected around 14 minutes after the ignition of the main vehicle, with the ignition zone being in the engine compartment. Important information was obtained for both the dynamics of the burning vehicle and the temperature field between the two cars.

1. Introduction

There are many reasons for the occurrence of a fire in road transport vehicles. Most often they are deducted to a road traffic accident, gaps/failures in the vehicle's systems or as a result of vandalism. The fire may occur both during vehicle movement and in a parked state.

The dynamics of the fire depends on both the fuel load of the vehicle and the environmental parameters. The development of fire in a burning vehicle in open space depends mainly on outside air parameters such as wind speed and temperature. In limited spaces (such as underground parking and garages), the development depends both on the available amount of oxygen in the air and on the availability of ventilation systems.

In literature, the research of burning vehicles in a limited space is common. An analysis of the reasons for the occurrence of a fire in a confined space is presented in [1, 2] and concrete measures to prevent the occurrence of a fire, as well as a risk assessment, are presented. Paper [3] presents a full scale flame fire experiment and a study of the possibilities for transfer of fire to a nearby vehicle. The radiant heat flux is measured with a heat flux gauge installed 5 m away from the car fire. The temperature of the flue gases in the vicinity of the vehicle has been measured while being calculated using the indirect calculation method. The influence of mechanical systems on the development of fires of vehicles in garages (indoor spaces) is presented in [4, 5]. Paper [6] is an analysis of the possible effect of passive fire protection techniques on the peak heat release rates and temperature inside the tunnel. The impact of the effects studied provides information on enhancing the fire safety in road tunnels.

In recent decades, there has been a growing interest in the investigation of fires in cars using numerical methods. In paper [7, 8] numerical approaches are presented in modelling of burning vehicles in garages. The studies show that affected by ventilation systems, the fire develops rapidly resulting in the most rapid oxygen consumption followed by early extinction of the fire. In the presence of fresh air supplied by the intake ventilation system re-ignition takes place at an accelerated rate. The sensitivity of the CFD results to the input parameters is highlighted. As a conclusion it is stated that in a 4MW vehicle the supplied ventilation rate should not exceed 1,1 m/s in order to limit air with a high concentration of oxygen to accelerate the combustion process.

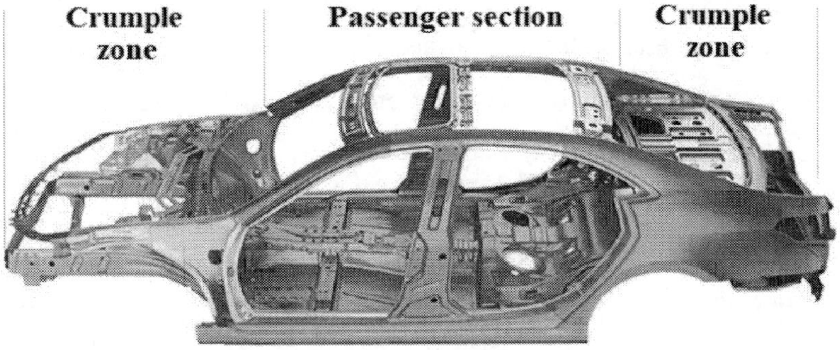

Fig. 1. Separated zones in the event of a fire [10]

The distribution of the fire in a car depends on a number of factors - the location of its ignition (engine compartment, passenger compartment, luggage compartment) (Figure 1), weather conditions, the presence of obstacles around the vehicle, the location of the vehicle (open spaces or Indoors), the characteristics of the vehicle, etc.

2. Experimental study

Experimental study is performed with a VW Golf Passat, fully-working, manufactured in the year 2000 (Figure 2). A "Peugeot" car was located at a distance of 0,8m away from the VW Golf. Based on the data presented in Table 1 regarding the combustion characteristics of the materials of the chosen vehicles it has been estimated that the combustion load is about 2,1 MW.

Fig. 2. Models and positioning of cars during the experiment

Table 1. Typical combustible material properties of the proposed cars

Combustible materials	**Basic Composite Material**	**Ignition point [°C]**	**Oxygen Index, [%]**
Engine compartment	Polypropylene	211	19
Seats	Polyurethane foam	205	17
Console shell	Polyvinyl chloride	238	20
Tires	Rubber	289	38

2.1. Measuring equipment used during the experiment

Various measuring equipment was used during the experiment. The multifunctional Testo 435-4 meter is used to determine the outside air parameters. It is equipped with sensors for measuring wind speed, temperature, humidity and atmospheric pressure. Table 2 provides information on the type and technical characteristics of the measuring equipment.

Table 2. Summary technical specification of a multifunction meter - Testo 435-4

Probe	**Parameter**	**Measuring Range**	**Accuracy, %**
Humidity probe	Humidity	0 to +100 [%rH]	0,1[%rH]
Pitot tube	Differential Pressure	0 to 25 [hPa]	0,01 [hPa]
Hot wire anemometer	Velocity	0 to 20 [m/s]	0,01 [m/s]
Hot wire anemometer	Temperature	-50 to +125 [°C]	± 0,4 [°C]

16 thermocouples have been installed to measure the temperatures during the fire in the passenger compartment. The outputs of the thermocouples were connected to two TC600 multichannel digital thermometers, with serial numbers 01648 and 01729 respectively. The thermocouple installation points are presented in figure 3. The operational range of the instruments is 0 ÷ 1200 °C. A calibration certificate with № 18A / 12.01.2016 and 19A / 12.01.2016 was issued by Interlabe AD.

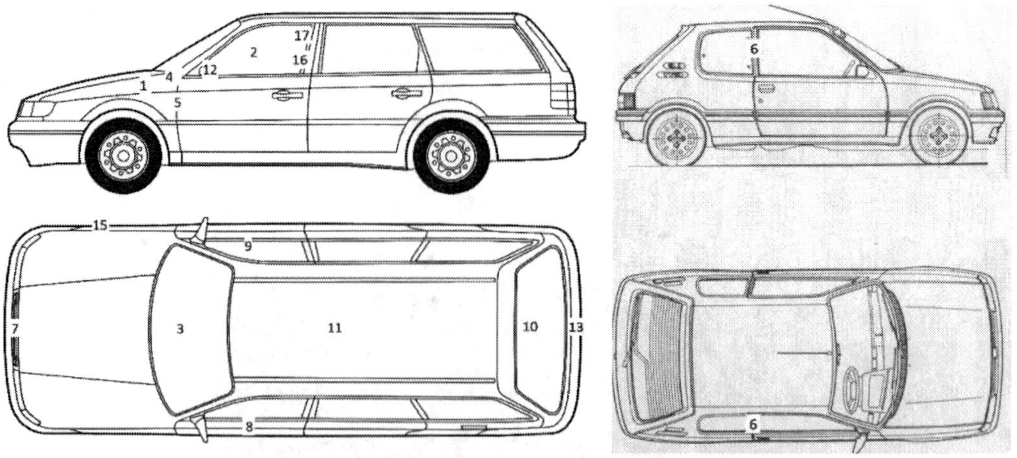

Fig. 3. Location of K type thermocouples: 1-engine compartment, 2-cup inner roof, 3-windshield, 4-fuel hoses, 5-fuse box, 6 - front left door to adjacent vehicle; 7-door hatch, 8-door front left, 9-door front right, 10-spoke glass, 11-compartment floor, 12-way dashboard, 13-door luggage compartment, 14-wheel front left, 15- wheel front right, 16-front left seat.

In addition, Infrared camera (FLIR E40) was used to capture the temperature field in front of the vehicles. The basic parameters of the measuring equipment used during the experiment are presented below:
- Resolution: 160 x 120 IR
- Spatial resolution: 2,72 mrad
- Thermal sensitivity: < 0,07 °C
- Temperature range L -20 to + 650 °C
- Spots: 33 boxes with min/max air
- Measurement of temperature difference -3,1 Mp digital camera and one LED light
- Bluetooth® / WiFi-1-2x digital zoom
- IR area on real image.

2.2. Experimental results

Experimental study was performed on 06 April 2017. The collected data during the experiment show the following: wind speed varies from 0 to 5 m/s with the average reading being 2,8 m/s. The change of temperature is in the range of 10 to 16 °C. Peak values are in times when the predominant wind direction is from the burning vehicle to the measuring probe. The humidity varies from 48 to 60%. Atmospheric pressure during the experiment changes in a narrow range of 925 - 926 hP. The change in outdoor air parameters during the experimental studies is shown in Fig. 4 a-d.

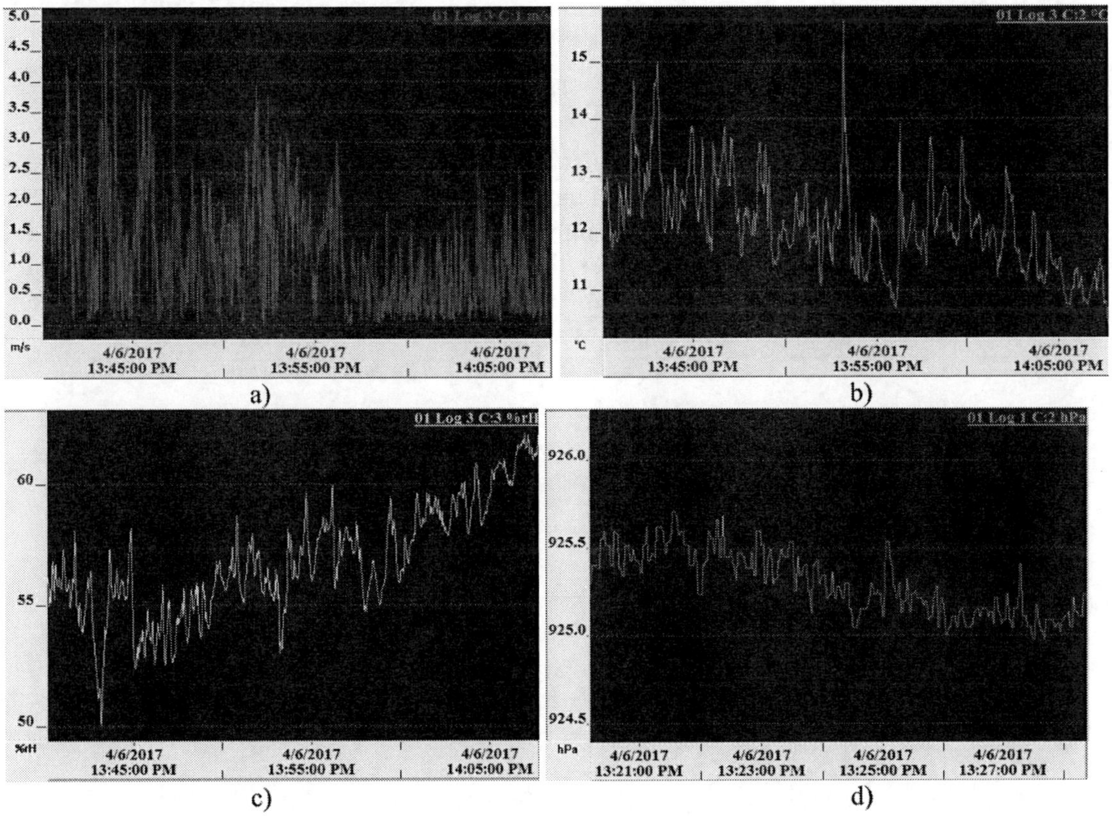

Fig. 4. Collected data for ambient air: a) air velocity; b) air temperature; c) relative humidity of the air; d) atmospheric pressure

a) b) c) d)

Fig. 5. Instant photographs during the fire: a) 9 min after ignition; b) 15 min after ignition; c) 22 minutes after ignition; d) after the test

Figures 6a and b show information about the temperature change during the engine fire and fuel system fire. These two areas are mentioned because they are directly in front of the infrared camera that shoots the fire. The impression is that after the fire occurs, the values recorded with the FLIR camera are lower than those measured with the thermocouples. This is due to the fact that the surface temperature is measured with the camera, whereas the values measured with the thermocouples are for the respective point.

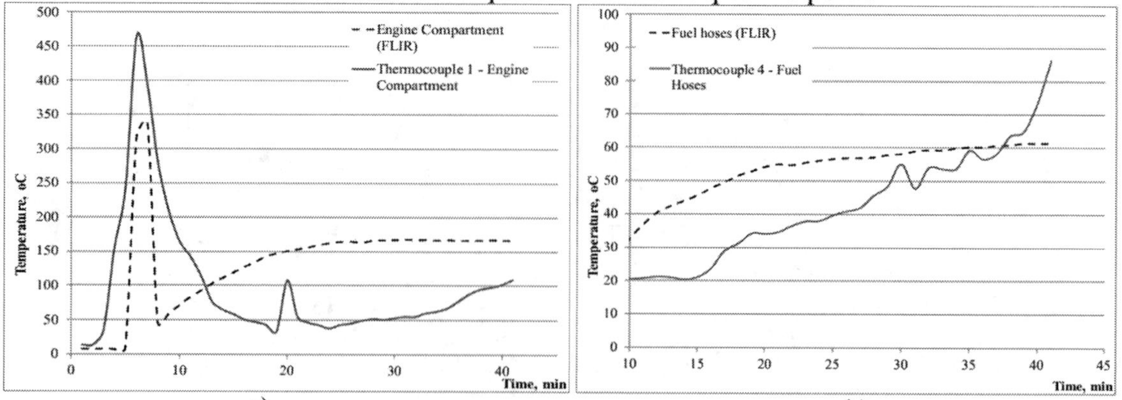

a) b)

Fig. 6. Temperature distribution a) zone Engine compartment; b) zone Fuel fuses

The results obtained by using the thermal imaging camera to study the fire provided enough detailed information about its development. The horizontal axis in Figure 7 represents the distance between cars respectively temperature profiles of the air between the two vehicles. Figure 7a the high temperature of the burning vehicle is clearly visible. A heat flux from the burning car is detected on the right side of the diagram. Figure 7b shows the temperature profile of the air between the two vehicles in the period of 20 minutes. Here, with the thermal imaging camera, a reversal of the temperature profile is detected as a result of a change in the flame direction due to wind. Figure 7c shows a temperature equalization between the two vehicles, respectively the ignition of the adjacent vehicle. Figure 7d shows an analogous situation like the one in Figure 7b (the case of a strong side wind), but unlike the previous case, here the velocity profile is considerably more aligned.

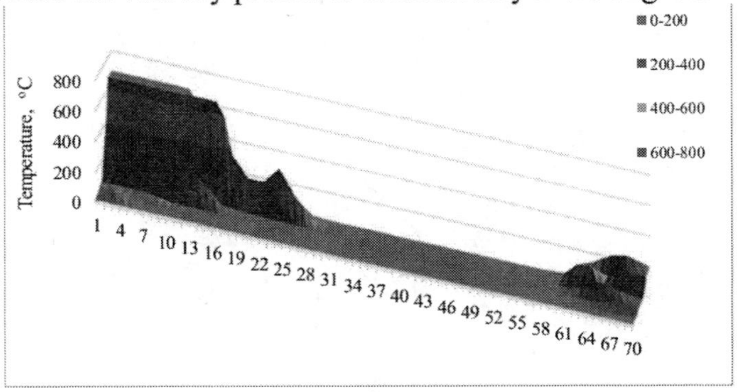

a)

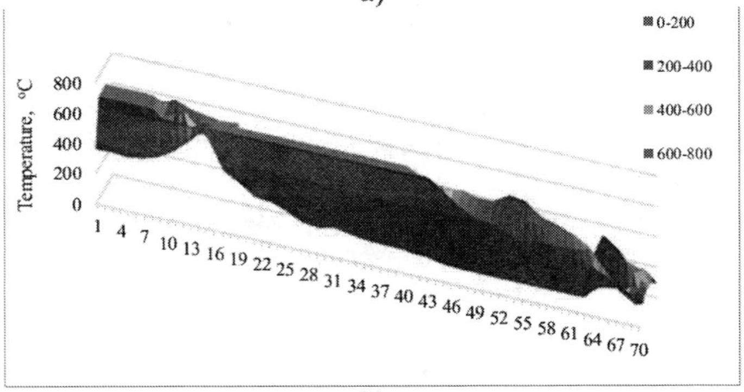

b)

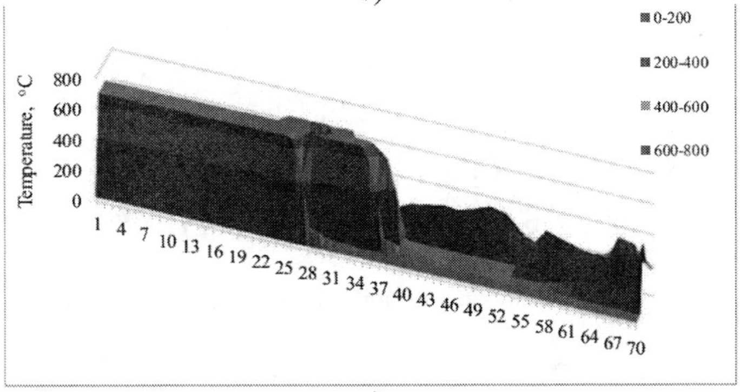

c)

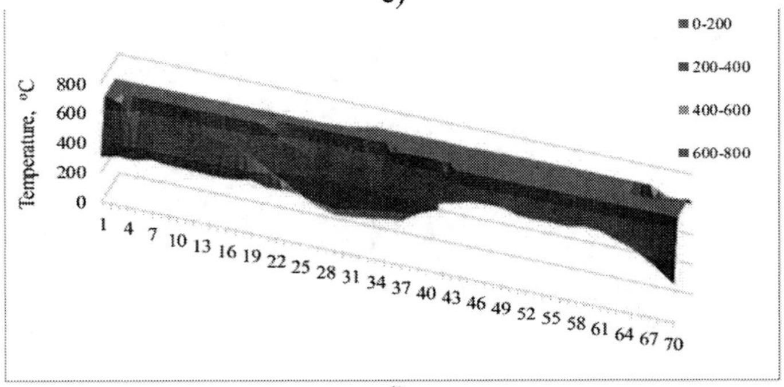

d)

Fig. 7. Temperature distribution a) zone Engine compartment; b) zone Fuel fuses

Fig. 8a-b give information from the thermal imaging camera for the distribution of the temperature field around the burning vehicle. The clearly identifiable are the temperature fields when the two different prevailing wind directions are registered.

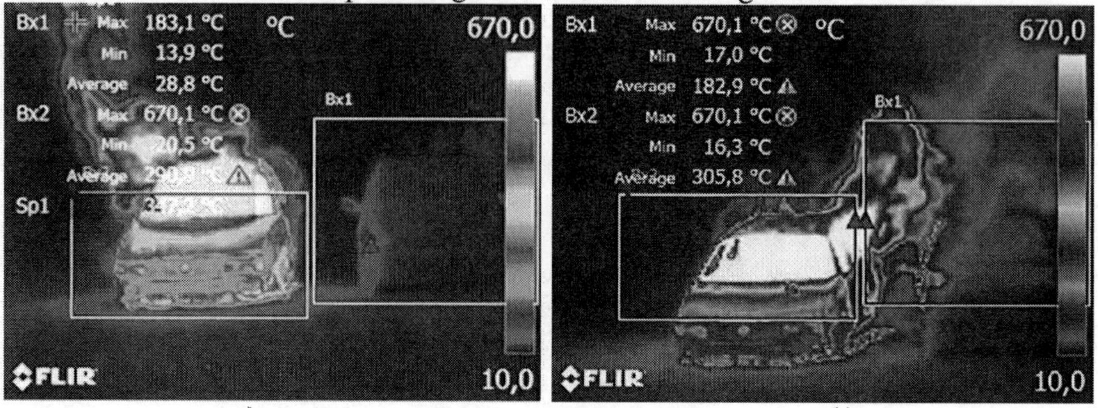

a) b)

Fig. 8. Temperature field around burning car captured with infrared camera: a) prevailing wind direction from east, b) prevailing wind direction from west

Figure 9 shows information about the temperature change of the two adjacent car doors (thermocouples 6 and 8). Once the fire starts in the engine compartment, it begins to take over the vehicle gradually. Initially, smoke begins to fill the vehicle's body and increase its temperature. The ignition of the individual elements begins after reaching their flash point. Thus, the door to the driver (made of acrylic fibers) is ignited at a temperature of 560 °C. A temperature gradient is apparent from the figure, which for a short time exceeds 800 °C. Upon reaching this peak, the temperature slowly began to decrease, with a slight increase 36 minutes later due to the cracking of the rear window of the vehicle and the entry of outside air with higher oxygen content. Burning of the left door of the vehicle points to a high temperature, which implies an increased heat transfer through radiation heat exchange. This is the reason for the sudden increase in the door temperature of the neighbouring vehicle. This rapid change is also due to the presence of wind, which further intensifies the heat transfer process (Figure 8b). After 14 minutes, there is a decrease in temperature of the left door of the burning vehicle and an increase in the temperature of the neighbouring door due to its ignition. The conclusion from the figure is that given the burning load of the vehicle, the ignition of the neighbouring car, when the distance is 0,8m, is in the order of 14 minutes.

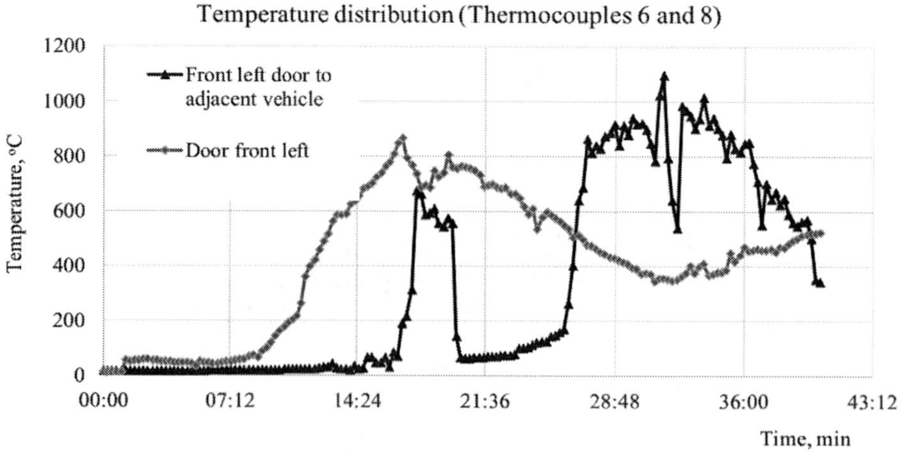

Fig. 9. Experimental data from the temperature change of the two adjacent car doors (thermocouples 6 and 8)

3. Conclusion

The dynamics of fire in vehicles positioned in open space differs significantly from the one in closed space (underground parking lots and garages). The environmental parameters play a major role in the development of the fire. The presence of wind significantly changes the behaviour of the combustion process as well as the direction of radiation heat exchange. In the presence of winds up to 5 m/s the ignition of the neighbouring vehicle may occur no later than 14 minutes, of course taking into account the fuel loading of the cars, which in this case is about 2,1 MW. The resulting distribution of the temperature field around the vehicle in a fire is important for analysing the dynamics of the fire and is extremely important when the vehicle is to be extinguished.

Acknowledgements
The author would like to thank the Research and Development Sector at the Technical University of Sofia for the financial support.

References

1. Moss PJ., et al, Incremental fire analysis (IFA) for probabilistic fire risk assessment. In: *Proc. 23rd Australasian Conference on the Mechanics of Structures and Materials (ACMSM23)*, 9–12 December 2014, Byron Bay, NSW, Australia, 2014, pp. 707–712
2. Shintani Y, et al, Experimental investigation of burning behavior of automobiles, *6th Asia-Oceania Symposium on Fire Science and Technology*, Korean Institute of Fire Science & Engineering, 17–20 March 2004, pp. 618–629.
3. Dayan Li, Guoqing Zhu, Hui Zhu, Zhichao Yu, Yunji Gao, Xiaohui Jiang, Flame spread and smoke emperature of full-scale fire test of car fire, Case Studies in Thermal Engineering, **10**, 2017, Pages 315-324
4. I. Antonov, R. Velichkova, K. Grozdanov, S. Antonov, M. Uzunova, I. Abbassi, Fire extinguishing system in large underground garages, *Advances in Science, Technology and Engineering Systems Journal*, **2**, No 3, pp. 221-226 (2017)
5. B. Merci, M. Shipp, Smoke and heat control for fires in large car parks: Lessons learnt from research?, *Fire Safety Journal*, **57**, 2013, Pages 3-10
6. Mukesh Singh Tomar, Shashank Khurana, Impact of passive fire protection on heat release rates in road tunnel fire: A review, Tunnelling and Underground Space Technology, **85**, 2019, Pages 149-159
7. Velichkova R., I. Antonov, K. Nikolov, K. Grozdanov, M. Uzunova, Modeling of the occurrence of fire in closed cars garages, *EFEA'2016*, Belgrade, Serbia, DOI: 10.1109/EFEA.2016.7748807
8. Victoria Kroll, David Crundall, Aperture judgement in fire-appliance drivers, Transportation Research Part F: *Traffic Psychology and Behaviour*, **63**, 2019, Pages 55-66
9. X. Deckers, S. Haga, N. Tilley, B. Merci, Smoke control in case of fire in a large car park: CFD simulations of full-scale configurations, *Fire Safety Journal*, **57**, 2013, Pages 22-34
10. https://carbrain.com/Blog/what-is-frame-damage

E3S Web of Conferences **112**, 01016 (2019) https://doi.org/10.1051/e3sconf/201911201016
TE-RE-RD 2019

Instantaneous field measurements of thermal bridge parameters in ground floor residential room

Martin Ivanov[1,*]

[1]Technical University – Sofia, FPEPM, Department:" Hydroaerodynamics and Hydraulic Machines", Sofia 1000, Bulgaria

Abstract. The "thermal bridges" are defined as an isolated building's areas, where the construction elements have higher thermal conductivity, compared with the rest of the building envelope. Thus, at cold winter conditions, a significant temperature difference may occur between neighbouring solid and air volumes within the construction. Moreover, it has been documented, that the heating energy demand of a building may be increased with more than 30%, due to the existence of thermal bridges and the increased heat losses from the indoors. Consequently, in the recent years, norms and standards have been developed, for avoiding thermal bridges during the building design process. But still, thermal bridges exist in the indoor environment, especially in older buildings, where no energy efficient measures have been applied. That is why, the presented study focuses on instantaneous field measurements of thermal bridge parameters in real existing ground floor residential room. The thermal bridge propagation is analysed relative to the indoor and outdoor air temperature and relative humidity, as well as with infrared thermal images of the affected external walls. The achieved results give valuable information about the generic conditions for thermal bridge existence, without considering the building envelope properties.

1 Introduction

In the developed countries worldwide, the building sector is responsible for about 40% from the total energy demand and consumption. [1] Also, to satisfy these energy needs, the building sector becomes indirectly responsible for about 40% from the CO2 and other greenhouse gases emissions in the atmosphere. [2] That is why, different strategies and norms were developed and adopted in the recent years, in order to assure sustainable reduction in the energy consumption in the residence buildings, as well as in the industrial sector. [3]

In the past years, building technologies and structure elements were continuously analysed and improved, in terms of thermal properties and contribution to the energy efficiency of the buildings. [4] All this research helped the countries, especially in Europe,

* Corresponding author: m_ivanov@tu-sofia.bg

to develop and adopt norms for significant heat loss control. But also, heat transfer paths, like thermal bridges, have been presented in the recent construction regulations. [5] Historically, when the requirements for thermal insulation of the buildings were moderate, the role of thermal bridges over the energy efficiency was considered as minor. But now, it becomes progressively important, especially for the demand for heat flow reduction trough the different well isolated construction elements, with lower thermal transmission values (U-values).

However, while the overall heat losses from the newly developed buildings have been decreased, it has been documented that, the portion of heat losses due to the thermal bridges has been increased. This is especially valid for the point thermal bridges, issuing from different modern façade technologies, which may contribute to more than 25% of the heat flows through the building envelope. [5] But, the impact of the linear thermal bridges on the heating energy need of the buildings may become even higher – above 30%. [6]

Another very important aspect of the thermal bridges is their impact over the indoor environment, especially in residential occupied spaces. The cold spots of the thermal bridges, particularly in winter conditions, may be a prerequisite for humidity related problems in the premises. The accumulated moisture over the thermal bridge surface, may support mold growth, which has direct affect over the occupants' health, productivity and performance.

In thermal bridge analyses, the building structure is considered as the most important parameter, for the heat flow distribution. [7, 8, 9] But, for the generic conditions for thermal bridge existence, the indoor and outdoor atmospheric conditions, are also very important factors, and may be analysed without considering the building structure. Sometimes, the characteristics of the building envelope structure are not available, especially in older or historic buildings. But, thermal bridges may still exist in these premises, and there is a necessity for mitigation measures of the corresponding negative effects.

That is why, a substantial need exists, for measurement data and analyses of the generic conditions for thermal bridge existence, especially for boundary and initial conditions in the numerical modelling of the related phenomena. This is the main purpose of the suggested research in this paper. The described experimental measurements are oriented in the field of non-ventilated residence rooms, which have the highest impact over the today's society. Also, special attention is carried on the case of ground floor residential rooms. In these rooms, the thermal bridge development potential is significantly higher, due to the direct contact with the solid ground.

2 Objectives of the study

The global objective of the presented study is to obtain and analyse the generic conditions for thermal bridge existence in ground floor residential room, by instantaneous field measurements of the thermal bridge parameters and the indoor and outdoor environmental conditions.

3 Methods of the study

The main approach in the presented study consists of field experimental measurements of the microclimate parameters, in real existing ground floor residential room, with confirmed thermal bridge in the floor area. The thermal bridge zone is confirmed by preliminary taken thermal images of the outer walls. The instantaneously measured parameters include: room air temperature, wall surface temperature, relative humidity, CO_2 concentration, outdoor air temperature and outdoor relative humidity. Also, infrared images of the thermal bridge area were taken in during the measurements, each morning and evening, in order to visualize the bridge propagation in the occupied space. The room is occupied, and the measurements were taken under uncontrolled occupant behaviour, which corresponds to the real room use. The effect of the occupant behaviour, over the generic conditions for the thermal bridge existence, will be studied later.

The measurements were taken in the winter period, with duration of 3 days and 3 nights, which corresponds to a short-term experiment. The winter period was selected, because the generic conditions of the thermal bridge existence are much better established and captured in the heating period of the year. Also, a weekend period was selected for the measurements – from 22:00h at 15. February 2019 (Friday), till 22:00h at 18. February 2019 (Monday). The reason was that during the weekend period, the selected residential room is mostly occupied, and the impact of the occupants over the thermal bridge propagation could be studied.

The physical measurements of the environmental parameters are performed by advanced wireless instrumentation (HOBO data loggers, made by ONSET Company), allowing recording of the changes in air temperature, relative humidity and carbon dioxide concentration in the room. The records were taken for each minute in each measurement point, for the entire duration period - 3 days and 4 nights. In addition, an FLIR C3 infrared thermal camera is used, to capture the surface temperature of the inner side of the outer walls in the thermal bridge area.

All obtained results are descriptively analysed and presented in terms of tables and graphics. It is considered that, this analysis gives valuable information about the generic conditions for thermal bridge existence, without considering the building envelope properties. The data will be used later, for initial and boundary conditions for the development of CFD (Computational Fluid Dynamics) based numerical model. This model will be able to predict the thermal bridge distribution in the affected occupied zones, but also a conjugated heat transfer could be included, in order to investigate the complex effect of the thermal bridge over the building structure itself.

4 Experimental set-up and instrumentation

On Fig. 1 are presented the ground floor residential room layout and the corresponding measurement points. The room is part of auxiliary guest house in a property, situated in the western part of Sofia, Bulgaria. The orientation of the room is South, and the window is situated in South. The floor area is 16 m^2, and the room height is 2.80 m, which corresponds to approximately 44.80 m^3 of free space. This room is basically used for a sleeping room and is normally occupied by two adult persons. The heating of the room is provided by a radiator, situated under the window, which is part of central heating system of the auxiliary building. During the experiments, the radiator was working constantly, to maintain the comfortable 20 °C – 21 °C in the room, and the heating set-point was not modified additionally.

As indicated on the figure, the confirmed thermal bridge zone is on the floor area in the corner, between the two external walls. Considering the building structure, it should be noted that both walls are well insulated, but the ground concrete floor is not. Also, in the thermal bridge area, there is a concrete support column, supporting the concrete roof. This structural layout is a prerequisite for thermal bridge existence itself, but still the generic conditions need to be additionally examined.

The measurement devices were divided in three separate systems, the position of which are indicated on Fig. 1. System 1 and System 2 consist of one HOBO data logger in combination with Telaire 7001 CO_2 sensors. These devices allow instantaneous measurements of air temperature, relative humidity and CO_2 concentration. The measurement points were selected to be away of the breathing zones of the occupants and away of the walls, in that way to cover the established measurement standards. The elevation from the ground was approximately one meter.

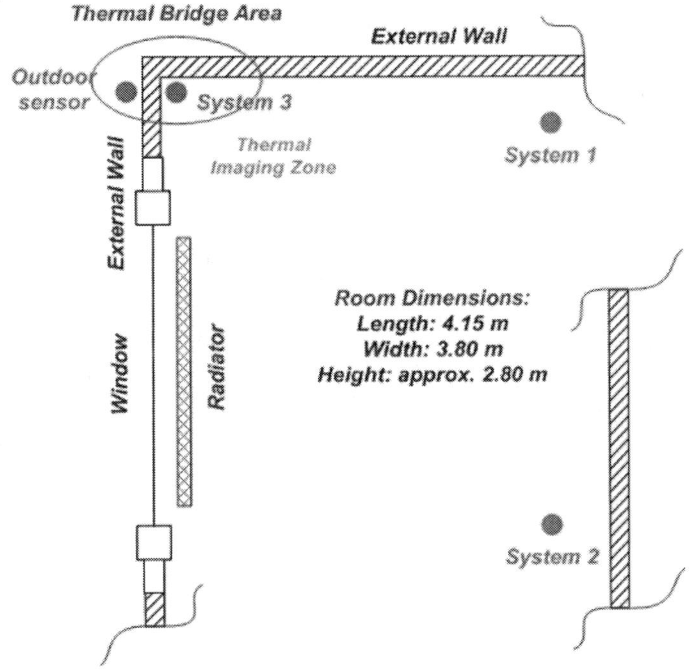

Fig. 1. Experimental room layout

System 3 consist of only one HOBO data logger for measurement of air temperature and relative humidity. This sensor was installed on the floor of the room, as close as possible to the external wall surface, in the thermal bridge zone. In that way System 3 measures the closest possible indoor parameters, affected by the thermal bridge existence. As it was mentioned, this is the zone for the thermal images as well. It should be noted, that when the thermal pictures were taken, the System 3 sensor was slightly removed from the place, for less than a minute, and then returned into place.

On the opposite side of System 3, over the outside wall surface, was placed the Outdoor sensor. This sensor provided the outdoor atmospheric conditions in terms of air temperature and air relative humidity. Thermal images in this zone were also taken, during the experimental period.

As it was mentioned, the room was occupied by two adults, during the experiments. Also, the period was attentively selected to be during the weekend, in order to stimulate the thermal load in the room. However, the occupant behaviour towards the indoor environment parameters, was not modified by any means. This will help in the understanding of the thermal bridge generic conditions.

4 Results and discussion

All obtained measurement data is processed and analysed. In this paper, only the results for the measured temperatures and relative humidity will be presented. It is considered that these values justify the instantaneous generic conditions for the thermal bridge existence in the ground floor residential room. The results are presented in terms of figures and tables, describing the temperatures and relative humidity distribution, within the entire experimental period.

The outdoor atmospheric conditions, in terms of air temperature and relative humidity, are presented on Fig. 2, for the entire measurement period. The summarized values, as well as the standard deviation are presented in Table 1. The minimum value of the measured temperature is -1.07 °C. The maximum measured value is 22.94 °C, which is not normal for February in Bulgarian climate, according to the national statistics. The average measured temperature is 5.88 °C for the entire experimental period, which is also higher than the normal statistical values for that month.

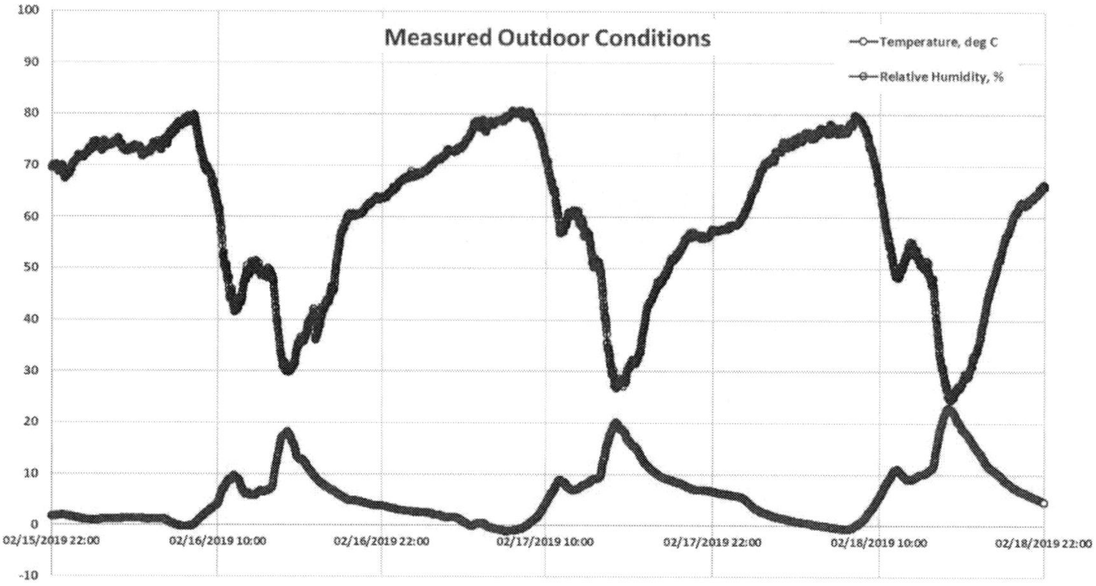

Fig. 2. Measured outdoor atmospheric conditions

Table 1. Summarized outdoor atmospheric conditions

Outdoor condition	min	max	average	SD
Temperature, °C	-1.07	22.94	5.88	5.46
Relative Humidity, %	24.55	87.92	60.46	15.03

The measured values of the relative humidity are as follows: minimum value is 24.55%, the maximum value is 87.92%, and the corresponding average value is 60.46%. The deviations of the relative humidity values clearly correspond with the increase and the decrease of the atmospheric air temperature. Both deviations happen distinctively during the day and night periods, as is clearly shown on Fig. 2. During the night period, the atmospheric air temperature is lower, and the relative humidity is higher. Obviously, these deviations will strongly affect the heat transfer trough the wall, in the thermal bridge area.

The indoor environment air temperature and relative humidity are presented on Fig. 3 and Fig. 4, for the measurement Systems 1 and 2. They are presented in different graphs, for more clear visualization of the measured values' distribution. The summarized values from the corresponding measurement points are presented in Tables 2 and 3, respectively after each graph.

The minimum value of the measured indoor temperature for System 1 is 18.39 °C. The maximum measured value is 21.32 °C, and the average measured temperature is 20.14 °C, with standard deviation of only 0.62 °C for the entire experimental period. The measured minimum value of the indoor relative humidity is 33.53%. The maximum value is 59.22%, and the corresponding average value is 47.29%.

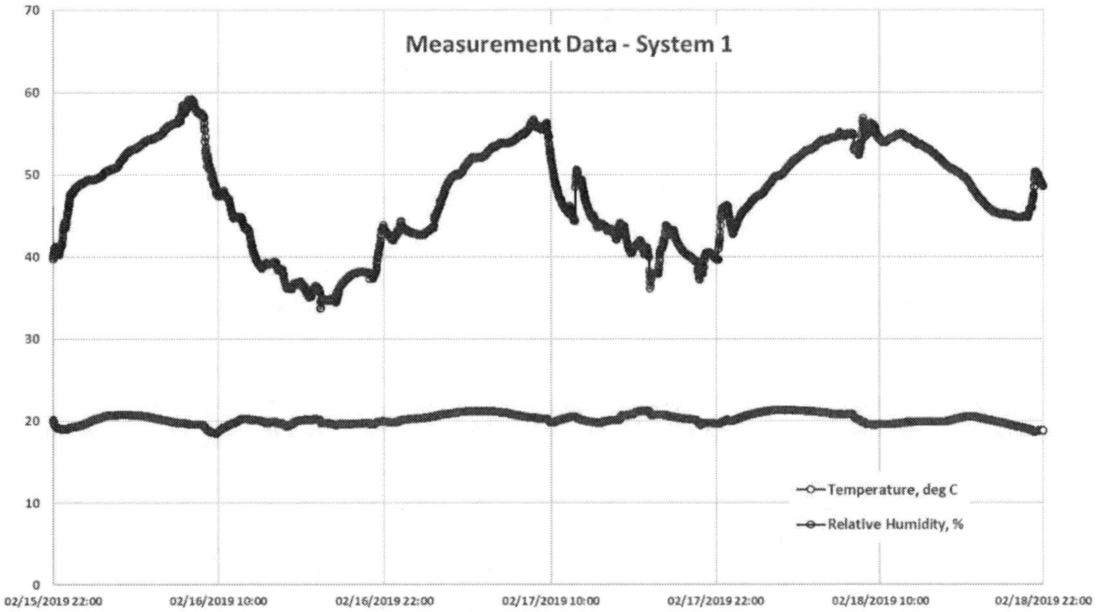

Fig. 3. Measured indoor data for System 1

Table 2. Summarized indoor data for System 1

Indoor data, System 1	**min**	**max**	**average**	**SD**
Temperature, °C	18.39	21.32	20.14	0.62
Relative Humidity, %	33.53	59.22	47.29	6.18

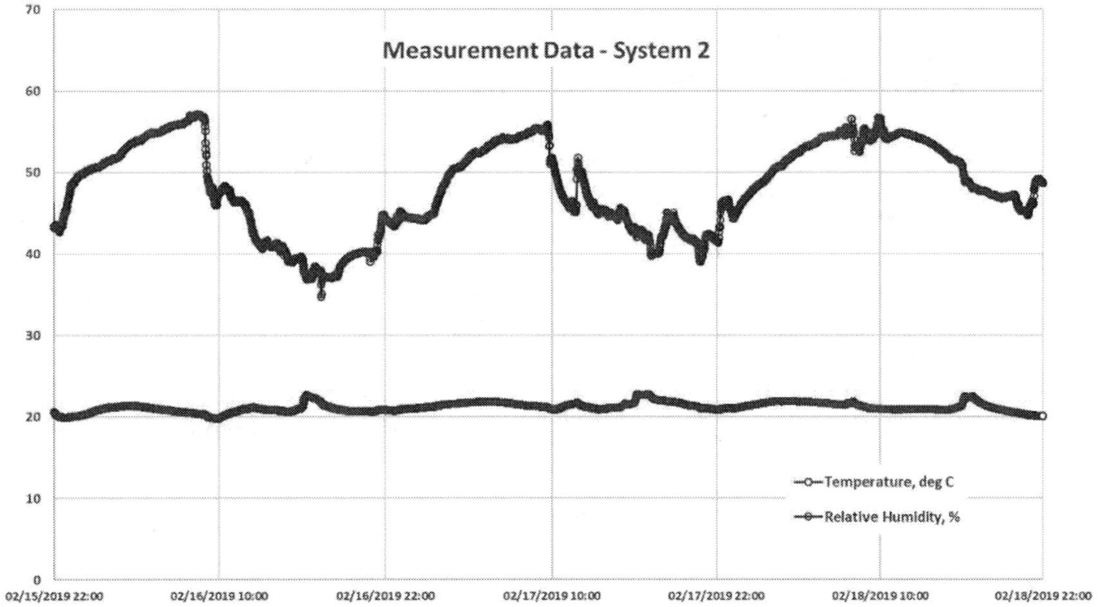

Fig. 4. Measured indoor data for System 2

Table 3. Summarized indoor data for System 2

Indoor data, System 2	min	max	average	SD
Temperature, °C	18.58	22.73	21.11	0.60
Relative Humidity, %	34.68	57.14	48.14	5.39

The corresponding minimum value of the measured indoor temperature for System 2 is 18.58 °C. The maximum measured value is 22.73 °C, and the average measured temperature is 21.11 °C, with standard deviation again of only 0.60 °C for the entire experimental period. The measured minimum value of the indoor relative humidity, according to System 2 is 34.68%. The maximum value is 57.14%, and the corresponding average value is 48.14%.

The summarized data for the two measurement systems shows relatively high degree of homogeneity in the indoor air temperature and relative humidity distribution in the experimental room. System 2 is closer to the inner wall and closer to the door of the room, which leads to adjacent indoor spaces. It is away from the thermal bridge zone, and that is why it shows relatively higher average values of the measured parameters. The proximity of System 1 to the thermal bridge area, explains the lower value of the average measured temperature. But, the difference of the relative humidity is mostly due to the non-perfect mixing of the room air and due to a decreased aeration rate in relatively large periods of time. The analyses of the CO_2 concentration measurements will give additional explanation of this effect and will be presented in another supplementary paper. The metabolic CO_2 concentration is a faster indicator for the aeration rate in the room, compared with the temperature and relative humidity, especially when there is a heating device installed. The longer periods with high CO_2 concentration, are indicator for low aeration or infiltration rate in the room. However, the deviations between the results in the two measurement points are considerably small, and the values itself are in the comfortable range for the indoor environment requirements for the winter season. This means that, the thermal bridge areas will be more accurately defined, and the generic conditions will be more precisely established.

The measured values of the air temperature and relative humidity in the thermal bridge zone, System 3, are presented on Fig. 5 and summarized in Table 4. The minimum value of the measured air temperature for System 3 is 11.93 °C. The maximum measured value is 20.29 °C, and the average measured temperature is 13.89 °C, with standard deviation of 1.06 °C for the entire experimental period. The measured minimum value of the relative humidity, in the thermal bridge zone is 39.67%. The maximum value is 83.77%, and the corresponding average value is 67.56%.

The obtained results for the thermal bridge zone clearly show significantly lower values of the air temperature and significantly higher values of the relative humidity, compared with the rest of the residence room. The graphics on Fig. 5 shows the predisposition of constant heat losses through the thermal bridge zone. The higher values of the relative humidity in this zone indicates the possibility for moisture related problems on the surface of the external wall.

These effects are confirmed by the thermal imaging results, presented in Table 5. The images show the surface temperature in the thermal bridge zone, captured each morning and each evening, during the experimental period. The camera marker is set to show the minimum measured temperature in the focus area. Thermal images of the outside wall conditions were also taken at the same time but will not be presented in this paper. The results here are concentrated on the impact of the thermal bridge over the indoor environment.

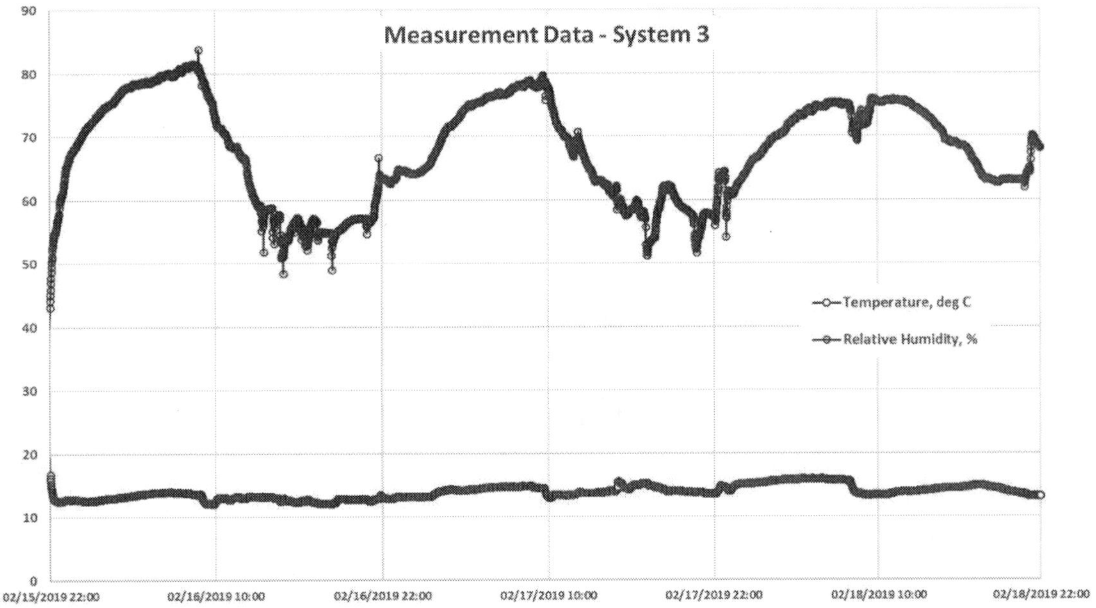

Fig. 5. Measured indoor data for System 3

Table 4. Summarized indoor data for System 3

Indoor data, System 3	min	max	average	SD
Temperature, °C	11.93	20.29	13.89	1.06
Relative Humidity, %	39.67	83.77	67.56	8.13

Table 5. Thermal images from the field experiments

2/15/19 22:00h	2/16/19 9:00h	2/16/19 22:00h	2/17/19 12:00h
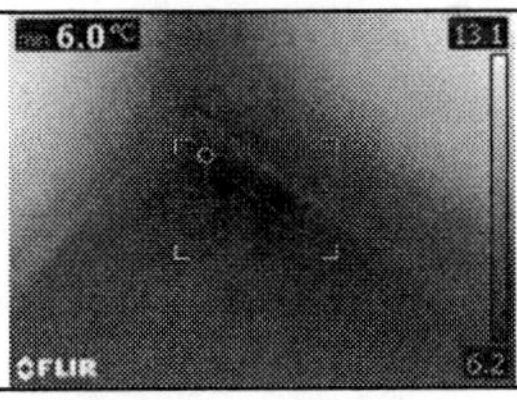	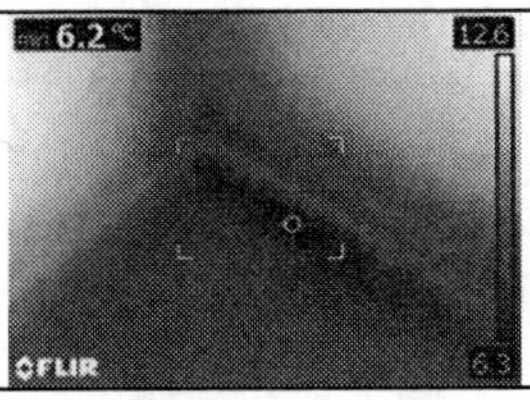	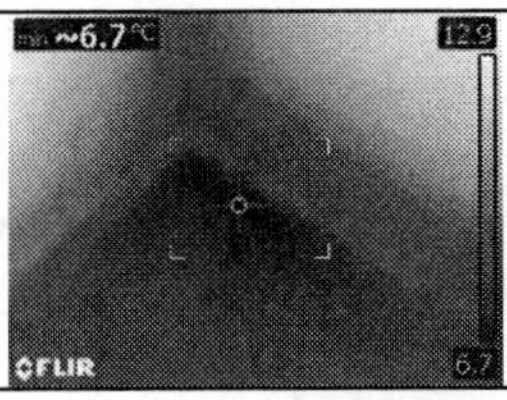	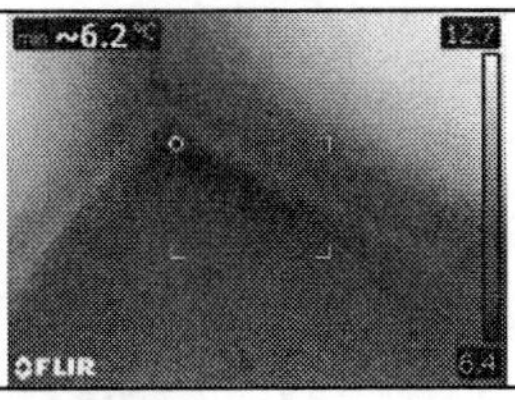
2/17/19 22:00h	**2/18/19 9:00h**	**2/18/19 22:00h**	
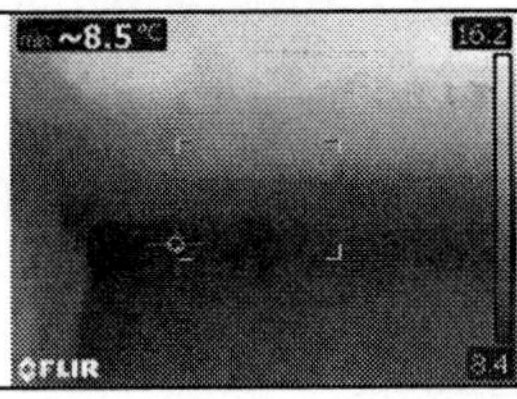	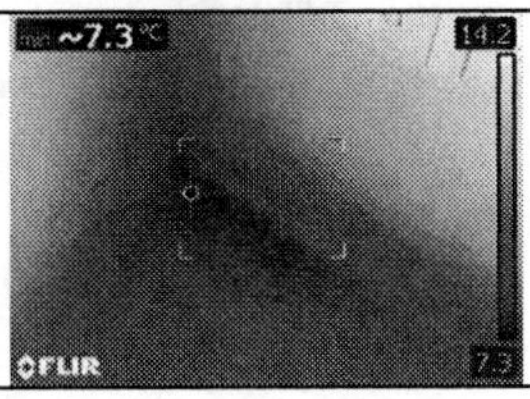	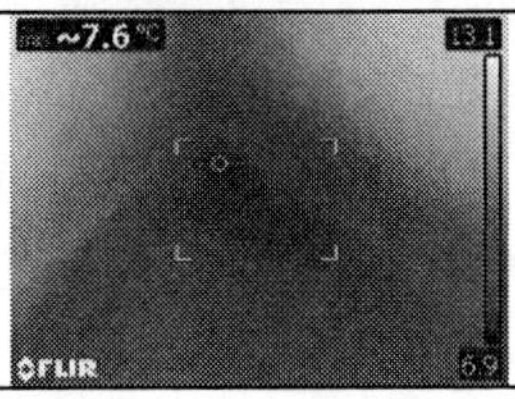	

For further analyses, this thermal imaging data is compared with the measured values of the air temperature form System 3 and it is presented on Fig. 7. As it can be seen, the measured surface temperature values are significantly lower than the one, measured with the logger. This could be expected, because the logger measures the air temperature in very close proximity with the thermal bridge zone, compared to the direct surface temperature, captured by the thermal camera.

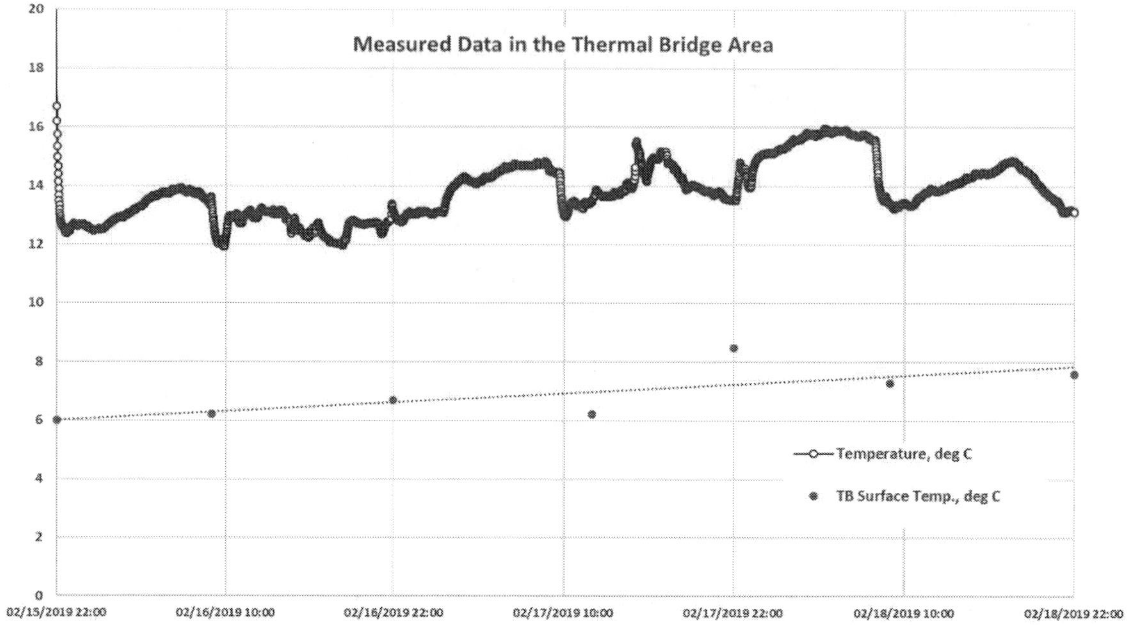

Fig. 7. Comparison of the measured data in the thermal bridge area

Obviously, the thermal camera may account better for the heat loss trough the solid surface of the wall. However, for precise modelling of the conjugated heat transfer in such experimental set up, the air temperature around the solid cold surface, is also important generic parameter for the thermal bridge existence. Collecting such boundary condition experimental data, seems to be innovative approach in the existing knowledge on thermal bridges and their impact on the indoor air quality.

Additional processing of the obtained data will also be accomplished. The next step will be to analyse the dew point temperature distribution in the occupied residential space, which will be an indicator for humidity related problems, over the thermal bridge surface area. Mitigation measures could also be suggested, in order to minimize the thermal bridge propagation zone. Depending on the further analyses, these measures may include not only structural isolation, but for example slight increase of the room temperature, or increase of the aeration rate, for lowering the humidity related problems in the thermal bridge area.

5 Conclusions

Instantaneous field measurements of thermal bridge parameters, in ground floor residential room, were performed in the presented study. The analyses of the results showed the generic conditions for the thermal bridge existence, without considering the building structure.

The achieved measurement results showed relatively high degree of homogeneity in the indoor air temperature and relative humidity distribution in the selected experimental room.

As it was occupied by two persons, the values of the temperature and relative humidity are in the comfortable range for the indoor environment requirements in the winter season.

However, the results showed that the room have distinctively confirmed thermal bridge zone, situated in the floor area between the two outer walls. The obtained results near the thermal bridge clearly showed lower values of the air temperature and higher values of the relative humidity, compared with the rest of the residence room.

This effect was confirmed by the thermal images and is considered as predisposition of constant heat losses through the thermal bridge area. The higher values of the relative humidity in this zone indicates the possibility for moisture related problems on the surface of the external wall.

The presented experimental results provide important information about the mechanism of interaction of the different microclimate parameters, relative to the generic conditions for thermal bridge existence. The data obtained will be used to create a numerical model, which might predict the heat losses as well as the appearance of humidity related problems in the premises.

Acknowledgements

The presented study is supported by the National Science Fund at the Ministry of Education and Science of Bulgaria, under the activities of: "Competition for financial support for projects of junior basic researchers and postdocs – 2018", with Contract № КП-06-М27/4, entitled: "Numerical assessment of effective measures for prevention of moisture accumulation on external walls with thermal bridge, in residential rooms without organized ventilation".

References

[1] Eurostat: *"Tables, Graphs and Maps Interface"*, https://ec.europa.eu/eurostat, (Accessed April 2019)

[2] International Energy Agency, *"Modernising Building Energy Codes"*, (2013)

[3] International Energy Agency, *"Transition to Sustainable Buildings: Strategies and Opportunities to 2050)"*, (2013)

[4] T. Theodosiou, K. Tsikaloudaki, S. Tsoka, P. Chastas, *"Thermal bridging problems on advanced cladding systems and smart building facades"*, J. Clean. Prod., 214, 62-69 Elsevier (2019)

[5] A. Papadopoulos, *"Forty years of regulations on the thermal performance of the building envelope in Europe: achievements, perspectives and challenges"*, Energy Build. 127. https://doi.org/10.1016/j.enbuild.2016.06.051, (2016)

[6] H. Erhorn-Kluttig, H. Erhorn, *"Impact of thermal bridges on the energy performance of buildings"*, Paper P148 of the EPBD Buildings Platform, (2009)

[7] ISO 10211, *"Thermal Bridges in Building Construction - Heat Flows and Surface Temperatures. Detailed calculations"*, International Organization for Standardization, (2017)

[8] ISO 10456, *"Building Materials and Products, Hygrothermal Properties, Tabulated Design Values and Procedures for Determining Declared and Design Thermal Values"*, International Organization for Standardization, (2007)

[9] EN ISO 6946, *"Building Components and Building Elements - Thermal Resistance and Thermal Transmittance - Calculation Method"*, International Organization for Standardization, (2012)

Dew point temperature analyses in ground floor residential room with existing thermal bridge

Martin Ivanov[1,*]

[1]Technical University – Sofia, FPEPM, Department:" Hydroaerodynamics and Hydraulic Machines", Sofia 1000, Bulgaria

Abstract. The presented study reveals a dew point temperature analyses in ground floor residential room with existing thermal bridge. The dew point temperature is analysed, based on field measurements of indoor air temperature and relative humidity in the residential room, without organized occupants' behaviour. Furthermore, the dew point temperature is cross-analysed with existing thermal bridge propagation in one of the outer walls of the room, via infrared thermal images. The results represent a valuable indicator for moisture accumulation over the thermal bridge zone, as well as an indicator for future mold growth and other humidity related problems. In the building design practice, the "thermal bridge" is defined as a distant zone, where construction elements have higher thermal conductivity, compared with the rest of the building envelope. These thermal bridges mostly affect the energy performance of the buildings, because of the higher heat losses from inside towards outside. But even more important, moisture build-up and considerable humidity related problems in the occupied areas are probable, due to the decreased surface temperature over the affected thermal zones.

1 Introduction

According to various studies, humidity indoors increases the risk for development of wide range of health problems in the occupants. [1] These may include: respiratory tract infections, asthma, shortness of breath, but also fatigue, weakness and headache. Some studies also suggest that prolonged occupation of humid environment is a risk factor for the development of hypersensitivity and the emergence of allergies. [2,3] However, except for the house dust mites, it is not yet definitely established which components of the indoor air associated with the humidity, are responsible for the existence of these problems. [4]

Nevertheless, it is recognized that excess humid indoor environment is dangerous for the occupants themselves. The moisture in the rooms can exist in the form of water vapor in the air or simply by water, and there are basically four main sources. The first are the external sources, due to the moisture from the ground, outside air and various gaps (leakages), due to rains or prolonged snow. The second are from the internal sources, most often due to the people (occupants), their pets and their activities, like cooking, cleaning, washing, and drying clothes etc. The third are the building sources, which cover the

* Corresponding author: m_ivanov@tu-sofia.bg

moisture contained in the building materials, like concrete walls and slabs, brick walls, different types of plasters, floor and wall coverings and others. The last are considered all accidental sources, which include the different kinds of leaks from water pipes, floods and others.

An interesting matter is the difference between moisture in the building structure and moisture in the indoor air. Moisture in the building structure can affect the construction materials, leading to microbiological and chemical processes that produce various emissions of odours, allergens and other respiratory irritants. In turn, the relative humidity of indoor air can cause moisture accumulation on cold surfaces and within the structure, which also assists the chemical processes in the development of mold and fungus. This effect my easily happen in thermal bridge areas, where construction elements have higher thermal conductivity, compared with the rest of the building envelope. These construction elements were continuously analysed and improved in the past decades, in terms of thermal properties and influence on the energy efficiency of the buildings. [5] But also, thermal bridges were recognized as important heat transfer paths, and have been presented in the recent construction regulations, in order to avoid their existence. [6]

In reality, thermal bridges still exist in the indoor environment, especially in older residential premises, and it is still important to study the moisture accumulation indicators in such indoor environment. That is why, the presented research is concentrated on the dew point temperature analyses, in a ground floor residential room with confirmed thermal bridge. The dew point temperature is the saturation temperature of the water vapor, which corresponds to the partial pressure of water vapours in the humid air under the specific conditions. Compared with the surface temperature over the thermal bridge area, a direct indicator for moisture accumulation problems, is attained.

Dew point temperature was also previously analysed in extensive long-term study in more than 300 residential rooms, in two of the major cities in Bulgaria [7]. In 32.5% of the observed rooms in one of the cities and in 28.8% in the other, the average dew point temperature is higher than 10 °C. These values are significantly high and suggest a high risk of moisture accumulation in the building structure, as well as mold growth over the outer enclosures of the rooms.

2 Objectives of the study

The global objective of the presented study is to analyse the dew point temperature, in ground floor residential room with existing thermal bridge, as an indicator for moisture accumulation problems in the indoor environment.

3 Methods of the study

The main approach in the presented study consists of field experimental measurements of the microclimate parameters, in real existing ground floor residential room, with confirmed thermal bridge in the floor area. The measured parameters include: room air temperature, wall surface temperature, relative humidity, CO_2 concentration, outdoor air temperature and outdoor relative humidity. Some of these measured values are used to calculate and analyse the dew point temperature in the occupied space.

In the environmental engineering practice, various parameters are used to characterize the amount of water vapor in the humid air. The most widely used is the relative humidity in % - the ratio of the partial pressure of the water vapor dissolved in the air under given conditions, relative to the saturation pressure of the water vapor (the maximum amount of water vapor that can be dissolved in the air) under these conditions. But, in order to assess the humidity related problems in the indoor environment, the dew point temperature in °C is also widely used. This parameter represents the saturation temperature of the water vapor, which corresponds to the partial pressure of water vapors in the humid air under the specific conditions.

The saturation pressure of the water vapor as a function of its temperature is calculated from the expression:

$$p_s = 100\exp\left(18.956 - \frac{4030.18}{t_r + 235}\right),\ Pa \tag{1}$$

where the temperature of the water vapor (and the humid air) t_r is in °C. This formula is derived in [7], based on ISO 7730 [8]. From this expression, after transformations and taking into account the definitions of relative humidity and dew point temperature [8], the following expression is reached:

$$t_{dp} = \frac{1}{\dfrac{1}{235 + t_r} - \dfrac{\ln\left(\dfrac{RH_r}{100}\right)}{4030.18}} - 235 \tag{2}$$

from which the dew point temperature (t_{dp}) of the air in a room is calculated as a function of its temperature (t_r) and relative humidity (RHr).

It could be approximated that, the dew point temperature of the air in a given room will be the temperature of the surroundings at which and below which a moisture accumulation will be observed. Typically, the coldest elements in the dwellings are the outer walls and the glass surfaces of windows and outer doors, and most often moisture accumulation is formed on them. But these are also the areas with thermal bridges as well.

The retention of moisture on these surfaces, leads to a decrease in their heat insulation capacity. Moreover, in the presence of suitable nutrient medium and temperature, it may lead to the development of molds of different types. The mold in the premises is recognized as a major respiratory tract irritant and causes a number of respiratory problems, including the development of allergies and asthma.

Instead of the thermal bridges, there are also other situations in which the temperature of the inner surface of the external enclosures becomes lower than the dew point temperature of the air in the room. For example, in homes without central heating, or where the heaters are switched on periodically, the construction elements of the building remain colder and the risk of moisture accumulation on the walls is much higher.

Considering the methods of the presented study, the CO_2 concentration measurements have also an important role. The CO_2 sources in the selected residence room are only the two occupants, via their metabolic and breathing physiological functions. At first, the expected distinguished build-up and decays in the metabolic CO_2 concentration measurements, will show the precise periods, when the room is occupied and not occupied. Second, at the build-up periods, due to the berthing process characteristics, the exhaled air is also with higher temperature and higher relative humidity. As it is mixed with the room air, the room relative humidity and temperature is also changed, which reflect over the dew point temperature in the occupied space. So, it is supposed that, the moisture accumulation over the thermal bridge area will be higher when the metabolic CO_2 concentration in the room is higher as well.

The metabolic CO_2 concentration is a faster indicator for the aeration rate in the room, compared with the temperature and relative humidity, especially when there is a heating device installed. The longer periods with high CO_2 concentration, are indicator for low aeration or infiltration rate in the room. At these periods, when a thermal bridge is established, the risk of moisture accumulation over the thermal bridge inside surface area is significantly high. Thus, at high CO_2 concentration levels, there exist an increased risk of harmful exposure of the occupants, to different mold growth and spores over the thermal bridges, which may lead to allergies, asthma and other respiratory problems.

The rest of the study methods are also described in [9], which is supplementary paper to the same research. The thermal bridge zone in the selected experimental room is confirmed by preliminary taken thermal images of the outer walls. The short-term measurements were taken in a weekend winter period, with duration of 3 days and 3 nights. In that way, the generic conditions of the thermal bridge existence are much better established, and the selected residential room is occupied by two adults in most of the time, which corresponds to the normal room use. Moreover, the occupant behaviour towards the indoor environment parameters, was not modified by any means.

4 Experimental set-up and instrumentation

The experimental room layout in schematically presented on Fig. 1. The room is used for a bedroom and is part of single floor, auxiliary guest house in a property, situated in the western part of Sofia, Bulgaria. The orientation of the widow-wall is South, the floor area is 16 m^2, and the room height is 2.80 m. The presented radiator was working continuously during the measurements, maintaining 20 °C – 21 °C in the room, without modifying the heating set-point.

The figure also shows a thermal image of the established thermal bridge on the floor area in the corner, between the two external walls. In fact, both walls are well insulated on the outside, but the ground concrete floor is not. Also, in the presented zone, there is a concrete support column, supporting the concrete roof, which is a structural a prerequisite for the thermal bridge existence.

The green points on Fig. 1, denoted by Systems 1, 2, 3 and Outdoor Sensor, represent the positions of the measurement devices. System 1 and System 2 consist of one HOBO data logger (manufactured by ONSET company) in combination with Telaire 7001 CO_2 sensors, allowing instantaneous measurements of air temperature, relative humidity and CO_2 concentration. According to the established measurement standards, the positions of these two systems were selected to be away of the breathing zones of the occupants and away of the walls, on approximately one meter from the ground.

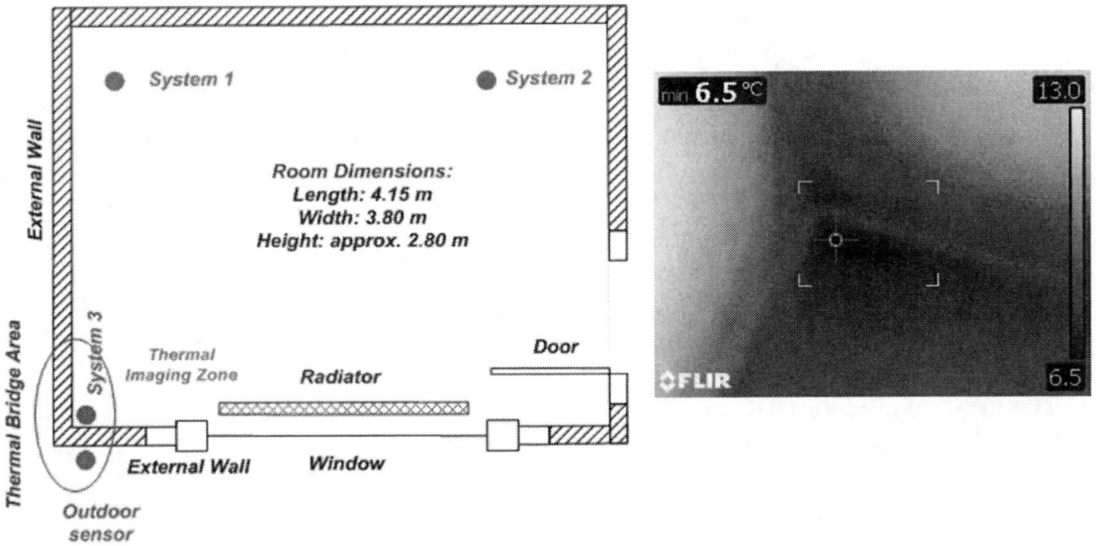

Fig. 1. Experimental room layout and thermal image of the established thermal bridge

System 3 consist of only one HOBO data logger, for measurement of air temperature and relative humidity, and was installed on the floor of the room, as close as possible to the thermal bridge zone. This is also the zone for the thermal images, taken each evening and each morning, during the experimental period. Over the outside wall surface, on the opposite side of System 3, was placed the Outdoor sensor, providing the outdoor atmospheric conditions in terms of air temperature and air relative humidity. Thermal images in this zone were also taken, during the experimental period.

4 Results and discussion

For the purpose of the presented study, only the results for the measured temperatures, relative humidity, CO_2 concentration and the calculated dew point temperature will be presented, in terms of figures and tables. The outdoor atmospheric conditions were also measured and presented in [9]. The corresponding parameters, measured in the two room locations, are presented on Fig. 2 and Fig. 3, for both measurement Systems 1 and 2. They are presented in separate graphics, for more clear visualization of the values' distribution. For cross-analyses, the summarized values are shown together in Table 1.

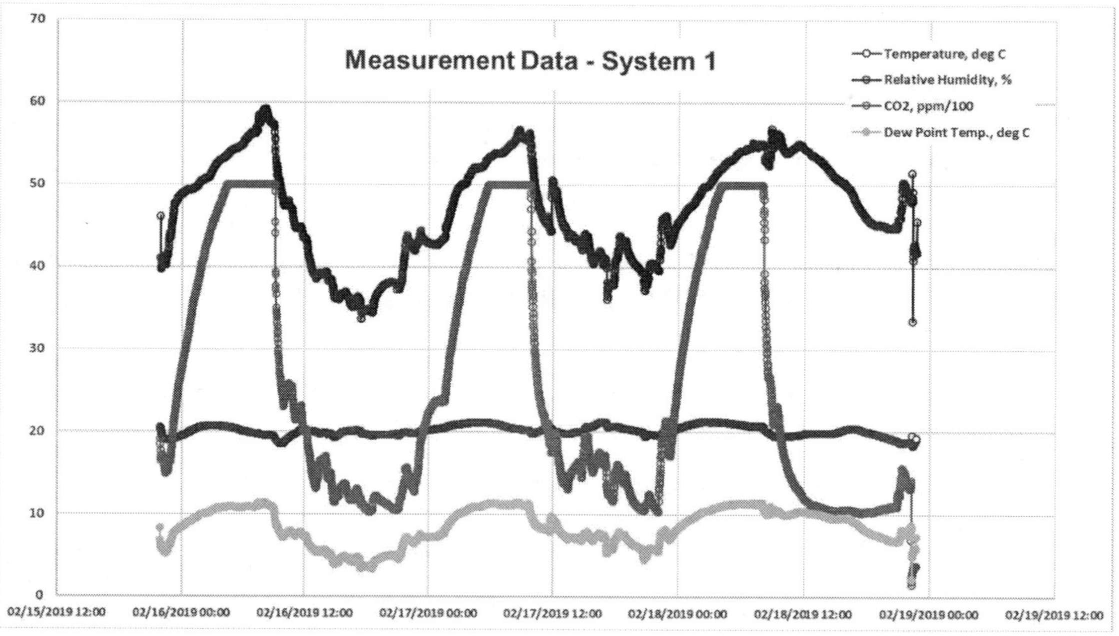

Fig. 2. Measured indoor data for System 1

The processed measurement data for System 1 shows, that the average room air temperature is 20.14 °C, with standard deviation of only 0.62 °C. The minimum and maximum values are 18.39 °C and 21.32 °C, respectively. The average value of the indoor air relative humidity is 47.29%, with minimum and maximum measured values of 33.53% and 59.22%, during the entire measurement period.

The measured indoor CO_2 concentration varies in large extend for both systems, and the build-up during the occupied night periods is well presented on both Fig. 2 and Fig.3. It should be noted, that the minimum measured values are not correct, since the outdoor environmental CO_2 concentration is around 400 ppm. There might be an incorrect reading of the sensor, at the end of the experimental period, as shown on the figures. However, the maximum values are correct, and as seen, the real concentration in the room exceeds the maximum measurement capabilities of the sensors – 5000 ppm. This result was unexpected and is a clear indicator for low aeration, especially when the room is occupied. The average registered concentrations are above 2500 ppm, for both system, which is an indicator for low infiltration and exfiltration, and probably for decreased level of indoor air quality.

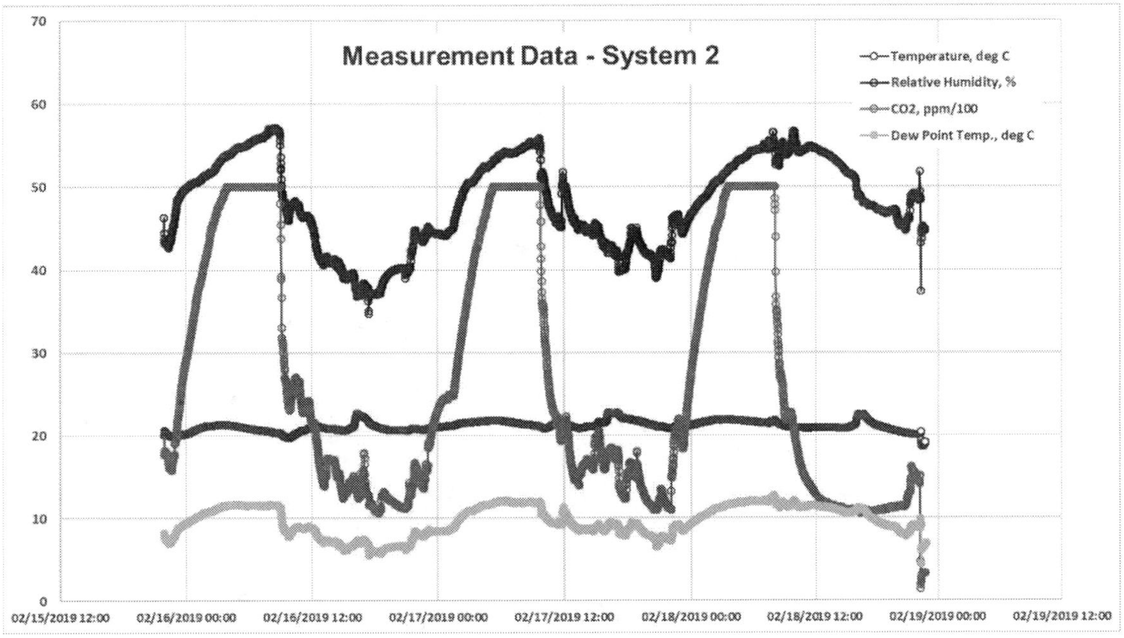

Fig. 3. Measured indoor data for System 2

Table 1. Summarized indoor data for System 1 and System 2

Indoor condition	min	max	average	SD
System 1 - Temperature, °C	18.39	21.32	20.14	0.62
System 2 - Temperature, °C	18.58	22.73	21.11	0.60
System 1 - Relative Humidity, %	33.53	59.22	47.29	6.18
System 2 - Relative Humidity, %	34.68	57.14	48.14	5.39
System 1 - CO_2, ppm	151.00	4998.00	2595.55	1530.21
System 2 - CO_2, ppm	140.00	4998.00	2661.90	1520.61
System 1 - Dew Point Temp., °C	2.20	11.50	8.48	2.17
System 2 - Dew Point Temp., °C	4.40	12.71	9.65	1.79

The average dew point temperature, calculated by the measurements of System 1, is 8.48 °C, with standard deviation of 2.17 °C. The minimum and maximum values are 2.20 °C and 11.50 °C, respectively. For the measurements of System 2, the calculated average dew point temperature is 9.65 °C, with standard deviation of 1.79 °C. The minimum and maximum values, for the same system, are 4.40 °C and 12.71 °C, respectively.

As seen form the figures, the dew point temperature deviates during the experimental period to a higher extend, compared with the room air temperature. The higher values are calculated with the highest relative humidity values, respectively. But also, it could be noted, that when the dew point temperature and relative humidity are higher, the CO_2

concentration is also higher. This happens especially during the night period, when the room is occupied. And since the room air temperature does not deviate largely, it can be concluded that the occupants themselves are the major factor for excess humidity and high dew point temperature in the room.

This is the point at which the thermal bridge parameters become important factor for future humidity related problems in the presented residence room. That is why, a comparison is made, between the calculated dew point temperatures in Systems 1 and 2, and the measured temperature by System 3. These results are presented on Fig.4. System 3 is situated in very close proximity to the thermal bridge zone, but it does not measure the surface temperature. This system shows the air parameters, near the thermal bridge zone, which are summarized in Table 2. The surface temperature is evaluated by the thermal camera, and the images are analysed in [9]. The minimum measured surface temperature values are also presented in Fig.4. As it can be seen, the measured surface temperature values are significantly lower than the one, measured with the logger, which is expected.

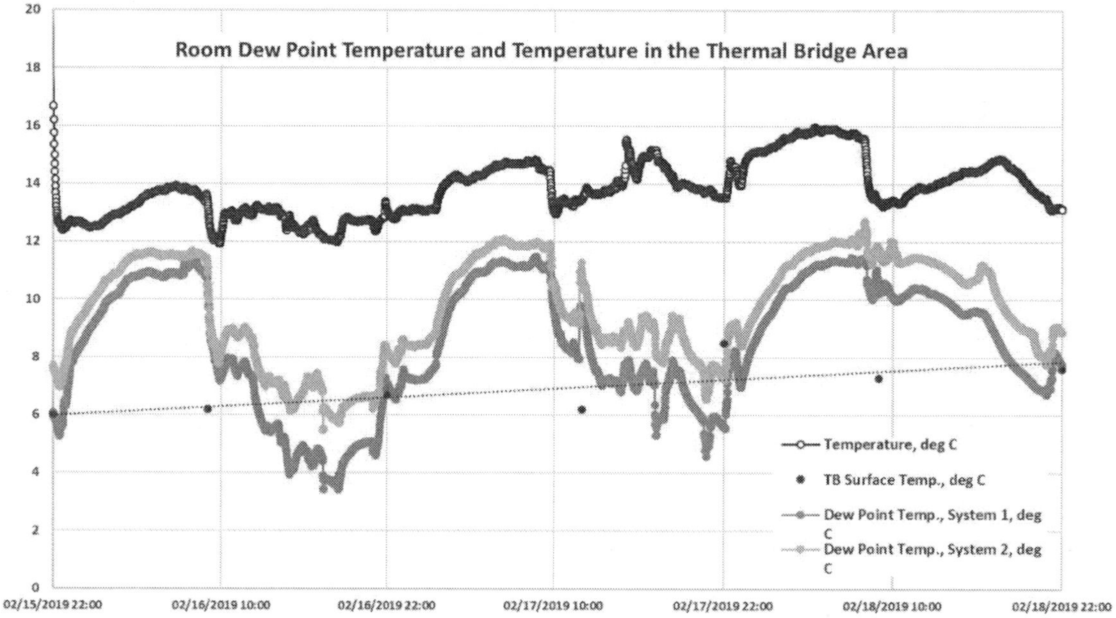

Fig. 4. Calculated room dew point temperatures and measured temperature for System 3

Table 2. Summarized indoor data for System 3

Indoor data, System 3	**min**	**max**	**average**	**SD**
Temperature, °C	11.93	20.29	13.89	1.06
Relative Humidity, %	39.67	83.77	67.56	8.13
Dew Point Temp., °C	1.73	11.46	7.90	2.18

As seen on Fig. 4 and in Table 2, the average measured temperature in the thermal bridge zone is 13.89 °C, with standard deviation of 1.06 °C for the entire experimental period. The minimum and maximum values are respectively 11.93 °C and 20.29 °C. For the relative humidity in the same zone, the measured average value is 67.56%. The minimum and maximum values are 39.67% and 83.77%, respectively. The values of the minimum surface temperature, measured by the thermal camera, are presented in Table 3, with the corresponding imaging time.

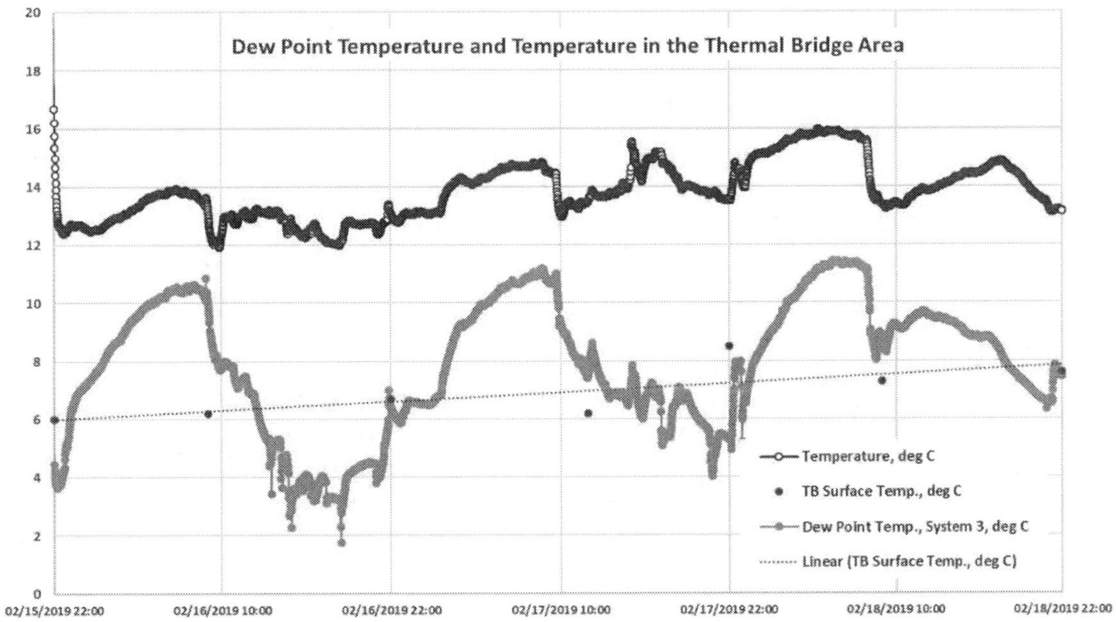

Fig. 5. Measured temperature and calculated dew point temperature with the values for System 3

Table 3. Minimum values of the surface temperatures in the thermal bridge zone

2/15/19 22:00h	2/16/19 9:00h	2/16/19 22:00h	2/17/19 12:00h
6.0 °C	6.2 °C	6.7°C	6.2 °C
2/17/19 22:00h	**2/18/19 9:00h**	**2/18/19 22:00h**	
8.5 °C	7.3 °C	7.6 °C	

Compared with the rest of the residence room, the presented results for the thermal bridge zone clearly show significantly lower values of the air temperature and significantly higher values of the relative humidity. This is a prerequisite for constant heat losses through the thermal bridge zone. The higher values of the relative humidity in this zone indicates the possibility for moisture related problems on the surface of the external wall. But, this effect is not so clear from the comparison between the dew point temperature distribution in the room and the values, measured near the thermal bridge zone. The two graphs go closer, but do not cross each other.

However, comparing the surface temperatures, measured by the thermal camera, it is clear that large time intervals exist, in which the dew point temperature is higher than the surface temperature in the thermal bridge zone. For the presented weekend period, there are distinguished three such periods. This result is well shown on Fig. 5, where the dew point temperature distribution is calculated with the values measured in the thermal bridge zone, by System 3. Obviously, in these peak periods, moisture accumulation over the thermal bridge surface is expected. In this case, the thermal camera may account better for the heat loss trough the solid surface of the wall and seem to be better way for establishing the cold spots of the thermal bridge areas.

In general, the summarized measurement data for the two systems shows relatively high degree of homogeneity in the indoor air temperature and relative humidity distribution in the experimental room. In most of the time, the measured values are in the comfortable range for the indoor environment requirements for the winter season. The huge exception is the measured CO_2 concentration values, and additional measures for aeration should be taken. The small differences in the measured values between the two systems show, that the mixing of the room air is not perfect. That is why, a use of a small fan for mixing the room air is recommended is such experimental studies. Nevertheless, the deviations between the results in the two measurement points are considerably small. This means that, the thermal bridge parameters are more accurately defined, and the measured generic conditions are more precisely studied.

It is considered, that all presented results would assist future numerical modelling of the related phenomena. For the accurate modelling of the conjugated heat transfer in similar set-ups, the measured and presented above air temperature around the solid cold surfaces, is an important generic parameter for the thermal bridge simulation. Collecting such experimentally measured values is considered as innovative approach in the field. The development of accurate numerical model for simulation of the thermal bridge distribution, will be the next step in the presented research study.

5 Conclusions

Dew point temperature analyses is performed, in ground floor residential room with existing thermal bridge. The experimental data is collected by conducting field measurements of indoor air and thermal bridge parameters, in real existing ground floor premises. The analyses of the results demonstrated the generic conditions for the thermal bridge existence, without considering the building structure.

The summarized measurement data for the two systems shows relatively high degree of homogeneity in the indoor air temperature and relative humidity distribution in the experimental room. In most of the time, the measured values are in the comfortable range for the indoor environment requirements for the winter season.

However, the measured CO_2 concentration values are significantly high, considering that there were only two occupants in the room. This is an indicator for low aeration and low infiltration and exfiltration, and probably for decreased level of indoor air quality.

The general results show, that during the experimental period, the dew point temperature deviates to a higher extend, compared with the room air temperature. The higher values are calculated with the highest relative humidity values, respectively.

The high levels of the dew point temperature and the relative humidity correspond with the high level of the CO_2 concentration. Thus, the occupants themselves are the major factor for excess humidity and high dew point temperature in the room, since the room air temperature does not deviate largely.

The results for the thermal bridge zone clearly show significantly lower values of the air temperature and significantly higher values of the relative humidity, compared with the rest of the rom. This is a prerequisite for constant heat losses through the thermal bridge zone. The higher values of the relative humidity in this zone indicates the possibility for moisture related problems on the surface of the external wall.

The comparison with the measured surface temperatures, performed by the thermal camera, clearly show large time intervals, in which the dew point temperature is higher than the surface temperature in the thermal bridge zone. In these peak periods, moisture accumulation over the thermal bridge surface is expected.

The global results show, that in such experimental set-ups, the thermal imaging camera may account better for the heat loss trough the solid surface of the wall and seem to be better way for establishing the cold spots of the thermal bridge areas.

It is believed, that all presented results would assist future numerical modelling of the related thermal bridge phenomena, which will be the next step in the presented research study. For the accurate modelling of the conjugated heat transfer in similar set-ups, the measured and presented above air temperature around the solid cold surfaces, is an important generic parameter for the thermal bridge simulation.

Acknowledgements

The presented study is supported by the National Science Fund at the Ministry of Education and Science of Bulgaria, under the activities of: "Competition for financial support for projects of junior basic researchers and postdocs – 2018", with Contract № КП-06-М27/4, entitled: "Numerical assessment of effective measures for prevention of moisture accumulation on external walls with thermal bridge, in residential rooms without organized ventilation".

References

[1] CG. Bornehag, J. Sundell, L. Hagerhed, S. Jason, and the DBH study group. *„Dampness in buildings and health. Dampness at home as a risk factor for symptoms among 10 851 Swedish children (DBH-step I)"*, Proc. to Indoor Air Conference 2002, 431-436, (2002)

[2] Institute of Medicine, *"Clearing the air. Asthma and Indoor Air Exposures"*, The National Academy Press, Washington, (2000)

[3] M. Ivanov, D. Markov, P. Stankov, *"On the research of the parameters of the indoor environment in Bulgarian homes, based on the experimental project ALLHOME"*, J. „Heat Tech. for Everyday Life", ISSN 1310-9405, Issue 4, pp. 13-17, (2010)

[4] M. Ivanov, D. Markov, *"Analyses and assessment of the indoor air quality in residence rooms in Bulgaria and comparison with the Bulgarian norms and foreign standards, based on the ALLHOME Project"*, Proc. to: „Int. Sc. Conf. UNITECH'10", Vol. III, pp. 558-564, ISSN 1313-230X, (2010)

[5] T. Theodosiou, K. Tsikaloudaki, S. Tsoka, P. Chastas, *"Thermal bridging problems on advanced cladding systems and smart building facades"*, J. Clean. Prod., 214, 62-69 Elsevier (2019)

[6] A. Papadopoulos, *"Forty years of regulations on the thermal performance of the building envelope in Europe: achievements, perspectives and challenges"*, Energy Build. 127. https://doi.org/10.1016/j.enbuild.2016.06.051, (2016)

[7] M. Ivanov, D. Markov, *"Analyses and assessment of the dew point temperature in Bulgarian homes, based on results of the ALLHOME project"*, Proc. to "Int. Sc. Conf. in Ruse'10", Vol.49, Series 1.2, pp. 14-18, (2010)

[8] ISO 7730, "Ergonomics of the thermal environment – Analytical determination and interpretation of thermal comfort using calculation of the PMV and PPD indices and local thermal comfort criteria", International Organization for Standardization, (2006)

[9] M. Ivanov, *"Instantaneous field measurements of thermal bridge parameters in ground floor residential room"*, Submitted for publishing in: Proc. of "8th Int. Conf. on Thermal Equipment, Renewable Energy and Rural Development – TE-RE-RD'19", (2019)

The initiation stage of Food Court Waste during air versus steam gasification processes

Raluca Nicoleta Tîrtea, *Cora* Bulmău*, *Gabriela* Ionescu, *Cosmin* Mărculescu

University Politehnica of Bucharest, Department of Energy Production and Use, Splaiul Independenţei 313, Romania

Abstract. Using different gasification agents: air and steam, two types of gasification process were performed into a batch reactor at temperature of 750°C and 850°C and atmospheric pressure. The only difference between of the two compared experimental configurations was represented by the gasification agent used in the process. The amount of oxygen introduced into the reactor for air gasification at an ER of 0.3 was computed. Therefore, in the steam gasification process, the same amount of oxygen was introduced, so establishing an unordinary steam to biomass ratio. In this way, the two processes, air vs. steam gasification, were compared, the rest of the process parameters being kept constant. This paper approaches the transitory regimes (initiation stage) of gasification process in order to observe the influence of process temperature and gasification agent on the process run. According to the experimental results, better gas quality is obtained if steam is used as a gasifying agent, yet the conversion and energy efficiencies decreases. By optimizing time residence in steam gasification, process efficiencies may be increased.

1 Introduction

Each year, the world's population generates more than 2 billion tons of waste [1]. By 2050, the world's population is expected to exceed 9 billion people [2]. So, if the society continues to keep the actual trend, in high-income countries, the waste generation will reach at 7 billion tons per year by 2025 [1]. In most of the countries the landfills continue to remain the main method for solid waste disposal [3, 4]. According to several studies, solid waste generation is increasing faster than any other environmental pollutant, including CO_2 [5-7].

Food waste represents a significant fraction of municipal solid waste, that requires proper management with minimum environmental impact. with According to Food and Agricultural Organization (FAO), food waste and the losses related from this waste segment accounts for 8% of global greenhouse gas emissions. Currently, in the EU, around 88 million tons of food waste are generated annually [6]. Therefore, there is a continuing interest in management waste plans development that includes drastically diminish waste

* Corresponding author: cora4cora@gmail.com

landfilling. Food waste has a great potential in the energy field as a feedstock and represents a serious advantage that implies energy recovery due to its notable energy content [7]. This type of waste could potentially be converted to various forms of useful energy through thermochemical processes such as gasification, solving the waste landfill binding targets. Gasification of waste is preferred over combustion because it provides gas product, syngas, that can be used in many energetic applications, as rival for hot combustion gases. In gasification technology, the gas is usually composed by carbon monoxide (CO), hydrogen (H_2), carbon dioxide CO_2 and methane (CH_4) [8, 9]. There are abundant examples of studies that aim to optimize the gasification conditions using thermal and catalytic treatments in order to improve the gas quality by reducing the tar content in the syngas [10, 11], along with the hydrogen content increase and the processing stages reduction. Starting from the analysis of the gasification transitory regime, the process steps can be delimited time approximately, the composition of the gas produced at each stage can be determined, and the efficiency of the material conversion and also the process can be estimated according to the studied parameters [12].

Since for batch gasification experiments in the transitory regime, the process is not complete, like in the case of a continuous gasification process, the results could not be used to predict the final composition of the gas resulted from the process and just a primary composition of it. Therefore, the proposed approach aims to investigate the initiation stage of the gasification processes and serves as a starting point for obtaining the optimal process parameters necessary to initiate the gasification of various materials studied, under specific conditions. For this, two types of experimental gasification surveys were conducted using a batch reactor at temperatures of 750°C and 850°C and atmospheric pressure, using air and steam. To assess the gasification process in terms of gas quality, the gas composition and its lower calorific value were analyzed. Based on the experimental results, the process efficiencies can be calculated, even if the experiments do not reproduce a continuous gasification process. Their values are significantly different comparing with those obtained for a typical gasification process, but these can be used to characterize the gasification processes and to observe the influence of parameters on process run.

2 Material and methods

The feedstock used in the current gasification processes was food court waste (FCW). The organic material has been collected from a food court area existing in a commercial center. Its composition keeps the standard values of the concentration for all the components and the characteristics of a FCW residue. More details about the collection stage, pretreatment and characterization were previously reported by the authors in [11, 13].

Experiments of air/steam gasification of FCW – transitory regime – were conducted at atmospheric pressure, while the process temperature was set at 750°C or 850°C. For the air gasification experiments an ER of 0.3 was considered, therefore the amount of oxygen introduced in the reactor was computed. The amount of water introduced in the reactor was calculated to ensure the equivalent oxygen as for the air gasification process. In order to compare the two processes, air versus steam gasification, all process parameters were maintained constant. The amount of oxygen introduced in the process, supplied either from air or from water, was kept constant. The description of the experimental setup procedure concerning gasification using air was already reported by the authors in another research study [11,21]. The experimental steps are like those depicted in Figure 1, except the gasifying agent, steam in this case. Therefore, water was introduced in the reactor at a constant and controlled flow rate, using an automatic syringe injection pump manufactured by Indiamart, model 300.

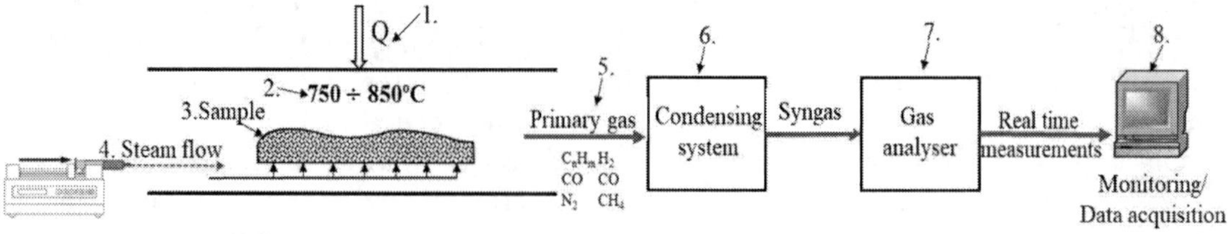

Figure 1. Air/Steam gasification experimental set-up (adopted after [11, 12, 14 -16])

1. Externally heated batch reactor; 2. Temperature controller (750°C - 850°C); 3. Food Courte Waste sample; 4. Syringe Pump (Steam flow); 5. Gas line outlet (primary gas); 6. Cooling systems with water-cooled ice; 7. Testo 350 XL gas analyser; 8. Data monitoring and acquisition: EASY- emission software, Testo, Inc.

In the following, the main parameters related to the process gasification. are defined and explained.

Carbon conversion efficiency represents the ratio between the amount of carbon in gas products and the amount of carbon introduced in the process by fuel [17, 18]:

$$CCE = (Carbon\ in\ gasification\ products\ /\ Carbon\ in\ fuel) \cdot 100\ [\%] \quad (1)$$

or it can be computed by considering the amount of carbon unburned [19, 20]:

$$CCE = (1 - Carbon\ in\ ash\ /\ Carbon\ in\ fuel) \cdot 100\ \ [\%] \quad (2)$$

For the CCE computation, data such as product gas composition, the amount of substance produced (expressed in kmol), the amount of carbon in gas products etc. must be known. For the determination of the amount of gas produced, two methods may be applied:

- First method (1): determining the quantity of gas produced in mass units by mass balances, and the gas molar mass;
- Second method (2): determining according to Basu, by Nitrogen conservation of mass [19].

To compute the CCE, it is imperative to know the amount of gas produced, expressed in kmol. Therefore, by applying the two methods described above, two values for CCE will result and they may be similar.

Cold-gas efficiency (CCE) is the ratio between the chemical energy of the produced gas and the chemical energy of the gasified material, thus [17-20]:

$$CGE = (LHV_{gas} \cdot M_{gas} / LHV_{fuel} \cdot M_{fuel}) \cdot 100\ [\%] \quad (3)$$

where LHV_{gas} - lower calorific value of gas [kJ/kg], LHV_{fuel} - lower calorific value of feedstock [kJ/kg], M_{gas} - mass of gas produced from the gasification of M_{fuel} [kilograms of fuel].

The gasification processes take place at high process temperatures; therefore, the gas temperature and its sensible heat are very high. If the gas produced through gasification is used in applications such as ovens or boilers, the gas sensible heat energy can be used. Therefore, hot-gas efficiency (HGE) can be defined, since, besides its gas chemical energy, it also takes into account the sensible heat of the gas produced [17, 18, 20]:

$$HGE = ((LHV_{gas} \cdot M_{gas} + M_{gas} \cdot c_{pgas} \cdot (T_{gas} - T_{fuel})) / (LHV_{fuel} \cdot M_{fuel})) \cdot 100\ [\%] \quad (4)$$

where c_{pgas} - specific heat of gas [kJ/kgK], T_{gas} - gas temperature at the gasification reactor outlet [°C], and T_{fuel} - fuel temperature entering the gasification reactor [°C].

3 Results and discussions

In the figure below, variations of CO and H_2 concentrations from the syngas function of the process parameters are presented in case of the air vs. steam FCW gasification. It can be observed that neither the process temperature, nor the gasification agent does not influence the concentration of CO in the analyzed temperature range. On the other hand, the H_2 concentration is influenced by both process temperature and gasification agent in case of air vs. steam gasification processes at a temperature of 850°C. Using steam in gasification processes leads to H_2 concentration increase by 3 percentage points, compared to air gasification at 850°C.

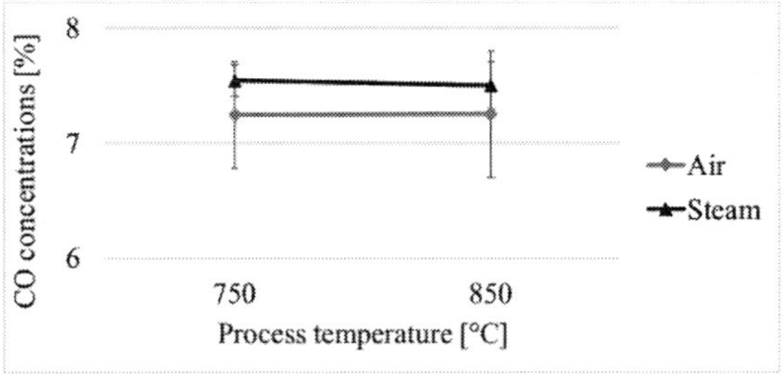

Figure 2. Variation of CO concentration – air vs. steam gasification of FCW

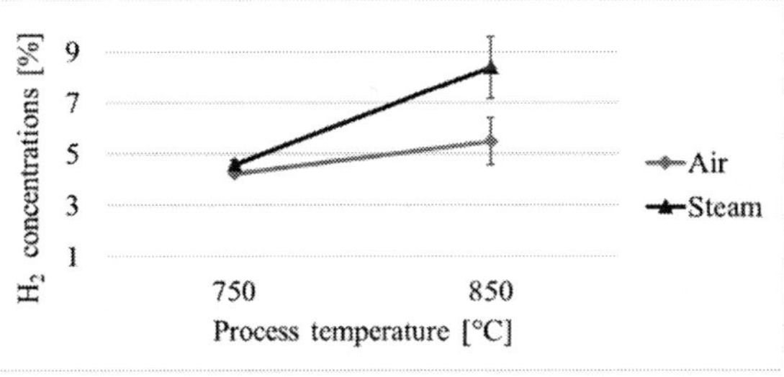

Figure 3. Variation of H_2 concentration – air vs. steam gasification of FCW

Figure 4 and Figure 5 illustrate the CCE variation for air vs. steam gasification processes of FCW function of the process temperature and type of oxidant. CCE has different values, depending on the method for the gas quantity determination. For the air gasification processes, CCE increases slightly with temperature rise and decreases slowly with temperature rise for the FCW steam gasification processes. Concerning the method applied for the gas quantity determination, the obtained values of the CCE are considerably lower when steam is used.

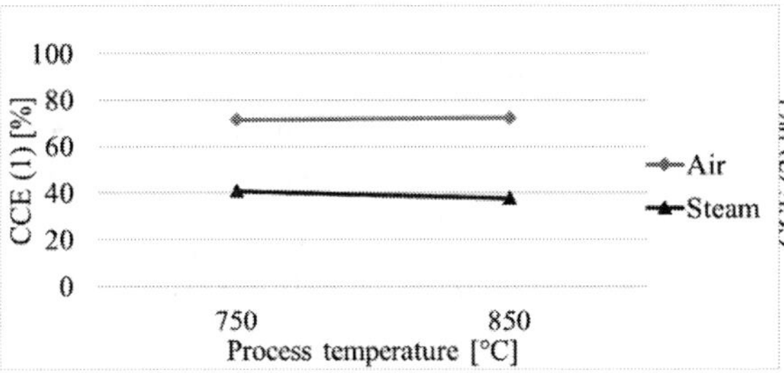

Figure 4. CCE (1) - air vs. steam gasification of FCW

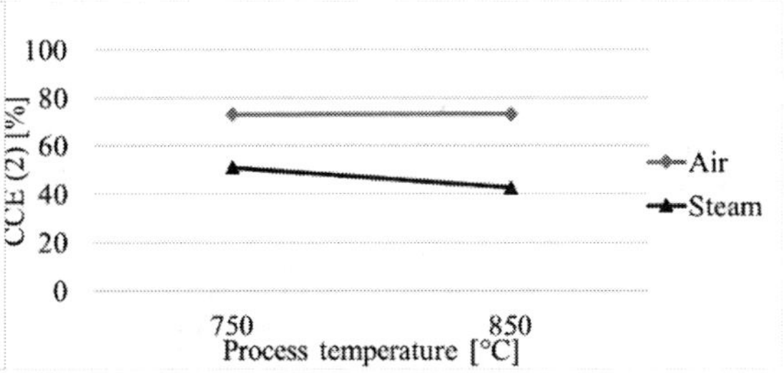

Figure 5. CCE (2) - air vs. steam gasification of FCW

In the figures below the variation of CGE and HGE function of the process parameters for air vs. steam gasification processes of FCW are presented. The calculated energy efficiencies are influenced both by process temperature and type of oxidant. Regarding the type of the oxidant used in the process, energy efficiencies increase with the process temperature rise, in the range considered. Similar to CCE, CGE and HGE decrease when steam is used as an gasification agent.

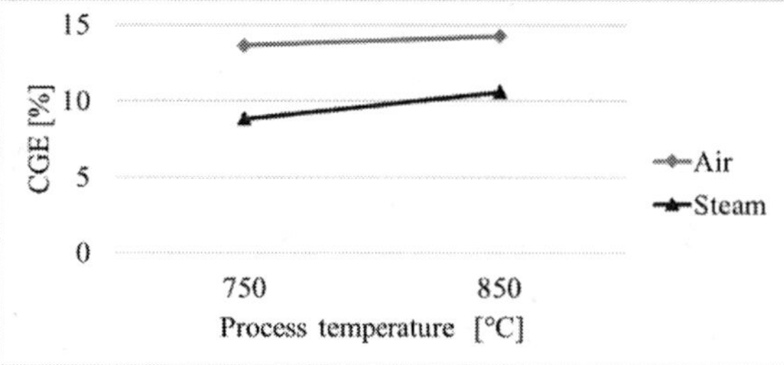

Figure 6. CGE - FCW air vs. steam gasification

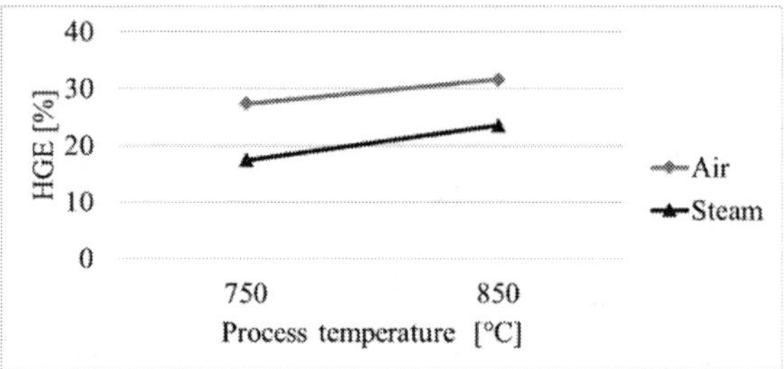

Figure 7. HGE - FCW air vs. steam gasification

CGE and HGE decrease by approximately 3-5, respective, by 8-10 percentage points, when steam is used in the process. The process temperature increase by 100°C leads to an optimization of CGE of maximum 2 percentage points, while HGE increases by 6-8 percentage points.

Char gasification step is the most important stage in gasification processes since heterogenous reactions between gas and solid phase takes place, therefore this step determines gasification process duration. The reaction between solid carbon and oxygen (air) is very fast, whereas oxygen is the most reactive oxidizing. Reaction between steam and solid carbon is 3 to 5 magnitude orders slower [17]. Therefore, in steam gasification processes, residence time must be increased, compared to air gasification process, to ensure enough time for reaction to take place. Since process duration was maintained constant in the air vs. steam gasification experiments conducted, the conversion efficiencies decreased in steam gasification.

According to mass balance, the amount of gas produced in air gasification of FCW varied between 95.19 and 96.93%, while the gas produced from steam gasification of FCW was 87.41-91.27%, depending of process temperature. The decrease in the amount of gas produced is also a consequence of the short residence time.

4 Conclusions

When the air is used as a gasification agent, the gas production exceeds 95%. The experimental results showed that the production of gas is intensively influenced by the process temperature, increasing with temperature rise. Also, high gasification temperatures favor the solid material conversion into gas.

Concerning the steam gasification processes, the gas production is sometimes less than 90%. The use of the steam as gasification agent caused also a decrease of the CCE. The results of the gasification experiments demonstrated that the CCE tends to increase when the temperature and air equivalence ratio are rising. Higher process temperatures provide a higher reaction rate, the greater amount of air introduced in the process producing more oxygen in the reactor, which can be used for carbon oxidation reactions.

The values obtained for CGE and HGE are smaller compared to those generated by the standard gasification processes, due to the type of the experiments performed in our study. So, a complete gasification process does not take place in the laboratory reactor used in this case. Thus, the energy efficiencies values are calculated based on the intermediate gas compositions, representing the efficiencies of some gasification process stages, not of the whole process. However, the calculated values of the energy efficiencies can be used to characterize the gasification processes and to observe the influence of the process parameters on the obtained gas. Usually, the difference between CGE and HGE in a typical gasification process is about 10-20 percentage points. This range is keeping also for the gasification processes performed for the already mentioned conditions.

Based on the experimental results of our study, we can conclude that better values for CGE and HGE are generated when the air is used as an gasification agent, but taking into account the gas composition, better gas compositions are obtained when the steam is used as a gasifying medium. Therefore, for better overall results, steam gasification of FCW at an increased time residence should be studied.

Acknowledgement

Funding provided through Romania's "Competitiveness Operational Programme 2014-2020" Priority Axis 1: Research, Technological Development and Innovation (RD&I) to Support Economic Competitiveness and Business Development Action 1.1.4. Attracting high-level personnel from abroad in order to enhance the RD capacity ID/Cod My SMIS: P_37_768/103651; No. Contract: 39/02.09.2016

References

1. Solid Waste Management in the World's Cities, *Water and Sanitation in the World's Cities* (UN-HABITAT. 2010). http://www.unhabitat.org/pmss/listItemDetails.aspx?publicationID=2918.

   Report, World Population (Population Reference Bureau, 2011). http://www.prb.org/ Population Reference Bureau. April 2011. World Population. http://www.prb.org/
2. Report, Global Food Waste Management, (World Biogas Association, 2018) http://www.worldbiogasassociation.org/wp-content/uploads/2018/05/Global-Food-Waste-Management-Full-report-pdf.pdf
3. E.C. Rada, L. Squazardo, G. Ionescu, A. Badea, UPB Sci. Bull. Serie D, **76**, 199-206, (2014).
4. G Ionescu., E. C. Rada,L. I. Cioca, Environmental Engineering & Management Journal (EEMJ), **14** (11) (2015)
5. Report, Estimates of European food waste levels, https://ec.europa.eu/food/safety/food_waste_en (EC, 2016)
6. A. N., Izaharuddina Paula, M. C., Theppitakb, S., Daib, X., & Yoshikawab, K., *Food waste gasification through hydrothermal carbonization pre-treatment*, (Joint Meeting the German and Italian Sections of the Combustion Institute, 2018)
7. AA. Erakhrumen, *Biomass gasification: Documented information for adoption/adaptation and further improvements toward sustainable utilisation of renewable natural resources*, ISRN Renewable Energ. **8** (2012).
8. R. N. Tirtea, , C. Mărculescu, A. Badea, University Politehnica Of Bucharest Scientific Bulletin Series C-Electrical Engineering And Computer Science, **79**(1), 315-322, (2017).
9. S. Ciuta, D. Tsiamis, , M. J. Castaldi, *Gasification of waste materials: technologies for generating energy, gas, and chemicals from municipal solid waste, biomass, nonrecycled plastics, sludges, and wet solid wastes*. Academic Press, (2017).
10. T. Wongsiriamnuay T, N. Kannang, N. Tippayawong, *Effect of operating conditions on catalytic gasification of bamboo in a fluidized bed,* I. J. Chem. Eng. **9** (2013).
11. R. Nicoleta Tirtea, G. Ionescu, C. Bulmau, C. Marculescu, *Experimental study of food court waste air gasification process performances in a batch reactor*, 8th International

Conference on Thermal Equipment, Renewable Energy and Rural Development (TE-RE-RD), 2019 (Under review)

12. R. N. Tîrtea, C. Mărculescu, A. Badea, 2015 International Conference on ENERGY and ENVIRONMENT (CIEM) (2015)
13. I. Stănciulescu, , C. Mărculescu, D. Boldor, 2018 IEEE International Conference on Environment and Electrical Engineering and 2018 IEEE Industrial and Commercial Power Systems Europe (EEEIC/I&CPS Europe), 1-4 (2018)
14. C. Bulmău, C. Mărculescu, S. Lu, Z. Qi, Journal of Geochemical Exploration **147**, 298-305 (2014)
15. C. Mărculescu, S. Ciuta, Renewable energy 57, 645-652 (2013)
16. R. N. Tirtea, I. Stanciulescu, G. Ionescu, C. Bulmau, C. Marculescu, D. Boldor, 2017 International Conference on ENERGY and ENVIRONMENT (CIEM), 390-393 (2017)
17. C. Stevens, *Thermochemical processing of biomass: conversion into fuels, chemicals and power*. Ed. Robert C. Brown. John Wiley & Sons, 2011
18. W. De Jong, J. Ruud van Ommen, *Biomass as a sustainable energy source for the future: Fundamentals of conversion processes*. John Wiley & Sons, 2014
19. C. Higman, M. van der Burgt, *Gasification*, f Professional Publishing, 2008
20. M. L. de Souza-Santos, *Solid Fuels Combustion and Gasification: Modeling, Simulation*. CRC Press, 2010
21. C. Bulmău, G. Ionescu and C. Mărculescu, *Bio-Gaseous Fuels from Agricultural Waste Pyrolysis (Part II)*, 9th International Conference on Manufacturing Science and Education (MSE) 2019, (Accepted for publication)

Experimental study of Food Court Waste air gasification process performances in a batch reactor

Raluca Nicoleta Tîrtea, *Gabriela* Ionescu*, *Cora* Bulmău and *Cosmin* Mărculescu

University Politehnica of Bucharest, Department of Energy Production and Use, Splaiul Independenței 313, Romania

Abstract. The paper presents the experimental results of Food Court Waste air gasification using a batch reactor. This type of waste is generated mainly in the food court areas of services spaces, shopping centers, airports and malls and is generally composed of paper & cardboard, plastic, organic, wood, metal and glass waste. The process operating parameters were: temperatures between 650°C and 850°C and an equivalent ratio of 0.25 and 0.4 respectively. The study focused on the influence of process parameters on the energy conversion rate considering the cold gas and hot gas efficiency correlated with the feedstock carbon conversion rate. The recorded instantaneous concentrations of sampled gas species were plotted in time dependent graphs for accurate variation curves of gases concentrations. The results can be used to isolate the startup stage of the gasification process and to establish the optimal process parameters for increased overall energy efficiency. In conclusion, considering the current setup, operational parameters and process energy efficiency, the optimal temperature for the air gasification of Food Court Waste is 850°C with an equivalent ratio of 0.40.

1 Introduction

Presently, the global municipal solid waste (MSW) generation accounts for 1.3-1.9 billion tonnes per year and is forecasted to increase to 2.2 billion tonnes, by 2025 [1]. The necessity of waste recovery in the form of secondary raw material or energy production represents a worldwide priority in the renewable sources' management field. The EU strategic long-term vision for 2050 to a novel, self-sustained conversion chain, and climate-neutral economy includes ascending common key targets such as 65% recycling of municipal waste and 70% of packaging materials by 2035, and 2030 respectively, correlated with a maximum landfilling rate of 10% of MSW [2]. However, the transition towards the circular economy concept and its design still accounts a demanding moving from theory to practical implementation, by adapting industrial processes.

The complexity of the treatment choice is dictated firstly by the waste structure, composition and properties, engaged as stand-alone fractions (e.g. biomass, plastic) or in

* Corresponding author: gabriela_ionescu@ymail.com

mixtures (e.g. MSW, solid recovered fuel) [3-5]. On average, the share of combustible materials in MSW is about 70-80% making it suitable for thermochemical recycling [6-8]. Combustion is the preferred waste treatment technology followed by gasification [9]. The chemical recycling, via contemporary methods (pyrolysis and gasification) offer prominent waste management with environmental benefits [10,11]. During thermal treatment, the syngas produced during the process can be used for heating and power generation. The conversion and energy efficiency of the gasification process can be evaluated based on the quality of the syngas obtained which depends mainly on: volumetric composition and flow rate, gasifying agent (the ratio of oxidant to fuel) and characteristics of the conversion process (process temperature and pressure, residence time, reactor type) [12-15].

The feedstock used in our experiments was food court waste (FCW). The waste is mainly produced in the food court areas of services spaces, shopping centers, airports and malls. The feedstock is composed of paper & cardboard, plastic, wood, metal and glass that represents the main source of packaging, along with the disposable cutlery and dishes. To all these, organic waste can be added. The composition of the FCW can be broadly different depending on the type of food court area where it is produced, season, weekday and time of the day. Over time, the combustible fractions of FCW (paper, cardboard, plastic, organic waste and wood) have been studied individually [16-24] as well as in various mixtures [25-28], as feedstock for thermochemical conversion purpose.

The aim of the current research is to determine the influence of operating parameters on the process energy conversion rate based on cold gas and hot gas efficiency corelated with the feedstock carbon conversion rate.

The recorded instantaneous concentrations of sampled gas species were plotted in time dependent charts for accurate variation curves of gases concentrations. The results can be used to isolate the startup stage of gasification process and to establish the optimal process parameters for increased overall energy efficiency.

2 Material and methods

2.1 Material

The material used in the present experimental study has a typical composition of the FCW feedstock. The details regarding the selection criteria and collection steps were previously reported in other already published papers of the research group [29].

Briefly, to establish an equivalent composition of waste to work with, waste samples were collected from three different locations during different times of day and week. For its energetic valorization, the waste was sorted by components, non-combustible fractions such as glass and aluminum cans packaging, have been removed, while the combustible ones have been dried, grounded and homogenized in specific proportions. The composition of FCW and its properties are presented in the next table.

Table 1. Properties of food court waste [29]

		Unit	Value
Waste average composition	Paper	%	29.29
	Cardboard	%	26.22
	Plastic	%	18.92
	Organic residues	%	25.18
	Wood	%	0.39
Primary analysis	Moisture	%	26.12^{as}
	Volatile matter	%	80.89^{db}
	Fixed carbon	%	7.97^{db}

	Ash	%	11.53^{db}
Ultimate analysis	C	%	51.09^{dab}
	H	%	7.03^{dab}
	O (by difference)	%	40.75^{dab}
	N	%	1.14^{dab}
High heating value		kJ/kg	20620as
Low heating value		kJ/kg	19173as

^{as} – *as received*
^{db} – *dry basis*
^{dab} – *dry ash free basis*

2.2 Methods

Air gasification experiments were carried out in a batch tubular electrically heated reactor. The reactor is specially designed to simulate thermochemical conversion experiments of solid organic material in a batch mode using a custom made tubular electric oven, produced by NABERTHERM, model RO 60/750/13.

The design of the FCW air gasification process in the batch tubular reactor is shown in Figure 1.

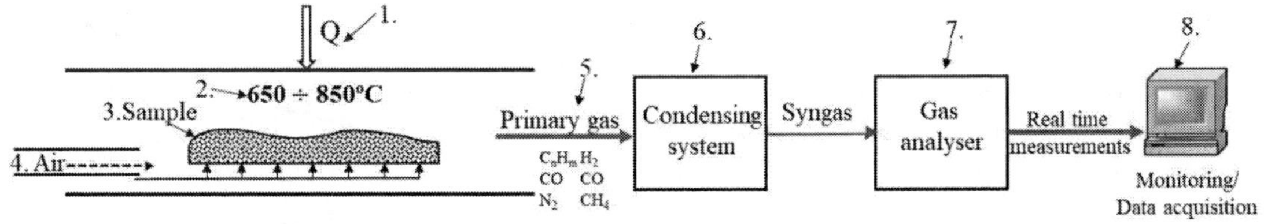

Figure 1. Air gasification process experimental set-up (adopted after 30-34)

1. Externally heated batch reactor; 2. Temperature controller; 3. Food Court Waste sample; 4. Inlet air flow meter; 5. Gas line outlet (primary gas); 6. Cooling systems with water-cooled ice; 7. Testo 350 XL gas analyser; 8. Data monitoring and acquisition: EASY-emission software, Testo, Inc.

A detailed presentation of the reactor main geometrical characteristics and potential configurations and runs, along with the transitory regime's mechanism are presented in previous papers of the authors [31-34]. Briefly, in the experimental runs the batch reactor (1.) was electrically heated at temperatures between 650°C and 850°C (2.). The FCW sample was distributed along the active zone of the crucible and then inserted in the preheated batch reactor (3.). The air was introduced in co-current flow and kept constant using a flowmeter (4.). To ensure a constant amount of air in the reactor and process duration, the amount of air needed to gasify 10 g of sample at an equivalent ratio (ER) of 0.25 was computed (15.88 l_{air}). Then, considering the amount of air resulted for the first case, the amount of sample was calculated so that the air/fuel ratio for an ER of 0.40 to be assured. The primary gases formed during the process (5.) were passed through a condensing system with water-cooled ice (6.). In order to establish the transitory regimes, the gases produced were analyzed in real time using the Testo 350 XL gas analyzer (7.) and monitored, registered respectively in a continuous mode with EASY- emission software, Testo, Inc (8.).

The conversion and energy efficiency of the gasification process have been calculated considering three factors: Carbon conversion efficiency – CCE; Cold-gas efficiency – CGE;

Hot-gas efficiency – HGE according to [13-15, 35,36]. The formula used to compute the three mentioned parameters are described in another research by the same authors [30]. Briefly the CCE was determined using two methods. In the first method the amount of syngas produced determined by mass balance is considered (1), while in the second one the amount of syngas produced determined by calculation based on the amount of nitrogen introduced into the gasification process and the concentration of nitrogen measured in syngas produced.

3 Results and discussions

To evaluate the process in terms of gas quality, the variation concentration of the most important combustible gas species (carbon monoxide and hydrogen) according to process parameters were analyzed. In figures 2 and 3, the CO and H_2 concentration as function of the FCW air gasification parameters (temperature and ER) are presented. According to the figures below, gas species concentrations increase with temperature rise, but decrease when ER increases. The reduction of the combustible gas species concentration from the primary syngas at higher ER could be a consequence of air quota increase in the process. As ER increases, more air is introduced in the reactor, therefore, more Nitrogen is present in the reactor, and thus in the produced gas.

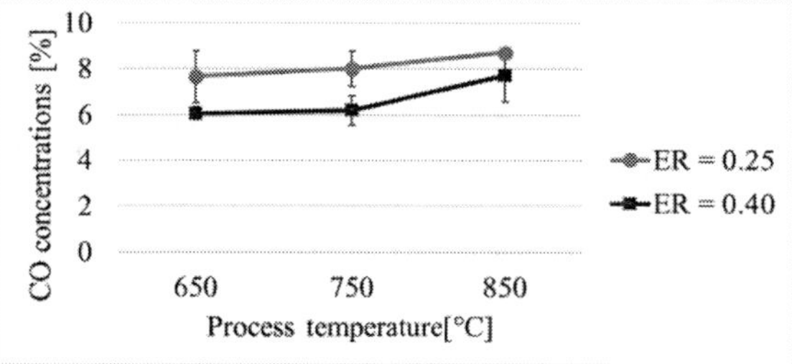

Figure 2. Variation of CO concentration - air FCW gasification

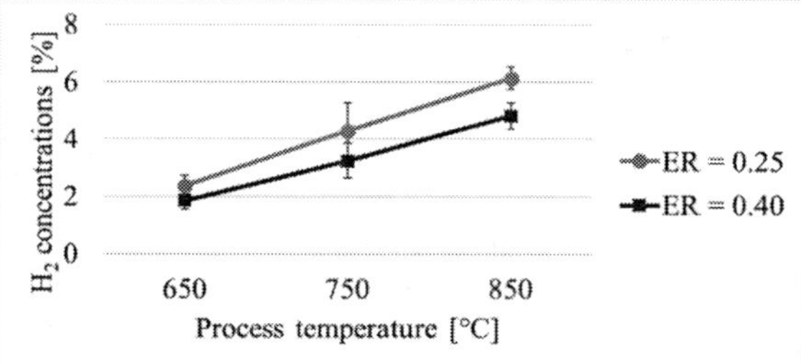

Figure 3. Variation of H_2 concentration - air FCW gasification

To evaluate the gasification processes in terms of conversion and energy efficiency, CCE, CGE and HGE were computed for each experimental configuration studied. Depending on computation method, two values of CCE resulted [13-15,36]. In Figure 4 and Figure 5 the variation of CCE in FCW air gasification as function of the process parameters is presented. In both cases, CCE increase with process temperature rise, for the studied range. Best results are obtained at 850°C and ER of 0.40. Depending on the calculation method applied, results may differ by 16 percentage points.

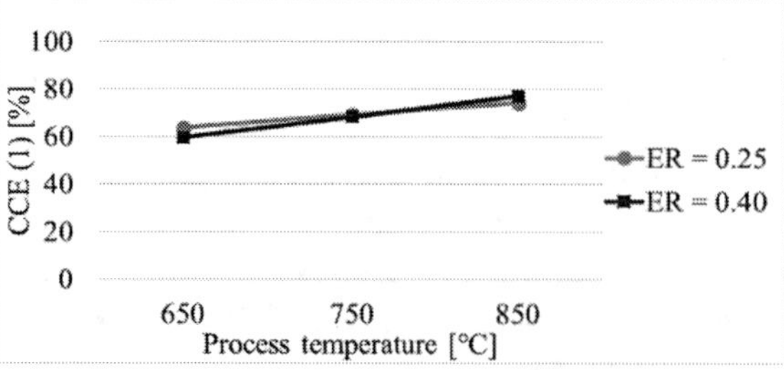

Figure 4. CCE (1) - FCW air gasification

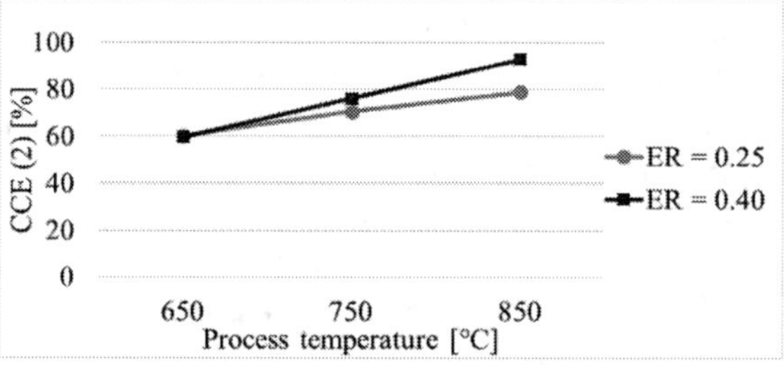

Figure 5. CCE (2) - FCW air gasification

In figures below are presented the values of process energy efficiencies obtained for air gasification of FCW at different process parameters. CGE and HGE increase with temperature and ER.

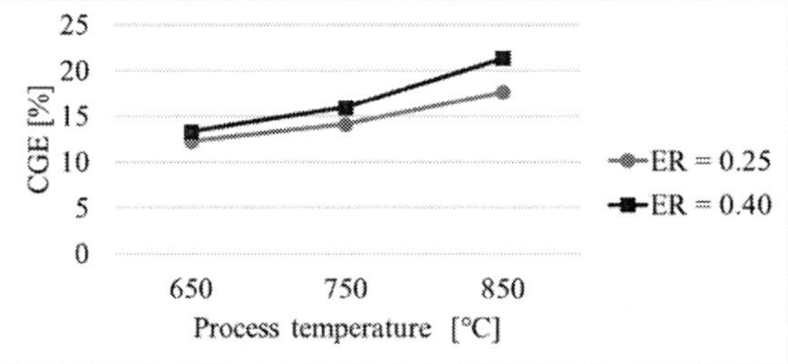

Figure 6. CGE - FCW air gasification

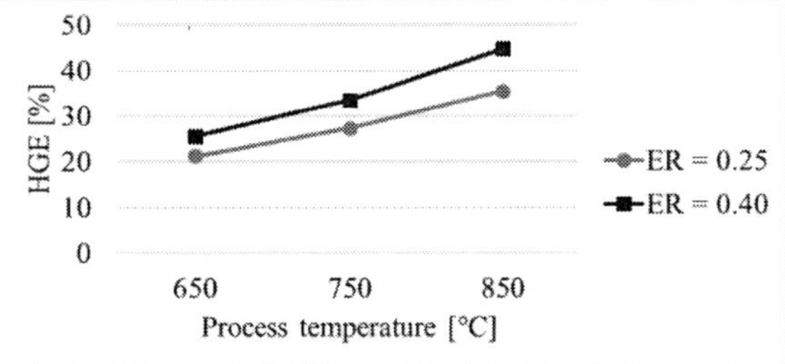

Figure 7. HGE - FCW air gasification

At the same process temperature, the increase of ER by 0.15 leads to CGE and HGE increase by up to 4 and 9 percentage points, while the process temperature increase by 200°C leads to energy efficiencies rise by 5-8 and 14-19 percentage points. Best CGE and HGE results are obtained at 850°C and ER of 0.40.

4 Conclusions

Based on the experimental results, the FCW air gasification leads to a syngas production more than 95%. The amount of the syngas produced is appropriate due to the easily flammable feedstock properties, that is mainly composed of lignocellulosic and polymer-based material accounting for 75% from the total amount of FCW analysed. As expected, the syngas production is strongly influenced by the process temperature. With the augmentation of the process temperature, the overall conversion efficiency of the feedstock into gas increases. In all the experimental results the CCE tends to increase and being directly proportional with the temperature and ER with their increase. The enhancement of process temperature leads to the augmentation of the reaction rates. The increase of the air used in the process assures more oxygen supplied in the reactor and therefore further carbon oxidation reactions might occur.

Due to the experimental setup performed in batch mode the CGE and HGE values obtained are small in comparison with standard gasification processes. Thus, in this case, the gasification process doesn't fully complete its cycle. Generally, the difference between CGE and HGE in a typical gasification process is about 10-20 percentage points [13,14]. However same trends are registered along the CGE and HGE values obtained.

Therefore, the values of the energy efficiency are determined considering the intermediate syngas composition that reflects certain sequences of the gasification process and not the entire one. The present energy efficiencies values can give a reliable and comprehensive view on the gasification process transition stages, corelated with temperature and ER influence on the obtained gas composition.

In conclusions, for the air gasification of FCW, considering the process energy efficiencies the optimal process temperature is 850°C and ER = 0.40.

ACKNOWLEDGMENT

Funding provided through Romania's "Competitiveness Operational Programme 2014-2020" Priority Axis 1: Research, Technological Development and Innovation (RD&I) to Support Economic Competitiveness and Business Development Action 1.1.4. Attracting high-level personnel from abroad in order to enhance the RD capacity ID/Cod My SMIS: P_37_768/103651; No. Contract: 39/02.09.2016

References

1. E. C. Rada, L. I. Cioca, G. Ionescu, MATEC Web of Conferences **121,** 8, (2007)
2. European Commission, https://ec.europa.eu/clima/policies/strategies/2050, Accessed in March 2019.
3. G. Ionescu, C. Bulmău, E3S Web of Conferences **85**, 6 (2019)
4. G. N. Sava, G. Ionescu, H. Necula, M. Scripacriu, M. Q. Duong, S. Leva, M. Mussetta, 2017 IEEE International Conference on Environment and Electrical Engineering and 2017 IEEE Industrial and Commercial Power Systems Europe (EEEIC/I&CPS Europe), 1-5 (2017)
5. E. C. Rada, G. Ionescu, F. Conti, U. I. Cioca, V. Torretta, Calitatea **19(167)**, 118-122 (2018)
6. E. C. Rada, M. Ragazzi, V. Torretta, G. Castagna, L. Adami, L. I. Cioca, AIP Conference Proceedings **1968**, 030050-1- 030050-6 (2018)
7. S. Chen, A. Meng, Y. Long, H. Zhou, Q. Li, Y. Zhang, Journal of the energy institute **88(3)**, 332-343 (2015)
8. E. Ranieri, G. Ionescu, A. Fedele, E. Palmieri, A. C. Ranieri, V. Campanaro, Waste Management & Research **35(8)**, 890-898 (2017)
9. H. A. Gabbar, M. Aboughaly, N. Ayoub, Journal of the Energy Institute **91(4)**, 481-488 (2018)
10. S. Ciuta, F. Patuzzi, M. Baratieri, M. J. Castaldi. Journal of Analytical and Applied Pyrolysis, **108**, 316-322, (2014).
11. M. S. Ciuta, C. Marculescu, , C. Dinca, A. Badea (2011). UPB Sci. Bull, **73**, 307-320.
12. R. N. Tîrtea, C. Mărculescu, A. Badea, University Politehnica Of Bucharest Scientific Bulletin Series C-Electrical Engineering And Computer Science **79(1)**, 315-322 (2017)
13. C. Higman, M. van der Burgt, *Gasification*, Gulf Professional Publishing, 2008
14. P. Basu, *Biomass gasification, pyrolysis and torrefaction: practical design and theory*. Academic press, 2013.
15. C. Stevens, *Thermochemical processing of biomass: conversion into fuels, chemicals and power*. Ed. Robert C. Brown. John Wiley & Sons, 2011.
16. P. A. Caton, M. A. Carr, S. S. Kim, M. J. Beautyman, Energy Conversion and Management **51.6**, 1157-1169 (2010)
17. C. Wu, P. T. Williams, Fuel **89.10,** 3022-3032 (2010)
18. R. Xiao, B. Jin, H. Zhou, Z. Zhong, M. Zhang, Energy Conversion and Management **48.3**, 778-786, (2007)
19. C. M. van der Meijden, H. J. Veringa, L. PLM Rabou, Biomass and bioenergy **34.3**, 302-311 (2010)
20. I. Ahmed, A. K. Gupta, Applied Energy **86.9**, 1732-1740 (2009)
21. I. Ahmed, A. K. Gupta, Applied energy **86.9**, 1813-1821 (2009)
22. P. N. Sheth, B. V. Babu, International Journal of Hydrogen Energy **35.19**, 10803-10810 (2010)
23. G. Lopez, M. Artetxe, M. Amutio, J. Alvarez, J. Bilbao, M. Olazar, Renewable and Sustainable Energy Reviews **82**, 576-596 (2018)
24. R. Muangrat, J. A. Onwudili, P. T. Williams, Bioresource technology **101.17**, 6812-6821 (2010)
25. I. I. Ahmed, N. Nipattummakul, A. K. Gupta, Applied energy **88.1**, 165-174 (2011)
26. I. Ahmed, A. K. Gupta, Applied Energy **86.12**, 2626-2634 (2009)
27. K. G. Burra, A. K. Gupta, Applied Energy **211**, 230-236 (2018)
28. F. Pinto, C. Franco, R. N. Andre, M. Miranda, I. Gulyurtlu, I. Cabrita, Fuel **81.3**, 291-297 (2002)

29. I. Stănciulescu, , C. Mărculescu, D. Boldor, 2018 IEEE International Conference on Environment and Electrical Engineering and 2018 IEEE Industrial and Commercial Power Systems Europe (EEEIC/I&CPS Europe), 1-4 (2018)
30. Tirtea R.N., Bulmau C., Ionescu G., Marculescu C., The initiation stage of Food Court Waste air versus steam gasification processes, TE-RE-RD 2019, eISSN: 2267-1242), under review.
31. R. N. Tîrtea, C. Mărculescu, A. Badea, 2015 International Conference on ENERGY and ENVIRONMENT (CIEM) (2015)
32. C. Bulmău, C. Mărculescu, S. Lu, Z. Qi, Journal of Geochemical Exploration **147**, 298-305 (2014)
33. C. Mărculescu, S. Ciuta, Renewable energy **57**, 645-652 (2013)
34. R. N. Tirtea, I. Stanciulescu, G. Ionescu, C. Bulmau, C. Marculescu, D. Boldor, 2017 International Conference on ENERGY and ENVIRONMENT (CIEM), 390-393 (2017)
35. M. L. de Souza-Santos, *Solid Fuels Combustion and Gasification: Modeling, Simulation*. CRC Press, 2010.
36. W. De Jong, J. Ruud van Ommen, *Biomass as a sustainable energy source for the future: Fundamentals of conversion processes*. John Wiley & Sons, 2014

Researches on the mixture limits of animal fats with liquid hydrocarbons for combustion at industrial level

Dana Andreya Bondrea[1*], *Lucian* Mihaescu[2], *Gheorghe* Lazaroiu[3], *Ionel* Pisa[4] and *Gabriel* Negreanu[5]

[1]University POLITEHNICA of Bucharest, Department of Hydraulics, Hydraulic Machinery and Environmental Engineering, Romania

[2,4,5]University POLITEHNICA of Bucharest, Romania Department of Thermotechnics, Engines, Thermal and Frigorific Equipment

[3]University POLITEHNICA of Bucharest, Department of Energy Production and Use

Abstract. The experimental research has highlighted the variety of possibilities of combustion of animal fat from bovine and swine mixed with liquid hydrocarbons. Previous research has established that the upper limit for an efficient combustion was 30 %. For a perfect mixing, the lower temperature limit was set to 40 °C. In the fuel laboratory, at the department TMETF was determined the viscosity of the mixtures for different proportions. The values obtained for various concentrations and preheating temperatures were close to the values for liquid hydrocarbons. The experimental researches have studied the combustion of the mixture using a mechanically spraying burner with constant pressure between 14 and 18 bar. The aspiration of the mixture is done from a specially designed tank; this tank is equipped with an electric heater, in order to maintain the mixture at a constant temperature between 40 °C to 50 °C. After that, the burner heats again the mixture with an integrated heating device up to 75 °C. The burner is also equipped with an air blower, pump and a calibrated nozzle. The combustion resulted from the experimental boiler with a power rated to 55 kW were monitored with a thermal vision camera and an exhaust gas analyser. This research has demonstrated the viability of using this type of mixtures in energetic burning equipment designed for liquid hydrocarbons.

1. Introduction

The method of combustion of animal fats (especially for bovine) was developed as a requirement for leather industry, because the animal fats are a waste which is hard to neutralize. As a result, the combustion of these animal fats is a solution which has energy and ecologically impact. The amount of animal fat waste in the leather industry is not very high, so it is planned to produce thermal energy through combustion in the boiler`s furnace [1-3]

The combustion research on a diesel engine is not part of this paper. In the current state of development, the leather industry produces annually 11.000 to 12.000 tons of waste for combustion neutralization, representing about 50 % animal fats and 50 % protein. If protein is studied in order to transform it into biogas through anabolism, the animal fats can be burned by mixing in liquid hydrocarbons. The method of burning animal fats in liquid hydrocarbons is the main subject of this paper [4-7].

Previous studies on the physical, chemical and energetic characteristics of animal fats have indicated:

- Elemental analysis;
- The melting temperature in the liquid phase, minimum 40°C.

Preliminary research has demonstrated the possibility of combustion for the 30 % concentration of animal fats in liquid hydrocarbons. For spraying and combustion studies it was necessary to determine the viscosity of the previously defined mixtures [8-11].

*Coressponding author: andreya.dana@yahoo.com

2. Determination of viscosity of the mixtures composed from animal fats and liquid hydrocarbons

The determinations were made in the University Politehnica of Bucharest, The Faculty of Mechanical Engineering and Mechatronics, department Thermal Systems and Equipment, Hydraulic and Pneumatic Systems and Equipment.
In figure 1 are presented the experimental stand with:

1. Process control thermometers;
2. Viscometer Engler;
3. System with electric heating stovetop for samples;
4. Balance for weighing the mixture;

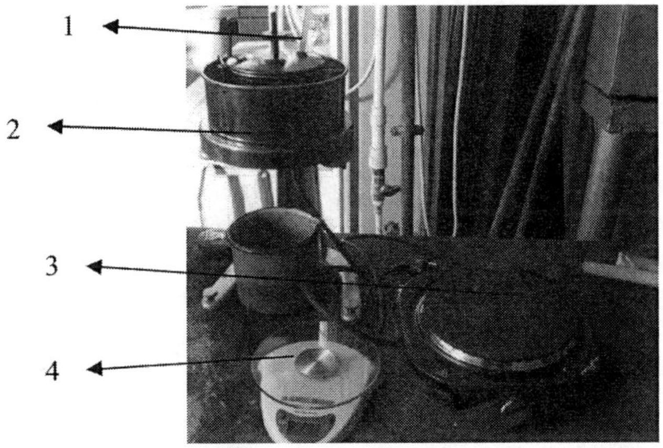

Fig.1. The equipment used to determine viscosity for animal fat mixed in liquid hydrocarbons

The experiments were done with fat from bovine and swine. The bovines and goats are a more important source for leather industry than swine. Figures 2 and 3 show the variation of the viscosity of the mixed liquid fuels with different animal fat concentration (10 %, 20 % and 30 %) depending on the temperature.

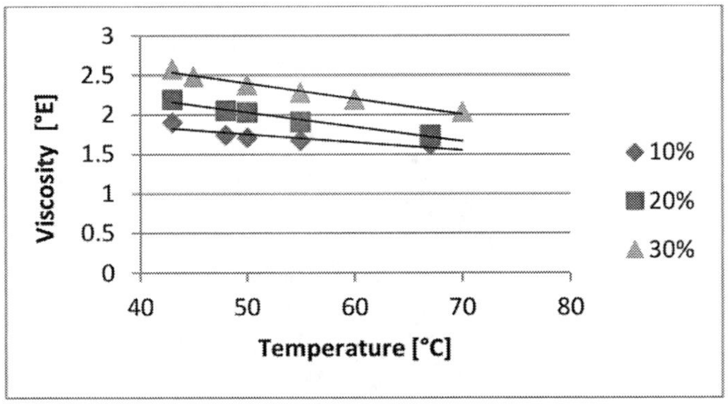

Fig.2. The viscosity variation of the mixture animal fats from bovine with liquid hydrocarbons

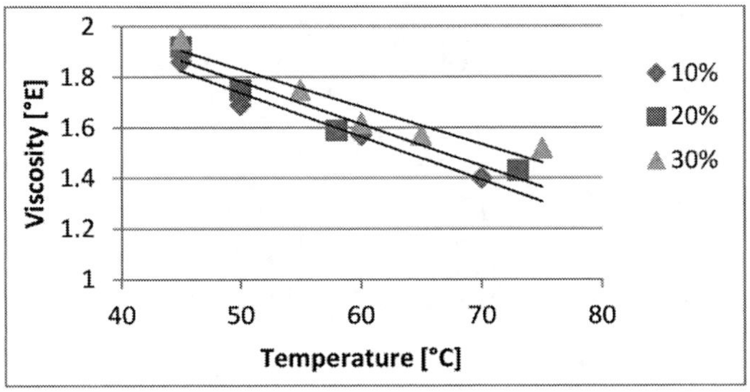

Fig.3. The viscosity variation of the mixture animal fats from swine with liquid hydrocarbons

In figure 4 and figure 5 are show the variation of the density of the mixture analysed.

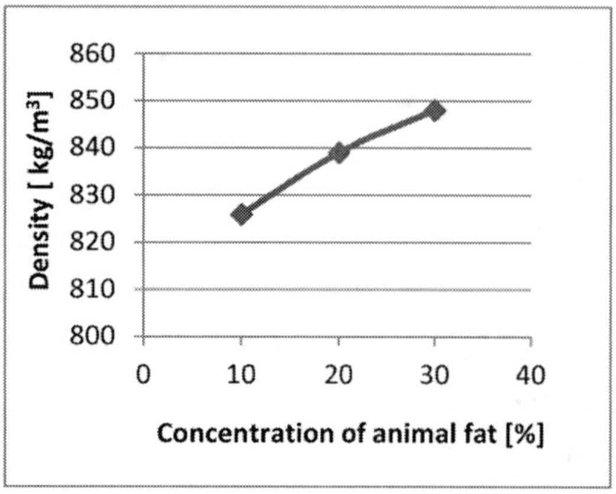

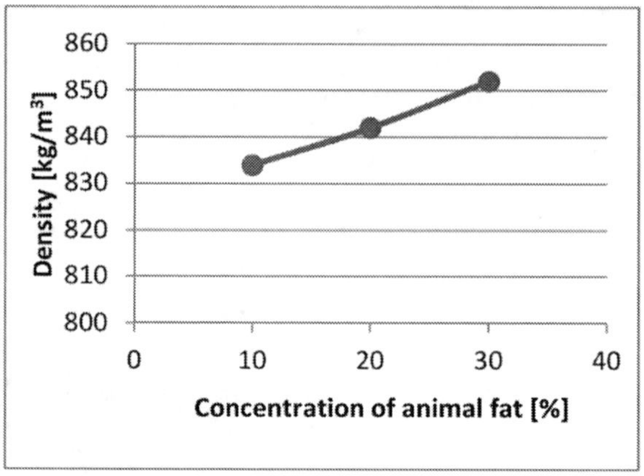

Fig.4. The variation of the density of the mixture of animal fats from bovine with liquid hydrocarbons

Fig.5. The variation of the density of the mixture of animal fats from swine with liquid hydrocarbons

For the viscosity and density characteristic, the experimental determinations have demonstrated a very large approximation for mixture with pure liquid hydrocarbons, thus confirming the results of an efficient combustion with non-modified burners for liquid hydrocarbons.

3. The results of experimental and simulation tests of combustion of mixture

The combustion tests occurred in a hot water boiler, the boiler has a power rated to 55 kW and it is equipped with a burner GB - Ganz type ANYO -12R-2-1-0. The burner is designed for burning liquid hydrocarbons. The experimental results confirm the successful use of these mixtures to small and medium sized hot water boilers. These boilers can be used to heat the space of production units in the leather industry or for industrial hot water generation [12-14].

In figure 6 can be seen the boiler installation together with a specially designed tank for the preparation of mixtures equipped with an electric heater located outside the burner.

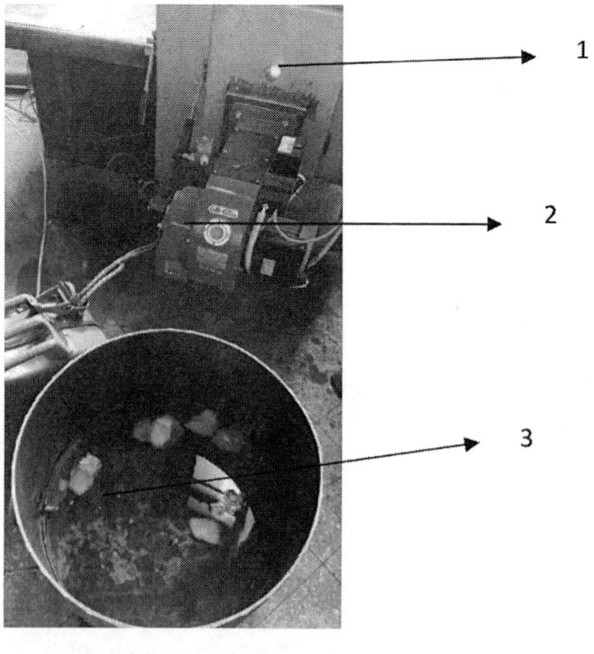

Fig.6. The experimental installation: 1) the boiler 2) the burner and 3) the specially tank

In order to achieve and control mixtures, it was used a free volume tank presented in figure 6, number 3. The tank is equipped with an electric heater and a contact thermocouple. The mixture is made in this tank with various concentrations by heating at temperatures between 40 °C to 50 °C; at these temperatures the animal fat is completely dissolved in liquid hydrocarbons (in this case: diesel oil) . The maximum temperature of 50 °C is required by the operation of the burner pump on the suction side [15].

- The burner has a mechanical spray pump with constant pressure between 14 bar to 18 bar, the spray quality is also achieved with an integrated heating device at 70 °C to 75 °C. The burner has a flame stabilizer mounted on the air outlet section, which generates a local swirl. The swirling is achieved through channels with 60 ° inclined air outlet. The first part of the flame develops into a circular channel with a 74 mm inner diameter, which fixes the tangential component, so that the flame will be narrow. The construction of the burner in the aerodynamic part of the flame is presented in the figure 7 [16].

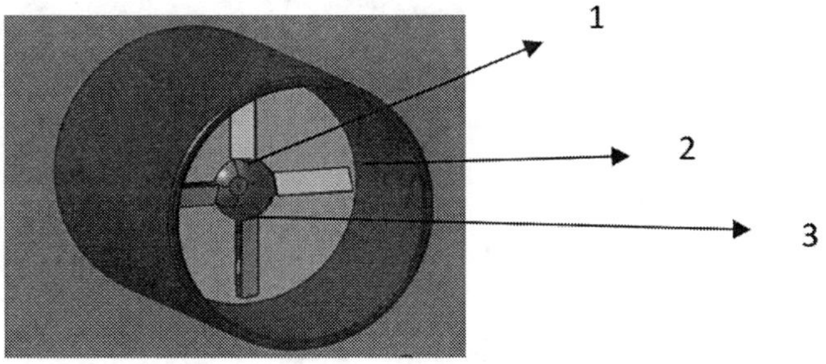

Fig.7. The construction details of the burner 1) spray nozzle 2) circular flame flow channel 3) tangential slits

The analytical calculation highlighted the next functional elements for the burner:

- Axial air velocity in the exit section: u_x=20 m/s;
- The tangential velocity: $u_\varphi = 37\ m/s$;
- The equivalent degree of swirl: n=27;
- The mass of fluid, m, entrained by the jet with the initial mass m_0, at a distance, x, from the diameter, D:

$$m = m_0(0.32 + 0.4n)\frac{x}{D} \tag{1}$$

- The average fuel particle diameter for the 18 bar pressure and animal fat concentration 30 %: d= 23μm
- The combustion time calculated according to Spalding model: $\tau = 0.23\ s$
- The length of the flame, L, calculated with the relation below, has a value confirmed by the figure 8:

$$L = u_x \cdot \tau = 20 \cdot 0.23 = 0.46\ m \tag{2}$$

In figure 8 is showed the aspect of the flame for the mixture with a concentration of 30 % bovine animal fat in liquid hydrocarbons. The measured flame was 0.468 m, this value confirming the calculations of the analytical models.

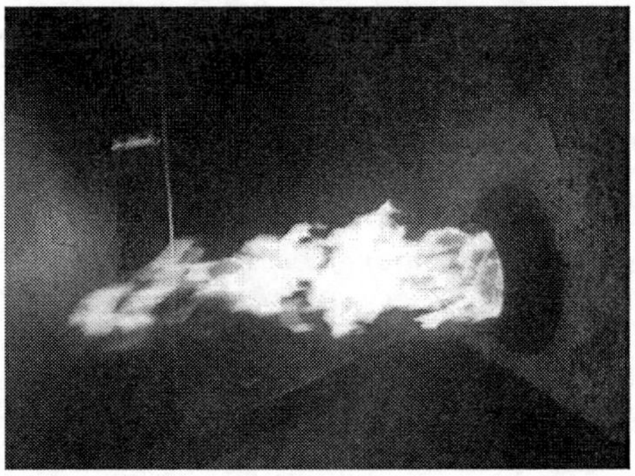

Fig.8. The aspect of the flame for 30 % mixture bovine animal fat

For a complete characterisation of the combustion processes, a numerical modelling of the processes by using the program ANSYS WORKBENCH 14.0 CFX was performed. The simulation of the flame was done in 2D. Aerodynamics of the flame jet regarding the velocity range from exit of the spray nozzle to the end of the boiler is shown in figure 9.

https://doi.org/10.1051/e3sconf/201911202001

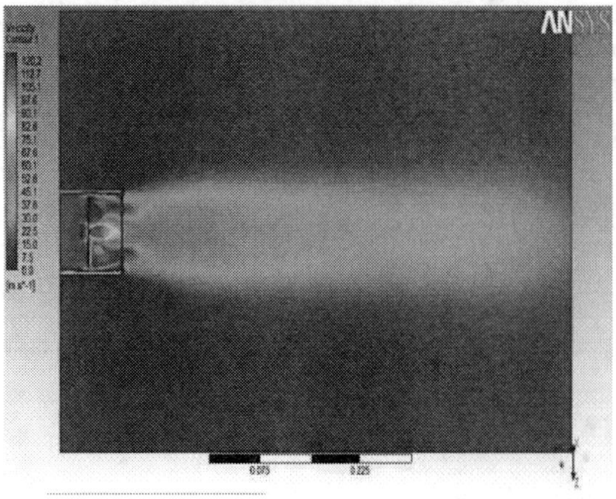

a)

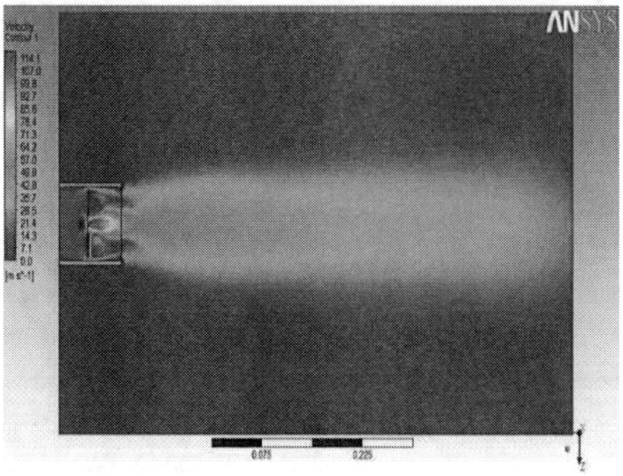

b)

Fig.9. The flame speed range a) for the mixture with a 30 % concentration of bovine animal fat; b) for the mixture with a 30 % concentration of swine animal fat

The aspect of the flame respects the results of the analytical study, the flame being long and thin. However, when using swine animal fat, the flame is 10 % thicker.

The heat range, defined by the flame temperatures, indicates for both mixtures, a core peak in the first flame combustion.

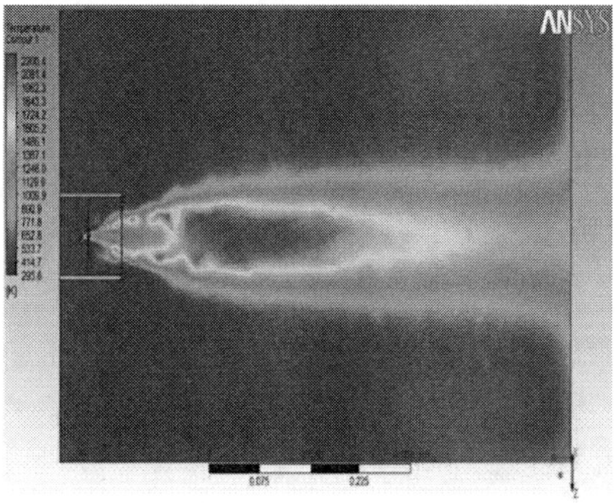

a)

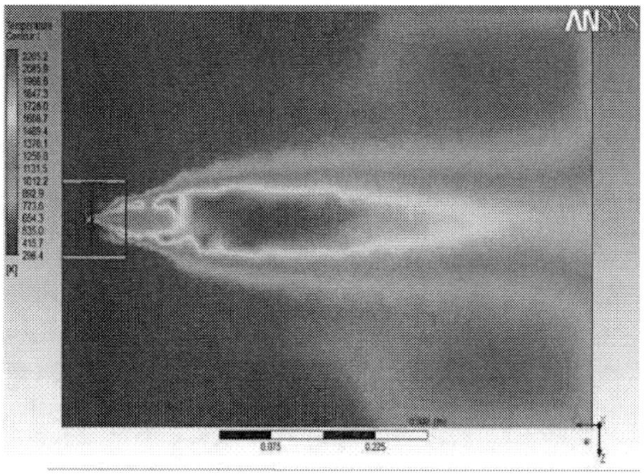

b)

Fig.10. The flame heat range a) for the mixture with a 30 % concentration of bovine animal fat b) for the mixture with a 30 % concentration of swine animal fat

The reverse flow characteristic of the swirling was not noticed due to the presence of the initial flame control tube. Combustion research comprised and determined the combustion gas emissions at evacuation to the chimney, the measurements being made with the MAXYLIZER NG. Excessive air and pollutant emissions were monitored. A good burning, no pulsation of the flame and no particle emission, was achieved with an excess of air, ranging between 1.74 and 3.11. As shown in the figure 11 and figure 12, the combustion gas emissions do not depend of the fat mixing in the tested range.

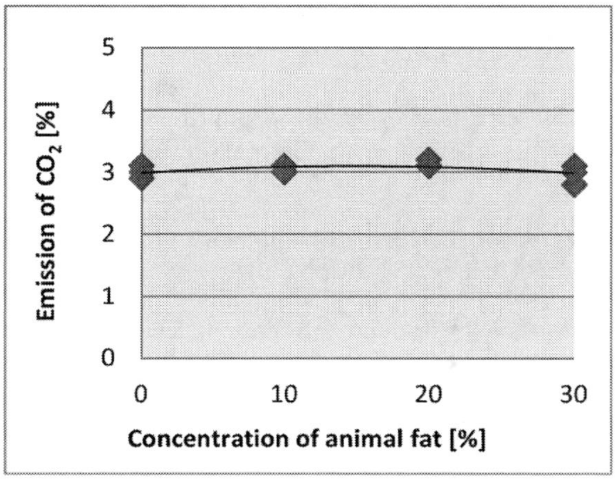

Fig.11. The variation of the emission of CO_2 with the concentration of bovine animal fat

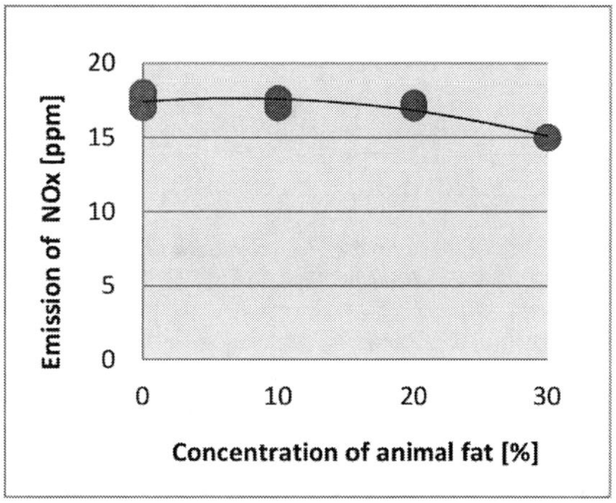

Fig.12. The variation of the emission of NO_x with the concentration of bovine animal fat

The set of research methods consisting of the analytical study, experimental data and numerical simulation, confirms through the data convergence, the viability of the possibilities of burning animal fat mixed with liquid hydrocarbons. Regarding air pollution, a special attention was given to NO_x emissions; this emission was measured in exhaust gases to the chimney. The measurements indicated very low values for nitrogen oxides, even observing a tendency to decrease when increasing the percentage of animal fat, bovine or swine - this tendency can be observed in figure 12.

4. Conclusion

Many researches on the combustion of mixed animal fats with liquid hydrocarbons have shown total applicability in the use of industrial burners with mechanical spraying; these are equipped with spray pumps with pressures above 14 bar and also with heating the mixture in two steps to a temperature of 75 °C. The data obtained by modelling and numerical simulation confirmed and completed the data obtained by analytical studies and experimental measurements. The experiments with 40 % mixture animal fat in liquid hydrocarbons showed a pulsating flame as in figure 13 for a spray pressure increased to 18 bar.

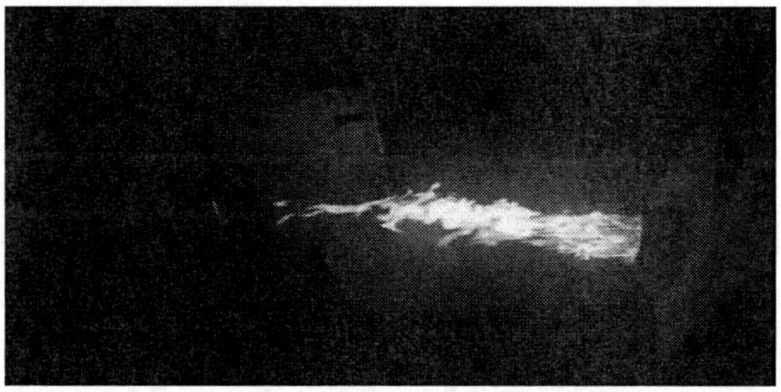

Fig.13. The aspect of the flame for 40 % mixture animal fat with liquid hydrocarbons

The research has not continued on other types of burners, with higher spray pressures, because the obtained flame already indicates an incompatibility. The results obtained imposed the maximum limit of mixture with animal fat up to 30 %. Pollution has not increased by mixing liquid hydrocarbons with animal fat.

Acknowledgment

This work was supported by a grant from the Romanian Ministry of Research and Innovation, CCCDI—UEFISCDI, project number PN-III-P1-1.2-PCCDI-2017-0404/31PCCD/2018,within PNCDI.

References

[1] D.A. Bondrea, G. Lazaroiu, *Analysis and simulation by ansys cfx of the aerodynamics flame for burning animal fats with liquid hydrocarbons*, UPB Scientific Bulletin, Series D: Mechanical Engineering, BDI, **80**, Issue 4, Pages 103-116, (2018).

[2] Environmental, Health, and Safety (EHS) Guidelines for Tanning and Leather Finishing (2010).

[3] A. İşler , S. Sundu, M. Tüter, F. Karaosmanoğlu, *Transesterification Reaction Of The Fat Originated From Solid Waste Of The Leather Industry*, Waste Management **30** (2010) pages 2631–2635, Elsevier.

[4] M.E. Mavrodin., G. Lăzăroiu, *Experimental research on combustion of biogas obtained through anaerobic fermentation of tanneries wastes,* UPB Scientific Bulletin, Series B: chemistry and materials science, **80**, Issue 3 Pages: 105-116, (2018).

[5] E. Onem, G. Gulumser, M. Renner, O. Yesil-Celiktas, *High pressure vegetable tanning of sheepskins using supercritical carbon dioxide,* The Journal of Supercritical Fluids, **104**, Pages: 259-264, b (2015).

[6] G. Lazaroiu, C. Pana, L. Mihaescu, A. Cernat, N. Negurescu, R. Mocanu, G. Negreanu, *Solutions for energy recovery of animal waste from leather industry*, Energy Convers. Manag., **149**, Pages: 1085-1095 (2017).

[7] G. Lazaroiu, E. Pop, G. Negreanu., I. Pisa, L. Mihaescu, A. Bondrea, V. Berbece, *Biomass combustion with hydrogen injection for energy applications*, Energy, **127**, Pages: 351-357 (2017).

[8] L. Mihaescu., G. Lazaroiu, G.P. Negreanu, I. Pisa, *Influence of the characteristics of biogas generated in the leather industry on combustion quality*, Therm.sci., **22**, Pages: S1349-S1357, Supplement: 5 (2018).

[9] D.A. Bondrea, G. Lazaroiu, R. Mocanu, A.N. Stan, L. Balanescu, L. Mihaescu, V. Berbece, *The energy method to protect the environment form waste in the leather industry*, 8th International Conference On Energy And Environment (CIEM), Pages:105-109, Energy & Fuels; Engineering, Book Series: International Conference on Energy and Environment (2017).

[10] G. Lazaroiu, R.C. Moacanu, A.D. Bondrea, E. M. Mavrodin, L. Balanescu, *Use of animal fat waste as a bio-fuel*, 17th International Multidisciplinary Scientific GeoConference SGEM 2017, **17**, Issue 42, Pages 675-682 (2017) , Section Renewable Energy Sources and Clean Technologies, Bulgaria, Albena.

[11] K.E. Lorber, *Waste Management and Treatment of Residues in the Leather-Tanning Industry* -University of Leoben, Austria, 2009, Middle Pomeranian Scies. Society. of the Environ.t Protection, Pages: 8-28.

[12] L. Mihăescu, G. Lăzăroiu, I. Pîşă, E. Pop, G. Negreanu, V. Berbece, A. Bondrea, *Analysis Of The Possibilities Of Combustion Of Animal Fat Mixed With Liquid Hydrocarbons In Boilers With Small Furnaces*, 6th International Conference on Thermal Equipment, Renewable Energy and Rural Development TE-RE-RD (2017), Proceeding, Romania, pages 83-86 (2017).

[13] *L. Mihăescu* „Arzătoare pentru hidrocarburi cu NO_x scăzut"(Burners for hydrocarbons with low NO_x), Ed. Printech, Bucureşti 2004, ISBN 973-718-039-9.

[14] J. Kanagaraj , T. Senthilvelan , R.C. Panda , S. Kavitha, *Eco-Friendly Waste Management Strategies For Greener Environment Towards Sustainable Development In Leather Industry: A Comprehensive Review*, Jour. of Clean. Prod. **89** (2015) Elsevier, pages 1-17, (2015).

[15] G. Lăzăroiu, L. Mihăescu, I. Pîşă, E. Pop, C.R. Mocanu, *Animal Fats Mixed With Liquid Hydrocarbons Combustion Efficiency*, Proceedings of University of Ruse, Bulgaria, 55, pages 12-16 (2016).

[16] G. Lăzăroiu, L. Mihăescu, G. Negreanu, C. Pana, I. Pisa, A. Cernat, D.A. Ciupageanu, *Experimental investigations of innovative biomass energy harnessing solutions*, Energies, **11**, Issue:12, (2018).

E3S Web of Conferences **112**, 02002 (2019) https://doi.org/10.1051/e3sconf/201911202002
TE-RE-RD 2019

Conceptual matrix for the biofuels of the future applicable to the Romanian economy

Gheorghe Lăzăroiu[1,*], *Lucian* Mihăescu[2] , *Gabriel* Negreanu[2] , *Viorel* Berbece[2] , *Elena-Adriana* Jarcu[3] , *Laura Alexandra* Stănescu[3]

[1]University Politehnica of Bucharest, Departament of Energy Production and Use, Adress: Splaiul Independenţei, no 313, sector 6, Romania

[2]University Politehnica of Bucharest, Departament Termotechnics, Engines, Thermal and Frigorific Equipment, Adress: Splaiul Independenţei, no 313, sector 6, Romania

[3]University Politehnica of Bucharest, Department of Hydraulics, Hydraulic Machinery and Environmental Engineering, Adress: Splaiul Independenţei, no 313, sector 6, Romania

Abstract. Romania is in a full process of development in regards to producing energy from renewable sources. Successes have been recorded in the wind and hydro sectors, however, challenges are still to be solved in the field of biofuels. In previous years, research and investments targeting facilities that make use of a larger range of renewable fuels and their respective incineration technologies were carried out. The experience thus gained allows for a redefinition of the range of fuels which can constitute cost-worthy power producing solutions. Defining the renewable fuels of the future is of utmost importance for a country with a significant agricultural potential, but also with a temperate climate. After eliminating some cultures specific to warmer climates and not taking into account the quantity/price ratio, the current paper highlights a synthetic approach (a matrix) to the fuels of the future applicable to the power sector in Romania. At the definition base for the matrix also stands data gained from the experience of the researchers in the respective domain.

1 Classification criteria of biofuels

The notion of conceptual matrix for biofuels comes as a necessity in the definition and classification of a significant number of such fuels. Grouping these fuels from the perspective of provenience characteristics and technological applicability for energy production was considered.

The economic criteria will have a primordial position in the following grouping, the climate being, from this standpoint, an essential selection element [1],[2].

At the current technological development level there is a multitude of biofuel categories. The products resulted from their physico-chemical processing are more feasible to be used in producing energy. There are cases where biofuels in a natural (pure) state can be used alongside their by-products, a representative case being wood [3],[4].

* Corresponding author: glazaroiu@yahoo.com

In order for a biofuel to be used it has to fulfil the following conditions [4]:

- an initial plant production for a certain level of use;
- possibility to collect and store efficiently;
- the existence of an economic processing branch;
- efficient final use.

The biofuels of the future will be those that in time will amass all these four criteria. The evolution of society, which includes agriculture, technological development but also the increased power requirement and environmental protection, may contribute to the addition or subtraction of some biofuel categories from those targeted by the future [7]. The ratio between calorific power and fuel unit represents a prime economic criteria. The ratio between pollution (polluting emissions) generated and fuel unit represents the second socio-economic criteria.

The evolution in time, which essentially contributes to the definition of a future biofuel, indicates the following trends (Figure 1):

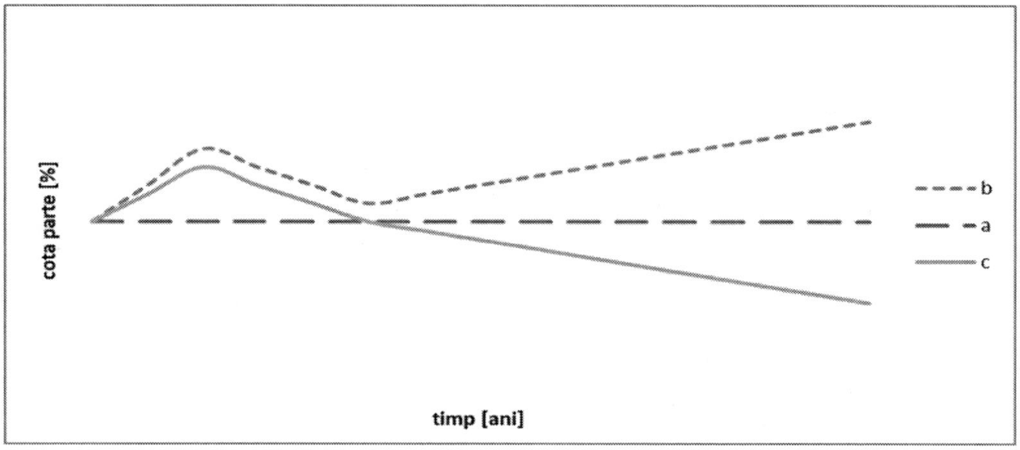

Fig 1. Evolution in time of biofuel consumption

The biofuels from the *a* and *b* categories will be considered the fuels of the future, as they exhibit a constant or increasing trend. Those from category *c* do not present future interest.

The initial increase characterises the start of use for a specific biofuel and is due to the technological momentum, but is corrected in time by real working conditions and costs.

The use of an alcohol fuel is highly desired due to the low or non-existent CO_2 emissions. This leads to the initial definition of fuel production potential, starting from biomass production.

Alcohol production essentially targets two products:

- ethanol (C_2H_5OH), with hydrolysis and fermentation as the main processes, including almost all agricultural products and by-products;
- methanol (CH_3OH), obtained through various physico-chemical processes, such as pyrolysis, gasification and chemical or biological synthesis, through methanogenic anaerobic fermentation.

Gasification leads to the formation of gaseous components represented by CH_4, CO, CO_2, H_2 and others, while anaerobic methanogenic fermentation leads to mainly CO_2 and CH_4.

Methanol can lead to easily obtained hydrogen, differing from the classic production branches based on methane and water dissociation. Methanol is also a synthesis source for ethylene.

In electrical energy production, methanol was imposed from an economic standpoint.

For transportation, alongside methanol, ethanol was also adopted, usually mixed with classic hydrocarbons from petroleum. Research has proved that the percentages of 15% methanol and 20% ethanol cannot be surpassed.

Oil type biomass leads to the formation of metal-esters (biodiesel) through chemical treatment with alcohol.

The state of aggregation (solid, liquid or gaseous) will represent an essential group in the fuel matrix, more so since some of the fuels change states through physico-chemical processing. Technologically, liquid and gaseous fuels present advantages in respect to storage and combustion, and if the price rises compared to the base starting fuel, it will have to be analysed against the price of fossil fuels.

An essential economy cycle includes the comparison between energy price and total production costs.

2 Primary and secondary usable biomass in Romania

According to HG 1844/2005 [5], biomass represents the biodegradable part of products, waste and agricultural waste (including related animal and industrial), forestry and its connected industries, as well as the biodegradable components of industrial and urban waste.

Alongside primary biomass, directly resulted from photosynthesis, secondary biomass has, in time, gained an increasing margin, due to the technological developments and environmental protection standards [6].

The residues can be classified as primary, secondary and tertiary residues.

The primary residues are products represented by plants or forestry elements. From the activity to which this class of residues is subjected to in order to obtain dietary products, the secondary class of residue is obtained.

Energy crops are steered towards starch, sugar or oil producing plants, or lignocellulosic crops [8].

From the perspective of energy producing biomass it should be mentioned that Romania, due to its significant agricultural potential and continentally temperate climate, is a large cereal and oil producer, the rest of the energy crops, including willow, being in the definition phase.

For Romania, through the Strategy for use of renewable resources, according to HG 1535-2003 [9], the share of biofuels will have to reach the value of 6.87% of power production by the year 2020.

Concerning the biofuels that can be obtained from the agricultural sector the following data regarding potential can be mentioned [8][10]:

• for cereal crops, the production of straw used in energy generation is $2*10^6$ t/ha leading to $7*10^6$ t/year being produced, which translates to $7.5*10^5$ toe/year;

• from a surface of about 2.5-2.7 million ha cultivated with corn, a quantity of stems of up to 17 mil. t/year and 2 mil. t/year of cobs can be produced. From an energy production perspective, around 60% of the total quantity can be used;

• from a surface of about 1 million ha cultivated with sunflower, around 1.8 mil. t/year of stems can be obtained. The shells, with a high calorific power, are also used for the purpose of energy production;

• grape plant vines resulting from maintenance operations carried out in viticulture amount to around $2.2*10^5$ t/year, and about $2.3*10^5$ t/year of biomass can be obtained from orchard maintenance.

It should be mentioned that in 2018, Romania was determined to be the largest producer of corn and sunflower in Europe.

Rapeseed is cultivated on around 0.75 mil. ha, the oil being almost exclusively dedicated to the transportation sector. It is worth acknowledging that in the past few years, the stems of the plants have also started being used for energy production.

The forest sourced biomass potential is also significant in Romania, an important percentage of the produced wood being directed towards direct combustion in the citizen's homes (around 65% of Romanians use fire for heating) but also towards power production.

In Figure 2, the evolution of wood type biomass harvesting for Romania can be observed, including the harvesting for direct burning.

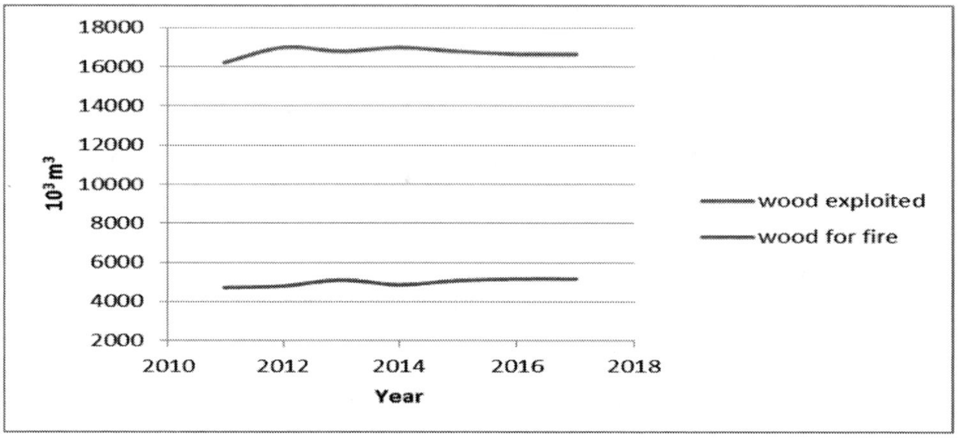

Fig 2. The evolution in time of wood production in Romania [10]

Although wood represents the oldest biofuel used, the requirement to limit its humidity before combustion will always remain a drawback. The simplest prerequisite limits the humidity to 30%, but as the demands of the various combustion technologies develop, this limit gains multiple nuances.

These humidity related aspects reflect also on the willow used for energy production purposes, relatively underdeveloped in Romania (around 2500 ha). Although the combustion technology for this type of willow has developed accordingly, being considered an agricultural crop it has to adhere to inhibiting regulations.

The criteria of calorific power is outlined as can be observed from the data in Table 1 by all the biofuels, representing the limitations provided by the humidity content (wood is especially affected). For reference, it should be mentioned that the calorific power of lignite variates between 9000-9500 kJ/kg.

Table 1. The calorific power of some biofuels, in kJ/kg

Biofuel	The calorific power in anhydrous state	Biofuel	The calorific power in anhydrous state
Sunflower seed shells	16200	Corn cobs	17200-17600
Sunflower stems	21000-21800	Wine plant vines	16500 (humidity 7%)
Wheat straw	17000-18500	Fruit tree vines and branches	15000 (humidity 7%)
Barley straw	15000-15500	Wood	15500
Corn stems	15500-16200	Willow for energy use	15000

The influence of humidity on the calorific power of wood and willow is given by the graph in Figure 3.

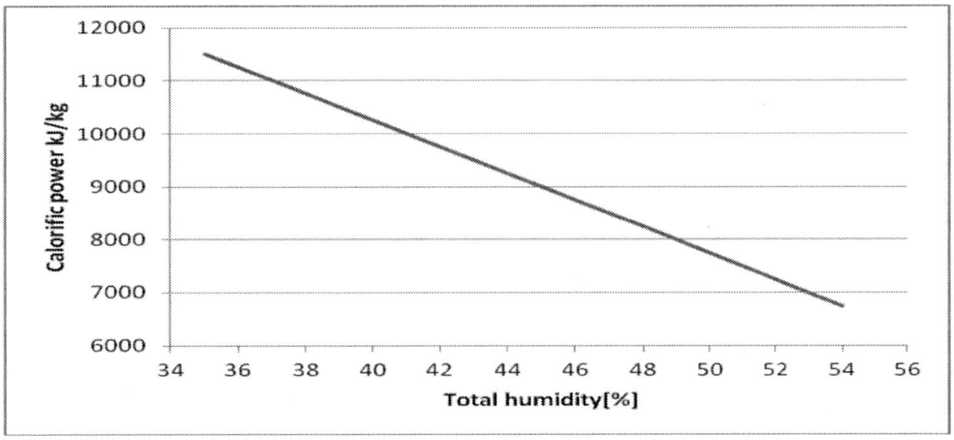

Fig 3. The importance of the relationship between total humidity and calorific power [10]

Starting from these aspects, the primary biomass nomenclature for the Romanian environment is presented in Table 2, from a current potential of use perspective.

Table 2. Future biofuels in Romania

Feedstock	Biofuels				
	Solid	Liquid			Gaseous
	Direct combustion	Bioethanol and biomethanol	Biodiesel	Pure vegetable oil	Biogas
A – Forest biomass					
A1 – Raw wood (firewood)	A1.1				
A2 – Industrial wood waste	A2.1				
A3 – Pellets and briquettes	A3.1				
A4 – Char, pyrolysis gas	A4.1	A4.3			
B – Agricultural products					
B1 - Cereals (barley, oats, rye, wheat)		B1.3	B1.2		
B2 - Sugarcane		B2.3			B2.3
B3 - Sorghum	Stems B3.1		B3.3	B3.3	B3.3
B4 - Rapeseed		B4.3	B4.2	B4.2	

B5 – Corn kernels, stalks, cobs	B5.1	B5.3		B5.2	B5.3
B6 - Sunflower (stems, shells, seeds)	B6.1	B6.3	B6.3	B6.1	
B7 - Potato		B7.3			
B8 - Soy			B8.2	B8.1	
B9 – Cereal straw (ballots, briquettes, pellets)	B9.1		B9.3		B9.3
B10 – Willow and poplar for power production (chips, briquettes, pellets)	B10.1				
C – Vinicultural sources					
C1 - Grapevine (cords)	C1				
D –Tree farm sources					
D1 – fruit trees, ornamental trees, leaves, branches	D1				
E – Animal fats	E1				
F – Animal protein					F3
G – Combustible municipal waste	G1				G3
H – Organic and combustible industrial waste	H1				H3
I – Water treatment					I3
J - Waste	J1				J3

Reducing the impact on the environment was also taken into account. The development of pellet production has recently lead to the use of a large range of shells, seeds and roots in products to be used in energy production [12], [14], [15].

The matrix has on the abscissa the provenience of the biofuels, and on the ordinate the state in which the fuel is used (Figure 4).

$$\begin{vmatrix} A1 & B1\ B2\ B3\ B4\ B5\ B7\ B8\ B9\ B10 & C & D & E & F & G & H & I & J \\ A1.1\ A2.1\ A3.1\ A4.1 & B3.1\ B4.1\ B5.1\ B6.1\ B8.1\ B9.1\ B10.1 & C1 & D1 & E1 & & G1 & H1 & & J1 \\ A6.2 & B1.2\ B4.2\ B3.3\ B8.2 & & & & & & & & \\ A5.3 & B1.3\ B2.3\ B3.3\ B4.3\ B6.3\ B7.3\ B9.3 & & & & F3 & G3 & H3 & I3 & J3 \end{vmatrix}$$

Fig 4. The matrix of future biofuels in Romania

From the biofuel matrix some fuels were removed, considered to not be economically feasible for Romania, the majority of which being unfit for the climate, amongst which:

- sugarcane
- artichoke (Cynara)
- elephant grass (Miscanthus)
- palm, coconut tree and macadamia
- panicum virgatum
- phalaris Arundinacea
- eucalyptus
- paulownia
- hemp
- cameline

3 Biofuel usage scale in Romania for energy production

In Romania, during year 2018, the installed power in electrical energy producing facilities from biofuels was 124.16 MW, in 28 units, according to data from ANRE [13].

Depending on the biofuel used and synthesised in the matrix presented before, the electrical energy producing units (some with cogeneration) can be grouped as following in the Table 3:

Table 3. Biofuels used in electrical energy production

Biofuel	Company	Installed electrical power, MW
A1	A6 IMPEX S.R.L. - Dej	9.73
A2	EGGER ROMANIA S.R.L. - Rădăuţi	14.5
	SAUCOLEMN S.R.L.	0.31
	FIRST BIOGAZ S.R.L.	1.49
	GENERAL ENERGETIC S.R.L. - Pângăraţi	6.5
	SORTILEMN S.A. - Gherla	1.44
	HOLZINDUSTRIE SCHWEIGHOFER S.R.L.	21.25
	EXPLOCOM GK S.R.L. – Lupeni - Harghita	0.82
	BIO ELECTRICA TRANSILVANIA S.R.L.	19.93
	BIOENERGY SUCEAVA S.R.L.	29.65
B1.2, B1.3 A4.3 B5.3 B6.3	AAYLEX PROD S.R.L.	2.26
	AGROTRUST S.R.L.	0.25
	ARMAN CONSTRUCTION S.R.L.	0.40
	COM ABM S.R.L.	0.99
	IRIDEX GROUP IMPORT	3.60

	EXPORT S.R.L.	
	MEVCER S.R.L.	0.50
	NEW LIFE ENERGY S.R.L.	0.50
	RENEWABLE POWER S.R.L	0.66
	SANA RA S.R.L.	1.60
	TEB PROJECT ONE S.R.L.	2.98
	UAT ORASUL SEINI	0.37
	BIOCARNIC ESCO S.R.L.	0.52
	GENESIS BIOTECH S.R.L.	1.06
I5	APASERV SATU MARE S.R.L.	0.35
	COMPANIA DE APA ORADEA S.A.	0.72
	COMPANIA DE UTILITATI PUBLICE S.A.	0.25
B10.1	BEST TEAM CONSULTING S.R.L.	0.25
	RIG BIOMASS S.R.L.	1.28

The presented data shows that the dominant biofuel is wood stock (85%), followed by biogas resulting from waste fermentation (12.7%). This repartition of biofuel use leads to the conclusion that agricultural biomass is almost fully reused in electrical energy production [11].

The quantity of forestry biomass used in energy production is a consequence of the large exploited volumes. As can be seen in Figure 2 the quantity of wood used for fire has to be taken into account in the thermal energy scale of the country. To this, the agricultural biofuels, including the power production used willow, will be added.

If an average annual quantity of 3.5 million tons of firewood is to be considered for thermal energy production, for a thermal power of 3.3 kWh/kg, the resulting produced thermal energy is $11.5*10^6$ MWh/an. For an efficiency of 70%, the actual thermal energy produced is around $8.08*10^6$ MWh/an.

Referring to the biofuels of agricultural provenience, the use of cereal straw, corn stems, sunflower and cobs, represents the largest potential of use, although in power production under 10%.

4 Conclusions

For energy production, both electrical and thermal, in Romania, in the biofuel balance the forestry sector has the largest margin. To be noted is the significant potential of agricultural biofuels.

Referring to the waste use, the range of applicability is extremely reduced, more so today, when the notion of circular economy is in full progress.

From the census of biofuel use in electrical and thermal energy production, an extremely reduced number of biofuels that are actually used can be observed.

References

1. G. Lazaroiu, L. Mihaescu, G. Negreanu, C. Pana, I. Pisa, A. Cernat, D-A. Ciupageanu, *Experimental Investigations of Innovative Biomass Energy Harnessing Solutions*, Energies, ISSN: 1996-1073, Volume **11** Issue:12, (2018)

2. M. Balat and G. Ayar *Biomass Energy in the World, Use of Biomass and Potential Trends*, Journal Energy Sources, Volume **27** Issue 10, Pages. 931-940, (2006)
3. I. Pisa, G. Lazaroiu, C. Radulescu, L. Mihaescu, *Experimental Studies on the Corrosion Occurrence During Biomass Combustion Process*, Environmental Engineering and Management Journal, ISSN: 1582-9596, EISSN: 1843-3707, Volume: **11**, Issue: 9, Pages: 1555-1560, (2012)
4. G. Lazaroiu, I. Oprea, L. Mihaescu, T. Prisecaru, I. Pisa, G. Negreanu, C.R. Mocanu, *Biomass Briquettes from Pitcoal-Wood: Boiler Test Facility Combustion Case Study*, Journal of Environmental Protection and Ecology, ISSN: 1311-5065, Volume: **13**, Issue: 2a, Pages: 1070-1081, (2012)
5. Hotărârea nr. 1844/2005 privind promovarea utilizării biocarburanţilor şi a altor carburanţi regenerabili pentru transport
6. T. Vintila, S. Neo, *Biogas in Romanian Agriculture, Present and Perspectives*, Scientific Papers: Animal Science and Biotechnologies 44, (2011)
7. Directive 2003/30/Ec of the European Parliament and of the Council of 8 May 2003 on the Promotion of the Use of Biofuels or Other Renewable Fuels for Transport, Official Journal of the European Union 17.5., L 123/42, (2003)
8. Run-Cang Sun, *Cereal Straw as a Resource for Sustainable Biomaterials and Biofuels*, Elsevier, ISBN: 978-0-444-53234-3
9. Hotărârea nr. 1535/2003 privind aprobarea Strategiei de valorificare a surselor regenerabile de energie
10. G. Lazaroiu, L. Mihaescu, T. Vatuiu, I. Pasa, D-A. Bondrea, G.P. Negreanu, E-A. Jarcu, V. Berbece, *Biomasa Solida Sursa de Energie Curata*, Volume **2**, (2018)
11. Study of Detection of Business Opportunities for Entrepreneurs and SMEs in Biomass Sector, https://ec.europa.eu/energy/intelligent/projects/sites/iee-projects/files/projects/documents/biobusiness_market_opportunities.pdf?fbclid=IwAR0kVR6_7XORug4ZTtOrLwUdDqpvSBIWdN7AyMRcglhHqtEpK-yf_Zbl92w
12. T. Amon, *New Trends on Energy Generation Using Biomass*, Simpósio Internacional sobre Gerenciamento de Resíduos Agropecuários e Agroindustriais – II SIGERA, Volume **1**, Pages 65-68 (2011)
13. Autoritatea Naţională de Reglementare în domeniul Energiei
14. F-P. Langue, *Biomass market trends and bioenergy sector views on EU policy*, AEBIOM, (2015)
15. M. Balat, N. Acici and G. Ersoy, *Trends in the Use of Biomass as an Energy Source*, Journal Energy Sources, Volume **1** Issue 4, Pages 367-378, (2006)

Energy passivation of a building. The ICSTM building case study

Otilia Nedelcu[1,*], *Ioan Corneliu* Salisteanu[1], *Traian* Ivanovici[1], and *Valentin* Dogaru[1]

[1]Valahia University of Targoviste, Department of Electronics, Telecommunication and Energy Engineering, Aleea Sinaia no.13, Targoviste, 130004, Romania

Abstract. The renewable energy sources are becoming more and more current in the energy supply of buildings, from residential buildings to institutional buildings. The renewable energy sources have proved to be a solution and an important element of the Romanian Energy System. We are starting from the idea of using the renewable sources to passivize a building, and we are relying on existing facilities in the Multidisciplinary Scientific and Technological Research Institute (ICSTM) from Valahia University of Targoviste (UVT) to produce the energy from its own sources. We are also considering the consumption data from previous year as well as the electric link to external power grid. This paper aims to propose new renewable energy units so that ICSTM become energetical independent. Currently, the ICSTM building consumes an amount of energy produced from renewable sources about three times less than that consumed from the external network. In order to determine the number of equipment, specifically CPV (concentrating photovoltaic) panels placed on trackers, it is desirable to create a 40% reserve above the installed power, taking in account the possible future development of the research laboratories.

1 Introduction

An energy-efficient construction is no more than a construction concept that can be reached and applied by anyone. [1]

Multidisciplinary Scientific and Technological Research Institute is part of the UVT, being built on an area with a ground footprint of 2240 m^2 and a total built area of 7250 m^2. The internal part of the building is formed by 35 laboratory spaces, 5 technological laboratories, 6 functional annexes, 7 administrative spaces and 4 dissemination spaces, with installations and equipment used for different activities of research. [2]

Multidisciplinary Scientific and Technological Research Institute is equipped with more sources which produce renewable energy: photovoltaic platform, wind platform, thermosolar platform.

All those sources cover only one part from necessary consumption for ICSTM, the other part is consumed from the National Energy System (SEN) network.

* Corresponding author: otilia.nedelcu@valahia.ro

2. The current state of the ICSTM supply with energy from renewable sources

The current ICSTM equipment consists of:
-The photovoltaic experimental platform which has an active surface of more 250 m^2, and the installed approximate power of 79,65 kWp;
-The wind experimental platform which has the installed approximate power of 25 kWp;
-The thermo-solar experimental platform which has the active surface of more 250 m^2, with an approximate capacity of 10000 litters.

2.1 The photovoltaic experimental platform

This platform is composed of many installations with different installed powers:

- *OnGrid installation,* which is located on the institute's terrace, it has PV panels with the orientation towards south and the inclination 45°. The installed power is 33.15 kWp. It is consisting of three type of panels: monocrystalline silicon panels, polycrystalline silicon panels and amorphous silicon panels. It is equipped with 6 inverters of SMA 5000TL-21kW type. [2]
- *OffGrid installation* located on the ICSTM terrace, it has PV panels with the orientation towards south and the inclination 30°. It has the installed power of 42.5 kWp and it is consisting of two types of PV panels, two types of invertors and 100 batteries.
- *The Off Grig installation of parasolar type.*

The ICSTM building is separated in three building bodies and the panels repartization had been made in the following way: the system in front of ICSTM consists of 120 photovoltaic panels of semi-transparent Altius AFP type.

PV of parasolar type consists of polycrystalline silicon panels with the 250 W nominal power.

- *The OffGrid installation of curtain type.* It consists of 51 Q.Cells Q.Pro-G3 BLK photovoltaic panels, polycrystalline silicon panels, 7 inverters of Victron Multiplus 5 kW, 2 inverters of Victron Multiplus 3 kW and 100 VRLA Gel Deep Cycle 220Ah and Victron VRLA Deep Cycle 220Ah batteries.
- *Fixed Tracker with 4kWp power*

The Tracker has two axes and the 25 m^2 surface. It consists of 15 Q.Cells Q.Pro-G3 BLK photovoltaic panels and 2 SMA SunnyBoy 2 kW inverters. The photovoltaic panels on the tracker are from polycrystalline silicon.

2.2 The wind experimental platform

The wind experimental platform has the wind installations with turbines which have vertical axis or horizontal axis: the OnGrid installation; the MagLev installation; the SmallWind installation.

- *20 kWp OnGrig installation* on the ground_has: one Aeolos HAWT wind turbine with 10 kW power and horizontal axis, 24m monopole; the Aeolos VAWT wind turbine with 10 kW and vertical axis, 260 rpm, 18m monopole; two ABB Wind inverters 15kW.
- *MagLev installation.* It is a wind turbine which function on the principle of magnetic levitation [3], it has vertical axis and is located on ICSTM terrace with 3 kWp power.
- *2 kWp SmallWind installation* has: one Aeolos HAWT wind turbine with 1 kW power, horizontal; one Aeolos HAWT wind turbine with 1 kW power, vertical.

2.3. The thermo-solar experimental platform

The thermo-solar experimental platform from ICSTM includes: the thermo-solar installation for heat input; the thermo-solar installation for domestic hot water.

➢ *26 kWp Thermo-solar installation* for heat input includes: 12 Westech SP-S58/1800A-20 EHPT thermo-solar panels with vacuum tube; 5 Alfa-Bit Beta 58/1800-12 thermo-solar panels; 10 Tesy SP 07 250 ASL flat-plate thermo-solar panels; Taconova Tacosol Circ ZR High efficiency pumping stations; Ferroli Ecounit 1000-2WB heat exchanger and storage boilers.

➢ *thermo-solar installation* for domestic hot water by 600 liters which includes: two SunSystem boilers of 300 litres -a thermosiphon system with flat catheters.

3 The own produced energy and total consumption of energy from ICSTM

Calculation were made for one calendar year considering the energy consumed from the produced energy by ICSTM's own equipment from and the energy consumed from national network and the cost for that moment.

Total energy produced by the photovoltaic platform in the studied period is 112417.22 kWh, it shown in the figure 1, the total produced energy by the wind platform is 21600.58 kWh, it shown in the figure 2, and the total produced energy by the thermo-solar platform is 21252.20 kWh.

Energy consumption in the same period from SEN was: the consumption of electrical energy 239908.4 kWh; the consumption of natural gas 259008.32 kWh, so a total consumption 498916.72 kWh shown in the figure 3, this value represents an annual daily average of approximatively 59.95 kWh. The value payed to supplier was 35470.60 EURO.

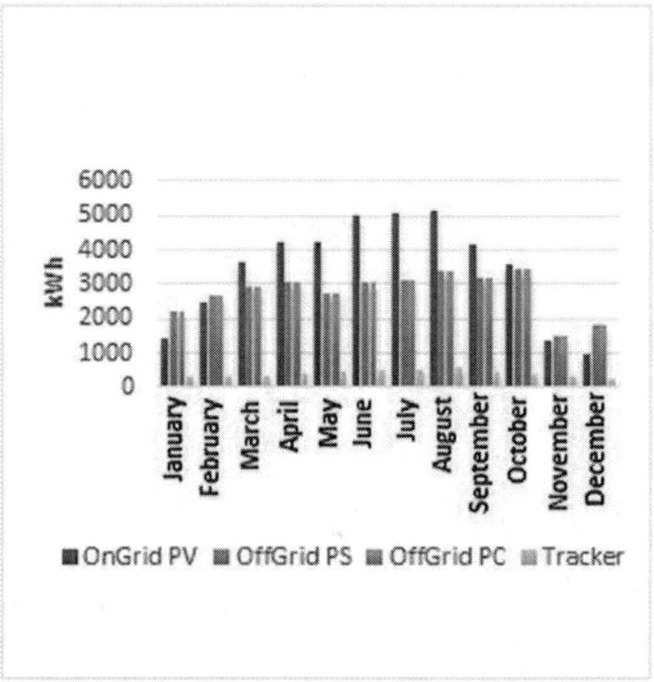

Fig. 1. The total produced energy by the photovoltaic platform

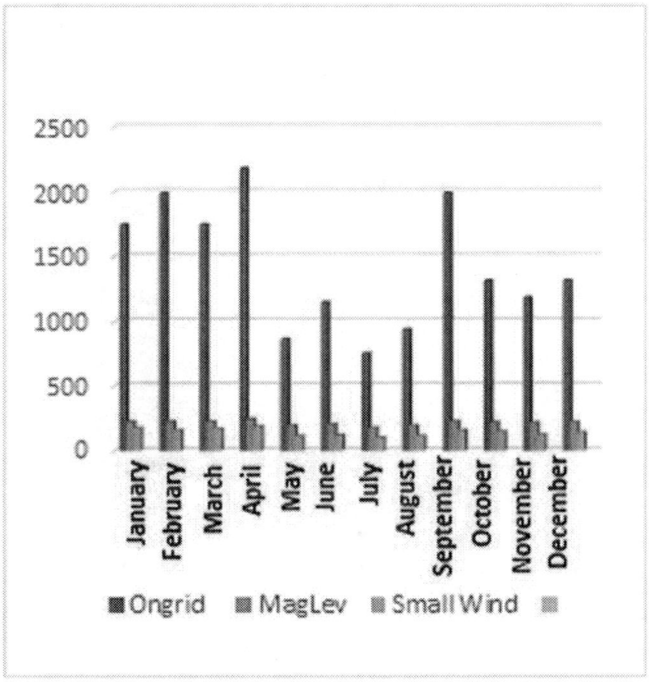

Fig. 2. The total produced energy by the wind platform

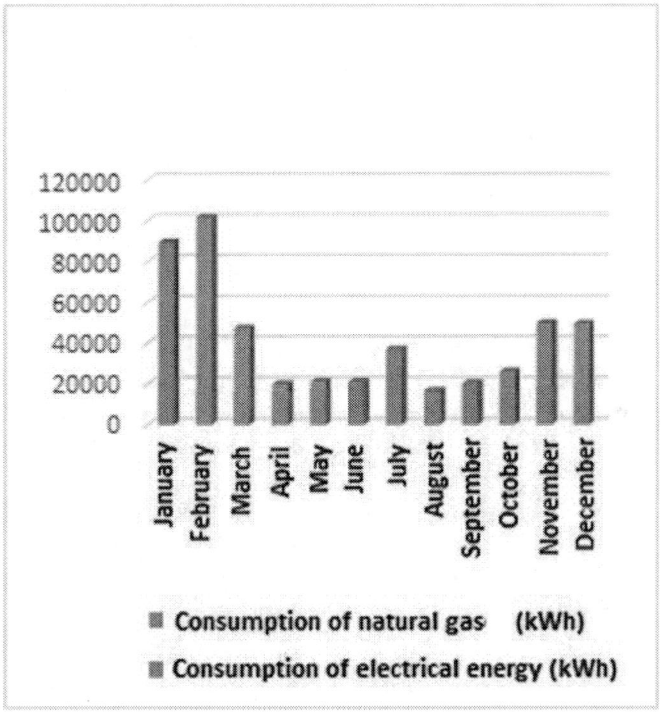

Fig. 3. The total consumption of energy from SEN

We find that the absorption of energy from the national network is about three times higher than the production of the developed energy with the currently own equipment as shown in figure 4.

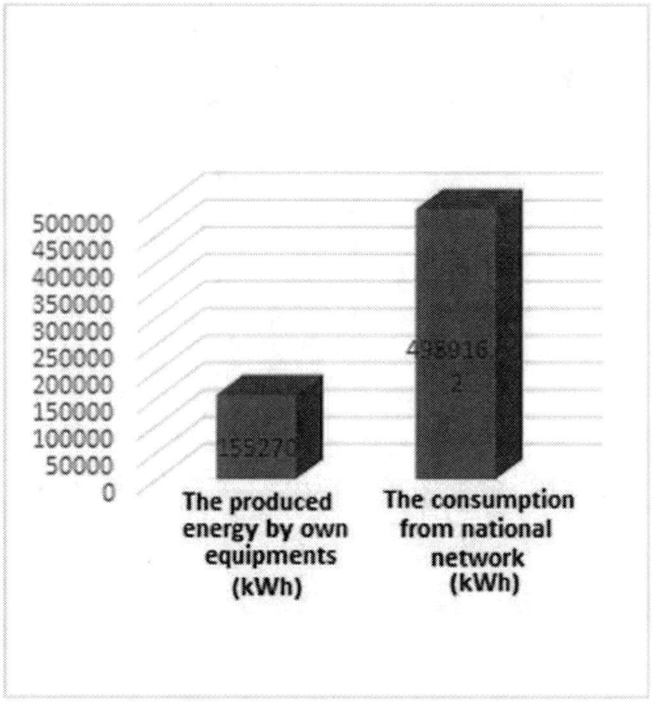

Fig. 4. The total energy (from national network and produced with own equipment)

4 Proposal for the ICSTM building power supply only with energy from own renewable sources

The purpose is to fully feed the ICSTM building whit renewable energy produced from own sources, because the amount of bills paid to the energy provider rises to very high values.

To cover all the energy consumption that ICSTM currently has but taking into account the possible development of ICSTM laboratories or the failure of current sources of renewable energy we need to consider a 40% surplus. This means that the necessary energy from renewable sources will increase to approximatively 689483.408 kWh a year. Thus, the annual daily average needed will be around 80 kWh.

Solution we propose to achieve the autonomy of ICSTM building is: installing ten systems of *concentrating photovoltaic panels* (CPV) Tracker type with two axes of 12.00 kWp.

Trackers will be located near ICSTM building, figure 5, on same line whit the existing tracker on the right of the alley and other six on the left of the alley, between the alley and the Greek Amphitheatre, parallel whit the ICSTM building.

Fig 5. The location of the trackers

4.1 The equipment chosen

➢ *Concentrating photovoltaic panels* (CPV) were chosen because these have the efficiency of 27.2% against the usual panels which have efficiency around 15%. These panels behave well in small but unshaded spaces in tracker systems. [4]

Panels chosen are Concentrix Solar 75 Wp CX-75 CPV Multi-junction, nominal capacity 75 Wp, voltage MPP 135 V, current MPP 0.55A, short circuit current 0.64 A, output circuit voltage 150 V, maximum voltage 1000 V, length 828 mm, width 428 mm, weight 9.5 kg.

➢ *Tracker – because the concentrating photovoltaic panels* (CPV) have smaller sizes than classic ones, and the tracker offers a more space than the existing one, approximatively 70.6 m^2, we will used 198 panels a single tracker. The nominal power is 8000 – 12000 Wp, the elevation angle is from 10° to 90° and the rotation angle Est-West of maximum 300°.

➢ *Inverter* —5 inverters SolarMax 20C type will be used, with recommended PV power 24000 W, Max. DC input voltage 900 V, Max. DC input current 48 A, AC nominal power 20000 W, maximum AC output power 20000 W, weight 275 kg.

➢ *Batteries* – To ensure the storage of produced energy by the ten trackers (1980 panels) is takes 4 modules of 40 batteries are needed which together give:

$$40\ batteries\ x\ 480\ V\ x\ 260\ A = 124.8\ kW \qquad (1)$$

Batteries will be Battery Rolls Solar 4000 - S260 type with the nominal capacity of 200Ah / C 20 – 266 Ah/ C 100, voltage 12 V. [5]

➢ *Regulator – the type chosen is REGULATOR* VICTRON ENERGY BLUESOLAR MPPT 150/60 – TR with the characteristics: auto-selector system voltage/battery 12/24/36/48 V, power of photovoltaic panels 12V-860W / 24V-1720W / 48V-3440W, the maximum voltage for photovoltaic panels 150 V when the temperatures is low / maximum 145 to start and normal operating, maximum charge current 60 A, maximum efficiency 98% and the operating temperature -30/+60°C.

➢ *PV electric cable* – 500 m

On a tracker with the chosen dimension are necessary 198 CPV panels and we obtain for each formed system 10 kWh. Given that we need 10 such systems, the total cost for the panels is 135229.79 EURO.

Table 1. The total costs on the investment.

Equipment	Number of equipment	Price (EURO)
Concentrix Solar 75Wp CX-75 CPV Multi-junction	1980	135229.79
Invertor - SolarMax - SolarMax 20C	5	7420.22
Battery Rolls Solar 4000 - S260	160 (4 modules)	48935.90
Tracker	10	122340.50
PV electric cable	500 m	590.00
REGULATOR VICTRON ENERGY BLUESOLAR MPPT 150/60 – TR	10	5000
Set cables/box connections/fuses/connectors		1900.00
The workmanship		17021.30
The total cost		**338469.90 EURO**

4.2 The amortization of the costs

The total cost of purchasing the sources used to obtain ICSTM energy passivation, is 338470 EURO. Regarding the values of the bills in one year, which ICSTM payed – 35470.563 EURO, we can approximate the amortization time for the investment costs.

Table 2. The amortization of the investment costs.

Initial Investment - 338470 EURO			
The produced energy on year - 698.483 MWh			
The year	**The produced energy MWh**	**The amortization of the investment (RON)/(EURO)**	**The remaining amortized value (RON)/(EURO)**
1	698.483	35470.563	302999.2
2	691.498	35470.563	267528.7
3	684.583	35470.563	232058.2
4	677.737	35470.563	196587.6
5	670.959	35470.563	161177
6	664.249	35470.563	125656
7	657.606	35470.563	90175.9
8	651.030	35470.563	54705.3
9	644.520	35470.563	19234.75

10	638.075	19234.75	0
11	631.694		0
12	625.377		0
13	619.123		0
14	612.932		0
15	606.803		0
16	600.735		0
17	594.727		0
18	588.780		0
19	582.892		0
20	577.063		0
21	571.293		0
22	565.580		0
23	559.924		0
24	554.325		0
25	548.782		0
Total	**15518.70**		

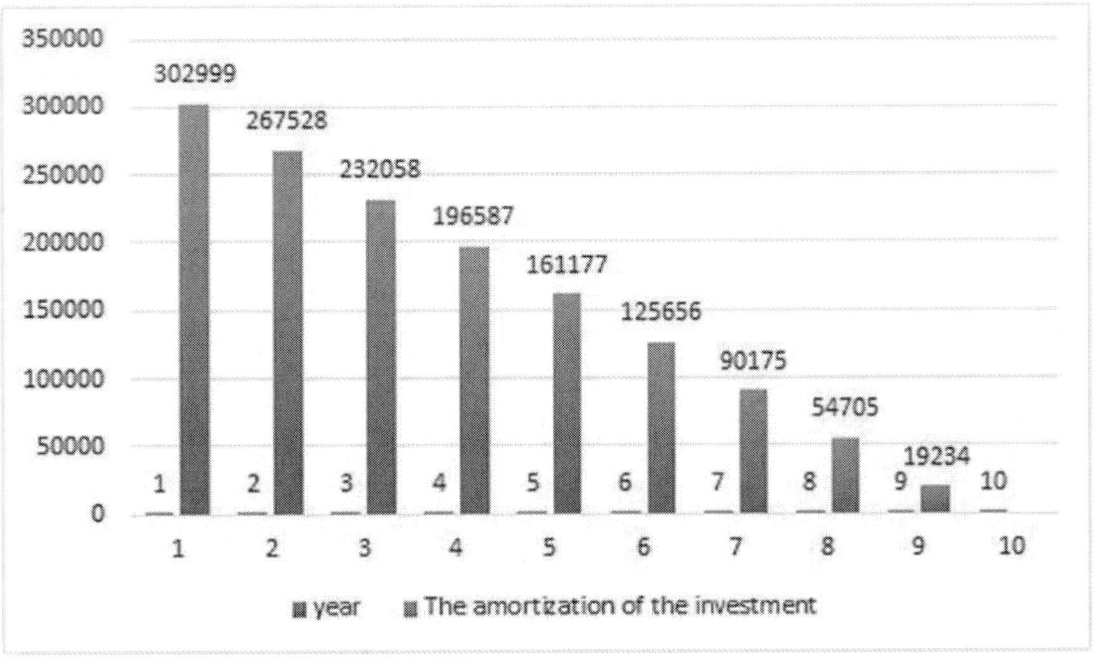

Fig. 6. The amortization of the investment

The amortization of the investment is made in 9 years and 7 months, and the lifetime of the equipment in 25 years, table 2, figure 6.

5 Conclusions

The ICSTM building is now partially powered from renewable sources and partially from national network (SEN) of energy. The quantity of energy produced from renewable sources is about three times less than the energy consumed from the national network.

The study drawn for the ICSTM building to became autonomous in terms of energy consumption shows that this is possible if we add to the current renewable sources, some others renewable sources.

The best solution resulted after several atempts, is the use of 1980 CPV panels mounted on 10 trackers with 10 charge regulators for 4 banks which contain each 40 batteries and 5 inverters – 24 kW toghether with necessary materials for instalation.
When we determined the number of equipment, a 40% reserve over the required power to install for the future development of the research laboratories was considered.

By comparing the cost of the investment with the value of the energy bills, it was found that the investment is depreciated within 9 years and 7 months, the lifetime of the equipment being 25 years.

References

1. I.R. Rancov, *Cercetari privind cresterea randamentului instalatiilor electrice fotovoltaice*, Editura Politehnica, Timisoara, (2014)
2. http://916.icstm.ro/content/rezultate-obtinute
3. B.K. Balankutty, A. Raju, H. A. Mammen. A.K. Abraham, *Design and Analysis of Maglev Vertical Axis Wind Turbine,* International Journal of Emerging Technology and Advanced Engineering Website: www.ijetae.com (ISSN 2250-2459, ISO 9001:2008 Certified Journal, Volume **4**, Issue 4, April (2014)
4. https://www.secondsol.com/de/anzeige/13085/photovoltaikmodule/konzentrierte-photovoltaik-cpv/soitec/concentrix-solar-75wp-cx-75-cpv-multi-junction
5. http://www.e-solare.com/produs/battery-rolls-solar-4000-t12-250

Experimental research on influence of functional parameters of combined installations designed at separating the impurities out of cereal seeds

Carmen Brăcăcescu[1,*], *Ioan* Ţenu[2] , *Costin* Mircea[1] and *George* Bunduchi[1]

[1]National Institute of Research-Development for Machines and Installations designed to Agriculture and Food Industry- INMA/ Bucharest, Romania;
[2]"Ion Ionescu de la Brad" University of Agricultural Sciences and Veterinary Medicine of Iasi - USAMV Iasi, Romania

Abstract. The paper presents the experimental installation, the methodology and measuring apparatus used for experimental research of qualitative indexes of impurities separation out of grain seeds for combined separating systems (according to specific weight and aerodynamic properties of seeds). The experimental installation used was composed of a gravity separator with mechanical shaker with unbalanced masses (mounted on the platform working surface) and an aspiration installation with fan. The experimental research has aimed at quantitative and qualitative influence on separation quality index of the following operating parameters: material flow rate of shaking separator, the air flow of aspiration installation, tilting work surface, work surface oscillation amplitude. Based on data obtained by measurements and qualitative indicators the separating process indexes have been determined, namely: degree of impurities separation, degree of good seeds loss, as well as, the index of technological separation effect for different types of combinations of separation installation parameters.

1 Introduction

After the harvesting process, grain seeds cannot be used directly for different purposes (storage, consumption, industrialization, sowing, etc.) because they contain impurities (vegetal remains, foreign bodies, etc.) and injured products. Prior to receiving a particular destination, it is necessary and obligatory for the harvested products to undergo cleaning and sorting, called as primary processing operations.

The preparation of cereal seeds before processing them is a technological process, which requires technical equipment and combined installations types designed to separate and remove the impurities existing in the cereal seed mass, whose construction depends on the different principle of separation [1, 2, 3, 4].

* Corresponding author: carmenbraca@yahoo.com

So, the researches in this area focused on the study of the phenomena that influence the impurities separation process, the aim being to minimize the impurities that reach the final processing stages of cereals seed [5,6, 7]

The modern conditioning units use complex installations carrying out the separation using combined principles, the most used being the differences between the specific weight and the aerodynamic properties (mainly, the air flow velocity) of the components of the seed mixtures [8,9,10].

In order to obtain the maximum quality with minimum material, electricity and labour consumption it is necessary a thorough knowledge of the constructive and functioning principles of the technical equipments specific to the technology as well as the regulation of the technical and functional parameters of these technical equipments.

The experimental model of gravity separator SG-00 design, manufactured and testing by INMA Bucharest is a innovative technical equipment representative for cereals seed cleaning.

The paper presents the experimental combined installation, the methodology and measuring apparatus used for experimental research and some results of experimental research, especially the index of technological separation effect for different types of installation's technical and functional parameters.

2 Material and method

In the following is presented the installation and the experimental determination used to determine the qualitative indices of the separation process of impurities from cereals seed in the case of combined separation systems (relative to the specific weight and the aerodynamic properties). The installation presented in the technological flow scheme in figure 1, consists of the gravity separator SG-00 (1) connected to a suction installation with air composed of the fan (2) and the cyclone (3).

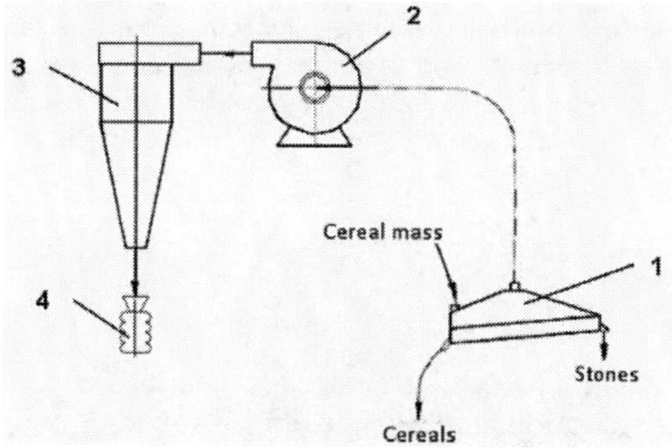

Fig. 1. Technological flow scheme of combined separation installation:
1 - gravity separator SG-00, 2 - fan, 3 - cyclone, 4 - bags for dust and light impurities

Fig.2. The stand (combined separation installation) used for experimental research
1 - gravity separator SG-00; 2 - fan; 3 - cyclone

The functional technical parameters of SG-00 are:

- capacity of processing, t/h
  - cleaning process min.2,5
  - precleaning process max.4
- necessary air flow rate, m^3/min. 150
- installed power, kW 0,7 (2x0.35)
- frequency of oscillations, osc/min. 960

By adjusting the motovibrators eccentric masses, the acting force F_{cf} of motovibrators is regulate, and the direction of oscillations can be adjusted by rotating the motovibrators fixing support (figure 3).

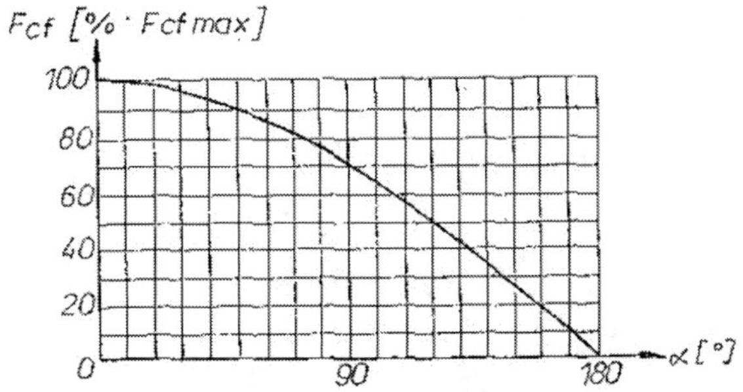

Fig. 3. Variation of centrifugal forces F_{cf} of unbalance masses depending on the angle α of overlap of the non-balanced eccentric masses [11]

Due to the combined effect of the vibrations of the working platform and the action of the air currents, the product mass is stratified on the sieve of working surface. Particles of impurities with a specific weight greater than grain seeds remain in contact with the inclined surface of the platform, while light and very light particles are arranged in upper layers. Thus, under the action of the aspirated air stream, the grain seeds are fluidized and float above the working surface, with which they do not come in constant contact, flowing on its inclined plane to the outlet connections located at the lower end of the gravity separator SG-00.

The material used at experimental researches was the cereals seed (mainly wheat) obtained from experimental plots of INMA Bucharest. The seed mixture mass was first introduced in the intensive suction separator SI 800, being subjected to separation by size.

For experimental determinations were used specific measurement devices for acquisition or registration [12] of the following technical and functional parameters:

- masses of products and impurities;
- material supplying flow rate;
- velocity of air currents in the pipe of suction installation, using an anemometer model Testovent 4000;
- inclination angle respect to the horizontal of the working surface (vibrating frame);
- oscillation amplitude of the vibrating working frame, using an inductive displacement transducer, model W10-Hottinger;
- frequency of oscillation of the two motovibrators, by determining their rotation speed;
- power consumption of motovibrators.

On entry and exit of product (wheat seeds) in the gravity separator SG-00 and at evacuation of impurities in the process of separation were determined the moisture content and hectolitric mass of product.

The coefficient of loss of good wheat seeds C_{ps} is calculated with the relation [4]:

$$C_{ps} = (m / M) \cdot 100 \ [\%] \tag{1}$$

 https://doi.org/10.1051/e3sconf/201911202004

where: m is the good wheat seeds mass which are found at the exit of SG-00 in the quantity of total impurities eliminated, kg; M - good wheat seeds mass at the entry into SG-00, kg.

The index of technological effect of the separation process E_{cs} represents the percent of impurities (foreign bodies) eliminated from the mass of processed product (wheat seeds) and is determined with the relation [4]:

$$E_{cs} = [(c_{csi} - c_{cse}) / c_{csi}]x100 \quad [\%] \tag{2}$$

where: c_{csi} is content impurities at the entrance of product in SG-00, %
c_{cse} - content impurities at evacuation of product from SG-00, %

Knowing the velocity v (m/s) recorded at the anemometer model Testovent4000 and the section S (m^2) of the suction pipe, the air flow rate required for aspiration was determined with the relation:

$$Q_a = 3600 \cdot v \cdot S \quad [m^3/h] \tag{3}$$

The quantity of electricity W consumed by the two motovibrators is calculated by the relation:

$$W = (P_a \cdot t) / 3600 \quad [kWh] \tag{4}$$

where: P_a is the power absorbed by the pair of motovibrators, kW, measured with the phase and frequency analyzer model CA 8334; t - operating time at tests, s.

The working capacity Q of the gravity separator SG-00 was determined by calculating with the relation:

$$Q = 3600 \cdot m/t \quad [kg/h] \tag{5}$$

where: m is the initial product mass, kg; t - time required for experimental determination, s.

The values resulting from processing determined parameters are mentioned in Table 1 and Table 2.

Table 1. Functional operating parameters of SG-00

Crt. iss.	Determined parameter	U.M.	Parameter value determined at tests
1.	Product supply flow rate (test versions)	kg/h	1000; 2000
2.	Sample mass	kg	1000
3.	Duration of test	s	2400; 1800
4.	The quality of the processed product at the entry into SG-00: humidity; hectolitric mass; impurities	 % kg/hl %	 11.93 78.72 2.41; 2.46
5.	Coefficient of loss of good seeds C_{ps}	%	0.01÷0.95
6.	Index of technological separation effect E_{cs}	%	max.98.57
7.	The suction air flow rates of the installation (test versions)	m^3/min.	100; 125; 150
8.	Air speed in the pipe at air flow rates of the suction installation (test versions)	m/s	17.36; 21.7; 26.04
9.	Power absorbed by motovibrators	kW	0.7
10.	Specific consumption of electricity	kWh/kg	0.00045

Table 2. Some results of the tests effected for the determination of optimal regime of the gravity separator SG-00

Wheat seed feeding flow (kg/h)	Air flow (m^3/min)	Angle of operation surface (°)	Amplitude of operation surface (mm)	Quantity of impurities separated at 1000kg of wheat processed (kg)
0	1	2	3	4
1500	100	5	1.5	35.5
			2.0	38.6
			2.5	41.2
		7.5	1.5	33.4
			2.0	35.1
			2.5	37.2
		10	1.5	31.6
			2.0	33.7
			2.5	34.8
	125	5	1.5	26.3
			2.0	29.2
			2.5	33.1
		7.5	1.5	24.5
			2.0	25.8
			2.5	27.6
		10	1.5	21.3
			2.0	22.8
			2.5	23.6
	150	5	1.5	29.7
			2.0	21.4
			2.5	23.8
		7.5	1.5	16.4
			2.0	18.2
			2.5	19.4
		10	1.5	15.6
			2.0	16.8
			2.5	17.9
2000	100	5	1.5	37.6
			2.0	40.2
			2.5	43.4
		7.5	1.5	35.6
			2.0	38.9
			2.5	39.9
		10	1.5	33.3
			2.0	35.4
			2.5	36.7
	125	5	1.5	28.1
			2.0	31.4
			2.5	34.9
		7.5	1.5	26.6
			2.0	27.4
			2.5	29.1
		10	1.5	23.1
			2.0	25.1
			2.5	26.0

Table 2. Some results of the tests effected for the determination of optimal regime of the gravity separator SG-00 (continuation)

Wheat seed feeding flow (kg/h)	Air flow (m^3/min)	Angle of working surface (°)	Amplitude of working surface (mm)	Quantity of impurities separated at 1000kg of wheat processed (kg)
0	1	2	3	4
2000	150	5	1.5	21.2
			2.0	23.0
			2.5	24.9
		7.5	1.5	17.8
			2.0	19.8
			2.5	21.6
		10	1.5	17.1
			2.0	18.3
			2.5	19.2

3 Results and discussions

For an analysis of the influence of constructive and functional parameters of the combined separation installation on the global technological separation index values E_{cs} graphics were drawn expressing the following technological dependency of the index values of the functional parameters (adjustment) of the gravity separator SG-00: the supplying flow rate with product *Qg* subjected to processing material (wheat seeds), the air flow rate *Qa* of the suction installation, the angle of inclination respect to the horizontal working surface of the gravity separator α and the oscillation amplitude *A* of the gravity separator working surface.
In fgures 4 – 6 are presented, for illustration, some representative graphics.

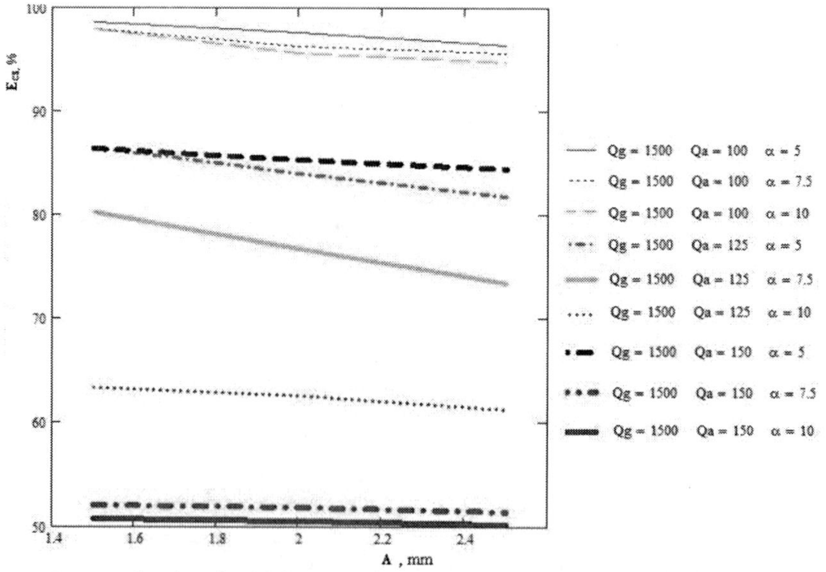

Fig. 4. The dependence of technological separation effect E_{CS} =*f(A)* for Q_g=1500kg/h, Q_a =100÷150m^3/min and α =5÷10°

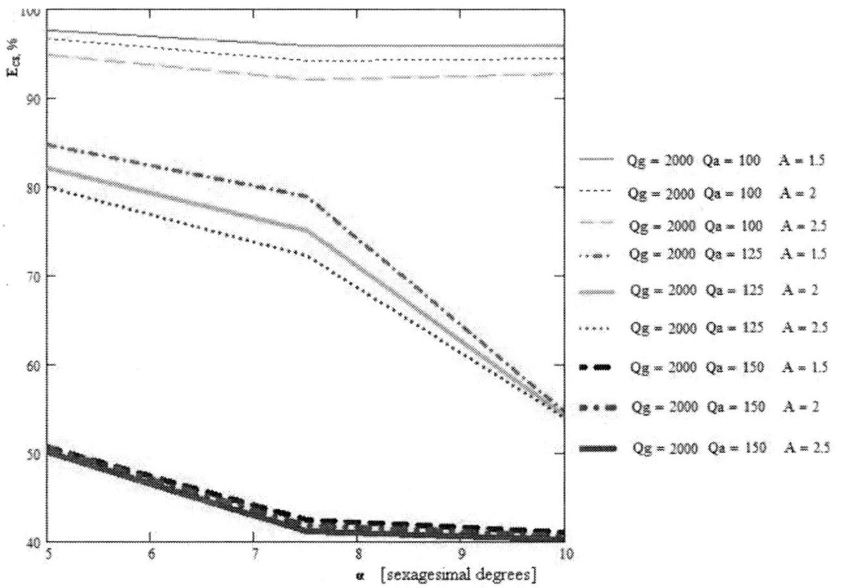

Fig. 5. The dependence of technological separation effect $E_{CS}=f(\alpha)$ for Q_g=2000kg/h, Q_a =100÷150m^3/min and A=1,5÷2,5mm

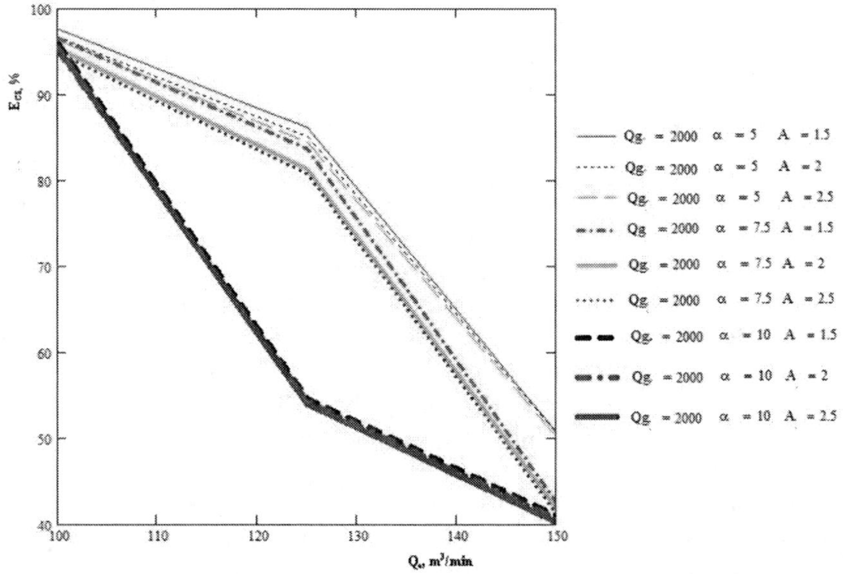

Fig. 6. The dependence of technological separation effect $E_{CS}=f(Q_a)$ for Q_g=2000kg/h, α =5÷10° and A=1,5÷2,5mm

The technological separation effect function E_{CS} is obtained by applying linear regression method at determined experimental data processing, expressed by the relation [13]:

$$E_{CS}(Q_g,Q_a,\alpha,A)= -0.15891Q_g -0.848267Q_a-3.256434\,\alpha -2.736311A+237.645462 \quad (6)$$

The correlation between the experimental data and the data calculated in the experimental points with the regression given by the relation (6) is 0.932.

 https://doi.org/10.1051/e3sconf/201911202004

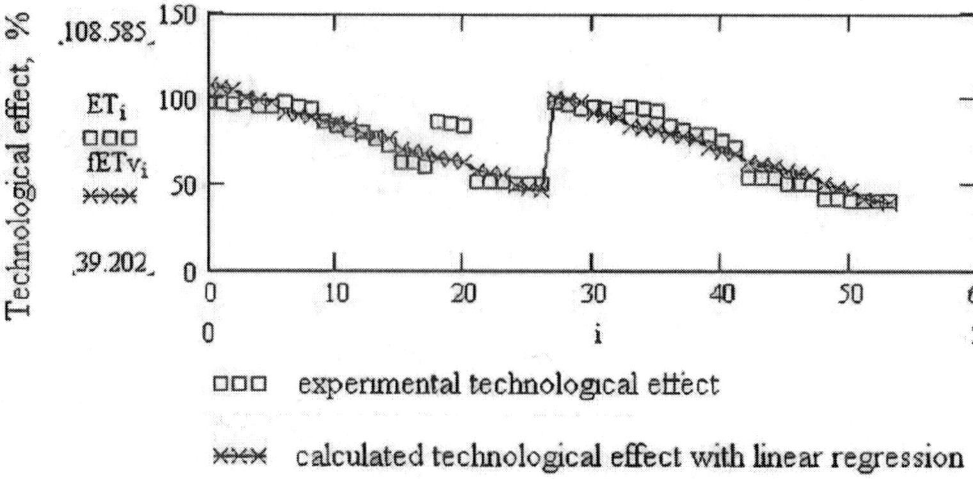

Fig.7. Comparison of the experimental data with those obtained by linear regression at the determination of the technological separation effect index

From the graphical representation in figure 4-7 and in correlation with formula (6) result that the index of global technological separation effect decreases with the increase of the supplying flow rate with material, of the working surface angle, of the suction air flow rate and of the working surface amplitude.

It was noticeable that, by increasing of the cereal seeds feed flow through the impurities pipes are eliminated a part of basis crop grain.

4 Conclusions

The working process of the gravity separator SG-00 is efficient and it achieves a high precleaning of the wheat seeds. The technological parameters obtained during tests depend on: the regulation of the working vibration surface (oscillations amplitude and inclination angle), the aspiration air flow as well as wheat feed flow.

By equipping with a pair of motovibrators, the oscillating movement induced by them allow to the gravity separator SG-00 a high output capacity and an increased separating efficiency.

From the analysis of data obtained by acquisition and processing the experimental measurements revealed the following conclusions:

1. the suction air current generated by the suction installation and introduced into the gravity separator SG-00 favors the stratification of seeds mixture uniform dimensionally with different specific weights, their movement on the working surface of the gravity separator being possible even under the minimum impulses;

2. by increasing the air flow rate between 100÷150 m^3/min the combined separating installations performances are enhanced, due to a better product layer structure achieved depending on the specific weight of particles (seeds and impurities);

3. the oscillations amplitude influence upon the working regime, establishing as optimum values between 2 ÷ 2.5 mm;

4. by increasing the tilting angle of the working surface between 5 ÷ 10° the wheat seeds quantity eliminated through the joint of impurities is reduced. Also, over 10° the gravity separator SG-00 have diminished their performances;

5. the technological separation effect resulted at a single passage of the product was of 98.57% for big foreign bodies eliminated, 91.35% for small foreign bodies eliminated, 95.3% for light foreign bodies eliminated.

The results of experimental researches obtained by direct measurement or determined by calculation reveal that the combined installation consisting of gravity separator model SG-00 and the suction installation comply with the requirements in terms of destination, functioning mode and multiple possibilities for adjustment and servicing.

Acknowledgements

This work was supported by:
- a grant of the Romanian Research and Innovation Ministry, through Programme 1 – Development of the national research-development system, subprogramme 1.2 – Institutional performance – Projects for financing excellence in RDI, contract no. 16PFE.
- a grant of the Romanian Research and Innovation Ministry CCDI - UEFISCDI, Project INNOVATIVE TECHNOLOGIES FOR IRRIGATION OF AGRICULTURAL CROPS IN ARID, SEMIARID AND SUBHUMID-DRY CLIMATE, project number PN-III-P1-1.2-PCCDI-2017-0254, Contract no. 27PCCDI / 2018, within PNCDI III.

References

1. T. Căsăndroiu, *Primary processing and agricultural products storing,* lithographed course, Politehnica University of Bucharest (1993)
2. I. Danciu, *Cereals cleaning*, University “Lucian Blaga” Publishing House, Sibiu, (2001)
3. Fl. Rus, *Separation operations in the food industry*, Publishing Transylvania University of Braşov (2001)
4. V.F. Didyh, *Improving the efficiency of drying agricultural plant materials: monograph*, Lutsk pp. 75-101 (2002)
5. A.D. Zewdu, *Aerodynamic properties of the grain and straw material*, Biosystems Engineering Journal, **98**, pp. 304 –309 (2007)
6. Z. Jing Z, *Design of New Grain Threshing Cleaning Device*. Journal of London University, 1, 17, London/UK (2013);
7. C. Brăcăcescu , Sorică C., D. Manea , Yao Guanxin, G.A. Constantin - *Theoretical contributions to the drive of cereal cleaning technical equipment endowed with non balanced vibration generating systems,* INMATEH - Agricultural Engineering, vol. 42, no. 1 / 2014, pp. 69 – 74, (2014)
8. A.V.Nesterenko, S.M.Leshchenko, O.M.Vasylkovskyi, D.I.Petrenko, *Analytical assessment of the pneumatic separation quality in the process of grain multilayer feeding*. INMATEH Agricultural Engineering, **53**, No. 3, pp.65-70, Bucharest/Romania (2017),
9. I. Costin, *Miller's Book*, Technical Publishing House, Bucharest, Romania (1999)
10. Chr. Geankoplis,*Transport processes and separation process principles*, Prentice Hall PTR, Fourth Edition, 14, pp.903-944, (2003)
11. O. Vasylkovskyi, K. Vasylkovska, S. Moroz, M. Sviren, L. Storozhyk, *The influence of basic parameters of separating conveyor operation on grain cleaning quality*, INMATEH- Agricultural Engineering, **57,** no.1/2019, pp 63-70, (2019),
12. V. Damian*, Manual of the Food Industry Engineer*, 28-Measuring and control devices, control and adjustment, pp.1234-1262, Technical publishing house, Bucharest (1998)
13. C. Brăcăcescu, *Theoretical and experimental researches regarding the optimization of working processes of gravimetrical separators aimed at cereals impurities*, PhD thesis (Transilvania University of Brasov, 2011)

Usage of biomass CHP for balancing of power grid in Ukraine

Oleksii Epik [1,*], *Vitalii Zubenko*[1]

[1] Institute of Engineering Thermophysics of National Academy of Sciences, Zhelyabova 2a street, 03067, P.O. box 66, Kyiv-67, Ukraine

Abstract. The article contains the first considerations of the problematic of Ukrainian grid balancing issues raised by a rapid increase of RES share in total electricity supply. The provision of balancing electricity with accent on biomass combined heat and power plants (CHP) usage is considered. Three technical concepts are proposed for engaging of existing and planned biomass CHP into balancing operation primary operating in baseload regimes, namely – greenfield biomass thermal power plant (TPP) and CHP working primary in baseload regimes and provide balancing electricity when needed (with and without steam accumulation). It is shown that there are no principle technical limitations for biomass CHP/TPP usage for grid balancing. The levelized cost of electricity (LCOE) of balancing electricity for proposed concepts are calculated and compared with the reference technology proposed by the national grid operator (gas-piston engines and/or gas turbine). According to the calculations performed the LCOE (EUR/MWh) of balancing electricity could be 77-88 EUR/MWh for biomass CHP primary operating in baseload and 216 EUR/MWh for greenfield biomass TPP against 206 EUR/MWh for gas-piston/gas-turbine for applied assumptions, prices and tariffs.

1 Introduction

Electricity generation by renewable energy sources (RES) in Ukraine during 2014-2018 demonstrated dramatic increase (see **Fig. 1**). According to [1] the essential part of grid electricity as for 1st quarter 2018 is supplied by solar PV (55%) and wind power (30%). Total installed capacity of RES-to-grid objects for the 1st quarter of 2018 has already achieved 1.7 GWel with the increment of almost 700 MWel for the last 3 years [1, 2]. The results of specific sectoral RES-to-power grid modelling for Ukraine [3, 4, 5] demonstrates that such tendency will be continued and the structure of RES-to-power grid supply with dominant role of solar PV and wind power will remain the same till at least 2035.

From the other hand, the national grid operator (dispatcher) – National power company "Ukrenergo" in their own comprehensive research [6] stipulates that "…present power grid balancing facilities (mainly large-scale coal-fired TPPs and at some extent large scale hydropower plants (HPPs) and hydro-accumulation power plants (HAPPs)) allows addition of installed capacity of non-stable RES-to-grid generation on the level of no more than 3 GW_{el}".

* Corresponding author: epik@biomass.kiev.ua

In other words, additional 0.7 GW_{el} (2.3 GW_{el} is already in the grid) of additional unstable power sources may be considered as critical of Ukrainian grid balance. Bearing in mind also the observed tendencies and mentioned results of dependent modelling forecasts, this figure may be achieved between 2020 and 2021. According to NPC "Ukrenergo", the excess of the 3 GW_{el} threshold will result in "...the non-regime frequency fluctuations and consequent rapid growth of the risk of grid failures and disbalance of the whole integrated power system" [6]. The only solving of this potential problem currently proposed by NPC "Ukrenergo" is to build up new gas-fired power facilities (based on gas-piston and gas turbine technology) which are technically most suitable for balancing operation due fast response time on the request of grid balancing operator.

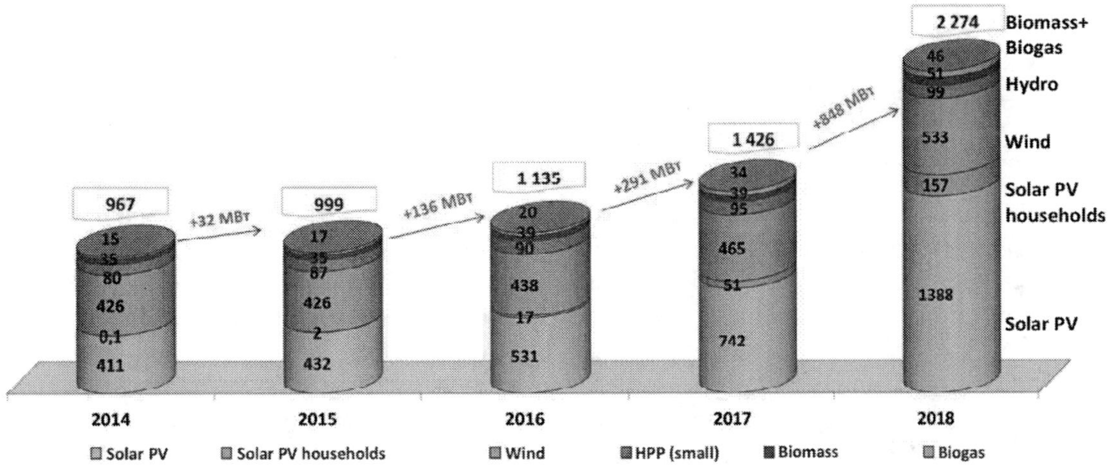

Fig. 1. Evolution of installed capacities from all RES types in Ukraine, 2014-2018, MW_{el}

However, for the past 20 years there were no case and no publicly announced plans by any active investor or Ukrainian government for building up new gas-to-power units in Ukraine. Moreover, existing gas-fired CHP operated in big cities mainly for heating purposes can't compete with other sources on electricity market, which is confirmed by the recently adopted special Resolution of Parliament # 324 dated 18 April 2018 [7] aimed on direct financial support of gas-fired CHP to let them survive on the new electricity market otherwise there is a risk of switch from cogeneration operation to the heat-only operation.

The alternative approach for grid balancing could imply use of biomass/biogas CHP/TPP. These facilities are actively implemented in Ukraine since 2015 when key problematic issues has been finally regulated in the newly approved Law on "green" tariff # 514-VIII [8] along with solar PV and wind power-to-grid generation facilities. As for the 31 December 2018 more than 46 MW_{el} of solid biomass power-to-grid capacities have been put into operation (see **Table 1**).

Table 1. The list of implemented solid biomass CHPs/TPPs in Ukraine

#	Name of CHP, company-owner	Fuel used	Sector	Capacity		Comis-sioned since
				MW_{el}	MW_{th}	
1	"Kirovogradoliya", PJSC	sunflower husk	Industry	1,7	33	2006
2	"Smilaenergopromtrans", LLC, Smliyanska CHP	wood chips	District heating	8,5	40	2010
3	"Kombinat Kargill", LLC	sunflower husk	Industry	2	20	2012
4	"Biogazenergo", LLC	wood chips	Power-only	6 (18)*	0	2013

5	Agroholding "Yevgroyl", LLC Mykolaiv CHP	sunflower husk	Industry	5	30	2013
6	"Clear Energy", LLC Koryikivka CHP	wood chips	Industry	3,5	10	2016
7	Rakhiv CHP	wood chips	District heating	6.0	10	10/2017
8	"SINGA ENERGIES", LLC	sunflower husk	Industry	5,12	25	04/2018
9	"Kriger" LLC, Kamyanetsk ORC CHP	wood chips, pellets, straw	District heating	1,2	40	07/2018
10	"Yougenergo", LLC Pereysalav CHP	wood chips	District heating	5	12	09/2018
11	Mebel-service, LLC, Radekhiv CHP (furniture)	wood chips/dust	Industry	2,2	10	09/2018
	TOTAL CAPACITY			46 (58)	230	

* - only first turn in operation

Additionally there are dozens of publicly available biomass CHP projects announced by investors, some of them already on commissioning stage. According to National Commission of Electricity Regulation of Ukraine (NERC), more than 20 biomass CHP/TPP projects are currently on different implementation stages and are to be commissioned in 2019-2021 [9] increasing total installed capacity to 200-250 MW_{el}. These facilities are focused on baseload electricity-to-grid supply by "green" tariff, some of them working in cogeneration regime.

The latest state-level programmatic documents like Energy Strategy-2035 (2017) [10], National Renewable Energy Action Plan (NREAP-2020) (2014) [11], Plan of actions for Energy Strategy-2035 (2018) [12] also consider substantial growth of biomass/biogas CHP/TPP input in total energy balance of Ukraine. For example, the Energy Strategy till 2035 foresees increasing of primary energy supply from biomass to 11 Mtoe in 2035. This means also the increasing of installed capacity of solid biomass CHPs/TPPs to at least 1.7 GW_{el} in 2035 [13]. NREAP-2020 contains the figure of 660 MW_{el} of solid biomass and biogas power capacities in 2020. Some portion of these capacities may be engaged not only in baseload operation but also in balancing operation.

At the same time, in Ukraine, despite the number of comprehensive public debates and available researches on grid balancing with large shares of unstable power-to-grid sources, the engagement of biomass/biogas technologies for these purposes has not yet been considered by somebody. So, there is a mature necessity to start considering these issues. The purpose of this article is to at least raise the problematics, drafting first assumptions for viable concepts of biomass-to-grid facilities usage for grid balancing in Ukraine.

2 Concept of technical realization and key assumptions

Biomass CHPs/TPPs are commonly considered as baseload capacities with low maneuvering possibilities, characterized by slow response rate on the request of grid operator for changing of power capacity. The typical biomass CHP in Ukraine and the EU-28 is operating in industrial or district heating sector with main purpose of baseload heat supply. The installed power capacity of typical biomass CHP/TPP is most commonly in range of 3-10 MW_{el}, often defined by the respective heat loads. There are only a few examples of biomass CHPs in the EU-28 with capacities more than 20 MW_{el} working in power-only generation regime [14].

On practice in the EU-28 and other countries, biomass CHP and TPP are actively used for grid balancing. The typical concepts of their engagement are different, and could be summarized as follows [14, 15, 16]:

1) Large-scale (>200 MW_{el}) existing fossil fuel (usually coal) TPP units with biomass co-firing (share of biomass 5-40%, most common 10%);
2) Large-scale (>30 MW_{el}) biomass-only TPPs;
3) Medium-scale (3-10 MWel) biomass-only TPPs/CHPs (industrial and district heating);
4) IGCC installations with biomass co-firing;
5) Biogas CHPs based on gas-piston and gas-turbine generation technologies
6) Usage of biomethane;
7) In perspective, medium-scale (5-15 MW_{el}) power plants on II-nd III-rd generation of liquid biofuels (for example bio-oil);

Biomass-to-power facilities are implemented in the vast majority of cases not for the purpose of balancing, but as baseload facilities aimed on gaining profits through electricity, heat and (in some cases) cold supply. Their operation in balancing regimes may be considered as bonus activities only, opposite to the specially designed maneuvering natural gas balancing capacities. It is extremely important to understand that main purpose of existing or planned biomass CHP/TPP in Ukraine and abroad is baseload operation, not balancing. That is why it is necessary to consider balancing operation regimes as supplementary to the permanent, primary activity of biomass facilities.

Another issue is the ratio between balancing capacities and capacities of unstable grid-connected facilities, "balancing MWel/unstable MWel". There are different views on this factor, it can vary in the range of 0.1…2.0 [14, 15, 16]. According to the NPC "Ukrenergo" each grid-connected Megawatt of unstable power sources shall be supported by same equivalent grid-connected Megawatt of balancing generation [6].

For the purpose of current work, the biomass CHP concept considered is based on steam Ranking cycle with condensing turbine. It is the most commercially sound technology for power generation from solid biomass in Ukraine proved by implemented (10 out of 11 are based on this technology) and planned projects as well as practical experience of the EU-28 and other countries. Power output capacity is taken as 10 MWel (on generator), respective heat capacity of the boiler is 40 MWel, 30 MWth is coming to condenser. These parameters are chosen as they are typical for medium-scale biomass CHPs in Ukraine. The reference case is gas-fired piston engines or gas turbine with the same power capacity according to views of NPC "Ukrenergo" which technically have fastest response time on power capacity increasing/decreasing request among other technologies and thus most suitable for the balancing operation regimes (progressing from cold state to the nominal capacity during 30 sec-5 min [6, 17, 18, 19]).

As the Concept #1 let's consider the greenfield biomass TPP based on the steam Ranking cycle with condensing turbine (see the simplified principal scheme at **Fig. 2**).

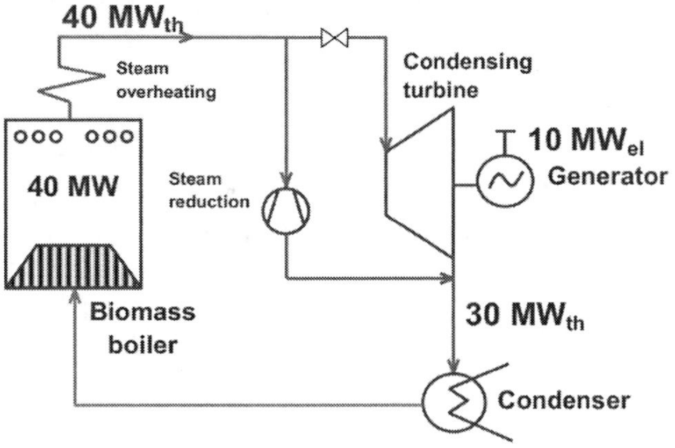

Fig. 2. Principal scheme of greenfield biomass TPP with condensing steam Ranking cycle used only for grid balancing (Concept # 1).

Such installation could technically operate in both baseload and balancing regimes (7,600 and 1,000 hours respectively). 1,000 hours are chosen according to daily operation on peak load for maximum 2.5-3 hours and corresponds to the daily peak load requested by the grid operator according to average summer/winter daily power graph of the Ukrainian power grid [20]. Further concepts will be based on Concept #1 supplemented with additional technical modifications.

3 Methodology and input data

With purpose to compare different technologies the common Levelized Cost of Electricity (LCOE) factor is used [21].

$$LCOE = \frac{\sum_{i=1}^{n} \frac{CAPEX_i + OPEX_i}{(1+r)^i}}{\sum_{i=1}^{n} \frac{E_i}{(1+r)^i}}, \quad (1)$$

where:
$CAPEX_i$ – Total capital cost of technology in the year *i*;
$OPEX_i$ – Total operational cost of technology in the year *i*;
E_i – useful electricity output in the year *i*;
r – discount rate. For the purpose of the calculation equals zero;
n – project lifetime, years. For the purpose of the calculation equals 25 years.

$$OPEX_i = F_i + M_i + W_i + C_i\,, \quad (2)$$

where:
F_i – cost of fuel in the year *i;*
M_i – cost of maintenance and materials in the year *i*;
W_i – cost of workforce in the year *i*;
C_i – cost of contingency in the year *i*.
The input data used for the calculation is presented in **Table 2**:

Table 2. Input data for the calculation

Parameter	Value	Unit
Power capacity of CHP/TPP	10	MW_{el}
Operational hours for balancing regime	1,000	hours/year
Operational hours for baseload regime	7,600	hours/year
Specific cost of biomass CHP/TPP [22]	3,000	EUR/kW_{el} installed
Specific cost of gas-fired piston engine/gas turbine [23]	1,000	EUR/kW_{el} installed
NCV of biomass	12	MJ/kg
NCV of natural gas	33	MJ/m^3
Cost of biomass w/o VAT	30	EUR/t
Cost of natural gas w/o VAT [24]	300	$EUR/1000m^3$
Average weighted operational electric efficiency of biomass CHP/TPP	25%	%
Average weighted operational electric efficiency of gas-fired piston engine/gas turbine CHP/TPP	30%	%

4 Main results and discussion

At first iteration it is necessary to calculate LCOE for the Concept # 1 – greenfield biomass TPP specifically designed and constructed exclusively for grid balancing.

According to the applied assumptions and performed calculations, the LCOE of balancing electricity for the Concept # 1 is 216.4 EUR/MWh, which is much higher than current wholesale electricity tariff in Ukraine (70 EUR/MWh) and "green" tariff (123.9 EUR/MWh). The respective LCOE for baseload generation of the Concept #1 is 63.4 EUR/MWh. For the reference case, LCOE for balancing electricity is 206.4 EUR/MWh (bearing in mind 300 EUR/1000m^3 natural gas price which is quite conservative and may be much higher depending on the region of Ukraine) and 139.4 EUR/MWh of baseload generation (see **Table 3**).

Table 3. LCOE of balancing electricity and baseload electricity for Concept # 1 and for the reference case – gas piston engine/gas turbine TPP

Technology/ Operation regime	CAPEX, 10^3*EUR (lifetime)	MWh_{el}/ year	Fuel, t or 10^3*m^3	Fuel cost annual, 10^3*EUR	Maintenance, materials, workforce, other contingency, EUR	LCOE, EUR/ MWh
Biomass TPP baseload	30,000	76,000	91,200	2,736	886,000	63.4
Biomass TPP balancing	30,000	10,000	14,500	435	529,425	216.4
Gas fired TPP baseload	10,000	76,000	27,636	8,290	1,903,891	139.4
Gas fired TPP balancing	10,000	10,000	4,136	1,240	423,390	206.4

Obviously, such high indicators of LCOE for balancing electricity means that Concept # 1 is not viable in balancing operation regime for current market conditions. Thus, there is a necessity to search a ways for decreasing of LCOE for biomass TPPs operating in balancing regimes.

One of the possible solutions is to partly engage in balancing operation regimes existing biomass CHPs or those designed for purposes other than balancing – Concept #2 (see **Fig. 3**). In this case, CHP is partly engaged into balancing operation by enhancing of electric installed capacity of condensing turbine to the level required by both balancing and baseload operation regimes. For the strict comparison with Concept #1, let's consider the same input heat and output power capacity levels but divide them between baseload operation and balancing operation. Also, the very important feature is that this is CHP working in cogeneration regime providing useful heat (in our example, for industrial purposes (see **Fig. 3**)). The principal scheme for this updated Concept #2 consists of 2 boilers 20 MW_{th} each and condensing steam turbine with extraction (to provide steam technological needs) with the same installed power capacity as for Concept #1 (10 MW_{el}). However, in Concept #2 the baseload capacity is 6 MW_{el}. Another part – 4 MW_{el} – is "reserved" for balancing operation regime only. The appropriate heat load for technological needs is adjusted in such a way that 6 MW_{el} baseload capacity on nominal is enough to provide 100% of these needs. 4 MW_{el} of balancing capacity is engaged only when needed (3 hours per day on nominal or 1,000 hours per year). 2 boilers of 20 MW_{th} each are specially designed to make regulation of the whole cycle more effective and to maximize the capacity factor (60-70% in baseload for each boiler).

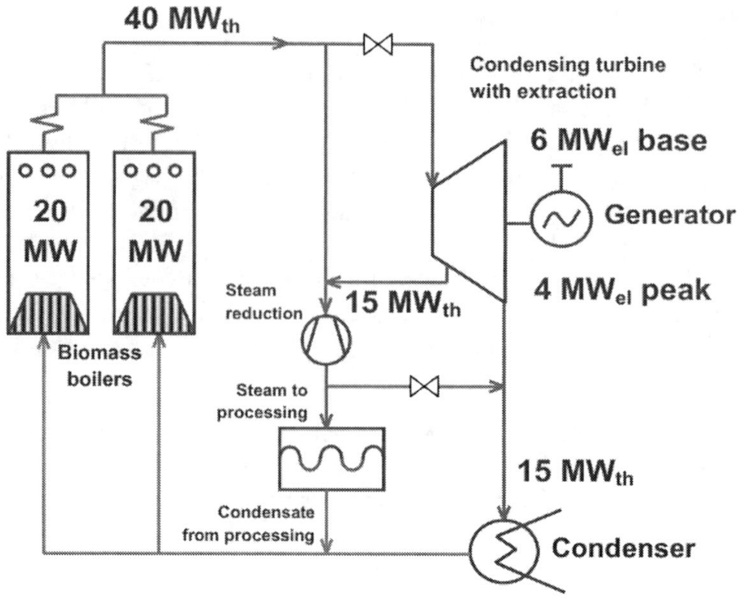

Fig. 3. Principal scheme of biomass CHP primary aimed for baseload operation and partly engaged for grid balancing (Concept # 2)

The calculation shows that for Concept #2, LCOE for baseload generation is on the level of 58.8 EUR/MWh. This figure is obtained for the case when all CAPEX and workforce associated with both baseload and balancing operation regimes are attributed to baseload operational regime only. In this case, according to the calculation, LCOE for biomass CHP baseload will be 58.8 EUR/MWh and the LCOE for balancing electricity from same biomass CHP will be 86.4 EUR/MWh.

Table 4. LCOE of balancing electricity for Concept #2

Operation regime	CAPEX, 10^3*EUR (lifetime)	$\frac{MWh_{el}}{MWh_{th}}$/ year	Fuel, t	Fuel cost annual, 10^3*EUR	Maintenance, materials, workforce, other contingency, EUR	LCOE, EUR/ MWh
Biomass CHP baseload	18,000	45,600 / 114,000	107,335	3,220	842,310	58.8
Biomass CHP balancing	12,000	4,000	6,161	184.8	160,646	86.4

For both Concepts #1 and # 2, the regime of dynamic increasing/decreasing of power capacity include the lag between response time required by grid balancing operator (delivery of necessary power in 30 seconds-15 minutes, depending on the reserve class of the installation, most usual is 5-10 minutes [6, 17]) and actual response time depending to maneuver possibilities of the solid biofuel installation to run from zero capacity to nominal (30 minutes – 2 hours, usually 1 hour [25, 26]).

To make the cycle more efficient, the Concept #3 has been developed (**Fig. 4**). It is similar to Concept #2, the only new feature is addition of steam storage tank aimed on accumulation of steam (heat) during baseload and during acceleration/ deceleration of biomass boilers. For current scheme, the installed capacity of such a tank is 10 MW_{th} and the volume is calculated according to [27] as 170 m^3.

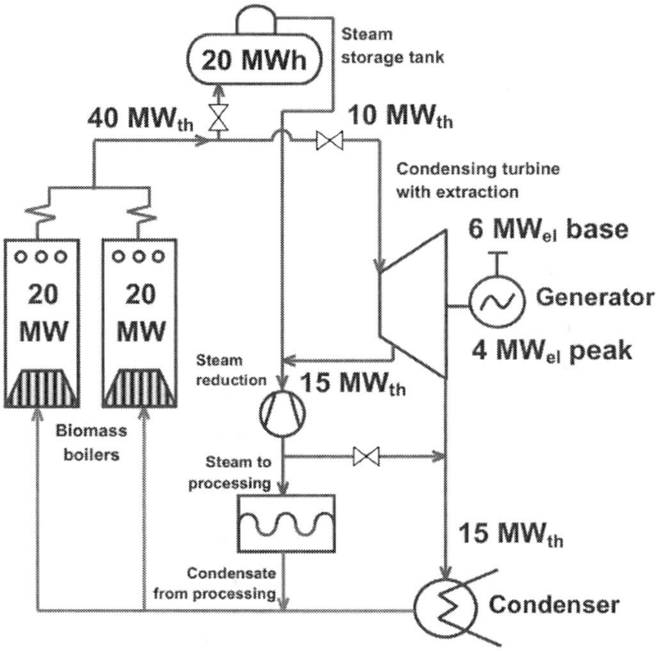

Fig. 4. Principal scheme of biomass CHP with steam accumulation primary aimed for baseload operation and partly engaged for grid balancing (Concept #3)

The calculation of LCOE for the Concept #3 shows that installation of steam storage tank may decrease it to 77.9 EUR/MWh with the conservatively estimated cost of such equipment (including all auxiliary, integration to the cycle and other contingency cost) of 1.2 million EUR.

Table 5. LCOE of "balancing electricity" for Concept # 3

Operation regime	CAPEX, 10^3·EUR (lifetime)	MWh_{el} MWh_{th}/ year	Fuel, t	Fuel cost annual, EUR	Maintenance, materials, workforce, other contingency, EUR	LCOE, EUR/ MWh
Biomass CHP balancing	13,200	4,000	4,800	144,000	167,520	77.9

With such indicators, electricity from biomass CHP operating in both baseload and balancing regimes could compete with other technologies on the balancing market (and reserve market), as the LCOE of balancing electricity is slightly higher than current wholesale electricity tariff in Ukrainian grid.

According to **Table 6**, which summarizes all analyzed scenarios, the lowest LCOE of balancing electricity is for the option "Biomass CHP with steam accumulation" and the highest LCOE is for both – biomass greenfield TPP or gas-fired TPP.

Table 6. LCOE for all concepts considered and gas-fired TPP as reference case

Technology/ Operation regime	LCOE, EUR/MWh(el)
Biomass TPP baseload	63.4
Biomass TPP balancing	216.4
Gas fired TPP baseload	139.4
Gas fired TPP balancing	206.4

Biomass CHP baseload	58.8
Biomass CHP balancing without steam accumulation	86.4
Biomass CHP balancing with steam accumulation	77.9

5 Conclusions

1) The gas-fired piston-engine based balancing facilities could be used for balancing purposes but are not the only alternative for coverage of non-balance peaks caused by fast RES share increasing in power system of Ukraine;
2) Usage of greenfield biomass TPP for grid balancing based on steam Ranking cycle (primary aimed for grid balancing only) have the LCOE of "balancing electricity" of 216 EUR/MWh$_{el}$, which is equal or slightly higher than for gas-fired balancing facilities (206 EUR/MWh$_{el}$) in current fuel prices;
3) Usage of biomass CHP based on steam Ranking cycle primary aimed for baseload operation (including existing ones) and modified for grid balancing have much lower LCOE of "balancing electricity": 78 and 86 EUR/MWh$_{el}$ with and without steam accumulation respectively;
4) For open electricity market starting from 1 July 2019 in Ukraine, there are no technical and economic barriers for the baseload biomass CHPs to partly operate as balancing facilities equally competing with other balancing electricity sources.

References

1. State Agency of Energy Efficiency and Energy Saving, Official issue, 2018. http://saee.gov.ua/sites/default/files/PR_EE_RE_4_years_30_08_2018.pdf
2. Ukrainian Association of Renewable Energy, Enhancing of RES production in Ukraine. Official newsletter, 2018.
3. O. Diachuk, M. Shepelev. R. Podolets, Yu. Oharenko at all., Transition of Ukrainian Energy Sector to 100% of RES till 2050. Issued by: Heinrich Boll Fund in Ukraine, Institute of Economics and Forecasting of NASU, 2017.
4. O. Tryboi, A. Epik, G. Boye Olsen et all., Transition of Ukrainian energy sector towards 100% RES in 2050. International project ASET, NGO "REA", INFORSE-Europe, 2017.
5. M. Child, D. Bogdanov et. all. Transition towards a 100% Renewable Energy System by 2050 for Ukraine. Issued by Lappeenranta University of Technology, 2016. http://www.neocarbonenergy.fi/wp-content/uploads/2016/02/3_Child.pdf
6. Report of NPC "Ukrenergo" On Estimation of Necessity (Adequacy) of Generation Capacities, 2017. https://ua.energy/wp-content/uploads/2017/10/Zvit-z-otsinky-vidpovidnosti-dostatnosti-generuyuchyh-potuzhnostej.pdf
7. Resolution of CMU # 324 "On Approval of the Temporary Support Provision for the Producers Performing Combined Heat and Power Production on CHP" as of 18 October 2018 (approved) http://zakon2.rada.gov.ua/laws/show/324-2018-%D0%BF
8. Law of Ukraine on Amendments to Regulations for Provision of Competitive Conditions of Electricity Generation from Alternative Energy Sources, 2015. http://zakon0.rada.gov.ua/laws/show/514-19
9. Registry of Power Generation Objects on Alternative Energy Sources by National Commission on Electricity Regulation of Ukraine, as for 1 October 2018:

http://www.nerc.gov.ua/data/filearch/litsenziini_reestry/REESTR_ALT.pdf
10. New Energy Strategy till 2035 "Efficiency, Security, Competiveness" (in effect, 18 August 2017): http://mpe.kmu.gov.ua/minugol/control/uk/publish/article?art_id=245230270&cat_id=35109
11. National Renewable Energy Action Plan of Ukraine till 2020 (in effect, 2014) http://saee.gov.ua/documents/NpdVE_eng.pdf
12. Plan of Measures for Energy Strategy of Ukraine till 2035 (in effect, 6 June 2018) https://www.kmu.gov.ua/ua/npas/pro-zatverdzhennya-planu-zahodiv-z-realizaciyi-etapu-reformuvannya-energetichnogo-sektoru-do-2020-roku-energetichnoyi-strategiyi-ukrayini-na-period-do-2035-roku-bezpeka-energoefektivnist-konkurentospromozhnist
13. A. Epik V. Zubenko, Forecast of heating sector transition to biomass-based generation till 2050. Proceedings of International Conference "World Sustainable Energy Days-2018", 28 February – 2 March 2013, Linz-Wels, Austria, 2018.
14. A. Arasto, D. Chiaramonti, J. Kiviluoma, et all., Bioenergy's role in balancing the electricity grid and providing storage options – an EU perspective. Published by IEA Bioenergy: Task 41, **P 6** (2017).
15. Renewable Energy Integration in Power Grids. Published by ETSAP-IRENA. 2015. http://www.irena.org/publications/2015/Oct/Renewable-energy-integration-in-power-grids
16. Integrating Renewable Electricity on the Grid. A Report by the APS Panel on Public Affairs. American Physical Society. https://www.aps.org/policy/reports/popa-reports/upload/integratingelec.pdf
17. Technical Requirements of response time of balancing capacities by Fingrid https://www.fingrid.fi/en/electricity-market/reserves_and_balancing/frequency-containment-reserves/
18. Integration of solar and wind power capacities into the Ukrainian power grid. Presentation of NPC "Ukrenergo", 2018. https://www.slideshare.net/Ukrenergo/ss-92759300
19. Problematic issues of power-to-grid production by RES in Ukraine. Presentation of NPC "Ukrenergo", 2018. https://www.slideshare.net/Ukrenergo/ss-90511879
20. Daily graphs of power grid operation in Ukraine https://ua.energy/diyalnist/dyspetcherska-informatsiya/dobovyj-grafik-vyrobnytstva-spozhyvannya-e-e/
21. LCOE methodology by EIA https://www.eia.gov/outlooks/aeo/pdf/electricity_generation.pdf
22. Renewable power generation costs in 2017. Published by IRENA. ISBN 978-92-9260-040-2, 2018.
23. Energy technology perspectives 2017. Pathways to a clean energy system. International Energy Agency, ISBN: 978-92-64-27597-3, 2017.
24. The tariff for end customers for Kyivgas as for 1 October 2018 https://energy.kyivgaz.ua/ofitsijna-informatsiya/tsini-ta-tarifi-na-gaz.html
25. V. Kovetsky M. Kovetskaya. Estimation of maneuver possibilities of power generation technologies for provision of proper electricity quality in the grid. System researches and complex problems of energy, **16** (2007).
26. Maneuver possibilities of different technologies http://energetika.in.ua/ru/books/book-3/part-2/section-2/2-8
27. M. Gurevych, P. Fedorov, Heat and power facilities of sugar industry. Published by State Publishing House of Technical Literature of USSR, 1962.

Biomass consumption in residential sector of Ukraine in 2007–2016

Viacheslav Antonenko[1,*] and *Sofiia* Levinska[2]

1 Researcher, PhD student, Institute of Engineering Thermophysics of NASU, 03058 2a Marii Kapnist str., Ukraine
2 Lead engineer, Master's degree, Scientific engineering center "Biomass" , 03058 2a Marii Kapnist str., Ukraine

Abstract. Residential sector plays a decisive role in the bioenergy sector growth in Ukraine. Larger half of biomass used by households comes from so-called self-production, which is problematic for the statistical monitoring. State Statistics do not publish detailed fuel mix reports on regional level. In this article we are providing a detailed data on fuel mix used by the households at regional level during 2007–2016 and determine the biofuel self-production amount. The facilities that are direct emission sources and are the final fuel consumers, including individual heating/hot water boilers/stoves out of district heating system and individual cookers are considered in detail.

1. Introduction

According to the International Energy Agency, in preventing the global temperature increasing on the Earth more than to 2°C, concerning to the preindustrial period, the most important role in reducing CO_2 emissions in the period up to the 2050 will be playing energy efficiency (40%) and renewable energy (30%). Determination of the biofuel self-production amount in households at regional level concerns both and that's why is actual nowadays.

According to the Energy Strategy of Ukraine until 2035 [1], Ukraine is going to reduce fossil fuels consumption and increase the share of renewable energy sources in the total primary energy production from 4,4% (in 2017) to 25% (in 2035). In recent years, biomass sector has accounted for about 80% of renewable energy in the country and is based mainly on utilization of woody biomass in heat generation: wood logs and wood pellets mostly for population and district heating, wood chips and residues mostly for industrial purposes, public heating and power production [2]. However, the Ukrainian corn stover has relatively high ash melting temperature defusing requirements for the boiler additional systems and steel quality. As a result, the Ukrainian corn stover potentially may be used within wood-fired boilers [3,4].

2. Objectives of the project

In recent years, a biomass consumption in Ukraine shows a rapid growth. Analysis of similar data for the period of 2010–2016 shows that the average growth rate of the bioenergy sector in Ukraine is 35% per year by the indicator of the "total supply of primary energy from biofuels and waste". For example, the total supply of primary energy from biofuels and waste was 3,046 ktoe in 2017 (against 2832 ktoe in 2016) [5]. And residential sector plays a decisive role in this growth: the final biofuel consumption of households was 1,506 ktoe in 2016 (against 1,097 ktoe in 2015). Larger half of biomass used by households comes from so-called self-production, which is problematic for the statistical monitoring. Furthermore, there are large differences between regions in the fuel mix. However, the State Statistics do not publish detailed fuel mix reports on regional level. The differences are related to mix of causes e.g. gas pipe network availability, coal and peat deposits, forest cover, population density, urbanization level, income level etc.

Therefore, the objective of the present paper is to provide a detailed data on fuel mix used by the households at regional level during 2007–2016 and to determine the biofuel self-production amount. The present study is focused on the facilities that are direct emission sources and are the final fuel consumers. That include individual heating/hot water boilers/stoves out of district heating system and individual cookers.

[1*] Corresponding author: viacheslavon@gmail.com

3. Methodology

The main data source of the present study is the State Statistics of Ukraine. The statistical office collects and process data on the gross final energy consumption. The monitoring covers 41 different fuel types and refinery products. These fuels are divided into 4 groups: coal (including peat); oil refinery products; natural gas; biofuel and waste.
The required data on specific fuel consumption in different subsectors on regional level can be found in statistical reports available by Request for Information only, namely:

- Report on Remains and Usage of Energy [6] (hereinafter, - 4-MTP) and
- Households Living Conditions Survey [7] (hereinafter, - HLCS).

Fig. 1. Data Flow Diagram

The Statistical Survey by Form 4-MTP applies to business entities of all types of economic activity, regardless of size, organizational and legal forms of management and subordination, which use and have in the balances energy materials and products of oil refining [8,9]. All indicators are based on the data of the primary documentation and accounting. The 4-MTP include data on the volumes of fuel sold to population by enterprises and organizations to satisfy their household needs. However, in reality population (residential sector) utilize more biofuels and wastes than reflected in standard 4-MTP statistical form. The substantial portion of this biofuels is coming from so-called "self-production" or "self-procurement" performed by the households themselves. The fuel self-production volumes are based on data from the HLCS.

3.1 Self-production Methodology Extensive Description

In rural settlements many people are searching for cheap and available firewood due to poverty. Many of them are involved in self-production of firewood. In order to estimate the self-production amount the State Statistics uses specific mathematical model developed on the basis of statistical sampling.

The HLCS is based on generally accepted international standards and corresponds to the current socio-demographic and economic situation in Ukraine [10].

The methodological foundations for the sample formation are given in the documents [11,12]. The HLCS is carried out on a separate network of respondents (non-institutional (private) households). The volume of the annual sample of households for the survey in 2014–2018 is 13,029 households.

3.2 Extrapolation (dissemination) on the general population

For the dissemination to the entire population of Ukraine (except temporarily occupied Autonomous Republic of Crimea and Sevastopol) the statistical weights are used. The statistical weights are taking into account the probabilities of including the households into the sample and the actual level of participation in the survey, coordination of the survey results with external data, including the data of demographic statistics (number, sex and age structure of the population, according to social statistics on the number and placement of the population, etc.).

3.3 Data Quality

In the analysis of the quality of The HLCS data, the replicate method of balanced repeat replications (BRR) is implemented, which is calculated in the standard WesVarPC program package [13].

Relative standard error (RSE) is often used as an indicator of data suitability.

- RSE ≤ 5% - then the score is considered reliable;
- 5% < RSE ≤10% - evaluation is suitable for quantitative analysis, but its reliability not high enough;
- 10% <RSE ≤ 25% - the assessment is suitable only for qualitative analysis and should be used with caution.

In general, for the regional level, the accuracy of the assessments of the indicators under consideration is satisfactory. For 13 regions out of 26 regions (which comprise 60% of population and 50% of territory), the relative standard error of the sampling does not exceed 5% (for all indicators), that is, the estimates are reliable. In other regions, the relative standard error of sampling for all indicators does not exceed 10%. In general, survey data can be used for statistical analysis at the regional level.

4. Results and discussion

4.1 Standard statistics

Household natural gas consumption shows a stable pattern, explained by the price of gas for households in real terms being almost constant from 2008 until 2013. However, between 2013 and 2016, total gas, coal and oil consumption decreased dramatically (See **Fig. 2**).

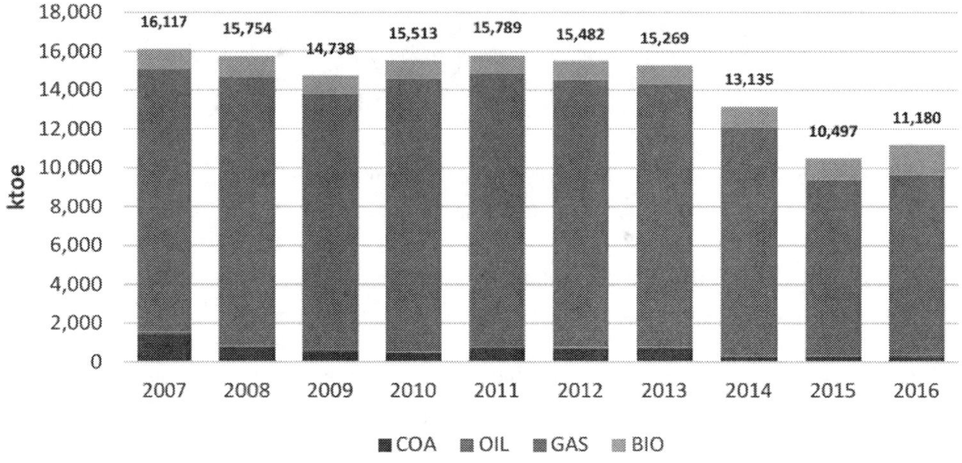

Fig. 2. Residential Fuel Mix of Ukraine 2007–2016

The key drivers of this decrease are the loss of control over 46.2 th. m^2 of the country`s territory (7%), 6 million people (13%) and 20% of GDP. At the same time, households experienced a 119% weighted average USD increase in the tariffs for natural gas while simultaneously experiencing a decline of average per capita income of almost 50% [14]. As a result, the hitherto relatively constant Ukrainian household consumption of natural gas has declined in 2016 by some 31% (from the 2013 level) (-4,228 ktoe), coal consumption shows 62% decline (-456 ktoe).

Since natural gas prices for households raised (up to 70% of the market level), a large share of gas consumers are no longer able to pay bills with new price of natural gas. Therefore, biomass-to-energy utilization became one of the economically reasonable alternative for heat supply of households. The biomass consumption pattern supports the judgement. While consumption of other fuel types show decrease in 2014–2016, the biomass consumption of households increased by 51% for mentioned period (+500 ktoe) regardless military action and statistical data gap (See **Fig. 3**). Thus, the share of biomass in total final fuel consumption of households increased from 8.1% in 2014 to 13.5% by the end of 2016 (excluding district heating and electricity).

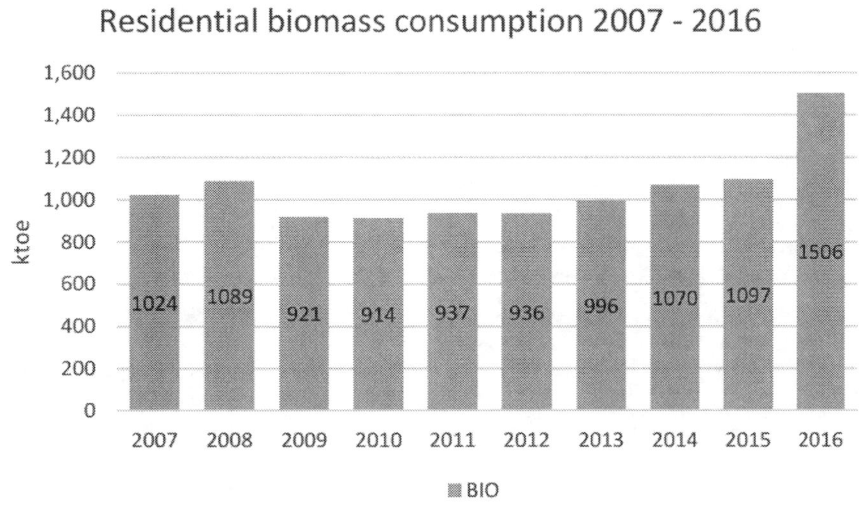

Fig. 3. Residential Biomass Consumption of Ukraine

4.2 Self-production

Along with firewood, a self-production of peat and coal takes place. However, according to data of previous observations the dominant fuel type is woody biomass (own expert judgment suggests that >95% of total self-production comes from biomass). Thus, self-production of the coal and peat may be neglected.

Since 2009, the biomass self-production demonstrates dramatic growth from 555 to 1,145 ktoe/yr corresponding to 62% - 76% of total biomass consumption of the residential sector (**Fig. 4**).

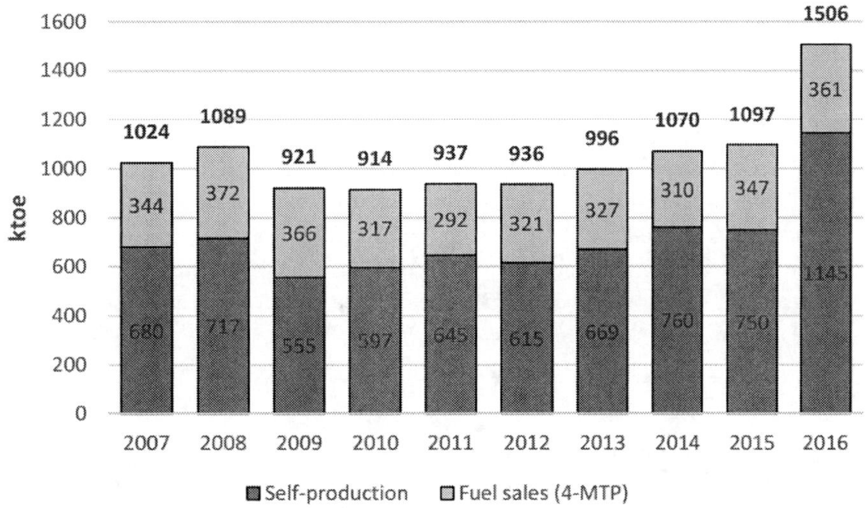

Fig. 4. Final biofuel consumption by households 2005–2016

Noteworthy, that in 2016 a considerable growth of self-production (+70%) coincided with gas consumption drop (-31%) comparatively to 2013 despite the data gap from occupied Donbass and Crimea.

Despite the biomass growth, natural gas remains the most common fuel for each region. However, biomass covers remarkable share in some forest-rich regions: Chernivtsi (32%), Chernihiv (29%), Zhytomyr (28%), Rivne (21%) etc. (**Fig. 5**). On the other hand, the coal mining regions show high coal consumption rates: Luhansk (34%), Donetsk (19%).

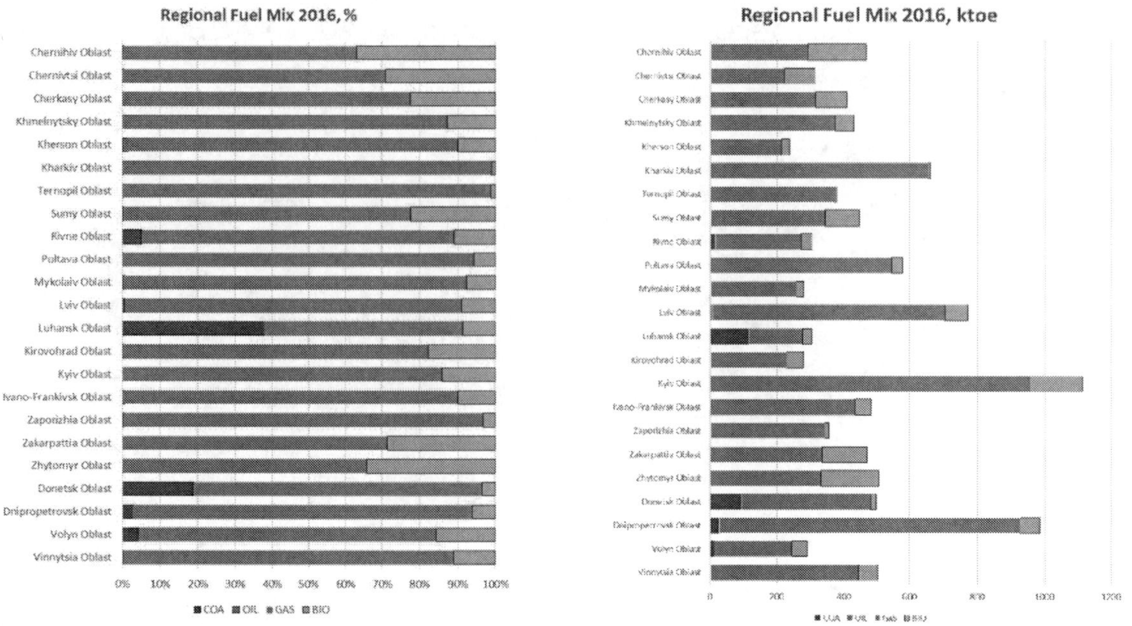

Fig. 5. Regional Fuel Mix*

* Biofuel breakdown is based only on 4-MTP "Fuel Sales" regional distribution factors. Assume the biomass self-production is around 70% (differs each year) of every region due to no details on the regional level.

At the same time, 6 regions: Zhytomyr, Kyiv, Sumy, Cherkasy, Chernivtsi, Chernihiv consume more than 52% of biofuel in residential sector of the country.

5. Conclusions

The paper gives a detailed review on fuel final use in the residential sector of Ukraine during 2007–2016 years at the regional level based on official statistical reporting. It gives a clear picture of Residential Sector final energy consumption record for the period 2007–2016 with deep breakdown by the fuel type on regional level.

The study shows that a self-produced biofuel is a main biomass source for residential sector of Ukraine. Biofuel self-production reached a menacing level providing 76% of total biofuel consumption of households. And this activity is illegal and problematic for control and monitoring. Therefore, a stricter control of biofuel procurement and trade may be needed.

References

1. International Energy Agency World Energy Outlook Special Report 2015: Energy and Climate Change
2. National Renewable Action Plan till 2020 in force since 1 October 2014 (Decree of Cabinet of Ministers # 902-r)
3. V.O. Antonenko, V.I. Zubenko, O.V. Epik, Fuel Properties of Ukrainian Corn Stover, *Industrial Heat Engineering* **40**, No3 // ISSN 0204-3602 https://doi.org/10.31472/ihe.3.2018.11 (2018)
4. V. O. Antonenko, Fuel Properties and Fertilization Cost of Ukrainian Corn Stover, *Proceedings of CYSENI 2017*, May 25-26, Kaunas, Lithuania. ISSN 1822-7554 (2017)
5. State Statistics Service of Ukraine http://www.ukrstat.gov.ua/
6. Report on Remains and Usage of Energy And Oil Recycling Products (Form 4-MTP (Annual), Table 3), State Statistics of Ukraine. Confidential.
7. Households Living Conditions Survey, State Statistics of Ukraine. Confidential
8. Instruction on filling in the form of state statistical observation № 4-mtp "Report on the residual and use of energy materials and products of oil refining", approved by the order of the State Statistics Committee of 15.09.2008 number 331 and registered with the Ministry of Justice of Ukraine 10.10.2008 for № 959/15650 .
9. Methodological Provisions From The Organization Of Public Statistics Observations On Use Fuel And Energy. Approved Order of the State Committee Statistics of Ukraine 03.10.2011 number 250.
10. Expenditure And Resources of Households Of Ukraine in 2015 Year. Methodological Explanations. State Statistics
11. Methodological Provisions on Statistics. Volume 2. State Committee of Statistics of Ukraine. - K .: ICC of the State Statistics Committee of Ukraine, 2006. - p. 456-483, p. 497-525.

12. Methodology For The Formation Of Sample Communities For Conducting Sample Surveys Of The Population (Households) In 2014-2018: Living Conditions Of Households, Economic Activity Of The Population And Agricultural Activity Of The Population In Rural Areas. State Statistics Service of Ukraine. - K .: 2013. - 204 p.
13. Methodological Position To Preparation Of Standard Reports On The Quality Of Government Sample Surveys Of The Population (Households), Approved By The Order Of The State Statistics Committee Dated 05.06.2008 No. 178 (With Amendments Approved By The Order State Statistics Service As Of 01.01.2017 № 107)
14. The Ukrainian residential gas sector: a market untapped ISBN 978-1-78467-060-3. https://www.oxfordenergy.org/wpcms/wp-content/uploads/2016/07/The-Ukrainian-residential-gas-sector-a-market-untapped-NG-109.pdf

Optimization of biogas composition in experimental studies

Radostin Dimitrov[1,*], *Zdravko* Ivanov[1], *Penka* Zlateva[2] and *Veselin* Mihaylov[1]

[1]Technical University of Varna, Department "Transport engineering and technologies", Varna, Bulgaria

[2]Technical University of Varna, Department "Thermal engineering", Varna, Bulgaria

Abstract. The article is focused on the potential and application of biogas, as an alternative fuel from Renewable Energy Sources, for use mainly in gas-generator stations. Biogas fuel is basically a mixture of methane and carbon dioxide. Its composition depends on the type of raw material used for its production. Methane concentration in biogas is between 50÷80%. To be possible engine to work with maximum efficiency with different biogas fuels, it is necessary to modify specific adjustment parameters depending on the concentration of methane in the mixture. This requires the creation of a biogas simulation system for different concentrations of the main components.

The aim is to investigate and determine the optimum and permissible biofuel blend concentrations and their impact on engine performance and fuel consumption. Biogas can be used as a fuel to produce electricity, heat or steam or as fuel for internal combustion engine, and its use will help to reduce harmful emissions into the atmosphere.

1 Introduction:

The study focuses on the use and potential of biogas as an alternative fuel from renewable energy sources for use in gas generators. Studies show that the content of particulate matters and nitrogen oxides is significantly reduced. In case of burning of methane - one of the components of biogas, CO_2 emissions in the atmosphere are lower. The aim is to study the impact of different concentrations of biogas on the performance and fuel consumption of automotive engines, to determine the optimum and permissible biofuel blend concentrations. For this purpose, it is necessary to create and build a system for mixing and maintaining the concentration of gases.

2 Results and discussion:

The use of alternative fuels is necessary due to the ever increasing standards for the content of harmful components in the exhaust gases as well as for the protection of the environment. Biogas is a gas mixture of methane and CO^2. Currently in Bulgaria biogas is

* Corresponding author: r_dimitrov@tu-varna.bg

not yet used as fuel for internal combustion engines, and this is a new perspective for us, while in some European countries (Germany, Sweden, Norway, etc.), Japan, China, USA and others, biogas for the drive of internal combustion engines has been used for a long time. The prospect of building biogas production and processing plants in the next ten years is becoming increasingly popular with automotive and internal combustion engines.
The composition of biogas depends on the type of raw material used for its production. Biogas is a combustible gas very similar in chemical composition to natural gas. The difference is mainly in the relative share of methane. In natural gas, the methane CH_4 is about 95-97%, while for biogas the relative share of methane is 50-80%, which also determines calorific value of the mixture. The biogas is colorless and odorless and besides methane it also contains carbon dioxide (CO_2), nitrogen, ammonia (NH_3), sulfur dioxide (SO_2), hydrogen sulphide (H_2S), and hydrogen, depending on the raw material used. It can be used as a fuel to produce electricity, heat or steam, or as a fuel for internal combustion engine, and also its use helps to reduce harmful emissions in the atmosphere.
Table 1 shows some of the main characteristics of biogas fuel depending from raw material used for its production.

Table 1. Parameters of biogas according to raw materials

Raw material	Methane content	CO_2 content	Other gases	Density	Lower heating value
	[%]	[%]	[%]	[kg/ m³]	[kJ/ kg]
Plant raw materials					
Corn silage	50,00	45,90	4,10	1,304	12724,00
	55,00	40,90	4,10	1,242	15690,00
Beet leaves	55,00	40,90	4,10	1,242	15690,00
Grass silage	55,00	40,90	4,10	1,242	15690,00
Substance from the processing industry					
Beer malt	55,00	40,90	4,10	1,242	14665,00
	60,00	35,90	4,10	1,180	18268,00
Waste water	50,00	36,90	13,10	1,250	13250,00
	60,00	26,90	13,10	1,126	19210,00
Forage waste	70,00	25,90	4,10	1,056	23812,00
	75,00	20,90	4,10	0,994	27105,00
Organic raw materials from household waste and slaughterhouses					
Household bio-waste	58,00	37,50	4,50	1,209	16965,00
	65,00	30,50	4,50	1,122	20889,00
Waste fat	60,00	35,45	4,55	1,184	18278,00
	72,00	23,45	4,55	1,034	25119,00
Sewage sludge from sewage treatment plants	60,00	27,50	12,50	1,126	19210,00
	72,00	15,50	12,50	0,999	25171,00
Manure					

Liquid beef manure	60,00	35,50	4,50	1,184	18278,00
Liquid pork manure	60,00	35,50	4,50	1,184	18278,00
	70,00	25,50	4,50	1,059	23806,00
Bird manure	60,00	35,50	4,50	1,184	18278,00

The table shows that biogas is essentially a mixture of methane and carbon dioxide. Other gases have a very small percentage and they do not affect the combustion process. Carbon dioxide, as it is a non-combustion, ballast gas, has a major impact on biogas fuel. Its higher content in biogas has a negative impact on the calorific value of the fuel. Biogas with the highest calorific value after processing is obtained from raw materials from manure, organic household waste and some raw materials from the processing industry, and the methane content varies depending on the conditions of production.

Biogas is a fuel from a renewable energy sources that can be used in pure form as an alternative fuel for propulsion of an internal combustion engine. It is necessary to study the possibility of the engine to work steadily with this type of alternative fuel or additional actions are needed to improve its effective performance. Because biogas has a high octane rating (RON=130), increasing its effective engine power may get through increase its compression ratio, which may reach to $\varepsilon = 13$. Using the optimal biogas fuel (depending on the raw material) for engine operation will reduce the costs incurred to get the most efficient engine performance. It is necessary to study the possibility of optimizing or determining the permissible fuel type that can be used to drive the engine and at the same time not to achieve a sharp reduction in its effective power and a significant increase in fuel consumption. The research is aimed at exploring the possibility of using not upgraded biogas to produce electricity from a gas generator station using a gas internal combustion engine. The study aims to determine the minimum acceptable concentration of methane and carbon dioxide where the engine can work steadily and provide the necessary torque needed to drive the generator. Biogas fuel upgrade technologies to reduce concentration of carbon dioxide are permissible but they are expensive, economically unprofitable and require a lot of energy.

Engines used in gas generating stations for power generation, need to be able to operate steady-state of stationary mode with a constant crankshaft rotation speed of typically 1500 min-1 or 3000 min-1, which allows the use of a lower energy fuels for engine operation. To carry out the experiment, a system is required to simulate different types of biogas fuels based on methane and carbon dioxide, and at the same time can provide gas with a constant concentration of elements in each engine operating mode.

To be possible to make experimental studies, it is necessary to create a methane and carbon dioxide mixing system. Figure 1 shows a schematic diagram of a gas mixing system, created in department "Transport engineering and technologies" in Technical University of Varna.

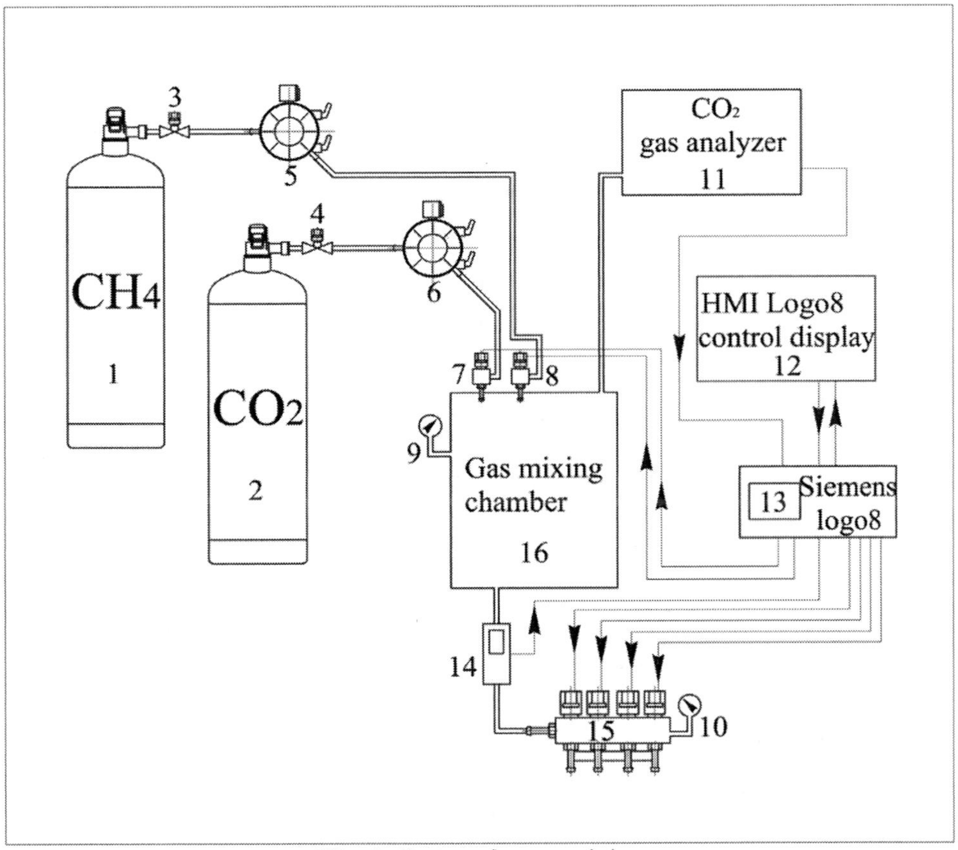

Fig. 1. System for gas mixing

The bottles for storage of methane and carbon dioxide are shown in positions 1 and 2. Positions 3 and 4 are electro-magnetic shut-off valves controlling the flow of gases. In positions 5 and 6 are pressure regulators which reduce the pressures of the two gases and feed them to the gas injectors 7 and 8 with the same pressure, so that they can be injected into the mixing chamber 16. The content of CO_2 in the chamber is measured from the gas analyzer – pos. 11. After measuring the value of the CO_2, programmable logic controller Siemens Logo8 - position 13 compares the measured value with the set concentration value and changes the injection time of the injector 7, to reach the desired concentration. A flow rate sensor - position 14 is installed at the outlet of the mixing chamber, which measures the flow rate of the gaseous mixture entering to the injector rail. Through the injectors - position 15, the gas mixture is fed to the intake manifold of the internal combustion engine with the required flow and pressure for the defined mode of operation. The programmable controller sets and manages the injection time of the injectors depending on the engine operating mode. The system has the ability to mix two different gases across a wide range and different concentrations. An equal pressure of the mixture, measured by the pressure gauge 9 and 10, which can be set by the control unit, is maintained. The mixing of the gases is precise and the accuracy of the percentage ratio depends only on the accuracy of the gas analyzer. The system has the ability to deliver fuel, satisfying engine performance requirements for different RPM and load performance.
Engine experiments are made on SI engine with volume 1,275 liters, compression ratio ε=9,75 and nominal power 46 kW at 5500 min^{-1}. The change of effective power and specific fuel consumption at external speed characteristic (maximum open throttle) at deferent concentration of methane and carbon dioxide was measured.

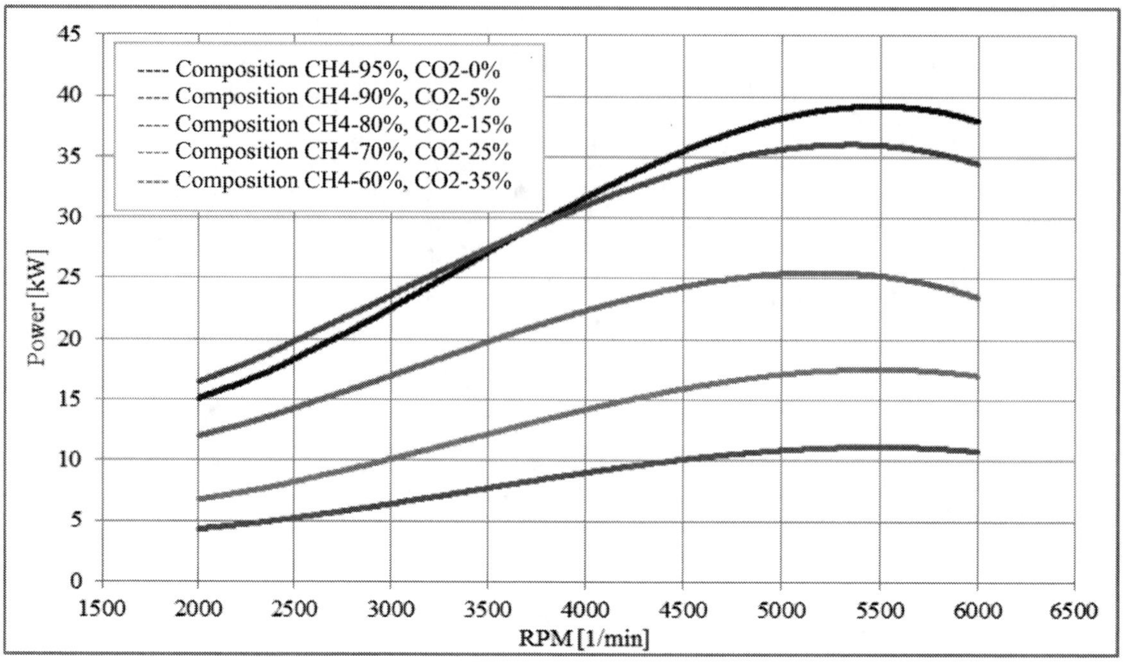

Fig. 2. Power indicators

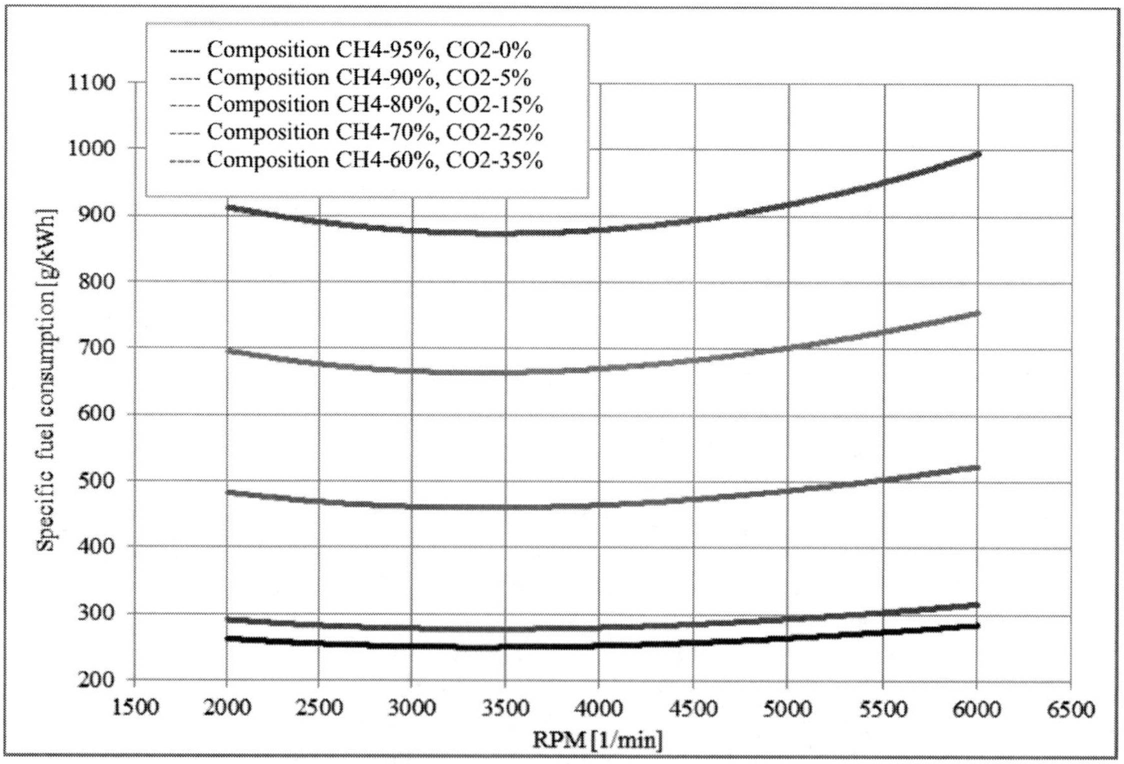

Fig. 3. Fuel consumption indicators

Variation of power and specific fuel consumption with different concentration of biogas mixtures has been investigated at wide-open-throttle operating characteristics. Figure 2 shows power variation at different concentration of biogas fuel – 95% CH_4 (upgraded biogas), 90% CH_4 and 5% CO_2, 80% CH_4 and 15% CO_2, 70% CH_4 and 25% CO_2, 60% CH_4 and 35% CO_2. Figure 3 shows specific fuel consumption variation at different concentration of biogas fuel – 95% CH_4 (upgraded biogas), 90% CH_4 and 5% CO_2, 80% CH_4 and 15% CO_2, 70% CH_4 and 25% CO_2, 60% CH_4 and 35% CO_2.

As the carbon dioxide content is greater in the composition of biogas, the calorific value of the fuel is lower. Optimum results are obtained when methane in biogas is about 75-80%. Biogas with this content can be obtained from Forage waste or from sewage sludge.
The use of biogas with lower methane concentrations is possible, but it is necessary to improve the efficiency of the engine.
The use of biogas as a fuel for internal combustion engines has great potential with important socio-economic benefits (biogas is an alternative energy source). Its use can significantly reduce emissions of exhaust gases from the engines release in the atmosphere. In addition, biogas is a major greenhouse gas and has a high methane and carbon dioxide content of about 60/40%. This makes mandatory the use of biogas for energy purposes in the places of its receipt (urban landfills, different farms and places of organic decay).

3 Conclusion:

•The system allows obtaining different concentrations of elements in the biogas composition. The mixing of the gases is precise and the accuracy of the percentage ratio depends only on the accuracy of the gas analyzer. The system has the ability to deliver fuel, satisfying engine performance requirements for different frequency and load performance.

•At reduction of concentration of methane with 15% from biogas composition, the nominal power of engine decrease with 37% from 39,7 kW to 24,9 kW. At the same ratio the minimum value of specific fuel consumption increase with around 85% from 243 g/kWh at 95% CH_4 to 448 g/kWh at 80% CH_4 and 15% CO_2.

• Optimum results are obtained when methane in biogas is about 75-80%. Biogas with this content can be obtained from Forage waste or from sewage sludge.

•The use of biogas with lower methane concentrations is possible, but it is necessary to improve the efficiency of the engine (for example: through increase of compression ratio of the engine).

References

1. Z. Ivanov, Dimitrov R., Mihaylov V., Petkov D.; System for regulating the composition of biogas used as fuel for internal combustion engines; PROCEEDINGS of University of Ruse, volume 57, book 4, 41-45; (2018)
2. R. Dimitrov, C. Ianasi, K. Bogdanov. Research of cycle by cycle variation of si engines working with methane as a fuel,; Annals of the "Constantin Brancusi" University of Targu-Jiu; Engineering series, 62-65, Romania, (2017)
3. R.Dimitrov, Zlateva P., Demirova S.. Investigation of the characteristics of biogas fuels and opportunities for their distribution in bulgaria; Acta technica corviniensis; Bulletin of Engineering Tome X, 69-72, Romania; (2017) http://acta.fih.upt.ro.
4. Z. Ivanov, Mihaylov V, Kolev A. Izsledovatelska sistema za opredelyane ekologichnite haracteristiki na avtomobilni dvigateli, XVIII Nauchno-tehnicheska konferenciya s mejdunarodno uchastie ECO-VARNA, 511-518, Bulgaria, (2012)
5. http://agrobio.elmedia.net/bg/2013-1/editorials/

Acknowledgements
The research received funding from the National Research Program "Low Carbon Energy for the Transport and Household (E+)" of the Bulgarian Ministry of Education and Science under grant agreement DO1-214/28.11.2018

Contributions to the development of a methodology for determining the flow rate

Cristian Purece[1], *Valeriu* Panaitescu[2], *Irina Alina* Chera Anghel[1]

[1]Energy Research and Modernization Institute – ICEMENERG, Bucharest, România
[2]University Politehnica of Bucharest, Power Engineering Faculty, Bucharest, România

ABSTRACT. The implementation of the requirements of the European Parliament Directive 2009/28 / EC requires efficient use of the water supply of a hydropower plants installation in order to obtain a higher amount of electricity by producing the same volume of water. In order to achieve efficient utilization of the energy of the water is necessary framing the operation of hydropower plants in the ranges of head course and the electric power so that the energy conversion is carried out in the optimum efficiency characteristic operation of the system. In order to determine as accurately the actual operating characteristics of a hydro- unit, in situ tests are required to determine the actual operating parameters of the hydro-unit. These parameters, the flow rate is the parameter that requires the most complex methodologies to determine. The paper presents a way of improvement but also a simplification of the methodology for in situ determination of the flow to small hydropower plants by using a mobile frame that has implemented a wireless data transmission system.

1 Introduction

The purpose of this paper is to obtain a modern methodology for in situ determination of the flow rate at small hydropower plants, but also in other locations such as open channels, etc. Depending on the in-situ flow rate, the actual performance characteristics of the turbine and the hydro-unit will be plotted.

The operation of the hydro-unit in the optimal operating areas established according to the actual performance characteristics will have the effect of increasing the hydro-unit efficiency.

The operation of the hydropower development and hydro-unit at any time is characterized by certain values of their characteristic parameters, Q flow rate, H head, P power, η efficiency, etc. The correlation between the values of these parameters within their entire range of variation, determined by the type, construction, and dimensions of the arrangement, is graphically defined by the so-called characteristic curves of the arrangement or of the hydro-unit (HA) equipping a hydropower plant (turbines, pumps, etc.).

Characteristic curves, areas or optimal operation lines are determined from the design process. Due to the cover reserves necessarily taken into the project and due to the specificity of the execution, the actual curves differ not only from the projected ones but also from the hydro-unit to the hydro-unit, even if they were executed by the same performer on the same object. The modern measurement theory and technique allows the individuality of each unit to be determined individually, which creates the necessary premises for knowing and using both the internal reserves resulting from the project coverage of the hydro-unit and the differences between hydro-unit of the same hydropower plant, optimal differential use of hydro-aggregates. The optimal operation of a hydro-unit, a power plant, or a cascade of power plants is equivalent to an increase in energy or the existence of some (theoretical) hydropower development that would have an additional 2% of additional electricity production.

In order to obtain a maximum of electricity production, it is necessary for a multi-group hydroelectric plant to operate at the optimum parameters.

This means that hydro-unit are operating in the maximum efficiency areas (which are usually not at the maximum power of the hydro-unit) and the hydropower works with minimum head losses (with a minimum flow) under the conditions of a power imposed on the plant. Of the parameters of a hydro-unit, the parameter whose determination poses the most difficult problems is the flow rate. Determining this parameter requires a complex measurement methodology.

The main methods of determination of the flow rate for hydropower plants are provided in the international standards for in situ measurements [1], [2], [3].

The current meter method is one of the most appreciated and precise methods of flow rate determination, and the only recommended method for determining the flow rate at a low head hydropower plant.

This paper aims to improve and simplify the methodology for determining the flow rate at a small hydropower plant by using an adjustable mobile frame.

Simplifying the methodology for determining the flow rate and energy performance of hydro-units in the hydropower plant will reduce the time required to conduct in situ determinations and lower the cost of performing energy performance tests.

2 Description of the measurement method used

The current meter method consists into place in a circular, rectangular or trapezoidal section of a number of measuring transducers (current meter, Pitot tubes or acoustic transducers) to determine the water velocity at those points, is known in the literature as "punctual speed". The actual flow profile in the measurement section can only be traced using the current meter or Pitot tubes and to a certain extent using the acoustic method, but only when multiple measurement paths are used. The flow rate in a trapezoidal or rectangular section for a *dx·dy* elemental surface can be written as [4]:

$$\mathrm{d}Q = v\,(x,y)\mathrm{d}x\mathrm{d}y \tag{1}$$

For the entire measuring surface the flow is calculated using the formula:

$$Q = \sum_i v_i \Delta x_i \Delta y_i \tag{2}$$

written in full form the flow calculation formula for the entire measuring surface is:

$$Q = \iint_S v(x,y)dxdy \tag{3}$$

The relation (7) can be expressed in the form of a simple integral in two ways:

$$Q = \int_0^x dx \int_0^y v dy \qquad \text{or} \qquad Q = \int_0^y dy \int_0^x v dx \tag{4}$$

The calculation of the integral (8) can be done with the spline or cubic spline interpolation function [5], with Lagrange interpolation polynomials or the Simpson integration method [6]. In order to trace the distribution of speeds in the border area it is taken into account that from the walls to the first moose the velocity can be expressed by the law of Kármán [7]:

$$v_x = v_a \left(\frac{x}{a}\right)^{1/n} \tag{5}$$

where v_x is the velocity in the extrapolation zone at a distance x from the section wall, - the velocity at the nearest point to the section wall, and n is a coefficient which takes into account the roughness of the wall, the flow conditions and apply near the walls and radiator, when there are no special contraindications, it can be considered $n = 7$. The extraction from the marsh on the free surface of the water is based on observing the continuity of the curve drawn. If the surface velocity is also measured with floats, the account is taken of its value.

3 Description of the measurement system used

The proposed mobile measurement system (Figure 1) will be used for the determination of the absolute flow rate and calibration of the differential pressure inlet on the spiral chamber of the hydraulic turbines in the low head hydropower plants [8] is presented below.

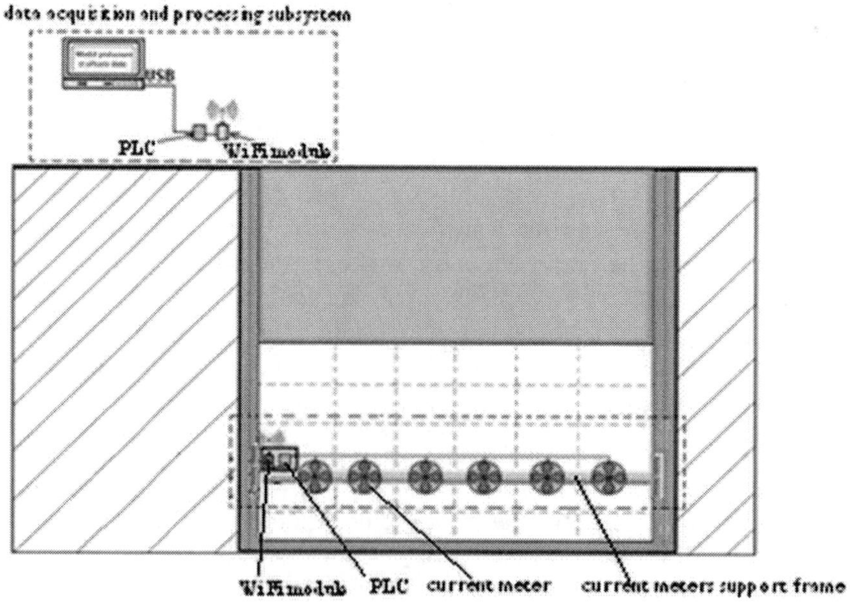

Fig. 1. Structure the mobile system of the flow rate in situ measurement

This mobile system will consist of a series of current meters, a modular mobile frame (necessary for the installation of current meters in safe conditions at points with fixed coordinates and their arrangement in the measurement section), a smart data acquisition interface represented by a PLC (automatically programmable), two wireless data transmission modules, a computing unit featuring a powerful laptop that will allow calculating and displaying in situ on the monitor of the computing unit of interest sizes.

On the monitor of the computing unit, the values of measured magnitudes measured by the current meters and the individual velocity, the calculated flow rate and average values in the measurement section, the graph of the velocity distribution in the measurement section, the information shall be stored in the the database (previous measurements, the parameters of the current meters and the turbine / hydro-units where the measurements were made, etc.).The mobile frame used to place the current meters is a modular frame (Figure 2) consisting of several segments and can be adapted for different dimensions of the flow measurement sections. Also, the modular mobile frame can be easily assembled and is easy to transport and handle, it is much easier than a classical hydrometric frame made up of a single component.

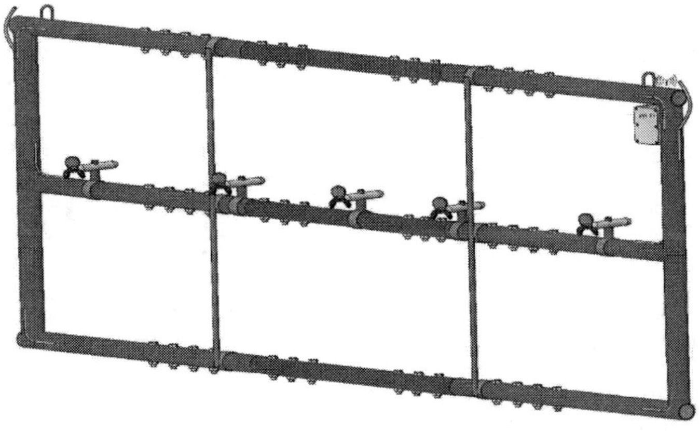

Fig. 2. 3D draw mobile frame support the current meters

Replacing the classic mode of multi-wire electric cable transmission of electrical signals purchased from current meters through the use of a wireless data transmission system is another advantage of the proposed measuring mobile system. Using wireless data transfer mode will allow you to reduce the working time required to mount the current meter on the support frame, thus reducing the downtime of the hydro unit. The location of the mobile system to measure in situ flow rate to low head hydropower is intake entrance in stop logs upstream of the hydraulic turbine.

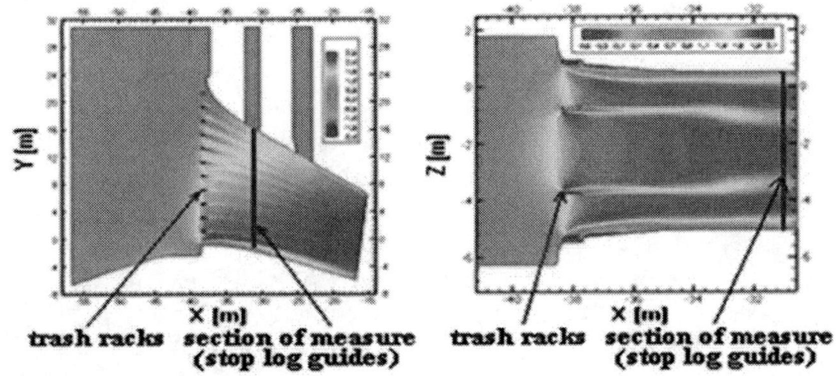

Fig. 3. The flow rate measurement section for low head hydropower plants

Figure 3 shows the location section of the flow rate measurement system for low head hydropower plants in the XY and XZ co-ordinates (9).

4 Results and discussions

In order to verify the operation of the flow rate measurement mobile system, both laboratory and in situ samples were performed.

The in situ samples were carried out at Zăneşti HPP, a low head hydropower plant located on the Bistriţa River.

Zăneşti HPP (figure 4) consists of two Kaplan turbines of the type 7.5-20 with a 20.2 m gross head, a power installed on the hydropower plant of 15.5 MW and a flow rate installed on the hydropower plant of 84 m3s-0, 5 and production of electricity of 81.5 GWh / year.

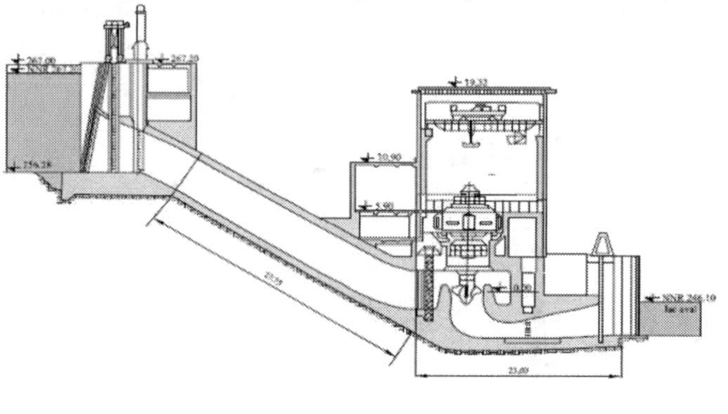

Fig. 4. Cross section through Zăneşti HPP

For the determination of the flow rate the current meter method was used, and the measuring section was the stop logs upstream of the turbine from the hydropower plants unit HA1 from Zăneşti HPP. Gross head and power on the plant were kept constant throughout the sample.

Samples were performed for a 3 MW power to the hydro-generator from the HA1 and a gross head of 19.1 m.

After the flow measurements in the stop logs upstream of the HA1, the following average velocities and flows were obtained:

- for the left stop logs an average velocity v_{m1} = 0.733 [ms-1] and an average flow rate Q_{m1} = 10.561 [m3s-1];
- for the right stop logs an average velocity v_{m2} = 0.583 [ms-1] and an average flow rate Q_{m2} = 8.391 [m3s-1];
- for HA1 an average velocity v_m = 0.658 [ms-1] and an average flow Q_m = 18.952 [m3s-1].

After performing of the measurements, a real picture of the velocity distribution in the two stop logs upstream of the HA1 turbine was obtained. 3D distribution of velocity measured with the mobil frame in the two of turbine stop logs according to the height (*H*) and the length (*L*) the measurement section is shown in Figures 5 and 6, [10].

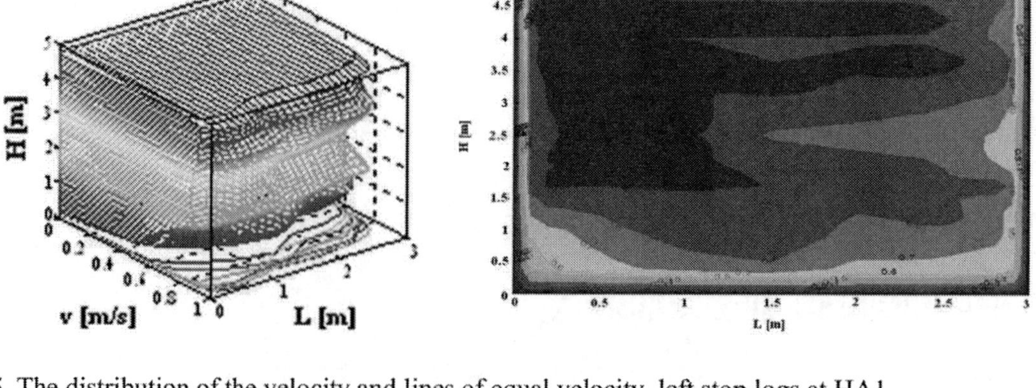

Fig. 5. The distribution of the velocity and lines of equal velocity, left stop logs at HA1

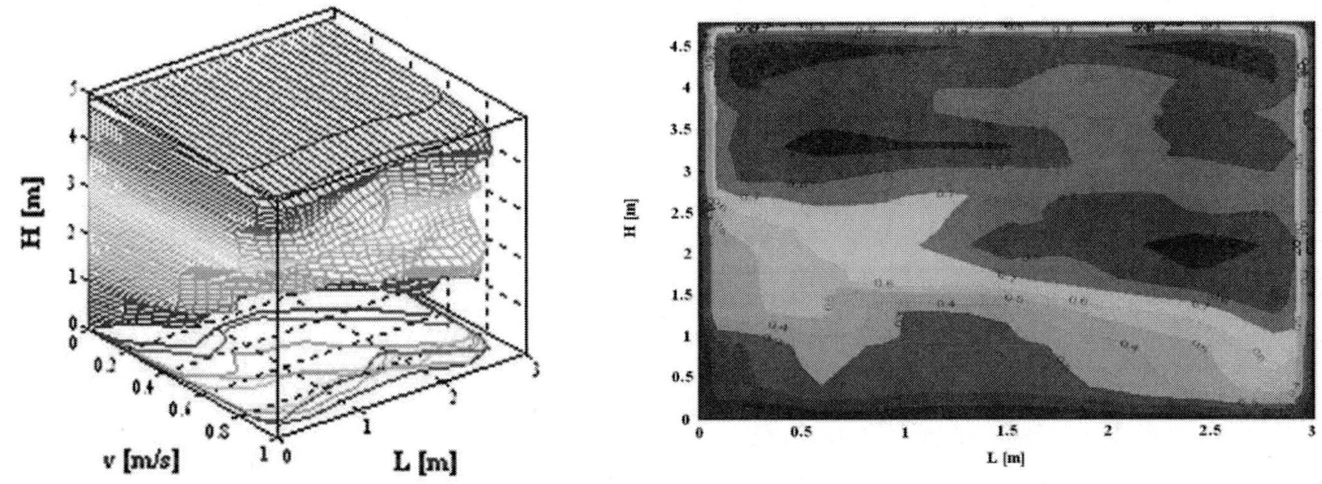

Fig. 6. The distribution of the velocity and lines of equal velocity, right stop logs at HA1

After performing the tests in situ, was calculated based on relationship (6) the turbine efficiency and were drawn the actual turbine operating characteristic $\eta_T = f(P_T)$ for HA1 (figure 7).

$$\eta_T = \frac{102\, P_T}{H_n Q} \tag{6}$$

where: P_T is the turbine power, η_T is the turbine efficiency, Q is the flow, H_n is the net head.

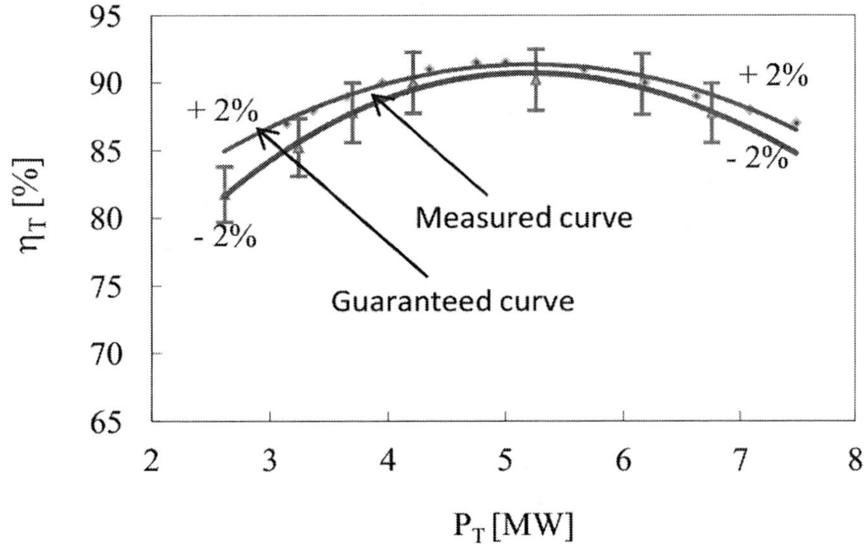

Fig. 7. The turbine efficiency characteristic according to turbine power $\eta_T = f(P_T)$ for HA1

As can be seen from Figure 7, the optimal operating range of the turbine from HA1 is between 3.75 MW and 6.70 MW, the field in which the turbine operates at a yield of at least 88%. The error of determination of the flow was 1.1% and the calculation error of the actual performance of the hydraulic turbine obtained by the current meter method was ± 1,4 %, lower than the limit of ± 2 % provided in the standard CEI 41/1991.

5 Conclusions

The aim of the paper was to simplify the methodology for in situ determination of the flow and to increase the accuracy of flow determination by using an absolute measurement method, current meter method.

The novelty of the methodology is to use in the mobile system the in situ measurement of the flow of a modular mobile frame for the location of current meter in the measurement section of a wireless data transmission system and a specialized flow calculation program. The use of the modular mobile frame instead of the fixed class frame and the Wireless system will reduce the time required to mount the current meter on the frame and reduce the number of current meters used. Also, with the help of the computing program, the flow will be calculated in situ using one or more calculation methods, which will increase the accuracy of the in-situ flow rate methodology. The modular mobile frame used in the mobile system is unique in Romania.

It has been designed to determine the flow rate for hydropower plants but can also be used in other locations (open channels, thermal power plants) and can be modified according to the measurement section. The mobile system for flow determination proposed in this paper can be used periodically for the determination of the flow rate and calibration of the pressure tape of the hydraulic turbine spirals in hydropower plants, in order to achieve a flow rate measurement permanent installation and as a benchmark for other methods flow measure, such as acoustic transit time method, acoustic scintillation method, Gibson method [11].

A general conclusion of this paper is that the simplification and improvement of the flow rate measurement methodology using the current meter is feasible. The solutions proposed in this paper are viable and perfectly applicable in the usual practice, resulting in the simplification the methodology for in situ flow determination and lower its cost price.

The research proposed in this paper can be continued by testing a mobile system for in situ flow rate determination using a mobile frame that has implemented a device capable of modifying the angle of location of the current meters depending on the flow spectrum in the measurement section, such that that the angle between the current meter axis and the tangent to the current lines of the flow spectrum to be max.5^0. The use of this device will eliminate the error introduced in the flow calculation by the angle between the axis of the current meters and the tangent to the current lines of the flow spectrum, error encountered in the flow measurements performed with the current meters placed on a mobile frame.

References

[1] * * * ASME PTC 18, American National Standard - Hydraulic turbines and pump-turbine performance test codes. New York, USA, ASME (2002) (consolidation of ASME PTC 18-1992 and ASME PTC 18.1-1978)

[2] * * * CEI 41/1991 - Field acceptance tests to determine the hydraulic performance of hydraulic turbines, storage pumps and pumps-turbines. Geneva, Switzerland, CEI 3rd edition, (1991)

[3] * * * CEI 62006/2010 - Hydraulic machines. Acceptance tests of small hydroelectric installations. Geneva, Switzerland, CEI 1st edition, (2010)

[4] A. Adamkowski, *Essential methods of SHP efficiency testing. Keynote address at the SHP Research& Development Actors Network Meeting*, The Szewalski Institute of Fluid-Flow Machinery, Gdansk, September 10-11th, (2009)

[5] Z. Krzemianowski, A. Adamkowski, *Use of Non-uniform Rational B-splines for Discharge Calculation in the Velocity Area Method.* Journal of Energy and Power Engineering **8**, pp. 911-919, (2014)

[6] * * * ISO 3354/2008 - Measurements of clean water in closed conduits. Velocity-area method using current-meters in full conduits and under regular flow conditions. ISO 3rd edition (2008)
[7] J. Steller, A. Adamkowski, A. Henke, W. Janicki, *Discharge measurement and performance tests of hydraulic units in low-head small hydropower installations*, Transactions of the Institute of Fluid-flow Machinery, No. **130**, pp. 3–29, (2015)
[8] * * * INCDE ICEMENERG, *Researches on the higher use of water resources by new solutions for measuring turbine flow at hydropower plants,* contr. MCI nr. 20 N/2016, Bucureşti (2016)
[9] G. Proulx, E. Cloutier E., *Comparasion of discharge measurement by current meter and acoustic scintillation methods at La Grande-1*, Proc. IGHEM, Lucerne Switzerland, (2004)
[10] * * * INCDE ICEMENERG, *Determination of performance parameters of the hydro units in Zăneşti HPP,* contr. Hidroelctrica nr. 18/2015, Bucureşti (2015)
[11] D. Gonzales, F. Murciaccia, *Turbine acceptance tests at Frieira HPP, Mino River, Spain with Acoustic Scintillation Flow Meters and current meters*. Proc. HydroVision, Denver USA (2013)

E3S Web of Conferences **112**, 02009 (2019) https://doi.org/10.1051/e3sconf/201911202009
TE-RE-RD 2019

Modelling and simulation of a wind energy system

Cristian-Gabriel Alionte[1] and *Liviu-Marian* Ungureanu[2,*]

[1] University Politehnica of Bucharest, Mechatronics and Precision Mechanics Department, RO-060042 Splaiul Independentei, No.313 room CH 402, Romania
[2] University Politehnica of Bucharest, Theory of Mechanisms and Robots Department, RO-060042 Splaiul Independentei, No.313 room JC111, Romania

Abstract: Nowadays, wind energy systems are the most efficient systems of all renewable energy production systems. Therefore, all sizes and types of this kind of systems are available, but there are few portable systems which can be temporarily mounted in certain isolated areas. We propose in this case study another solution as an alternative: an adaptable mechatronic system which can withstand even in very difficult and challenging weather conditions due to their active surfaces that have reconfigurable automatic structure.

1 Introduction

Wind energy systems have been known since ancient times [1], the sail boat is considered the earliest use of wind power (approx. 1000 B.C.) which has developed in sail-type windmills which was used to power the grain-grinding and water-pumping. The first vertical system was made in Persia about 500-900 A.D. but without documents it's not known the application type. Also, in 1219 A.D. another vertical-axis windmills have been documented by the Yehlu Chhu-Tshai a Chinese statesman. Another application of a wind power is the using of hundreds of sail-rotor windmills for water pumping of crops and livestock on the Crete island. In western Europe the first documented windmill is in 1270 A.D. and was a horizontal-axis configuration. In 1390, in Holland had been developed the tower mill design, which in the later design can be oriented into the wind manually, and in this way, the energy of the windmill and the power output was optimized, and the mill was protected from damage. In the beginning of the 19th century the development of the wind energy turbines has decrease because of the using of the steam engines until the 20th century when finding new energy sources, especially non-pollutant and regenerable, has become a main issue.

In principle, a wind energy generating system consists of the following components, see Fig. 1 where the most important part is the wind runner. Now, exist many shapes and dimensions, and can be with horizontally axis (HAWT) or vertically axis (VAWT). We focused on the VAWT because the weight did not introduce any force and can be achieved a perfect balance of the mechanical system.

* Corresponding author: ungureanu.liviu.marian@gmail.ro

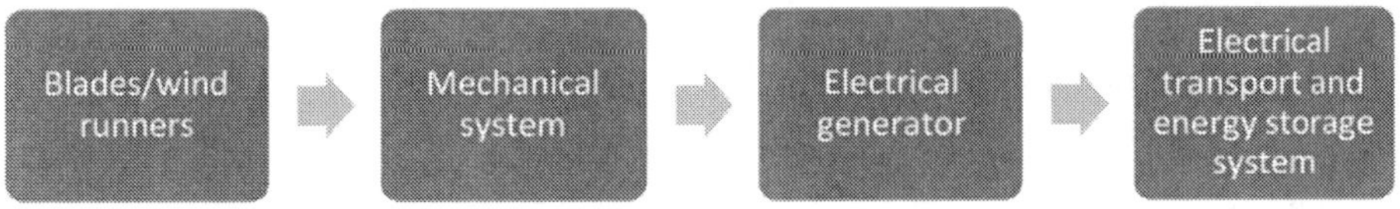

Fig. 1. Components of the wind energy generating system.

Typical wind runners of vertical axis used are Darrieus, helical, drag cup, Savonius, and H type, presented in fig. 2.

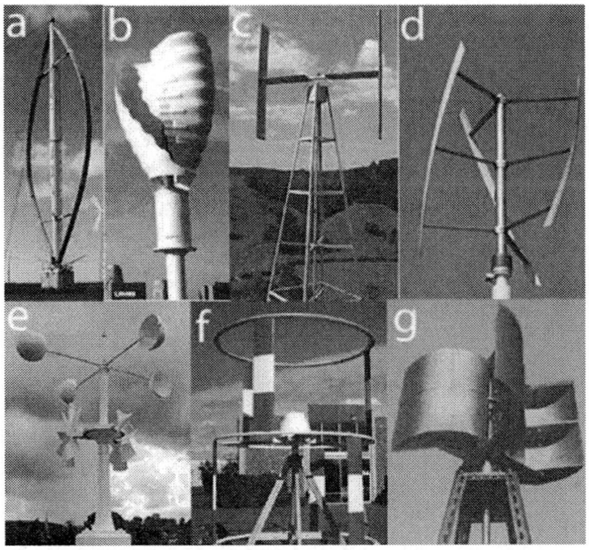

Fig. 2. Different types of vertical axis wind runners : a) Darieus [2]; b) Helix geometry[3]; c) H geometry [4]; d) Turby [5]; e) Drag cup [2]; f) Hybrid vertical wind turbine [2]; g) Savoniuos [6].

The design of the wind runner is important because influence the efficiency of the all-wind energy system and the protection of the system in case of powerful wind or storms. This is why we suggest making a vertical wind energy system with adaptable position of runners.

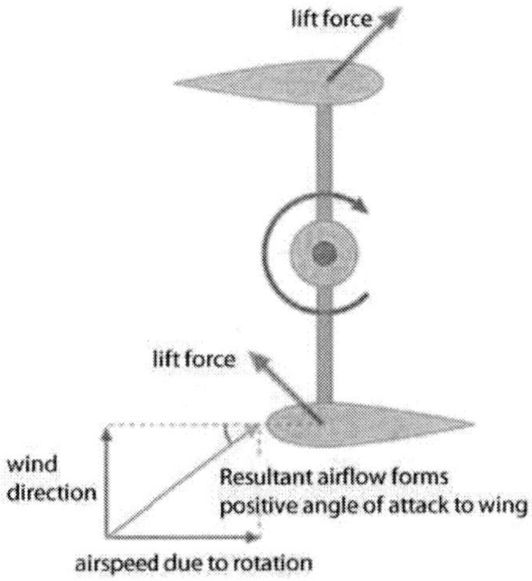

Fig. 3. Working principle of vertical axis wind runner [2].

The mechanical power generated by the vertical axis wind runner is:

$$P=\omega \cdot T \quad (1)$$

Where
ω - represents the angular velocity of the axis
T – represent the torque

Usually, the wind energy system is more efficient if the power coefficient (C_P) is higher. It is defined as the ratio of the power converted from the rotor power to the wind power flowing, it is a concept of aerodynamic energy conversion efficiency [8] and is also influenced by the power transmission efficiency (η) of the mechanical system used for transforming and transmitting the mechanical power to the electrical generator. This coefficient has a maximum value at a specific tip speed ratio (λ) and for example, Darrieus runner has a maximum power coefficient of about 0.4 at a tip speed ratio of 5 and all turbines has an optimum tip speed ratio with maximum power and is related it the speed of the wind (V). The tip speed ratio is the ratio of the wind runner tip speed and the wind speed at which the wind runner tip moves with rotation, is related with power coefficient, and can be described as:

$$\lambda = R \cdot \omega \cdot V^{-1} \tag{2}$$

The power output (P) is:

$$P = \rho \cdot R \cdot H \cdot V^3 \cdot \eta \cdot C_p \tag{3}$$

Where
ρ- is the density of air;
R- is the radius of the runner;
H- is the height of the wind runner.

The radius (R) multiplied by the height (H) represents the wind-swept area (A_ω). The bigger A_ω the bigger P become. If the R is longer, the generated torque is higher, but the solidity (σ) is lower and the generated torque is lower. If the H is longer than the rotational speed of the rotor should be increased to obtain the same power output. The solidity is defined as the ratio of total projected area of the rotor wind runner (NC, which is the number of wind runner multiplied by projected area) to the rotational area of the wind runner rotor [7].

So, by modifying the angle of the vertical runner with the according with the wind direction the angular velocity is increasing so the output power is increasing. In this way the system is efficient even at low wind speed, and this is why we chose to analyze such system by modeling and simulation considering a prototype design with four wind runners presented in Fig. 4 which is modifying the angle according with the wind direction.

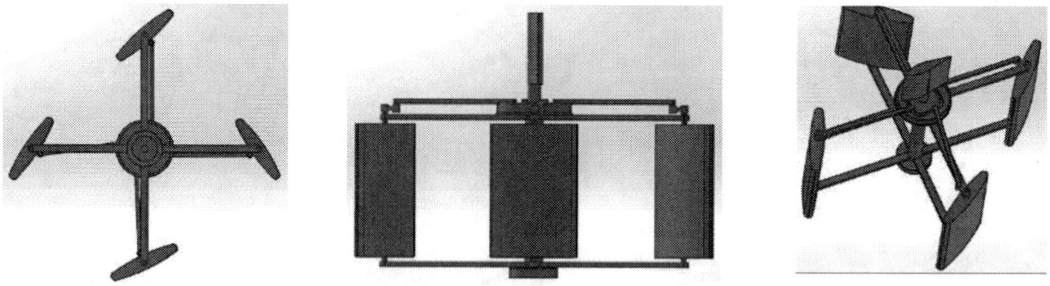

a) b) c)

Fig. 4. The prototype design of wind runner with four wind runners: a) Front view; b) Top view; c) Isometric view.

2 Method for calculation for the wind runners orientation system used in a vertical wind turbine

For calculation [9, 10] for the wind runners orientation system, we started to analyses one wind runner (as can be seen in Fig. 5) because our design is axial symmetric. There for the method can be applied starting for two or four wind runners as presented in our case or more.

The vertical wind turbine system consists of a quadrilateral mechanism (ABCD) and in the joint A is fixed the wind runner at the middle of its width. The element 3 (CE) has a ball bearing for the friction reduction, which it is fastened in the joint E, and it is rotated by the roller which has a role of follower for the cam 4 having a translation movement. For the determination of the kinematic parameters of the quadrilateral mechanism, we can use equations 1 and 2.

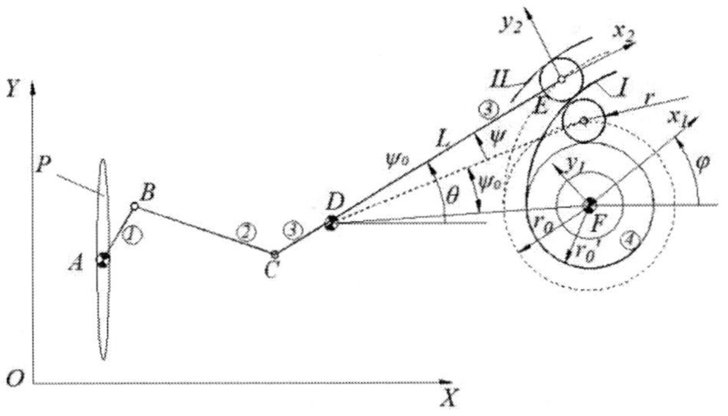

Fig. 5. Structure diagram for analyses.

The position of the mechanism is given by the value of the angle of the crank 1 (starting from 30^0) and corresponds to the minimum angle made by the vector DC with Ox axis, considering the positive direction (elements 1 and 2 are in the extension). Determining this position [9] is given by the equation:

$$\varphi = \arccos\left(\frac{XD}{AD}\right) + \arccos\left(\frac{AD^2+(AB+BC)^2-CD^2}{2AD(AB+BC)}\right) \quad (3)$$

where

$$AD=\sqrt{XD^2+YD^2} \quad (4)$$

For the synthesis of the rotating cam mechanism, which acts as a cylindrical active rotating surface, we can use the circle family envelope equations, which is moving with the center on the theoretical profile, are

$$\left(x-x_{1i}\right)^2+\left(y-y_{1i}\right)^2-r^2=0 \quad (5)$$

$$\left(x-x_{1i}\right)\frac{dx_{1i}}{d\varphi_i}+\left(y-y_{1i}\right)\frac{dy_{1i}}{d\varphi_i}=0 \quad (6)$$

where:
i= step

$$x_{1i}=\left(XE-XF\right)\cos\varphi_i+\left(YE-YF\right)\sin\varphi_i+L\cos\left(\theta_i-\varphi_i\right) \quad (7)$$

$$y_{1i}=-\left(XE-XF\right)\sin\varphi_i+\left(YE-YF\right)\cos\varphi_i+L\sin\left(\theta-\varphi_i\right) \quad (8)$$

 https://doi.org/10.1051/e3sconf/201911202009

$$\frac{dx_{1i}}{d\varphi_i}=-\left(XE\text{-}XF\right)\cos\varphi_i+\left(YE\text{-}YF\right)\sin\varphi_i+L\sin\left(\theta_i\text{-}\varphi_i\right)\frac{d\theta}{d\varphi_i} \quad (9)$$

$$\frac{dy_{1i}}{d\varphi_i}=-\left(XE\text{-}XF\right)\cos\varphi_i-\left(YE\text{-}YF\right)\sin\varphi_i-L\cos\left(\theta_i\text{-}\varphi_i\right)\frac{d\theta_i}{d\varphi_i} \quad (10)$$

By making the replacements in the above relationships, we obtain:
- the profile *I*:

$$x_{1I}=x_{1_i}+\frac{r\frac{dy_{1i}}{d\varphi_i}}{\sqrt{\frac{dx_{1i}}{d\varphi_i}^2+\frac{dy_{1i}}{d\varphi_i}^2}},\qquad y_{1I}=y_{1_i}-\frac{r\frac{dx_{1i}}{d\varphi}}{\sqrt{\frac{dx_{1i}}{d\varphi_i}^2+\frac{dy_{1i}}{d\varphi_i}^2}} \quad (11)$$

- the profile *II*:

$$x_{1II}=x_{1_i}-\frac{r\frac{dy_{1i}}{d\varphi_i}}{\sqrt{\frac{dx_{1i}}{d\varphi_i}^2+\frac{dy_{1i}}{d\varphi_i}^2}},\qquad y_{1II}=y_{1_i}+\frac{r\frac{dx_{1i}}{d\varphi_i}}{\sqrt{\frac{dx_{1i}}{d\varphi_i}^2+\frac{dy_{1i}}{d\varphi_i}^2}} \quad (12)$$

3 Results and conclusions

The joint C of the mechanism is fixed to the frame of the vertical wind runner. It can be seen in Fig. 5 that on the cam 4 (which has the joint F) is fastened a system, which establishes the wind direction, and which will move the cam 4 on this direction. The rotation movement of the AB element is limited on $\varphi = 100^0$ deg. Starting from the above relationships, a calculation program developed in SciLAB has as input date the angle φ in steps of 10^0 starting from 30^0 until 130^0. In fig. 6 we can see some results from the program.

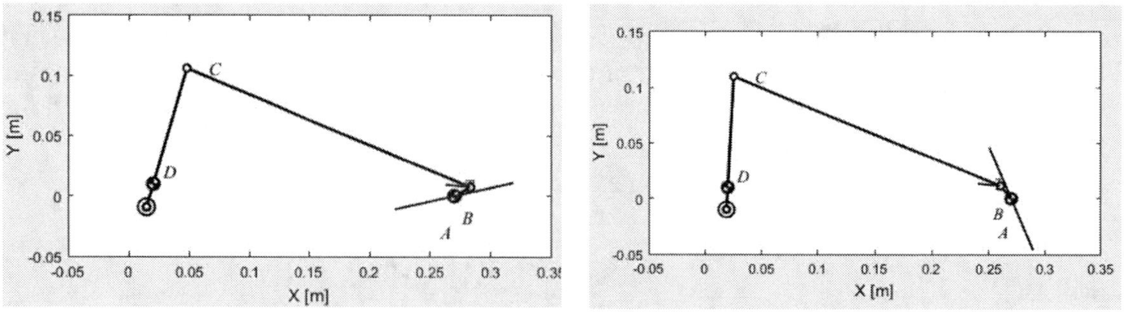

a) b)

Fig. 6. Graphical representation of the movement of the system at given angles: a) $\varphi = 30^0$; b) $\varphi = 130^0$.

Numerical values obtained φ are given in radian and are presented in Table 2.

Table 1. Numerical values obtained using the calculation method.

φ [rad]	ψ [rad]	θ [rad]	$\dot{\psi}$ [rad/s]	$\dot{\theta}$ [rad/s]	$\ddot{\psi}$ [rad/s^2]	$\ddot{\theta}$ [rad/s^2]
0.5230	2.7447	1.2889	0.4105	1.2012	-3.7176	8.8836
0.6975	2.7513	1.3112	0.3416	1.3398	-4.1558	6.9619
0.8721	2.7566	1.3355	0.2659	1.4429	-4.5063	4.8136
1.0466	2.7606	1.3613	0.1848	1.5068	-4.7791	2.4771
1.2211	2.7631	1.3879	0.0995	1.5285	-4.9815	-0.0043
1.3957	2.7640	1.4144	0.0113	1.5061	-5.1165	-2.5815
1.5702	2.7634	1.4402	-0.0787	1.4382	-5.1810	-5.1972

1.7447	2.7613	1.4643	-0.1691	1.3248	-5.1646	-7.7829
1.9193	2.7575	1.4862	-0.2584	1.1672	-5.0493	-10.2559
2.0938	2.7523	1.5048	-0.3446	0.9680	-4.8109	-12.5177
2.2683	2.7455	1.5197	-0.4254	0.7321	-4.4212	-14.4562

A wind runner according with the wind speed rotates the cam and the quadrilateral mechanism (ABCD) makes a complete rotation motion. Using the aforementioned calculation formulas in the same software, we verified the functioning of the vertical axis wind runner. In the fig. 7 we present the system in four distinct positions at $0^0, 90^0, 180^0, 270^0$.

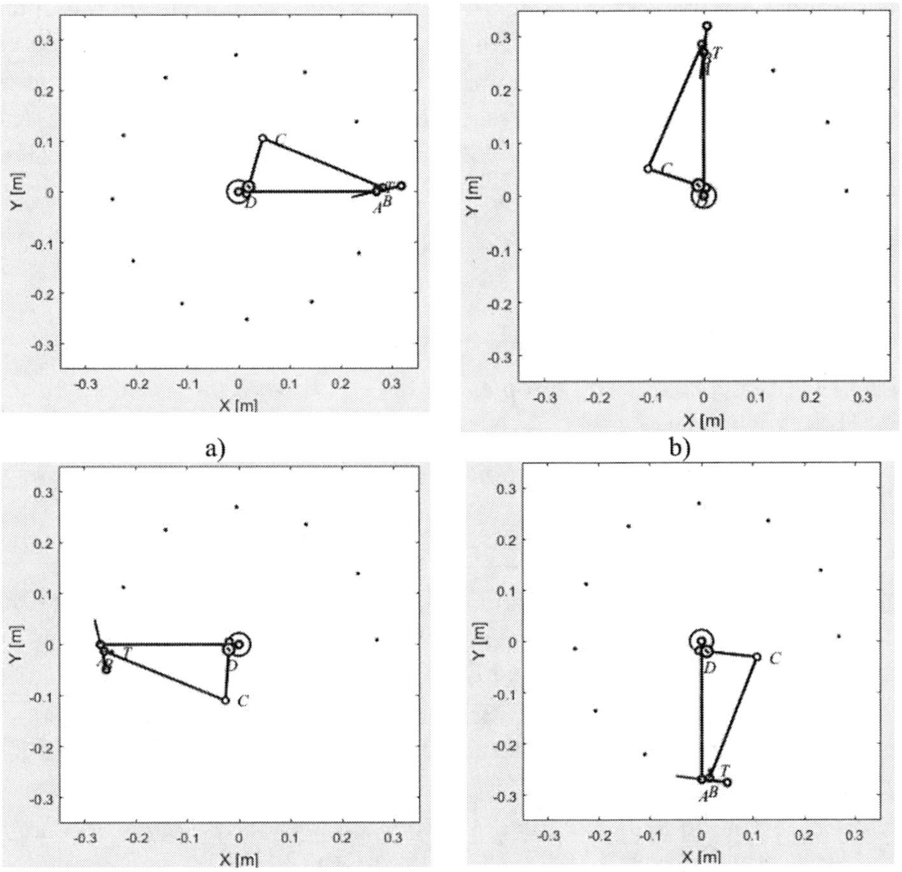

c) d)

Fig. 7. Graphical representation of the movement of the system at given angles: a) 0^0; b) 90^0 ; c) 180^0; d) 270^0.

The calculated numerical values for the quadrilateral mechanism are presented in table 2 when element 4 cam is on a fixed position by the wind energy.

Table 2. Numerical values for the quadrilateral mechanism obtained using the calculation method.

φ [rad]	ψ [rad]	θ [rad]	$\dot{\psi}$ [rad/s]	$\dot{\theta}$ [rad/s]	$\ddot{\psi}$ [rad/s^2]	$\ddot{\theta}$ [rad/s^2]
0.0000	2.7448	1.2890	9.6718	9.0386	-2.3803	5.6816
0.5236	3.2520	1.7768	9.5834	9.3316	-1.6277	7.6240
1.0472	3.7756	2.3004	9.5065	9.2083	-2.2839	10.6977
1.5708	4.2992	2.8239	9.5065	9.2083	-2.2294	10.7852
2.0944	4.8457	3.4056	9.6765	8.7299	-3.6974	6.3863
2.6180	5.3776	4.0937	10.2033	8.8150	-4.6204	-7.9994
3.1416	5.8938	4.6465	10.3269	9.0834	-4.3555	-11.1466

3.6652	6.4174	5.1701	10.3269	9.0834	-4.3555	-11.1466
4.1888	6.9410	5.6937	10.3269	9.0834	-4.3555	-11.1466
4.7124	7.4737	6.1768	10.1605	8.7450	-4.6554	-6.8581
5.2360	7.9990	6.6240	9.9061	8.5515	-4.4639	0.1447
5.7596	8.5162	7.0952	9.7483	8.6324	-4.0200	4.4663
6.2832	9.0279	7.5722	9.6112	8.8612	-3.2971	8.0979

The results presented in table 1 and table 2, proves that the wind energy system does not have singularities and the combination between two mechanisms cam with roller follower and quadrilateral mechanism, which has the positions in Fig. 7, are representing a feasible solution. In conclusion, our proposed wind energy system can modify the angle of the runners according with the wind direction if we use a passive mechanism presented in Fig.4 or with the wind direction and speed when we use an active mechatronic which comprise a speed sensor for sensing and a dc motor for actuation. However, the disadvantage is the possible contact wear in the area where the follower and cam come in contact. At next step we will focus our work on integration of this type of wind runner in different sound absorption or safety autonomous systems.

References

1. https://www.vertogen.eu/wind-turbine-history/
2. M. Ragheb, Vertical axis wind turbines (2015)
3. https://www.treehugger.com
4. https://www.hurricanewindpower.com
5. https://www.renugen.co.uk
6. https://vertogen.eu
7. H. Dowon, G. H. Young, J. C. Nak, H. N. Sang, H. C. Kyoung and C. K. Kyung, Design, Fabrication, and Performance Test of a 100-W Helical-Blade Vertical-AxisWind Turbine at Low Tip-Speed Ratio, Energies, **11**, 1517 (2018)
8. F. Niculescu, A. Săvescu, and C. Borzea, Analogue flow control of a network of centrifugal air compressors. At International Conference on Applied and Theoretical Electricity, Craiova, Romania (2018).
9. V. Moise, I. Simionescu, M. Ene, A. Rotaru, *Analysis of plane mechanisms with hinged bars. Applications in MATLAB*, Ed. Printech (2015)
10. V. Moise, I. Simionescu, M. Ene, *Optima Synthesis of mechanisms with plane cam. Applications in MATLAB*, Ed. Printech (2018)

Compressed Air Energy Storage Installation for Renewable Energy Generation

Claudia Borzea[1,*], *Iulian* Vlăducă[1], *Dan* Ionescu[1], *Valentin* Petrescu[1], *Filip* Niculescu[1], *Cristian* Nechifor[1], *Gabriel* Vătăşelu[2], and *Mihai* Hanek[2]

[1]Romanian Research and Development Institute for Gas Turbines COMOTI, 061126 Bucharest, Romania
[2]ICPE, Energy Centre, 030138 Bucharest, Romania

Abstract. Compressed Air Energy Storage (CAES) installations are used for storing electrical power, under the form of potential energy from compressed air. The heat generated during compression can be stored to improve the efficiency of compression-expansion cycle. The solution presented consists of a 100 kW screw compressor driven by a 110 kW asynchronous three-phase motor. The compressor supplies air into vessels which store it until a high electrical energy demand arises. At that time, the compressed air is released into a 132 kW screw expander whose shaft spins a 132 kW asynchronous generator, producing electric power and supplying it into the electrical grid. Before expansion, the air must be preheated in order to avoid the freezing of expansion equipment. If the heat generated during compression is used for air preheating before expansion, the process is adiabatic. A demonstrative model of the installation is currently being developed, with the expander part being completed so far. The maximum power to be produced was calculated to be around 100 kW. During expander commissioning tests with air supply from a 250 kW high pressure compressor, a maximum generated power of 49.7 kW was attained, expected to be higher when releasing air from the reservoirs.

1 Overview

Global concerns on climate changes and sustainability have led to recent worldwide approaches towards generating electricity from renewable sources. The use of renewable energies as a response to the European Union targets defined for 2030 Climate Change and Energy is increasing [1]. Non-dispatchable and intermittent renewable energy sources like wind and solar power cannot meet the necessary supply and demand, which can also cause certain problems in the grid. Some of the major limitations of renewable energy sources are represented by their low power density and intermittence, largely depending upon local site and unpredictable weather conditions.

Compressed air energy storage (CAES) is a technique for supplying electric power to meet peak load requirements of electric utility systems [2], being suitable for wind and solar sources. A CAES power generation facility uses electric motor-driven compressors to

* Corresponding author: borzeaclaudia@gmail.com; claudia.borzea@comoti.ro

 https://doi.org/10.1051/e3sconf/201911202010

inject air into a reservoir, later releasing the compressed air to turn turbines and generate electricity back into the grid. High capacity storage media can include underground salt dome caverns, depleted oil/gas reservoirs, underground aquifers, or in certain cases abandoned underground hard rock mines [3]. CAES power plant concept dates back more than 40 years ago [2,4,5], but environmental concerns sustained by technological advances have encouraged their research and development interest to increase worldwide [6-11].

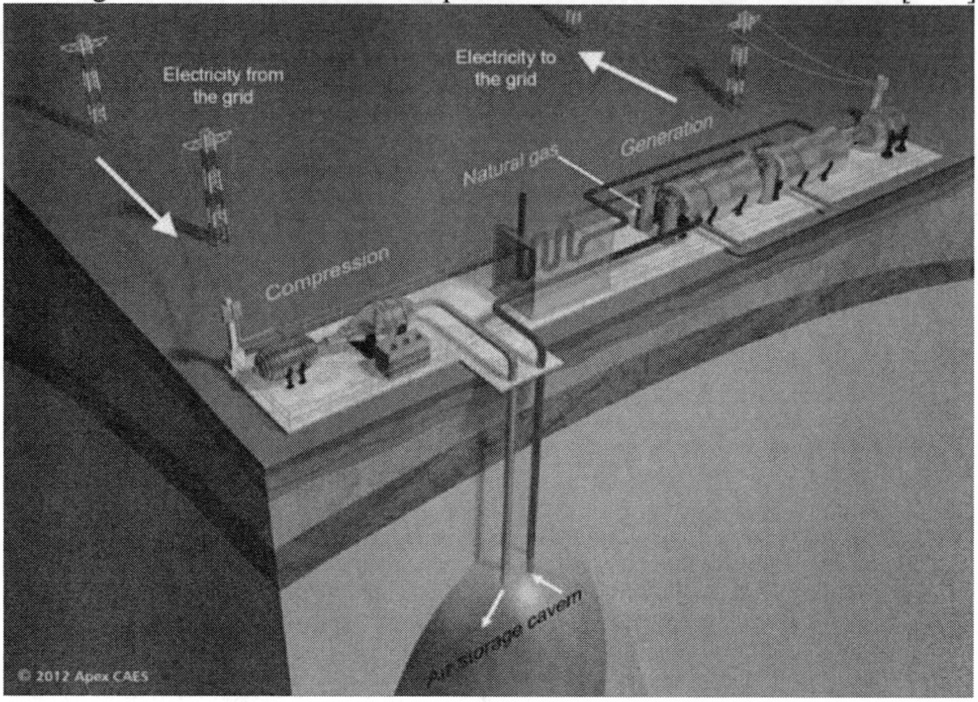

Fig. 1. Design of a CAES power generation facility [3]

In a CAES power generation plant, compression normally occurs during off-peak hours, when power prices are low. During hours of high electricity demand, compressed air is withdrawn from the cavern, preheated, and introduced into one of the several combustors. In the combustors, natural gas is fired to further heat the air, and the hot expanding gases are used to drive expansion turbines. These turbines are connected to a generator, which produces electricity for the grid.

The earliest commercially operated compressed air energy storage system was constructed in Huntorf, Germany, in 1978, with an installed capacity of 290 MW. The Huntorf plant is located on a 300,000 m^3 salt dome, in which compressed air is stored, originally to capture excess nuclear power production.

Fig. 2. CAES power plant in Huntorf, Germany [10]

Another successful CAES system was built in McIntosh, Alabama in USA, in 1991, and can store compressed air in the abandoned salt mines 450 m underground, providing compressed air to a 110 MW turbine for 26 consecutive hours. The construction of CAES plants is restricted by terrain and has special requirements for the geological structure [11].

Large-scale CAES plants are very inefficient, being commercially viable only in places where electricity price varies dramatically. Compressing and decompressing air introduces energy losses, resulting in an electric-to-electric efficiency of only 40-52%, compared to 70-85% for pumped hydropower plants, and 70-90% for chemical batteries [12]. Nevertheless, wind turbines plants relying on the intermittent nature of wind power can make CAES an attractive backup solution. In such systems, energy is stored during periods of low demand and is then converted back into electricity when the demand is high.

The low efficiency is mainly caused by air heating during compression. This waste heat, which holds a large share of the energy input, is dumped into the atmosphere. A related problem is air cooling during decompression, lowering electricity production and possibly freezing water vapours in the air. To avoid this, large-scale CAES plants heat the air prior to expansion using natural gas fuel, which further deteriorates the system's efficiency and makes renewable energy storage dependent on fossil fuels [12].

This paper presents the concept and design of the first Romanian CAES system. A prototype is currently being developed within ROCAES project.

2 ROCAES installation design and layout

The design presented is a relatively low scale model, using vessels to store the compressed air and hot water, without gas turbine engines or dynamic compressors (centrifugal or axial) which use fossil fuels. The prototype uses a twin-screw compressor and a twin-screw expander, actuated by three-phased asynchronous electric machines. The block diagram of ROCAES station with integrated SCADA (Supervisory Control and Data Acquisition) automation system is shown in figure 3 below.

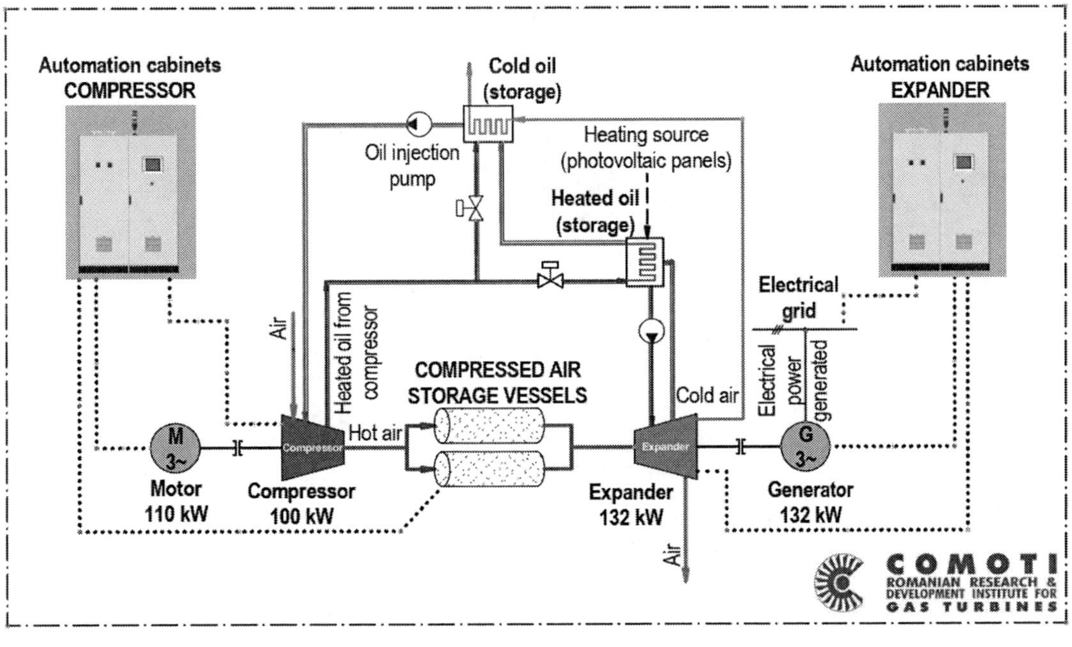

Fig. 3. Block diagram of ROCAES station

The system consists of a 100 kW twin-screw compressor that supplies compressed air into two storage vessels, and a 132 kW expander. The compressor is driven by a 110 kW asynchronous three-phase motor. The expander is driven by the compressed air released from the reservoirs, and spins a 132 kW asynchronous three-phase generator.

The compressor and the expander are not designed for simultaneous operation. The compressor supplies air into the storage tanks in the periods with low cost energy, until the maximum pressure of 26 bar absolute is attained in the vessels. The pressure is permanently monitored by a pressure transducer which transmits information to the PLC (Programmable Logic Controller) in the automation cabinet assigned to compressor and air storage. The compressed air is released from the storage tanks when an increased demand of electrical energy arises. The compressed air turns the expander, whose shaft spins the electrical generator, providing electrical power to the grid.

As presented schematically in figure 4, showing the total scale 1:1 of the 2D view, the overall area of the ROCAES power plant is 285 m^2, a relatively small scale CAES facility. The ROCAES plant is a pilot station, and the functional time for demonstration is about half an hour. The overall layout consists of four main areas, as designed, namely:

- ***Zone 1***: Compression skid – 1500 m^3 air screw compressor + compressed air storage vessel – 150 m^3 at 26 bar absolute air pressure.
- ***Zone 2:*** Heating water storage vessel – 1 bar absolute at 90°C water storage vessel, with two heat exchangers for oil to water heat transfer.
- ***Zone 3:*** Expander skid – 1500 ÷ 4500 m^3 air screw expander.
- ***Zone 4:*** Automation and control room – enclosure for automation cabinets, station computer and human operator controls.

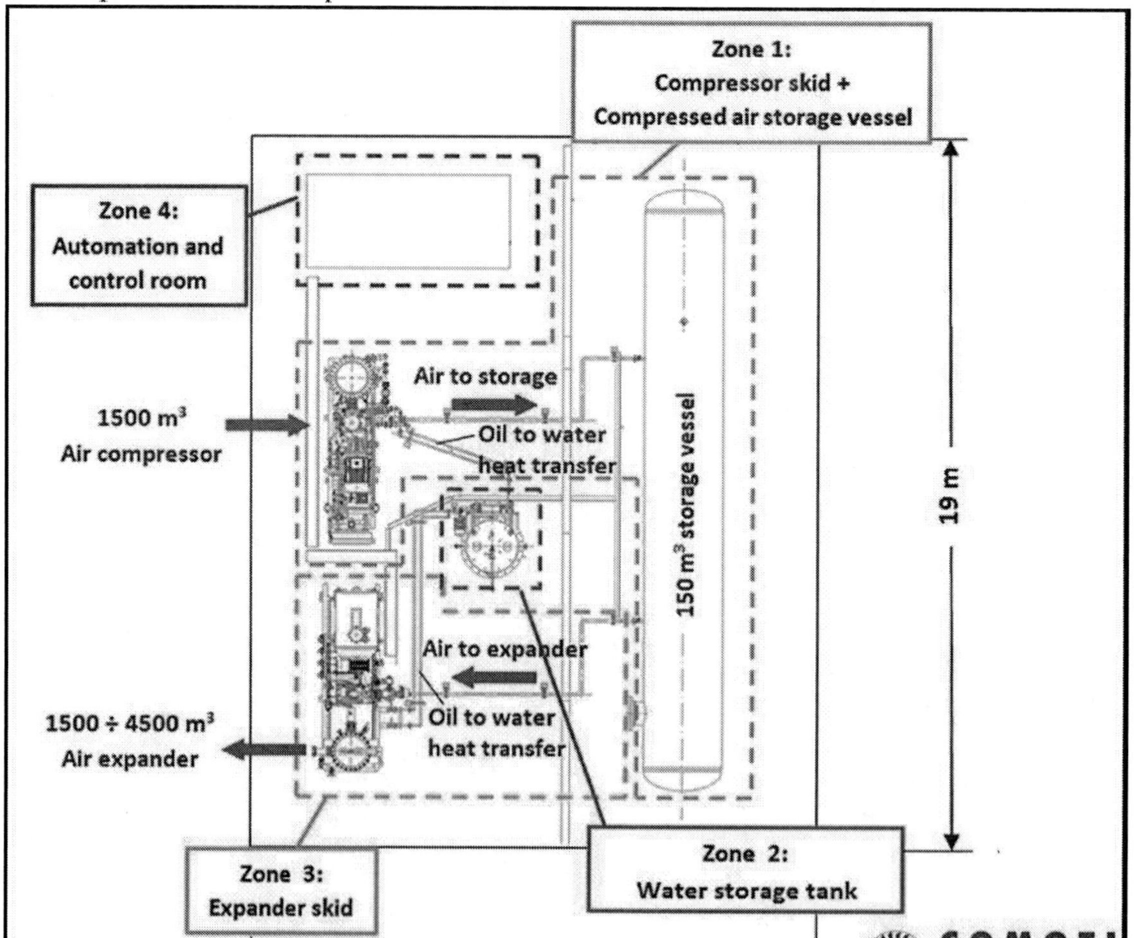

Fig. 4. ROCAES station 2D layout

During expansion, air temperature can drop below freezing point (down to -10 ÷ -15°C or even lower). When the compressor is functioning, hot air from compression is accumulated in the storage tanks. The lubrication and cooling oil tray passes through the thermal conditioning unit (water storage tank) and heats the thermal agent transferring heat to the oil circuit of the expander. The oil injection pump of the expander recirculates the oil on the bypass circuit configured by a three-way motor operated valve, ensuring oil heating. After compressor finishes operation, oil recirculation is also stopped for thermal energy conservation in the storage tank, maintaining the lube oil at an optimal temperature.

Fig. 5. Expander skid with automation cabinets in the assembly hall

A long period of time, neither the compressor nor the expander will function. The operation of the installation depends on the optimal timing. During idle periods, it is preferable that the compressed air vessels to be full, waiting for air to be released into the expander. If the air inlet motor operated valve is open, a time of approximately 20s is estimated until starting to generate electrical energy.

3 Electrical and automation considerations

The purpose of the automated SCADA system represented by the drives and automation cabinets is to monitor the operation parameters of thermal, mechanical and electrical processes, to regulate and keep these ones under control, in order to ensure maximum safety conditions.

On compressor and expander skids, transducers are mounted for monitoring the important parameters of the installation. Local indication is realized by means of manometers and thermometers. For the important sizes that need to be transmitted to PLC for a safe operation, pressure transducers are mounted in several points for monitoring air and lube oil pressures. The air flow before inlet is monitored by means of differential pressure cells with orifice plate, which constitute analogue inputs into PLC. Temperature data is acquired using PT100 resistance temperature detectors (RTD) for air temperature on inlets and lube oil temperature in the storage tank. The information from the transducers is transmitted to two PLCs, so as the expander and the compressor to be able to operate independently of one another. The analogue I/O (inputs/outputs) signals for varying parameters are transmitted in unified current signal 4-20 mA. The digital I/O (open/close, on/off type) are transmitted in 0-24VDC voltage signal.

On expander skid, the automation system controls two motor operated valves for air inlet and oil bypass, a solenoid valve with rapid response for emergency shutdown, and the opening angle of three electro-pneumatic valves (providing feedback for air flow regulation through expander, for flow through the bypass circuit, and for outlet air flow).

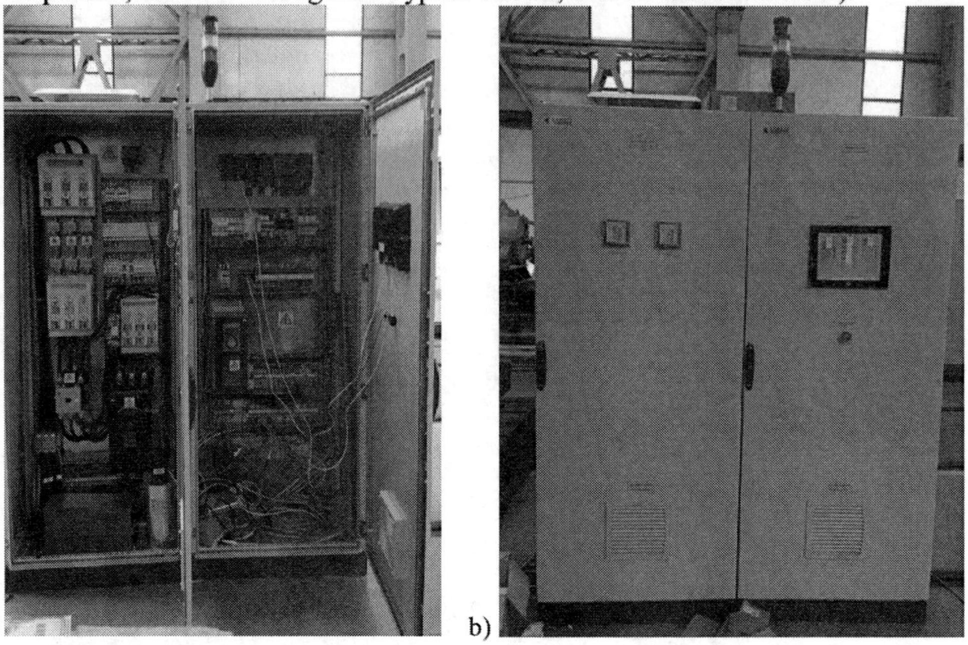

a) b)

Fig. 6. Automation cabinets assembly: a) fully equipped, with open doors; b) during operation

The asynchronous three-phase machine of the expander is started in motor regime, using a softstarter, until reaching no load speed (~1470 rpm). The compressed air released turns the expander, which spins the shaft of the electric machine above the synchronous speed of 1500 rpm, this one entering automatically in generator regime and supplying electrical power into the grid. The speed of the machines is maintained constant, thus being possible to use softstarters instead of variable speed drives, and therefore reducing costs.

An asynchronous generator is an electric induction motor driven at speeds above its synchronous speed. It has a solid armature (squirrel cage), which acts as an electrical short circuit. When power supplied, the machine will start turning like a motor at a speed slightly below the synchronous speed of stator's rotating magnetic field. If the rotor is driven at exactly the synchronous speed (1500 rpm), the magnetic field rotates at the same speed as the rotor, causing no induction phenomena in the rotor, which will not interact with the stator. If the speed is increased above the synchronous speed, the rotor moves faster than the rotating magnetic field of the stator, and the stator induces a current in the rotor. The more mechanical power that is delivered to the rotor, the more power will be transferred as electromagnetic force to the stator, and consequently converted into electricity, which is supplied into the electrical grid.

The choice of using asynchronous induction machines in the detriment of synchronous machines brings significant advantages:

- No additional manoeuvres for coupling to the electrical grid;
- Asynchronous electric machines are able to operate without requiring any human operator intervention, rendering the start-up - loading - connection to electrical grid cycle to be completely automatic;
- High dynamic load capacity;
- High start-up capacity.

However, besides providing a lower efficiency than synchronous machines, the asynchronous generator also has a reactive power component. During operation, a continuous, oscillating reactive power exchange occurs between the generator and the electrical grid. For overcoming this disadvantage, a three-phase capacitor has been ensured. The capacitive reactance provides the reactive power component to the electric machine, thus avoiding inducing power fluctuations in the electrical grid.

The apparent power of an asynchronous generator consists of an active power component and a reactive power component and is measured in VA (Volt Ampere). The active power (at shaft) is the actual mechanical power measured in W (Watts).

4 Estimated power computations

This section presents the computations made for the power expected to be produced by the generator. The expander works at a suction pressure p_{suc} = 20 bar absolute (2 MPa). At 20°C and 60% relative humidity, the mass air flow rate is $M_{air,exp}$ = 1500 ÷ 4500 m^3/h and the volume flow rate is $Q_{air,exp}$ = 0.487 ÷ 1.461 kg/s [13]. The computations considered the minimum and maximum volume flow rates in [kg/s].

The discharge pressure p_{dis} is 2 bar absolute. The expansion ratio is:

$$\pi_{exp} = \frac{p_{suc}(abs)}{p_{dis}(abs)} = 10 \tag{1}$$

The temperature in the storage vessel will increase adiabatically from compressor output temperature, according to [14]. If t_{comp} = 90°C and the initial pressure in the storage vessel $p_{st,in}$ = 2 bar absolute, the compressed air temperature in the storage vessel is calculated, at a maximum final pressure attained of $p_{st,fin}$ = 26 bar absolute:

$$T_{st,fin} = (t_{comp} + 273.15) \cdot \frac{k}{1+(k-1)\cdot\frac{p_{st,in}}{p_{st,fin}}} = 493.23\ \text{K} \Rightarrow t_{st,fin} = 220.08°\text{C} \tag{2}$$

The nominal operating regime of the system corresponds to the following parameters:

- p_{suc} – air pressure at the inlet of the expander: 20 bar (abs);
- p_{dis} – discharge air pressure of the expander: 2 bar (abs);
- $\dot{m}$ – air mass flow rate: 1500 ÷ 2500 m^3/h, at 20°C and 60% humidity;
- Q – air volume flow rate: 0.487 ÷ 0.812 kg/s, at 20°C and 60% humidity;
- k – adiabatic exponent: 1.4;
- η_{iz} – isentropic expander efficiency: 0.7;
- R – gas constant: 287 J/(kgK).

Due to the oil injected screw turbine, the oil will be heated/cooled, which will cause a constant temperature at the compressed air inlet in the turbine. The suction temperature will be maintained constant at maximum temperature supported by the expander of t_{suc} = 135°C (T_{suc} = 408.15 K absolute value).

The temperature difference $\Delta T \cong 85°C$ is used to compensate the storage heat losses, which can oscillate depending on insulation, from 3.5 to 5°C/h. The calculations were performed for different storage tank diameters and insulation thicknesses, for an external storage vessel of 150 m^3. The specific heat c_p at constant pressure is:

$$c_p = k \cdot R/(k-1) = 1.010\ \text{kJ/(kgK)} \tag{3}$$

The power of the expander is calculated using the following relation:

$$P_{ex} = \dot{m} \cdot \eta_{iz} \cdot c_p \cdot T_{suc} \cdot \left[1 - \left(\frac{1}{\pi_{exp}}\right)^{\frac{k-1}{k}}\right] = 73.8 \div 112.9\ \text{kW} \tag{4}$$

Considering generator efficiency $\eta_{gen} = 0.9$, the estimated power generated will be:

$$P_{gen} = P_{exp} \cdot \eta_{gen} = 66.4 \div 101.6 \text{ kW} \quad (5)$$

5 Commissioning tests for expander-generator system

The functionality of the expander-generator system was tested with the air inlet supply connected to a 250 kW screw CHP (Compressor of High Pressure) designed by INCDT COMOTI for attaining high pressures up to 40 bar. The electric machine of the expander was started up in motor regime, with expander air inlet valve open. The system start-up must be realized only with the inlet valve open because otherwise vacuum is created during expansion, inducing large vibrations in the equipment when vacuum is discharged at the outlet, into a higher external atmospheric pressure.

When the compressed air pressure and flow became sufficient to turn expander's helical screw rotors, the shaft spinning above the synchronous speed determined the generator to produce electrical energy. The power consumption as motor, as well as the power generated and supplied into the grid are displayed by the power analysers. The consumed power values are positive and the generated power is displayed as negative (figure 7).

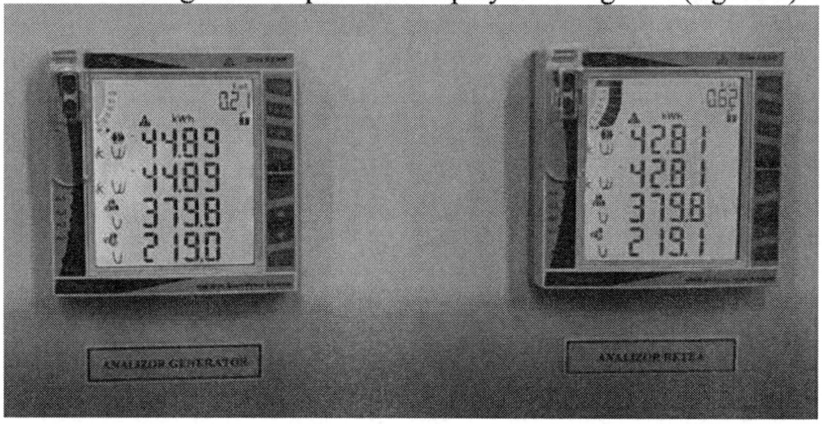

Fig. 7. Power generated (left) and power supplied into grid (right) displayed on the analysers

The power analysers transmit the data to the PLC using Ethernet communication protocol, also displaying the information on the control panel, along with the important operational parameters (figure 8).

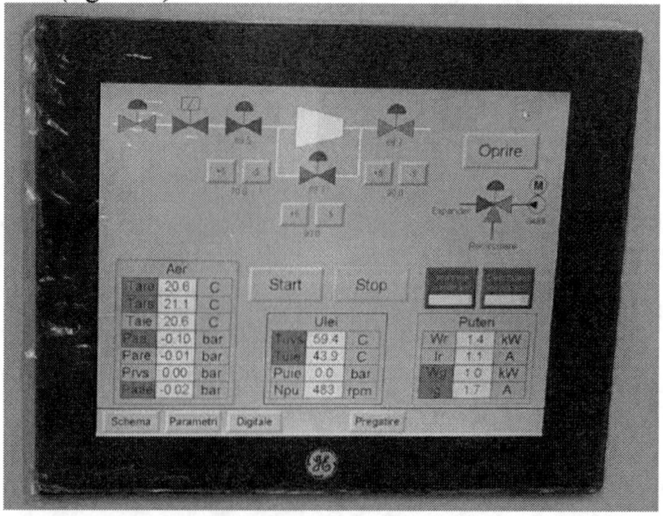

Fig. 8. Control panel displaying technological diagram screen and important parameters

Data acquisition (DAQ) was performed during commissioning tests. Parameters values were acquired through the program realized in Proficy Machine Edition software, by connecting the computer on which the program was ran to an Ethernet module enabling the communication with the PLC. Relying on the DAQ, the subsequent graph was represented.

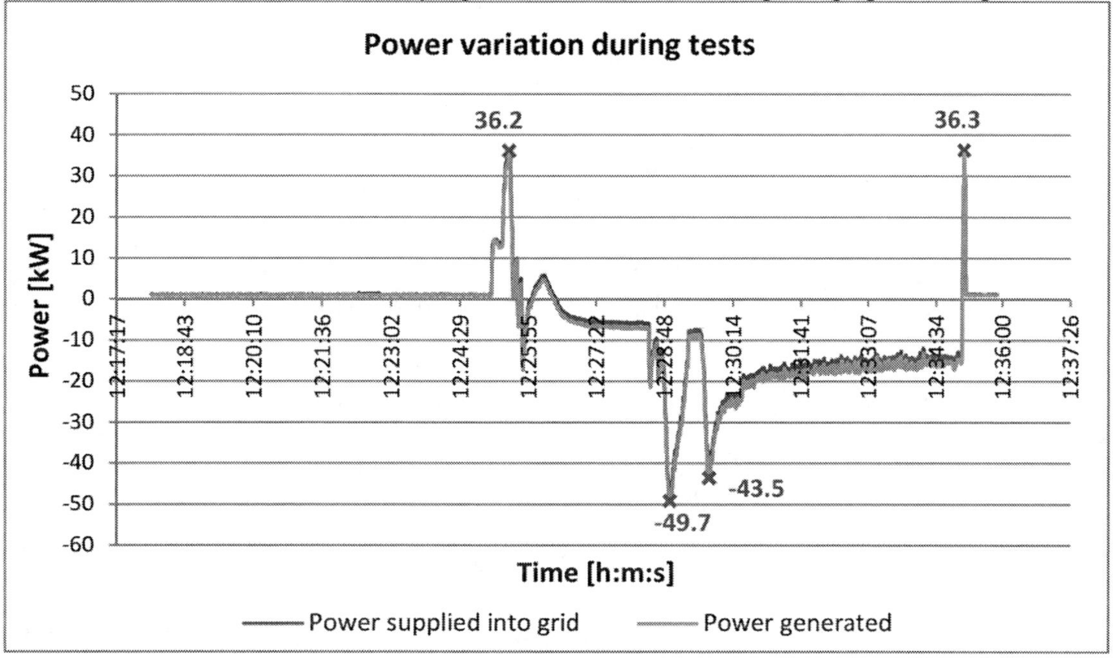

Fig. 9. Power variation chart drawn with acquired data

As it can be observed from the graph above, the power supplied into the grid is slightly lower. The reason is that the 4 kW motor of the oil pump and the electrical equipment in the automation cabinets also have to be supplied from the generated power.

The CHP provided a pressure of 12-14 bar, air entering the expander at around 4 bar. By increasing the air flow at expander inlet, a maximum generated power of 49.7 kW was recorded, the power supplied into the grid being of 48.4 kW.

A peak of consumed power of 36.2 kW from the electric grid occurs at rated speed as motor. Another peak of 36.3 kW power consumption occurs when stopping the expander-generator system, due to suddenly cutting off the compressed air supply. In the complete ROCAES application, the compressed air released from the storage vessels is expected to provide a smooth operation stop, as air pressure will decrease gradually in the reservoirs.

6 Conclusions

The paper proposes a CAES installation with air storage in vessels. The solution presented is a low scale model, suited for electrical power backup on small areas such as villages. ROCAES system can be used for ensuring a continuous electrical energy distribution when interruptions occur in the power generated by wind turbines or solar panels.

The advantages of ROCAES installation design are:

- high product efficiency due to an almost isothermal expansion – by oil injection;
- high flexibility for variations of gas parameters at the expander suction;
- small scale model, able to be replicated easily;
- eliminating terrain restrictions and constraints regarding special geological structures to provide compressed air storage, such as caverns and salt mines. ROCAES power plant can basically be placed in any area respecting the environmental conditions for which it has been designed.

https://doi.org/10.1051/e3sconf/201911202010

The paper presents the concept and design of the first CAES power plant developed in Romania. A maximum generated power of 49.7 kW was attained during commissioning tests of expander-generator system, with the air inlet connected to a high pressure screw compressor. ROCAES is an ongoing project, only the expander part being completed so far. The overall installation has not yet been functionally tested. The mathematical calculations estimate that the facility will be able to produce up to 100 kW.

The estimated operation time is about 30 minutes, being limited by storage vessels capacity. The power and the operation time are not that high, but are enough to validate the conceptual model with low costs. An underground cavern is currently being searched for increasing the air volume stored and therefore the operation time.

Future research will focus on optimising the oil temperature conditioning circuit by introducing photovoltaic panels for oil heating. Using a system for oil thermal energy storage (TES) has also been discussed. Solutions for reducing pressure drops and losses will also be studied.

The research work presented herein was funded by Competitiveness Operational Program 2014-2020, Priority Axis 1 - Research, Technological Development and Innovation in Support of Economic Competitiveness and Business Development, Action 1.2.3, Partnerships for Knowledge Transfer, contract no. 79 / 08.09.2016, carried out with the support of Romanian Ministry of European Funds.

References

1. A. Cristescu, C.G. Alionte, *Echipamente Termice Clasice şi Nucleare ETCN* (2008).
2. W.V. Loscutoff, Ener. Stor., 467-471, (Elsevier, 1980).
3. http://www.apexcaes.com/caes.
4. K. Vosburgh, J. of Ener., **2**, no. 2, 106-112 (1978).
5. O.K. Mawardi, *Energy Storage for Electric Utilities* (1981).
6. M. Masoum, E. Fuchs, *Power Quality in Power Systems and Electrical Machines*, 961-1084 (2015).
7. L. Rublack, O. Warweg, P. Bretschneider, S. Freund, M. Bieber, *2016 IEEE PES Innovative Smart Grid Technologies Conference Europe (ISGT-Europe)* (2016).
8. S. Donadei, G. Schneider, *Storing Energy*, 113-133 (2016).
9. J. Wang, K. Lu, L. Ma, J. Wang, M. Dooner, S. Miao, J. Li, D. Wang, Ener. **10**, no. 7, (2017).
10. T.M.I. Mahlia, T.J. Saktisahdan, A. Jannifar, M.H. Hasan, H.S.C. Matseelar, Renew. and Sust. Ener. Rev. **33**, 532-545 (2014).
11. K. Ding, J. Zhi, *Large-Scale Wind Power Grid Integration*, 211-232 (2016).
12. https://www.lowtechmagazine.com/2018/05/ditch-the-batteries-off-the-grid-compressed-air-energy-storage.html.
13. N. Toma, V. Petrescu, M. Niţulescu, *Huitième édition du COlloque FRancophone en Energie, Environnement, Economie et Thermodynamique, COFRET'16* (2016).
14. V. Radcenko, N. Alexandrescu, E. Ionescu, M. Ionescu, *Calculul şi proiectarea elementelor şi schemelor de automatizare* (1985).

Noise assessment of the small-scale wind farm

Cristian-Gabriel Alionte[1,*] and *Daniel-Constantin* Comeaga[1]

[1]Faculty of Mechanical Engineering and Mechatronics, University Politehnica of Bucharest, RO-060042, Street Splaiul Independenței no.313 , sector 6, Bucharest, ROMANIA

Abstract. The importance of renewable energy and especially of eolian systems is growing. For this reason, we propose the investigation of an important pollutant - the noise, which has become so important that European Commission and European Parliament introduced Directive 2002/49/CE relating to the assessment and management of environmental noise. So far, priority has been given to very large-scale systems connected to national energy systems, wind farms whose highly variable output power could be regulated by large power systems. Nowadays, with the development of small storage capacities, it is feasible to install small power wind turbines in cities of up to 10,000 inhabitants too. As a case study, we propose a simulation for a rural locality where individual wind units could be used. This specific case study is interesting because it provides a new perspective of the impact of noise on the quality of life when the use of this type of system is implemented on a large scale. This option, of distributed and small power wind turbine, can be implemented in the future as an alternative or an adding to the common systems.

1 Introduction

Energy production systems based on wind power have had a rather rapid development in the last 50 years being a direct result both of the gradual growth policy of oil prices [1] and pollution reduction policies.

Starting with the time of the energy crisis in 1973 due to the OPEC embargo, developed countries that have a deficit of classical energy resources have been oriented to the renewable energy resources. As a high-success example, Denmark was one of the first countries to develop and implement energy production systems based on wind energy.

Another important argument of wind energy is that it is one of the most environmentally friendly, the energy is cheap, and the investment is recovered much faster than using nuclear energy or coal fired power plants [2].

Globally, currently, there is an acceleration of new wind farms, as can be seen in Fig.1, due to several factors such as the improvement of new technologies, becoming more performant in terms of costs, pollution reduction and production capacity.

So far, priority has been given to very large-scale systems connected to national energy systems, wind farms whose highly variable output power could be regulated by large power systems. At present, with the development of small storage capacities, it is feasible to

* Corresponding author: cristian.alionte@upb.ro

install small wind turbines in cities up to 10,000 inhabitants[4]. Few possibilities exist, starting from wind-powered generators obtained by scaling down the most common systems, with huge dimensions, to new devices, with vertical helicoidal blade [4]. A new idea has emerged recently in regions with high noise traffic, reduced with sound absorbing systems, through use of small turbines placed on the top of sound barriers.

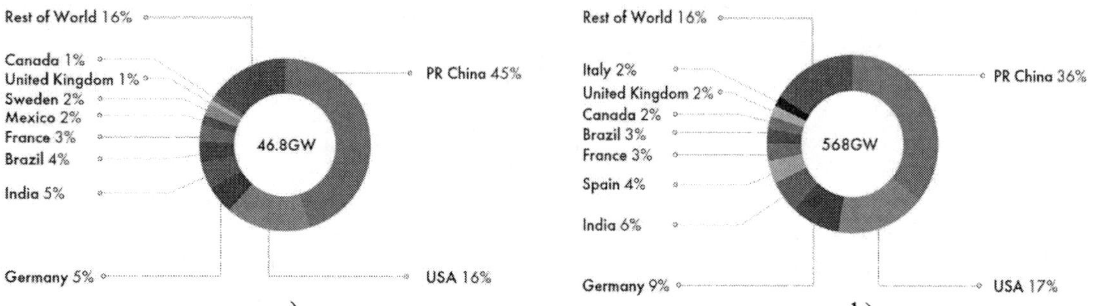

a) b)

Fig. 1. Development of wind energy sources: a) new installations of onshore wind systems in the year 2018; b) total number of onshore wind systems [3].

The noise analysis produced by a wind farm is based on the national and European legislation, mainly the EU directives.

At European level we can emphasize the Directive 2002/49/EC of the European Parliament on the assessment and management of ambient noise, in which the main elements of European noise assessment methods and the mechanisms for reporting the results from national bodies to the European Environmental Agency, with authority in the field, are specified. In the Directive 2002/49/EC it is specified for acoustic planning and noise zoning the obligation to use of L_{DEN} and L_{night} as noise indicators that can be obtained by calculation or measurements. These indicators are linked with two human factors, according World Health Organization: annoying noise (with L_{DEN}) and sleep disturbance noise (with L_{night}).

But for predictions only the method of calculation can be used, using the formula:

$$L_{DEN} = 10\lg 1/24(12\cdot 10^{(Lday/10)} + 4\cdot 10^{((Levening+5)/10)} + 4\cdot 10^{((Lnight+10)/10)}) \quad (1)$$

where:

L_{day} is the average long-term weighted acoustic pressure level for the day period, estimated or measured as an average value during a year, as defined in ISO 1996-2:1987;

$L_{evening}$ is the average long-term weighted acoustic power level for the evening period, estimated or measured as an average value during a year, as defined in ISO 1996-2:1987;

L_{night} is the average long-term weighted acoustic power level for night period, estimated or measured as an average value during a year, as defined in ISO 1996-2:1987.

In Romania, the Directive was implemented at national level by the Government Decision no. 321/2005 on ambient noise assessment and management, updated in which the standard for calculating the noise parameters is specified, namely SR ISO 9613-2: "Acoustic-Attenuation of sound spread outdoors, Part two: General Method of calculation." Also, as is beforementioned the noise assessment has to use calculations methods which is cheaper than using the monitoring stations.

In this article a small locality, Cochirleanca in Buzau County, Romania, was chosen as a model and several scenarios of wind-producing devices were studied, the output parameters being selected according to the aforementioned directive and national regulations.

The locality was chosen because it is located in an area with wind potential and is representative for rural localities covering a significant area and placed along traffic routes, which adds a significant exposure to road noise.

The locality Cochirleanca has in WGS system the coordinates 45°12′30"N,27°2′20"E and has a population of 5805 inhabitants (website: http://www.edrc.ro).

2 Input data

The calculation method has to include the minimal following elements [6]:

1. Acoustical and physical characteristics of sound pollution sources: type (punctual, linear, volume), size, height, characteristic data of noise emission. In our case we defined two types of sound pollution sources:

-wind turbine which is a punctual source;

-road traffic which is a linear source.

2. Acoustical and physical characteristics of the land type (the existing reflections from the source to the receiver) and especially to green spaces, lakes, parking, asphalt areas. It is also including the topography of the land (height), both the emission points and the receptors (the altitude of each point in the field component is used as a parameter). If the proportion of a type of land is very low (less than 10% of the total land) for simplification it can be considered either the land is fully reflective or absorbent. In our case we have:

- data for land type, (Fig.2.a), from Open Street Maps (website: https://www.openstreetmap.org);

- data for land model, (land height points) from Shuttle Radar Topography Mission (website: https://lpdaac.usgs.gov).

3. Acoustical and physical characteristics of obstacles (dimensions, shape, material), buildings, bridges, walls. In our case we defined:

- data for buildings (in Fig.2.b)- Open Street Maps (website: https://www.openstreetmap.org);

- we made a parapet on both sides of the road, on we considered that a vertical wind turbine is built.

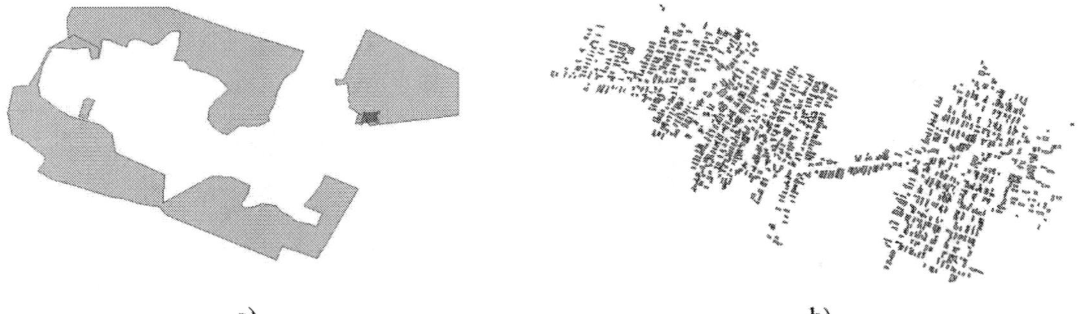

a) b)

Fig. 2. GIS data used in the model : a) land; b) buildings.

4. Receivers characteristics. The receivers can be receptors distributed:

- evenly over the entire surface of the calculation model (as we present in the calculation maps in this article);

- at the front level (relevant only for high-rise buildings with a population distributed vertically);

- evenly distributed inside buildings (to determine the affected population).

For modelling and calculation, NoiseModelling software was used [7] where all the elements above mentioned make a GIS database (Geographic Information System) comprising the geometric elements and attributes of each element. The calculation model includes:

-Weather statistical data for the whole year. For our model all weather statistical data (temperature – Fig.3a), humidity, wind speed and wind direction Fig. 3 b) are from Meteoblue website (website https://www.meteoblue.com).

-calculation method (for industry or for road);

-the grid size (10mx10m in our case);

-the number of reflections of the acoustical ray with the model objects (2 in our case);

-the day intervals (Day from: 07h-19h, Evening:19h-23h and Night:23h-07h);

-the tracer method(fast ray, ray and beam).

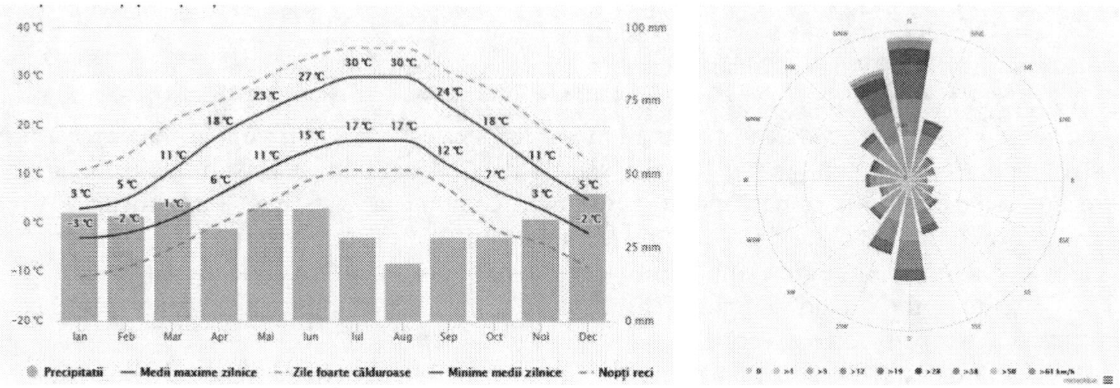

a) b)

Fig. 3. Weather data used in the model (website https://www.meteoblue.com): a) temperature and humidity; b) wind speed and direction

3 Noise assessment

In the article, we have analyzed the following scenarios, depending on the size and distribution of eolian generating systems:

- Scenario 1 - Large size wind turbines - producing 750kW with a 45m rotor diameter (including blades);

- Scenario 2 - Small size wind eolian farm – consisting of 70 independent systems producing each 10 kW with a 3m rotor diameter. The distance between the turbines is 300m and has been considered the maximum number that could be distributed in the calculation area.

- Scenario 3 - Small horizontal wind turbines – which produce under 1kW with 1m rotor diameter. There were considered 750 wind systems to reach the minimum energy level generated in Scenario 1, uniformly distributed on the area.

- Scenario 4 - Small vertical wind turbines – producing less than 1kW located at the road parapets.

For each scenario, calculations were made and the values of noise indicators are graphically displayed in the form of noise maps, in which the calculation area is divided into equal areas of 10m x 10m each and for each noise level band of 5dB a unique color is used (the color code is recommended in the regulations).

3.1 Scenario 1 - Large size wind turbines

In the first scenario was considered the case where a single large wind is used. The wind turbine is assumed to be on a 50m tower and can be acoustically modeled as an omnidirectional source with the sound power level of 102 dB(A) [5]. The location of the wind system was considered more than 500m from the nearest dwelling. As can be seen in the Fig. 4, the impact is relevant at the buildings level in the area with immission over 75dB.

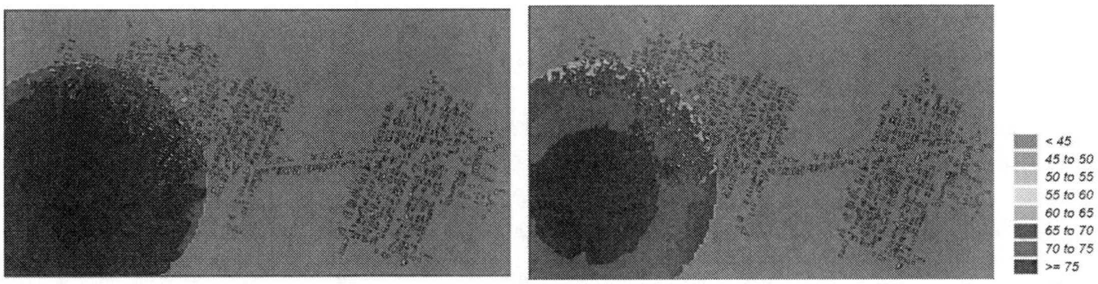

a) b) c)

Fig. 4. Noise Map considering the first scenario: a) L_{DEN} indicator level; b) L_{night} indicator level ; c) graphical correspondence using the color code, the values are in dB

3.2 Scenario 2 - Small size wind eolian farm

In this case we consider a wind farm consisting of Bergey BWC Excel 10 (website: https://en.wind-turbine-models.com) type systems, acoustically modeled as omnidirectional source with the sound power level of 100dB(A) (website: http://www.wind-works.org). The farm is split in two, one location is in the South West of the map and the second is in North East, because there is a minimum distance limitation between the turbines. In this case you can see that the impact is much stronger than the one in Scenario 1.

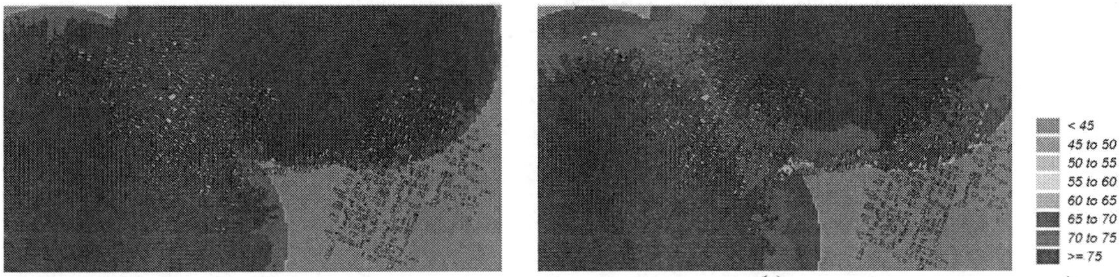

a) b) c)

Fig. 5. Noise Map considering the second scenario: a) L_{DEN} indicator level; b) L_{night} indicator level ; c) graphical correspondence using the color code, the values are in dB

3.3 Scenario 3 - Small Wind individual eolian turbines

In this case we consider that each dwelling has an individual wind turbine, Bergey BWC Excel 10 (website: https://en.wind-turbine-models.com) type, acoustically modeled as an omnidirectional source with the sound power level of 60.79 dB(A)[4]

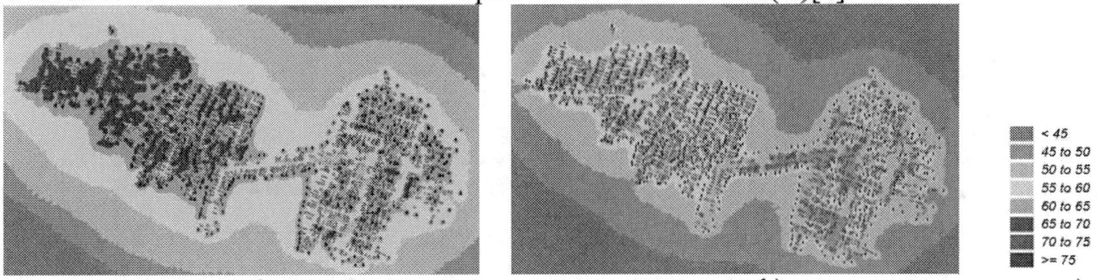

a) b) c)

Fig. 5. Noise Map considering the third scenario: a) L_{DEN} indicator level; b) L_{night} indicator level ; c) graphical correspondence using the color code, the values are in dB

3.4 Scenario 4 - Small Wind vertical eolian turbines

As final scenario, we have considered that we can use some vertical wind turbines which are located on the parapet of the road. They are put in motion by air currents generated by the vehicles moving along the road.

Due to the relevant road traffic through the locality, the inclusion of the population in the legally accepted immission intervals requires the use of noise reduction solutions. The possibility of using acoustic insulating panels, modified according to new concepts, that propose the recovery of energy of air currents produced in the circulation of motor vehicles through the integration of micro-turbines in panels has been analyzed. Although with a high manufacturing complexity (it is necessary to use a construction that integrates the turbines but does not allow the propagation of the sound, the easiest placement of the turbines being in the upper area of the panels, where usually are placed devices for reducing sound diffraction) the solution is interesting from the energy point of view and allows faster recovery of the investment. At the same time, by placing the walkways, you can remove the speed restrictions through the locality. In this case, according to Government Decision No. 321/2005 updated the method for calculating the noise parameters "NMPB Routes-96 (SETRA-CERTU-LCPC-CSTB)", a French national method of calculation, part of the French standard XPS 31-133, was used for assessment of noise generated by car traffic.

From the street network of the locality, we considered only the main street (DJ220). The input data collected on April 12, 2019, is:

- The traffic data
  - Small vehicles (under < 3, 5kg) Day-75, evening-40, night-10;
  - Large vehicles (> 3.5 kg), day-30, evening-15, night-5;
  - Vehicle speed – 50km/h;
- the surface is asphalt without pits;
- traffic is considered to be pulse flow type.

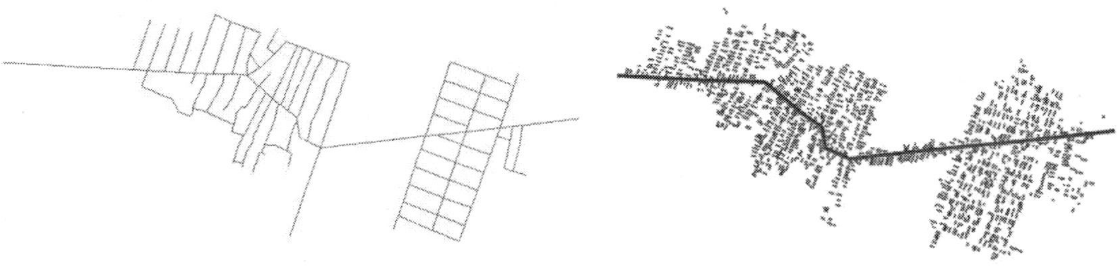

a) b)

Fig. 6. The road network of the locality : a) full road network; b) the calculation model with DJ220

As could be seen in Figure 7, the impact of the noise produced by the road is an important one, requesting solutions for protecting the dwellings and population.

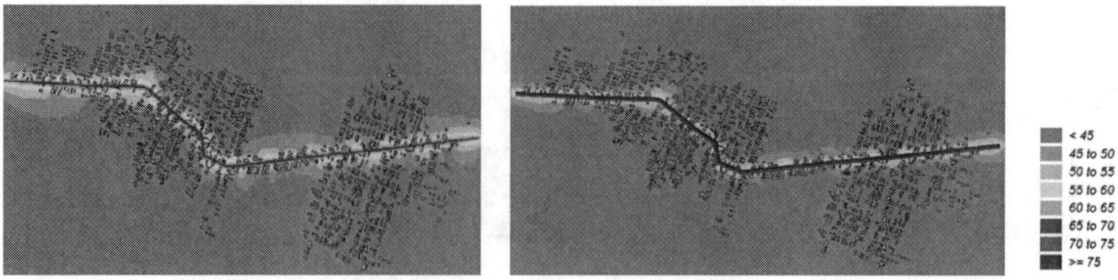

a) b) c)

Fig. 7. Noise Map considering the road noise: a) L_{DEN} indicator level; b) L_{night} indicator level ; c) graphical correspondence using the color code, the values are in dB

As a result of the application of the panels that support micro-turbines, a reduction in the noise produced by road traffic may be observed (Fig. 8).

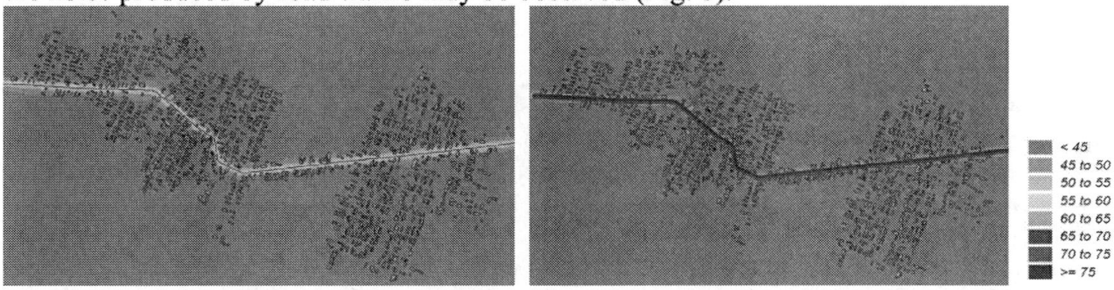

a) b) c)

Fig. 8. Noise Map considering the road noise and panels: a) L_{DEN} indicator level; b) L_{night} indicator level ; c) graphical correspondence using the color code, the values are in dB

The microturbines were modeled separately as a network of omnidirectional noise sources of under 40dB [8] placed at height on the panels. A separate noise map has been calculated for these (Figure 9).

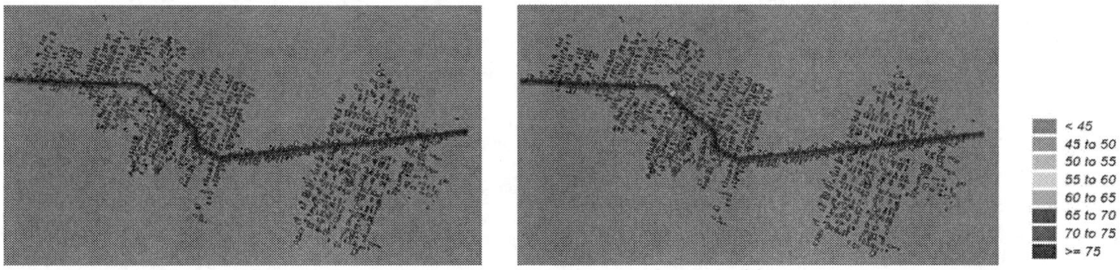

a) b) c)

Fig. 9. Noise Map considering the micro-turbines: a) L_{DEN} indicator level; b) L_{night} indicator level ; c) graphical correspondence using the color code, the values are in dB

Noise maps for road traffic and microturbines were summed up and the exposed population estimated (Figure 10).

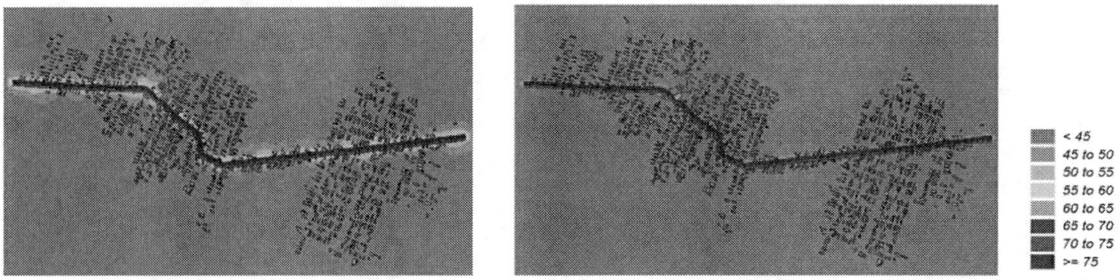

a) b) c)

Fig. 10. Cumulated Noise Map considering the road noise with panels and microturbines: a) L_{DEN} indicator level; b) L_{night} indicator level ; c) graphical correspondence using the color code, the values are in dB

4 Results and conclusions

The maximum noise values for Romania, set according to the Ministry Order MMDD/MT/MSP/MIRA Nr. 152/558/1119/532/2008 (MO No. 531/15.07.2008), are presented in table 1:

Table 1. Noise limit values

Noise source	Maximum Allowed Values, computed on a 24-hour basis L_{DEN} dB(A)	Maximum Allowed Values during Night L_{night} dB(A)
Streets, roads and highways	65	50
Industrial Areas	60	50

Using the calculation method, we have obtained the affected population, divided in bands of 5dB range of noise, as presented in table 2. As can be seen, for common wind-turbines, in terms of 24 hours noise the weakest variant consists of Scenario 2 - the L_{DEN} level; scenario 3 is the weakest night values – the L_{night} level. We can say that using only one wind turbine is the best case when we take into consideration the efficiency, the cost and the pollution.

Table 2. Comparison of affected inhabitants

Noise interval (dB)	Affected inhabitants Scenario 1		Affected inhabitants Scenario 2		Affected inhabitants Scenario 3	
	L_{DEN}	L_{night}	L_{DEN}	L_{night}	L_{DEN}	L_{night}
<45	4449	4449	1408	1408	0	11
45 - 50	0	1	0	8	0	577
50 - 55	1	11	0	0	284	2097
55 - 60	5	16	8	8	1611	2490
60 - 65	11	160	8	33	2690	604
65 - 70	74	632	24	536	1168	25
70 - 75	450	480	250	1824	52	0
75>	814	55	4107	1988	0	0
Population above the limit	1349	1354	4389	4389	3910	5216

We must consider the fact that the acoustic level of the eolian system is valid at strong wind exceeding 50km/h, which is considered the most disadvantageous case.

The scenario 4 uses only secondary the wind because the currents are produced by road traffic. This is an advantage because this is independent of the weather conditions. Another advantage is a small noise pollution, as can be seen in table 3 and, even more, the panel is isolating a part of the noise generated by traffic. The disadvantage lies in a reduced efficiency and therefore in a higher energy production price.

Table 3. Comparison of affected inhabitants

Noise interval (dB)	Affected inhabitants without panel		Affected inhabitants with panel and micro-wind turbines	
	L_{DEN}	L_{night}	L_{DEN}	L_{night}
<45	3947	4759	4387	4890
45 - 50	547	252	405	670
50 - 55	301	766	164	230
55 - 60	456	28	656	15
60 - 65	555	0	192	0
65 - 70	0	0	0	0
70 - 75	0	0	0	0
Population above the limit	555	0	192	0

In the case of small cities with limited areas, analyzing the affected population, the costs of implementation and from the practical point of view, we can say that the use of large turbines is the most efficient of existing turbine solutions. An interesting solution especially for cities along high traffic roads that produce high pollution is the use of acoustic insulating panels incorporating micro-turbine solutions which can be implemented in different architectural cityscapes for a better visual impact. It achieves both a strong reduction in street noise and the possibility of removing speed restrictions, so a reduction of pollution from thermal motors, as well as a generation of electricity. These micro-turbines are much easier to install and maintain than large turbines without the need for specialized personnel.

These types of micro-turbines will be the subject of additional studies, analyzing the possibility of combining with asphalt-mounted energy recovering systems (using piezoelectric materials).

References

1. V. Lapčík, *Wind Farm - Technical Regulations, Potential Estimation and Siting Assessment* (InTech 2011)
2. S. Mathew, *Wind energy. Fundamentals, Resource Analysis and Economics* (Springer 2006)
3. M. Dyrholm, *Global wind report – 2018* (GWEC 2018)
4. J. Taylor, C. Eastwick, C. Lawrence, R. Wilson, Ren. En., **55** (2013)
5. A.L. Rogers, J. F. Manwell, S. Wright, *Wind Turbine Acoustic Noise* (White paper, 2006)
6. G. Licitra, *Noise Mapping in the EU – Models and Procedures* (CRC Press 2013)
7. E. Bocher , G. Guillaume, J. Picaut, G. Petit, N. Fortin, ISPRS Int. J. Geo-Inf. **8**, 3 (2019)
8. J. D. M. Botha, H. Rice, *A novel method of vertical axis wind turbine noise prediction*, (EuroNoise 2015)

Irradiance characteristic of a small-scale solar simulator for testing thermal collectors

Adina-Teodora Gheorghian[1,*], *Valentin* Apostol[1], *Tudor* Prisecaru[1], *Camelia* Stanciu[1] and *Horaţiu* Pop[1]

[1] Politehnica University of Bucharest, Department of Engineering Thermodynamics, 313 Splaiul Independentei, 060042, Bucharest, Romania

Abstract. This paper describes the development of a small-scale solar simulator for research and educational purposes. The main goal is to provide a uniform and stable distribution of irradiation from a set of four floodlight halogen lamps, over a targeted area of 40x40 cm^2. The floodlights are placed perpendicularly to each other, on the top end of a light tube. Two types of halogen lamps were characterised and the total irradiation of the solar simulator was measured over a 5x5 grid. The results showed that the two arrangements of lamps are able to achieve a level of 1000 W/m^2 on the 40x40 cm^2 targeted area, but the non-uniformity is unsatisfactory. In order to reach a non-uniformity level of 10% using the proposed lamps arrangements, the targeted area should be reduced to 31x31 cm^2.

1 Introduction

Thermal collectors are often tested indoors using a solar simulator that provides uniform and stable irradiance at certain levels.

The development of a solar simulator for indoor testing and the certification procedure for photo-thermal solar devices is discussed in [1]. The solar simulator consists of 14 Philips no. 13989 R quartz halogen lamps (1000 W, 25000 lm, 3200 K) mounted in a Philips NV 51 luminaire with built-in reflectors and provide a testing area of about 1x1 m^2. The power of the lamps can be modified by means of adjustable transformers. Thus, the radiation at the testing area level can be adjusted between 400 and 1500 W/m^2. The irradiance distribution over the testing area has been recorded using a precision Eppley pyranometer.

The light source and production of an economical large-scale solar simulator is presented in [2]. The solar simulator consists of 188 lamps (400 Watt, metal halide gas discharge type), mounted on a steel frame (4.5x3.88 m) and four variable speed elevators, which are either manually or computer controlled. Measurements of the irradiance indicates that the simulator conforms to Class B of the ASTM E927 (2005) and 60904-9 (2007) standard in regard to spectrum match, irradiance uniformity and stability. The average irradiance on the targeted area can be adjusted from 150 to 1100 W/m^2 by varying the number of lamps and/or the lamp-to-area distance.

* Corresponding author: adina_gheorghian@yahoo.com

Paper [3] describes an empirical method for evaluating a solar simulator testing facility based on an experimental performance. For testing evacuated solar collectors, a sunlight simulator comprising of an array of 30 halogen floodlights (400 W, 3200 K) covering an area of 2.32 m^2 has been assembled and analysed. The halogen lamps were divided into groups of 10 pieces. The three groups of halogen floodlights were connected to the grid via a three-phase adjustable transformer, which allowed continuous adjustment of the power consumption and, implicitly, of the irradiation intensity. The authors showed that the higher the radiation intensity, the greater the non-uniformity of irradiation. Maximum non-uniformity based on experimental data was 9.1%, less than the 15% limit set in British standards for solar simulator performance.

A simple solar simulator which consists of one 300 W halogen lamp, a Luxmeter (Lutron LX-101A), and solar panel holders placed on a closed light box with the dimension of 40×40×40 cm^3 is described in [4]. The spatial distribution of irradiation inside the light box was theoretically estimated with the Monte Carlo method and experimentally validated using a luxmeter. It was found that the theoretical and the experimental data are in good agreement and the irradiation uniformity was acceptable inside the light box.

Paper [5] presents a development of a solar simulator for education and research purposes. The simulator was built using eight 500 W/220 V halogen lamps. Two groups of four lamps were connected in series, and then the two groups were connected in parallel. The lamps were placed in the nodes of a 2x4 equidistant grid, perpendicularly to each other. The irradiation produced by the eight lamps was measured with a pyranometer at different distances between reflectors and different distances up to the target area. Experiments have shown that the uniformity and intensity of irradiation are appropiate when the distance between the reflectors is 15 cm and the distance to the target plane is between 20 and 40 cm.

A solar simulator using an internally reflecting light tube has been designed and built for testing small size (0.5 x 0.5 m) flat plate thermal collectors [6]. Four halogen floodlights were placed at the top end of the light tube to ensure uniform illumination. Ray-tracing simulations were used to choose the tube dimensions and maximum allowable distance from the lower end of the light tube to the collector. Simulations carried out with this solar simulator show that the efficiency of the solar collector based on the pyranometer's readings is approximately 1% higher than it would have been obtained when using an AM1.5 spectrum.

Paper [7] is aimed at discussing in detail the development, characteristics and calibration of a simple low cost solar simulator using 16 halogen lamps (50 W, 110 V) mounted on an aluminium plate of 380mm×350 mm as light sources. The non-uniformity of irradiance values were investigated at different distances D between the light surface simulator and solar collector's surface and the maximum non-uniformity was found to be ~12% (at D = 32 cm).

This paper presents the development and characterization of a solar simulator using four halogen floodlights and a light tube for testing flat plate thermal collectors with a maximum targeted area of 400x400 mm^2. The design of the solar simulator is similar to those described in paper [6].

2 Experimental set up and procedure

2.1 Irradiance distribution of the floodlight

During the first stage, the intensity and distribution of radiation produced by a single floodlight was determined experimentally over a plane located one meter below. The measurements were carried out with two different types of halogen lamps, in order to determine which one ensures the maximum required irradiation level.

The shape and appearance of the two lamps used in the experiment are shown in Fig. 1.

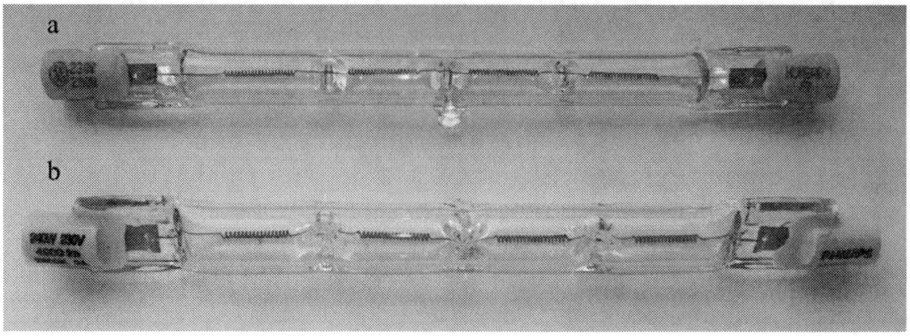

Fig. 1. The selected lamps for experiment: a - Halogen Linear OT K9 C200W 230V R7S 117MM GE General Electric; b - Plusline ES Small 118mm 2y 240W R7s 230V 1BB Philips

The comparison between the lamps with respect to performance and electrical data is presented in Tables 1, 2 and 3.

Table 1. Halogen lamps - Product data

Brand	Product code	Bulb shape	Bulb diameter	Overall Length	Cap/Base
			mm	mm	-
General Electric	64968	tubular	8	119	R7s
Philips Plusline ES Small	87279008 5380300	tubular	12	117.6	R7s

Table 2. Halogen lamps - Performance data

Brand	Rated efficacy	Energy efficiency	Rated Life	Color temp.	Nominal lumens
	lm/W	-	h	K	lm
General Electric	20	C	2000	3000	4000
Philips Plusline ES Small	20.4	C	2000	2900	4900

Table 3. Halogen lamps - Electrical data

Brand	Switching cycles	Coil type	Dimming Capability	Nominal power	Nominal voltage
	-	-	-	W	V
General Electric	8000	CC-8	yes	200	230
Philips Plusline ES Small	8000		yes	240	230

The selected lamps were inserted into a symmetrical model 0201 HEPOL floodlight (Figure 2).

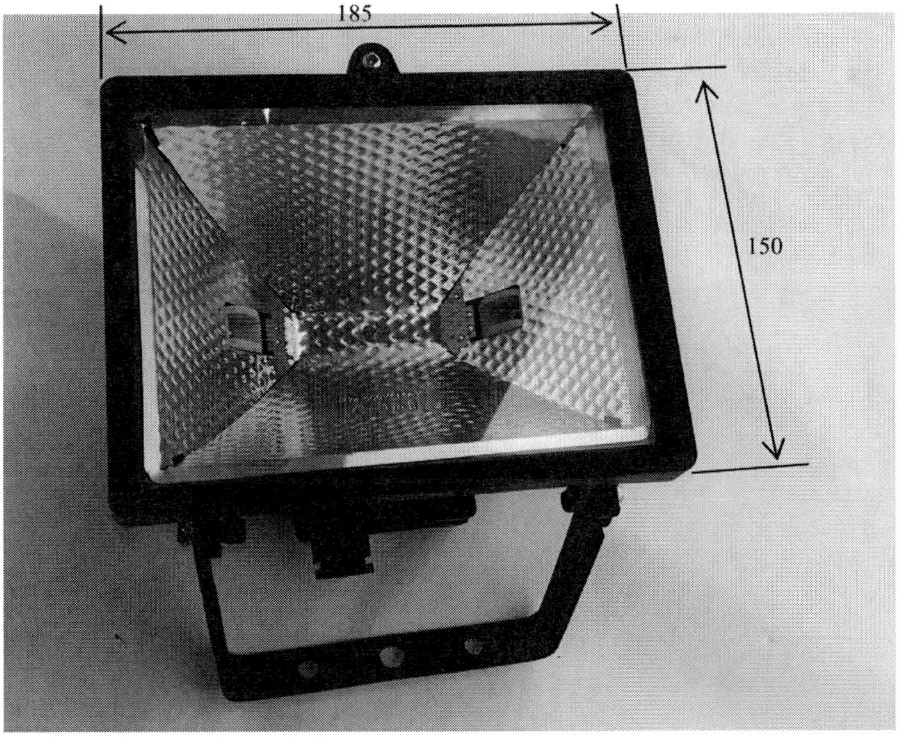

Fig. 2. Symmetrical floodlight model 0201 HEPOL

The intensity of the irradiation produced by the floodlight over the targeted plane was measured using a pyranometer at several equidistant points along the two perpendicular axes, xx and yy, defined as shown in the scheme in Figure 3.

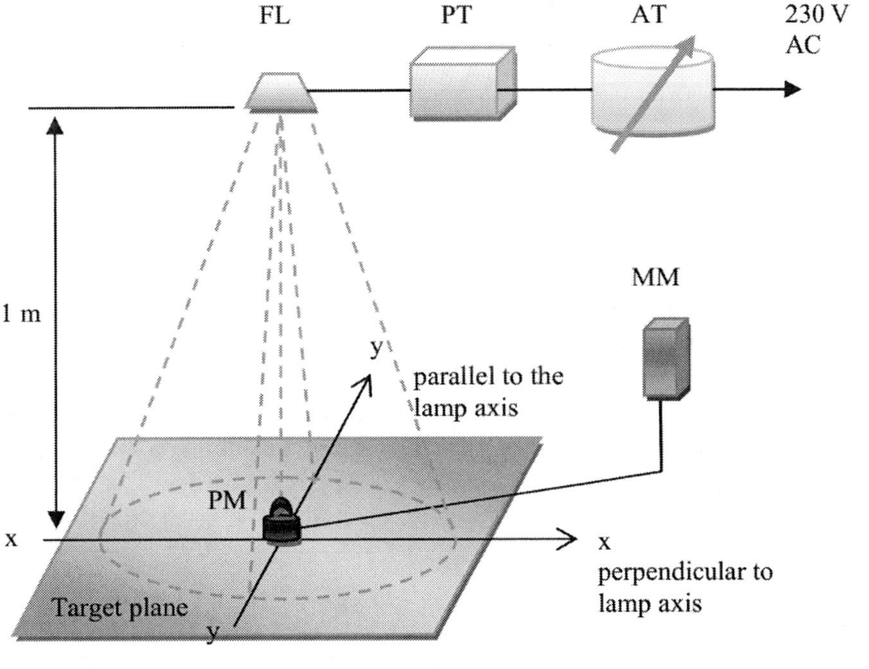

Fig. 3 Schematic of irradiation measurement setup: FL – HEPOL floodlight; PT – SIRAX BT5400 power transducer; AT – HSN METREL adjustable transformers; MM – FLUKE 87 V multimeter; PM – EKO MS-60 first class pyranometer

The power input of the lamp was monitored on the power transducer display. During measurements, the input power of the lamp was kept constant and equal to the nominal power of the lamp by means of the adjustable transformer. The irradiance was read on the multimeter's display and recorded manually, from 5 to 5 centimetres, along the axis. Figure 4 reveals the irradiation distribution of each tested lamp.

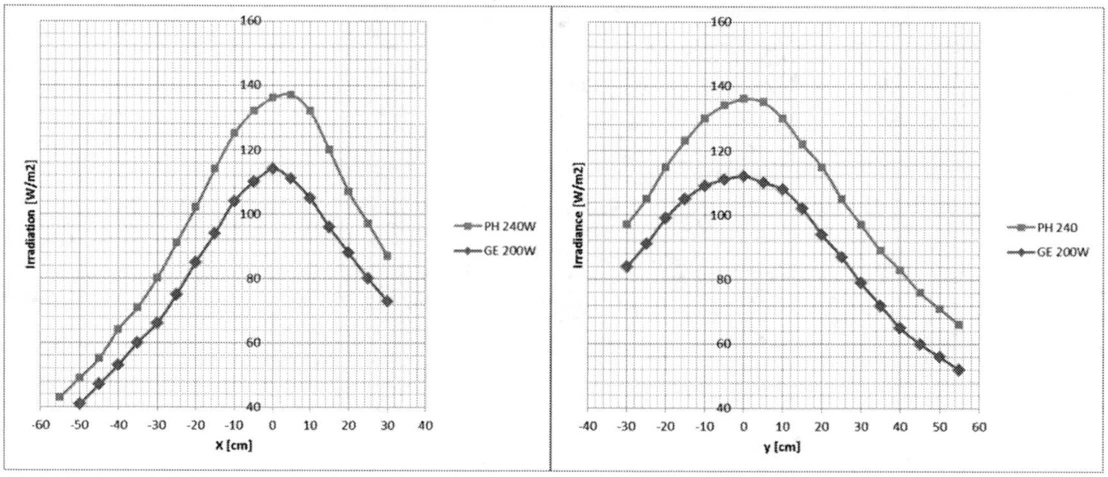

Fig. 4. Distribution of lamp irradiation perpendicular to lamp axis (left) and parallel to lamp axis (right)

As expected, the 240 W Philips lamp is brighter than the 200 W General Electric lamp by about 24 W/m^2 in the centre and by only 10 W/m^2 at the edges. Comparing the two charts, it can be noticed that irradiance along the x axis is stepper than the one along the y axis, which is due to the cylindrical (linear) bulb shape.

2.2 Irradiance distribution of the solar simulator

To reduce irradiation's non-uniformity over the target plane, the four floodlights of the solar simulator were placed equidistantly on a circle and rotated 90° apart from each other on the top end of the light tube, as shown in Figure 5.

The walls of the light tube were made of 10 mm OSB plywood, reinforced with 24x48 mm timber laths on the edges. An aluminium foil was previously glued using BISON Colla™ spray adhesive, on the inner surface of the tube walls, in order to get a proper reflective surface for the floodlights.

The power transducer and the multimeter were attached on the outer surface of the light tube (see Figure 6).

Similarly to the previous experiment, the input power of the four floodlights was kept constant and equal to four times the nominal power of the lamp, while the irradiance was measured by reading the multimeter's display in 25 locations on a 5x5 grid (the distance between two adjacent points of the grid on x or y direction was 10 cm, so the targeted area was 40x40 cm^2).

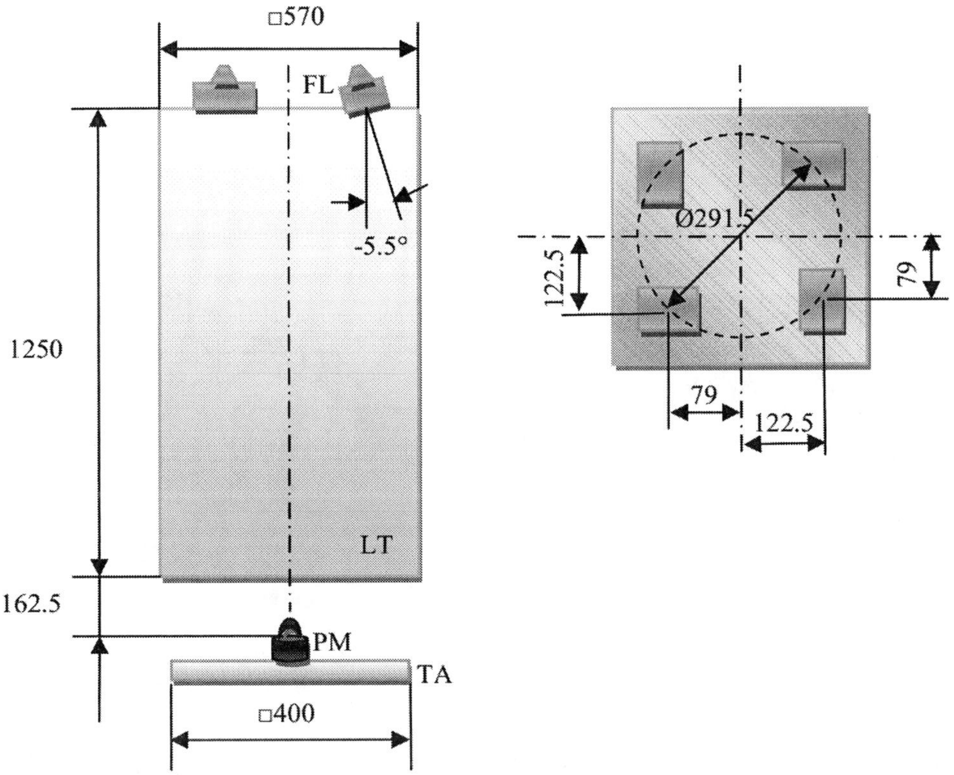

Fig. 5 The arrangement of the floodlights on the top of the light tube: FL – floodlight; LT – light tube; PM – pyranometer; TA – tageted area

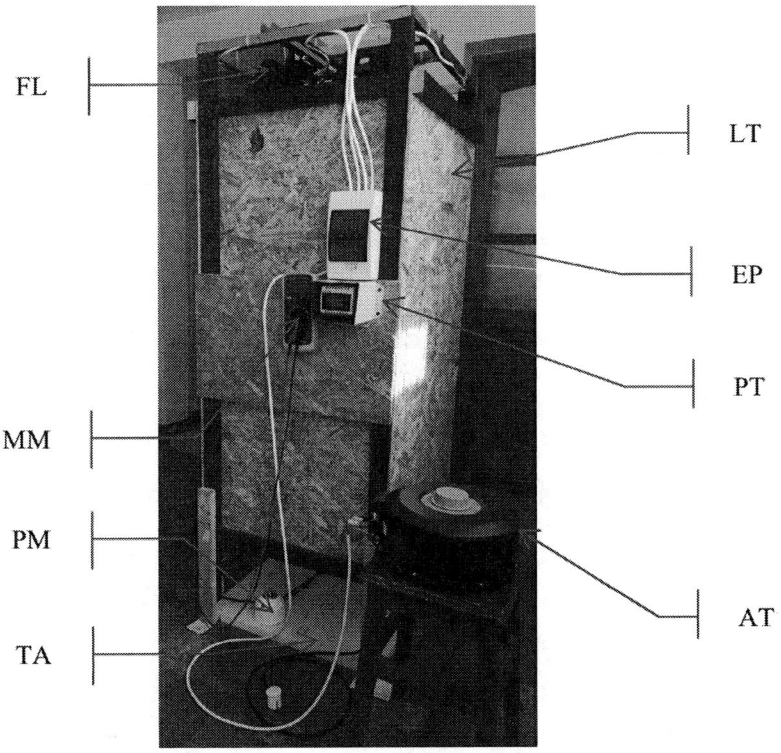

Fig. 6 View of the solar simulator: FL – floodlight; EP – electric panel; PT- power transducer; LT – light tube; AT – adjustable transformer; MM- multimeter; PM – pyranometer; TA – targeted area

3 Results and discussion

Figure 7 shows the results for the four floodlights arranged on the top of the light tube as presented in Fig. 5.

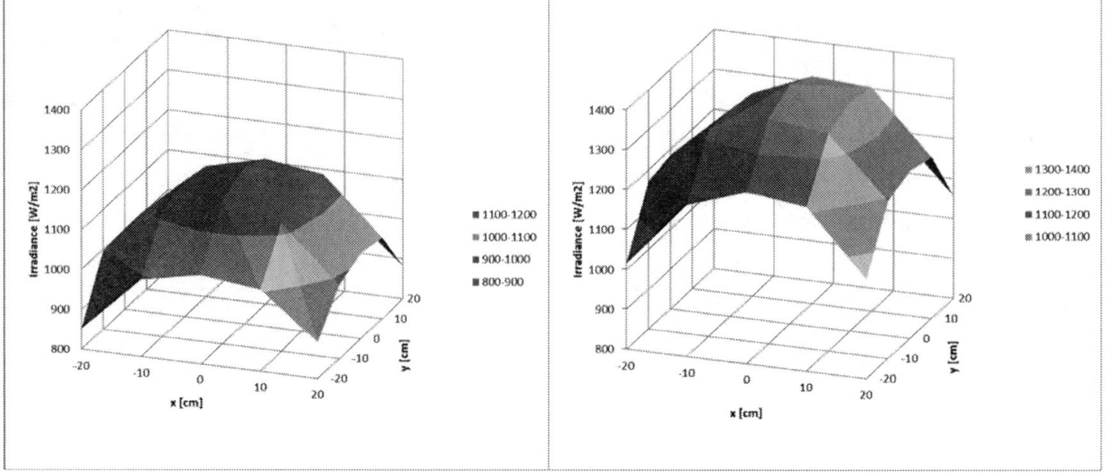

Fig. 7 The irradiation distribution over the targeted area: 4x200W General Electric lamps (left) and 4x240W Philips Plusline ES Small lamps (right)

The statistics of the two sets of experimental data are summarized in Table 4.

Table 4. Irradiance statistics

Lamp	Mean	Min	Max	SD	Non uniformity
-	W/m^2	W/m^2	W/m^2	W/m^2	%
4x200W General Electric	1035.6	852.0	1194.0	98.1	16.7
4x240W Philips Plusline ES Small	1226.2	1000	1403.0	115.7	16.8

According to the IEC 60904-9 international standard, the non-uniformity of the irradiance on the test plane is

$$non\_uniformity = (max_{\mathrm{irr}} - min_{\mathrm{irr}})/(max_{\mathrm{irr}} + min_{\mathrm{irr}}) \cdot 100\,\%, \quad (1)$$

where the maximum and minimum irradiance were measured using the pyranometer over the tested area.

The average area mean irradiation computed by integration on the 40x40 cm^2 targeted area was 1076.76 W/m^2 for 4x200 W General Electric lamps and 1273.67 W/m^2 for 4x240 W Philips Plusline lamps, therefore both arrangements are able to ensure a mean irradiance of 1000 W/m^2. However, as one may notice from Figure 7 and the Table 4, both arrangements display a low level of irradiance in the corners of the 40 x40 cm^2 targeted area. The non-uniformity of the irradiation is significantly lower than 10% as provided ASTM and IEC standards for C class simulators.

The additional calculations show that using these arrangements, the surface of targeted area should be reduced to 31x31 cm^2 in order to achieve a C class non-uniformity.

4 Conclusions

Based on the experimental study presented in this paper, the following concluded remarks have been made:

- The main purpose of this study was to develop a solar simulator capable of producing a stable and uniform irradiation with adjustable intensity between 400 and 1000 W/m^2, over a targeted area of 40x40 cm^2, for testing prototypes of thermal plane collectors;
- The study was carried out for research and educational purposes, using two types of halogen lamps;
- The four floodlights are placed on the top end of a 57x57x125 cm light tube, with a clearance of 16.2 cm, and wired to an electrical circuit which enables accurate measurements and adjustments of input power and irradiance levels;
- The two lamps arrangements are able to achieve a level of 1000 W/m^2 on the 40x40 targeted area, but the non-uniformity has high values;
- In order to achieve a non-uniformity level of 10 % using the studied arrangements of lamps, the targeted area should be reduced from 40x40 cm^2 to 31x31 cm^2.

This research work was supported by Politehnica University of Bucharest within the program National Grants GNaC 2018 ARUT, contract no ME15-18-01.

References

1. H. P. Garg, A. R. Shukla, Indrajit, Madhuri, R. C. Agnihotri and S. Chakravertty, *Development of a Simple Low-Cost Solar Simulator for Indoor Collector Testing*, Applied Energy **21** 43-54, (1985)
2. Qinglong Meng, Yuan Wang, Linhua Zhang, *Irradiance characteristics and optimization design of a large-scale solar simulator*, Solar Energy **85** 1758–1767, (2011)
3. Mahmoud Shatat, Abdulkarym Mayere, Saffa Riffat, *A standardized Empirical Method of Testing Solar Simulator Coupled with Solar Tube and Concentrator Collectors*, Int. J. of Thermal & Environmental Engineering **5**, No. 1 13-20, (2013)
4. Rahmat Awaludin Salam, Casmika Saputra, Elfi Yuliza, Irwan Meilano, Muhammad Miftahul Munir, Hamzah Latief, and Khairurrijal, *Development of a Simple Low-Scale Solar Simulator and Its Light Distribution,* 2016 International Conference on Instrumentation, Control and Automation (ICA), Institut Teknologi Bandung (ITB), Bandung, Indonesia, August 29-31, (2016)
5. Hazim Moria, *Radiation distribution uniformization by optimized halogen lamps arrangement for a solar simulator*, Proceedings of the International Conference on Industrial Engineering and Operations Management Rabat, Morocco, April 11-13, (2017)
6. R.W. Moss, G.S.F. Shire, P.C.Eames, P. Henshall, T.Hyde and F.Arya, *Design and commissioning of a virtual image solar simulator for testing thermal collectors*, Solar Energy **159**:234–42, (2017)
7. Erkata Yandri, *Uniformity characteristic and calibration of simple low cost compact halogen solar simulator for indoor experiments*, International Journal of Low-Carbon Technologies, **13**, 218–230, (2018)

Monitoring the State of Small Power Equipment

Andrei Dragomirescu[1,*], *Crăița* Carp-Ciocârdia[2], *Carmen-Anca* Safta[1], *Nicolae* Orășanu[2], and *Ioan* Magheți[2]

[1]University Politehnica of Bucharest, Department of Hydraulics, Hydraulic Machinery and Environmental Engineering, 313 Splaiul Independentei, Bucharest, Romania
[2]University Politehnica of Bucharest, Department of Mechanics, 313 Splaiul Independentei, Bucharest, Romania

Abstract. The paper presents analyses of vibrations measured on power equipment that can be encountered in rural spaces. A centrifugal pump that can be used in a small supply water network or in an irrigation system and a small hydraulic turbine designed for small hydropower plants were tested. Power spectra obtained with the Fast Fourier Transform for the vibrations measured at the pump and at the turbine are presented and discussed and defects that affect the normal operation of the machines are identified. The results highlight the importance of vibration diagnosis in the process of monitoring the operation of the machines, as a proactive maintenance method.

1 Introduction

For the last two decades, the Romanian rural economy has been undergoing a continuum process of development aimed at improving the quality of life. Important investments in infrastructure, reflected in new facilities for electric power and thermal energy production, drinking water supplies, public sewerage systems, and waste management, have changed rural areas, which cover 87% of the country [1]. The Rural Development Programme for Romania, which is available for the period 2014-2020, supports investments in projects related to water efficiency and renewable energy in order to improve the competitiveness of agriculture [2].

Irrigation and water supply systems as well as systems for harvesting the renewable energy of small rivers require specific hydraulic machines, namely pumps and water turbines. The efficiency of these machines depends not only on their design and operating conditions, but also on the maintenance management. Vibration monitoring and analysis are nowadays widely used for the predictive maintenance [3], since they allow to obtain valuable information on the state of different types of equipment. In case of hydraulic machines, such as pumps or water turbines, the vibrations have a significant impact on the safety and reliable operation of these machines [4]. There are various sources of vibrations in hydraulic machines: the complexity

*Corresponding author: andrei.dragomirescu@upb.ro

of the flows inside the runners, the mechanical and hydraulic imbalances of the rotating parts, the wear of the bearings, the misalignment of the shafts, etc. The complexity of the flow causes turbulence and boundary layer separations that are usually accompanied by pressure fluctuations which are sources of vibrations. The mechanical imbalance can be caused either by added masses, e.g. small debris that can remain stuck between two runner blades or the uneven distribution of the discharge inside the runner, or by subtracted masses, e.g. pieces of material removed by abrasive or cavitation erosion. The hydraulic imbalance is usually caused by the interactions between rotating parts – the runners – and fixed parts – stay vanes, wicket gates, diffuser vanes, or scroll casing tongs. The wear of the bearings occurs naturally, its rate being influenced by the operating regimes, loads, and maintenance. The misalignment of the shafts is usually caused by improper assemblies.

Identifying the causes of vibrations is a complex problem due to the features of the signals in the frequency spectrum, that characterize each particular cause. A frequent mechanical cause of vibrations is the malfunction of the bearings due to their wear that leads to defects.

The present paper presents analyses of the results obtained on two experimental models – one of a pumping station, the other of a small hydropower plant. Of all the causes of vibrations highlighted previously, the paper focuses on the identification of those frequencies in the vibrations spectra, that are caused by the defects due to the wear of the bearings of the machines that operate in the investigated installations.

2 Theoretical considerations regarding the mechanical vibrations caused by the wear of rolling bearings

Since many machines that have rotating parts are equipped with rolling bearings, an important cause of vibrations are the defects of the bearings. A defective bearing causes excessive vibrations that are transmitted to the casing or to the chassis of the machine, where they can be measured with accelerometers. In this way, a global evaluation of the bearing can be carried out. The deterioration of a bearing is usually a lengthy process that takes place in several stages:

- in the first stage, early minor defects can occur, being indicated by vibrations having ultrasonic frequencies;
- in the second stage, small defects can appear, stimulating the natural frequencies of the bearing elements – these frequencies are usually found in the range 500–2000 Hz;
- in the third stage, as the wear evolves, frequencies corresponding to the defects of the rolling bearings and their harmonics appear, being also possible to identify side bands around them – these frequencies are usually below 500 Hz;
- in the fourth stage, the deterioration becomes significant and leads to an increase in the clearings of the bearing accompanied by the disappearance of the usual frequencies of the bearing and the appearance of random high frequency bands; in the same time, the frequency of rotation becomes dominant and the number of its higher harmonics increases.

The excitation harmonics are correlated with the frequencies that appear in the last stage and, consequently, many spectral components become visible, which hinders the interpretation of the measured signals.

The frequencies that are specific to the defects of the rolling bearings depend on the frequency of rotation of the shaft, $f_r = n/60$, where n is the shaft speed in rpm, without being integral multiples of it. It can be stated that the defects of the rolling bearings are among the few sources of vibrations that generate harmonics which are not multiples of the frequency of rotation. When in the spectrum no frequencies that are specific to defects appear, it means that no defects exist. On the contrary, the appearance of such frequencies should be regarded as an evidence of failure. These frequencies and the relationships that allow to calculate them are the following [5–7]:

- fundamental train frequency (for a defect on the cage or separator):

$$FTF = f_r \frac{1}{2}\left(1 - \frac{D_b}{D_m}\cos\beta\right) \quad [\text{Hz}]; \tag{1}$$

- ball passing frequency inner race (for a defect on the inner race):

$$BPFI = f_r \frac{Z}{2}\left(1 + \frac{D_b}{D_m}\cos\beta\right) \quad [\text{Hz}]; \tag{2}$$

- ball passing frequency outer race (for a defect on the outer race):

$$BPFO = f_r \frac{Z}{2}\left(1 - \frac{D_b}{D_m}\cos\beta\right) \quad [\text{Hz}]; \tag{3}$$

- ball spin frequency:

$$BSF = f_r \frac{D_m}{2D_b}\left[1 - \left(\frac{D_b}{D_m}\cos\beta\right)^2\right] \quad [\text{Hz}] \tag{4}$$

- ball fault frequency (for a defect on a rolling element):

$$BFF = 2BSF = f_r \frac{D_m}{D_b}\left[1 - \left(\frac{D_b}{D_m}\cos\beta\right)^2\right] \quad [\text{Hz}]. \tag{5}$$

In the above formulas, f_r is the frequency of rotation of the shaft, Z is the number of rolling elements, that can be either balls or rollers, β is the contact angle, D_m is the pitch diameter (the average diameter of the bearing), and D_b is the diameter of the rolling element.

The relationships presented above were used to calculate coefficients that multiply the frequency of rotation in order to obtain those frequencies that indicate a particular defect. Such coefficients are usually tabulated in catalogs provided by producers of rolling bearings [8].

3 Experimental setups

As mentioned in the Introduction, the vibrations were measured on two experimental models: one of a standard pump station and the other of a small hydropower plant.

The experimental setup of the pump station, schematically depicted in Figure 1, contains a free surface water tank, a suction pipe having the inner diameter of 128 mm, a centrifugal

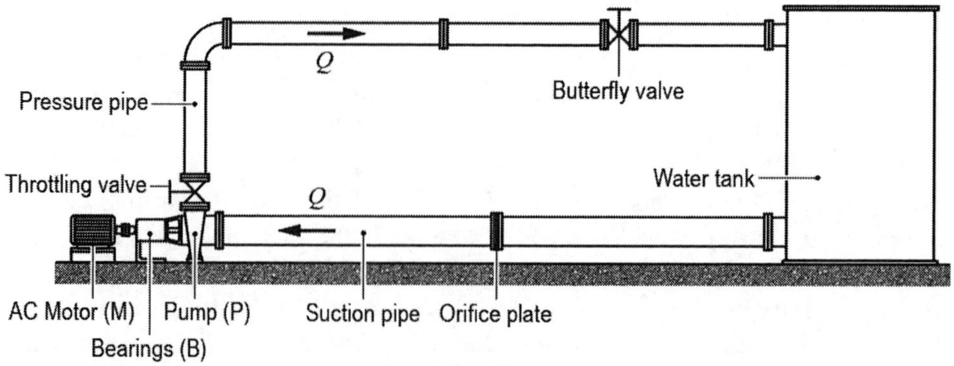

Fig. 1. Schematic of the experimental setup of the pump station.

pump driven by a three-phase AC motor, and a discharge pipe having the inner diameter of 104 mm. The AC motor has four poles, a rated power of 4 kW, and a nominal speed of 1460 rpm, to which a frequency of rotation, f_r, of 24.3 Hz corresponds. Another frequency which is important for turbomachines, i.e. machines provided with a bladed impeller or runner, is the blade passing freqency, f_b, defined as follows:

$$f_b = Z_b f_r, \tag{6}$$

where Z_b is the number of impeller or runner blades. The pump impeller has 5 blades, hence the blade passing frequency is of 121.5 Hz. The pump shaft is supported by two ball bearings of series 6310 and is connected to the motor shaft by a pin and bush coupling. At the pump outlet a gate valve for throttling the discharge is installed. The discharge is measured with an orifice plate installed in the suction pipe. The rather unusual installation of the orifice plate is due to the fact that this experimental setup can also be used to investigate the cavitation caused by a butterfly valve installed in the discharge pipe [3, 9]. Both the throttling valve and the butterfly valve can be used to simulate the variable demand of consumers supplied with water by the pump. To measure the differential pressure caused by the orifice plate, an U-tube differential manometer with mercury was used. At zero head, the pump delivers a maximum flow rate, Q_{max}, of roughly 43 l/s. This approximate value was obtained by extrapolating the head curve obtained experimentally.

Not all the data required by the relationships (1) to (5) were available for the bearings of series 6310. For this reason, the failing frequencies of these bearings were calculated by analogy with a similar bearing of series 310S from the MRC catalog [8].

The experimental model of the small hydropower plant is schematically depicted in Figure 2. The water turbine is of Banki type and is fed with water by a pumping station through an upstream pipe. Under the experimental conditions, the maximum power that can

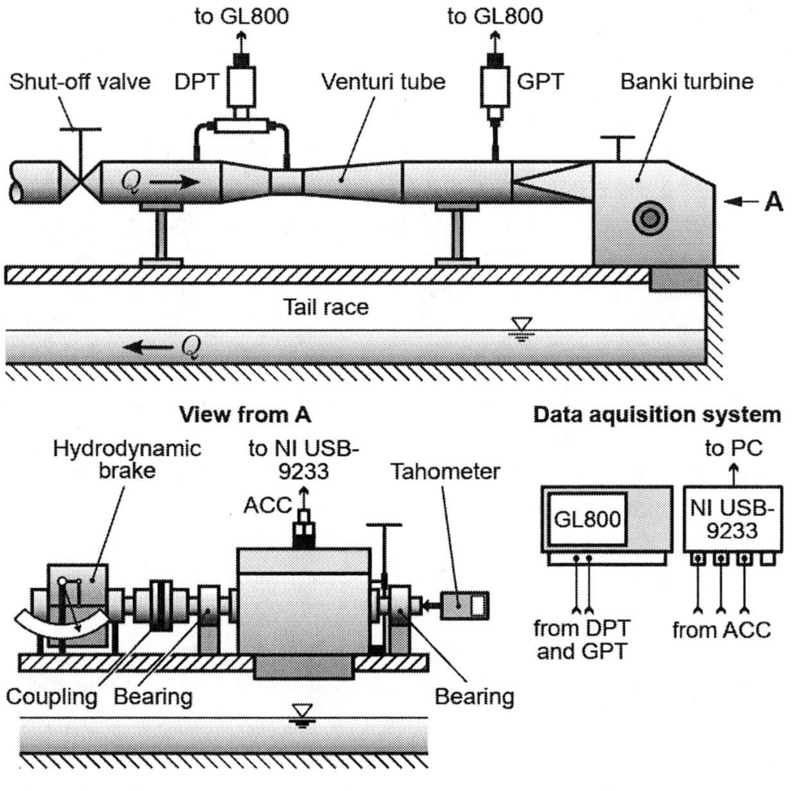

Fig. 2. Schematic of the experimental model of the small hydropower plant.

be attained by the turbine is of roughly 1.1 kW, although the turbine was initially designed for a higher power. The pipe has the inner diameter of 300 mm and is provided with a shut-off valve and with a welded Venturi tube for discharge measurement. The turbine is coupled by a pin and bush type coupling to its load represented by a hydrodynamic brake with oscillating stator, which replaces the generator. The brake allows to easily simulate different loads and to measure the corresponding torque (or shaft moment). It should be mentioned that, due to its operating principle, the hydrodynamic brake does not allow to slow down the turbine below a threshold speed, since the efficiency of the brake decreases rapidly as the shaft speed decreases. Nevertheless, the hydrodynamic brake has the important advantages of not generating additional shocks and vibrations and of allowing to accurately measure the torque.

The theory, construction and operation of Banki turbines are thoroughly described by Mockmore and Merryfield [10]. It should be mentioned here that these turbines are provided with an inner flap that allows the adjustment of the discharge. Banki water turbines are nowadays increasingly used in small hydro power plants due to their advantages, such as simple construction and maintenance, low price, and flat efficiency curve.

The runner of the tested turbine has 19 blades, a breadth of 300 mm, and the outer and inner diameters of 250 mm and 165 mm, respectively. The runner shaft sits on two double-row self-aligning radial ball bearings of series 2213, having the following design parameters: $Z = 18$ balls, $\beta = 0°$, $D_m = 93$ mm, and $D_b = 14$ mm.

The differential pressure at the Venturi tube and the static pressure at the turbine inlet were measured with a differential pressure transducer (DPT) and a gauge pressure transducer (GPT), respectively. The DPT and GPT have the measuring ranges of 0...100 mbar and 0...2.5 bar, respectively. Both transducers have accuracies of 0.5% FS (i.e. of their measuring ranges). The transducers were connected to a GL800 data logger for data acquisition.

At both experimental setups, the vibrations levels were measured with three identical Brüel & Kjaer DeltaTron accelerometers (denoted ACC in Figure 2) provided with built-in preamplifier and transducer data sheet (TEDS), that were connected to a laptop via a portable bus-powered USB carrier from National Instruments. The reason for using three accelerometers is that measurements were carried out simultaneously either at three locations or in three directions, as it will be detailed below. The accelerometers have the reference sensitivity of 20 m/s^2 RMS at 159.2 Hz, the measuring range of ±7000 m/s^2 peak, the frequency range 10% of 0.1 Hz to 6 kHz, and the phase response ±5° of 0.5 Hz to 5 kHz [9].

At the pump, the vibrations levels were measured for the shaft speed of 1460 rpm at zero flow rate (i.e. with the throttling valve fully closed) and at two flow rates close to the best efficiency point. For each flow rate the measurements were carried out simultaneously only in vertical direction on the motor, on the bearings housing and on the pump. The three measuring points will be denoted M, B, and P, respectively. The sampling frequency was of 50 kHz. The high value of the sampling frequency was chosen in order to catch the effects of the hydrodynamic interaction inside the pump between the runner blades and the tongue of the scroll casing and of the pressure waves that could propagate back and forth between the pump and the partly opened gate (or disk) of the throttling valve.

At the turbine, the vibrations levels were measured for a flap opening a_0 of 25% of the maximum opening at four decreasing shaft speeds that were obtained by progressively filling the hydrodynamic brake with water: 448.8 rmp, 397.7 rpm, 322.8 rpm, and 294.8 rpm. Under the measurement conditions, the chosen flap opening allowed to obtain the highest overspeed and the largest speed interval. The measurements were carried out only on the turbine casing in three directions simultaneously: transverse direction, i.e. the direction of the upstream pipe that feeds the turbine, vertical direction, and axial direction, i.e. the direction of the turbine shaft axis. These three directions will be denoted T, V, and A, respectively. The reason for measuring in three directions simultaneously was to identify the direction which better highlights the dominant frequencies of the vibrations. The sampling frequency was of 2 kHz. The value of the sampling frequency was chosen significantly lower than that used at the pump since no high frequency hydrodynamic phenomena was expected inside the turbine due to its operation at atmospheric pressure and only partly filled with water.

After the measurements, all data sets delivered by the accelerometers were filtered with a digital high-pass filter using the software dbTrait [11]. The cut-off frequency was chosen at 2 Hz to remove the low frequency random noise coming from outside of the experimental setups.

4 Results and discussion

The failing frequencies of the turbine bearings were calculated with the relationships (1) to (5). As mentioned previously, in the absence of some parameters, the failing frequencies of the pump bearings were calculated by analogy with corresponding bearings from the MRC catalog. The results along with the frequencies of rotation and the blade passing frequencies are presented in Table 1. The calculated values were used to identify the failing frequencies in the frequency spectra of the signals obtained from the vibrations measurements.

The values of the failing frequencies obtained by analyzing the frequency spectra of the measured vibrations are presented in Table 2 together with the errors relative to the calculated values from Table 1. In the table, *exp.* means *experimental value*, while *err.* means *error relative to calculated value*. It can be seen that all relative errors are below 5%, which indicates that the theoretical and experimental failing frequencies are in good agreement.

The Figures 3 to 5 present FFT spectra of the vibrations measured at the pump – on the motor, on the bearings housing, and on the pump itself – for the three operating regimes described previously. The frequency of rotation, f_r, and its second and third-order harmonics, $2f_r$ and $3f_r$, respectively, as well as the blade passing frequency, f_b, can be easily identified.

Table 1. Speeds, frequencies of rotation, blade passing frequencies of the runners, and calculated failing frequencies of the bearings in the experimental setups.

Parameter		Pump	Banki turbine			
n	(rpm)	1460	448.8	397.7	322.8	294.8
f_r	(Hz)	24.33	7.48	6.63	5.38	4.91
f_b	(Hz)	121.7	142.1	126.0	102.2	93.29
FTF	(Hz)	9.27	3.18	2.82	2.28	2.09
BFF = 2*BSF*	(Hz)	96.38	49.68	44.02	35.73	32.63
BPFI	(Hz)	120.5	77.45	68.64	55.71	50.88
BPFO	(Hz)	74.15	57.19	50.67	41.13	37.56

Table 2. Failing frequencies obtained experimentally for the investigated bearings.

Parameter	Pump		Banki turbine							
	n = 1460 rpm		n = 448.8 rpm		n = 397.7 rpm		n = 322.8 rpm		n = 294.8 rpm	
	exp. (Hz)	err. (%)	exp. (Hz)	err. (%)	exp. (Hz)	err. (%)	exp. (Hz)	err. (%)	exp. (Hz)	err. (%)
f_r	24.41	0.33	7.44	0.53	6.84	3.17	5.37	0.19	4.88	0.61
FTF	9.16	1.19	3.31	4.09	2.93	3.90	2.32	1.75	2.08	0.48
BFF = 2*BSF*	96.64	0.27	50.17	0.99	44.68	1.50	36.01	0.78	32.84	0.64
BPFI	121.05	0.46	77.15	0.39	68.6	0.06	55.54	0.31	50.42	0.90
BPFO	74.77	0.84	57.86	1.17	51.39	1.42	41.5	0.90	37.84	0.75

https://doi.org/10.1051/e3sconf/201911202013

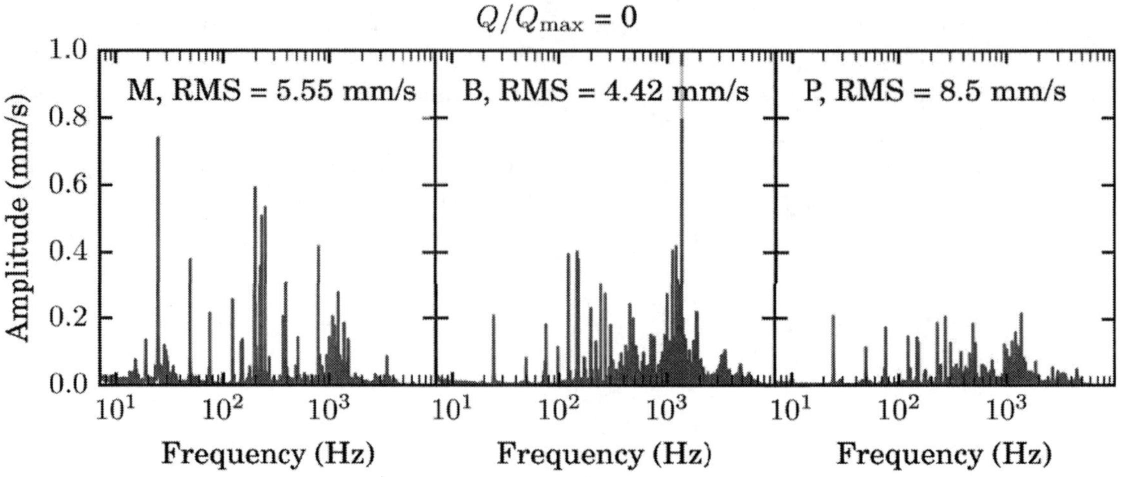

Fig. 3. FFT spectra of vibrations measured at the centrifugal pump at $Q/Q_{\max} = 0$ (zero flow rate).

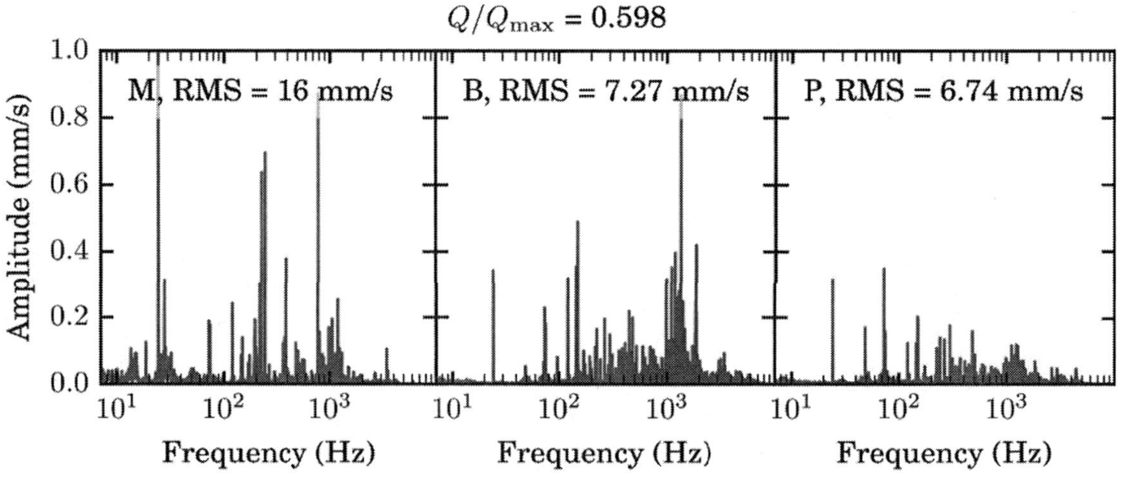

Fig. 4. FFT spectra of vibrations measured at the centrifugal pump at $Q/Q_{\max} = 0.598$.

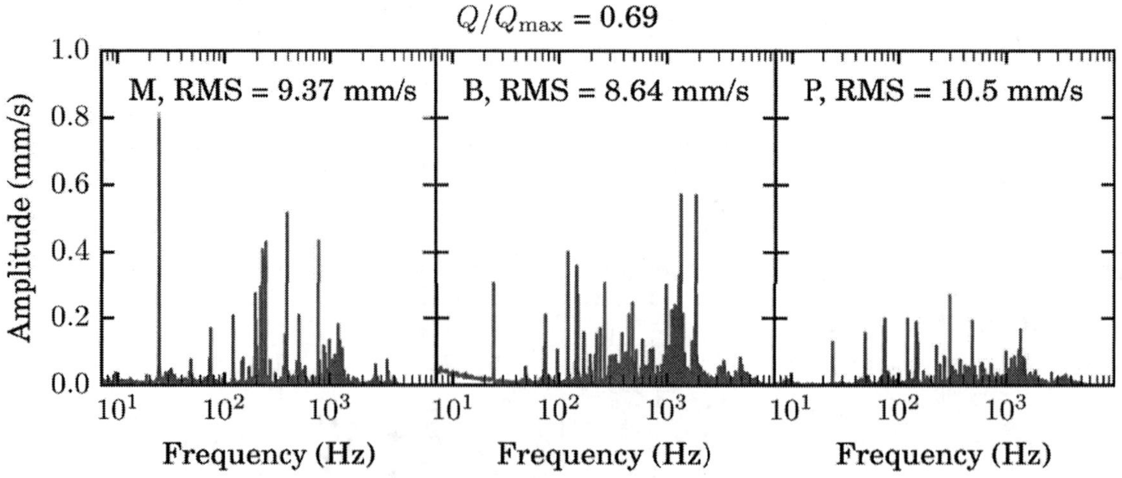

Fig. 5. FFT spectra of vibrations measured at the centrifugal pump at $Q/Q_{\max} = 0.69$.

In case of the motor, the frequency of rotation is always dominant in the spectra. On the pump, the amplitudes at f_r are close to the maxima in the spectra. On the bearings, the amplitudes at f_r are close to the average values in the spectra. The *BPFI* and at least its second-order harmonics (2*BPFI*) is also visible on all spectra – even at the motor. The same remark is valid for the *BPFO* and its harmonics. It should be noted that, for the measurements on the bearings, the level of the spectral components associated to these frequencies is dominant in the specific frequency domain, i.e. below 500 Hz. The same remark remains valid for the pump as well. This behavior is rather natural when seen in relation to the experimental assembly.

The analysis of the experimental data obtained at the Banki turbine for the four operating speeds (Figures 6 to 9) reveals that in all cases the FFT spectra of the vibrations measured in axial direction are more relevant than those corresponding to the transverse and vertical directions. This could be explained by the fact that the vibrations are dampened in transverse direction due to the rigid assembly of the turbine with the upstream pressure pipe, while in vertical direction the dampening is caused by the rigid mounting of the turbine on the concrete laboratory floor. Nevertheless, some important frequencies can also be observed on the spectra obtained for the transverse and vertical directions. Independent of the speed, dominant amplitudes at the frequencies of rotation, f_r, are clearly visible on all spectra. Only the first-order harmonic, $2f_r$, can be clearly observed at the highest speed (Figure 6), while

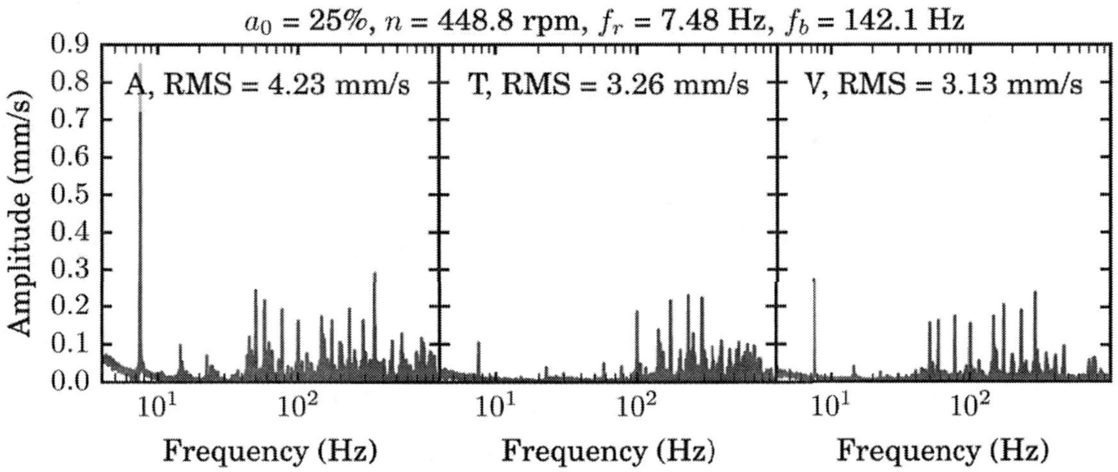

Fig. 6. FFT spectra of vibrations measured on the Banki turbine at n = 448.8 rpm.

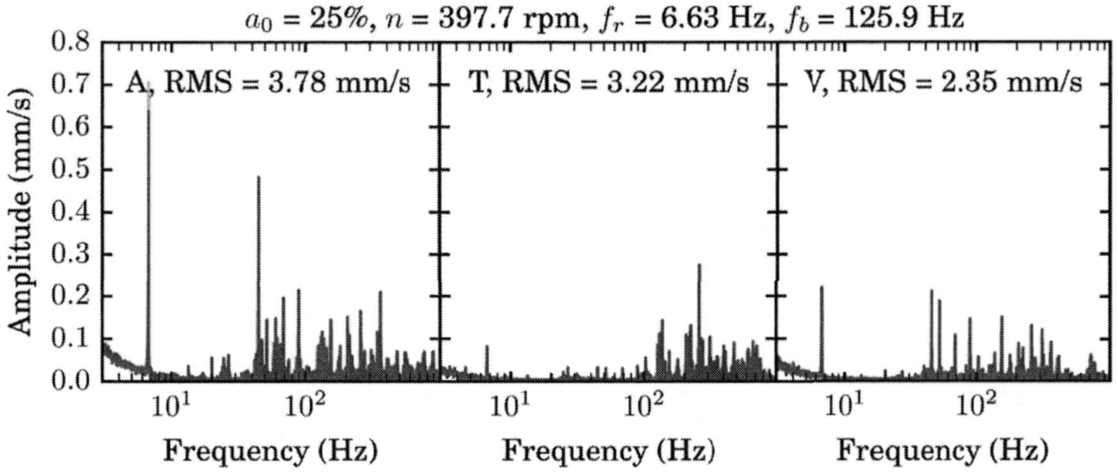

Fig. 7. FFT spectra of vibrations measured on the Banki turbine at n = 397.7 rpm.

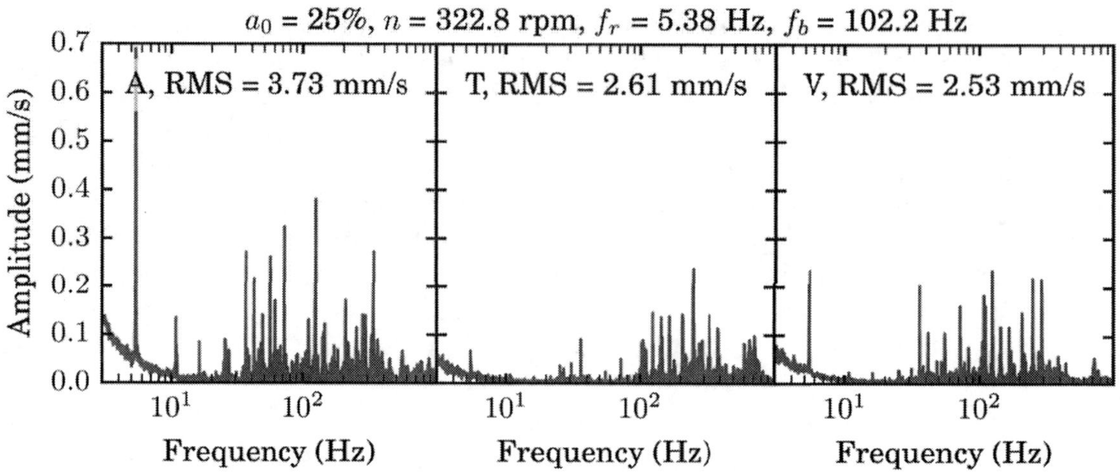

Fig. 8. FFT spectra of vibrations measured on the Banki turbine at n = 322.8 rpm.

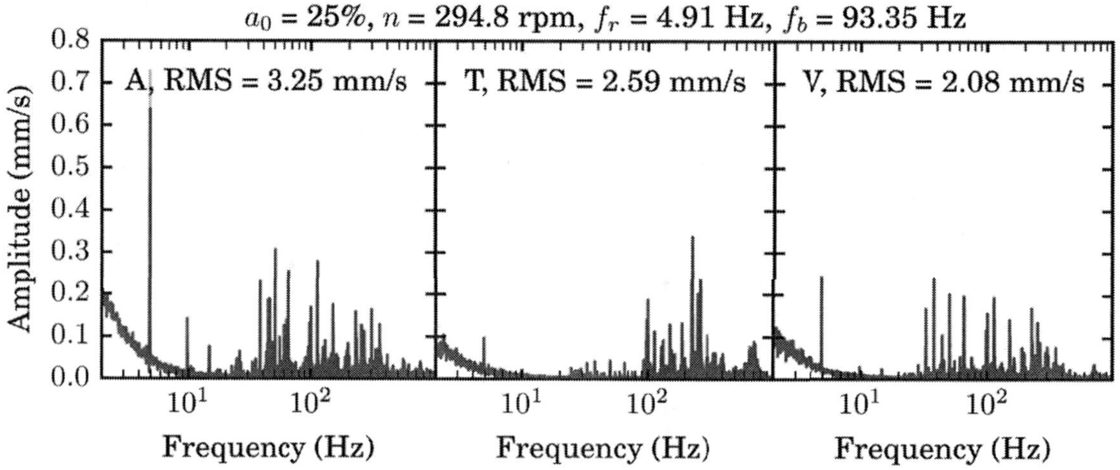

Fig. 9. FFT spectra of vibrations measured on the Banki turbine at n = 294.8 rpm.

two harmonics, at $2f_r$ and $3f_r$, are visible at the other speeds (Figures 9 to 7). It is interesting to note the the blade passing frequency can be easily identified at the maximum speed of 448.8 rpm (Figure 6), although its corresponding amplitude is relatively low, and becomes less visible as the speed decreases (Figures 7 to 9). For all four speeds, significant peaks are visible at the corresponding *BPFI*s. The second and third order harmonics of the *BPFI*s are also visible. Peaks can be noticed at the *BPFO*s and, in some cases, at their second-order harmonics, although they are smaller than those at the *BPFI*s. At higher speeds, local maxima are visible at the *BFF*s (Figures 6 to 8). It should be noted, however, that all the measurements were carried out on the turbine casing, where other factors – mostly hydraulic phenomena – could influence the vibrations levels as well.

5 Conclusions

In this paper, the vibrations measured on two experimental models of a pump station and of a small hydropower plant were analyzed in order to identify in their frequency spectra those frequencies that are specific to defects caused by the wear of the bearings.

The results obtained suggest that both the bearings of the pump and those of the turbine have defects on their inner and outer ring raceways. These defects are indicated by the occurrence of the corresponding failing frequencies and of their harmonics.

In case of the pump, the ball fault frequency can also be identified, although it is very attenuated. Additionally, a high global RMS value can be noticed, both at the bearings housing and at the pump. Considering this high RMS value and the number of fault frequencies harmonics that can be identified in the spectra, it can be concluded that the pump bearings are affected by severe wear. According to the limit values for Small Machines of Class I, having powers below 15 kW, from ISO 10816-1 [12], the bearings are at the upper limit of the Unsatisfactory (alert) zone, which has RMS values in the range 1.8–4.5 mm/s, and in the Unacceptable (danger) zone, characterized by RMS values above 4.5 mm/s. This indicates that the long-term operation of the pump should be restricted of even avoided, since the alarm values are exceeded and the shutdown values are almost attained.

In case of the Banki turbine, the occurrence of the ball fault frequency, even though not at all speeds, indicates the existence of an early failure of the rolling parts. In this case, the global RMS values place the turbine bearings in the Unsatisfactory (alert) zone according to ISO 10816-1 [12]. It should also be considered that the occurrence of the frequencies corresponding to defects of the raceways could indicate a poor lubrication or an incorrect loading of the bearings.

The results presented in this paper illustrate how tracing the fault frequencies in the spectra obtained during the operation of pumping systems or small water turbines and comparing the RMS values of the vibrations with those indicated by norms is an efficient method in monitoring the operation of such machines and in providing maintenance for them.

References

[1] E. Sima, Strategii Manageriale **4**, 515 (2014)

[2] European Comission, *Factsheet on 2014-2020 Rural Development Programme for Romania*, (https://ec.europa.eu/agriculture/sites/agriculture/files/rural-development-2014-2020/country-files/ro/factsheet_en.pdf)

[3] G. Băran, I. Catană, I. Magheți, C.A. Safta, M. Savu, IOP Conference Series Earth and Environmental Science **12**, 012100 (2010)

[4] Y. Wu, S. Li, S. Liu, H.S. Dou, Z. Qian, *Vibration of Hydraulic Machinery* (Springer, Dordrecht, 2013)

[5] M. Gafițanu, S. Creu, B. Drăgan, *Vibro-acoustic diagnostics of machines and equipment* (Technical Publishing House, Bucharest, 1989)

[6] A. Bilosova, J. Bilos, *Vibration Diagnostics, Investments in Education Development* (Report CZ.1.07/2.2.00/15.0132, Ostrava, 2012)

[7] M. Radeș, *Dynamics of Machinery III* (Printech, Bucharest, 2008)

[8] MRC Bearing Services, *Engineering Handbook* (SKF USA Inc., 2005)

[9] A. Dragomirescu, C.A. Safta, L. Mândrea, N. Orășanu, I. Magheți, *Preventive Maintenance of Hydraulic Installations Based on Monitoring Cavitation-Induced Vibrations*, in *International Conference and Exposition on Electrical and Power Engineering* (2018), pp. 883–888

[10] C.A. Mockmore, F. Merryfield, Oregon State College Engineering Experiment Station Bulletin **25** (1949)

[11] https://www.01db.com/activity-area/noise-and-vibration-monitoring/

[12] *ISO 10816-1:1995 Mechanical vibration – Evaluation of machine vibrations by measurement on non-rotating parts – Part 1: General Guidelines* (1995)

Fish Ladder Geometrical Sizes and Hydraulic Performances. Experimental Approach

Cristina-Catalina Petica[1], *Carmen-Anca* Safta[1,*], *Irina* Pincovschi[1] and *Lucian* Mandrea[1]

[1]University Politehnica of Bucharest, Faculty of Power Eng., 313 Splaiul Independentei, 060042, Bucharest, Romania

Abstract. The paper presents the hydraulic experimental results from different types of fish ladders heaving pools and baffles with submerged orifice or vertical slot. The goal of the experiments is to develop a database of velocities and power densities for each type of fish ladder with the baffles in different arrangements, for different lengths of the pools. So, based on the fish ladder type, the hydraulic characteristic database and the experimental laboratory model size provided by this study, one can improve the design of a fish ladder. The laboratory experimental model is designed with respect to Froude similarity criterion.

1 Introduction

Fish ladders must facilitate the passing of aquatic biota along the river from downstream to the upstream, or vice versa, when river cross-section is blocked by hydraulic structures. The transverse obstacles could be weirs, intake weirs, check dams, barrages, or small hydropower plants (SHP). According to the Romanian Water Law (107/1996) the newly built obstacles that require fish passes are the ones with heights higher than 0.5 m, up to 13.6 m.

Fish ladders (or fish passages, or fish ways) are hydraulic structures of slope channel type usually having the length divided by baffles into pools. Through the baffles the water flows over weirs, through slots or submerged orifices.

The Romanian Water Law (107/1996) [1] stipulates in the article 53, item 4 that the new barrage or water intake located across the watercourses must be provided with facilities to insure the longitudinal continuity and a servitude flow rate (which is the sum of the ecological flow and the discharge necessary to supply the downstream users, as it is defined in section 26 of annex no. 1). As appropriate, the facility should include construction works to ensure the migration of aquatic fauna, especially of the ichthyofauna, in order to achieve the environmental objectives (stipulated in art. 2^1, in compliance with the provisions of art. 2^4). In the same law it is also highlighted that the applied technical solutions must be feasible and without "disproportionate costs".

The Romanian energy and rural development policies from the last decade encouraged the development of renewable energies. Small hydropower plants usually built on small

* Corresponding author: safta.carmenanca@gmail.com

rivers within rural areas represent an important renewable energy source. More than 450 SHPs have been inventoried in Romania by 2016 [2].

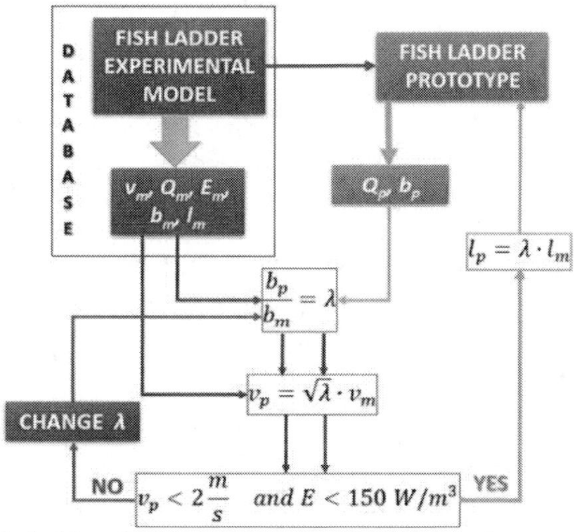

Fig. 1. Flowchart of the fish ladder design process developed based on experimental hydraulic characteristics included in the experimental database; where *v, Q, E, l* and *b* are symbols for velocity, flow rate, power density, length and width, respectively; λ is the scale factor and the subscripts *p* and *m* indicate the prototype and the model.

The presence of river cross-section hydraulic structures without fish passages may have a strong negative impact on the longitudinal migration of aquatic biota. To protect the environment and to connect the upstream and downstream reaches along the stream a fish ladder must be properly designed and built.

In the study published by Luca E. [3] it is specified that "15% from the fish ladders in Romania are inefficient". The non-efficiency of the ladders means their hydraulic parameters and water biological conditions (temperature, biological dissolved oxygen, light intensity etc.) do not comply with fish migration conditions and therefore fish will not pass it. This is due to a poor design that does not provide these optimal conditions suitable with the behaviour of fish species. Fish ladder performances depend mainly on water velocity, depth, degree of turbulence, hydraulic losses, the geometrical sizes of the pools, channel bottom roughness, flow direction [4].

The objective of this paper is to analyse experimentally the hydraulic behaviour of different types of fish ladders with pools and baffles having submerged orifice or vertical slot and to gather all results in a data base which could be used in the fish ladders design.

In Figure 1 is presented the flowchart of the design process on a knowledge data base of fish ladders hydraulic behaviour.

The hydraulic parameters of velocity and power density of different types of fish ladders were compared with those measured for different arrangements of the baffles (in line or zigzag) and different length of the pools.

For the experiments a flume was designed for geometrical sizes and flow rate which are specific to small hydropower plants (SMH) having fish ladders with flow rates, usually between 40 l/s to 400 l/s.

2 Fish ladder hydraulic characteristics

The review of scientific papers and technical reports regarding fish ladder design [4÷9] mentions that the design of fish ladders has a systemic approach and the overall compliance of the biological, hydraulic, thermic, hydrological and economic conditions must be considered.

There are different types of fish ladders according to different classification criteria, mainly: pools and weirs (including the submerged orifices), vertical slots, Denil, close-to-nature and culvert [6÷8]. These types of fish ladders are widespread in association with the small hydropower plants (mainly run of river - RoR). Other types of fishways are fish locks and lifts, which are usually used on larger rivers and/or with high dams [7].

The DVWK technical report [6] states that depending of the fish species and size, the hydraulic requirements for baffle fish ladders are: flow velocities in the orifices (of fish ladders with submerge orifices) or in the vertical slots (v_s in Figure 1) should not exceed 2.0m/s. Also, the flow power density, *E*, in terms of its volumetric dissipation, should not exceed 150 W/m^3 for most fish species, even 200 W/m^3 for salmonids, so that the low-turbulence flow regime in the pool should be ensured [6].

With the notations from Figure 2, the flow velocity through the submerged orifice and the power density are calculated with [6, 8]:

$$v_s = \sqrt{2g\Delta h} \quad , \tag{1}$$

$$E = \frac{\rho g \Delta h Q}{b h_m (l_b - d)} \tag{2}$$

where Q is the flow rate with $Q = Q_s + Q_a$, where Q_s and Q_a are the flow rates through the submerge orifice and the slot, respectively.

In relation (1) v_s is the theoretic velocity. The real velocity through on a submerged orifice depends on the coefficient of velocity, $= \sqrt{\frac{1}{1+\varsigma}}$, caused by friction loss (with ς loss coefficient [9]). If the flow is in the turbulent rough regime (with Reynolds number **Re**>200,000) $\varphi = 1$ [9, pp. 230] and v_s is calculated with equation (1), else the velocity is calculated with formulae $v_s = \varphi\sqrt{2g\Delta h}$, and $\varphi = \varphi(\mathbf{Re})$.

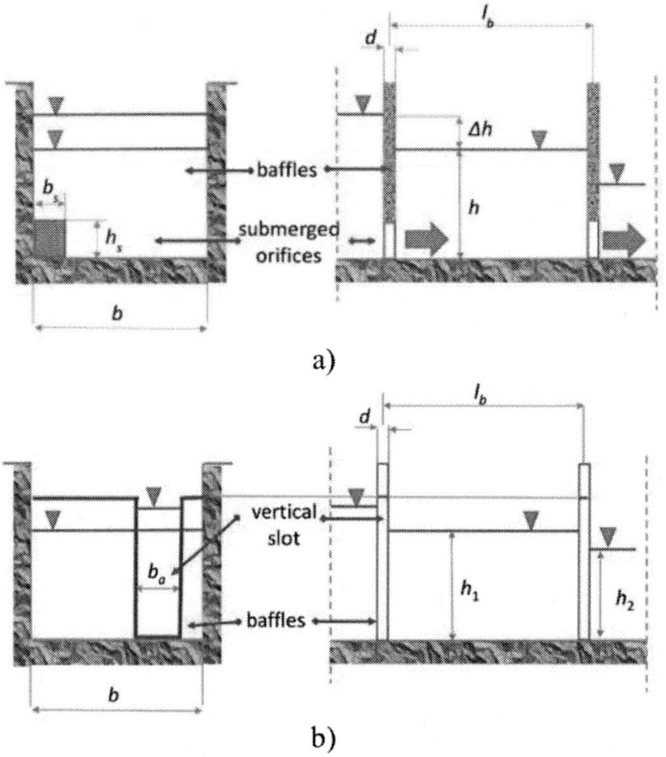

Fig. 2. Longitudinal and cross-section profiles through fish ladders with submerged orifices (a) and slots (b).

The discharge in the orifice is $Q_s = \psi \cdot b_s \cdot h_s\sqrt{2g\Delta h}$ with the discharge coefficient ψ = 0.65 to 0.85 [6]. The flow rate through the slot is:

$$Q_a = \frac{2}{3} C_s b_a \sqrt{2g}(h_1)^{3/2} \quad , \tag{3}$$

where h_1 is the mean level of the flow in the pool upstream of the baffle; and C_s is the discharge coefficient for slots given by Villmonte as:

$$C_s = \beta_0 \left[1 - \left(\frac{h_2}{h_1}\right)^{1.5}\right]^{0.385} . \tag{4}$$

The coefficient β_0 depends on the geometry of the slot, pool dimensions, and the discharge equation used [8]. It was considered that $\beta_0 = 0.705$ for a flow rate 0.455 m^3/s as it was measured [8, 10]. In equation 4, h_2 is the mean level of the flow in the pool downstream of the baffle, Fig. 2.b.

In the literature, the geometrical sizes of the pool and the slot width must be proportional with the fish size. Also, according to USFWS technical report, the pool length should be ten times the orifice width and the pool width should be eight times the orifice width (b_s or b_a, Fig. 2) [11].

The above indications were considered in the dimensioning of baffles (with submerged orifice or slot) and pools fish ladder experimental model built in the laboratory.

3 Experimental setup

The measurements of the hydraulic characteristics of different types of fish ladder, were carried out on the experimental setup in Figure 3.

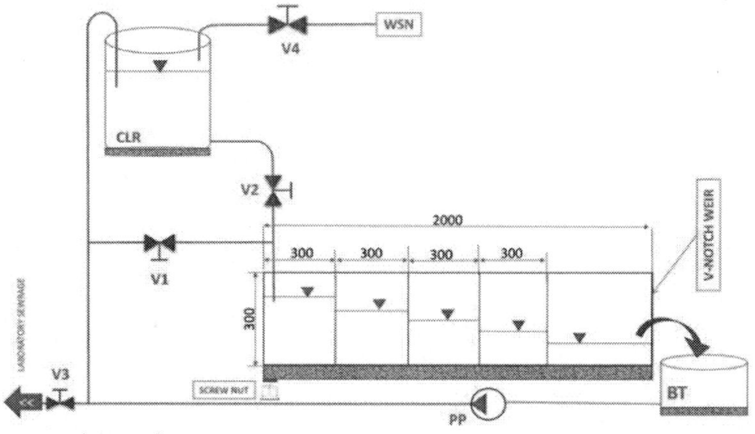

Fig. 3. Sketch of the experimental setup.

The testing facility has a hydraulic flume which is 2 m long, 0.24 m wide and 0.3 m high; a constant level reservoir (CLR, 0.1 m^3 capacity); a buffer tank (BT, 0.1 m^3 capacity); a radial pump (PP, Grundfos type UPS 15-50 CES 87); four valves (V1-V4, 1 inch diameter), and flexible pipes (1 inch diameter, except for the supply pipe of the pump which is 1-1/2 inch diameter) [12].

The flume was designed to comply with the Froude hydrodynamic similarity criterion ($Fr = V^2/(g \cdot L)$) so that the prototype discharge should range between 0.04-0.4 m^3/s, which is usually common for the fish ladders of SHP RoR. The scale factor $\lambda = L_p/L_m$ was chosen $\lambda = 25/1$, to avoid scale effects caused by the viscous forces, as Chanson recommend [13]. In the definition of scale factor λ, the subscripts p and m are for prototype and model.

The flume is divided into pools with adjustable length, by using removable baffles (Figure 4, Figure 5) with different geometrical configurations, specific for fish ladders with pools and weirs. The baffles can have a submerged orifice or a slot, placed laterally or central, across the section (Figure 6). More details regarding the manufacture of the flume are in reference [12].

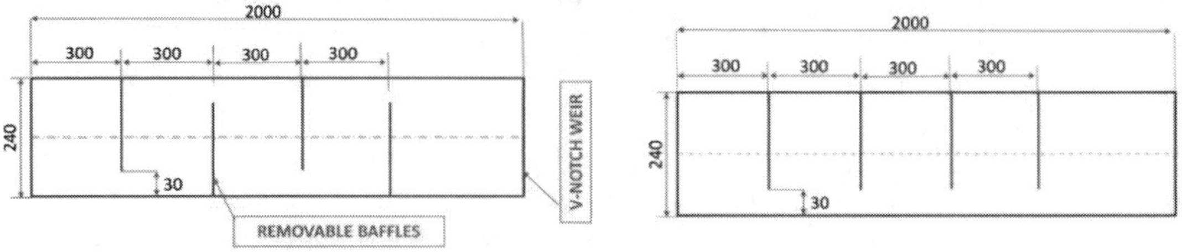

a) b)

Fig. 4. The fish ladder with the removable baffles in different arrangements: (a) zigzag; (b) in line.

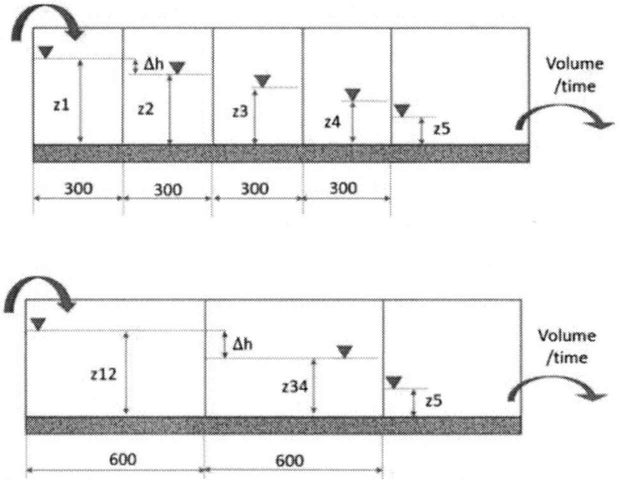

Fig. 5. Sketches of the fish ladder with baffle arrangement for different pool lengths

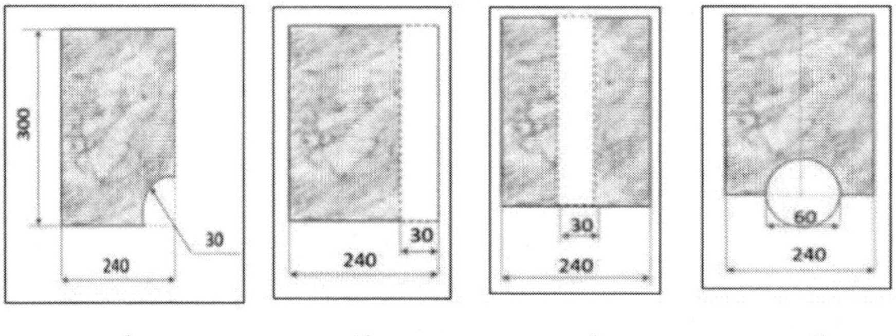

a) b) c) d)

Fig. 6. Different types of baffles: (a) with lateral submerged orifice, LSO; (b) with lateral vertical slot, LVS; (c) with vertical slot placed in the middle of the baffle, MVS; (d) with submerged orifice placed in the middle of the baffle, MSO.

At the downstream end of the flume there is a 90° V-notch (with a 100 mm wall height). The weir was calibrated by measuring the discharge volumetrically (volumetric flow rate $Q = V/t$) and the water level (h) above the crest, leading to the rating curve with the equation $Q = 0.5353\ h^{2.3788}$.

The flume can operate with slopes from zero to 5% by adjusting a screw nut at the upstream end of the channel (as in Figure 3). The experiments were carried out on the flume with zero slope because only a structure without slope can provide the non-uniformity flow conditions between pools [8, 10, 14, 15]. A fish needs to swim faster than the velocities in the slots when moving upstream and for this reason the uniform flow conditions in the design of fish ladders are not recommended [14, 15].

In this experimental study the roughness of the flume was not considered because only by using a distorted model the bed geometry and the roughness could be included [13].

The ratios between corresponding hydrodynamical parameters in prototype and model namely length (L), time (t), velocity (V), flow rate (Q), force (F), pressure (p) and power density (E) with respect to Froude similarity are shown in Table 1, considering the same fluid ($\rho_p = \rho_m$, where ρ is the fluid density).

Table 1. Rations between corresponding parameters in prototype and model.

Ratio	L_p/L_m	t_p/t_m	V_p/V_m	Q_p/Q_m	F_p/F_m	p_p/p_m	E_p/E_m
Froude $Fr = V^2/(g \cdot L)$	λ	$\lambda^{1/2}$	$\lambda^{1/2}$	$\lambda^{5/2}$	λ^3	λ	$\lambda^{1/2}$

The fish ladders prototypes are considered to have the flow rates in the range of $(0.04 \div 0.4)$ m^3/s, which is specific for the RoR SHPs.

4 Results and discussions

Using the experimental set up presented above, the hydraulic parameters of velocity and power density (v, E) for different types of fish ladder, having a simplified vertical slot baffle (Fig. 6), were measured and calculated.

Twelve different scenarios were considered (in Table 2) by changing the baffle type and the pool length in the flume. A short pool is 300 mm long, and a long pool is 600 mm, as in Figure 5. Each baffle (Fig. 6) was placed in multiple possible arrangements (Fig. 4):

- zigzag in short pools, for LVS and LSO baffles (Fig. 6.a and Fig. 6.b);
- in-line in short pools, for LVS and LSO baffles;
- zigzag in long pools, for LVS and LSO baffles;
- in-line in long pools, for LVS and LSO baffles;
- central openings in short pools, for MVS and MSO baffles (Fig. 6.c and Fig. 6.d);
- central openings in long pools, for MVS and MSO baffles.

Table 2. Scenarios considered in relation with baffles type from figure 6

Pool length / Baffles arrangements / Baffles Type	Short pools		Long pools	
	Zigzag	In-line	Zigzag	In-line
LVS (lateral vertical slot)	Scenario 1	Scenario 2	Scenario 3	Scenario 4
LSO (lateral submerged orifice)	Scenario 5	Scenario 6	Scenario 7	Scenario 8
MVS (middle vertical slot)	Scenario 9		Scenario 10	
MSO (middle submerged orifice)	Scenario 11		Scenario 12	

The flow rate can be adjusted by using a variable flow centrifugal pump in close circuit and by operating with the constant level reservoir (CLR) in an open circuit [12]. For each flow rate the levels in the centre of pools were measured.

The hydraulic characteristics (velocity in the orifice/slot and the power density) were calculated for each level measured between the pools. The results obtained on the model were transposed on the prototype by using the relations in Table 1.

The results obtained on the prototype are plotted in Figures 7 and 8.

In all scenarios the values of power density are much less than the maximum limit given in the literature, of 150 W/m^3 [6]. However, in terms of velocity, one can notice that the limit of 2 m/s is exceeded in the case of baffle with the submerged orifice in the shape of a quarter of circle, placed in the corner of the baffle (Scenarios 5, 6 7 and 8). This observation is valid for short and long pools, as well as in-line and zigzag arrangements (Figure 7.a and Figure 8.a).

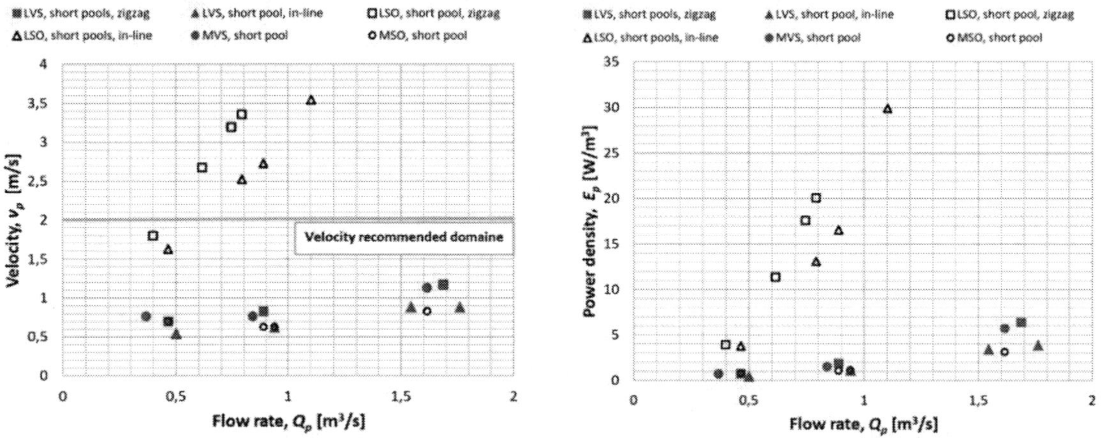

a) b)

Fig. 7. Short pools scenarios: (a) velocities vs. flow rates; (b) power density vs. flow rates.

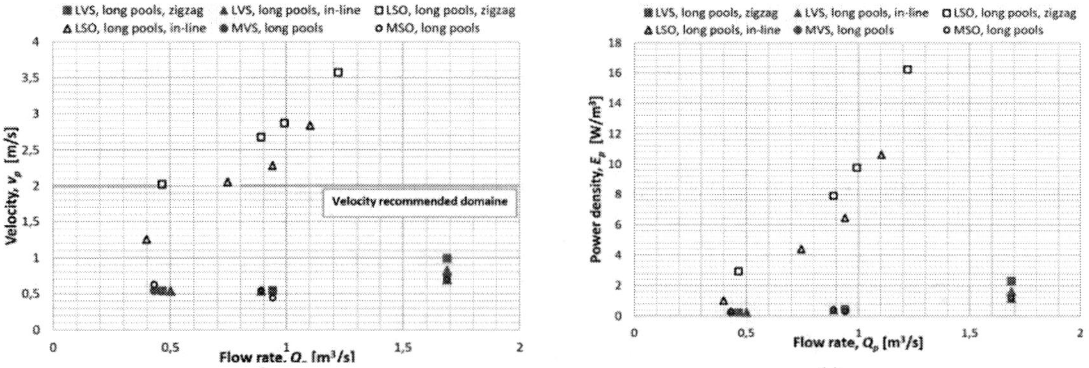

a) b)

Fig. 8. Long pools scenarios: (a) velocities vs. flow rates; (b) power density vs. flow rates.

The hydraulic behaviour of the vertical slot baffles is comparable with baffles having the submerged orifice in the middle of the baffle, MSO. The velocities are in the range of 0.5 to 1.2 m/s and power density is less than 10 W/m^3 in the case of short pools, Fig. 7.

In the case of long pools, the velocities are not very well individualized for each scenario even if the range of velocity remains approximately the same (0.5 to 1.0 m/s). Because the volume of water increases in the pool, the power density is lower in long pool scenarios than in short pool scenarios.

The velocities obtained on the fish ladder with vertical slot and the slot placed in the middle (MVS case - Scenario 9) are comparable with the velocities through the lateral vertical slot arranged in zigzag (LVS case - Scenario 1) but less than in the case of LVS arranged in-line, Scenario 2. This trend is found in the scenarios with long pools, too.

5 Conclusions

The paper presents the experimental results of different types of fish ladders having a vertical slot baffle or a submerged orifice, which were carried out on an experimental laboratory model. The test facility was designed to allow experiments on baffle fish ladders with pools and submerged circular orifices, or pools and vertical slots in different arrangements and different pool lengths.

The measurements were performed in twelve distinguished scenarios for which the velocity through the orifices and the power density in the pools were calculated. The results obtained on the model were transposed on the prototype considering Froude similarity. The scale factor was chosen 25 ($\lambda = 25$) to avoid scale effects.

The results show that the fish ladder having a vertical slot baffle has a hydraulic behaviour close to the fish ladders with submerged orifices placed in the middle. For this type of vertical slot, the baffles arrangements, in zigzag or in-line, are not very important, the velocities obtained being approximately the same.

For a fish ladder design, it is important that the proportion of geometrical sizes of the pools (length and width), which depend on the dimension of the submerged orifice or the slot width should be kept (length =10 × orifice or slot width and width = 8 × orifice or slot width). Otherwise the fish ladder could become inefficient and therefore unattractive for the fish.

The goal of the study was to obtain an experimental database of hydraulic characteristics of different types of fish ladders which can be used in fish ladder design. The experimental results obtained on the experimental model should be compared with the ones available in the literature, so a further development of the study is required.

References

1. Romanian Water Law (107/1996) amended and supplemented with the Government Emergency Ordinance OUG no. 78/2017 published in the Monitorul Oficial no. 885 from the 10th of Nov. 2017, http://www.monitoruloficial.ro)
2. B.Popa, *Microhidroenergia astazi, in Romania*. Romanian HydroPower energy summit, februarie 2016, http://govnet.ro/uploads/files/11_ARmHE GOVNET.pdf
3. E. Luca, *Practici mondiale privind solutii de asigurare a migrarii ihtiofaunei si posibilitatea aplicarii acestora in Romania*, Institutul National de Hidrologie si Gospodarirea apelor, Conferinta stiintifica anuala, 1-3 nov. 2011, https://www.scribd.com/doc/222635287/Practici-Mondiale-Privind-Asigurarea-Migrarii-Faunei-Piscicole
4. J. O'Connor, M. Mallen-Cooper, I. Stuart, Performance, operation and maintenance guidelines for fishways and fish passage works, Technical Report Series No. 262 for the Water and Catchments Group, Department of Environment, Land, Water and Planning. Arthur Rylah Institute for Environmental Research, Department of Environment, Land, Water and Planning, Heidelberg, Victoria, Arthur Rylah Institute for Environmental Research, (2015), https://www.ari.vic.gov.au/__data/assets/pdf_file/0024/39453/ARI-Technical-Report-

262-Performance-operation-maintenance-guidelines-for-fishways-and-fish-passage-works.pdf

5. T. Linnansaari, B. Wallace, R.A. Curry, G. Yamazaki, (2015). *Fish Passage in Large Rivers: A Literature Review*. Mactaquac Aquatic Ecosystem Study Report Series 2015-016. Canadian Rivers Institute, University of New Brunswick v + 55 p, https://www.researchgate.net/publication/308558211_FISH_PASSAGE_IN_LARGE_RIVERS_A_LITERATURE_REVIEW/download (10/04/2019)
6. DVWK, *Fish passes – Design, dimensions and monitoring*, Published by the Food and Agriculture Organization of the United Nations in arrangement with Deutscher Verband für Wasserwirtschaft und Kulturbau e.V. (DVWK) Rome, (2002), http://www.fao.org/3/y4454e/y4454e00.htm
7. EPRI, Upstream and Downstream Fish Passage and Protection Technologies for Hydroelectric Application: A Fish Passage and Protection Manual, EPRI, Pala alto, CA: (2002). 1005392, http://hydrovolts.com/wp-content/uploads/2012/02/Upstream-and-Downstream-Fish-Passage-Technologies-for-Hydroelectric-Application.pdf
8. J. F. Fuentes-Pérez, A. García-Vega, F. J. Sanz-Ronda, A. Martínez de Azagra Paredes, *Villemonte's approach: a_general method for modeling uniform and non-uniform performance in stepped fishways*. Knowl. Manag. Aquat. Ecosyst., 418, **23**, (2017).
9. D. Cioc, Hidraulica, Ed. Didactica & Pedagogica, Bucharest, (1983).
10. N. Rajaratnam, C. Katopodis, S. Solanki, *New designs for vertical slot fishways*. Can J Civil Eng 19: 402–414, (1992).
11. USFWS (U.S. Fish and Wildlife Service), Fish Passage Engineering Design Criteria, (2017), https://www.fws.gov/northeast/fisheries/pdf/USFWS_R5_2017_Fish_Passage_Engineering_Design_Criteria.pdf
12. C-C. Petica, C-A. Safta, I. Pincovschi, L. Mandrea, *Experimental Laboratory Model of a Fish Ladder Used in the Small Hydropower Plants*, The 11th International Symposium on ATEE, March 28-30, (2019), Bucharest, Romania, 978-1-7281-0101-9/19/$31.00 ©2019 IEEE (to be published).
13. H. Chanson, *Hydraulics of Open Channel Flows: An Introduction*, 512 pp., Chaper 14, Eduard Arnold, London, (1999).
14. J.F. Fuentes-Perez, A.T. Silva, J.A. Tuhtan, A. García-Vega, R. Carbonell-Baeza, M. Musall, M. Kruusmaa, *3D modelling of non-uniform and turbulent flow in vertical slot fishways*, Environmental Modelling & Software 99 (2018), pp. 156-169.
15. Fuentes-Pérez JF, Sanz-Ronda FJ, Martínez de Azagra-Paredes A, García-Vega A, Martínez de Azagra A, García-Vega A. 2016. *Nonuniform hydraulic behavior of pool-weir fishways: a tool to optimize its design and performance*. Ecol Eng 86: 5–12.

E3S Web of Conferences **112**, 03001 (2019) https://doi.org/10.1051/e3sconf/201911203001
TE-RE-RD 2019

Drones and windbreaks

Petru Cardei[1,*], *Sebastian* Muraru[1] , and *Mihai* Matache[2]

[1]INMA Bucharest, Computerized Engineering Department, 013813 Bucharest, Romania
[2]INMA Bucharest, Testing Department, 013813 Bucharest, Romania

Abstract. The article presents a possibility of applying the techniques specific to the drones (UAV) in order to estimate the efficiency, and optimize the windbreaks (vegetative barriers, forest curtains).In the first part the article presents a variant of determination of the dynamic and optical porosities of forest protection curtains using one or two drones simultaneously. In the second part, ascertaining some language inaccuracies, it observes that the issue of the efficiency of the forest curtains became a problem of mitigation of wind intensity. Finally, the last proposed application shows that using the drones, you can estimate the protected volumes behind the curtains, basically giving up the optical porosity. In order to achieve these applications, there is an essential condition: records of the magnitude of the wind speed and of the coordinates must not be excessively affected by the wind.

1 Introduction

The definition of the windbreak will be accepted after [1] or [2] (row of trees or a fence, wall, or screen, which provides shelter or protection from the wind). Synonymous terms, as used in this article, are vegetative barriers (usually formed by high herbs) or forest curtains or shelterbelt. We will not detail the definition of the windbreak, but will take over the definition and we will summarize the information from [2]. A windbreak (shelterbelt) is a planting usually made up of one or more rows of trees or shrubs planted in such a manner as to provide shelter from the wind and to protect soil from erosion. Other benefits include contributing to a microclimate around crops, providing habitat for wildlife, and, in some regions, providing wood if the trees are harvested, [4]. The importance of vegetal barriers is also highlighted in the case of the protection of communication ways, [5].

We followed the path described by the authors [3] until to the method of obtaining the experimental data. Vegetative barrier effectiveness is commonly measured by its sheltering efficiency leeward of the barrier via reduced wind speed, [3].The aerodynamic (α_p) and the optical (β_p) porosities of the vegetative barrier are defined in [3]. The aerodynamic porosity is defined by the formula:

$$\alpha = \frac{\int_0^H \bar{v} dz}{\int_0^H \bar{v}_0 dz} \tag{1}$$

* Corresponding author: petru_cardei@yahoo.com

where H is the height of the barrier, in m, $\bar{v}$ is the leeward wind speed (bleed wind speed), in m s^{-1}, and v_0 is the mean windward wind speed, in m s^{-1}. The optical porosity is defined by the formula, [3]:

$$\beta = \frac{(Background\ area)}{(Total\ area)} \tag{2}$$

According to [3], the background area (representing gaps outside the area occupied by the tree including leaves, branches, stems, and trunk), is obtained from image analysis of digitized photographs. Although β_p is convenient to use because it does not require wind measurement, a relationship between α_p and β_p is necessary, especially for deciduous trees that experience porosity variations throughout the year, as it is specified in [3].
Specific objectives of the study from article [3], were to: *determine optical porosity* using image analysis, determine aerodynamic porosity and drag coefficients during leaf-on and leaf-off conditions using wind profile measurements, and *measure the effects of vegetative barrier leaf* condition on wind speed reduction.
This article proposes a new approach, with new tools and concepts, for the same problem of forest curtains. *Using the drones to measure parameters* and *more accurately defining a wind-dampening factor* were the main objectives of this study.

2 Material and methods

For the data acquisition experiments, as in [3], it is primarily necessary to have an area containing vegetal barriers, with the properties required in [3]. At the beginning of the proposed activity, it is advisable to take into account the indications from [3], the site must be chosen based on the following criteria: (a) flat terrain to avoid topographical effects on wind speed patterns close to the vegetative barrier; (b) absence of obstructions such as buildings near the vicinity to avoid adverse effects on wind speed patterns along the vegetative barriers; (c) similar vegetation height on both sides of the vegetative barrier to maintain constant surface roughness; and (d) minimum length : height ratio of 10:1 to avoid edge and vortex effects at the ends of the vegetative barrier.
The description of the geographic areas chosen for experimentation, the climatic conditions and the vegetation type will be made when the experiments will be performed and the results will be presented, as in [3].

2.1 Wind Profiles

This is the first step of the method, which will be different from [3]. To determine the wind speed profiles, the authors [3] used two towers on which they placed several anemometers. In the experiences which are proposed in this article, one or more drones will be used. For measuring wind speed, drones (UAVs) will be equipped with anemometric sensors. These sensors will measure the speed and direction of the wind. Such sensors, for drone, exist in commerce, for example [6], [7], [8] and [9].

2.2 Porosity determination and vegetative barrier drag coefficient

To determine optical porosity, the same methods as [3] can be used, but using images obtained with a proper camera, installed on the drone (UAV). The linear force density along the forest curtain and the drag coefficient, can be calculated according to formula (3) and (4) from [3], neglecting the atmospheric pressure in the two parts of the curtain:

$$D_b = \int_0^z \left[\left(\frac{1}{2}\rho_{air}u_{up}^2\right) - \left(\frac{1}{2}\rho_{air}u_{down}^2\right)\right] dz\,,\, C_D \;=\; \frac{2D_b}{(\rho_{air}\bar{u}^2 H)} \tag{3}$$

2.3 Meteorological data and data analysis

As in [2], one meteorological station will be used on board each of the drones used. These stations will record: atmospheric temperature, humidity and pressure. All parameters will be reported to the geographic coordinates. In this paper, we are only predicting the structure of the data because we have not yet done such experiments with drones in the vicinity of the windbreaks. The data recorded during the flight of the drones in the neighbourhood of the windbreaks will contain the following parameters:

Table 1. Parameters to be recorded in each flight plan location.

No.	**Data**	**Parameters**	**Notation**	Units
1	Geographic Coordinates	Latitude	λ	degree
		Longitude	θ	degree
		Altitude	h	m
		Local abscissa	x	m
		Local coordinate	y	m
2	Meteorological data	Atmospheric temperature	T	Celsius degree
		Atmospheric pressure	p	Pa
		Humidity in the atmosphere	u	%
		Average[†] wind speed	v	ms^{-1}
		Direction of the maximum wind speed	ν	grade
3	Photographic images	Windbreaks images	If	-
4	Local time		t	Hours, min, sec

All data from the Table 1 will be recorded simultaneously in the designated data acquisition locations.

2.4 Setting up the drones route and storing data

To make the experiments as effective as possible, it's a good idea to set the drones' route before the start of the experiments. Also, the data sampling locations will be pointed out. The software will specify the sampling points of the data that will be automatically stored. An equidistant point network is recommended at varying distances (0.5 to 30 times the windbreak height of the longitudinal axis of the windbreak and at set heights up to 0 to 6 times the height of the windbreak). The record list will take the form:

$$\{\lambda_i, \theta_i, h_i, T_i, p_i, u_i, v_i, \nu_i, If_i, t_i\}, i = 1, \ldots, N \tag{4}$$

where N is the number of recordings made. The geographic coordinate array can be converted to local Cartesian coordinates using the national coordinate system (e.g. STEREO70 or WGS84). Is obtain the next string: $\{x_i, y_i, z_i, T_i, p_i, u_i, v_i, \nu_i, If_i, t_i\}; i = 1, \ldots, N$, and the conversion errors are negligible for the purposes (assuming $z_i = h_i$).
Therefore, in the data processing process, the following points of the fields (time-space dependent parameters) are obtained: maximum speed $\{v_i, \nu_i, \lambda_i, \theta_i, h_i, t_i\}; i = 1, \ldots, N$, or in

[†] When is working with a special anemometer (e.g. multi-parametric sonic anemometer, [9], [17], one can measure the three components of the wind velocity at a given time.

Cartesian coordinates $\{v_i, v_i, x_i, y_i, z_i, t_i\}; i = 1, \dots, N$, and the pressure, $\{p_i, \lambda_i, \theta_i, h_i, t_i\}; i = 1, \dots, N$ and in Cartesian coordinates $\{p_{i,} x_i, y_i, z_i, t_i\}; i = 1, \dots, N$ and similarly for the humidity and atmospheric temperature field, but these fields do not directly interest in the first stage of calculation. There is also an important field of computation, namely that of the photographic images, which are processed, and generates, according to formula (2), a field of optical porosity (variable in space and time): $\{\beta_i x_i, y_i, z_i, t_i\}; i = 1, \dots, N$ is generated. Operating with two drones simultaneously and collecting the data, on the each part of the forest curtain, in symmetrical network points in report to the axis of the windbreak, the next vectors of data will be obtained:

$$\{\lambda_{0i}, \theta_{0i}, h_{0i}, T_{0i}, p_{0i}, u_{0i}, v_{0i}, v_{0i}, If_{0i}, t_{0i}\}; i = 1, \dots, N \quad (5)$$

for the semi-space in which the wind enters the windbreaks and (4) for the semi-space where the wind comes out of the windbreaks. In this case, using the formula (1), and eventually a process of interpolation of the maximum speed field, an aerodynamic porosity field is obtained. Next, this field is averaged in a volume area given on the each side of the windbreaks. It can also proceed by calculating the average velocity and then directly calculating the aerodynamic porosity (1).

When working with at least two drones at the same time, each of them on different part of the forest curtain in the vectors (4) and (12) will be satisfied (by the working mode, or even by interpolation), the equality: $t_i = t_{0i}$. If the work is not done with two drones simultaneously, but only one, then you have to use the hypothesis that the wind has not undergone significant direction and intensity changes during the data collection process (which is physically unlikely in many cases).

2.5 Data processing

Data stored in the forms (4) and (5) can be processed to obtain aerodynamic and optical porosities. Data processing can be done in the same way as in [3], but it can also be used in a more complex way to processing. This option will briefly describes in order to be prepared for then when the data will be available.

Data is collected in the form (4) or (4) and (5) and then, by interpolation, the next functions will be obtained: $v, v_0, u, u_0, v, v_0, p, p_0, T, T_0$ in geographic or, in local Cartesian coordinates, x, y, z, t.

Using the definitions of the wind velocity field (vector) can be defined in this way:

$$\vec{v}(t) = v_x(t)\vec{\iota_x} + v_y(t)\vec{\iota_y} + v_z(t)\vec{\iota_z} \quad (6)$$

atmospheric pressure and temperature being scalar fields. The direction of maximum wind speed is established using the wind speed in Cartesian coordinates. The wind speed fields from the front and behind the windbreak has to be inserted as:

$$v_{wx} = v_x(x, y, z, t), v_{wy} = v_{wy}(x, y, z, t), v_{wz} = v_{wz}(x, y, z, t)\ , for\ y > \varepsilon$$

$$v_{lx} = v_x(x, y, z, t), v_{ly} = v_y(x, y, z, t), v_{lz} = v_z(x, y, z, t), for\ y < -\varepsilon \quad (7)$$

and in the same manner, for the other parameters: $u_w, p_w, T_w, u_l, p_l, T_l$, where ε is a constant of the windbreak, equal to, for example, the distance at which the drone can safely approach the curtain axis (due to the crown of the trees). The constant ε is also the limit of approach of the drones to the axis of the windbreak, so that the parameters fields will not measure in the intense turbulence area. The temporal mean of the physical parameters from (7), over a given time interval of length Θ, is defined by (8):

$$\bar{v}_{wx} = \frac{1}{\Theta}\int_0^{\Theta} v_x(x,y,z,t)dt\,, y \in [\varepsilon, nH] \tag{8}$$

and in the same manner for the other parameters, see [18]. Similarly, spatial average is defined by integrating the physical fields involved in the process, on the considered atmospheric volumes, using the triple integral, see [18].

2.6 Aerodynamic porosity

Dynamic porosity, as defined in [3], and in this paper in (1), is interpretable. The two speeds involved in the calculation formula (1), are defined as average speeds, but a way to calculate these mean values is not defined. Wind speed is a vector field that depends on the temporal variable and has three components, (15).Be:

$$v = \sqrt{v_x^2 + v_y^2 + v_z^2} \tag{9}$$

the intensity of wind speed. Be θ the angle between the Ox axis and the projection of the vector velocity on the plane xOy, and β the angle between the wind speed vector, and its projection on the plane xOy. Using these notations, the components of the velocity vector can be written as:

$$v_x = v\cos\beta\cos\theta, v_y = v\cos\beta\sin\theta\,, v_z = v\sin\beta \tag{10}$$

The system of equations (10) connects the two manners to give the wind velocity vector: by coordinates, respectively, by velocity and angular coordinates (polar coordinates). Expressing the intensity of the velocity, as a function of the angular polar coordinates, and by annulling the derivatives, the directions of maximum wind intensity is obtained:

$$\theta = arctg\frac{v_y}{v_x}, \beta = arctg\frac{v_z}{\sqrt{v_x^2+v_y^2}}, \theta \in \left[-\frac{\pi}{2}, \frac{\pi}{2}\right], \beta \in [0,\pi]. \tag{11}$$

Therefore, using the relations (11), it is possible to switch from the wind sensor information that gives the velocity vector through the Cartesian components to the information that gives the velocity vector through the maximum speed value. Next, the problem that needs to be clarified is that of estimating the effect of attenuation of the wind intensity produced by the forest curtains. Since the average velocity value can be done in time and space, a calculation method applicable to the wind speed field must be established both in front and behind the protection curtain and for any type of protection (having the axis, line right or curve, right line on portions, single tree line or more, possibly from several species). For example, we the average time speed of the wind can be defined, like in [10]:

$$\overline{v} = \overline{v}(x,y,z,t) = \frac{1}{T}\int_t^{t+T} v(x,y,z,t)dt, T > 0 \tag{12}$$

If the estimation of efficiency is done on a finite length of the forest curtain, then, for the calculus of the aerodynamic porosity, the average spatial value of the temporal mean value can be proposed:

$$\widetilde{\overline{v}_f} = \frac{1}{L(h-\varepsilon)H}\int_0^L\int_\varepsilon^h\int_0^H \overline{v}(x,y,z,t)\,dx\,dydz \tag{13}$$

and in the same manner, with changing of the integral limits after Oy axis for the volume in front of the windbreak. Thus, a measure of the aerodynamic porosity of the windbreak can be defined:

$$\alpha = \frac{\widetilde{v_b}}{\widetilde{v_f}}. \tag{14}$$

The ratio (14) is rather a measure of the wind intensity attenuation by the forest curtain. Conventionally we admit that this ratio can be a measure of the dynamic porosity of the forest curtain. In terms of spatial mediation, it should be noted that the mean values (13) are calculated on parallelepiped volumes only when studying areas of small length of a windbreak. In the case of estimating the efficiency of a windbreak over its entire length, the average velocity in front of the windbreak can maintain formula (13), but for the average velocity behind the windbreak, the integration volume changes, being less than that considered in the above considerations. In order to understand the volume delimitation mode on which the average wind speed value is calculated behind the windbreak, the readers are sent to [11].

2.7 Mode of operation in the field

According to the issues problems discussed from [3], the working process of the drones should aim at measuring the aerodynamic porosity and the optical porosity of the windbreaks. The working process may only include determining the wind speed field if the objective is to calculate the efficiency, and the optical porosity estimation is not necessary. By computing, as in other trials, a relationship will be established between these two porosities. The relationship between the two porosities may depend on the season, see [18], wind speed, type of the trees, atmospheric humidity. These dependencies can only be determined by repeated measurements under different environmental conditions. To make accurate measurements, it is recommended to operate with two drones, each doing the same type of measurements simultaneously, on different parts of the windbreak.

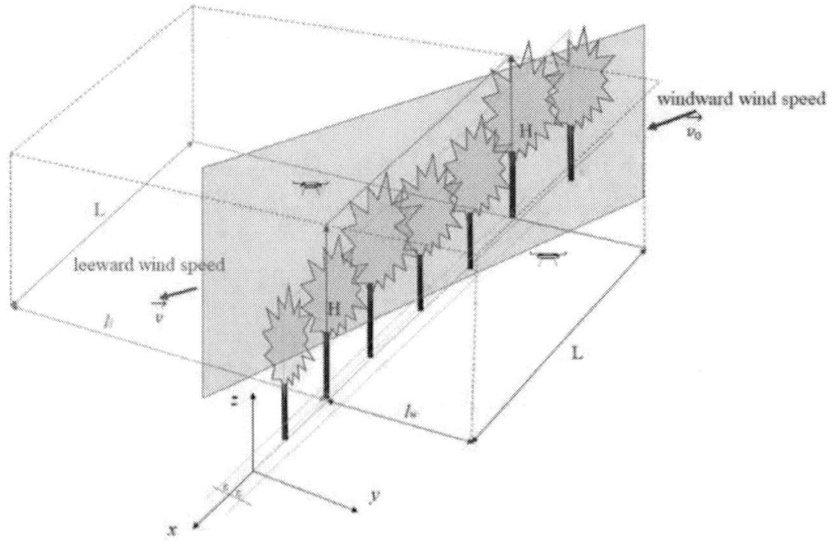

Fig. 1 The schema of the process of estimating the efficiency of a windbreak, using two drones simultaneously.

When working with two drones, one on each side of the forest curtain, they will sweep the chosen test volumes: $[0, L] \times [\varepsilon, h] \times [1.5, 2H]$, in the front of the windbreak, respectively$[0, L] \times [-h, -\varepsilon] \times [1.5, 2H]$, behind the forest curtain, see Figure 1. Lower and upper vertical limits are conventionally chosen by the operator team to cover the area of interest. Also, the constant ε, will be chosen so that no measurements are made in the area of intense turbulence near the windbreak. Depending on the experimental availability, the depth of the domain, h, can be chosen up to $10H$ - $20H$, [3]. The mode of operation of each of the two drones consists in stabilizing it in a measuring location for a minimum time interval to achieve wind speed recording and mediation. The following data vector is recorded in each measuring position:

$$\left(x_i, y_i, z_i, \bar{v}_{xi}, \bar{v}_{yi}, \bar{v}_{zi}, t_i, Im_i\right) \tag{15}$$

where $z_i > \varepsilon$ for the drone which is working in front of the windbreak, respectively $z_i < -\varepsilon$, for the drones which working behind the windbreak. The symbol Im_i is the image taken by the video camera of the drone, at the record with the index i.
If you are only interested in the intensity of the wind and possibly the direction of maximum intensity, you can move from the data structure (9) to the data structure (10):

$$\left(x_i, y_i, z_i, \bar{v}_i, \bar{\theta}_i, \bar{\beta}_i, t_i, Im_i\right) \tag{16}$$

using the Cartesian transformation and the formulas, (10), (11). The index i, varies from 1 to N_f, in front of the windbreak and from 1 to N_b behind the windbreak. The average wind speed values, which appear in (15) and (16) are calculated using the formula:

$$\bar{v}_i = \frac{1}{T}\sum_{j=1}^{M} \frac{v_{ij+1}+v_{ij}}{2}\left(t_{j+1} - t_j\right), 0 = t_1 < t_2 < \cdots < t_{M+1} = T. \tag{17}$$

where v_{ij} , is the speed recorded in the location i at the time t_j. If the recording time step at each location is constant, then the average (17) becomes the arithmetic average. For the other average data components measured, the formulas are similar, see [18].
In order to make a correct evaluation of the dynamic porosity, the reference speeds (6) - (7), averaged in relation to the spatial coordinates and thus located in the centre of the considered parallelepiped volume, should be reported. We can accept the hypothesis that in front of the forest curtain, the wind has small variations in relation to those behind the windbreak, so that it will be considered $N_f << N_b$. This means that the fluctuations in relation to the spatial coordinates are small in front of the windbreak in relation to the fluctuations of the wind speed behind the windbreak, without a narrow zone located along the forest curtain. For a first approximation of spatial mediation, one can consider the following arithmetic averages:

$$\hat{v}_f = \frac{\sum_1^{N_f} \bar{v}_i}{N_f}, \hat{v}_b = \frac{\sum_1^{N_b} \bar{v}_i}{N_b} \tag{18}$$

the first sum contain all the temporal average values of the maximum wind speed measured by the drones sensors in the front of the windbreak, and the second sum containing all time-averaged speeds determined by the drones behind the windbreak. A more precise variant of the spatial average can be obtained if there is availability of a drone's action plan, so that the spatial volumes in front and behind the windbreaks include at least 27 measurement points, in Cartesian network in each volume. In this case, the formulas (13), approximated by Riemann sums, according to the theory of the triple integral, for example [12], can be used for spatial average. Another possible option for spatial average of the wind speed calculus is the polynomial interpolation of experimental data, or other type of interpolation of temporal-span speeds, followed by the analytical exact calculus, using the formula (13).

3 Effectiveness of forest curtains - a mitigation problem

Since, generally, behind the windbreak, at a certain distance and height, the wind speed returns to the value of the one before the windbreak, the measure (14) of the forest curtain efficiency, cannot be correct unless it specifies the distance and the height at which it is made determining the wind speed behind the windbreak. The efficiency of the windbreaks can also be thought in other terms. In the design of vegetal barriers, the optimal height of their operation and the maximum limit of the attenuated speed, depending on the objective or the protected objectives, are specified, [5], [13].

In [14] there are results on the wind intensity behind the forest curtain at different distances and heights. Using Figure 6 from [14], we obtained data required to be able to estimate the distribution of wind intensity and to set a criterion for the efficiency of the barrier. To express the effectiveness of the forest curtain as a non-dimensional parameter, it is possible to divide the measured efficiency by the wind velocity return distance to the average value in front of the forest curtain, through the height of the curtain. We obtain the expression of a dimensionless efficiency, which can also be expressed in percentages: $f_a(h) = \delta(h)/H$ in which the distance $\delta(h)$, is the one at which the wind intensity, behind the windbreak, at height h (to the ground), is equal to the average wind speed in front of the forest curtain. Conventionally, δ can be a distance to the windbreak, for which the average speed behind the windbreak reaches the average value from the front of the same windbreak (or a fraction thereof) at any height, h, $\delta = \delta(h) = kH$. In this case, obviously, $f_a = k$.

Another way of defining the factor of attenuation of the wind intensity for a forest curtain is inspired by the graphical representation of the wind speed increase behind the windbreak [14]. It is possible to define the attenuation factor of the forest curtain, starting from the wind velocity distribution behind the curtain, in relation to the intensity of the wind speed in front of the curtain (considered as a mean constant value over time, but depending on spatial coordinates). Either the velocity field behind the curtain (Figure 1):

$$v_b = v_b(x, y, z), y < -\varepsilon, \tag{19}$$

It is considered a narrow section of the forest curtain so that the intensity variation (31) is negligible in relation to the variable across the windbreak, x (Figure 1). The above considerations also lead to more complex ways of defining the wind speed attenuation by a windbreak. Suppose, for example, using a drone, the time-averaged speed, more precisely the mean time value of the vector wind velocity is measured at some points behind the curtain. Because there aren't any own experimental data, the experimental data displayed as graphics is taken from [15]. Using the data published in [15] for an orange windbreak for an average wind speed in front of the 7 m/s barrier and an average barrier height 8.2 m, some estimation of the wind speed distribution behind the windbreak can be obtained.

Using the experimental information from [15], a distribution of the velocity field can be given in a plane section perpendicular to the axis or line of the forest curtain. Distribution can be obtained by interpolation. Such a distribution can be obtained by collecting the intensity of the wind speed in several locations behind the forest curtain. Using a wind distribution map, we can formulate different type of estimators of mitigate of wind intensity in the presence of a windbreak. The attenuation factor is defined as the ratio between the area of the field in which the velocity is less than the given speed and the total area of the same domain. If the maximum allowable speed is defined as a percentage of the average wind speed in front of the windbreak, then the maximum allowable speed is: $v_a = k_a v_f$. Maps of the velocity distribution in the atmospheric spaces in front and behind the vegetal barrier, can also be obtained by numerical simulation, see [18]. For comparison with the maps given in [18], [16] can be consulted, for example.

Starting from the surface definition of a plan area using Darboux sums, we calculate the area of the domain where the wind speed is below the maximum allowable speed:

$$A_u = \sum_{i=0}^{n_y-1} \sum_{j=0}^{n_z-1} (y_{i+1} - y_i)(z_{j+1} - z_j)\theta(w_{i,j}) \tag{20}$$

where:

$$\theta(w_{i,j}) = \begin{cases} 1, if\ w_{i,j} \leq k_a v_f \\ 0, if\ w_{i,j} > k_a v_f \end{cases} \tag{21}$$

In (20) and (21),

$$w_{i,j} = v(x_s, y_i, z_j),\ i = 0, \dots, n_y, j = 0, \dots, n_z \tag{22}$$

The coordinate along the windbreak line, x_s, is fixed, and $\{(y_i)_{i=1,\dots,n_y}\}, \{(z_j)_{j=1,\dots,n_z}\}$, are meshing of the edges of the volume considered, along the Oy axis, respectively Oz. The function v from (22) is obtained by interpolation from data collected by the drone. For the total area of the domain the dimensions of the rectangle can be directly used, or the formula:

$$A_T = \sum_{i=0}^{n_y-1} \sum_{j=0}^{n_z-1} (y_{i+1} - y_i)(z_{j+1} - z_j) \tag{23}$$

With these data, the attenuation factor will have the expression: $f_a = A_u/A_T$. For the dimensions of the section by 10 x 300 m and the characteristics of the case exposed in [15], taking $k_a = 0.95$, $f_a = 0.679$ is obtained. It can be seen that this attenuation factor depends on the length of the domain considered and its height. It also depends on the state of the crown of the windbreak and its type. In the formula (20), factors that take into account the sizes of the field used for agricultural crops can be introduced. For example, if the calculation is limited to a distance of 200 m behind the windbreak, the attenuation factor increases to 0.757. The limitation can also be made in height. The statement that the agricultural crop behind the windbreak needs protection up to a height of 3 m has been supposed. In this case the domain is restricted to this height (we keep the domain length is kept at 200 m), and, perhaps contrary to the expectations, the attenuation factor is lower than in the previous case, its value being 0.683. The explanation for this result can be found in the structure of the windbreak in the area of the stems, where, among the trunks of the trees, a wind of less attenuated intensity enters. In the case of complex windbreaks with double or triple rows of trees, obviously the situation will change.

4 Conclusions

The applications of the drones in the field of agricultural curtains aim at the following problems:
A1) General inspection of protection curtains for proper maintenance;
A2) Estimating the effectiveness of protection curtains according to the season and the type of vegetation used;
A3) Surveillance of the effect of forest protection curtains on protected crops (the effect of uneven baking in the vicinity of the curtain, harvest losses, humidity, exposure to light and temperature, etc.);
A4) Research activities on the relationship between optical porosity and dynamic porosity, depending on the type of vegetation used, the wind speed, the season, the vegetation response to some pest attacks.
For more details, see the source [18].

Acknowledgement: This work was supported by a grant of the Romanian Research and Innovation Ministry, through Programme 1 – Development of the national research-development system, subprogram 1.2 – Institutional performance – Projects for financing excellence in RDI, contract no. 16PFE, and by a grant of the Romanian Minister of Research and Innovation, CCCDI – UEFISCDI, project number PN-III-P1-1.2-PCCDI-2017-0560 "Eco-nanotechnologies and intelligent equipment for soil properties mapping and evaluating the dynamics of the plant in order to improve agricultural production and environmental protection", within PNCDI III.

References

1. Meriam - Webster dictionary, [Online]. Available: https://www.merriam-webster.com/dictionary/windbreak , (2018)
2. wikipedia\windbreaks, 17 July 2018. [Online]. Available: https://en.wikipedia.org/wiki/Windbreak , (2018)
3. H.B. Gonzales, M.E. Casada, L.J. Hagen, J. Tatarko, R.G. Maghiarang, C.J. Barden, Porosity and drag determination of a single-row vegetative barrier, Plant, Animal, &Facility Systems Community of ASABE, 2017, (2017)
4. C. G. Bates, WINDBREAKS: Their influence and value, Washington: *Government Printing Office*, (1911)
5. I. Musat, Gh. Guiman, Perdelele forestiere - modalitate eficienta de protectie a cailor de comunicatie si transport, Bucuresti: Academia Romana, www.acad.ro/com2014/doc/d0515/08Musat_Ilie.pps, (2014)
6. Wind-Sensors/Products/FT205, [Online]. Available: https://fttechnologies.com/Wind-Sensors/Products/FT205, (2018)
7. Applications/Applications/Drones-and-UAVs, [Online]. Available: https://fttechnologies.com/Applications/Applications/Drones-and-UAVs, (2018)
8. G. Giebel, U. Schmidt Paulsen, J. Bange, A. la Cour-Harbo, J. Reuder, S. Mayer, J. Mølgaard, Autonomous Aerial Sensors for Wind Power Meteorology - A Pre-Project Danmarks Tekniske Universitet, Risø Nationallaboratoriet for Bæredygtig Energi. Denmark. Forskningscenter Risoe. Risoe-R, (2012)
9. Applied Technologies, Inc., 2016. [Online]. Available: http://www.apptech.com/products/ultrasonic-anemometers/trisonica-mini/ , (2016)
10. Gh. D. Ionescu, Introducere in hidraulica, Bucuresti: Editura tehnica, (1977)
11. J.R. Brandle, Sh. Finch, "EC91-1763-B How Windbreaks Work, University of Nebraska, Lincoln, (1991)
12. M. Rosculet, *Analiza Matematica*, Bucuresti: Editura Didactica si Pedagogica, (1984)
13. Kika de la Garza Plant Materials Center, Vegetative Barriers For Erosion Control, Texas State Soil and Water Conservation Board, United States Environmental Protection Agency, Texas, (2013)
14. D. Rehacek, T. Khel, J. Kucera, J. Vopravil, M. Petera, Effect of Windbreaks on Wind Speed Reduction and Soil Protection against Wind Erosion, Soil & Water Res., (2016)
15. H.B. Gonzales, *Aerodynamics of wind erosion and particle collection through vegetative controls*, Manhattan, Kansas: Kansas State University, (2015)
16. G. Li, R.G. Maghirang, Numerical Simulation of Airflow and Particle Collection by Vegetative Barriers, Engineering Applications of Computational Fluid Mechanics Vol. 6, No.1, p. 110–122, (2012)
17. "GEONICA". [Online]. Available: http://english.geonica.com/prod/79/137/Wind-Speed-&-Direction/index.html. (2018)
18. P. Cardei, Drones and windbreaks, preprint, DOI: 10.13140/RG.2.2.15716.83843, https://www.researchgate.net/publication/329356139_Drones_and_windbreaks

Researches on the capitalization of some medicinal plants to obtain essential oils by using the performant innovative installation EUV 2x1000 type

Ion Grigore[1,*], *Andreea Iulia* Grigore[1], *Laurentiu-Constantin* Vladutoiu[1], *Mario* Cristea[1], *Elena* Sorica[1], *Dragos Dumitru*[1], *Florin Sirbu*[1], *Gabriel Badea*[3], *Marga Gradila*[2], *Marius Raischi*[3], *Denisa Duta*[4] and *Roxana Andreea Sandu*[5]

[1]INMA, TestingDepartment, Ion Ionescu de la Brad Blv. No. 6, Sector 1, Bucharest, Romania

[2]ICDPP Bucharest, Romania

[3]INCDPM Bucharest, Romania

[4]IBA Bucharest, Romania

[5]INCDTP Bucharest, Romania

Abstract. The article includes research on the production of essential oils from medicinal plants by applying water vapour distillation, the method most commonly used for the extraction of essential oils from plants. The results consist of extracting the essential oil from 2 types of medicinal plants: lavender and mint with the help of the EUV installation. The paper highlights the important aspects of obtaining quality products, on economically advantageous terms, addressing both specialists, farmers in larger farms and family associations. The article presents the experimental research results on obtaining essential oils from medicinal plants. Oils obtained by this method contain only volatile components, while wax and volatile resins are not found in the finished product.

1 Introduction

Medicinal plants are a real source of health. They have healing properties since ancient times and even today they enjoy special attention. Aromatic medicinal plants are found in nature as well as in cultures set up by small producers. Modern medicine recommends them as a natural remedy easy to obtain, cheap, effective and harmless. This is why in our country there is a large campaign for harvesting and capitalizing of medicinal plants, regardless of the season [1, 2]. Through the richness of terpenes and phenols, in alcohols and aldehydes, flavoring medicinal plants have demonstrated, over time, therapeutic antiseptic effects, so volatile plant oils administered in useful therapeutic doses have low toxicity, do not produce side effects and do not have contraindications (if administered

* *Ion* Grigore: ionica_grigore2001@yahoo.com

 https://doi.org/10.1051/e3sconf/201911203002

correctly). From a chemical point of view, volatile oils are complex mixtures of aliphatic and aromatic hydrocarbons, aldehydes, alcohols, esters and other constituents, but predominantly terpenoid compounds. In the past, they were called etheric oils, essential oils or aromatic essences [3, 4, 5]. Essential oils have been used for thousands of years for their healing and purifying effects on the body. They have always held an extreme value for doctors and ancient healers who have used them for aromatherapy, consumption and application on the skin. Research has now revealed its remarkable healing properties [6, 7].

Essential oils are concentrated by strong antioxidants. They come from raw plant material such as flowers, leaves, wood, bark, roots, seeds and peels [8, 9, 10]. Although they are called oils, these substances do not contain fatty substances: a drop of essential oil placed on a sheet of paper will not leave any trace, as opposed to a vegetable oil.

From the very wide range of medicinal and aromatic plants, we will focus on two of the most appreciated and used plants in our country, with high essential oil content: lavender (Lavandula angustifolia) (Figure 1) and mint (Mentha piperita) (Figure 2). Lavender is a perennial plant, considered a small bush, because the branches are wooded. It has evergreen leaves, even in winter, usually the leaves being green-silver, but there are also varieties of green or silver leaves. They reach maturity in 2-5 years, and can live without problems for 20-30 years. In our country, the most common kind of lavender is the narrow-leaved, 3-4 mm wide and 30-40 mm long, called Lavandula angustifolia. To establish a lavender plantation on a 1ha area, are needed about 18,000-20,000 seedlings that reach maturity after 3 years and production of green flower (fresh) can reach 4,000-4,500 kg / ha. If summer harvesting is done properly, you can also perform an autumn harvest from which you could get another 1,500-2,000 kg of lavender flower. Lavender essential oil is very used in medicine, cosmetics and perfumery.

Between the plants whose chemical composition can generate a significant amount of volatile oils and floral water, lavender occupies a leading position [1, 3, 4].

Fig. 1. Lavender (Lavandula angustifolia) [11]

Mint (Mentha piperita) (Figure 2) is one of the oldest known medicinal herbs. "Papyrus Ebers" mentions that mint was used with 1550 years before our time in Egypt. In Europe it was first cultivated in England (1760), where the first quantities of volatile oil were obtained [12]. The largest volatile oil production (3/4 of world production) is provided by the United States. To the world production of volatile oil of mint competes England, France, Italy, Bulgaria, Russia and Japan. Mentha piperita is a sterile hybrid, which in agricultural practice is propagated exclusively by vegetation, using the underground stools. Sometimes, for scientific purposes, mint is propagated by aerial stools, by the rooting of stem or even leaves [13]. Unlike many other plants and essential oils, the benefits of mint oil and mint have been studied and confirmed by the scientific community. Mint oil contains many minerals and nutrients, including iron, manganese, magnesium, calcium, folic acid, potassium and copper. It also contains omega-3 fatty acids, vitamin A and C. Mint harvesting can be done when the leaves reach 5-6 cm and a 50% inflorescence, time when the essential oil quantity is optimal. The ideal time for harvesting is in the morning between 10-14 o'clock in sunny weather, without dew, wind or fog [14].

Fig. 2. Mint (Mentha piperita) [15]

Removing a flavor from its natural environment is not an easy operation. It must be isolated or concentrated with a minimum of loss and compositional changes and without introducing new impurities.

Numerous factors can interfere and influence the quality of the distillation and the oils obtained. The operator performing the distillation must documents itself upon:

- the choice of distillation method for the chosen plants;
- the rules to be respected for the distillation of the retained plant;
- the quality of the flower which has effects on efficiency and quality, generally.

If the flower is collected at maturity, then the yield of distillation as well as the quality of the oils will be better. It can be seen an increase in yield and quality of up to 25%.

The time of collection, the quality of the transport materials (aerated bags made of natural materials), the storage time, the time between collection and distillation, the duration of distillation, the distillation temperature, the degree of cleaning of the alembic, the quality of the water required for the steam, the condensation temperature, all influence the yield and quality of the oils obtained.

The user must test the installation with his harvested plants to determine the best distillation parameters.

2 Materials and Methods

Technologies for obtaining volatile oils

The industrial extraction of volatile oils from various aromatic plants is performed according to different methods depending on their characteristic properties.

Each method is important and has its place in the production process of essential oils.

The main methods of obtaining volatile oils are [16]:

- Water vapor distillation;
- Extraction of volatile oils with organic solvents;
- Extraction with animal fats: enfleurage and maceration;
- Extraction with liquefied gases (supercritical CO2);
- Extraction by pressing;
- Adsorption on an adsorbent material.

Water vapor distillation is the most commonly used method for extracting volatile oils from most aromatic plants; it is used for industrial scale extraction, but equally effective can be used at laboratory level.

Using the EUV installation (Figure 3), the steam produced by the steam generator enters the vegetative mass, undergoes distillation, destroys the olefin glands cover, volatilizes the oil and then mixes with it. The mixture of water vapor and oil vapor passes into the refrigerant (condensation vessel), where it turns into a mixture of water and volatile oil. The maximum pressure inside the distillation vessel should not exceed 0.2 bar. A clean steam at a pressure of 0.18 - 0.20 bar, maximum, it is used. Distillation is generally ideal when the cooling water temperature is maintained between 24 and 35 °C. This mixture reaches the florentine vessel (separation vessel) where the separation occurs, namely, the volatile oil, being easier, deposit into a film above the floral water.

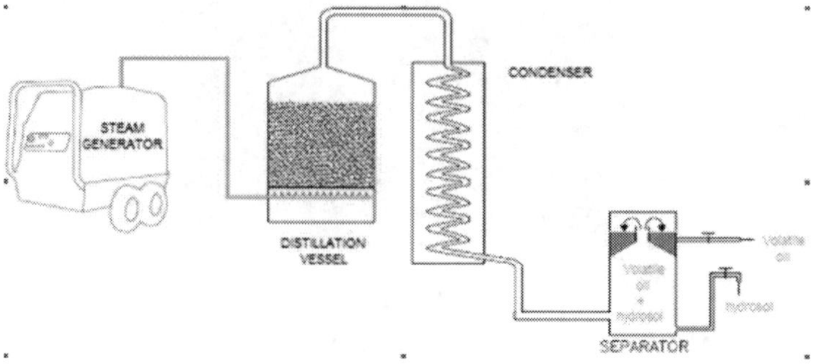

Fig. 3. Technological scheme of EUV distillation equipment

List of components of the EUV installation

- Stable metal supports for distillation vessels and condenser (cooling vessel);
- Steam generator with pressure regulator;
- One or two 1000 liter distillation vessels equipped with steam inlet/outlet sockets, drain tap and two support sieves for plant;
- Cover for the distillation vessel fitted with: 0.5 bar safety valves and a thermo-manometer;
- Condenser (cooling vessel with coil);
- Essences Separator (Florentine vessel);
- Food grade hose - Vaporcord Alim DN or food grade INOX pipe;
- L-type water tap for changing the direction of the flow (EUV-2X1000) (Figure 5).

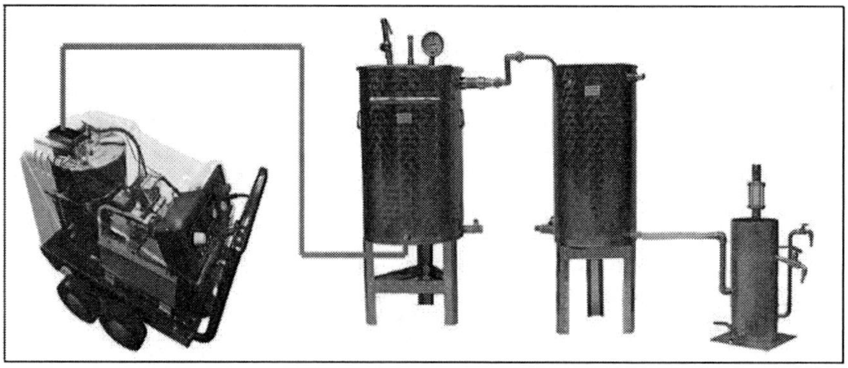

Fig. 4. The water vapor distillation installation EUV-1000

The EUV installation can work with a single distillation vessel fig. 4 or alternatively with two vessels Figure 5. In the case of two distillation vessels, the technical equipment is additionally equipped with two three-way stainless steel L-type water taps, with hoses (stainless steel pipe) for connection to the distillation vessels and the condenser. A water tap shall be installed at the outlet of the pressure regulator fitted on the steam generator so that by changing the outlet path, alternate steam supply through the hoses (stainless steel pipe) of the distillation vessels will be carried out. The second water tap is mounted at the inlet of the cooling vessel and by changing the direction of the flow through the L-type water tap, the alternate passage from one distillation vessel to the other towards the condenser (cooling vessel) is made. All metallic components of the installation are made of food grade INOX.

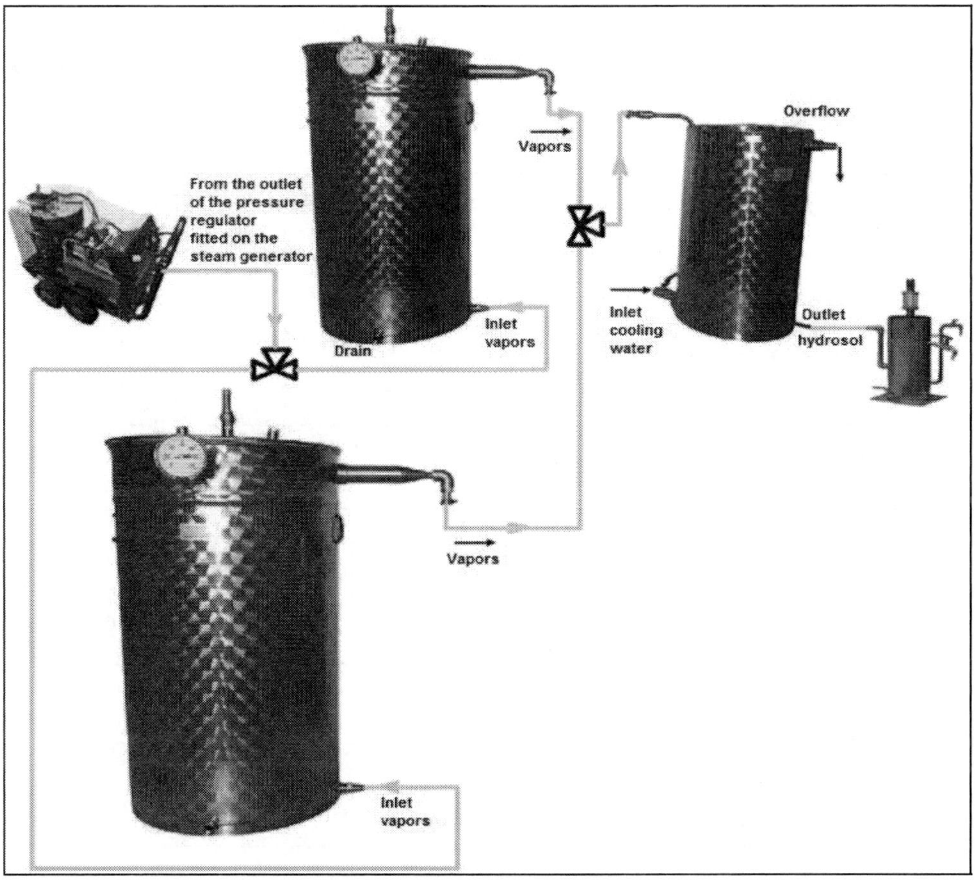

Fig. 5 The water vapor distillation installation EUV-2x1000

3 Results

The trials were carried out in Alba county with plants harvested from two lavender plots, respectively mint. Following the trials made, both with freshly harvested plants and dried plants in different amounts, the following results were obtained: with dry lavender, using between 30-90 kg, there were obtained between 0.08-0.27 l of oil (Table 1); lavender freshly harvested, using between 30-120 kg, there were obtained between 0.10-0.45 l of oil (Table 2); dried mint, using between 30-95 kg, there were obtained between 0.12-0.41 l of oil (Table 3) and fresh oil, using between 30-125 kg, there were obtained between 0.17-0.78 l of oil (Table 4). The amount of oil that can be obtained from distillation is greatly influenced by the quality of the plants under distillation.

Table 1 Dry lavender

Raw material	Quantity (kg)	Temperatures °C						Time (min)	Floral water (l)	Essential oil (l)
		Air	Cooling water	Steam		Interior vessel	Inlet cooling vessel			
				Outlet generator	Inlet vessel					
dry lavender	30	24	22	124	114	105	89	115	30	0.08
dry lavender	60	24	26	125	114	105	89	115	30	0.16
dry lavender	90	26	23	125	114	104	89	120	30.5	0.27

Table 2 Lavender freshly harvested

Raw material	Quantity (kg)	Temperaturi °C						Time (min)	Floral water (l)	Essential oil (l)
		Air	Cooling water	Steam		Interior vessel	Inlet vas cooling vessel			
				Outlet generator	Inlet vessel					
lavender freshly harvested	30	18	18	125	114	105	89	115	30	0.10
lavender freshly harvested	60	20	24	125	114	104	89	120	30.5	0.21
lavender freshly harvested	90	23	25	125	114	104	88	120	30.5	0.33
lavender freshly harvested	120	28	26	125	114	104	87.5	120	31	0.45

Table 3 Dry mint

Raw material	Quantity (kg)	Temperatures °C						Time (min)	Floral water (l)	Essential oil (l)
		Air	Cooling water	Steam		Interior vessel	Inlet cooling vessel			
				Outlet generator	Inlet vessel					
dry mint	30	19	19	125	114	105	89	120	30.5	0.12
dry mint	65	23	20	125	114	105	89	115	30.5	0.25
dry mint	95	26	24	125	114	104	88	120	31	0.41

Table 4 Mint freshly harvested

Raw material	Quantity (kg)	Temperatures °C						Time (min)	Floral water (l)	Essential oil (l)
		Air	Cooling water	Steam		Interior vessel	Inlet cooling vessel			
				Outlet generator	Inlet vessel					
mint freshly harvested	30	21	24	125	114	105	89	115	30	0.17
mint freshly harvested	65	25	24	125	114	104	88	120	30.5	0.37
mint freshly harvested	95	25	25	125	114	104	87.5	120	31	0.56
mint freshly harvested	125	26	23	125	114	104	87.5	125	31.5	0.78

The productivity differences recorded for lavender and mint, as a result of the use of both dry material and fresh material, are shown in Figures 6 and 7.

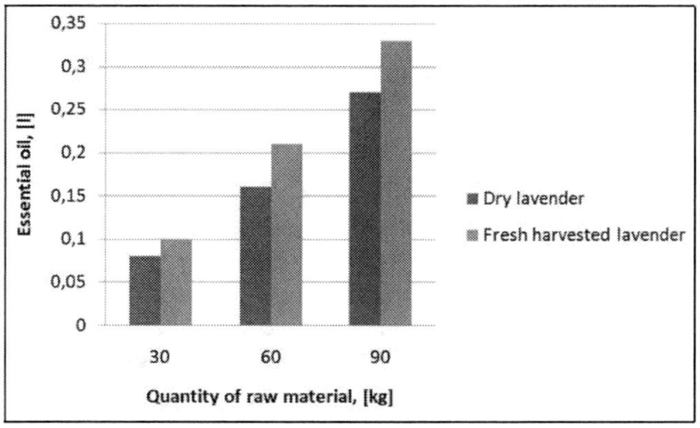

Fig. 6 The productivity of distillation process for lavender

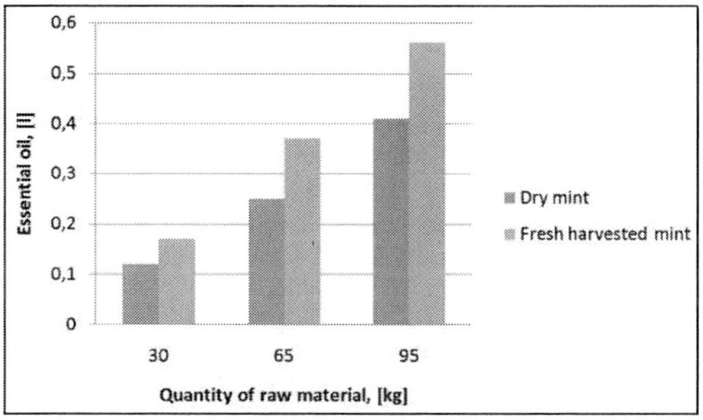

Fig. 7 The productivity of distillation process for mint

In order to prove the effectiveness of the installation with alternative operation EUV-2X1000 in comparision with EUV-1000 installation, tests were performed for two days, 10 hours a day, with each variant of the installation. The amount of distillation material was set at 120 kg freshly harvested lavender for each sample. Thus on the first day tests were made with EUV-1000. The duration of a sample was about 145-150 min, of which 25-30 min represents the times for cooling, evacuation of the meterial, cleaning the vessel and refilling it with the raw material for the next sample. Under these conditions, 4 samples were obtained within 10 hours. The next day with the EUV-2X1000 installation, 5 samples were made with the same amounts of material / sample. All tests were performed at an inlet pressure of 0.2 bar in the distillation vessel.

4 Conclusions

After the tests, the following aspects were identified:
- the amount of oil obtained depends largely on the quality of the plants subjected to distillation;
- the distillation time is approximately equal, irrespective of the quantity of plants in the vessel;
- from the freshly harvested plants subjected to distillation, about 25-30% more essential oil is obtained compared to the quantity obtained from dried plants;
- by using the EUV-2X1000 installation, a larger quantity of plants with about 20% / day can be distilled, comparing with the EUV-1000, due to the elimination of the distillation

vessel refueling times, this being done during the time in which the material from the first vessel is distilled, the distillation stream being continuous.

Acknowledgement

This work was supported by a grant of the Romanian Research and Innovation Ministry, through Programme 1 – Development of the national research-development system, subprogramme 1.2 – Institutional performance – Projects for financing excellence in RDI, contract no. 16PFE.

References

1. C. Popescu, A. Pruteanu, D. Cujbescu, I. Grigore, M. Matache, I. Voicea, V. Vlăduţ, E. Marin, C. Popescu, *Physical-chemical and microbiological characterization of lavender water obtained as secondary product from lavender (lavandula angustifolia L) essential oil,* International Symposium ISB-INMA TEH' 2015 Agricultural and Mechanical Engineering, 30 – 31 October, Bucharest, ISSN 2344-4118, Pages 335-342, (2015);
2. http://interferente.ro/plante-medicinale-izvor-de-sanatate.html;
3. C. Popescu, I. Voicea, V. Vlăduţ, D. Cujbescu, S. Bungescu, *Physical and chemical characterization of the lavender water resulted as waste at the obtaining of the lavender volatile oil, by driving with water vapour (Part 1),* Fifth International Conference "Research People and Actual Tasks on Multidisciplinary Sciences", Section 2 - Social and human sciences. Economic sciences. Ecology and Environment protection, Land & Water Use and Environment, Physics, Chemistry, Biology, Lozenec, Bulgaria, 24 – 28 June, Pages 153-160, (2015);
4. C. Popescu, I. Voicea, V. Vlăduţ, D. Cujbescu, I. Găgeanu, M. Duţu, *Physical-chemical characterization of lavender water resulted as waste when obtaining volatile oil from lavender, by engaging with water vapors (Part 2),*Fifth International Conference " Research People and Actual Tasks on Multidisciplinary Sciences", Section 2 - Social and human sciences. Economic sciences. Ecology and Environment protection, Land & Water Use and Environment, Physics, Chemistry, Biology, Lozenec, Bulgaria, 24 – 28 June, Pages 161-168, (2015);
5. http://www.creeaza.com/familie/medicina/ULEIURILE-VOLATILE-ESENTIALE127.php;
6. C. Popescu (Popiniuc), C. Popescu, St. Manea, V. Vlăduţ, I. Voicea, C. Covaliu, H. Abbas, A. Dune, D. Lupuleasa, *The influence of extraction solvent on the active principles content of Portulaca Oleracea native species*, Revista de Chimie, Volume **69** (10)/2018, Pages 2682-2692, (2018);
7. C. Popescu (Popiniuc), C. Popescu, St. Manea, V. Vlăduţ, I. Caba, C. Covaliu, H. Abbas, A. Dune, D. Lupuleasa, *Study on Portulaca Oleracea native species as vegetal source of Omega-3 and Omega-6 fatty acids*, Revista de Chimie, Volume **69** (11)/2018, Pages 2973-2980, (2018);
8. C. Popescu, I. Voicea, V. Vlăduţ, *Influence of breakage process on increasing the extraction yield of medicinal plants bioactive substances,* Romanian Biotechnological Letters, Volume **20**, Issue 4, Pages 10561-10571, (2015);
9. A. Pruteanu, C. Popescu, V. Vlăduţ, I. Găgeanu, *Biochemical analysis of some vegetal extracts obtained from indigenous spontaneous species of (Thymus serpyllum L.)*, Romanian Biotechnological Letters, Volume **23**, Issue 5, Pages 14013-14024, (2018);
10. https://viataverdeviu.ro/proprietatile-vindecatoare-ale-uleiurilor-esentiale/;
11. https://tse3.mm.bing.net/th?id=OIP.lB6JAuMdraQ1LEk8TSL5yAHaEs&pid=15.1&P=0&w=285&h=181\;
12. Solacolu T, Kollo C, Revista sanitara militara, (No 8, 1938);
13. http://plantemedicinale.crestere-melci menta-cultivarea-prelucrarea-si-utilizarea-mentei.ro/27/;
14. https://dietamediteraneana.wordpress.com/tag/recoltarea-mentei/;
15. https://tse4.mm.bing.net/th?id=OIP.-siBwpUFwAVvozT51WSObAHaD2&pid=15.1&P=0&w=326&h=171;

16. http://agro.afacereamea.ro/wp-content/uploads/floricultura/Metode-de-Obtinere-Si-de-Prelucrare-a-Uleiurilor-Volatile.p.

Theoretical and experimental research on sowing rate optimization of grassland drills

Dragoş Manea[1,*], *Eugen* Marin[1], *Marinela* Mateescu[1], and *Andrei* Dumitraşcu[1]

[1]National Institute of Research - Development for Machines and Installations designed to Agriculture and Food Industry - INMA, Bucharest 013813, Romania

Abstract. In this paper, sowing rate optimization of grassland drills consisted in finding an optimum combination between the seed flow of the seed metering device and the drill working speed, in order to achieve the sowing rates required by agro-technique for herbs seeds. It was developed a mathematical model that describes the seed flow variation depending on the working speed of the grassland drill and the active length of the groove cylinders. The experimental research was conducted using a grassland drill with eight metering devices and an automated installation, which simulates the working speed of the drill, collects and weighs seeds from each individual metering device. By using MathCad 14.0, it have been drawn the variation diagram of seed flow depending on active length of the groove and working speed. The comparative study of the variation diagrams based on experimental data and the variation diagrams obtained from mathematical modelling highlighted that the downward trend of the seed flow was observed in 66,67% of the results. Based on the mathematical model presented in this paper, it might be developed an algorithm enabling for fast obtain the optimal combination between the adjustment parameters, in order to apply a desired sowing rate.

1 Introduction

Improving and diversifying grasslands drills is the task of great importance to scientific research in the field of power and machinery systems. Development of the seed metering devices as main parts of drills was studied in some scientific works [1, 2].

Several authors [3,4] performed researches regarding optimization of drills seed flow, concluding that the cylinder dimension, flutes number, flutes length and revolution had important effects on the uniformity of seeds distribution.

Guler [5] investigated the seed flow distribution from drills meter rollers, concluded that the meter roller diameter should be between 6...8 mm and the flutes length should be between 15..25 mm at a rotation speed of 20...40 rotation per minute.

Gervais et al. [6] developed a method for test the performance of seed metering devices. This method was based on the metering roller flow continuity during operation and included an experimental procedure using testing devices.

* Corresponding author: manea_dragos_05@yahoo.com

Maleki et. al. [7] focused on how to increase the uniformity of seed placement by optimising the auger configuration, design parameters being derived to configure the auger groove depth and width, number of grooves, auger outer diameter and rotation speed.

Song et. al. [8] used *Festuca arundinacea* seeds and analyzed the effects of combing teeth space, rotation speed, inclining angle of flutes on the rate and uniformity. They observed that the variation of uniformity coefficient was based on the above four influencing factors.

The seeds metering and distributing process, has a decisive influence on the main quality indices of the drill: seed flow respectively of sowing rate uniformity, distribution uniformity on the working row. Therefore, the improvement of grasslands drills has been focused primarily on improving distribution devices, which are one of the main working part of the drill.

Theoretical and experimental researches worldwide performed, relating to grassland drills metering devices, have been more focused on their role in providing a wide range of flow rates that fully meet agro-technical requirements and less on concrete influence on the structural and functional parameters on the working process.

In this paper, optimization of the seeding rate of grassland drills, consisted in finding an optimum combination between the seed flow of the seed metering device and the drill working speed, to achieve the properly sowing rates required by agro-technique for herbs seeds.

2 Theoretical research

The main adjustment index of the metering roller is the seed discharge, which is given by the flutes active length in combination with the working speed of the grasslands drill. The seed flow (*F*) of the groove cylinder may be calculated with the following equation [9]:

$$F = \frac{V \cdot n_m \cdot \gamma_s}{60}\text{, (kg/s)} \quad (1)$$

where: V is the seeds volume discharged at one rotation, in m^3/rot; n_m - the flute cylinder rotation speed, in rpm; γ_s – seeds specific mass, in kg/m^3.

$$V = 1{,}65 \cdot S \cdot l \cdot z \cdot \psi\text{, (m}^3\text{/rot)} \quad (2)$$

where: S is the flute section area in m^2; l - flute active length, in m; z – number of flutes; ψ - flute filling coefficient ($\psi = 0{,}93 \div 0{,}98$ for small seeds).

According to [10], the relation between the seed flow (F) of the metering device and the sowing rate (N) is:

$$F = V \cdot n_m \cdot \gamma_s = \pi \cdot d \cdot n_w \cdot W \cdot \frac{N}{10^4} \quad (3)$$

where: d - diameter of the drive wheel, in m; n_w – the driving wheel rotation speed, in rpm; W - drill working width, in m.

The rotation speed of metering device was calculated with formula:

$$n_m = \frac{60 \cdot s}{\pi \cdot d \cdot i_t}\text{, (rad/s)} \quad (4)$$

where: s is the drill working speed, in m/s; $i_t = n_w / n_m$ - the transmission ratio between driving wheel and metering device.

By introducing the equations (2) and (4) into equation (1), were obtained the equation of the theoretical seed flow (F_t), which is:

$$F_t(l,s) = \frac{1{,}65 \cdot S \cdot z \cdot \psi \cdot \gamma_s}{\pi \cdot d \cdot i_t} \cdot l \cdot s\text{, (kg/s)} \quad (5)$$

The equation (5) shows that if we maintain parameters S, z, ψ, γ_s, W, d and i_t constant, then the theoretically seed flow depends directly proportional on the flute active length and working speed. From the experimental researches, authors observed for increasing working speed the seed flow decreases, which the theory does not show.

Following these observations, was used for modelling the approximation equation (6):

$$F_t(l,s) = \sum_{i=0}^{n} \sum_{j=-k}^{m} x_{ij} l^i s^j \qquad (6)$$

Data approximation by rational functions is an option dictated by the accuracy with which these functions can model the experimental data [11]. In equation (6) have been taken negative integer exponents for working speed, in order to modelling seed flow decrease.

The method of calculation is the method of least squares [11,12], form of approximation (6) leading to the following minimized function:

$$f(x_{ij}) = \sum_{p=1}^{n} \left(\sum_{i=0}^{n_l} \sum_{j=-k}^{n_s} x_{ij} l_p{}^i s_p{}^j - F_{tp} \right)^2 \qquad (7)$$

The equations system obtained by cancelling the partial derivative is linear and leads then to obtain the coefficients x_{ij}. The function (7) was defined on a infinite variety of functions. In this paper we considered terms of maximum 2 in *l* and minimum -2 in *s*. Using this method, we found the rational approximation function which minimizes the function (7):

$$F_t(l,s) = x_0 \cdot l + x_1 \cdot \frac{1}{s} + x_2 \cdot l^2 + x_3 \cdot \frac{1}{s^2} + s_4 \cdot \frac{l}{s} + x_5 \cdot \left(\frac{l}{s}\right)^2 \qquad (8)$$

The values of coefficients x_{ij} have been calculated with MathCad 14.0 program. The function (8) becomes:

$$F_t(l,s) = 2.5 \cdot 10^3 \cdot l - 62.5 \cdot \frac{1}{s} - 2,6 \cdot 10^4 \cdot l^2 + 0,7 \cdot \frac{1}{s^2} + 5,4 \cdot 10^3 \cdot \frac{l}{s} - 4,3 \cdot 10^4 \cdot \left(\frac{l}{s}\right)^2 \qquad (9)$$

Figure 1 shows the comparative chart between the theoretical seed flow values and the seed flow values obtained by applying the equation (8) to theoretical values for *l* and *s*.

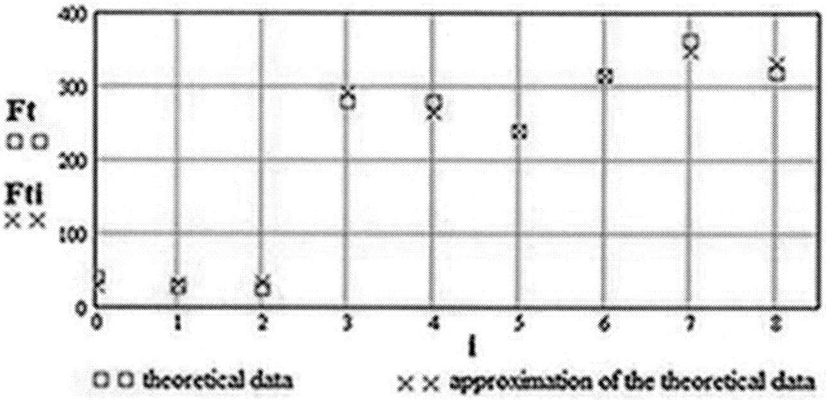

Fig. 1. Comparative chart between theoretical seed flow values and the ones obtained with the rational approximation function $F_t(l, s)$

Analyzing the graphs in figure 1, was observed a good approximation of the theoretical data by the function $F_t(l, s)$.

3 Experimental research

The aim of the experimental researches was to validate the theoretical model, in order to optimize the working process and to obtain mathematical relations helpful in making optimal adjustment of the grassland drills.

The experimental researches were performed at National Institute of Research - Development for Machines and Installations designed to Agriculture and Food Industry - INMA Bucharest. Tests were conducted on the test bench presented in Figure 2, which is made of: the grasslands drill MSP (*1*), having eight metering devices; an electric gear motor (*2*) with variable speed continuously adjustable, to simulate the working speed of the drill; two bands provided with seed collection boxes (*3*), powered by two electric motors; an electronic scale (*4*) for measuring quantity of seeds collected from each individual metering roller and the entire seeds quantity; an automated control and monitoring panel, having an PLC (programmable logic controller) and four frequency converters. Transmission of the motion to the metering devices axis is made through a gearbox with camshaft mechanism and with a chain transmission.

Fig. 2. Experimental installation used for tests

The seed used was *Trifolium repens,* with the main characteristics: 98 % purity ; 2.4 g mass of one thousand seeds; 1.04 g/ cm3 specific mass . Tests were performed in five repetitions for 3 seed flows (minimum, medium and maximum) and 3 working speeds (1.94 m/s; 2.36 m/s; 2.77 m/s). Data obtained from tests operations were processed by using MathCad 14.0.

In figure 3 there are presented variation diagrams of average seed flow obtained experimentally (F_e) versus theoretically seed flow (F_t).

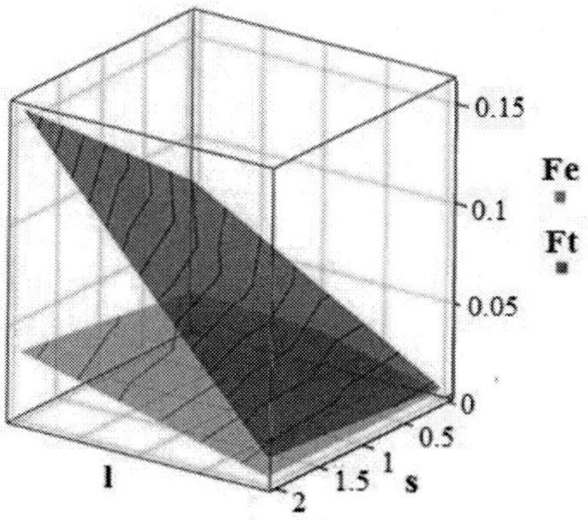

Fig. 3. The variation of experimentally and theoretically seed flow

Analyzing the charts in Figure 3, it was noted that the surfaces generated by experimental data have an aspect similar to the surfaces generated theoretically, and by plotting the curves of constant seed flow, could be observed that those obtained experimentally are very close to the theoretical.

In order to validate the experimental data, were compared on the same diagram, the variation of theoretically and experimentally seed flow, for the three values of the active length of grooves and working speed (Figure 4).

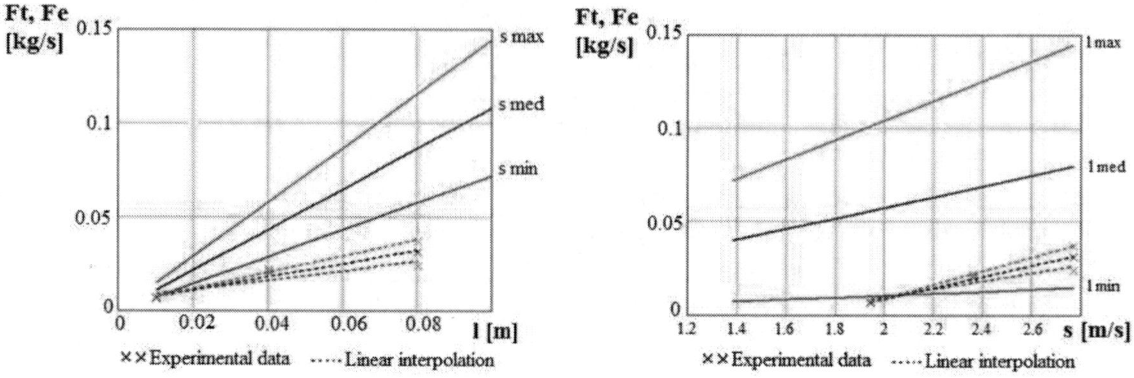

Fig. 4. Seed flow variation, theoretically and experimentally, in terms of: active length of grooves (*left*); working speed (*right*)

The theoretical values for the working speed and active length of grooves were given by theoretical limits of the variation intervals chosen for the study, as well as the average values located in the middle ranges. By linear interpolation of the experimental data, was observed similarity with the theoretical data.

The *deviation in relation to the average of probes (U)* was calculated with the eqaution:

$$U = \frac{U_{max} - U_{min}}{\overline{U}}, [\%] \tag{10}$$

where: U_{max} is the maximum value of seed flow at five repetitions; U_{min} – the minimum value of seed flow at five repetitions; $\overline{U}$ - average of seed flow at five repetitions.

The calculated values for the deviation in relation to the average of probes, are presented graphically in Figure 5.

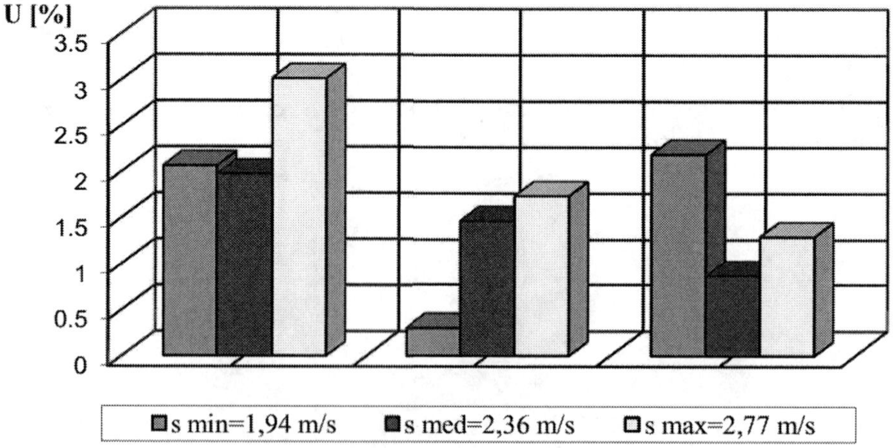

Fig. 5. Deviation in relation to the average of probes, for minimum, medium and maximum working speed

According to the requirements of Romanian National Standard [13], deviation in relation to the average of probes must not exceed 5% for grasses seeds. Experimental data presented comply with the requirements of the national standard, in conclusion the seed flow test was validated.

In order to validate the mathematical model of the seed flow, the rational approximation function chosen in the previous section, noted with $F_t(l, s)$ was applied to experimental data obtained for the seed flow. The equation describing the seed flow dependence on the grooves active length and working speed is:

$$F(l,s) = -65.3 \cdot l + 1.4 \cdot 10^{-5} \cdot \frac{1}{s} + 0.7 \cdot 10^{3} \cdot l^{2} + 0.2 \cdot \frac{1}{s^{2}} + 1.4 \cdot 10^{3} \cdot \frac{l}{s} - 2 \cdot 10^{4} \cdot \left(\frac{l}{s}\right)^{2} \quad (11)$$

Figure 6 are plotted the experimental values for seed flows and the values obtained by applying the function (11) on experimental values for active length of grooves and working speed. It can be seen a good approximation of experimental data by function (11).

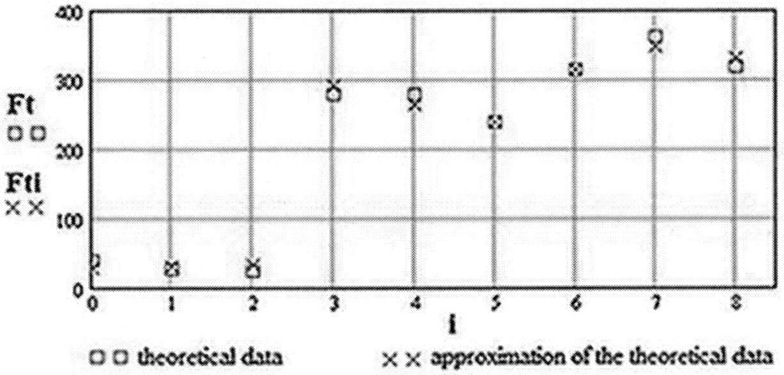

Fig. 6. Comparative chart between experimental seed flows and seed flows obtained with the approximation function

The seed flow variation, by using the equation (11), in terms of each adjustment parameter (working speed and active length of grooves) is presented in Figure 7.

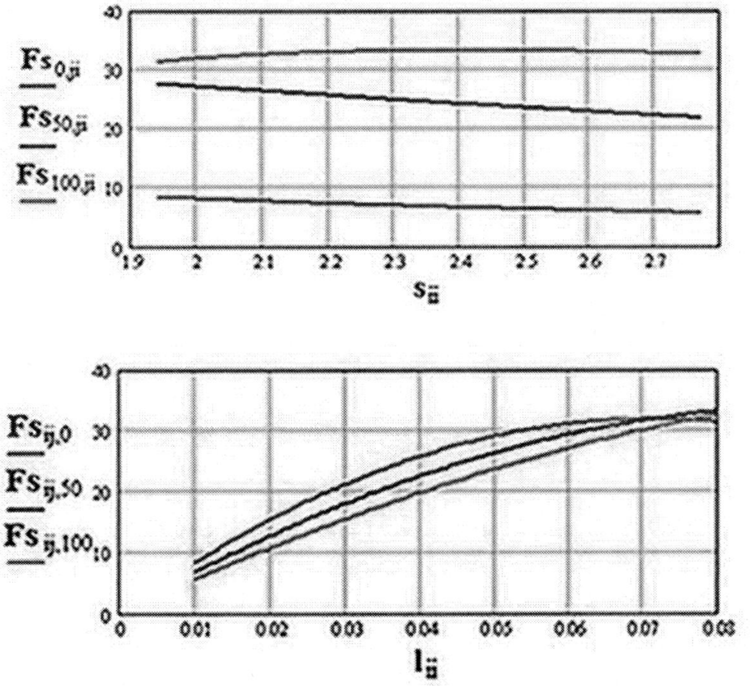

Fig. 7. Seed flow variation in terms of working speed (*up*) and active length of grooves (*down*)

Analyzing the graph in Figure 7, we have found up that at minimum and medium values of the active length of grooves, the seed flow decreases with increasing working speed, it

 https://doi.org/10.1051/e3sconf/201911203003

has a slight increase for the maximum value of groove active length, then returns to the downward trend.

In order to verify the theoretical hypothesis according to which the seed flow decreases with increasing working speed and to validate the mathematical model, were studied in comparison, the variation diagrams of seed flow depending on working speed, based on experimental data and the variation diagrams of seed flow depending on working speed obtained by applying theoretical mathematical model on experimental data (Figure 8).

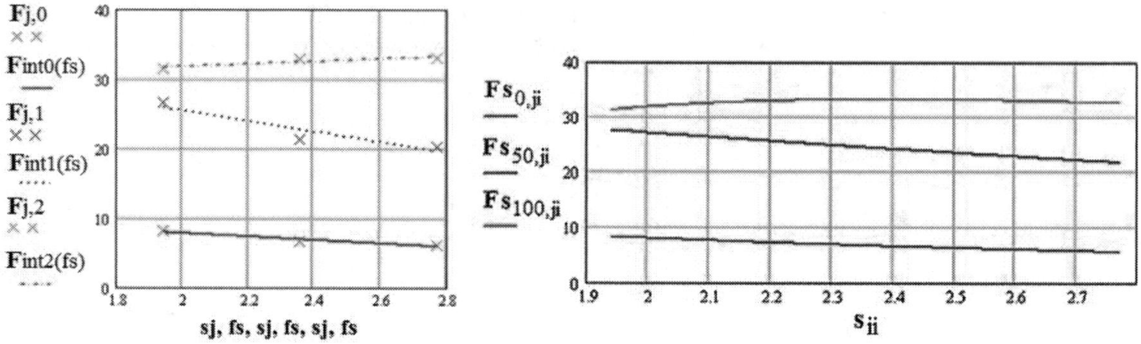

Fig. 8. Graphical comparison between seed flow based on experimentally data (*left*) and on mathematical model (*right*)

Observations which can be drawn from the comparative study are:

- the downward trend of the seed flow with increasing working speeds is observed in 66,67% of the results obtained both experimentally and by mathematical modelling;
- the remaining of 33,33 % covers the maximum seed flow, which have a tendency to increase with increasing working speed.

4 Conclusions

Seed flow of grassland drills decreases with increasing working speed. Starting from this hypothesis have been defined an approximation function.

It have been drawn the variation diagram of seed flow depending on flute length and working speed, using the approximation function derived from mathematical modelling applied on experimental data. The seed flow of the metering device determined experimentally for three working speeds and three active lengths of grooves, confirms theoretical variation law, and the deviation in relation to the average of samples comply with the requests of national standard.

The initial hypothesis has been verified and the theoretical mathematical model of seed flow has been validated, experimental data confirming its usefulness and correctness.

Were obtained the adjustment curves for the seed flow. These curves provide a clear image on the possibilities of achieving the desired seed flows or sowing rates for each type of grass seed, by the combination of the two adjustment parameters (grooves active length and working speed).

Based on the mathematical model presented in this paper, it might be developed an algorithm enabling for fast obtain the optimal combination of the adjustment parameters (grooves active length and working speed), of the grassland drills, in order to apply desired sowing rates.

This work was supported by a grant of the Romanian Ministery of Research and Innovation CCDI - UEFISCDI, Project INNOVATIVE TECHNOLOGIES FOR IRRIGATION OF AGRICULTURAL CROPS IN ARID, SEMIARID AND SUBHUMID-DRY CLIMATE, project number PN-III-P1-1.2-PCCDI-2017-0254, contract no. 27PCCDI / 2018 and Projects for financing excellence in RDI, contract no. 16PFE, within PNCDI III.

References

1. V. D. Crişan, V. Roş, Bulletin UASVM CN, **69**, 1 (2012)
2. V. D. Crişan, V. Roş, M. Gheres, *Proceedings of the 40th International Symposium, Actual Tasks on Agricultural Engineering*, **40**, 243-251 (2012)
3. I. Ozturk, Y. Yildirim, S. Hinislioglu, B. Demir, E. Kus, *Scientific Research and Essays*, **7**(1):78-85 (2012)
4. N. Turgut, I. Ozsert, M. Kara, Y. Yildirim, *16th National Congress on Agricultural Mechanization*, Bursa, Turkey, 529-537 (1995)
5. I. E. Guler, *Journal of Applied Sciences*, **5**(3):488 - 491 (2005)
6. J. Gervais, S. Noble, *XVIIth World Congress of the International Commission of Agricultural and Biosystems Engineering (CIGR)*, pp. 1-10 (2010)
7. M. R. Maleki, J. F. Jafari, M. H. Raufat, A. M. Mouazen, J. De Baerdemaeker, Biosystems Engineering, **94** (4): 535–543 (2006)
8. Y. Song, Z. Shumin, Int J Agric & Biol Eng **8**(1): 9—16 (2015)
9. Fl. Rus, *Agricultural machinery for soil and crop maintenance* (University Transilvania from Braşov, 1987)
10. V. Scripnic, P. Babiciu, *Agricultural machines* (Ceres Publishing House, Bucharest, 1979)
11. M. Iosifescu, C. Moineagu, V. Trebici, E. Ursianu, *Small Statistical Encyclopedia* (Scientific and Encyclopedic Publishing, Bucharest, 1985)
12. V. Bobancu, N. Mihăileanu, Ş. Gheorghiţă, A. Brezuleanu, A. Ştefănescu, T. Bălănescu, *General Mathematics Dictionary* (Romanian Encyclopaedic Publishing, Bucharest, 1974)
13. ***SR 13238 – 2. *Agricultural Machines. In Row Drills. Technical requirements.*

Experimental research on influence of functional parameters of combined installations designed at separating the impurities out of cereal seeds

Carmen Brăcăcescu[1,*], *Ioan* Ţenu[2] , *Costin* Mircea[1], and *George* Bunduchi[1]

[1]National Institute of Research-Development for Machines and Installations designed to Agriculture and Food Industry- INMA/ Bucharest, Romania;
[2]"Ion Ionescu de la Brad" University of Agricultural Sciences and Veterinary Medicine of Iasi - USAMV Iasi, Romania

Abstract. The paper presents the experimental installation, the methodology and measuring apparatus used for experimental research of qualitative indexes of impurities separation out of grain seeds for combined separating systems (according to specific weight and aerodynamic properties of seeds). The experimental installation used was composed of a gravimetric separator with mechanical shaker with unbalanced masses (mounted on the platform working surface) and an aspiration installation with fan. The experimental research has aimed at quantitative and qualitative influence on separation quality index of the following operating parameters: material flow rate of shaking separator, the air flow of aspiration installation, tilting work surface, work surface oscillation amplitude. Based on data obtained by measurements and qualitative indicators the separating process indexes have been determined, namely: degree of impurities separation, degree of good seeds loss, as well as, the separation quality index for different types of combinations of separation installation parameters.

1 Introduction

After the harvesting process, grain seeds cannot be used directly for different purposes (storage, consumption, industrialization, sowing, etc.) because they contain impurities of any kind (vegetal remains, foreign bodies, etc.) and injured products. Prior to receiving a particular destination, it is necessary and obligatory for the harvested products to undergo cleaning and sorting, called as primary processing operations.

The preparation of cereal seeds before processing them is a technological process, which requires technical equipment and combined installations types to eliminate foreign bodies existing in the mass of harvested seeds, whose construction depends on the different types of separation [1, 2, 3, 4].

* Corresponding author: carmenbraca@yahoo.com

Thus, research in this area focused on the determination of the physico-chemical properties of the components of a seed mixture that influence the separation process, with the aim of eliminating as much as possible the impurities that reach the final stages of grain processing [5,6, 7]

The modern conditioning units use installations for separation the impurities out of seed mass using a lot of separation principles, such as the differences between the volumetric weight and the air flow velocity of the components of the seed mixtures [8,9,10].

In order to obtain efficient separation equipment with a low consumption of material, electricity and labor, it is necessary to have a thorough knowledge of the technology, constructive principles, operation and maintenance of these technical equipments.

The experimental model of gravimetric separator SG-01 is a innovative technical equipment representative for cereals seed cleaning, design, manufactured and testing by specialists from INMA Bucharest.

The paper presents the experimental combined separation installation, the methodology and measuring apparatus used for experimental research and some results of experimental research, in order to optimize the functional parameters.

2 Material and method

The experimental stand presented in the technological flow scheme in figure 1, it is composed of the gravimetric separator (1) connected to a suction installation with air composed of fan (2) and the de-dusting cyclone (3).

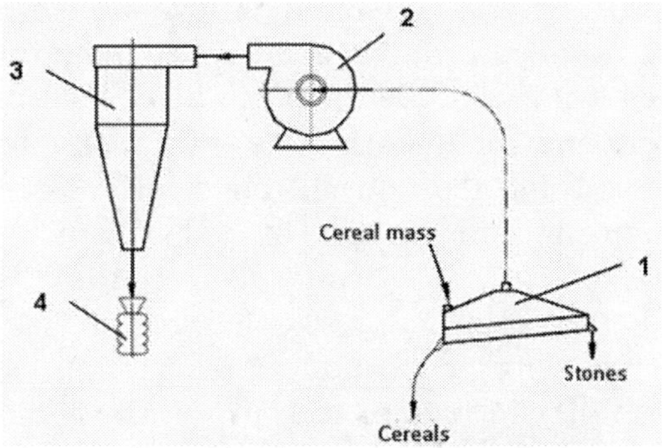

Fig. 1. Technological flow scheme of combined separation installation:
1 - gravimetric separator, 2 - fan, 3 – de-dusting cyclone, 4 - bags for dust and light impurities

Fig.2. The stand (combined separation installation) used for experimental research
1 - gravimetric separator SG-01; 2 – suction fan; 3 – de-dusting cyclone

The functional technical parameters of SG-01 are:

- capacity of processing, t/h
  - cleaning process min.2,5

- precleaning process max.4
- necessary air flow rate, m^3/min. 150
- installed power, kW 0,7 (2x0.35)
- frequency of oscillations, osc/min. 960

The centrifugal drive force F_{cf} of electrical motovibrators is regulate by adjusting the unbalanced masses, and the direction of oscillations can be adjusted by rotating the fixing support (figure 3).

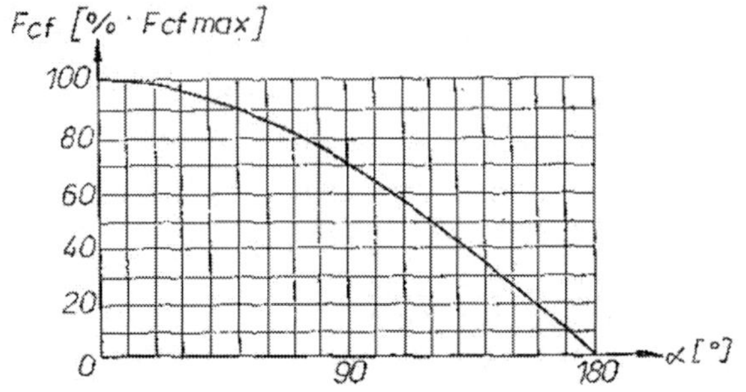

Fig. 3. Variation of centrifugal forces F_{cf} of unbalance masses depending on the angle α of overlap of the non-balanced eccentric masses [11]

The vibration movement of the working platform is imprinted on the impurities or particles in contact with it after a trajectory allowed by the elastic support springs. Due to the combined effect of the vibrations of the working platform and the action of the air currents, the product mass is stratified on the sieve of working surface. Particles of impurities with a specific volumetric weight greater than grain seeds remain in contact with the vibrating working platform, while light foreign bodies are distributed in upper layers. Thus, under action of the aspirated air stream, the grain seeds are fluidized and float above the working surface, with which they do not come in constant contact, flowing on its inclined plane to the outlet connections located at the lower end of the gravimetric separator SG-01.

The material used at experimental researches was the cereals seed (mainly wheat). Specific measurement devices for acquisition or registration were used for experimental determinations [12] of the following parameters:

- masses of products (seeds and impurities);
- material supplying flow rate;
- velocity of air currents in the pipe of suction installation, using an anemometer model Testovent 4000;
- inclination angle of the working surface (vibrating frame);
- oscillation amplitude of the vibrating working frame, using an inductive displacement transducer, model W10-Hottinger;
- frequency of oscillation of the two motovibrators, after determination of rotation speed;
- energy consumption after determination of power consumption of motovibrators.

Also were determined the moisture content and hectolitric mass of product on entry and exit in the gravimetric separator SG-01 and at evacuation of impurities in the separation process.

The loss rate of good wheat seeds C_{ps} is calculated with the relation [4]:

$$C_{ps} = (m / M) \cdot 100 \quad [\%] \tag{1}$$

where: m is the mass of good wheat seeds which are found in the mass of eliminated impurities at the exit of SG-01, kg; M - mass of good wheat seeds at the entry into SG-01, kg.

The separation quality index E_{cs} was determined at a single pass of the product through the separator, this indicator representing the percent of impurities (foreign bodies) eliminated from the mass of processed wheat seeds and is determined with the relation [4]:

$$E_{cs} = [(c_{csi} - c_{cse}) / c_{csi}]x100 \quad [\%] \tag{2}$$

where: c_{csi} is content impurities at the entrance of product in SG-01, %
c_{cse} - content impurities at evacuation of product from SG-01, %

Knowing the velocity v (m/s) recorded at the anemometer model Testovent4000 and the section S (m^2) of the suction pipe, the air flow rate required for aspiration was determined with the relation:

$$Q_a = 3600 \cdot v \cdot S \ [m^3/h] \tag{3}$$

The electricity consumed by the two electric motovibrators W is determinated by the relation:

$$W = (P_a \cdot t)/3600 \quad [kWh] \tag{4}$$

where: P_a is the power absorbed, kW, measured with the phase and frequency analyzer; t - operating time at tests, s.

The capacity of processing Q of the gravimetric separator SG-01 was calculated with the relation:

$$Q = 3600 \cdot m/t \quad [kg/h] \tag{5}$$

where: m is the initial product mass, kg; t - time required for experimental determination, s.

In Table 1 and 2 are mentioned the values resulting from processing determined parameters.

Table 1. Functional operating parameters of SG-01

Crt. iss.	Determined parameter	U.M.	Parameter value determined at tests
1.	Product supply flow rate (2 test versions)	kg/h	1000; 2000
2.	Duration of test versions	s	2400; 1800
3.	The initial quality indices of product at the entry into SG-01: - humidity; - hectolitric mass; - impurities	 % kg/hl %	 11.93 78.72 2.41; 2.46
4.	Loss rate of good wheat seeds C_{ps}	%	0.01÷0.95
5.	Separation quality index E_{cs}	%	max.98.57
6.	The suction air flow rates of the installation (3 test versions)	m^3/min.	100; 125; 150
7.	Air speed in the pipe at test versions of suction air flow rates	m/s	17.36; 21.7; 26.04
8.	Power absorbed by motovibrators	kW	0.7
9.	Specific consumption of electricity	kWh/kg	0.00045

Table 2. Some tests results for determination of optimal regime of the gravimetric separator SG-01

Wheat seed feeding flow (kg/h)	Air flow (m^3/min)	Angle of inclination of working surface (°)	Amplitude of working surface (mm)	Quantity of impurities eliminated at 1000kg of wheat seed processed (kg)
0	1	2	3	4
1500	100	5	1.5	35.5
			2.0	38.6
			2.5	41.2
		7.5	1.5	33.4
			2.0	35.1
			2.5	37.2
		10	1.5	31.6
			2.0	33.7
			2.5	34.8
	125	5	1.5	26.3
			2.0	29.2
			2.5	33.1
		7.5	1.5	24.5
			2.0	25.8
			2.5	27.6
		10	1.5	21.3
			2.0	22.8
			2.5	23.6
	150	5	1.5	29.7
			2.0	21.4
			2.5	23.8
		7.5	1.5	16.4
			2.0	18.2
			2.5	19.4
		10	1.5	15.6
			2.0	16.8
			2.5	17.9
2000	100	5	1.5	37.6
			2.0	40.2
			2.5	43.4
		7.5	1.5	35.6
			2.0	38.9
			2.5	39.9
		10	1.5	33.3
			2.0	35.4
			2.5	36.7
	125	5	1.5	28.1
			2.0	31.4
			2.5	34.9
		7.5	1.5	26.6
			2.0	27.4
			2.5	29.1
		10	1.5	23.1
			2.0	25.1
			2.5	26.0

Table 2. Some results of the tests effected for the determination of optimal regime of the gravimetric separator SG-00 (continuation)

Wheat seed feeding flow (kg/h)	Air flow (m^3/min)	Angle of working surface (°)	Amplitude of working surface (mm)	Quantity of impurities separated at 1000kg of wheat processed (kg)
0	1	2	3	4
2000	150	5	1.5	21.2
			2.0	23.0
			2.5	24.9
		7.5	1.5	17.8
			2.0	19.8
			2.5	21.6
		10	1.5	17.1
			2.0	18.3
			2.5	19.2

3 Results and discussions

After the analysis of the influence of constructive and functional parameters of the combined separation installation on the separation quality index E_{cs} graphics were drawn expressing the following technological dependency of the indices values of the functional parameters (adjustment) of the gravimetric separator SG-01: the product feed rate of the gravimetric separator *Qg*, the air flow rate *Qa* of the suction installation, the angle of inclination of the working surface α and the oscillation amplitude *A* of the gravimetric separator working surface.
In figures 4 – 6 are presented, for illustration, some representative graphics.

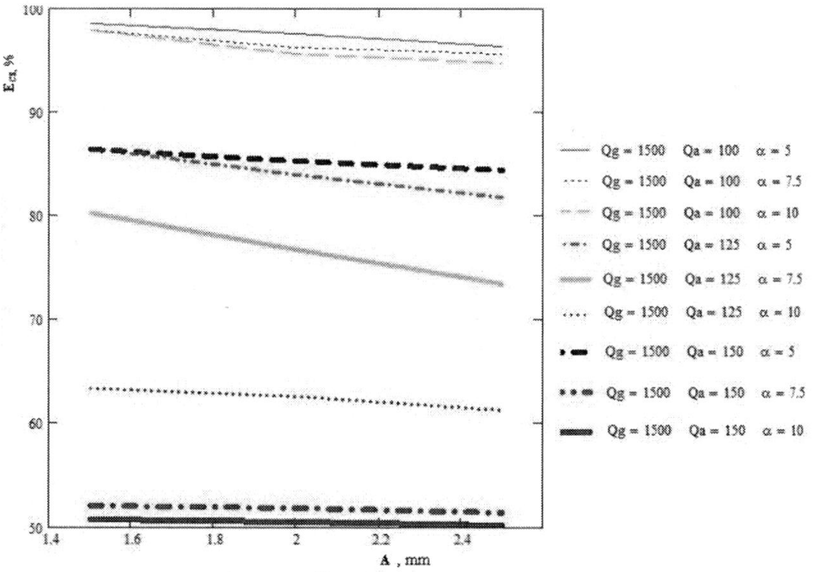

Fig. 4. The dependence of separation quality index $E_{CS} = f(A)$ for Q_g=1500kg/h, Q_a =100÷150m^3/min and α =5÷10°

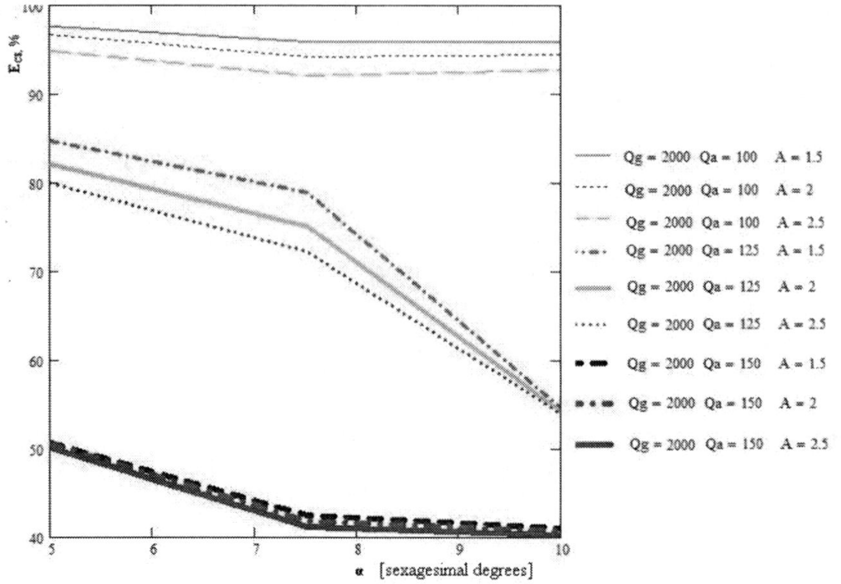

Fig. 5. The dependence of separation quality index $E_{CS}=f(\alpha)$ for Q_g=2000kg/h, Q_a =100÷150m³/min and A=1,5÷2,5mm

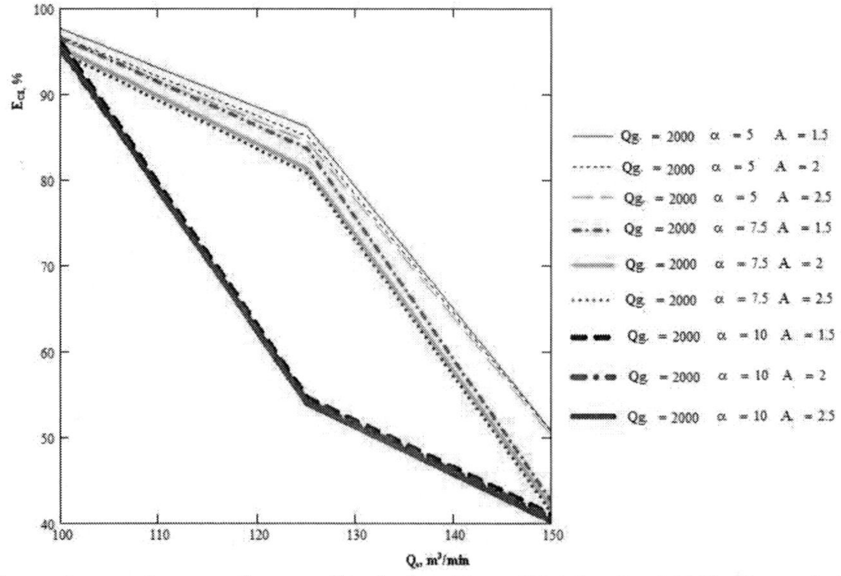

Fig. 6. The dependence of separation quality index $E_{CS}=f(Q_a)$ for Q_g=2000kg/h, α =5÷10° and A=1,5÷2,5mm

By applying linear regression method to determined experimental data processing, the separation quality index E_{CS} is obtained expressed by the following relation [13]:

$$E_{cs}\left(Q_{g,}Q_a,\alpha,A\right) = -0.15891Q_g - 0.848267Q_a - 3.256434\,\alpha - 2.736311\,A + 237.645462 \quad (6)$$

The correlation between the experimental data and the data calculated in the experimental points with the regression given by the relation (6) is 0.932.

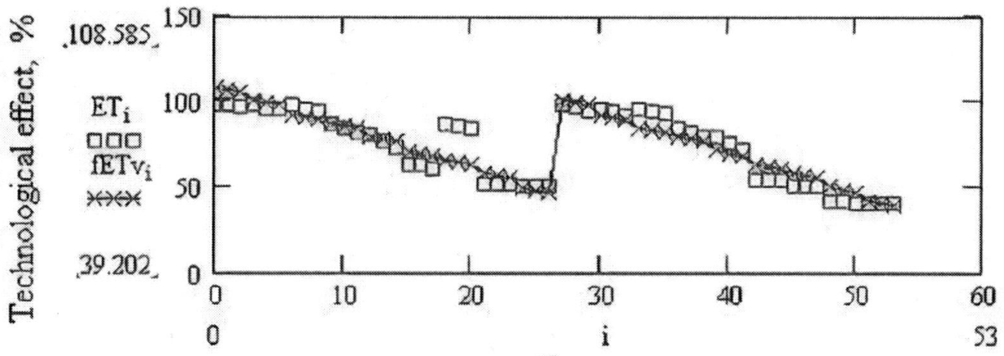

Fig.7. The separation quality index (technological effect) comparison of the experimental data with those obtained by linear regression

From the correlation of graphical representation in figure 4-7 with relation (6) result that the separation quality index E_{CS} decreases with the increase of the supplying flow rate with product, of the working surface inclination angle, of the suction air flow rate and of the working surface amplitude.

It was noticeable that, by increasing of the cereal seeds rate flow through the impurities pipes are eliminated a part of good seeds and the loss rate of good wheat seeds is above normal limits.

4 Conclusions

The working process of the gravimetric separator SG-01 achieves a high precleaning of the product (wheat seeds). The technological parameters obtained during experimental tests depend on: the regulation of the working vibration platform (oscillations amplitude and angle of inclination), the suction air flow as well as the wheat seed feeding flow.

By equipping with a pair of electric motovibrators, the oscillating movement induced by them allow to the gravimetric separator SG-01 a high output capacity and an increased separating efficiency.

The experimental measurements revealed the following conclusions:

1. the ascending airflow generated by the suction installation favors the stratification of seeds mixture uniform dimensionally with different specific volumetric weights, their movement on the working surface of the gravimetric separator being possible even at the minimum amplitudes of oscillation;

2. by increasing the air flow rate between 100÷150 m^3/min the combined separating installations performances are enhanced, through a better product layer structure achieved depending on the specific volumetric weight of grain seeds and other impurities;

3. the oscillations amplitude influence upon the working regime, establishing as optimum values between 2 ÷ 2.5 mm;

4. by increasing the tilting angle of the working surface between 5 ÷ 10° the wheat seeds quantity eliminated through the joint of impurities is reduced. Also, over 10° the gravimetric separator SG-01 have diminished their performances;

5. the separation quality index resulted at a single passage of the product was of 98.57% for big foreign bodies eliminated, 91.35% for small foreign bodies eliminated, 95.3% for light foreign bodies eliminated.

Therefore, the results of experimental researches obtained by direct measurement or determined by calculation show that the combined installation consisting of gravimetric separator model SG-01 and the suction installation comply with the requirements in terms of security and functioning mode offering multiple possibilities for adjustment and servicing.

Acknowledgements

This work was supported by:
- a grant of the Romanian Research and Innovation Ministry, through Programme 1 – Development of the national research-development system, subprogramme 1.2 – Institutional performance – Projects for financing excellence in RDI, contract no. 16PFE.
- a grant of the Romanian Research and Innovation Ministry CCDI - UEFISCDI, Project INNOVATIVE TECHNOLOGIES FOR IRRIGATION OF AGRICULTURAL CROPS IN ARID, SEMIARID AND SUBHUMID-DRY CLIMATE, project number PN-III-P1-1.2-PCCDI-2017-0254, Contract no. 27PCCDI / 2018, within PNCDI III.

References

1. T. Căsăndroiu, *Primary processing and agricultural products storing,* lithographed course, Politehnica University of Bucharest (1993)
2. I. Danciu, *Cereals cleaning*, University "Lucian Blaga" Publishing House, Sibiu, (2001)
3. Fl. Rus, *Separation operations in the food industry*, Publishing Transylvania University of Braşov (2001)
4. V.F. Didyh, *Improving the efficiency of drying agricultural plant materials: monograph*, Lutsk pp. 75-101 (2002)
5. A.D. Zewdu, *Aerodynamic properties of the grain and straw material*, Biosystems Engineering Journal, **98**, pp. 304 –309 (2007)
6. Z. Jing Z, *Design of New Grain Threshing Cleaning Device*. Journal of London University, 1, 17, London/UK (2013);
7. C. Brăcăcescu , Sorică C., D. Manea , Yao Guanxin, G.A. Constantin - *Theoretical contributions to the drive of cereal cleaning technical equipment endowed with non balanced vibration generating systems,* INMATEH - Agricultural Engineering, vol. 42, no. 1 / 2014, pp. 69 – 74, (2014)
8. A.V.Nesterenko, S.M.Leshchenko, O.M.Vasylkovskyi, D.I.Petrenko, *Analytical assessment of the pneumatic separation quality in the process of grain multilayer feeding*. INMATEH Agricultural Engineering, **53**, No. 3, pp.65-70, Bucharest/Romania (2017),
9. I. Costin, *Miller's Book*, Technical Publishing House, Bucharest, Romania (1999)
10. Chr. Geankoplis,*Transport processes and separation process principles*, Prentice Hall PTR, Fourth Edition, 14, pp.903-944, (2003)
11. O. Vasylkovskyi, K. Vasylkovska, S. Moroz, M. Sviren, L. Storozhyk, *The influence of basic parameters of separating conveyor operation on grain cleaning quality*, INMATEH- Agricultural Engineering, **57,** no.1/2019, pp 63-70, (2019),
12. V. Damian, *Manual of the Food Industry Engineer*, 28-Measuring and control devices, control and adjustment, pp.1234-1262, Technical publishing house, Bucharest (1998)
13. C. Brăcăcescu, *Theoretical and experimental researches regarding the optimization of working processes of gravimetrical separators aimed at cereals impurities*, PhD thesis (Transilvania University of Brasov, 2011)

Rain Use Efficiency in Southern Romania after Using the Device for Soil Modelling in Interrupted Furrows for Weeding Crops - DMBC-5

Remus Marius Oprescu[1], *Iulian Florin* Voicea[1*], *Iulian* Dumitru[1], *Ioana* Dragan[2], *Costel* Biolan[2], *Nicoleta* Ungureanu[3], *Emil Adrian* Neagu[4], *Bara* Norbert[5], *Daniel* Jaloba[6] *and Vlad Nicolae* Arsenoaia[7]

[1]National Research - Development Institute for Machines and Installations Designed to Agriculture and Food Industry, 6 Ion Ionescu de la Brad Blv., 013813 Bucharest, Sector 1, Romania
[2]AQUAPROIECT, 294 Splaiul Independentei, Bucharest, 060031, Bucharest, Romania
[3]University "POLITEHNICA" Bucharest, Faculty of Biotechnical Systems Engineering, 313 Splaiul Independentei, Bucharest, 060042, Romania
[4]SCDA Mărculeşti, Calarasi, Romania (Agricultural Research and Development Resort Marculesti, Calarasi, Romania)
[5]INCDPM, Bucharest, Romania (The National Institute for Research and Development in Environmental Protection, 294 Splaiul Independentei, 6th District, 060031, Bucharest, Romania)
[6]ICDPP, Bucuresti, Romania (Research - Development Institute for Protection Plant, 8 Ion Ionescu de la Brad Blv., 013813 Bucharest, Sector 1, Romania)
[7]Ion Ionescu de la Brad" University of Agricultural Sciences and Veterinary Medicine, Iasi, Romania (3, Mihail Sadoveanu Alley, Iasi, 700490, Romania)
*Corresponding author: *voicea_iulian@yahoo.com*

Abstract: The global impact of climate change, with the potential to affect agriculture through changes in temperature, rainwater distribution and amount, leads to the need to develop integrated technologies that increase rain use efficiency and support soil and environmental quality, ensuring higher agricultural outputs with lower costs. Soil modeling in interrupted furrows is a very efficient rainwater harvesting system, but has not yet been scientifically assessed for the conditions in Southern Romania as the other soil working systems have been. For this purpose, a soil modeling device consisting of a frame with triangle for coupling to the tractor, wheels for adjusting and limiting the working depth, soil loosening knives, ridge ploughs and furrow compartmenting equipment was compared with the conventional technology for sunflower culture in two localities in Southern Romania: Crânguri in Giurgiu County and Mărculeşti in Calarasi County. The device for soil modeling in interrupted furrows for weeding crops provides a viable option that has positive effects on soil properties and ensures increased crop yields compared to conventional works as well as the opportunity to increase the efficiency of water recovery in agriculture by harvesting rainwater.

Keywords: soil modelling; interrupted furrows; water conservation in soil; rain use efficiency.

1 Introduction

In the Southern part of Romania, characterized by a temperate continental climate, there is a need to preserve "in-situ" soil moisture due to insufficient rainfall for agriculture. Cultivated agricultural areas can lose a major part of the rainwater by surface leakage and large amounts of soil through erosion [1]. Rainwater harvesting has the potential to reduce soil erosion and improve the productivity of these areas. Rainwater harvesting is a general term used to describe the collection and concentration of surface leakage for various uses, including for agricultural and domestic use [2].

"In-situ" systems are the simplest and cheapest approaches to harvesting rainwater and can be practiced in many farming systems. Also called water conservation systems, they involve the use of methods to increase the amount of water stored in the soil profile by capturing or retaining precipitation water [3]. A concept on "in-situ" harvesting of rainwater, involving different techniques, is known as "Reservoir tillage" (RT). This approach has been developed taking into account the fact that soil work can increase the storage capacity of surface water and can be one of the most effective means of combating both surface leakage and soil erosion. RT creates basins or cavities to retain water, allowing it to infiltrate the soil and thus prevent leakage [4]. RT was defined by as a system where many small surface depressions are formed to collect and retain rainwater to prevent surface leakage. At present, RT is predominantly used to combat soil erosion in large annual rainfall environments with low intensities. In environments with low and rapid rainfall, as well as with an increase in droughts, interrupted furrows are used to collect precipitation water [5,9,10,11]. Through these harvesting works, rainwater is collected in mini-basins, allowing more time for infiltration, which reduces leakage and the potential for erosion and transport of soil particles [6]. In this method, the large infiltration surface created by the depressions and the depth of the water in them lead to higher infiltration rates and thus to a decrease in losses generated by surface leakage and evaporation [7,12].

Current scenarios predict that climate change will increase the water deficit in Southern Romania [8]. Climate change has the potential to affect agriculture through changes in temperature, rainwater distribution and amount. Precipitation changes will be one of the most critical factors that will determine the global impact of climate change. This issue indicates the need to develop integrated technologies that increase rain use efficiency and support soil and environment quality, ensuring higher agricultural outputs with lower costs [11,13].

The objective of this study was to develop and evaluate a new technology for soil working with interrupted furrows for weeding crops in the southern area of Romania. The technology includes a soil modeling device to ensure soil loosening and the creation of interrupted furrows on the soil surface for "in situ" harvesting of rainwater.

2 Materials and Method

The device for soil modeling in interrupted furrows for weeding crops, simultaneously on 5 intervals, DMBC-5 (Fig.1) is designed for soil modeling in interrupted furrows for weeding crops sown at 70-80 cm, for the accumulation of rainwater and the achievement of a uniform water distribution at the soil surface, on land with a slope of up to 6%, on soils ploughed to a minimum of 250 mm, with light, medium or heavy texture, at a humidity at which the soil does not stick to the active part.

Constructive description of the device

The device for soil modeling in interrupted furrows for weeding crops, simultaneously on 5 intervals, DMBC-5 consists of the following main parts:

- frame with triangle for coupling to the tractor;

- wheels for adjusting and limiting the device working depth;
- soil loosening knives before the ridge ploughs;
- ridge ploughs;
- furrow compartmenting equipment.

Fig. 1. Device for soil modeling in interrupted furrows for weeding crops, simultaneously on 5 intervals, DMBC-5

The ***frame*** of the device is made of two square cross-section pipes, parallel to each other, joined to the ends by welded cross-bars. The three-point tractor coupling triangle (ISO II) is welded to the front of the frame. On the sides of the frame are mounted some removable bars that will increase the working width of the device from 3600 mm to 5200 mm. Frame clearance will be adjusted according to the plant height by means of supports of loosening knives, respectively of ridge ploughs.

The ***wheels for adjusting and limiting*** the working depth are of the metal or rubber tyre type, currently used for soil-working machines, according to the attached documentation or serial production. The wheels have a diameter of 500 mm and are mounted on the sides of the frame. The working depth of the active parts is continuously adjusted by raising or lowering the wheels by means of a spindle screw actuated by a crank. The lateral removable bars and support wheels are positioned on the frame, respectively on the interval between the plant rows.

A ***workstation*** consists of a ***loosening knife***, a ***ridge plough*** and an ***impeller*** with the control equipment.

The 3, 4 or 5 ***loosening knives*** are fastened by means of straps and special cases to the front square pipe. The knives are chisel type and can penetrate into the soil at a depth of maximum 20 cm. The cases allow the knife supports to slide vertically; they can be adjusted to the plant height. Fixing the knife in a certain position is achieved by turning a screw.

The 3, 4 or 5 ***ridge ploughs*** have their supports fastened by means of straps and cases, on the second square pipe, behind the loosening knives and are mounted at varying distances on the frame of the device, according to the furrow pattern. The ridge plough consists of a removable knife, a ridge plough bearing support, two breasts and the vertical support. The ridge plough may have mobile or fixed breasts, making a variable or fixed opening. The bearing support, which adjusts the angle of ridge plough penetration into the ground and which is the overload safety element, will be fastened by means of two screws to the ridge plough support. The ridge ploughs are made of properly treated manganese sheet. The ridge plough angle of penetration into the ground can be adjusted by rotating the bearing support against the vertical support. The vertical support, which is a rectangular cross-section bar, is fixed to the second rear bar of the frame with two straps and a case. Adjusting the vertical support in the frame fastening case allows the frame to pass over the plants in the weeding crops. When the agricultural machine passes, the plants must not exceed 60 cm in height in order not to damage them. The active parts (the loosening knives and the ridge plough tip) are hardened and the edges have ensured self-sharpening during operation.

Furrow compartmenting equipment, mounted on the rear bar of the frame, consists of a camshaft, mechanisms for movement transmission to the bolts that lock the impeller blades and 3, 4 or 5 impellers. The rotation movement of the right-side wheel is transmitted by a drive chain to a shaft positioned parallel to the second square bar (rear), shaft that will have some cams positioned next to each work section. During rotation, the camshaft will actuate the lever/cable mechanism from each section in the direction of unlocking the blade with the locking bolt, and by rotating the blade the interruptions will be made; by cyclical unlocking of the rotor, depending on the distance travelled by the device, interruptions will be made at predetermined distances. The rotors will have 4 or 3 trapezoidal blades. The rotors are mounted behind the ridge plough in a frame that allows rotation and oscillation in the vertical plane and they are pressed on the bottom of the furrow with two side-mounted springs or springs that press vertically according to the attached documentation. While the rotation of one of the pallets will be locked by a bolt, it will scrape the bottom and side walls of the furrow, mobilizing a quantity of soil in front of it. When the eccentric cam actuates the lever mechanism, the bolt retracts, releasing the impeller, which rotates one step (the distance between two blades), leaving a soil plug on the furrow, with a base width between 20 and 40 cm and height equal to the depth of the furrow. While the impeller rotates 90^0, the bolt, released by the cam action, returns to the previous position, locking the next rotor blade, and then the cycle restarts. The device provides the modeling of the interrupted watering furrows on the intervals between plant rows in two ways:

- alternatively, a furrow interval and a furrowless interval or
- consecutively, on each interval according to the sowing scheme, the soil type and the root zone of the plant.

The adjustment of the furrow compartmenting equipment allows the creation of soil plugs along the furrow at distances of 1.5; 3 or 6 m.

Dimensional and functional characteristics

• aggregation tractor, HP	55-80
• optimal height of plants in the soil modelling phase, cm	30-60
• transport clearance, mm	min. 250
• frame operational clearance, mm	min. 500
• working speed, km/h	1.8-3
• device length, mm	1500-1700
• device width, mm	3600 or 5200
• device height, mm	1320
• mass, kg	800

Experiments under exploitation conditions of the DMBC-5 device were carried out in 2018 in two localities (Crânguri in Giurgiu County and Mărculești in Calarasi County), on an alluvial chernozem, respectively on vermic chernozem, in sunflower culture (in corn-wheat-sunflower rotation) and comprised a plot where the soil modeling work with interrupted furrows was applied and a control plot worked according to conventional technology. The lands were ploughed in autumn at 30 cm and in the spring the preparation of the seedbed was done by a superficial work with the combinator and the sowing with a SPC6 seeder. The soil modeling work in interrupted furrows was done when the sunflower crop height in the two localities, Crânguri and Mărculești, was 45.6, respectively 42.8 cm. In the tests, it was intended to make the interrupted furrows in an area with irregular surface and the adjacent surface was used as a control area. The machine was tested in a mechanically hoed area I, on 30 consecutive rows of 866 m long, 3.5 m working width, the worked area on a pass was 0.3031 ha/pass, the depth was variable 12 -15 cm depending on the height of the plant, the working speed and the degree of soil compaction. The adjacent control area was also 30 rows, a mechanically hoed zone II and the surface was as irregular as the area tested. The machine was tested in 3 areas representing repetitions of the experiment.

During tests under exploitation conditions, the procedures specific to the soil modeling devices were used and the following metrologically verified measuring and control devices and equipment were used: digital penetrometer with Fieldscout SC900 cone, HH2 portable humidometer with Theta Probe ML2X sensor, mechanical measuring tape, mechanical timer, forrow-meter, rulers, set squares, poles. Soil moisture and compaction at the time of the tests and qualitative working indices for the dimensions of the furrow made by impeller (Fig. 2) were determined.

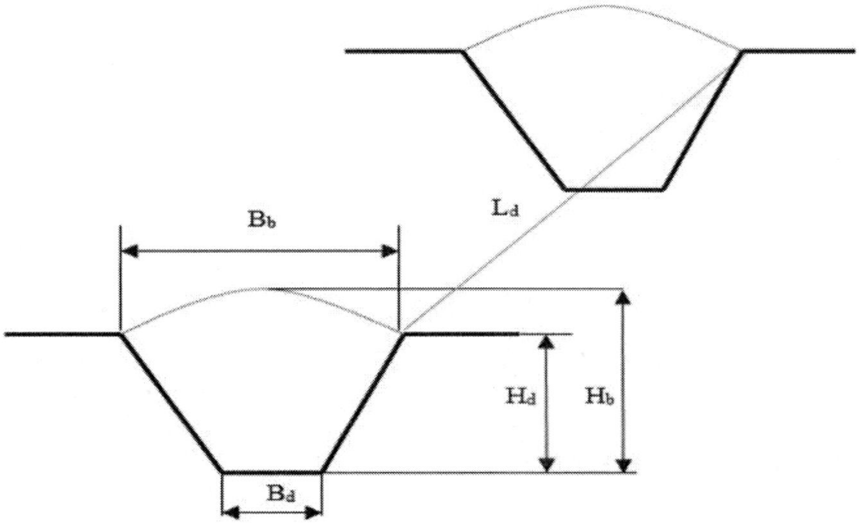

Fig. 2. Dimensions of the furrow made by impeller (Hb – depth of watering furrows, Bd – width of furrow bottom, Bb – furrow upper width, Hd – soil plug height, Ld – plug spacing)

The sunflower crop was harvested at the beginning of the maturation phase when the seed humidity dropped below 15%. Harvesting was done using a harvesting machine equipped with RFS-8 equipment. Production was reported at standardized humidity of 12%. For the assessment of rain use efficiency (RUE), grain production was reported to the precipitation recorded during the vegetation period (April to August) and expressed in kg ha^{-1} grains per mm precipitation.

3 Results

3.1 Soil moisture and compaction at the time of testing

From the analysis of the measured values for compaction (expressed by penetration resistance) and soil moisture it results that in the horizons 0-10 cm and 10-20 cm they recorded in Crânguri a significant increase from 100.7 to 153.3 daN cm^{-2} and from 11.3 to 19.1% (Fig. 3A).

In Mărculeşti the penetration resistance indicated a similar evolution, registering on average 74.1 daN cm^{-2} in the 0-10 cm horizon and 151.5 daN cm^{-2} in the 10-20 cm horizon. This is a significant difference. Humidity recorded an insignificant difference, 11.0% in the 0-10 cm horizon and 12.2% in the 10-20 cm horizon (Fig. 3B).

Soil compaction and moisture determined in the lands of the experimental plots of the two localities fall within the agrotechnical requirements of soil modeling works.

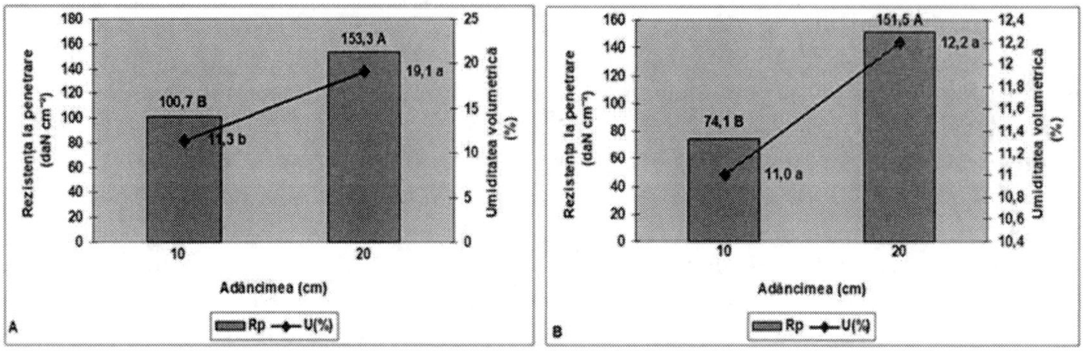

Fig. 3. Land characteristics of experimental lots from Crânguri (A) and Mărculești (B). Different letters indicate significant differences ($p < 0.05$).

3.2 Qualitative working indices for the dimensions of the furrow made by DMBC-5 device impeller

Following the experiments made according to the standards in force, in the two localities, it resulted that the qualitative working indices achieved fall within the agrotechnical requirements of soil modelling works (Fig. 4).

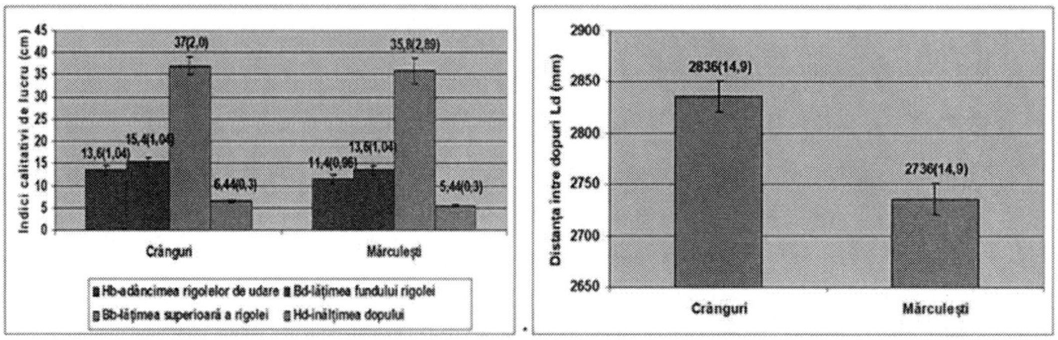

Fig. 4. Qualitative working indices obtained on the sunflower surfaces cultivated in Crânguri and Mărculești in 2018. The values in brackets represent the standard deviations

Thus, for the depth of watering furrows (Hb), furrow upper width (Bb), width of furrow bottom (Bd) and soil plug height (Hd) the coefficients of variation (CV) recorded values below 10%, while for plug spacing (Ld) the CV were less than 5% according to the requirements of the interrupted furrow works.

3.3 Sunflower production and rain use efficiency

In both localities where experiments were carried out, sunflower production was significantly influenced by the soil tillage technology used. The interrupted furrow technology (IFT) provided, compared to conventional technology (CT), an increase of 265 kg ha^{-1} in Crânguri and 319 kg ha in Mărculești (Fig. 5).

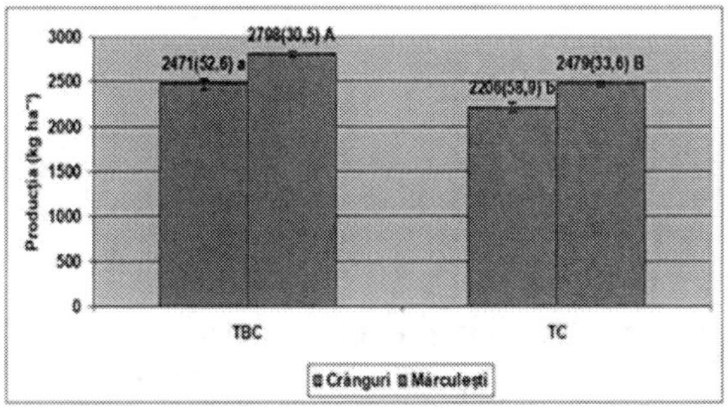

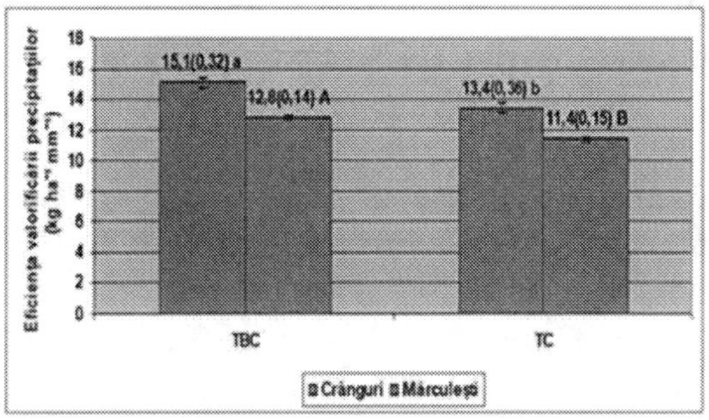

Fig. 5. The sunflower production and rain use efficiency obtained in the two technologies in Crânguri and Mărculeşti in 2018. The values in brackets represent the standard deviations. Different letters indicate significant differences ($p < 0.05$).

Figure 5 shows that rain use efficiency (RUE) increased clearly significantly in the two localities in IFT compared to CT. Thus, in Crânguri RUE increased from 13.4 kg ha^{-1} mm^{-1} in CT to 15.1 kg ha^{-1} mm^{-1} in IFT, respectively in Mărculeşti from 11.4 kg ha^{-1} mm^{-1} to 12.8 kg ha^{-1} mm^{-1}.

4 Conclusions

By promoting this technology with the new device for soil modeling in interrupted furrows for weeding crops (DMBC-5), it is ensured:

- water conservation in the soil by reducing evaporation in the root zone;
- elimination of the second hoeing;
- weed control in the area of the row between plants, and
- increase of agricultural production by over 10% as well as of rain use efficiency by over 1,4 kg ha^{-1} for every mm of precipitation.

Acknowledgement

This work was supported by a grant of the Romanian Research and Innovation Ministry, through Programme 1 – Development of the national research-development system, subprogramme 1.2 – Institutional performance – Projects for financing excellence in RDI, contract no. 16PFE.

References

[1]. F.L. Duley, Soil Sci. Soc. Am. **4**, 60 (1940).
[2]. FAO, Soils Bulletin **69**. Rome (1993)
[3]. G. Brhane, C.S. Wortmann, M. Marno, H. Gebrekidan, A. Belay, Agron. J. **98**, pp.124-134 (2006)
[4]. E.W. Rochester, D.T. Hill, K.H. Yoo, Trans. ASABE **37**, pp.1183-1192 (1994)
[5]. M. Kronen, Soil Tillage Res. **32**, pp.71-123 (1994)
[6]. E. Ventura, L.D. Norton, K. Ward, M. Lopez-Bautista, A. Tapia-Naranjo,.ASAE Annual Meeting ASABE (2005)
[7]. C. Patrick, C. Kechavarzi, I.T. James, M.O. Dogherty, R.J. Godwin, Soil Use Manage. **23**, pp.185-197 (2007)
[8]. A.C. Marica, A. Busuioc, Romanian J. Meteor. **6**, pp.50-57(2004).

[9] T. Sampathkumar, B. Pandian, M. Rangaswamy, P. Manickasundaram, Influence of deficit irrigation on growth, yield and yield parameters of cotton–maize cropping sequence, Agricult. Water Manag., **130**, pp. 90-102 (2013).
[10]. N. Korresa, R. Nilda, T. Burgosallias, T. Vurroc, K. Gitsopoulosd, K. Vijaya, O. Stephen, P. Dukef, C. Brabhama, E. Christopher, R Rouseh, P. Salas, New directions for integrated weed management: Modern technologies, tools and knowledge discovery, Adv. in Agron., **4**, pp.245-258 (2019).
[11]. R Oprescu., S. Şt. Biriş, E. Marin, C. Sorică, N. Ungureanu, E. Sorică, I. Dumitru, Considerations on mechanically active equipment for open intrrerupted furrow used in technology of weeding plant cultures fruit and viticultural plantations., , pag. 215 – 220, ISB-INMA TEH - International Symposium - Agricultural And Mechanical Engineering, Print: ISSN 2344 – 4118, pp. 215-220 (2017).
[12]. R. Oprescu, S. Şt. Biriş, I. Voicea, D. Cujbescu, C. Persu, I. Găgeanu, V. Vlăduţ, I. Dumitru, Considerations on the construction and operation of a device for soil modelling in interrupted furrows for weeding crops, 7 th International Conference on Thermal Equipment, Renewable Energy and Rural Development TE-RE-RD, ISSN 2457 – 3302, pp. 309-315 (2018).
[13]. R. Oprescu, S.Şt. Biriş, I. Voicea, V. Vlăduţ, Considerations regarding the construction and operation of an equipment designed to model the soil in compartmented furrows in vineyards and orchards, ISBINMATEH - International Symposium - Agricultural And Mechanical Engineering, Print: ISSN 2344 – 4118, pp. 431-436 (2018).

Modelling behaviour of a partially loaded road tanker with agro-food liquids during the breaking manoeuver

Gabriela Milian[1,*], *Sorin-Ştefan* Biriş[1], and *George* Ipate[1]

[1]University "Politehnica" Bucharest, 313, Splaiul Independentei, 060042, Romania

Abstract. Longitudinal and lateral liquid sloshing in a partially filled tanker affect the braking performance of the tankers. The movement during braking of the liquid load is determined by the change in the position of the centre of gravity of the load inside the vehicle. Becoming an undesirable phenomenon when analyzing the performance of the vehicle braking manoeuver. In this paper modelling of the behaviour of the cross-section circular tank with baffles, partially filled with liquid agro-food products, was performed, which performs a braking manoeuver. Two different fluids have been used to simulate the sloshing effect of agro-food liquid cargo and study the behaviour of the tank. In to order generate the geometric model of the tank was done by using SolidWorks, and simulating the behaviour of the tank and cargo with agro-food products during the ANSYS-CFX braking manoeuver. A sequence of numerical experiments has been carried out to estimate the pressure developed over the tanker wall and the free surface displacement of the fluids for different fill levels. Simulations with the two liquids and the results obtained were compared to be able to analyze the behaviour of the tank partially filled into a braking manoeuver.

Keywords: sloshing liquid, braking, agro-food liquid, ANSYS, baffles

1 Introduction

Sloshing has been described by the researchers as different: as a dynamic load acting on the tanker structure due to the movement of a fluid with a limited free surface inside the tank [1], or a strong movement of fluid from a moving tank [9]. The movement during braking of the liquid load is determined by the change in the position of the centre of gravity of the load inside the vehicle. In modeling the behavior of tankers, account is also taken of tank geometry; fill level; lateral and longitudinal force during typical maneuvers and dynamic interactions of the liquid structure [2]. For example, the ADR standard [3] specifies the detailed requirements for different tankers, as the structure shall of the tanks must resist an acceleration of 2g in vertical and longitudinal direction and 1g in lateral direction.

Over the years, the problem of partially filled tanker behavior with liquid products inside of which sloshing occurs has led of researchers to use different study methods and techniques. Salem et al. (2000) demonstrated that partially filled tankers undergoing turns or lane change maneuvers have a lower rollover threshold than any other type of vehicle [4].

* Corresponding author: gabi_isb@yahoo.com

Akyildiz et al.(2006) have been found by numerical and experimental research pressure variations and three-dimensional effects on liquid sloshing loads in a moving partially filled rectangular tank [9]. Liu et al. (2008) accomplished numerical model to study three-dimensional (3D) liquid sloshing in a tank with baffles [8]. Godderidge et al (2012) investigated using CFD code to study sway-induced sloshing flow in rectangular tank. They conducted a comparative study between homogeneous and inhomogeneous multiphase model for fluid density and viscosity. After a comparison between the results obtained numerically and experimentally shows that the homogeneous model gives less accurate result for peak pressure up to 50% as compared to inhomogeneous multiphase model [4]. Sarabi et al. (2012) studied the effect of ground motion on sloshing inside rectangular tank, used CFD simulation tool Open FOAM (Open Filed Operation and Manipulation). Results shows that the intensity of sloshing and pressure loads depend on the tank geometry, fill level, amplitude, frequency of excitation [4]. Di Nardo et al. (2012) studied the behavior of liquid fuel storage tank of cylindrical shape when subjected to an earthquake. They used VOF technique to assure an accurate description different filling levels, for 7 different acceleration inputs. Results shows that the sloshing of fluid depends on excitation frequency and filling levels [4]. Zheng et al. (2012) studied in-house two-phase fluid flow model by solving Navier-Stokes equations for cubic tank with different baffle configurations under the harmonic motion excitation [6]. Hasheminejed et al. (2014) studied two-dimensional transient sloshing in non-deformable baffled horizontal circular cylindrical vessels, filled with inviscid incompressible fluids to arbitrary depths, and subjected to arbitrary time-dependent lateral accelerations [9]. Babu et al. (2017) studied of sloshing impact on the tanker wall for different fluids with varying accelerations using VOF method for elliptical tanker. [5-8]. Otremba et al. (2018) studied more models to simulate the effects of the sloshing phenomenon on braking efficiency and the load transfer rate of a partially filled road tanker. They using CFD simulation.

2 Materials and methods

In this study the tanker is used to storage and transport agro-food liquid cargo under atmospheric pressure: crudes oil (sun-flower) and water for varying velocity (14m/s, 20m/s, 25m/s) and for different fill levels (40%, 80%). The tank model is presented in Figure1. It consists of a 3-dimensional fluids storage, with length 3840mm and height 1300mm, with circular cross-section tank, and two baffles (height 1000mm). The study of influence of liquid motion on tank structure was using by FSI simulation in ANSYS Workbench simulation environment by using ANSYS CFX and Static Structural codes (static and transient). The geometrical design with advanced modeling techniques SolidWorks software is presented.in Figure 2. CFD model presented on Figure 3 was prepared by ANSYS mesh tool and consist of 21673 nodes, of tetrahedral cell elements 114915.

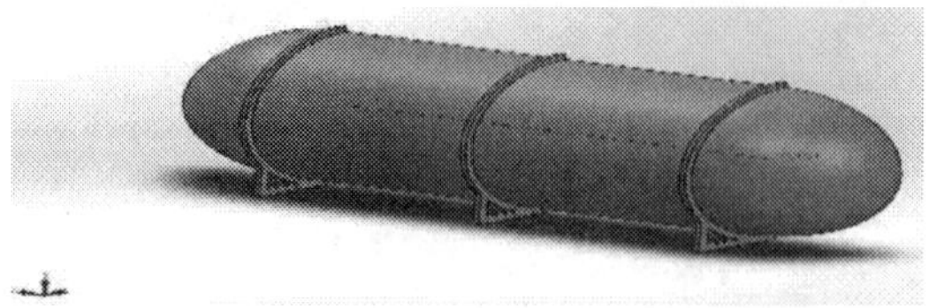

Fig.1. Physical model tanker

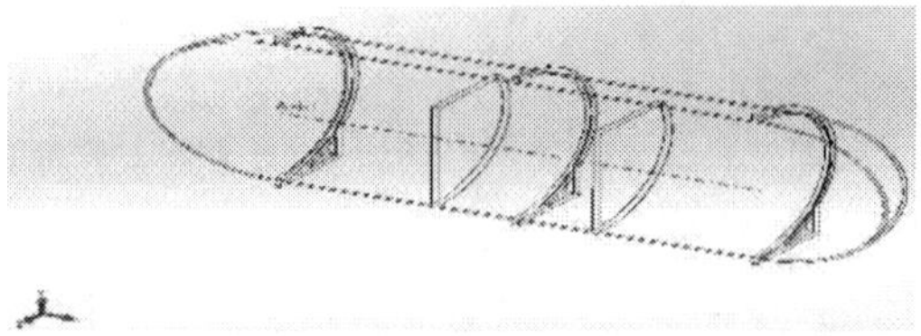

Fig.2. Geometrical design of the tanker

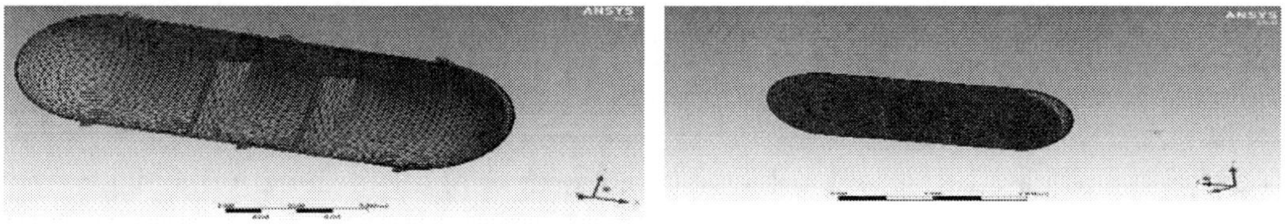

Fig.3. Mesh tanker and fluids

In this paper for both fluids it is studied the liquid behavior and simulated the modeled structure of the tank with the fill level of 80%, at the imposed velocity that decreases at 14-0m/s, 20-0m/s in 10 sec. And for the fill level of 40% at the imposed velocity that decreases at 25-0 m/s in 10 sec. Thus in these cases the deceleration simulates a sudden braking and the stop of the tanker. The tanker accelerates for first 5sec, such that the velocity goes up to 50km/h, 80km/h respectively, and last 5 sec deceleration (braking). The properties the fluids like water and oil sun-flower are investigated to find the slosh impact over the wall are given in Table 1.

Table 1. The specifications of the fluid agro-food simulation

S.NO	NAME OF FLUID	NAME OF PROPERTY	VALUE	UNITS
1	Water	Density	998.2	Kg/m^3
		Viscosity	0001003	$Kg/m^{-1}s^{-1}$
2	Oil sun-flower	Density	920	Kg/m^3
		Viscosity	0004956	$Kg/m^{-1}s^{-1}$

Inside the tank there are two baffles mounted 0.8m at a height of 1m and thickness 3 mm.

In present work, water and oil sun-flower fill level in tank is 40% and 80% of total height of tank and the rest part is occupied with air. The simulation with CFD is based on Volume Fraction (VOF) formulation when two or more fluids (or phases) don't penetrate each other. Having the following expression:

$$\sum_{p=1}^{n} \alpha_p = 1 \qquad (1)$$

where : α_p - volume fraction for those two fluids,

n - no of phases present in the computational cell,

p- fluids phase.

The value of α_p ranges between 0 to 1. 0 represent the cell is empty, 1 represents the cell is full with the fluids p. An intermediary value between 0 to 1 shows that there is an interface of one or more fluids.

VOF model can be analyzed from Figure 4. The red color represents the water and oil sun-flower phase, and the rest portion represents air from tanker. Having the following expression:

$$\alpha_{water} + \alpha_{air} = 1 \qquad (2)$$

$$\alpha_{oil\ sun\text{-}flower} + \alpha_{air} = 1 \qquad (3)$$

where : α_{water} - volume fraction for fluid phase (water),

$\alpha_{oil\ sun\text{-}flower}$ - volume fraction for fluid phase (sun-flower),

α_{air} - volume fraction for fluid phase (air).

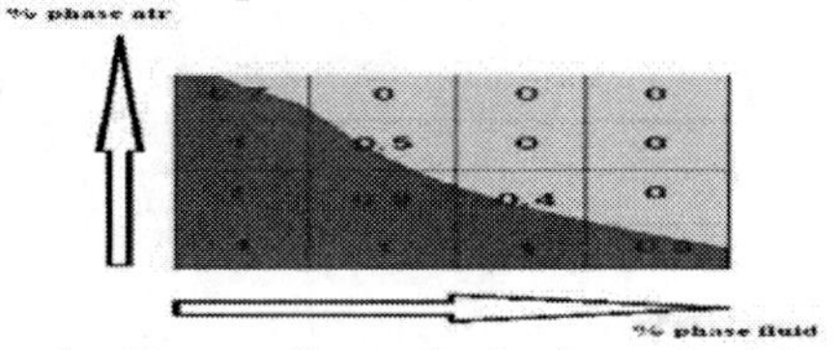

The fluid flow (water and oil sun-flower) during sloshing can be represented in the form of mathematical equations using continuity equations, the Navier-Stokes equations and turbulence modeling.

Flow simulation for those two fluids agro-food (water and oil sun-flower) was conducted for the following conditions in ANSYS:

- liquids is water and oil sun-flower with properties at normal ambient temperature (20°C); pressure based transient solver is used with explicit formulation and gravitational field is enabled;
- multiphase model with volume of fluid (VOF) method is used, and turbulent model is considered.
- flow is two phase (water and air, oil sun-flower and air respectively), homogenous with standard model of free surface and tank walls are ideal smooth with thickness 3mm;
- model is in thermal equilibrium, there is no heat exchange between fluids and environment and simulation is in transient conditions with acceleration peak 1g in longitudinal direction;
- operating conditions: pressure (101325Pa); gravitational acceleration: g_x= 0m/s^2, g_y= -9.81m/s^2, g_z= 0m/s^2; density air (1.225kg/m^3);
- non-iterative relaxation factor: pressure (0.8); momentum (0.6); turbulent kinetic energy:1; turbulent dissipation rate:1.
- tanker material is carbon steel in according SR EN 10216-2 [10].

3 Results and discussion

In this chapter a sequence of numerical experiments carried out to estimate the pressure developed over the tanker wall and the free surface displacement of the fluids for different fill levels and velocity. Simulations with the two fluids and the results obtained are compared to be able to analyze the behavior of the tank partially filled into a braking maneuver.

In the first stage the simulations for all three velocities (v=14m/s, v=20m/s, v=25m/s) were given, the tank being partially filled 80% (1m) with oil sun-flower. The simulation of the fluid being made only in the longitudinal direction gives the following pressures, and the behavior of the fluid in the tank can be seen in the Fig.5, 6 - a), b).

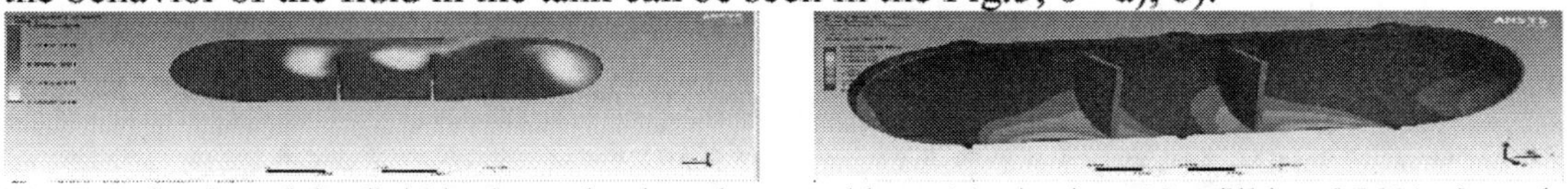

Fig.5. Behavior of the fluid in the tank when the speed is v=14m/s, time 10s, fill level 80%, the color white is air, blue is oil sun-flower respectively
a)Sloshing liquid motion in tanker and volume fraction fluid; b) total deformation static

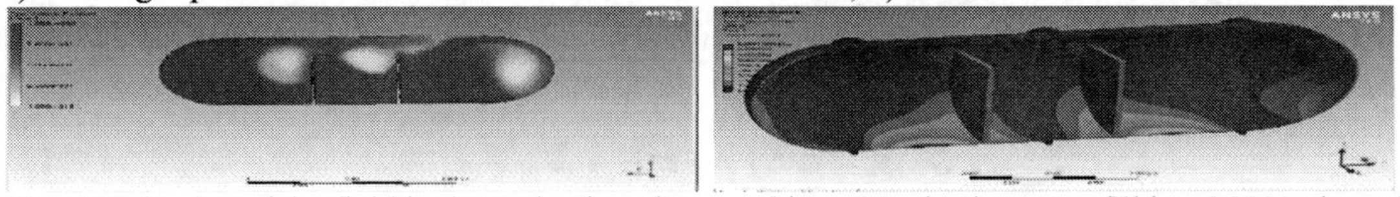

Fig.6. Behavior of the fluid in the tank when the speed is v=20m/s, time 10s, fill level 80%, the color white is air, blue is oil sun-flower respectively
a) Sloshing liquid motion in tanker and volume fraction fluid; b) total deformation static structural carbon steel

Then simulation is repeated and for a filling level of 40% and velocity v=25m/s^2 (Figure7 a), b)).

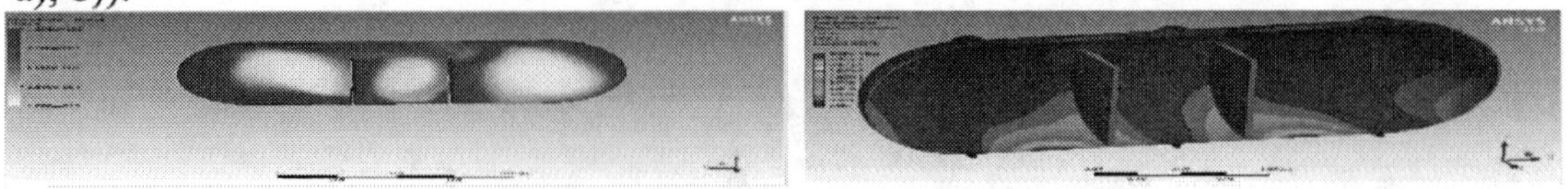

Fig.7. Behavior of the fluid in the tank when the speed is v=25m/s, time 10s, fill level 40%, the color white is air, blue is sun-flower respectively
a)Sloshing liquid motion in tanker and volume fraction fluid; b) total deformation static structural carbon steel

In the second part of the paper, the behavior of the tank with water for the three speeds is simulated, having a filling level of 80% (1m). Then for partially filled tank of 40%(0.55m). These simulation results are shown Fig.8, 9, 10 - a), b).

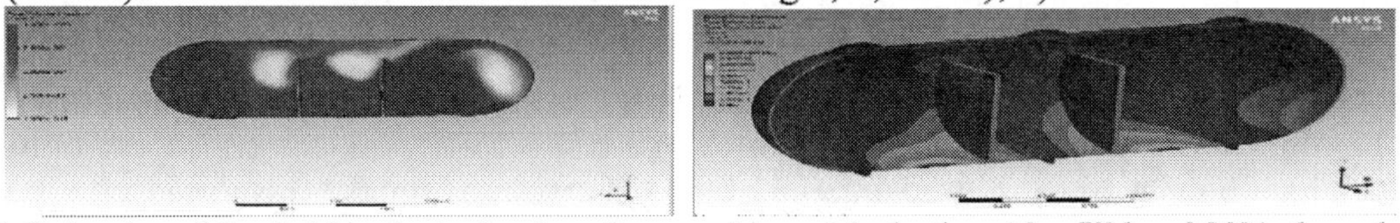

Fig.8. Behavior of the fluid in the tank when the speed is v=14m/s, time 10s, fill level 80%, the color white is air, blue is water respectively
a) Sloshing liquid motion in tanker and volume fraction fluid; b) total deformation static structural carbon steel

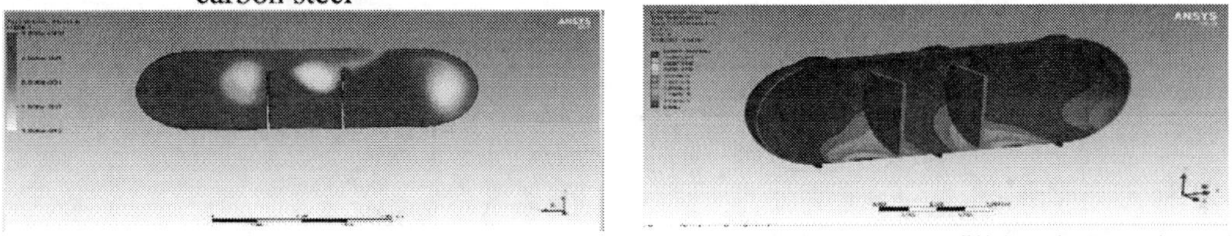

Fig.9. Behavior of the fluid in the tank when the speed is v=20m/s, time 10s, fill level 80%, the color white is air, blue is water respectively
a) Sloshing liquid motion in tanker and volume fraction fluid; b) total deformation static structural carbon steel

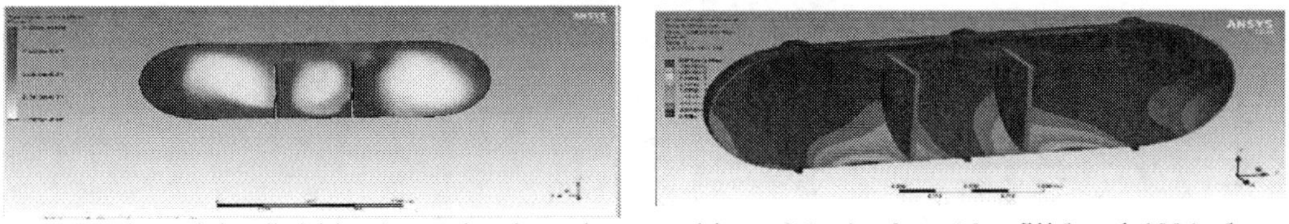

Fig.10. Behavior of the fluid in the tank when the speed is v=25m/s, time 10s, fill level 40%, the color white is air, blue is water respectively
a) Sloshing liquid motion in tanker and volume fraction fluid; b) total deformation structure carbon steel

The action of the pressures exercised by the two fluids (water and oil sun-flower) on the tank walls was analyzed both in static and transient structural for all speeds and filling levels.
Using FSI simulation could by highlighted the action and values of liquid velocity distributions, pressures, stress distribution (10^5Pa) as well as the shape of the free surface shown in Fig.11 a),b) both for static structural as well transient structural.

Comparing these simulations, we can see that the stress distribution (equivalent stress von-Mises) is much higher when the tank is filled in 40 % compared to 80% for varying velocity.

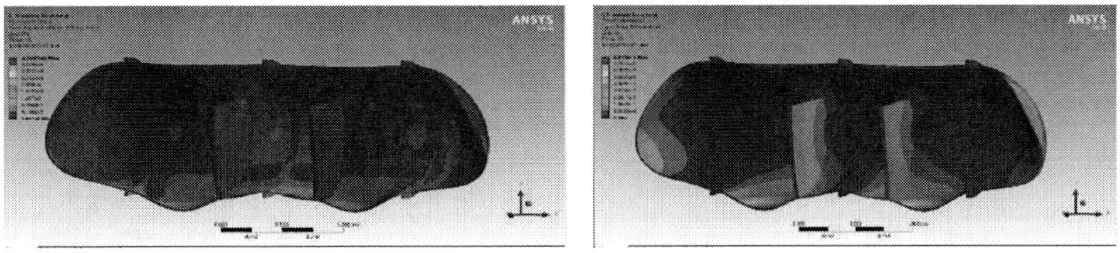

Fig.11. Using FSI simulation show the action and values of liquid velocity distributions, pressures, stress distribution

a) Equivalent stress in transient structural ; b) total deformation transient structural

4 Conclusion

The study may be a starting point to make for an analysis looking and determining the dynamic behaviour of fluids agro-food in tanker during braking manoeuver.

By simulating with FSI simulation using ANSYS CFX and ANSYS Structural codes enable to obtain peak pressure value, deformation value respectively, under defined excitation and check its influence on tank structure and of the behaviour of fluids transported.

Through simulating different imposed speed and for different filling levels it has led to a better understanding and analysis of fluid behaviour inside a tanker with baffles.

By using baffles there has been a liquid stabilization and a good reduction in the sloshing effect.

The braking maneuver is highly dependent upon the fill volume, presence of baffles, and severity of braking input.

5 References

[1] D. Chaudhari, D. Deshmukh, *Review on various roadways tanker cargo for controlling sloshing behavior*, *IJIRST*, **vol 2**, pp. 80-85 (2014)

[2] M. Toumi et al., *Impact of liquid sloshing on the vehicle tank design*, *Sciences &Techn.*, **n°28**, pp. 29-34 (2008)

[3] M. Damagala, E. Lisowski, *Interaction of liquid motion on mobile tank structure*, *JKPT*, **vol.18**, pp.66-71 (2011)

[4] G. Kishore et al., *Calculing Sloshing Impact on Tanker Walls for Fluids at Varying Acceleration*, *IJSR*, **vol.6**, pp.785-795 (2015)

[5] R. Babu, R.K. Gotru, *Study of Sloshing Impact on the Tanker Wall for Different Fluids with Varying Acceleration*, *IJIRST*,**vol.4**, pp.10-16 (2017)

[6] P.K. Panigrahy, U.K. Saha, D. Maity, *Experimental studies on sloshing behavior due to horizontal mouvement of liquids in baffled tanks*, *Oc. Eng.*, **vol.36**, pp.213-222 (2009)

[7] V. Singal, J. Bajaj, N. Awalgaonkar, S. Tibdewal, *CFD Analysis of a Kerosene Fuel Tank to reduce liquid sloshing*, *Pro. Eng.*, **vol.69**, pp.1365-1371 (2014)

[8]D. Liu, P. Lin, *Three-Dimensional liquid sloshing in a tank with baffles*, *Oc. Eng.*, **vol.36**, pp. 202-212 (2009)

[9] G. Brar, S. Singh, *An Experimental and CFD Analysis of Sloshing in a Tanker*, *2nd ICIAME*, pp.490-496 (2014)

[10] SR EN 10216-2

Experimental results on the solid organic fertilizer machine MG 5

Vasilica Ştefan[1*], *Raluca* Sfîru and *Lucreţia* Popa[1]

[1]INMA Bucharest, 6 Ion Ionescu de la Brad, Bucharest, Romania

Abstract. This paper is the first of a series of works devoted to the general issues of organic fertilizer spreading machines. Although the working process of solid organic fertilizer machines does not contain a large number of random constituents, however, the basic component, the material administered, gives the random characteristic of the process. In addition, another random component that influences the working process of the machine is the interaction of the work organ that performs the tearing and distribution of the scattered material with this material, as well as the movement of the fragments of material onto the surface of the device. A third random component is the movement of the material scattered into the atmosphere. The fragments of scattered material exhibit shape irregularities due to fragments during dislocation from the mass of feed material and due to particle collisions while moving into air until the contact with the soil. Therefore, the phenomenon of spraying the organic fertilizer on the layer from the soil surface is, as a whole, a random process. Consequently, it is logical to have experimental research prioritized in the study of the work on the distribution of solid organic fertilizers. The role of this article is to present the experimental results, some conclusions and to prefigure several theoretical or theoretical-empirical research directions. The results contained in this article are basics in all the other essays related to the topic machine.

1 Introduction

Processes of administration of some substances on agricultural land are an important component of most agricultural crop technologies. Through these processes, nutrients are ensured for the crops concerned, protection against pests or parasitic vegetation, the agricultural potential of the land is restored.

Mechanized processes for the administration of organic fertilizers are an important component of soil fertilization for agricultural use (and not only), especially in the context of a current (perhaps shy but existent) return to organic farming.

Most studies referring to precision farming applications address to chemical fertilizer. The motion of the particle on spinning disks equipped with pitched straight or curved vanes are described in [1,2,3,4]. In [5,6,7] the manure distribution process is studied (the interaction between manure particle and spreading equipment, the influence of the working parameters

* Corresponding author: *valle_vali@yahoo.com*

overs the distribution uniformity, numerical model for the flow in land application). In [8] the authors studied the relative motion over a rough surface of a rotary vertical axis helicoid and the movement equation were given and solved by numerical methods.

Generally, the organic fertilizers are found in the form of conglomerates in different physical and chemical states and contain many organic substances with very different properties, so that a characterization as acceptable in terms of physical (especially chemical) is very difficult to obtain. The first condition required to obtain an even distribution of the product on soil surface is to correlate manure properties with the characteristics of the land application equipment.

It is obvious that an attempt to estimate the coefficients of friction and viscosity of the contact between the particles of such a material and the metal of the fragmentation and throwing device is an extremely demanding test. The dependence of these properties on the chemical-physical composition of the material (organic fertilizer: compost or fermentation manure, for example) implies a multidisciplinary approach: chemistry, physics, mechanics, biology. It is difficult to carry out such a complex research to describe the dependence by the parameters mentioned above on the material humidity, ambient temperature, atmospheric pressure, concentrations of the main components of the material being administered. In other words, scattered scraps of scrap do not allow a good assimilation of them with the material points of Newtonian mechanics. Fragments of scattered material have appreciable areas of contact with the air and, in addition, in a large range of values, which makes the friction force with air an important component in their displacement into the atmosphere. The interaction between the particles thrown by the fragmentation and scattering apparatus is a very difficult factor to consider in a deterministic model. Even for a stochastic model, the movement of fragments of organs discharged by the organic fertilizer machine is very difficult and, given the very probably, the few effectively results, the effort required is hard to sustain.

Under these circumstances, the initial experimental study appears to be the only realistic solution, since that the theoretically-empirical and theoretical developments may occur later when some basic elements of the process will be established.

2 Material and method

To perform the experiments, the solid organic fertilizer machine MG 5 was used, fig. 1 and 2, the scraper of which consists of four vertical rotors, arranged at an angle of 10 ° to the vertical, in order to allow the material of the dump to be easily entrained according to the angle of its natural slope, fig. 1. Each rotor is made up of a central shaft on which there are 3 spirals at the end of which there are provided cutting and grinding blades of the material, and at the bottom there are provided centrifugal blades which allow the scraping of the remaining material in the conveyor or material squeegees missed by the first spiral of the rotor.

Fig. 1. Organic fertilizer spreader MG 5, rear view.

Fig. 2. Organic fertilizer spreader MG 5, lateral view during work.

The scraper is driven by power take-off (PTO) and has a transmission ratio of 1: 1.

The exploitation tests were carried out with compost and semi-fermented manure, distributed on stubble. The amount of fertilizer per hectare that can be dispensed by the MG-

5 has varied depending on the aggregate displacement speed, conveyor speed, and fertilizer characteristics.

The rotors speed in the distribution device component is correlated with the engine power of the power source. Three rotational speeds of 360, 440 and 530 rpm were used, corresponding to 1500, 2000 or 2200 rpm respectively of the tractor engine.

Three feed rates for compost (weighing 510 kg / m3, 52% humidity) were chosen, namely 6, 15 and 26 kg / s corresponding to three speeds of the scraper conveyor, namely 0.006, 0.0143 and 0.025 m / s. For semi-fermented manure we used material with a mass density of 800 kg / m3, humidity 75%, the working speeds at the same conveyor speeds were 10, 23 and 40 kg / s.

For collecting the samples, geotextile material of dimension L x l = 18x1 m, transversely arranged in the direction of movement of the machine, was used, Fig. 4.

After work shift, the distributed material was collected in a transverse plane on 1 m2 surfaces using a square metal frame with the 1 m side, fig. 3, 4. The samples were collected in plastic bags, numbered and weighed to the precision 1 gram.

The results obtained under the conditions shown have served to calculate the qualitative working indicators.

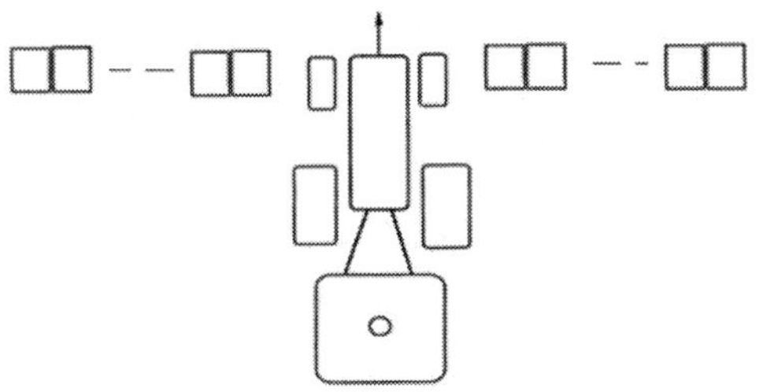

Fig. 3. The sketch of the work of the fertilizer machine, and of the sampling

Fig. 4. The fertilizer machine in working time

2.1 Statistical analysis of experimental data

Experimental data analysis is useful not only to make some appreciation of the machine performance but also to get useful suggestions in mathematical modelling, especially interpolation of experimental data.

In Table 1 are specify the main machine parameters, which have been varied in the experimental research done physically. It can be noticed that the parameters of the garbage spreader geometry did not change: the characteristics of the rotors (length, spiral pitch, radius), spatial arrangement.

Table 1. The list of parameters and their physical dimensions

Name	Notation	M	L	T
The coordinate of the measurement point*	x	0	1	0
Rotors speed	ω	0	0	-1
The fertilizer machine flow rate	q	1	0	-1
The angle of inclination of the rotors	β	0	0	0
Density of scattered material	ρ	1	-3	0
Linear density of scattered material	σ	1	-2	0
Working speed of the car	v	0	1	-1

*The coordinate of the measuring point on the axis perpendicular to the working direction, originating in the projection on the ground of the intersection between the axis of the working direction and the axis perpendicular to it.

Of the parameters considered in the experimental research (and obviously, varied), the parameter σ is, from a qualitative point of view, the most important. This parameter gives the linear distribution of the material scattered by the scraping device with which the machine is equipped. The measurement procedure is described in Chapter 1.The parameter σ is called linear density of distributed material and is measured in a 1 m wide rectangular band behind the machine over a 13 m long, measured in a transverse plane to the trajectory of the fertilizer machine, and is measured in kg / m^2.The variation of the linear density of the distributed material, in the plane of the application surface, must be less than 15% (see SR ISO 5690-1: 95) from the mean. Using this function is can estimate the effective working width, and the optimal working regime of the machine.

2.2 The descriptive properties of the distribution (descriptive statistics, estimators)

The elementary statistical analysis includes the values of the reference estimators of the linear density distribution of the scattered material. The experimental data containing 629 records in the form shown in Fig. 5.

	A	B	C	D	E	F	G	H
1	Mass	Coordinate X	Coordinate Y	Rotational speed	Feed flow	Tractor speed	Inclination angle	Bulk density
2	σ	x	y	ω	q	v	ß	γ
3	kg	m	m	rot/min	kg/s	m/s	°	kg/m3
4	0.114	-4.5	-0.5	360	6	1	15	510
5	0.254	-3.5	-0.5	360	6	1	15	510
6	0.687	-2.5	-0.5	360	6	1	15	510
7	0.897	-1.5	-0.5	360	6	1	15	510
8	0.782	-0.5	-0.5	360	6	1	15	510

Fig. 5. Sequence of records stored in the Excel database

The values of the calculated estimators are given in the Table 2.

Table 2. Statistical estimators

Statistical estimators†	Values
Minimum value	0.015
Maximum value	7.740
Arithmetic mean	1.843
Geometric mean	1.142
Harmonic mean	0.505
Mode	1.325
Median	1.382
Kurtosis	0.774
Skewness	1.107
Square root of the population variance	1.565
Square root of the sample variance	1.566
Variance	2.449
The sample variance	2.452

The large difference between the minimum and maximum values of the linear density of the scattered material is due to the fact that in the experimental and measuring process the extreme areas from which the material was distributed in insignificant amounts were taken

†For estimators definitions see, for example [10, 11] .

into consideration, areas which are not included in the actual work width of the machine. The average value of the linear density of the scattered material gives an approximation of the mean value applicable with this car, the most probable value of the studied parameter, under the given conditions, [12].In addition to this mean value, even higher and lower values are possible, depending on the needs, and these can be achieved using of operable adjustments for varied parameters in the experiment (Table 1).

Harmonic mean is more useful when looking for the estimation of a phenomenon rate and the parameter has of the specified physical size. Geometric mean is used when more variables occur in multiplicative combinations. The mode, is the most likely value of the random variable studied. Median is the so-called middle value (for which the number of samples with higher values is equal to the number of samples with lower values).Kurtosis is an estimator of the distribution form, which for normal distribution has zero value, being positive for higher and sharper distribution distributions than normal and negative distribution otherwise [13,14].In the case of the distribution analyzed in this paper, the distribution is leptokurtic. Skewness (asymmetry) refers to the distribution symmetry and has positive or negative values, as the mode is located to the left of the median and the middle to the right of the fashion, respectively the fashion is to the right and the average to the left of the median,[15]. Also, in this case, the asymmetry has the value of 1.107, so the mean distribution is asymmetric with deviation to the left of the middle of the fashion. The average square deviation and the sampled average square deviation (a better estimate of population variation), [13], differs very few. The average square deviation is slightly lower than the average (arithmetic mean) of the population. As a variant, the standard mean deviation is a measure of spreading the distribution values around the average of the same distribution, [12].

The distribution of the linear density samples of the scattered material after the sample mass is given numerically in fig. 6 and in fig. 7 in percentage.

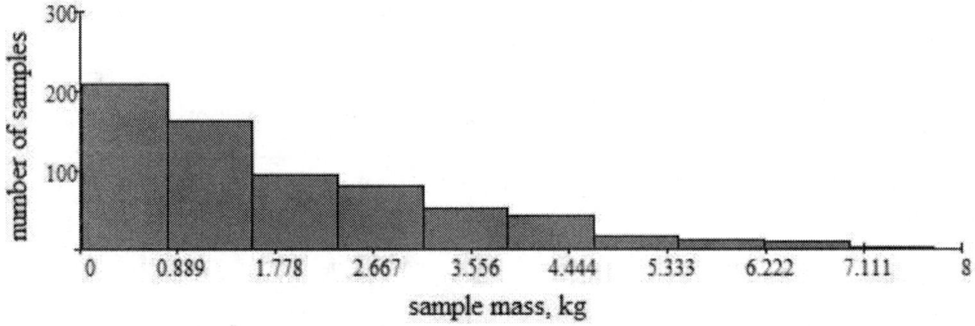

Fig. 6. Histogram of distribution σ, after sample mass.

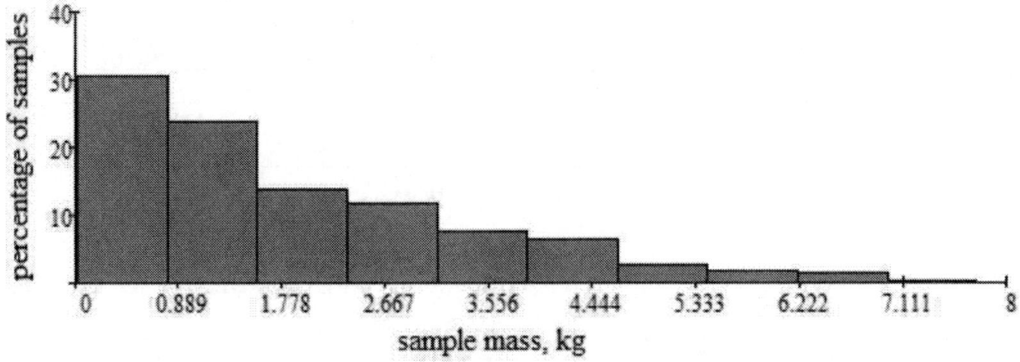

Fig. 7. Percent histogram of the distribution σ, after mass of samples

2.3 Correlations

The relations between the distribution of the linear density of the scattered material and the various experimentally varied parameters results from Table 3, which contains the values of covariance and correlation. The covariance and correlation are calculated according to [13]. Covariance and correlation (called correlation coefficient and correlation coefficient in some publications) are statistical parameters that highlight the relationship of dependence between two random variables [10]. The two statistical sizes show, from our point of view, the same phenomenon: a tight connection between the linear density of the scattered material and the *mass flow rate provided by the machine*, q, respectively, the *density of the scattered material*, ρ. The other parameters that were taken into account in the experimental work are insignificantly related to the main quality parameter of the machine, the linear *density of the scattered material*, σ.

This result shows that any attempt to model the mathematical link between σ, as the main quality parameter of the fertilizer spreader machine and the process input and control parameters, must first take into account the two parameters highlited by the covariance and correlation: q, ρ.

Table 3 Values of the covariance and correlation of linear density of the scattered material with experimentally varied parameters

	x	ω	q	v	β	ρ
Covariance σ with:	-0.297	-0.256	9.894	-0.024	-0.006	61.230
Correlation σ with:	-0.055	-0.023	0.565	-0.053	-0.057	0.270

It can be seen that the average linear density of the scattered material by the fertilizer machine is 1.843 kg/m^2, approximately 18 t/ha. For the fertilizers used in the experiments, the literature provides for the application rate of 10-40 t/ha, for compost which have density 500 kg/m^3 and 20-80 t/ha, for the waste which have density of 800 kg/m^3 [16,17].

2.4 Granulometric quality of scattered material

Among the performance of the organic fertilizer spreader, an important place is the granulometric quality of the scattered material, that is, what in terms of specialty is called the degree of crushing. This aspect of the organic fertilizer spreading process has been studied partially in [18]. The main results are summarized in Table 4.

Table 4 Values of spread width and shredding degree

Material type	Rotative speed of spreading apparatus [rpm]	Maximum distribution width [m]	Total mass of sample [kg]	Mass of particles sized of< 6cm [kg]	Shredding degree [%]
Fermented manure	365	8.5	14.08	10.78	76.56
	445	10	15.95	12.52	78.49
	530	13	20.05	17.56	87.58
Compost	365	9	17.79	16.85	94.71
	445	11.5	19.23	18.12	94.22
	530	14	24.11	24.11	100

The particle distribution in the scattered material is given in Table 5.

Table 5 Particle size distribution of tested materials

Material	Rotational speed [rpm]	Total sample mass [kg]	Particle size distribution of each material fraction [kg]					
			>6 cm	4-6 cm	2-4 cm	1-2 cm	0,5-1 cm	<0.5 cm
Fermented manure	530	3.45	0.250	0.300	0.446	0.954	0.492	1.008
Compost	530	4.82	0	0.245	0.520	0.745	1.360	1.950

By using the data in Table 4, information on the maximum spread width and the degree of crushing can be obtained, depending on the speed of rotation of the spreader. In fig. 11 and 12 are given the dependences, of the maximum spread width, respectively of the degree of crushing, by the speed of rotation of the apparatus. Because the experimental points available were only three, we preferred to calculate only the linear regression (linear dependence). If b_{max} is the maximum working width, and g_m, is the degree of crushing of the scattered material, and ω, the angular rotation speed of the scattering apparatus, then the linear dependence equations whose graphical representation is given in fig. 11 and 12, are written as follows:

$$b_{max}(\omega) = 0.261\omega - 1.719 \quad (1)$$

in case the scattered material is fermented manure,

$$b_{max}(\omega) = 0.289\omega - 2.031 \quad (2)$$

in case the scattered material is the compost,

$$g_m(\omega) = 0.642\omega + 50.858 \quad (3)$$

in case the scattered material is fermented manure,

$$g_m(\omega) = 0.34\omega + 81.823 \quad (4)$$

in case the scattered material is the compost. The functions (1) - (4) are plotted in fig. 8 and 9. In the same figures are also show the distribution of experimental data (measured directly).

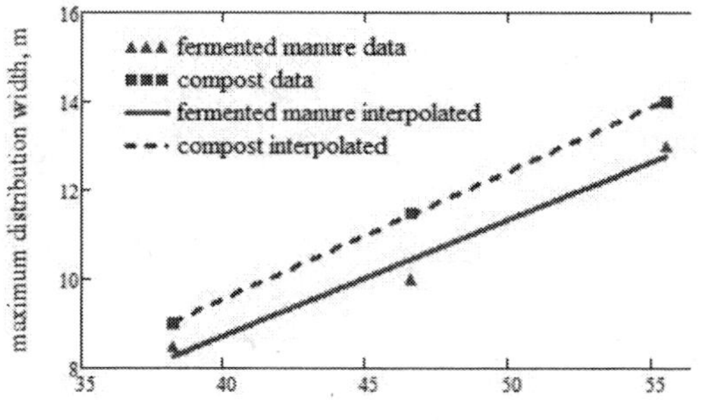

Fig. 8 The dependence of the maximum spread width, by the rotation speed of the appliance

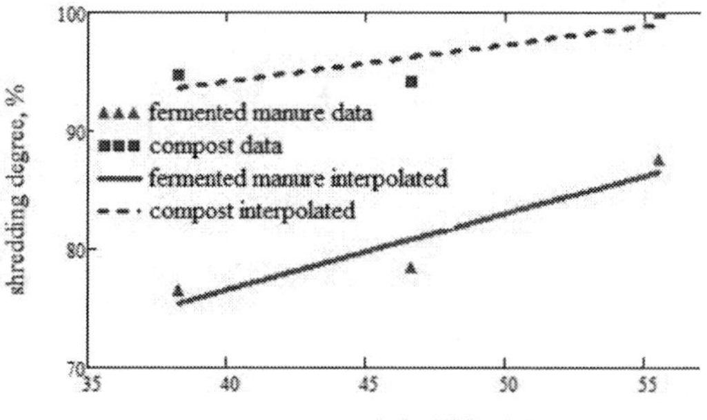

Fig. 9 The dependence of the degree of crushing of scattered material, by the speed of rotation of the rotors

As can be seen from the tabular data and graphs presented, the maximum spread width and the crushing rate increase, together with increasing the rotational speed of the spreader. Increasing the rotation speed of the scatter means also increasing the energy consumption of the machine.

3 Conclusions

Some of the conclusions have already been expressed immediately after the results are presented. The most important will be repeated in this chapter.

The minimum and maximum values of the scattered material quantities on the measured work width of 0 to 13 m allow the range of MG 5 fertilizer machine application rate to be set between 10 and 80 t / ha.

The variance of the linear distribution density of the scattered material is approximately 2.45 kg / m^2. This value is not yet relevant to the quality of the machine's uniformity of administration because it is calculated for all the work regimes achieved (v, q, β, ρ). The variance will be calculated as a measure of the uniformity of spreading material for each working regime. The lowest value of the variance relative to the mean value (desirable is that it is the standard indicated by scientific criteria) of the spreaded material amount denotes satisfactory uniformity, but a limit of their ratio is only conventional.

An information on the uniformity of the distribution of the material also gives the kurtosis estimator. For the uniformity to be the best, the value of this estimator must be negative (with a maximum ceiling below that of the normal distribution). For the time being, but only for the whole lot of experimental recordings, kurtosis has a small but positive value, therefore, on the perpendicular section on the fertilizer machine feed axis, the linear distribution density of the administered material shows a fairly visible peak.

Also, useful information on the uniformity of the linear density of the scattered material is given by its histograms. These must be interpreted in relation to the imposed rule, and then the surface of the field where the amount of material administered is excess and where it is less than what is required can be estimated. It is also necessary to take into account the tolerances admitted by the farmers specialists. In any case, it should be taken into account that the non-uniformity of the linear density of the material administered may be the cause of the crop irregularity.

The skewness factor has a positive value and denotes a deviation to the left of the median of the distribution pattern σ in the transverse direction of the machine. This is information that suggests better adjustment of the drive mechanisms as well as the geometry of the rotors placement.

The values of the correlation between the linear density of the administered material and the other recorded parameters show that its most intense dependence is on the feed rate and the bulk density of the material administered.

The quality of the work of the organic fertilizer machines is also expressed by the degree of crushing of the material used. The small study in this paper about this feature shows that the degree of crushing increases with the rotation speed of the rotors, which increases also the maximum working width of the machine (which is not, however, the actual work width).

In view of these results, separate analysis is planned on work regimes, and firstly to develop statistical mathematical models on which to estimate and optimize an effective working width of the fertilizer machine. Also, based on these models, a study of the factors involved influence is possible, as well as an economic study (work capacity and energy consumption) are made.

Acknowledgement

This work was supported by a grant of the Romanian Research and Innovation Ministry, through Programme 1 – Development of the national research-development system, subprogramme 1.2 – Institutional performance – Projects for financing excellence in RDI, contract no. 16PFE.

References

1. Xie S.Y., Yu T., Chen T.H., Yang Z.R., Yang L., Yang M.J., Design and experimental study of a rapeseed broadcast spreader of screw-drop type, INMATEH Agricultural Engineering Journal, 55, 19-26, 2018.
2. Popa L., Ştefan V., Andrei L.V., Guran G., Research regarding the construction of a chemical and amendments fertilizer machine MA 3,6, Scientific Papers (INMATEH), I, 54-59, 2009.
3. Kobets A.S., Naumenko M.M., Ponomarenko N.O., Kharytonov M.M., Velychko O.P., Yaropud V.M., Design substantiation of the three-tier centrifugal type mineral fertilizers spreader, INMATEH Agricultural Engineering Journal, 53, 13-20, 2017.
4. Popa L., Pirnă I., Vlăduţ V., Ştefan V., Researches Concerning the achievement of a machine for spreading low rates of chemical fertilizers", The 8th International Symposium "PROSPECTS FOR THE 3rd MILLENNIUM AGRICULTURE", University of Agricultural Sciences and Veterinary Medicine, Cluj-Napoca, 66, 441-447, 2009.
5. Stefan V., Cârdei P., Popa L., Experimental results on the solid organic fertilizer machine MG 5, April 2019. [Online]. Available: https://www.researchgate.net/publication/332112021_Experimental_results_on_the_solid_organic_fertilizer_machine_MG_5. [Accessed 22 april 2019].
6. Landry H., Piron E., Agnew J.M., Lague C., Roberge M., Performances of conveying systems for manure spreaders and effects of hopper geometry on output flow, Applied Engineering in Agriculture, 21, 159-166, 2005.
7. Landry H., Thirion F., Lague C., Roberge M., Numerical modelling of the flow of organic fertilizers in land application equipment, Computers and Electronics in Agriculture, 51, 35-53, 2006.
8. Duhovnik J., Benedičič J., Bernik R., Analysis and design parameters for inclined rotors used for manure dispersal on broadcast spreaders for solid manures, Transactions of the ASAE, 47, 1389-1404, 2004.
9. Pylypaka S.F., Klendii M.B., Klendii O.M., Particle motion over the surface of a rotary vertical axis helicoid. INMATEH Agricultural Engineering Journal, 51, 15-28, 2017.
10. V. Bobancu, N. Mihaileanu, S. Gheorghita, A. Brezuleanu, A. Stefanescu, T. Balanescu. *Dictionar de matematici generale*. Bucuresti: Editura Enciclopedica Romana. (1974).
11. MathSoft Engineering & Education, I. Mathcad User's Guide with Reference Manual Mathcad 2001 Professional. Cambridge: MathSoft Engineering & Education, Inc. (2001).
12. M. G. Bulmer. Principles of Statistics. Dover: doverpublication. (1979).
13. Statistic How To. Retrieved from https://www.statisticshowto.datasciencecentral.com/probability-and-statistics/statistics-definitions/kurtosis-leptokurtic-platykurtic/ .(2019, March 1).
14. Engineering Statistic Handbook. (2019, March 13). Retrieved from https://www.itl.nist.gov/div898/handbook/eda/section3/eda35b.html.
15. D. Jain. Retrieved from https://codeburst.io/2-important-statistics-terms-you-need-to-know-in-data-science-skewness-and-kurtosis-388fef94eeaa/ .(2019, March 13.
16. V. Scripnic, P. Babiciu. *Masini Agricole*. Bucuresti: Editura Ceres. 1979.
17. R. Majdar, V. Davidescu. *Agrochimie*. Bucuresti: USAMV. 2009.
18. V. Ştefan, R. Ciupercă, L. Popa, A. Nedelcu, G. Lazăr, A.S. Petcu, A. Zaica, *The influence of physical characteristics of solid organic fertilizers on quality of land spreading*. INMATEH - Agricultural Engineering, 46, 77-84, 2015.

Researches on ecological weed control in vegetable crops

Adriana Muscalu[1,*] , *Cristian* Sorică[1], *Cătălin* Persu[1], *Ana-Maria* Andrei[2], *Angela* Dorogan[3] and *Cătălina* Tudora[1]

[1]INMA, 6 Ion Ionescu de la Brad Blvd., Bucharest, 013811, Romania
[2]ICDPP, 8 Ion Ionescu de la Brad Blvd., Bucharest, 013813, Romania
[3]INCDTP, 16 Lucretiu Pătrăşcanu., Bucharest, 030508, Romania

Abstract. In recent years, intense concerns about the increasing resistance of weed populations to herbicide treatments, the low availability of bioactive ingredients for vegetable crops, as well as the expansion of organic farms for these niche crops have stimulated the development of new non-chemical methods of control or new approaches to those already used. In general, vegetable crops exhibit increased sensitivity to weeds compared to other crops. Therefore, weed control in organic vegetable crops is a real challenge for farmers, mainly using the so-called physical combat methods. Therefore, weed destruction is accomplished by using manual, mechanical, thermal or mulching means. The paper presents the preliminary experimental results obtained under laboratory conditions, referring to an innovative model of equipment for the ecological weed control in vegetable crops. It will achieve the destruction of weeds, combining the thermal method based on using hot water with the mechanical one, by equipping the machine with *environmentally friendly* active parts. The results obtained constitute an important premise for developing efficient equipment for the ecological weed control that can be used within integrated control systems, specific for organic onion, bean and bell pepper crops.

1 Introduction

The quality of agricultural products is a topic that has raised the interest of the general public, especially in recent years, when specialty studies have demonstrated the importance and beneficial effects of healthy eating. This involves a certain way of life as well as the consumption of healthy foods, especially "bio products". According to market studies conducted in EU countries, the consumption level of these products ranges between 3% and 13% for permanent consumers and between 23% and 53% for occasional consumers [1].

This context as well as the indications of nutritionists recommending higher consumption of vegetables has led to the development of organic farming for these species, for which the food and therapeutic value increases in these circumstances. [2]. Organic cultivation of vegetables ranks between so-called "minor" or niche crops, which provide

* Corresponding author: amuscalis@yahoo.com

productions representing more than 20% of the total value of agricultural production in the European Union [3].

In general, vegetable crops have increased sensitivity to weeds compared to other crops [4]. Weed control in vegetable crops without synthetic herbicides, according to the rules of organic farming, is one of the greatest challenges faced by vegetable growers [5].

The development of new non-chemical control methods or new approaches to those already used has been stimulated by weed populations' increase in resistance to herbicides and by the low availability of specific bioactive ingredients used for these crops [6]. Non-chemical weed control of organic vegetable crops is performed with mechanical and thermal methods, using characteristic equipment, which represent the traditional physical means [7,8].

Cultivation of vegetables in organic farming attaches great importance to soil works, crop rotation, as well as to soil fertilization. Some of these are even preventive weed control measures. To these are added the crop maintenance works (curative control measures) made with specialized equipment, intensely developed in the last period. Their action was intensified by the use of new active parts (brush, fingers, basket type, etc.) or intelligent equipment capable of recognizing and controlling weeds in vegetable crops [3,6].

The method of thermal control is based on heat transfer between an installation and harmful plants through foliar contact. For each type of crop, different techniques of thermal weed control should be created / adapted, ensuring the protection of cultivated plants [9]. Most of the researches in this regard have used treatments with: infrared radiation, flame, hot air, steam, hot water, saturated steam, etc. Short-term, sudden heating causes cellular protein distortion and protoplasm expansion, which leads to the elimination of water from weed tissue, causing damage to/destruction of the vital functions of the plant parts above the ground [9-11]. Applying the methods based on moist heat (steam, hot water, saturated steam with boiling water, etc.) is more convenient, being a little more efficient and less harmful to the environment. Hot water treatments can improve the bio-herbicidal potential of some pathogens (fungi). [5,9,10,12].

The paper presents an experimental model of weed control equipment for organic onion, bean and bell pepper crops, combining the thermal and mechanical methods. By using an electrical energy source, supplemented by the recovered energy, the environment will be protected. Finding a clean method of controlling weeds in vegetable crops and implicitly of lowering the prices of maintenance works for organic onion, bean and bell pepper crops is aimed at.

2. Material and method

The experimental model of organic weed control equipment for vegetable crops uses hot water as control agent, the action of which is completed by that of environmentally friendly active parts, of finger weeder type.

2.1 Organic weed control equipment

2.1.1 Description and operating principle

Considering that hot water flows gravitationally compared to steam and hot air, with a very good thermal conductivity, it can provide a higher heat transfer (approximately 23-27 times higher) compared to the other heat agents presented in Table 1, but in a slightly longer time [10, 13].

Table 1. Physical properties

Control agent (T=100^{0}C, p=1bar)	**Specific heat** kJ/(kg K)	**Thermal conductivity** W/(m K)
Hot air	1.01	0.03
Hot water	4.18	0.682
Steam	2.08	0.025

Controlling weeds with non-chemical methods depends on some technical-economic peculiarities of vegetable crops [3]. Organic control equipment (Figure 1) is intended for use on both shaped and flat land. The mobile frame (b) as well as the fastening systems of the nozzle holders (5), respectively of the active parts (d), allow their correct positioning reported to the soil and plants, according to the planting schemes for each culture. Nozzle holders also allow adjusting the distance between the insulated sidewalls, which are designed to adapt the thermal agent action zone to the spacing between plant rows. Hot water distribution nozzles are grouped by two, one after the other, to accentuate the control effect of hot water on weeds.

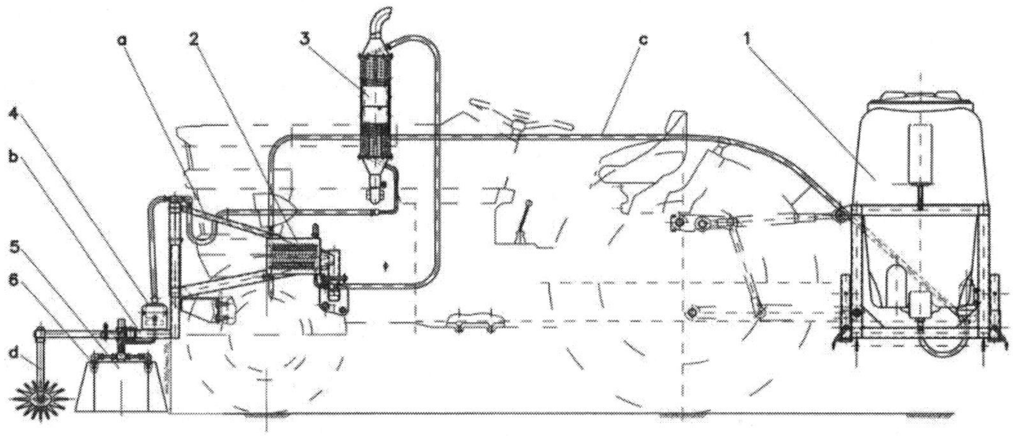

Fig.1 Experimental model of organic control equipment
a - fixed frame; b - mobile frame; c - heating and distribution installation (1 - tank, 2 - preheating system; 3 - heat recovery system; 4 - boiler type electrical resistance; 5- nozzle holders; 6 - nozzles); d – finger weeder

To be used as a control agent, the water in the tank needs to be heated. The process takes place step by step, being initiated in the pre-heating system, where the control agent takes over the heat of the water from the tractor radiator, which washes the spiral from the copper pipe through which it flows. The heating process continues in the heat recovery system, which is an exhaust pipe with heat recovery of exhaust gas, resulting from the tractor engine. It replaces the classic exhaust pipe of the tractor, consisting of a cylindrical body, to which the upper and lower caps are attached with screws, in the area of which there is a copper pipe spiral, through which the preheated water circulates. Flue gas, with a very high temperature at the engine exit, “washes” the spiral. [14,15]. Control agent heating is completed by means of a boiler type electrical resistance. Then, the hot water is distributed over the weeds, using the nozzles [16]. The equipment is fitted according to the scheme in figure 2, with two electrical resistances, the operation of which is controlled by electrical valves.

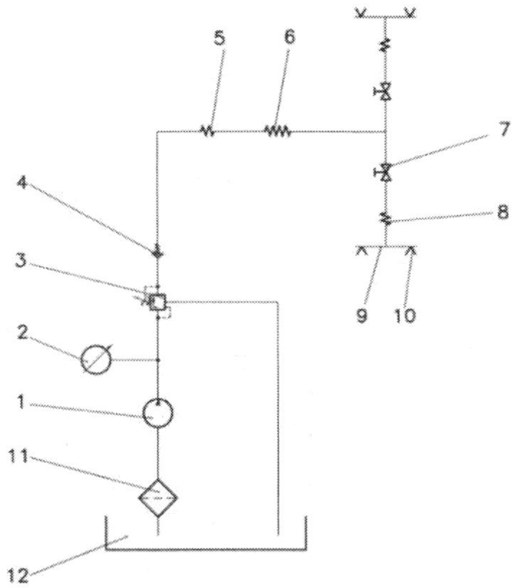

Fig. 2 Conventional scheme of hot water heating and distribution installation
1. Electric pump, 2. Manometer, 3. Pressure regulator, 4. Check valve, 5. Pre-heating system, 6. Heat recovery system, 7. Electrical valve, 8. Boiler type electrical resistance, 9. Nozzle holders, 10. Nozzle.

2.1.2 Technical characteristics

The main technical and functional characteristics of the experimental model of organic weed control equipment for vegetable crops are: pump supply voltage: 12 V dc; fluid pressure in the installation: max. 2 bar; water tank capacity: 300 l; number of operation intervals: 1-2 spaces between the plant rows; number of nozzle holders: 2 pcs .; number of nozzles: 4 pcs .; working width: 215-600 mm; mobile frame adjustment height: 400 mm; no. active parts: 2/4; active parts working depth: 2-4 cm.

2.2 Testing method

Laboratory experiments aimed at monitoring and determining some characteristics of the thermal agent, which are important to the control process. These refer to hot water temperature at the outlet of the nozzle, respectively the flow rate and the working pressure. Also, the uniformity of thermal agent volume distribution through the nozzles was followed, determining a coefficient of variation.

First, were connected to the tractor power source the pump and the boiler type electrical resistance, the latter requiring a pre-heating period of 60 seconds to reach the maximum temperature. After contact with the resistance, the temperature of the water from the tank was determined, at the outlet of the nozzle. Thus, it has been found that the temperature variation of the thermal agent is ΔT = 20-23^{0}C, under the conditions of its initial temperature of T_i = 11^{0}C.

Previously, we showed that the thermal agent heating process takes place step by step, depending on the operation of the tractor thermal engine afferent to the equipment. For this reason, it was necessary a period of 10-15 minutes of operation in advance, in order to reach the operating temperature. After the necessary preparation periods, the experimental determinations were carried out, under the operating conditions of the heating and distribution installation, in order to obtain the temperatures of the thermal agent, the pressure and the working flow. For this purpose, specific equipment such as: "Ama-digit 20th" digital thermometer, mechanical timer, graduated cylinder, etc. was used.

To determine the coefficient of variation for volume distribution of the liquid sprayed through nozzles, the following equation was used:

$$CV_{flow} = 100 \cdot \frac{\sigma_{flow}}{V_m}, \quad (1)$$

where σ_{flow} - standard deviation of the volumes collected from nozzles in one minute;
V_m - average volume collected from nozzles in one minute.

The standard deviation of the volumes collected from nozzles was determined using the following equation:

$$\sigma_{flow} = \sqrt{\frac{\sum (V_i - V_m)^2}{n-1}}, \quad (2)$$

where V_i - volume of liquid collected from the nozzle „i" in one minute;
n – number of nozzles.

The coefficient of variation for volume distribution of the liquid sprayed through nozzles, is a quality indicator for the spraying process and represents the ratio between the standard deviation and the average, expressed in percent. It points out the homogeneity of the values obtained and also the level of dispersion towards the average. Practically, this indicator shows if the spraying nozzles are properly calibrated by the producer and homogeneously distribute the liquid.

3 Results and discussions

Table 2 shows the mean values of control agent main characteristics, which influence the process, determined under laboratory conditions.

Table 2. Control agent characteristics (hot water).

Determination	**Working pressure** (bar)	**Nozzle flow** (l/min)	**Output temperature** (^{0}C)
1	2	0.91	68.3
2	1.8	0.87	77.6
3	1.6	0.82	80.4
4	1.4	0.78	84.9

The minimum temperature of the thermal agent should be as high as possible since, in order to direct heat to the plant, the highest possible temperature gradient is necessary. For effective control, the temperature of the control agent at the outlet of the nozzle must be as close to the boiling point of water as possible, so it must reach values of at least 95^{0}C. The recorded temperatures are below these values, generated both by the low temperature of water in the tank, at the time of the experiments, and by the heat losses recorded along the route.

Table 3 shows the average values of hot water volumes, distributed through each nozzle in one minute, obtained from three repetitions for each of them, corresponding to each considered working pressure.

Table 3. Qualitative indices related to the volume of liquid sprayed by nozzle in one minute

Qualitative indices	Working pressure (bar)	Nozzle no.			
		1	2	3	4
Average volume per nozzle V_i (l)	2.0	0.91	0.92	0.88	0.90
	1.8	0.87	0.88	0.85	0.86
	1.6	0.82	0.82	0.80	0.81
	1.4	0.78	0.79	0.77	0.78
Average volume V_m (l)	2.0	0.90			
	1.8	0.86			
	1.6	0.81			
	1.4	0.78			
Coefficient of variation for the volume distribution CV_{flow} (%)	2.0	1.44			
	1.8	1.34			
	1.6	1.23			
	1.4	1.04			

Figure 3 shows the average volume of control agent sprayed in one minute and the coefficient of variation for its volume distribution, depending on the working pressure.

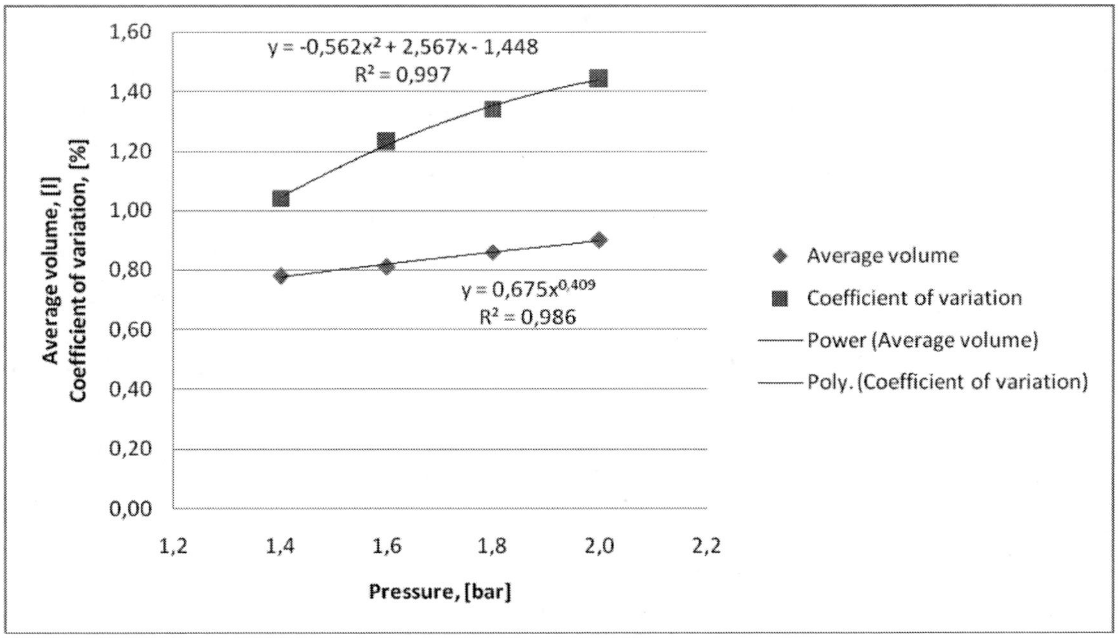

Fig. 3. Dependence of the average volume of control agent sprayed in one minute and of the coefficient of variation of its volume distribution on the working pressure

Linear regression, performed using Excel, allowed the identification of a power function which estimates the variation of the average volume of liquid sprayed through nozzles within one minute depending on the working pressure, with a coefficient of determination equal to 0.986. Also, a second-degree polynomial function has been identified, which estimates the coefficient of variation depending on the working pressure, with a coefficient of determination equal to 0.997.

4 Conclusions

The coefficient of variation for volume distribution of the liquid sprayed through nozzles has a maximum value of 1.44 % which is less than the statistical conventional limit of 5 %, below this limit the distribution being considered very homogeneous and the average value of volume is representative. These results show that the spraying nozzles are properly calibrated by the producer and homogeneously distribute the liquid through nozzles.

To improve the efficiency of the water heating and distribution installation, boiler type electrical resistances with greater power will be used. Experimental research on weed control will be continued under real operating conditions to assess the process effect on weeds in organic vegetable crops.

The equipment combines two methods of non-chemical weed control, not using any other source of environmentally harmful extra energy, except for the particulate matter generated by fuel combustion required for the self-driving of the tractor on which the equipment is mounted.

The results obtained are an important starting point for the production of efficient weed control equipment, which contributes to the development of organic onion, bean and bell pepper crops and of integrated protection systems, specific to these crops.

Acknowledgement

The work was supported by a grant of the Romanian Ministry of Research and Innovation, CCCDI - UEFISCDI project number PN-III-P1-1.2-PCCDI-2017-0659 / Nr.11 PCCDI/2018, within PNCDI III and a grant of the Romanian Research and Innovation Ministry, through Programme 1 – Development of the national research-development system, sub-programme 1.2 – Institutional performance – Projects for financing excellence in RDI, contract no. 16PFE.

References

1. D. Matei, I.S. Brumă, L. Tanasă, International Journal of Communication Research, **6(1),** ISSN 2246-9265, Iaşi, 7, (2016);
2. P. Skerret, W. Willett, J Midwifery, Women's Health, **55 (6)**, 10 (2010)
3. E. Pannaci, B. Lattanzi, F. Tei, J. Crop Protection, **96**, 14 (2017)
4. M. Fontanelli, M. Raffaelli, L. Martelloni, C. Frasconi, M. Ginanni, A. Peruzzi, Spanish Journal of Agricultural Research **11 (3)**, 10 (2013),
5. A.S. El-Sayed, G. M. El-Hameed, The Role of Agricultural and Biological Engineering for Advancement of the Agricultural Sector, Proceedings of 5th International Conference for Agricultural & Bio-engineering, 25-27 September, Cairo, Egypt, 30 (2017)

6. A. Peruzzi A., L. Martelloni L., C Frasconi, M. Fontanelli, M. Pirchio, M. Raffaelli M. Journal of Agricultural Engineering, XLVIII, **583** (14), (2017)
7. D.L. Shaner, H.J. Beckie H.J. Pest Management Science **70 (9)**, 11 (2014)
8. O.Z. Bundza, O.O. Nalobina, V.G. Nikitin, O.L. Tkachuk, INMATEH - Agricultural Engineering, **52(2)**, 6 (2017)
9. J. Nadzeikienė, A.P. Sirvydas, R. Čingienė, R. Vasinauskienė, P. Kerpauskas, J. Agronomy Research, **7(I),** 5 (2009)
10. P. Kristoffersen, A. M. Rask, S. U. Larsen, Weed Research, **48(1),** 7 (2008)
11. J. Ascard, P.E. Hatcher, B. Melander, M.K. Upadhyaya, Thermal weed control. In: Upadhyaya, M.K., Re Blackshaw, R.E. (Eds.), Non-chemical Weed Management - Principles, Concepts and Technology. CABI Press, Oxfordshire, UK, 21 (2007)
12. C.D. Boyette, R.E. Hoagland, K.C. Stetina, American J., Plant Sciences, **9**, 10 (2018)
13. F.T. Tanasescu, V. Stanciu, S. Nitu, *Technical Agenda*, Technical Publishing House, Bucharest, (1990)
14. Gh. Socaciu, M. Matache, M. Marian, A. Muscalu, M. Bîrsan *Patent application* No. A - 00504/21.07.2017 *Exhaust pipe with heat recovery of exhaust gas* (2017)
15. A. Tiwari, H. Vadnani, N. Kumari, M. Labana, IJRSAE, **6(9)**, 16 (2017)
16. A. Muscalu, C. Perşu, I. Ganea Hristu, G. Bolintineanu, C. Tudora, *Patent application* No. A - 0002/16.01.2019 *Equipment for weed destruction in vegetable growing* (2019).

E3S Web of Conferences **112**, 03009 (2019) https://doi.org/10.1051/e3sconf/201911203009
TE-RE-RD 2019

Concaves influence on the milling process of cereal seeds in hammer mills

Anişoara Păun[1,*], *Gheorghe* Stroescu[1], *Dumitru* Milea[1], *Mihai* Olan and *Mariana* Epure[1]

[1]National Institute of Research - Development for Machines and Installations designed to Agriculture and Food Industry - INMA, Bucharest 013813, Romania

Abstract. Obtaining quality fodder that meets the requirements of fodder receipts makes necessary the use of components that have to respect certain dimensions. The milled products must respect the granulation imposed by the receipt because they are very well assimilated by animals and the homogenization process is also very good. The milling process is a complex one, influenced by a number of factors that depend on the technical and constructive solution of the milling chamber and on the physical and mechanical properties of the processed raw material (seeds). In the hammer mills with open chambers and sieve surface up to 180 degrees, the presence of concaves (grooved plates) within the milling chamber is a must because it will lead to the acceleration of the seed milling process and implicitly to increase the productivity and reduce electricity consumption. The paper follows the theoretical and experimental presentation of an open chamber mill that is provided with these grooved plates.

1 Introduction

The production of concentrated fodder for livestock sector is an important concern both at global and national levels.

Fodder use in animal feed, in its natural state, does not meet all the requirements of the animal body and therefore it must be processed before consumption according to a particular technological scheme [1]. The milling of concentrated fodder, as a component of feed preparation technological processes, is one of the basic operations. Hammer mills improvement has been the focus of many researchers, but in most of the published works there are only partial approaches to constructive and functional parameters that influence the operation regime.

The working process of the mill must be regarded as a stochastic process described by pseudo-static methods [2,3,4, 5]. In this case, the established operation regime of the mill is a continuous, incidental, stationary process. Following the milling process, a multitude of small particles are formed, although milling can be characterized as a process of producing new surfaces of material particles. The size of the entire mass of the powdery material as a static mass is evaluated by the comprised content of classes (fractions), of certain dimensions, namely, by granulometric composition or granulation. Almost the entire

*Corresponding author: ani_paun@yahoo.com

crushing and milling of inorganic materials involves a reduction from a one-dimensional distribution to a smaller distribution.

For this reason, many crushing theories are based on the dimensional connection from one distribution to another.

The milling process, in the case of hammer mills, is influenced by the following factors: technological, mechanical, constructive, functional [6].

The main mechanical factors influencing the milling process in hammer mills are: hammers impact on fodder particles, loss of deformation energy during shock, hammers peripheral speed, the movement speed of the fodder layer in the milling chamber, the fan effect produced by the hammer rotor, hammer rotor's moment of inertia.

The constructive factors that influence the milling process are: the size of the milling chamber, the constructive shape of the hammers, the size of the distance between hammers and sieve, the constructive shape of the grooves, the method of feeding the machine, the way of grain milling discharge.

Seeds milling in hammer mills can be considered as being made up of three phases [7, 8, 9]:

1- shredding of fodder particles due to collision with the hammers of the rotating rotor when entering the milling chamber;

2- shredding of particles in non-uniform movement resulting from the first collision and projected by hammers on the grooved plates and also by the edges of the sieve holes, which border the milling chamber;

3- shredding of fodder particles as a result of mutual collisions in the milling chamber.

2 Theoretical research

The open chamber hammer mill, MC-3, Figure 1, is a mill where fodder is tangentially fed by free flow through the entire milling chamber length [2, 3, 6]. Fodder feeding is made through an opening at the peripheral side; at the entrance to the milling chamber grain seeds are subject to strong hammer blows.

Due to these blows the grain seeds are shred, and the resulting particles are projected by the grooved plates (counter knives), and there the second shredding occurs. From these plates, the resulting particles rebound and reach the hammer area again, after which the phenomena repeat.

The shredding phenomenon also continues in the sieve area due to the shocks and friction between the fodder particles and the edges of the sieve holes. The sieve is more like a system of grain milling separation (removal)[7,8,9].

Particles smaller than the sieve holes, under the action of the air flow created by a fan (mounted outside the mill) and centrifugal forces, pass through the sieve and reach the discharge and transport duct that connects to the separation and collection systems.

In these mills, fodder milling is made especially due to collisions between the hammers and the fodder particles, respectively the concaves.

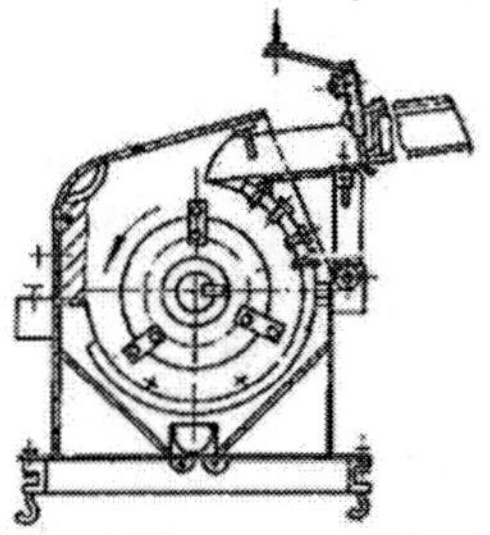

Fig. 1. Hammer mill with open chamber, MC-3

https://doi.org/10.1051/e3sconf/201911203009

The concave is the fixed part in the milling chamber participating in the seed shredding operation and which may be of different shapes: cylindrical, semi-cylindrical or cylindrical with groove, teeth or hole active surface [6, 7].

There are situations where the concave is missing, its role being taken over by the hole sieve. The use of groove or teeth concave leads to the best results in terms of shredding process quality.

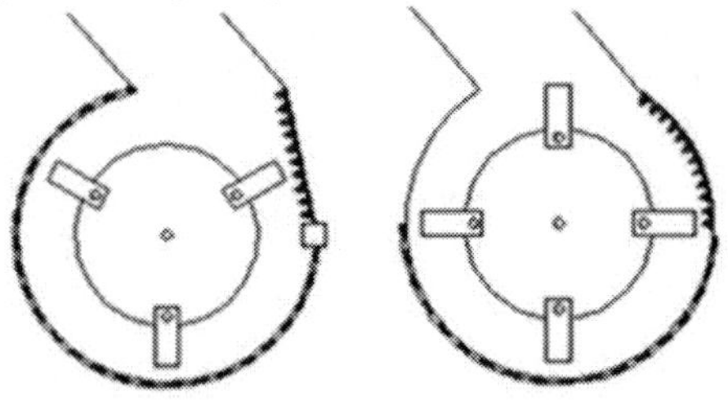

a b

Fig. 2. Constructive variants of sieves and concaves used in the open chamber hammer mill

a - *the sieve with the coverage angle of the milling chamber of 255°, b - the sieve with the coverage angle of the milling chamber of 180 ° with another type of counter knife*

We consider that the particle moves in the milling chamber in the area between the hammer rotor and the concave and strikes the concaving point A at the speed v_p (particle speed impacted by the hammer) at angle α having the normal at the point of contact, Figure 3.

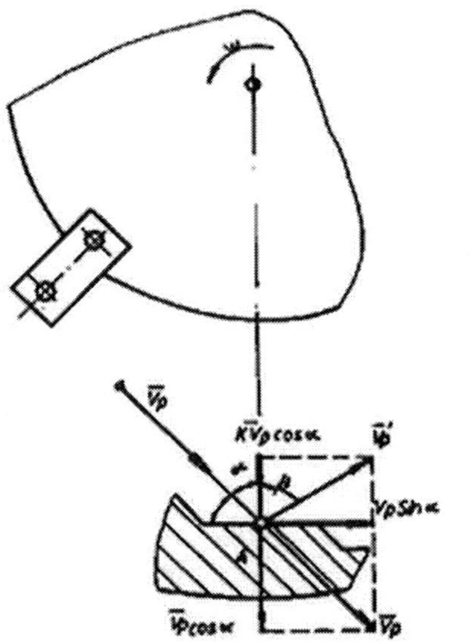

Fig. 3. Scheme of speeds at particle collision with the concave in shredding process [2]

The angle between the normal in the point of contact and the direction of the particle velocity after the collision is noted β

The concave is fixed, it results that its speed before and after the collision is null.

The case to be studied can be assimilated with:

- *the oblique collision, with no friction, of two spheres*, in which a sphere is represented by the concave's wall;
- *the oblique collision, with friction, of two spheres.*

As frictions are neglected, the tangential components of speeds will not change. The component $v_p \sin\alpha$ following the direction of the concave's wall does not change; the other component $v_p \cos\alpha$ changes its sense and as value it becomes $kv_p \cos\alpha$, since $v_{pn} = kv_p$.

After collision, the particle speed will be $v_p^{'}$ and will have the following expression:

$$v_p^{'} = \sqrt{v_p^2 \sin^2 \alpha + k^2 v_p^2 \cos^2 \alpha} = v_p \sqrt{\sin^2 \alpha + k^2 \cos^2 \alpha} \le v_p \text{, (m/s)} \qquad (1)$$

Angle β can be expressed in terms of particle speed before collision as follows:

$$tg\beta = \frac{v_p \sin\alpha}{k v_p \cos\alpha} = \frac{1}{k} tg\alpha \; (2)$$

As in the studied case the coefficient of restitution is $k \le 1$, from relation (2) it results that $\beta \ge \alpha$.

By analyzing relations (1) and (2), it is found that if there is collision of a sphere (fodder particle) with a fixed wall, the particle speed after collision $v_p^{'}$ decreases and the speed vector deviates from the normal.

After the collision with the concave, the particle rebounds. In order for the particle, after hitting the concave, to return to the hammer action area, it is necessary that the normal component of the speed after the collision is maximum. This situation occurs for:

$$\begin{cases} \alpha = \beta = 0 \\ v_p^{'} = v_{n\max}^{'} = -kv_p \\ k = 1 \end{cases} \; (3)$$

In concaves construction, the grooves are designed so that the incidence angle *i* has the maximum value $i = \frac{\pi}{2}$, Figure 4.

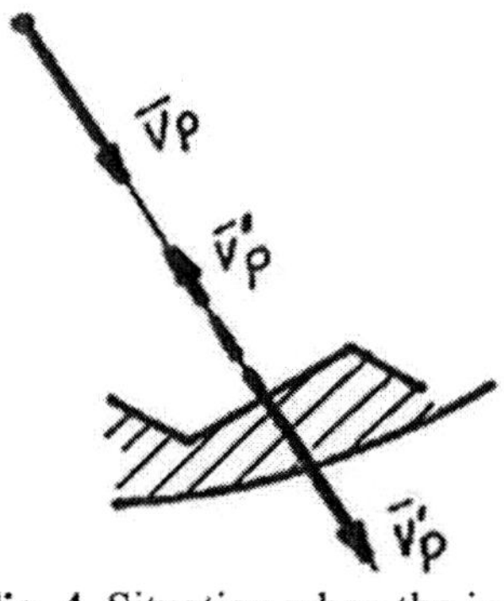

Fig. 4. Situation when the incidence angle has the maximum value [2]

The intensity of the blows (shocks) between the fodder particles and the grooved surface is also influenced by the radial distance between the circle described by the top end of hammers and the grooved surface, which must be as small as possible.

In the case of oblique collision, due to the friction forces, the relative speed and the rebounded particle movement direction correspond only partially to the laws of elastic collision, Figure 5 a and b [10].

In Figure 5 a, it is noted the case when the granule rebounds at an angle smaller than the angle of collision with the hammer.

In the case of breaking, the granules rebound as a dispersion cone with an angle bigger than the collision one (Figure 5b).

After the collision with hammers, the granules obtained hit the inner surfaces of the mill, undergoing a new shredding process.

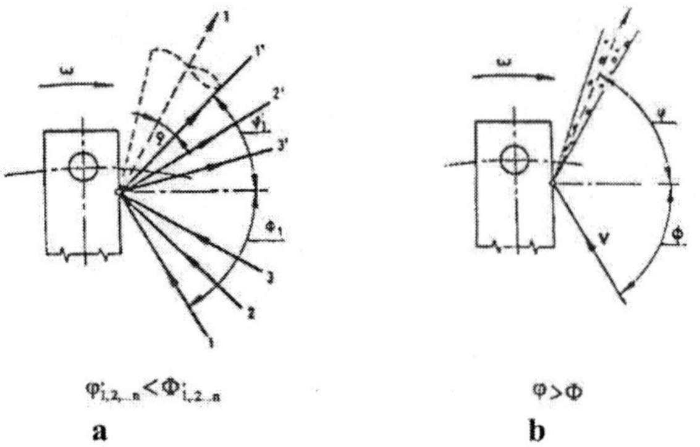

Fig. 5. Oblique collision–rebounding of granules [10]
a) without granule breaking; b) with granule breaking

3 Experimental research

The main objective of the experimental research was to verify the influence of the concaves on the grinding modulus and the power consumption in the process of milling the maize and barley seeds. Experimental research was carried out at the National Institute of Research - Development for Machines and Installations Designed to Agriculture and Food Industry - INMA Bucharest.

Fig. 6. Open chamber mill MC 22

The mill in Figure 6 consists of the following main sub-assemblies: frame, lower housing, upper housing, collecting chamber, assembled rotor, fan, tensioner, 22 kW electric motor.

The parameters that influence the performance of hammer mills are:

Q – feed rate of the mill, kg/h; v_c – peripheral speed of the hammers, m/s; D – diameter of the hammer rotor, m; v_{ds}–movement speed of the seed bed in the milling chamber, m/s; z - no. of hammers; b – hammer width, mm; e - distance between hammers top end and sieve, mm; d - diameter of sieve holes in the milling chamber, mm; β –milling chamber's angle of coverage, degrees; l –hammer length, mm; q_a- the amount of air required to remove the material from the milling chamber, m^3/min ; A –total area of the sieve holes, mm^2;

M –grain milling fineness modulus, mm; G –grain milling granulation, mm; F- seed crushing resistance, kgf/cm^2;E –seed elastic modulus, kgf/cm^2; ρ -density of the material to be milled, kg/m^3;

- the dynamic system rotor-hammers, the concave (grooved surface)

Because not all of the parameters have an equally large influence, determinations have only been made for some of them.

To demonstrate the influence of the concave (grooved surface in the milling chamber) on the cereal grain crushing process, tests were performed on open chamber hammer mill, shown in Figure 6, equipped with concave using the following seeds:

1) Seeds with moisture content:
-maize 13.6%;
- barley13.3%;

2) Mill constructive characteristics:
- sieve with ø3 mm holes;
- sieve with ø5 mm holes;
- peripheral speed 77 m/s;
- angle of coverage 180° and 255°;
- distance between hammers and sieve 10 mm

Grain milling granulation was determined on the basis of the samples taken:100 g grain milling using a sieve battery having the size of the holes: 0.5; 1.0; 1.6; 2.0; 2.5.

Determination of grain milling granulometric composition was achieved by sifting method. This method is most widely used, it is also called the Makarov method quoted by [5] and underlies the determination of the arithmetic average index of particle size. This index is also called the *grain milling modulus M.*

The sifting method consists in passing a defined quantity of particles, in a defined time, on multi stage sieves.

For the correct evaluation of the experiment results, the characteristics of the seeds subjected to the milling operation were determined in the laboratory. The values obtained are shown in Table 1. Experiments were performed for sieves with 180° and 255° angle of coverage. The results of the experiments are shown in chart 1. The tests were carried out with maize and barley seeds.

Table 1. The main characteristics of cereal seeds subject to milling

Seed type	Moisture [%]	Hectolitre mass / [kg/hl)	The weight of 1000 grains [g]	Impurities [%]
Maize seeds	13.6	70.8	284.6	0.85
Barley seeds	13.3	56.9	38.2	13.8

Following the experiments with the above-mentioned equipment, the results presented in the chart in Figure 7 were obtained.

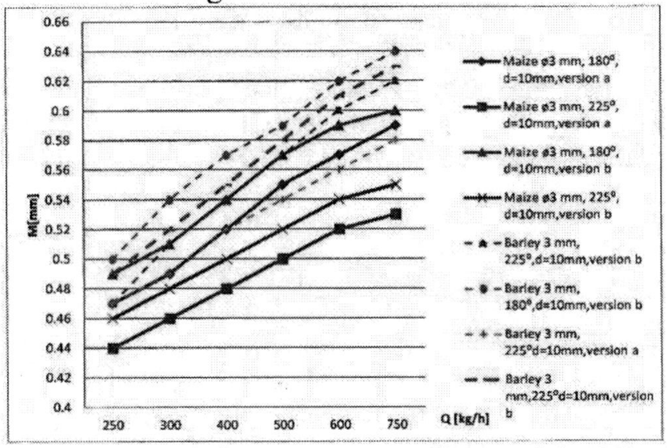

Fig.7. Concaves influence on *grain milling modulus M(mm)* in hammer mill grinding process

Analyzing the chart in Figure 7 it is noticed that:

- the most advantageous mill equipping system is, for maize, a sieve with 255° angle of coverage with concave, version a, and for barley, the sieve with 255° angle of coverage, version a.
- modulus M is better for grinding maize because it has lower crushing resistance than barley seeds.

Following the experiments conducted by various Russian and German researchers it was determined that barley seeds had the greatest resistance to pressing. As a result, it was decided that barley would be the standard for the comparative assessment of the other cereals.

A second series of tests was carried out to determine concaves effect on power consumption, presented in the graph in Figure 8.

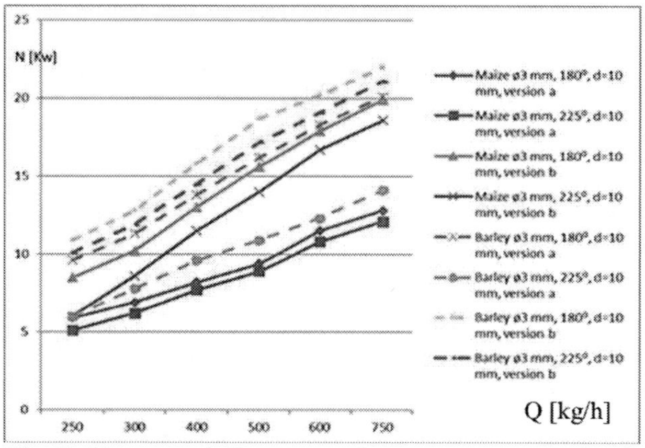

Fig. 8. Concaves influence on the power consumption in hammer mill grinding process

Analyzing the chart in Figure 8 it is noticed that:

- the most advantageous mill equipping system is, for maize, a sieve with 255° angle of coverage with concave, version a, and for barley, the sieve with 255° angle of coverage, version a.

4 Conclusions

The following conclusions can be drawn from the research carried out and the results obtained:

- seed hardness has a particular influence on the ability to form intermediate products, but also a criterion for estimating energy consumption for shredding.
- knowing that cereal seed fodder represent elastic, plastic and viscous bodies, but in the same time complicated colloidal, capillary-porous bodies where each part of the seed behaves differently, the determination of the resistance force for shredding under different working regimes, allows properly setting the constructive and functional parameters of the hammer mills.
- because of the complexity of the experimental determinations in the milling mechanism theory, the static resistance indicators are used, although they are insufficient to appreciate the dynamic working regime of the machine. In order to assess grains resistance to the milling as a whole, both due to the compressive and shear forces of their anatomical parts, the cereal hardness characteristic is used.
- hardness shows to what extent the seeds break, accidentally or when they are shred.

- the most advantageous mill equipping system is, for maize, a sieve with 255° angle of coverage with concave, version a, and for barley, the sieve with 255° angle of coverage, version a.
- it is recommended to enlarge the working surface of the sieve as this influences the grain milling granulation.
- to increase the milling process intensity in the event of hitting the concave, the particles must rebound in the hammer action area;
- concave's grooves must be constructively chosen so that the particles angle of incidence has the maximum value $\frac{\pi}{2}$;

Acknowledgements

This paper was financed by grant of the Romanian Research and Innovation Ministry, through Programme 1 – Development of the national research-development system, sub-programme 1.2 – Institutional performance – Projects for financing excellence in RDI, contract no. 16PFE.

References

1. V. Haţieganu,*Basics of animal nutrition and veterinary control of feed* (Didactic and Pedagogical Publishing House, Bucharest, 1978)
2. A. Păun, *Study on concentrated fodder milling process in hammer mills*, PhD thesis(Polytechnic University of Bucharest, 2004)
3. A.Păun,*Hammer mills. Study on concentrated fodder milling process*, (Terra Nostra, Publishing House, Iasi, 2008)
4. M-V Muntean, O. Marian, O. Ranta, I. Drocas, G. M. Catunescu, *The Influence of Hammer Type Used in Grinding Mills on Grist Fineness*, Bulletin UASVM Food Science and Technology **70**(1), 53-57, (2013)
5. V. Bulgakov, I. Holovach, V. Bandura, S. Ivanovs, *A theoretical research of the grain milling technological process for roller mills with two degrees of freedom*, INMATEH-Agricultural Engineering, **52** no.2, 99-106 (2017),
6. M. Segărceanu, T. Căsăndroiu, *Agricultural mechanics studies and research*, V, (3) (1971)
7. A.Păun, R.Ciupercă, B.Ivancu, Gh.Ioniţă, A.Zaica,*INMATEH* Agricultural Engineering,**43**, 445-451(2014)
8. Gh.Voicu, L. David, V. Safta, *Machine Construction Journal*, 4-5, 23-25, (1998)
9. G.Moiceanu, Gh.Voicu, G.Paraschiv, V.Vlăduţ,P.Cardei, M. Dincă, *Proceedings of the47th International Symposium,Actual Tasks on Agricultural Engineering*, **47**, 399-409 (2019)
10. A. Mikecz,*Azallattenyesztesgepei, MezogazdasagiKiado*, (Budapesta, 1985)

E3S Web of Conferences 112, 03010 (2019) https://doi.org/10.1051/e3sconf/201911203010
TE-RE-RD 2019

Innovative technical equipment for weed control in leguminous crops

Eugen Marin[1,*], *Marinela* Mateescu[1], *Ioan* Roşca[2], *Ana-Maria* Andrei[2], *Vasile* Jinga[2] and *Marga* Grădilă[2]

[1]National Institute of Research-Development for Machines and Installations designed to Agriculture and Food Industry-INMA Bucharest 013813, Romania
[2]Research and Development Institute for Plant Protection-ICDPP Bucharest, 013813, Romania

Abstract. Through its rural development policy, the EU aims to help rural areas cope with the many economic, social and environmental challenges that the 21st century brings. Weeds are a major scourge for farmers who cultivate organic leguminous crops in rural areas, which often have to invest in many weed control methods in order to increase crop yields. The paper presents innovative technical equipment for weed control through a process involving the blasting of young weeds with small fragments of organic material by means of air compressor and granular material distribution system that degrades in depth both the strain and the leaves of the weeds. Organic materials that can be used in weed blasting are walnut shells, granulated corn cobs, glauconite (also found in Romania) and soy flour. An additional advantage of the equipment is the potential for farmers in rural areas to use organic fertilizers, such as soy flour, as blasting material.

1 Introduction

The role of organic farming is to produce food appropriate to the metabolism of the human body in close connection with the development of the environment, in respect for nature and its laws [1].

Combating weeds under the conditions of an organic farming, especially among vegetable crops, must be approached with great care, because it can affect soil quality, degrade organic matter and accelerate the erosion, which leads to influencing production productivity and profitability [2].

The strategies and technologies used by organic farmers vary according to crops, soils, climate and other factors. Currently, vegetable growers based largely on traditional farming practices, such as crop rotation, fertility management, high density cultivation and the use of cover crops [3].

Some of the farmers, vegetable growers in organic farming, manages weeds by combating weed seeds in the critical period of the early season or by physical suppression of weeds with plastic or natural mulch [4].

* Corresponding author: marin_43eu@yahoo.com

Weed blasting is a new weed control technology, which has the potential to reduce the number of agricultural works in the context of organic farming. Some of the vegetable growers are interested in the possibility of using technical equipment that can apply organic fertilizers as abrasive materials to combat weeds and supplement the nutrition of the crops in a single pass [5].

A study based on research carried out over two consecutive years of abrasive weed control in a corn crop with granulated corn cobs has shown that the biomass of weeds can be substantially reduced and can help to maintain production, due to the fact that the cob particles (Fig.1) are valuable first because they are very absorbent, biodegradable, renewable and abrasive [6].

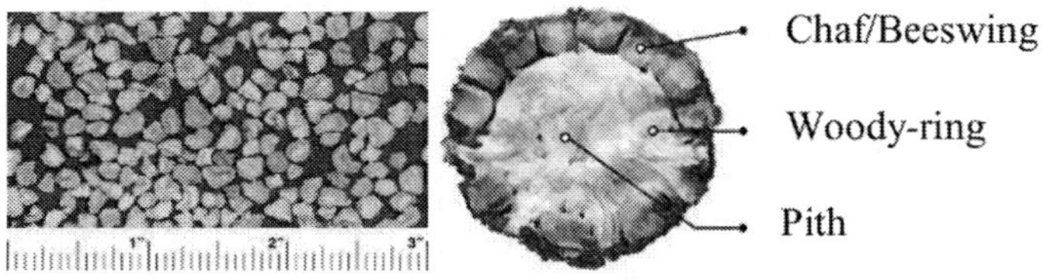

Fig. 1. Granulated corn cobs [7]

In some research papers, it is highlighted that by using FEA, which examines the behaviour of the product in contact with forces, considerable advantages are gained in dramatically reducing the costs of experimental testing and redesign. [8].

Seeking to respond to the demands of farmers who want to take appropriate measures for the correct application of natural, ecological products for combating weeds in vegetable crops through the blasting method, currently INMA in collaboration with ICDPP is conducting under a research contract with the project "*Integrated system environmental management (SIM) of phytosanitary risks by complex methods of sustainable management of agroecosystems*", which led to the design of a blasting equipment.

2 Methodology

For designing the technical equipment for weed blasting, which consists of a blasting machine connected to a compressed air compressor mounted on a frame worn on the three-point suspension mechanism of a 45-hp tractor on wheels, it was used the SOLIDWORKS 3D CAD software, developed by Dassault Systèmes SolidWorks Corporation [9].

The PhotoView 360 tool from the SOLIDWORKS 3D CAD menu allowed you to view the product a step further and make a photorealistic image directly from the 3D CAD model (Figure 2).

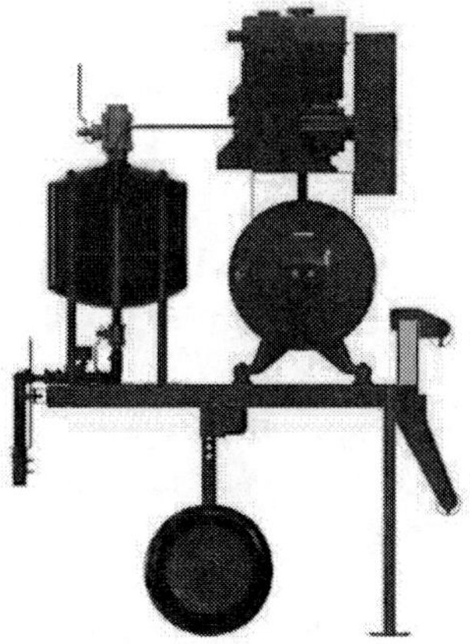

Fig. 2. Equipment for blasting weeding weeds- image taken with PhotoView 360 tool

In the 3D geometric 3D modelling, SOLIDWORKS 3D CAD allowed both basic operations - Extrude, Loft, Rotate, Sweep - in boss and cut variants (addition and cutting) as well as more complex operations such as Wrap, Free Form, Radiate, Surface - Offset, Face - Replace, Heal, or Spline on Surface. SOLIDWORKS 3D CAD is also very well developed from the point of view of geometric references (lines, points, planes, coordinate systems, etc.) that helped the operator both as landmarks and for the correct visualization of the 3D model. Testing in the virtual environment of the assemblies in terms of static load resistance, finite element analysis (FEA) was performed using the SOLIDWORKS® Simulation structural simulation application, which involved importing the geometry of the model made with SOLIDWORKS 3D CAD, defining the material of each component part, defining appropriate meshing restrictions and running linear stress analysis under the effect of an applied load to determine stresses, displacements, safety coefficient and result visualization [10].

Figure 3 shows a sequence from the comparison of the study analysis and results, which appears on the screen in the form of the von Mises voltage distribution, the displacement distribution, the image of the shape deformation and the safety coefficient for the welded frame of the technical abrasive weeding equipment.

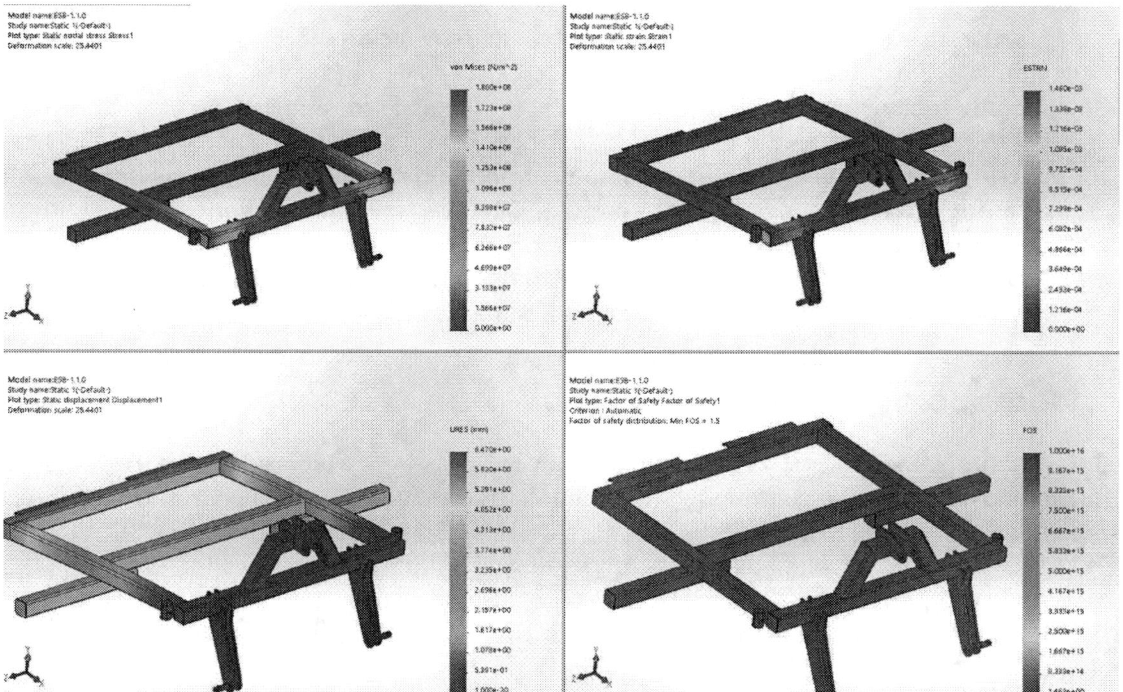

Fig. 3. Sequence that appears on the SOLIDWORKS® Simulation application screen when comparing the analysis study and results for the welded frame

At the same time, in order to carry out research on the experimental lots of SCDL Buzau in transplanted vegetable crops, which are substantially larger than weeds emerged at the time of application, the main component was subjected to the theoretical research to optimize the blasting process.

For uniform distribution of pressure and organic air-material mixture, to be used in weed blasting, theoretical researches were carried out to achieve a Venturi nozzle which has an extensive mark of the blasting jet and an increased speed of the abrasive by up to 100% at a given pressure.

Nozzle conceived is characterized by the flow geometry that produces a compact, controlled distribution beam in the form of a parabola of the sandblast jet and insensitive to pressure fluctuations. To perform the simulation has been used the SOLIDWORKS® Flow Simulation module, which is an intuitive Dynamics Fluid Dynamics (CFD) solution embedded in SOLIDWORKS 3D CAD [11]. Initially, was made the 3D model of the nozzle, after which were analysed the flux conditions. The finite element method was initiated by establishing the computing field, fluid subdomains, boundary conditions, and the networking. After running the simulation, it was possible to visualize the pressure and flow velocity through the nozzle (Figure 4).

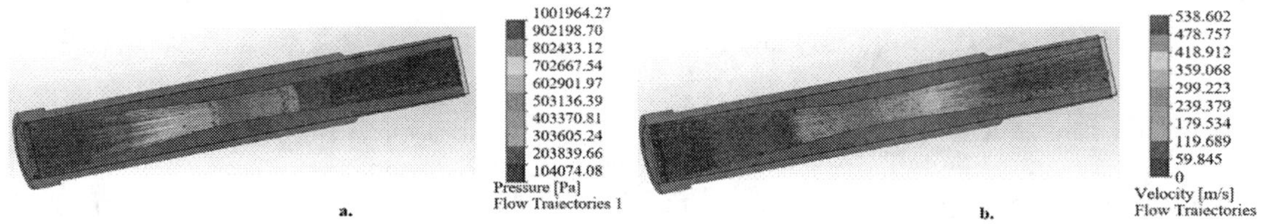

Fig. 4. Sequences of pressure (a) and speed (b) of organic air-material mixture in the nozzle

3 Conclusions

Studies conducted, by using CAD-assisted design software and CAE-assisted simulation, have allowed optimal design of the sub-assemblies and component parts of the weed abrasive weeding equipment.

This work can be developed using other optimization algorithms and other modelling techniques.

This work was supported by a grant of the Romanian Ministry of Research and Innovation, PCCDI – UEFISCDI, project number PN-III-P1-1.2-PCCDI-2017-0301/28PCCDI/2018.

Also, this work was supported by a grant of the Romanian Research and Innovation Ministry, through Programme 1 – Development of the national research-development system, subprogramme 1.2 – Institutional performance – Projects for financing excellence in RDI, contract no. 16PFE.

References

1. M. Lähdesmäki, M. Siltaoja, H. Luomala, P. Puska, S. Kurki, Journal of Rural Studies, **65**, 152-160, (2019)
2. Jerkins, D, and J. Ory, National Organic Farmer Survey and Listening Sessions. Organic Farming Research Foundation (www.ofrf.org), 128, (2016)
3. B.P. Baker and C.L. Mohler, Renewable Agriculture and Food Systems **30**, 418–427, (2015).
4. B. Brown and E.R. Gallandt, Renewable Agriculture and Food Systems, 1-7, (2017)
5. S. E. Wortman, Crop Protection, **77**, 157-162 (2015)
6. M. Erazo-Barradas, C.N. Friedrichsen, F. Forcella, D. Humburg, and, S.A. Clay, Renewable Agriculture and Food Systems 1–8, (2017)
7. *** http://www.greenproducts.com/greentru/corncob-products/
8. S.Şt. Biriş, N. Ungureanu, D. Cujbescu, INMATEH Agricultural Engineering, **57**, 261-270, (2019)
9. *** https://www.solidworks.com/product/solidworks-3d-cad
10. G. Gheorghe, C. Persu, I. Gageanu, D. Cujbescu, Proceedings of the 46th International Symposium ATAE, 79-86, (2018)
11. *** https://www.solidworks.com/product/solidworks-flow-simulation

Researches on the separation of impurities from a seed flow based on the aerodynamic principle

Gheorghe Stroescu[1,*], *Anişoara* Păun[1], *Gheorghe* Voicu[2], *Carmen* Brăcăcescu[1], *Radu* Popa[1] *and Vlad* Popa[1]

[1]National Institute of Research - Development for Machines and Installations designed to Agriculture and Food Industry - INMA, Bucharest 013813, Romania
[2]University Politehnica of Bucharest/Faculty of Biotechnical Systems Engineering, Romania

Abstract. Ensuring for production the necessary quantities of biological material of valuable varieties and hybrids to maintain their initial characteristics requires the permanent application of measures to be carried out on a scientific basis, depending on the biological particularities of the variety or hybrid. The production of seeds for leguminous plants is of great importance because the vast majority of species propagate that way. Due to the fact the yield obtained depends on the quality of the seeds, it is necessary to use seeds of varieties and hybrids with superior qualities and from higher biological categories without foreign bodies (impurities). Knowing the particularities of leguminous plant seed is of special practical importance because the conditioning technologies and machinery are based on it and also the culture technologies that are applied both in the field and in protected areas. The paper presents some theoretical aspects regarding the separation of seed mixtures based on the aerodynamic principle (resistance of the seed mass layer to the air passage, seed floating capacity, air flow working rates, total air flow, physical characteristics of seeds, etc.) which were the basis for designing a conditioning module for vegetable, flower and cereal seeds.

1 Introduction

Producing seeds for leguminous plants is of a major importance, as the majority of species propagate that way. Due to the fact the yield obtained depends on the quality of the seeds, it is necessary to use seeds of varieties and hybrids with superior qualities and from higher biological categories.

Knowing the particularities of leguminous plant seed is of special practical importance because it is the basis of the conditioning technologies and also of the culture technologies (that are applied both in the field and in protected areas). Seed quality is a rather relative concept, due to the fact that some species seeds can have different destinations: food or fodder consumption, sowing or industrialization [1, 2].

* Corresponding author: *gigistroescu62@gmail.com*

Knowing the aerodynamic characteristics of grains (terminal velocity, drag coefficient) is significant for the construction and operation of machines, which treat substances with air flow, and in all cases when substances are moved in the air [3,4,5]. Some studies showed that aerodynamic properties of an agricultural product (in the form of seeds) also changes according to its mass, volume, equivalent diameter, shape and projected area of the product that many researchers have studied for different products at different moisture content [6, 7].

There are various researches on the aerodynamic properties of agricultural product, in the form of non-spherical and spherical particles, which considered factor of its form and volume and stated the drag coefficient as a function of Reynolds number [8].

The paper presents some theoretical relevant aspects related to the separation of seed mixtures based on difference in aerodynamic properties and the movement of the particles in an air flow, aspects which underlay the elaboration of an modern technology for obtaining the seed material and the design of a Conditioning Module for vegetable, flower and cereal seeds, efficient world-wide, intended for sorting and cleaning seed mixtures [9].

2 Materials and methods

During separation, a mass particle *m* is introduced in the area affected by the air flow (Figure 1) and it moves in this air flow under the action of gravity force *G* =*mg* and resistance force *R*.

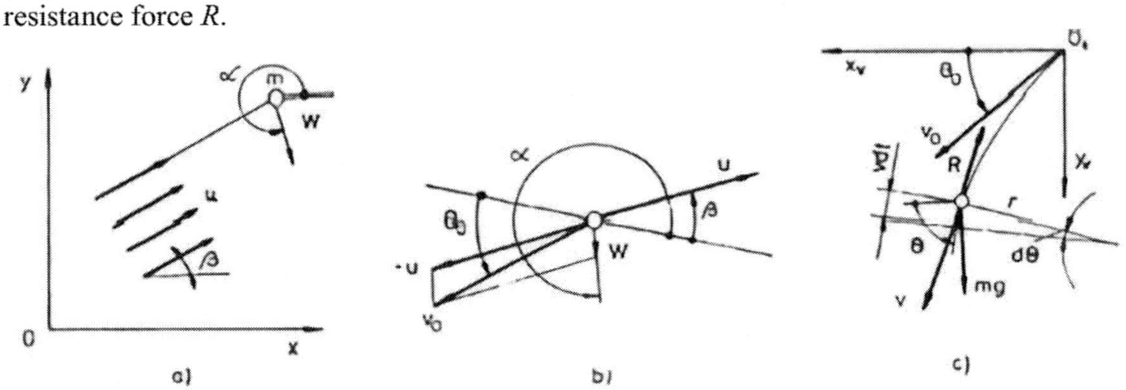

Fig. 1. Scheme of particles movement in a uniform air stream:
a) the introduction of the mass particle *m* in the interaction zone with the air flow; b) estimating the relative velocity of particle movement *v* by introducing it in the area affected by the flow; c) the effect of the flow and particle interaction on the relative movement trajectory

The resistance force direction is opposite to the moment of relative velocity *v* and its value is proportional to the second power of velocity kv^2. When the particle enters the air flow, it moves to the relative velocity *v*, the value and direction of which are described as the resultant of the parallelogram constructed using the vectors - $\bar{u}$ si $\bar{\omega}$.

From Figure 1 it results that:

$$v_0 = \sqrt{u + \omega^2 - 2u\omega\cos(\alpha - \beta)}, \text{ m/s} \tag{1}$$

Considering the law of sines, the following can be obtained:

$$\theta = \beta - arcsin\left[\frac{\omega}{v_0}\sin(\alpha - \beta)\right] \tag{2}$$

Assuming that the axes of the Cartesian coordinate system move with the air flow, the equations of the particles relative movement become:

$$\frac{dv}{dt} = gsin\theta - kv^2 \tag{3}$$

$$\frac{v^2}{r} = gcos\theta \tag{4}$$

where: r – radius of curvature of trajectory, θ - angle between axis x^v and the tangent to the relative velocity *v*.

Dividing equation (3) by (4) and considering that $vdt = r\, d\theta$, we get the velocity equations:

$$\frac{dv}{v^3 dv} = \frac{tg\theta}{v^2} - \frac{k}{gcos\theta}$$

$$-\frac{2dv}{v^3 cos^2\theta} = \frac{k}{g}\left(\frac{2d\theta}{cos^3\theta}\right) \tag{5}$$

where: k – coefficient of velocity.

For an air flow velocity equal to the particles critical velocity we can accept $k =$ constant.

Integrating equation (5) we get:

$$\frac{1}{v^2 cos^2\theta} + c_2 = \frac{k}{g}\left[\frac{sin\theta}{cos^2\theta} + lntg\left(\frac{\theta}{2} + \frac{\pi}{4}\right)\right] \tag{6}$$

The significance of constant c_2 can be found based on some initial conditions that are: $t=0$, $v=v_o$, $\theta=\theta_o$ as follows:

$$c^2 = \frac{k}{g}\left[\frac{sin\theta_0}{cos^2\theta_0} + lntg\left(\frac{\theta_0}{2} + \frac{\pi}{4}\right)\right] - \frac{1}{{v_0}^2 cos^2\theta_0} \tag{7}$$

From equations (6) and (7) we obtain a formula that helps to get the relative velocity *v* according to the angle θ.

$$v = \frac{1}{cos\theta\sqrt{\frac{k}{g}\left[\frac{sin\theta}{cos^2\theta} + ln\ \left(\frac{\theta}{2} + \frac{\pi}{4}\right)\right] - c_2}}, \text{ m/s} \tag{8}$$

Since the relative velocity *v* is a quantity of function θ, it results:

$$t = \int_{\theta_0}^{\theta} \frac{d\theta}{gcos^2\theta\sqrt{\frac{k}{g}\left[\frac{sin\theta}{cos^2} + lnt\ \left(\frac{\theta}{2} + \frac{\pi}{4}\right]\right] - c_2}} \tag{9}$$

The relative movement is described by the expression:

$$S_v = \int vdt = \frac{1}{2k} ln\left\langle\left[\frac{sin\theta}{cos_\theta^2} + lntg\left(\frac{\theta}{2} + \frac{\pi}{4}\right)\right] - c_2\right\rangle - c_3 \tag{10}$$

where

$$c_3 = \frac{1}{2k} ln \frac{1}{v_0^2 cos^2\theta_0} \tag{11}$$

At this stage we can estimate the particle transition during the relative movement with reference to the Cartesian coordinate system as follows:

$$x_v = \int vcos\theta\, dt = \int_{\theta_0}^{\theta} \frac{d\theta}{gcos^2\theta\left\langle\frac{k}{g}\left[\frac{sin\theta}{cos_\theta^2} + ln\ \left(\frac{\theta}{2} + \frac{\pi}{4}\right)\right] - c_2\right\rangle} \tag{12}$$

$$y_v = \int vsin\theta dt = \int_{\theta_0}^{\theta} \frac{tg\theta d\theta}{gcos^2\theta\left\langle\frac{k}{g}\left[\frac{sin\theta}{cos_\theta^2} + lntg\left(\frac{\theta}{2} + \frac{\pi}{4}\right)\right] - c_2\right\rangle} \tag{13}$$

Thus, the equations of relative movement of the particles under unrestricted air flow and resistance forces can be described as follows:

$$\frac{dv}{dt} = -kv^2 \tag{14}$$

$$\frac{v^2}{r} = 0 \tag{15}$$

From equation (14), after integration, we get the expression for *v*:

$$v = \frac{v_0}{1+v_0 kt} \text{, m/s} \quad (16)$$

But the value of v_o has to be calculated from equation (1).

From equation (15) it results that $r=\infty$ and it means that θ=cons and that the trajectory of the particle relative movement is a straight line. If the coefficient of volatility k is constant., we can easily obtain the expression that estimates the projection of the particle movement in absolute movement:

$$\frac{dx}{dt} = ucos\beta - \frac{v_0 cos\theta}{1+v_0 kt} \quad (17)$$

$$\frac{dy}{dt} = usin\beta - \frac{v_0 sin\theta}{1+v_0 kt} \quad (18)$$

where:

$$\theta = \beta - arcsin\left[\frac{\omega}{v_0}\sin(\alpha - \beta\right]$$

By integrating equations (17) and (18) we obtain:

$$x = utcos\beta - \frac{cos\theta}{k} ln|1 + v_0 kt| \quad (19)$$

$$y = utsin\beta - \frac{sin\theta}{k} ln|1 + v_0 kt| \quad (20)$$

3 Results and discussions

Equations (19) and (20) can be used in pneumatic separation processes, especially when calculating the movement parameters of the various components in the initial seeds mass, although air flow velocity may be several times higher than the critical velocity of the particles to be separated and sorted. It should be emphasized that those mathematical expressions have non-linear interdependencies and is virtually impossible to finding an analytical solution. Thus, in order to solve them practically, the transformation into computerized modelling is necessary.

It can be concluded that in order to carry out the separation process by aerodynamic proprieties of a mixture of solid particles (seeds in particular) one has to take into account a series of factors that have a major impact on the way of its development, through which it is highlighted [1, 10, 12,15]

- properties of solid particles subjected to aerodynamic separation process;
- air flow direction and flow regime;
- mode of realization of the biphasic gas-solid mixture (G-S).

The resistance of the product mass layer to the passage of air shall be calculated with the relation:

$$R = A \cdot h \cdot W^{n} \text{, mmH}_2\text{O} \quad (21)$$

where: h - thickness of product layer, in m;

W - conventional air velocity relative to the whole section of the seed layer, in m/s;

A, n - coefficients determined experimentally according to the physical characteristics of the seeds.

The aerodynamic properties of the seeds have not yet been determined with sufficient precision since they depend on a number of factors, the mutual connection of which cannot be taken into account with sufficient precision.

The floating velocity depends on the specific weight, the surface condition, the shape and geometric dimensions of the particles of the seed mixture.

Table 1 shows, according to the literature [2, 3, 6], the aerodynamic properties for several types of agricultural products (in the form of seeds or grains).

Table 1. Aerodynamic characteristics for several types of agricultural products

Product	Aerodynamic properties		
	Specific weight (kg/m³)	**Aerodynamic drag coefficient**	**Floating velocity (m/s)**
Wheat	1.216	0.184 – 0.265	8.9 – 11.5
Rye	1.210	0.16 – 0.22	8.3 – 9.89
Barley	1.200	0.191 – 0.272	8.08– 9.11
Maize	1.238	0.16 – 0.24	12.48 – 14.03
Soybean	1.092	0.115-0.152	17.25-20.16
Pea	1.264	0.190-0.229	15.50-17.50
Bean	1.220	0.159-0.259	12.48-14.03

For precleaning or cleaning, the air flow can be determined with the relation:

$$Q_{ap} = k_u \cdot S_{cp} \cdot v_{a1} \text{ , m}^3\text{/s} \tag{22}$$

while for sorting, it can be determined with the relation:

$$Q_{as} = k_u \cdot S_{cs} \cdot v_{a2} \text{ , m}^3\text{/s} \tag{23}$$

where: k_u is the coefficient of the air flow on the sieve (k_u=0.65);
v_{a1} and v_{a2} – air stream velocities, m/s;
S_{cp} and S_{cs}- sections of channels for the transport of separated fractions, m² that are determined by relations:

$$S_{cp} = \frac{Q\cdot(100-q_{cp})}{100\cdot q_s} \text{ , m}^2 \tag{24}$$

$$S_{cs} = \frac{Q\cdot(100-q_{cp}-q_{cs})}{100\cdot q_s} \text{ , m}^2 \tag{25}$$

where: Q is the feed flow rate of the installation, m³/s;
q_{cp} and q_{cs}- flow rates of the fractions that must be eliminated;
q_s- product (seed mixture) specific feed rate for an installation that is supplied with an air flow rate of 1 m³/s.

The total air flow rate (m³/s) required by suction cleaning separator can be determined by the relation:

$$Q_a = \sum_{i=1}^{n} Q_{si} + \sum_{i=1}^{n} Q_{ti} \text{ , m}^3\text{/s} \tag{26}$$

where:

Q_{si} is the air flow rate required for impurities separation in step i (for an installation with n separation steps);

Q_{ti}- the air flow rate required to transport the product (seed mass) from step i to the next step.

Seed conditioning module for leguminous species MCSL, designed and made by INMA Bucharest can perform both the cleaning of vegetable seeds from light impurities and the sorting of the product, according to the technological flow diagram in Figure 2.

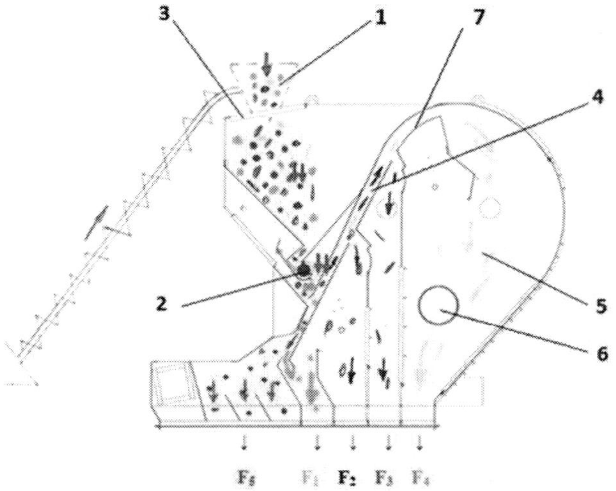

Fig. 2. Technological flow scheme of seed conditioning module MCSL

The product (seed mass) is brought into the feed funnel **(1)** from where it is distributed evenly through the sorting column **(4)** by means of cylinder **(2)** and the feed flap **(3)**. Under the influence of the ascending air flow produced by the fan **(6)** within the floating velocity limits, the heavier bodies fall to the bottom obtaining fraction F_1, while the lighter ones rise to the first orifice where fraction F_2 is separated, to the second orifice where the fraction F_3 is separated or leaves the column at the top entering the decanting chamber **(5)** where F_4 fraction of light bodies (impurities) is deposited. The sucked air from the decanting chamber is fed by the fan into the second decanting chamber where dust and other very light impurities are separated forming the fraction F_5. The adjustable wall **(7)** is designed to modify the sorting column section **(4)** to influence the ascending air flow velocity for seed sorting through four separation windows.

Fig. 3. 3-D representation of the sorts discharge system

4 Conclusions

In order to achieve the separation of a solid particles (heterogeneous mixture) in good conditions it is necessary to know, for each component of the mixture, their properties;
The sorting of solid particle mixtures according to the aerodynamic properties of the components is used in order to increase the purity degree of the basic product;
The aerodynamic separation process is a complex process due to the fact that it is influenced by a number of factors:

- the physico-mechanical characteristics of the solid particles contained in the heterogeneous mixture subjected to the aerodynamic separation process;
- air flow direction;
- constructive and functional characteristics of the installation performing the aerodynamic separation.

The MCSL module made by INMA Bucharest is designed to improve the seed conditioning process, an operation performed only by using air flow.

Acknowledgements

This work was financed with:

- the support of the Romanian Ministry of Research and Innovation, Programme P1 – Development of the National R & D system; Subprogram 1.2. Institutional Performance Institutional Development Projects - Complex Projects Realized in RDI Consortia, project: FUTURE FARMING - ENSURING ENVIRONMENTAL SUSTAINABILITY BY INCREASING IMPROVEMENT COMPETITIVENESS FOR ORGANIC FARMING, Contract: 14PCCDI/06.03.2018), project component 4 "Making efficient the technology of producing ecological seeds for legume species by achieving a module of seed conditioning"- INMA Bucharest,

- a grant of the Romanian Ministry of Research and Innovation, through Programme 1 – Development of the national research-development system, subprogramme 1.2 – Institutional performance – Projects for financing excellence in RDI, contract no. 16PFE.

References

1. N. Bucurescu, D. Roman, *Seed and its preparation for sowing*, Ceres Publishing House, Bucharest (1992)
2. L. Ioancea – Conditioning *and superior capitalisation of plant raw materials for food purposes. Technology and installation*, Ceres Publishing House, Bucharest (1988)
3. B Y. Gorial, J R. O' Callaghan, Journal of Agricultural Engineering Research, **46**(3), (1990).
4. R.K. Gupta, G. Arora, R. Sharma, , Journal of Food Engineering, **79**, (2007)
5. A.D. Zewdu, Biosystems Engineering, **98**, (2007)
6. P. H. Mortensen, H. I.Andersson, J. J. J. Gillissen and B. J Boersma., Physics of fluids, **19**, (2007)
7. E. Moşneguţu, *Contributions to the aerodynamic sorting of agricultural products*, "Gh. Asachi" Technical University of Iaşi (2006)
8. NN. Mohsenin, *Physical Properties of Plant and Animal Materials* (Handbook). (2nd ed.). Gordon & Breach (1986)
9. A. Păun, C. Brăcăcescu, D. Milea, G. Bunduchi, Proceedings of 16-th International Scientific Conference Engineering for Rural Development, Jelgava, Latvia (2017)

 https://doi.org/10.1051/e3sconf/201911203012

Researches on the capitalization of some medicinal plants to obtain essential oils by using the performant innovative installation EUV 2x1000 type

Ion Grigore[1,*], *Andreea Iulia* Grigore[1], *Laurentiu-Constantin* Vlăduţoiu[1], *Mario* Cristea[1], *Elena* Sorica[1], *Dragos* Dumitru[1], *Florin* Sirbu[1], *Gabriel Badea*[3], *Marga* Grădilă[2], *Marius* Raischi[3], *Denisa* Duţă[4] and *Roxana Andreea* Sandu[5]

[1]INMA, TestingDepartment, Ion Ionescu de la Brad Blv. No. 6, Sector 1, Bucharest, Romania

[2]ICDPP Bucharest, Romania

[3]INCDPM Bucharest, Romania

[4]IBA Bucharest, Romania

[5]INCDTP Bucharest, Romania

Abstract. The article includes research on the production of essential oils from medicinal plants by applying water vapour distillation, the method most commonly used for the extraction of essential oils from plants. The results consist of extracting the essential oil from 2 types of medicinal plants: lavender and mint with the help of the EUV installation. The paper highlights the important aspects of obtaining quality products, on economically advantageous terms, addressing both specialists, farmers in larger farms and family associations. The article presents the experimental research results on obtaining essential oils from medicinal plants. Oils obtained by this method contain only volatile components, while wax and volatile resins are not found in the finished product.

1 Introduction

Medicinal plants are a real source of health. They have healing properties since ancient times and even today they enjoy special attention. Aromatic medicinal plants are found in nature as well as in cultures set up by small producers. Modern medicine recommends them as a natural remedy easy to obtain, cheap, effective and harmless. This is why in our country there is a large campaign for harvesting and capitalizing of medicinal plants, regardless of the season [1, 2]. Through the richness of terpenes and phenols, in alcohols and aldehydes, flavoring medicinal plants have demonstrated, over time, therapeutic antiseptic effects, so volatile plant oils administered in useful therapeutic doses have low toxicity, do not produce side effects and do not have contraindications (if administered

* *Ion* Grigore: ionica_grigore2001@yahoo.com

correctly). From a chemical point of view, volatile oils are complex mixtures of aliphatic and aromatic hydrocarbons, aldehydes, alcohols, esters and other constituents, but predominantly terpenoid compounds. In the past, they were called etheric oils, essential oils or aromatic essences [3, 4, 5]. Essential oils have been used for thousands of years for their healing and purifying effects on the body. They have always held an extreme value for doctors and ancient healers who have used them for aromatherapy, consumption and application on the skin. Research has now revealed its remarkable healing properties [6, 7].

Essential oils are concentrated by strong antioxidants. They come from raw plant material such as flowers, leaves, wood, bark, roots, seeds and peels [8, 9, 10]. Although they are called oils, these substances do not contain fatty substances: a drop of essential oil placed on a sheet of paper will not leave any trace, as opposed to a vegetable oil.

From the very wide range of medicinal and aromatic plants, we will focus on two of the most appreciated and used plants in our country, with high essential oil content: lavender (Lavandula angustifolia) (Figure 1) and mint (Mentha piperita) (Figure 2). Lavender is a perennial plant, considered a small bush, because the branches are wooded. It has evergreen leaves, even in winter, usually the leaves being green-silver, but there are also varieties of green or silver leaves. They reach maturity in 2-5 years, and can live without problems for 20-30 years. In our country, the most common kind of lavender is the narrow-leaved, 3-4 mm wide and 30-40 mm long, called Lavandula angustifolia. To establish a lavender plantation on a 1ha area, are needed about 18,000-20,000 seedlings that reach maturity after 3 years and production of green flower (fresh) can reach 4,000-4,500 kg / ha. If summer harvesting is done properly, you can also perform an autumn harvest from which you could get another 1,500-2,000 kg of lavender flower. Lavender essential oil is very used in medicine, cosmetics and perfumery.

Between the plants whose chemical composition can generate a significant amount of volatile oils and floral water, lavender occupies a leading position [1, 3, 4].

Fig. 1. Lavender (Lavandula angustifolia) [11]

Mint (Mentha piperita) (Figure 2) is one of the oldest known medicinal herbs. "Papyrus Ebers" mentions that mint was used with 1550 years before our time in Egypt. In Europe it was first cultivated in England (1760), where the first quantities of volatile oil were obtained [12]. The largest volatile oil production (3/4 of world production) is provided by the United States. To the world production of volatile oil of mint competes England, France, Italy, Bulgaria, Russia and Japan. Mentha piperita is a sterile hybrid, which in agricultural practice is propagated exclusively by vegetation, using the underground stools. Sometimes, for scientific purposes, mint is propagated by aerial stools, by the rooting of stem or even leaves [13]. Unlike many other plants and essential oils, the benefits of mint oil and mint have been studied and confirmed by the scientific community. Mint oil contains many minerals and nutrients, including iron, manganese, magnesium, calcium, folic acid, potassium and copper. It also contains omega-3 fatty acids, vitamin A and C. Mint harvesting can be done when the leaves reach 5-6 cm and a 50% inflorescence, time when the essential oil quantity is optimal. The ideal time for harvesting is in the morning between 10-14 o'clock in sunny weather, without dew, wind or fog [14].

Fig. 2. Mint (Mentha piperita) [15]

Removing a flavor from its natural environment is not an easy operation. It must be isolated or concentrated with a minimum of loss and compositional changes and without introducing new impurities.

Numerous factors can interfere and influence the quality of the distillation and the oils obtained. The operator performing the distillation must documents itself upon:

- the choice of distillation method for the chosen plants;
- the rules to be respected for the distillation of the retained plant;
- the quality of the flower which has effects on efficiency and quality, generally.

If the flower is collected at maturity, then the yield of distillation as well as the quality of the oils will be better. It can be seen an increase in yield and quality of up to 25%.

The time of collection, the quality of the transport materials (aerated bags made of natural materials), the storage time, the time between collection and distillation, the duration of distillation, the distillation temperature, the degree of cleaning of the alembic, the quality of the water required for the steam, the condensation temperature, all influence the yield and quality of the oils obtained.

The user must test the installation with his harvested plants to determine the best distillation parameters.

2 Materials and Methods

Technologies for obtaining volatile oils

The industrial extraction of volatile oils from various aromatic plants is performed according to different methods depending on their characteristic properties.

Each method is important and has its place in the production process of essential oils.

The main methods of obtaining volatile oils are [16]:

- Water vapor distillation;
- Extraction of volatile oils with organic solvents;
- Extraction with animal fats: enfleurage and maceration;
- Extraction with liquefied gases (supercritical CO2);
- Extraction by pressing;
- Adsorption on an adsorbent material.

Water vapor distillation is the most commonly used method for extracting volatile oils from most aromatic plants; it is used for industrial scale extraction, but equally effective can be used at laboratory level.

Using the EUV installation (Figure 3), the steam produced by the steam generator enters the vegetative mass, undergoes distillation, destroys the olefin glands cover, volatilizes the oil and then mixes with it. The mixture of water vapor and oil vapor passes into the refrigerant (condensation vessel), where it turns into a mixture of water and volatile oil. The maximum pressure inside the distillation vessel should not exceed 0.2 bar. A clean steam at a pressure of 0.18 - 0.20 bar, maximum, it is used. Distillation is generally ideal when the cooling water temperature is maintained between 24 and 35 °C. This mixture reaches the florentine vessel (separation vessel) where the separation occurs, namely, the volatile oil, being easier, deposit into a film above the floral water.

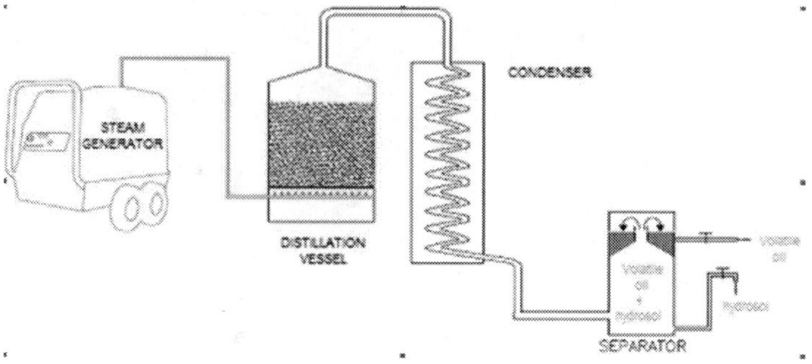

Fig. 3. Technological scheme of EUV distillation equipment

List of components of the EUV installation

- Stable metal supports for distillation vessels and condenser (cooling vessel);
- Steam generator with pressure regulator;
- One or two 1000 liter distillation vessels equipped with steam inlet/outlet sockets, drain tap and two support sieves for plant;
- Cover for the distillation vessel fitted with: 0.5 bar safety valves and a thermo-manometer;
- Condenser (cooling vessel with coil);
- Essences Separator (Florentine vessel);
- Food grade hose - Vaporcord Alim DN or food grade INOX pipe;
- L-type water tap for changing the direction of the flow (EUV-2X1000) (Figure 5).

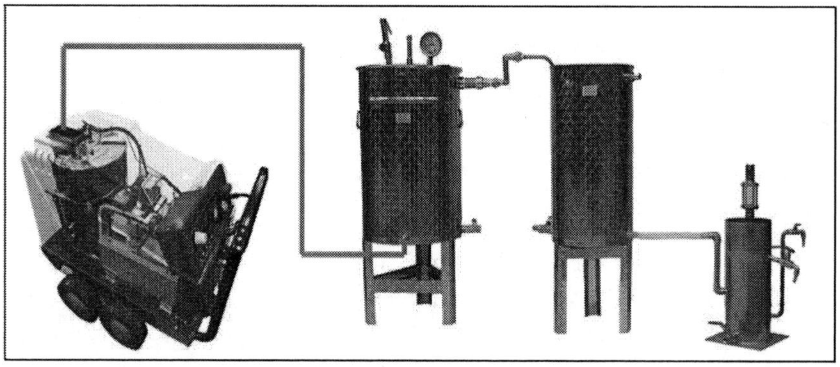

Fig. 4. The water vapor distillation installation EUV-1000

The EUV installation can work with a single distillation vessel fig. 4 or alternatively with two vessels Figure 5. In the case of two distillation vessels, the technical equipment is additionally equipped with two three-way stainless steel L-type water taps, with hoses (stainless steel pipe) for connection to the distillation vessels and the condenser. A water tap shall be installed at the outlet of the pressure regulator fitted on the steam generator so that by changing the outlet path, alternate steam supply through the hoses (stainless steel pipe) of the distillation vessels will be carried out. The second water tap is mounted at the inlet of the cooling vessel and by changing the direction of the flow through the L-type water tap, the alternate passage from one distillation vessel to the other towards the condenser (cooling vessel) is made. All metallic components of the installation are made of food grade INOX.

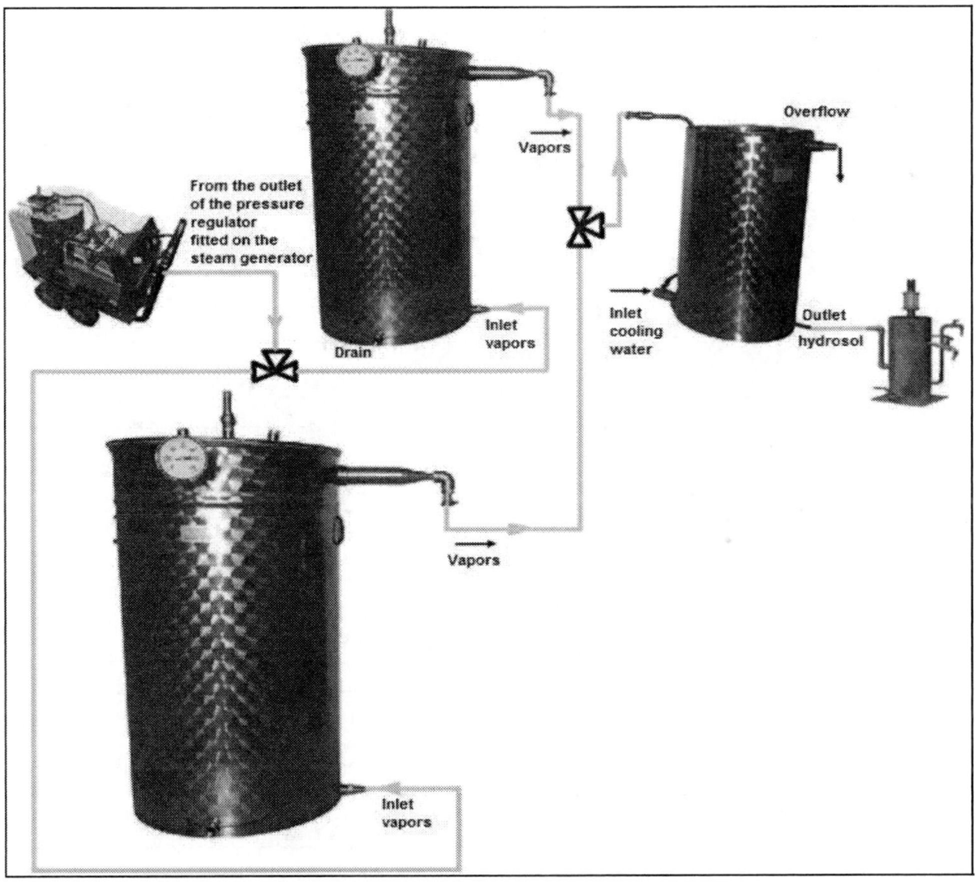

Fig. 5 The water vapor distillation installation EUV-2x1000

3 Results

The trials were carried out in Alba county with plants harvested from two lavender plots, respectively mint. Following the trials made, both with freshly harvested plants and dried plants in different amounts, the following results were obtained: with dry lavender, using between 30-90 kg, there were obtained between 0.08-0.27 l of oil (Table 1); lavender freshly harvested, using between 30-120 kg, there were obtained between 0.10-0.45 l of oil (Table 2); dried mint, using between 30-95 kg, there were obtained between 0.12-0.41 l of oil (Table 3) and fresh oil, using between 30-125 kg, there were obtained between 0.17-0.78 l of oil (Table 4). The amount of oil that can be obtained from distillation is greatly influenced by the quality of the plants under distillation.

Table 1 Dry lavender

Raw material	Quantity (kg)	Temperatures °C						Time (min)	Floral water (l)	Essential oil (l)
		Air	Cooling water	Steam		Interior vessel	Inlet cooling vessel			
				Outlet generator	Inlet vessel					
dry lavender	30	24	22	124	114	105	89	115	30	0.08
dry lavender	60	24	26	125	114	105	89	115	30	0.16
dry lavender	90	26	23	125	114	104	89	120	30.5	0.27

Table 2 Lavender freshly harvested

Raw material	Quantity (kg)	Temperaturi °C						Time (min)	Floral water (l)	Essential oil (l)
		Air	Cooling water	Steam		Interior vessel	Inlet vas cooling vessel			
				Outlet generator	Inlet vessel					
lavender freshly harvested	30	18	18	125	114	105	89	115	30	0.10
lavender freshly harvested	60	20	24	125	114	104	89	120	30.5	0.21
lavender freshly harvested	90	23	25	125	114	104	88	120	30.5	0.33
lavender freshly harvested	120	28	26	125	114	104	87.5	120	31	0.45

Table 3 Dry mint

Raw material	Quantity (kg)	Temperatures °C						Time (min)	Floral water (l)	Essential oil (l)
		Air	Cooling water	Steam		Interior vessel	Inlet cooling vessel			
				Outlet generator	Inlet vessel					
dry mint	30	19	19	125	114	105	89	120	30.5	0.12
dry mint	65	23	20	125	114	105	89	115	30.5	0.25
dry mint	95	26	24	125	114	104	88	120	31	0.41

Table 4 Mint freshly harvested

Raw material	Quantity (kg)	Temperatures °C						Time (min)	Floral water (l)	Essential oil (l)
		Air	Cooling water	Steam		Interior vessel	Inlet cooling vessel			
				Outlet generator	Inlet vessel					
mint freshly harvested	30	21	24	125	114	105	89	115	30	0.17
mint freshly harvested	65	25	24	125	114	104	88	120	30.5	0.37
mint freshly harvested	95	25	25	125	114	104	87.5	120	31	0.56
mint freshly harvested	125	26	23	125	114	104	87.5	125	31.5	0.78

The productivity differences recorded for lavender and mint, as a result of the use of both dry material and fresh material, are shown in Figures 6 and 7.

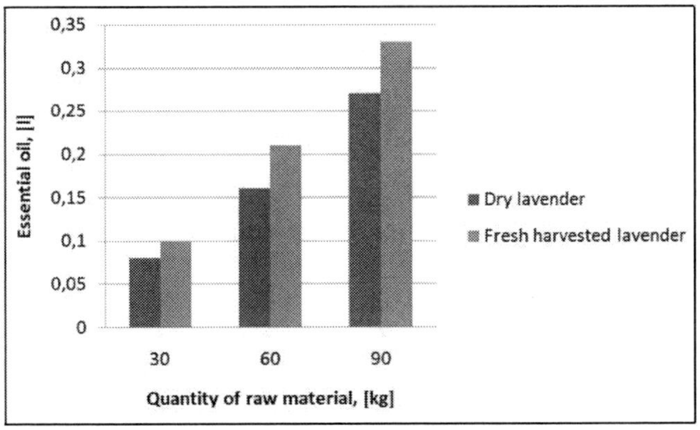

Fig. 6 The productivity of distillation process for lavender

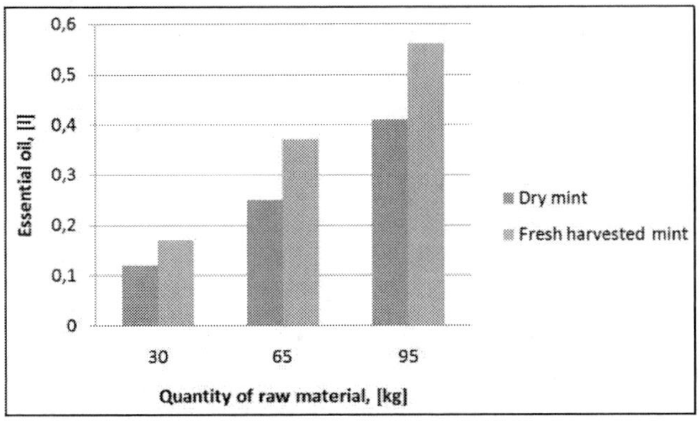

Fig. 7 The productivity of distillation process for mint

In order to prove the effectiveness of the installation with alternative operation EUV-2X1000 in comparision with EUV-1000 installation, tests were performed for two days, 10 hours a day, with each variant of the installation. The amount of distillation material was set at 120 kg freshly harvested lavender for each sample. Thus on the first day tests were made with EUV-1000. The duration of a sample was about 145-150 min, of which 25-30 min represents the times for cooling, evacuation of the meterial, cleaning the vessel and refilling it with the raw material for the next sample. Under these conditions, 4 samples were obtained within 10 hours. The next day with the EUV-2X1000 installation, 5 samples were made with the same amounts of material / sample. All tests were performed at an inlet pressure of 0.2 bar in the distillation vessel.

4 Conclusions

After the tests, the following aspects were identified:
- the amount of oil obtained depends largely on the quality of the plants subjected to distillation;
- the distillation time is approximately equal, irrespective of the quantity of plants in the vessel;
- from the freshly harvested plants subjected to distillation, about 25-30% more essential oil is obtained compared to the quantity obtained from dried plants;
- by using the EUV-2X1000 installation, a larger quantity of plants with about 20% / day can be distilled, comparing with the EUV-1000, due to the elimination of the distillation

vessel refueling times, this being done during the time in which the material from the first vessel is distilled, the distillation stream being continuous.

Acknowledgement

This work was supported by a grant of the Romanian Research and Innovation Ministry, through Programme 1 – Development of the national research-development system, subprogramme 1.2 – Institutional performance – Projects for financing excellence in RDI, contract no. 16PFE.

References

1. C. Popescu, A. Pruteanu, D. Cujbescu, I. Grigore, M. Matache, I. Voicea, V. Vlăduţ, E. Marin, C. Popescu, *Physical-chemical and microbiological characterization of lavender water obtained as secondary product from lavender (lavandula angustifolia L) essential oil,* International Symposium ISB-INMA TEH' 2015 Agricultural and Mechanical Engineering, 30 – 31 October, Bucharest, ISSN 2344-4118, Pages 335-342, (2015);
2. http://interferente.ro/plante-medicinale-izvor-de-sanatate.html;
3. C. Popescu, I. Voicea, V. Vlăduţ, D. Cujbescu, S. Bungescu, *Physical and chemical characterization of the lavender water resulted as waste at the obtaining of the lavender volatile oil, by driving with water vapour (Part 1),* Fifth International Conference "Research People and Actual Tasks on Multidisciplinary Sciences", Section 2 - Social and human sciences. Economic sciences. Ecology and Environment protection, Land & Water Use and Environment, Physics, Chemistry, Biology, Lozenec, Bulgaria, 24 – 28 June, Pages 153-160, (2015);
4. C. Popescu, I. Voicea, V. Vlăduţ, D. Cujbescu, I. Găgeanu, M. Duţu, *Physical-chemical characterization of lavender water resulted as waste when obtaining volatile oil from lavender, by engaging with water vapors (Part 2),*Fifth International Conference " Research People and Actual Tasks on Multidisciplinary Sciences", Section 2 - Social and human sciences. Economic sciences. Ecology and Environment protection, Land & Water Use and Environment, Physics, Chemistry, Biology, Lozenec, Bulgaria, 24 – 28 June, Pages 161-168, (2015);
5. http://www.creeaza.com/familie/medicina/ULEIURILE-VOLATILE-ESENTIALE127.php;
6. C. Popescu (Popiniuc), C. Popescu, St. Manea, V. Vlăduţ, I. Voicea, C. Covaliu, H. Abbas, A. Dune, D. Lupuleasa, *The influence of extraction solvent on the active principles content of Portulaca Oleracea native species*, Revista de Chimie, Volume **69** (10)/2018, Pages 2682-2692, (2018);
7. C. Popescu (Popiniuc), C. Popescu, St. Manea, V. Vlăduţ, I. Caba, C. Covaliu, H. Abbas, A. Dune, D. Lupuleasa, *Study on Portulaca Oleracea native species as vegetal source of Omega-3 and Omega-6 fatty acids*, Revista de Chimie, Volume **69** (11)/2018, Pages 2973-2980, (2018);
8. C. Popescu, I. Voicea, V. Vlăduţ, *Influence of breakage process on increasing the extraction yield of medicinal plants bioactive substances,* Romanian Biotechnological Letters, Volume **20**, Issue 4, Pages 10561-10571, (2015);
9. A. Pruteanu, C. Popescu, V. Vlăduţ, I. Găgeanu, *Biochemical analysis of some vegetal extracts obtained from indigenous spontaneous species of (Thymus serpyllum L.)*, Romanian Biotechnological Letters, Volume **23**, Issue 5, Pages 14013-14024, (2018);
10. https://viataverdeviu.ro/proprietatile-vindecatoare-ale-uleiurilor-esentiale/;
11. https://tse3.mm.bing.net/th?id=OIP.lB6JAuMdraQ1LEk8TSL5yAHaEs&pid=15.1&P=0&w=28 5&h=181\;
12. Solacolu T, Kollo C, Revista sanitara militara, (No 8, 1938);
13. http://plantemedicinale.crestere-melci menta-cultivarea-prelucrarea-si-utilizarea-mentei.ro/27/;
14. https://dietamediteraneana.wordpress.com/tag/recoltarea-mentei/;
15. https://tse4.mm.bing.net/th?id=OIP.-siBwpUFwAVvozT51WSObAHaD2&pid=15.1&P=0&w=326&h=171;

 https://doi.org/10.1051/e3sconf/201911203012

16. http://agro.afacereamea.ro/wp-content/uploads/floricultura/Metode-de-Obtinere-Si-de-Prelucrare-a-Uleiurilor-Volatile.p.

E3S Web of Conferences **112**, 03013 (2019) https://doi.org/10.1051/e3sconf/201911203013
TE-RE-RD 2019

Utilization of the thermohydric stress in the psamosols area in Southern Oltenia through the cowpea culture

Reta Draghici[1*], *Iulian* Draghici[1], *Aurelia* Diaconu[1], *Mihaela* Croitoru[1], *Alina Nicoleta* Paraschiv[1], *Milica* Dima[1] and *Mircea* Constantinescu[1]

[1]Research Development Station for Plant Culture on Sands Dabuleni, 207170, Călăraşi, Dolj, Romania

Abstract. The area of psamosols in southern Oltenia is characterized by thermohydraulic stress, accentuated by the thermal resources recorded and by the poor hydrophysical properties of the soil. Higher capitalization of these lands has been an important concern both in the world and in our country, and the choice of plant species with increased drought resistance and low water consumption is a measure to counteract the negative effects of the drought. The humidity analysis of a psamosol, on the 0-125 cm profile, indicates a momentary supply of reduced water and a water deficit compared to the field capacity of - 34.4 ... -65.5 m^3 / ha. In order to obtain economic results on these soils, it is necessary to cultivate low-water plants, such as a cowpea (*Vigna unguiculata L. Walp)* that consumes about 4160 m^3 / ha, compared to soybeans that have almost double consumption. The results obtained in 4 genotypes of cowpea highlight the daily average values of the plant's foliar sweating, in the blooming phase, ranging between 4.05 - 5.23 mmol H_2O / m^2 / s, which correlated negatively with the level of grain production.

1. Introduction

Cowpea (*Vigna unguiculata (L.) Walp*), is a drought-resistant plant, with little pretentions to soil conditions and an important source of protein for climatic-deprived areas [7, 10]. In most of the crop areas, the plant ensures its water needs mostly from precipitation and less from irrigation, being more tolerant to drought compared to soybeans [11]. In some tropical areas of developing countries, such as the areas of Sahelian Africa and the northeast region of Brazil, legumes are cultivated only during the rainy season, and drought stress is the most important threat for the production of biomass for many small farmers [15, 16]. For people in these regions, cowpea is an important source of protein.

In Central Africa and West Africa, 64% of the total cultivated area is the culture of cowpea. In the tropical and subtropical regions of the sub-Saharan African region, cowpea is cultivated for biomass and grains, being an important resource for the economy. According to the Food and Agriculture Organisation (FAO), cowpea is cultivated on at least 12.5 million hectares [2]. Cowpea is a specie with reduced requirements to the soil, and from this point of view areas with sand and sandy soils provide favourable conditions for the growth and development of the plant. As a leguminous plant, the cowpea forms on the roots numerous nodozities, which develops the bacterium that secures atmospheric nitrogen [5].

The nitrogen in the atmosphere that is biologically fixed by leguminous plants, compared to mineral nitrogen, has advantages, because fossil energy is not consumed for its production and is environmentally friendly. The research carried out in 1988 by Gheorghe D. and Şuteu G.E., quoted by Draghici Reta [13], by using the isotope nitrogen to fertilize the cowpea, shows that the cowpea can synthesize with the help of symbiotic bacteria of the genus Rhizobium cca. 78,2-80,6% of the nitrogen required for production. From the chemical fertiliser applied, the plant uses only 0,53-2.71%, with a very low coefficient of use of the fertiliser (3.98-4.56%), due to its leaching in the soil.

2. Methodology

The research was carried out at Research Development Station for Plant Culture on Sands Dăbuleni (RDSPCS), and aimed at the response of the cowpea plant to the thermic and hydric stress in southern Oltenia. The study was conducted on a low natural fertility psamosol, characterized by a reduced supply of organic carbon (0.28-0.45%) nitrogen (0,039-0.08%), a medium-sized supply of phosphorus (21.8-30.5 ppm) and potassium (75-91,5 ppm). The PH of the soil on which the experience was placed showed values of 4.18-6.25, which shows a strong acidic to moderately acidic reaction.

The Analysis of soil fertility was carried out by harvesting of soil samples which were recorded and conditioned in the laboratory, from which the following determinations were carried out:

- Total Nitrogen – Kjeldahl method;
- Extractable phosphorus (P-AL) – Egner – Riem Domingo method, whereby phosphates are extracted from the soil sample with a solution of acetate – ammonium lactate at pH – 5.75, and the extracted phosphate anion is determined as-blue molybdenum;
- Replaceable potassium (K-AL) – Egner – Riem Domingo method whereby the hydrogen and ammonium ions of the extraction solution replace the potassium ions in the changeable form of the soil sample, which are thus passed in the solution. The Dosing of potassium in the solution thus obtained is made by flame emission photometry.
- Organic carbon – wet oxidation method and titanium dosing (after Walkley – Blak in the Gogoaşă modification);
- Soil pH, potentiometric method

To establish the irrigation regime of the cowpea, soil samples were harvested monthly on the depth of 0-125 cm and the soil humidity was determined, using the gravimetric method, by evaporation of water physically linked to the soil at 105^{0}C, with the help of the oven. The following hydrophysic indices of the soil were calculated: the momentous supply of water, the reserve against the wilting coefficient, the deficit to the field capacity and the water supply in the soil, to which the irrigation was applied.

In this respect, the following calculation formulas were used:

- Water momentan supply = (100 * H * Da * W), where,:
  W = soil humidity (%)
  H = depth of soil (m)
  Da = apparent soil Density
- Wilting coefficient (CO) = 100 * H * Da * CO
- Field Capacity (CC) = 100 * H * Da * CC
- Minimum ceiling of water (30% of the active humidity range, on the depth of 50 cm) = CO + 1/3 (CC-CO)
- Water Reserve = momental Supply-Minimum cap
- Water Deficit = Momental supply – Field Capacity

Hydrophysic Indices of the sandy soil:

Da = 1.38-1.46 g/cm^3 (depending on the depth of the soil)
CO = 2; CC = 9

Experience aimed at the morpho-physiological behavior of 4 cowpeas genotypes (*Jiana, Aura 26, Ofelia, Doljana*) cultivated in a 3-year rotation: cowpea-rye-sorghum. The Jiana variety is a late variety, with a 103 day vegetation period, reddish brown bean and a crude protein content of about 21.8%. This variety develops a rich vegetative mass, being recommended as a green fertilizer for sands, or as feed for animal feed. The other varieties of cowpea, respectively, *Aura 26* with a vegetative period of 91 days and grain white and hil brown, *Ofelia* with a vegetative period of 95 days and grain white and hil black and *Doljana*, with a growing period of 92 days and white grain are especially recommended for human food in the form of beans with a high protein content in grains of 22.2-22.9 % or green pods due to the high vitamin C content (over 35 mg / 100 g of fresh substance). During the vegetation of the cowpea varieties, they were carried out the determinations of biometrics, physiology (dry substance, water formes in the plant, photosynthesis, transpiration) and variety-based productivity were performed.

The forms of water were determined gravimetrically with the help of the oven, and the photosynthesis rate, the transpiration rate and the stomatic conductance were performed with the help of the device: LC Pro +.

3. Rezults and discusions

The climatic conditions recorded in the year 2018 (Table 1) were favourable to the growth and development of the cowpeas plants, with an average air temperature recorded during the plant vegetation period (May-August) of 22.97 °C, with a deviation in addition to 2 °C, compared to the same multiannual period. The thermal regime of the area, created by the accumulation in the air of approx 2825.31 °C, was favorable for the normal conduct of biological and physiological processes in the culture of cowpea, the literature mentioning a necessity of approx 2000 °C [3]. Rainfall during this period exceeded the multiannual average by 262.21 mm, but were unevenly distributed to the requirements of the cowpea plant.

Table 1. Maximum Air (^{0}C) and precipitation (mm) in the period April-September 2018, recorded at Dabuleni RDSPCS weather Station

Month/decade	April	May	June	July	August	May - August
Monthly average monthly Temperature (^{0}C)	17.8	20.7	22.5	23.6	25.1	22.97
Average monthly multi-annual temperature (^{0}C) (1956-2017)	11.8	16.8	21.6	23.1	22.4	20.97
Temperature deviation from multi-annual average(^{0}C)	6	3.9	0.9	0.5	2.7	2
Rainfall (mm)	17	106.6	195.2	148.7	30	480.5
Sum of multiannual monthly rainfall (mm) (1956-2017)	47.52	62.39	68.46	54.24	33.2	218.29
Rainfall deviation from multi-annual average (mm)	-30.52	44.21	126.74	94.46	-3.2	262.21

Thus, from the analysis of the daily distribution of rainfall (Figure 1), it is noted that between 10 April and 10 May there were 12.6 mm, insufficient quantity for a uniform emergence of plants, being necessary to complement the water deficit with irrigation. The 2nd water shortage for cowpea was recorded in the flowering period of the plant, which coincided with the last decade of June. During the period of the occurrence of florals-flowering (16-25 June) were recorded just 14.4 mm precipitation, insufficient amount for the conduct of the physiological processes of the plant, and it was also necessary to irrigate the culture.

The research results obtained in the sandy soils in southern Oltenia show that the cowpea has high requirements for the water, for the imbibition water and germination of the seeds of cowpeas, and if in this interval the edaphic drought occurs in the superficial layer (0-20 cm) an emergence watering is required with a reduced watering quota (150-200 m^3 water/ha) [6]. In the first 30-40 days the cowpea has a slow growth rate, during which the radicular system develops. In this interval, the application of watering is not required. The need for water rises after the formation of the 4th and 5th leaves, recording the maximum in the period of formation of the floral organs, flowering and formation of pods. The studies aimed at the water consumption and its sources of coating in the cowpea (Table 2) in the conditions of the sandy soils have highlighted the drought resistance of the cowpeas compared to soybeans. The results obtained show that the cowpea provides 4% from the soil reserve, 60% from rainfall and 36% from irrigation, compared to soybeans that provide the largest amount from irrigation water (60%).

Table 2. Water consumption and supplying sources to some plant species cultivated on sandy soils

Culture	*Average production (kg/ha)*	*Water consumption (m^3/ha)*	*Sources of coverage*					
			Soil reserve		*Rainfall*		*Irrigation*	
			%	m^3/ha	%	m^3/ha	%	m^3/ha
Cowpea	2570	4160	4	160	60	2500	36	1500
Soybean	3020	7340	6	450	34	2500	60	4480

The results obtained on sandy soils, showed that the minimum soil humidity ceiling, were we must intervene with irrigation for the cowpea is at 1/3 in the range of humidity accessible on the depth of 50 cm, compared to soybeans in which the application of the irrigation is made at minimum ceiling of water of 2/3 in the range of humidity accessible on the depth of 50 cm [6]. .

Starting from these studies and in order to ensure optimum water consumption of the cowpea plant, soil samples were harvested on the 25th day of each month from April to July, to determine the soil humidity on different depths (0-125 cm), and in function of the results to determine the amount of water supplemented by irrigation. The results achieved, underlined the existence of two moments with moisture deficiency during the vegetation period of the cowpea (Tables 3, 4).

The first moisture deficiency compared to the field capacity was recorded during the preparation of the germinal bed (25 April) at the level of-203-49 m^3 water/ha, when the water supply of water in the soil (-20.2 m^3 water/ha) was lower than the minimum ceiling of water in the soil, on 0-25 cm soil profile in which the seeds of cowpea germinate and the young plant develops. This moisture deficiency was accentuated in the period following the planting of the varieties of cowpea (27 April), when for a period of 10 days there were no precipitation, it was necessary to apply an emergence watering of 250 m^3 water/ha. During the branching phases of the plant and the initiation of the generative organs water consumption was ensured from the recorded rainfall.

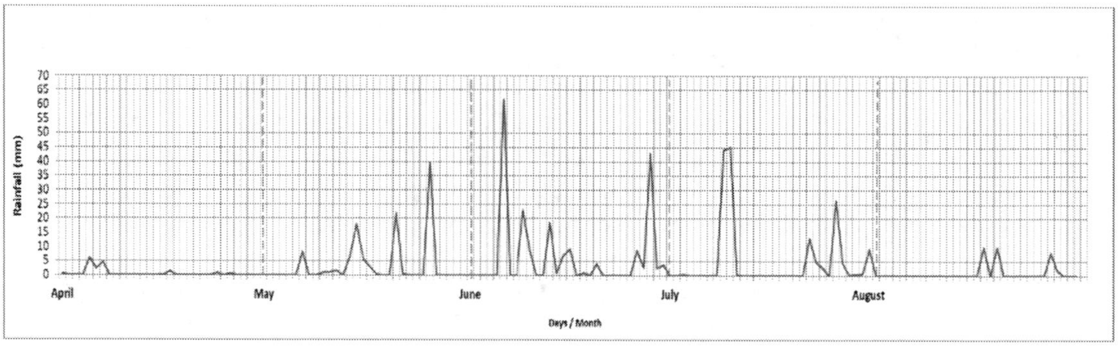

Fig. 1. Evolution of daily rainfall recorded from April to August 2018

Table 3. The evolution of some hydrophysic indices of soil (momental supply, deficit versus capacity field) on the soil profile during the water consumption period of the cowpea plant

The depth of the soil (cm)	25 April		25 May		25 June		25 July	
	Water momental supply (m^3/ha)	*Water deficit versus field capacity (m^3/ha)*	*Water momental supply (m^3/ha)*	*Water deficit versus field capacity (m^3/ha)*	*Water momental supply (m^3/ha)*	*Water deficit versus field capacity (m^3/ha)*	*Water momental supply (m^3/ha)*	*Water deficit versus field capacity (m^3/ha)*
0-25	165	-203	393	44	263.6	-34.4	545	239
25-50	158	-217	410	59	224.2	-59.8	487	178
50-75	216	-162	423	76	233.6	-45.4	553	240
75-100	347	2	429	93	212.5	-65.5	444	130
100-125	391	49	465	141	228.7	-47.3	442	127

Starting with the stage of the emergence of florals – flowering (16-25 June) the water consumption of the plant has increased, and the water resources existing in the soil have been insufficient, because over a period of 10 days were registered only 14.4 mm (Figure 1), the plants of cowpeas beeing exposed to the thermic and hydric stress specific to sandy soils. The humidity determinations carried out on June 25th (plant reproductive phase) on the soil profile 0-125 cm, revealed a moisture deficiency versus the field capacity contained within the limits-65,6...-34.4 m^3 water/ha (Table 3).

The water supply in the soil, calculated on the depth of 50 cm in this phase was less than 163.4 m^3 water/ha than the minimum moisture limit of the soil (Table 4), it is necessary to complement the water deficit by irrigation for the purpose of normal conduct of the metabolism of the plant.

The results achieved by in South Africa showed that the reproductive stage is most susceptible to water scarcity, resulting in a reduction of at least 50% of the efficiency of water use and seed production [3]. Research carried out in France by F. Monteiro De Paula and colab [4], showed that in the case of a moderate shortage of water, the content of lipid in the leaf decreased to the varieties of susceptible cowpea and increased for the resistant ones. At severe stress, the lipid content has been drastically reduced for both susceptible and resistant varieties. Studies in Senegal have highlighted the implication of moisture deficiency over the quality of the plant and have shown that water deficiency in the flowering phase has significantly decreased the starch content of the leaf, the number seeds in the pod and the number of pods per plant [7]. The same study shows that cawpea varieties with better drought tolerance are less affected by water shortages compared to sensitive varieties. Most of the areas where the cowpea are grown are prone to drought and it is necessary to solve this problem, because water scarcity becomes a major concern in agriculture.

Table 4. Development of hydrophysic indices of soil (minimum limit, water supply in soil) on the soil profile during the water consumption period of the cowpea plant

The depth of the soil (cm)	25 April		25 May		25 June		25 July	
	Minimum ceiling of water (m^3/ha)	*Water supply in soil (m^3/ha)*	*Minimum ceiling of water (m^3/ha)*	*Water supply in soil (m^3/ha)*	*Minimum ceiling of water (m^3/ha)*	*Water supply in soil (m^3/ha)*	*Minimum ceiling of water (m^3/ha)*	*Water supply in soil (m^3/ha)*
0-25	185.2	-20.2	181	212	164.67	98.93	149.5	196.88
25-50	189.87	-31.87	182.33	227.67	160.93	63.27	151.67	210.23
50-75	191.33	24.67	182.33	240.67	158.2	75.4	153.83	242.35
75-100	178	169	177.33	251.67	157	55.5	156	256.92
100-125	178.53	212.47	177.33	287.67	157.47	71.23	157.5	274.86

The growth and development of plants and implicitly their production in a particular area, constitute the result of the complex interaction between climatic factors and edafic factors. In this interaction of the existing factors in the area of the sandy soils of Southern Oltenia, the cowpea harnessis with good results the biological potential of the area, due to drought resistance and reduced claims towards the natural fertility of the soil [12]. The drought resistance of the cowpea is given both by the very powerful root system, with a high absorption power, the waxy layer on the leaves, and the number of stomata in the leaf that ranges between 280-327 stomata/mm^2 on the upper face. The closure of the stomates is a common strategy used by different genotypes of cowpea to avoid dehydration of the foliar apparatus, and the genotypic variance in the conductance of the stomates increases considerably in drought conditions.

In order to assess the use of thermic and hydric stress conditions in the year 2018, the physiological behavior of 4 varieties of cowpea in the flowering phenophyasis of plants was analysed. The results obtained, on the forms of water in the plant and the dry matter content, underlined a different behaviour of the genotypes, depending on the structural particularities of the foliar apparatus and the climatic conditions currently present (Table 5).

Table 5. The dry matter content and the forms of water recorded in the cowpea plants (flowering phase)

Genotype	Dry substance (%)	Total water (%)	Free water (%)	Linked water (%)
Jiana	21.31	78.69	75.40	3.29
Aura 26	15.44	84.56	81.61	2.95
Ofelia	17.85	82.15	79.46	2.69
Doljana	18.43	81.57	79.43	2.14

Thus, the dry substance in the leaves varied in the range between 15.44-21.31%, noting that the maximum value was reached by the *Jiana* variety. The percentage values of total plant water were between 78.69% (in the *Jiana* variety) and 84.56% (in the *Aura 26* variety). The free water recorded the highest values in the *Aura 26* variety (81.61%), and the lowest in the *Jiana* variety (75.40%). Under the appearance of linked water from the leaves, an indicator emphasising drought resistance, it was highlighted by a good behavior at drought for the *Jiana* variety, at which the highest percentage of linked water was recorded. The speciality literature mentions that under unfavourable environmental conditions, when the vital activity of the plant is reduced, the amount of free water decreases, and that of the linked water increases, which causes greater resistance of plants to drought [9]. The higher the amount of linked water in comparance to the free water, the cells are better resistant to high temperatures [11]. The authors underline that the drought resistance of cowpea is due to the properties of cell protoplasm that allows the preservation of the plants life during drought and the resumption of vital processes, with increased intensity, when the soil is supplied with water again. The stress tolerance of the cowpea depends on the genotype and the thermic and hydric regime, and it was determined in the flowering phase of the plant by physiological determinations on the transpiration and photosynthesis of the plant. The obtained results revealed a diurnal variation in the physiological processes in the plant, depending on the variety and climatic conditions (Figure 2). The maximum rate of photosynthesis (37.69 μmol $CO_2/m^2/s$) was performed by the *Doljana* variety at 9 o'clock, when it was on average 30.5 ^0C at the leaf level, and the active radiation in photosynthesis was 1560 μmol/ m^2/s. The increase in temperature and active radiation level in photosynthesis resulted in a reduction in the intensity of the plant photosynthesis. Determinations regarding the rate of transpiration of the plant have highlighted a positive correlation with the temperature at the leaf level. The values of the foliar transpiration oscillated between 2.23-4.86 mmol $H_2O/m^2/s$ at 9 o'clock, when the temperature at leaf level was 30.5 ^0C, between 5.39-6.38 mmol $H_2O/m^2/s$ at 12 o'clock, when the leaf temperature was 33.41 ^0C and between 4.05-5.23 mmol $H_2O/m^2/s$ at 15 o'clock, when the temperature at the leaf level was 33.3 ^0C (Figure 2).

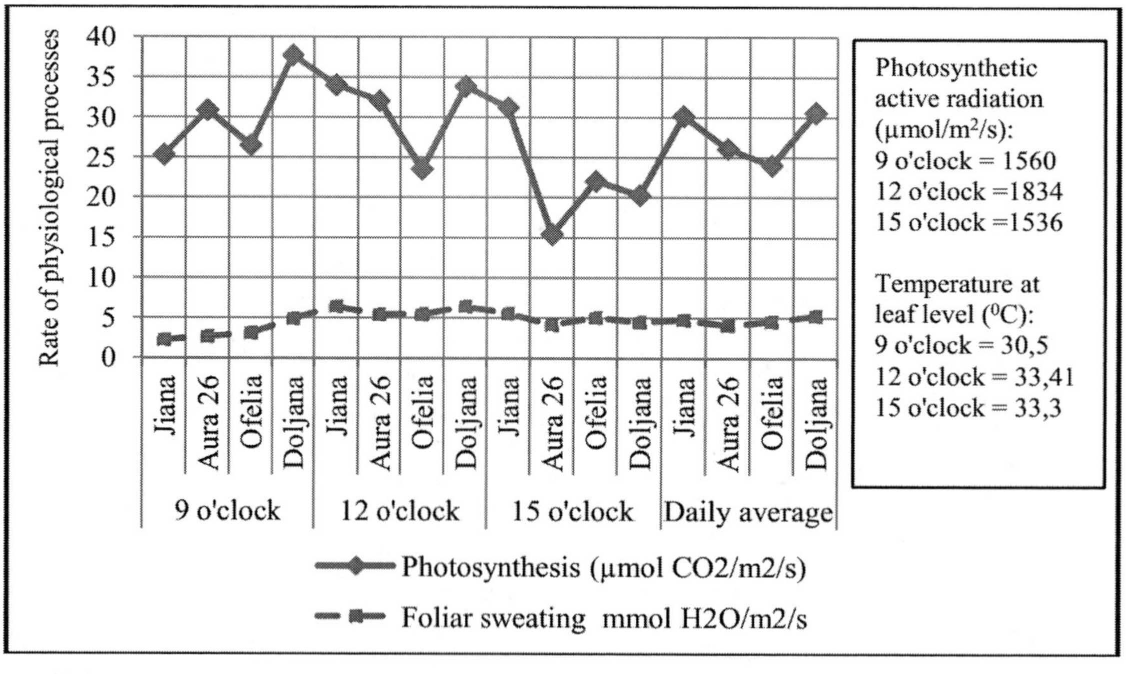

Fig. 2. The diurnal variation of physiological processes in the cowpea according to genotype

The Varieties *Jiana* and *Doljana*, although they recorded high values of the transpiration of the plant, best exploited the water lost by transpiration, as a result of the accumulation of high average values of CO_2 through the photosynthesis process (30.17-30.6 μmol/ m^2s). The daily average of the transpiration of the plant in the 4 cowpea genotypes, determined in the flowering phase, highlighted values ranging from 4.05 to 5.23 mmol $H_2O/m^2/s$, which correlates negatively to the level of the grain production (Table 6).

Table 6. Biological qualities in some varieties of cowpea studied in the conditions of sandy soils in Romania

Variety	Plant vegetation period		Plants height (cm)	Foliar Index	Photosynthesis rate (μmol/m^2/s)	Foliar sweating mmol $H_2O/m^2/s$	No. pods / Plant	Production of grains kg/ha
	No. day	Σ^0C						
Jiana	115	2648.5	140.6	7.25	30.17	4.71	6	1273.8
Aura 26	103	2354.5	100	5.9	26.1	4.05	13	2194.4
Ofelia	102	2329.8	112	6.05	24.03	4.51	11.2	2309.5
Doljana	101	2336	105	6.72	30.6	5.23	10.4	1718.2
Correlation between the foliar index and the photosynthesis rate of the plant				$Y = -3.7012 x^2+52.942x-158.84$; $R^2 = 0.7846$; $r = 0.885$				
Correlation between foliar index and grain production				$Y = -744.7x+6699.7$ $r = -0.980$*				
Correlation between the number of pods/plants and the production of grains				$Y = -4.4609x^2+226.81x+61.039$; $R^2 = 0.8173$; $r = 0.904$				
Correlation between foliar sweating and the production of grains				$Y = -529.34x+4322.2$; $r = -0.544$				

The drought tolerance of plants is a genetic feature and is highlighted by the drought index, which represents the ratio of the yields obtained on the supplied and unprocessed parcels of water. The value of this indicator varies depending on the genotype between 0.22 and 0.9 [8] The results obtained from the cowpea culture revealed the differentiation of morphological and productive qualities by genotype (Table 6). The vegetation period of the cowpea genotypes was between 101-115 days, with positive correlation with the thermal resources accumulated during this interval of each variety. In the study conditions, the leaf area index (L.A.I.) in the four varieties of cowpea presented values ranging from the range 5.9-7,25. It was noted by the high values of the L.A.I. varieties *Jiana* and *Doljana*, which also recorded a high intensity of the photosynthesis process. The analysis of the production of grains obtained at harvest reveals with the best results (2194.4-2309.5 kg/ha) the varieties *Aura 26* and *Ofelia* and underlines positive correlation with the number of pods/plants and negatives with the index of the foliar surface and the rate of foliar sweating.

4. Conclusions

1. The culture of cowpea has successfully exploited the thermic and hydric stress of southern Oltenia.
2. During the preparation of the germinating bed for cowpea sowing (25 April), the first humidity deficit was recorded in relation to the field capacity at the level of -203-49 m^3 water/ha, registering on the soil profile 0-25 cm a reserve of water in the soil 20.2 m^3 water/ha, reserve less than the minimum moisture level in the soil.
3. In the maximum consumption phase of the cowpea plant (plant reproductive phase) a moisture deficiency has been determined against the capacity of the field contained within the limits -65,6...-34.4 m3 water/ha.

4. In the reproduction phase of the plant the water supply from the soil on the depth of 50 cm was less than 163.4 m^3 water/ha versus the minimum moisture level of the soil, which led to the completion of water deficit by irrigation, for the purpose of normal performance of plant metabolism.
5. The results obtained revealed a diurnal variation in the physiological processes in the plant, depending on the variety and climatic conditions
6. The analysis of the production of grains obtained at harvest reveals the best results (2194,4-2309,5 kg/ha) in the *Aura 26* and *Ofelia* cowpea varieties.
7. The production of grains obtained from the varieties of cowpea has been positively correlated with the number of pods/plants and negative with the index of the foliar surface and the rate of foliar sweating.

Acknowledgement

This work was supported by a grant of the Romanian Ministery of Research and Innovation CCDI - UEFISCDI, Project INNOVATIVE TECHNOLOGIES FOR IRRIGATION OF AGRICULTURAL CROPS IN ARID, SEMIARID AND SUBHUMID-DRY CLIMATE, project number PN-III-P1-1.2-PCCDI-2017-0254, Contract no. 27PCCDI / 2018, within PNCDI III.

References

1. A. Zăvoi, *Contributions on biology, improvement and agrotechnique of cowpea - Vigna sinensis (Torn) Endl.* PhD Thesis, Cluj (1967).
2. C. Egashira, T. Yamauchi, Y. Miyamoto, T. Yuasa, Y. Ishibashi, M. Iwaya-Inoue, *Physiological Responses of Cowpea (Vigna unguiculata (L.) Walp) to Drought Stress during the Pod-filling Stage*. Cryobiology and Cryotechnology, vol. **62,** no. 1, pp. 69-75, 2016, Japan (2016)
3. F. E. Ahmed, A. S. H. Suliman, *Effect of water stress applied at different stages of growth on seed yield and water-use efficiency of cowpea*. Agriculture and Biology Journal of North America, ISSN : 2151-7517, vol.**1**, No.4, p.534-540, (2010)
4. F. Monteiro De Paula, A.T. PhamThi, J.Vieira De Silva, A.M. Justin, C. Demandre, P. Mazliak, *Effects of water stress on the molecular species composition of polar lipids from Vigna unguiculata L. Leaves.* Plant Science, Vol. **66**, Issue 2, https://doi.org/10.1016/0168-9452(90)90203-Z, France, pp. 185-193, (1990).
5. Gh. Eliade, I. Ghinea, Gh. Ştefanic, *Soil microbiology.* Ed. Ceres, Bucharest (1975)
6. Gh. Marinică, 1994. *Research on the irrigation regime for cowpea (Vigna sinensis), cultivated on the sandy lands of southern Olteniei.* Workers. Sc. SCCCPN Dabuleni, vol. **8**, pp.43-53, (1994).
7. H. Falalou, Z. Gérard, G. Sita, D. Omar, N.D. Ndèye, B. Serge, *Physiological, biochemical and agromorphological responses of five cowpea genotypes (Vigna unguiculata (L.) Walp.) to water deficit under glasshouse conditions*, Biotechnol. Agron. Soc. Environ, Base, vol. **11**, pp. 225-234, Senegal (2007).
8. I. Burzo, *Climate change and effects on horticultural plants*, Ed. Sitech, Craiova, pp. 42-91, (2014).
9. I. Burzo, A. Dobrescu, *Thermodynamic stress on plants - High temperature and drought-*, Ed Ceres, Bucharest, pp.28-50, (2011).
10. O. Düzdemir, A. Ünlükara, A. Kurunç, *Response of cowpea (Vigna unguiculata) to salinity and irrigation regimes*. New Zealand Journal of Crop and Horticultural Science. ISSN: 0114-0671 (Print) 1175-8783 (Online), vol. **37:3**, DOI: 10.1080/01140670909510273, pp. 271-280, Turkey, (2010).

11. R. B., Dadson, F. M. Hashem, I. Javaid, J. Joshi, A. L Allen, T. E Devine, *Effect of Water Stress on the Yield of Cowpea (Vigna unguiculata L. Walp.) Genotypes in the Delmarva Region of the United States.* Journal of Agronomy and Crop Science, vol. **191**, pp: 210–217. doi: 10.1111/j.1439-037X.2005.00155.x, USA, (2005).
12. R. Drăghici, *Contributions to the establishment of the cultivation technology of the cowpea Vigna unguiculata (L.) Walp on the irrigate sands of southern Olteniei.* Doctoral Thesis. University of Craiova (1999)
13. R. Draghici, *Cowpea – the plant of sandy soils.* Ed. Sitech, Craiova, ISBN 978-606-11-6587-2, (2018).
14. R. Draghici, I. Draghici, A. Diaconu, M. Croitoru, M. Dima, *Significant progress achieved in cowpea breeding in Romania*. www.nordsci.org/proceeding2018, ISBN 2603-4107, ISSN 978-619-7495-01-0, DOI 10.32008/B2/**V1/34**, pp. 321-328, (2018).
15. R. J. Donohue, M. L. Roderick, T. R. McVicar, G. D. Farquhar, *Impact of CO_2 fertilization on maximum foliage cover across the globe's warm, arid environments.* Geophysical Research Letters, vol. **40**, http://dx.doi.org/10.1002/grl.50563, Australia, pp 3031-3035, (2013)
16. R. Rivas, H.M. Falcão, R.V. Ribeiro, E.C. Machado, C. Pimentel, M.G. Santos, *Drought tolerance in cowpea species is driven by less sensitivity of leaf gas exchange to water deficit and rapid recovery of photosynthesis after rehydration.* South African Journal of Botany, vol. **103,** pp. 101–107, (2016).

Results on bioenergetic potential of some sweet sorghum hybrids cultivated under psamosols conditions in Southern Oltenia

Iulian Draghici[1*], *Reta* Draghici[1], *Aurelia* Diaconu[1], *Mihaela* Croitoru[1], *Alina Nicoleta* Paraschiv[1], *Milica* Dima[1], *Elena* Ciuciuc[1] and *Daniel* Ciuciuc[1]

[1]Research Development Station for Plant Culture on Sands Dabuleni, 207170, Călăraşi, Dolj, Romania

Abstract. *Sorghum bicolor L. Moench* is a herbaceous annual C4-type plant, with a high ecological plasticity because it can be grown in tropical, subtropical, temperate and semi-arid regions, harnessing soils with reduced natural fertility, from the category of sandy soils. Studies on 5 sweet sorghum hybrids studied under psamosols in southern Oltenia highlight a bioenergy potential in the milk-wax phase of the grain, ranging from 60847.4-89338.65 Mcal / ha, with a maximum in the hybrid Shaşm 2, which achieved a distinctly significant difference in production, compared to the average production of hybrids. The results recorded in sugar sorghum production were positively correlated, significantly distinct with leaf index determined during the flowering of the plants (r = 0.9854**).

1. Introduction

Sorghum (*Sorghum bicolor L. Moench*) is an annual herbaceous C4 plant with a high ecological plasticity, as it can be cultivated in tropical, subtropical, temperate and semi-arid regions, harnessing the soils with low natural fertility, from the category of sandy soils. Due to the high content of sweet juice in the stems and the high proportion of starch from the grains, sweet sorghum (*Sorghum bicolor var. saccharatum* or simplified *Sorghum saccharatum*) is one of the most important agricultural sources used for production of biomass that can be processed and converted relatively easy into a type of fuel (bioethanol, synthesis gas), vegetable proteins, food for animals or green manure [1, 3, 5]. Renewable energy is a critical source of energy that contributes to energy security, reducing dependence on fossil fuels and greenhouse gas emissions. Studies in India show a need of more than 6.3 billion litres of ethanol to achieve its target of replacing 20% of oil fuel with biofuel consumption (bioethanol and biodiesel) by 2017, according to Ethanol Blending Program (EBP) [10]. It shows the importance of the sweet sorghum for renewable energy in India, stating that the plant is a promising biological raw material for dry areas, which is addressing favourably the problem of supplying with superior fuels. The widespread cultivation of sweet sorghum is achievable if cultivation with increased yield of sugar with multiple tolerance to biotic and abiotic stress are available. Sweet sorghum is widely

*Corresponding author: iuliandraghici54@yahoo.com

acknowledged as a highly promising energy culture for biomass with special potential to complement the production of sugar cane in diversified culture systems [6]. The Research carried out at the Agricultural Centre of the Louisiana State University, shows that the percentage of sugar in the strains was different depending on the hybrid taken in the study, and varied in the range 12.7%-15.1%. The results obtained in a study at Giza Agricultural Research Station, Giza Governorate, Egypt, highlighted the use of various forms of sweet sorghum syrup for the manufacture of cakes [13]. The chemical properties indicated that the syrup and the lyophilised juice from the sweet sorghum were rich sources in both sugars and minerals (K, Ca and Fe). Due to its genetic variability, with regard to the sugar content of the stem, such as total soluble sugars, the yield of green stem, the quantity of juice and the yield of grains, various research institutes in the country and abroad have developed varieties and hybrids of the culture of sweet sorghum. In our country, the research undertaken by I. Antohe, [8], on the chernoziom of Fundulea, highlighted that the average sugar productions in the strains were between 4.6-6.3 t/ha in the *Carmen* variety and 6.8-8.5 to the simple hybrid *Doina*.

Under the current conditions of aridization of the climate, the identification of agricultural practices with the potential to mitigate the impact of climate change on the security of agricultural production is increasingly more interesting. In this context, the orientation towards drought-resistant crops is shaped as an alternative solution. Sweet sorghum is one of the cultures with high adaptability to slightly favourable ecological conditions (poor soils, arid climate), due to the high capacity to efficiently exploit natural resources and it's increased tolerance to drought.

2. Methodology

The research carried out at Research Development Station for Plant Culture on Sands Dăbuleni (RDSPCS Dabuleni), aimed at the conduct of 5 sweet sorghum hybrids in the conditions of the sandy soils of southern Oltenia. The sowing was made in the first decade of May, at 70 cm between the rows, ensuring a density of 15-20 germinable grains/m^2. The experience was placed after the method of randomized blocks, in three repetitions, on a sandy soil with a accentuated non-uniformity from a chemical point of view. The soil quality analyses where the experience was located highlighted a nitrogen content of between 0.035% and 0.13%, values indicating a low soil supply state, an extractable phosphorus content between 59 ppm and 115 ppm, values, that characterizes the soil as well stocked in phosphorus, and a percentage of 24 ppm to 62 ppm in replaceable potassium, which shows a low-medium supply state. Organic Carbon presented values in the range 0.19%-0.65%, indicating a state of supply of soil in low organic matter, with portions to the middle, and the soil pH oscillated between 4.83 and 6.13, values showing a moderately strong acid reaction to poorly acidic.

Morphological, biometric, physiological and quality measurements of the biomass obtained have been carried out during the plant vegetation period. From a physiological point of view, determinations were made in the flowering phenophyasis of plants, concerning:

- PAR = photosynthesis active radiation (µmol/m^2/s), determined with LC PRO +
- T = temperature at the level of the analyzed leaves (^{0}C), determined with LC PRO +
- F = Photosynthesis rate (µmol CO_2/m^2/s), determined with LC PRO +
- E = foliar sweating rate (mmol H_2O/m^2/s), determined with LC PRO +
- The average foliar Surface of a leaf was determined using the Area Metter AM 300, in the flowering phase of the plant
- The sugar concentration in the sorghum plants was determined with a Hand Refractometer, in the milk wax of the grain phase of the plant. The Refractometer has a 0 – 32% BRIX scale and temperature compensation in the range 10-30 degrees Celsius.

In the phase of milk wax of the grain, determinations were made on the size of the plant, the diameter of the stem and the production of biomass.

3. Results and discusions

Analyzing the characteristics of the plant at 5 hybrids of sweet sorghum studied on the sandy soils, a plant height was revealed within the range of 227-299 cm and a diameter of the stem of 18.4-22.08 mm (Table 1). They were distinguished by above-average values, the hybrids of *Shaşm 1* and *Shaşm 2.* The measurements of sugar in the strains, carried out in the wax-milk phase of the grain, underlined values of 14.8-16.7%, with a maximum at the hybrid *Es Athena.* Out of the 5 hybrids, 3 registered above-average values in terms of sugar content in the strains. The results obtained in Romania [4, 5], highlighted a high-soluble solid content in some hybrids of sweet sorghum cultivated in suthwestern Romania. Thus, the total quantity of sugar ranged between 14.02% and 15.87%, of which a higher percentage 67.74% to 79.43% is the non-reducing sugar. The leaf area index (Table 1) recorded values of 8.28-9.85, with an average of 9.12, highlighting the hybrids of *Shaşm 2, Es Athena* and *Shaşm 1.* Numerous researches have resulted in a positive correlation between the height of the sweet sorghum plant and the accumulation of sugar. Since the physiological mechanisms underlying the accumulation of sugar in the sweet sorghum are not well understood, it is not known whether the apparent association between the accumulation of sugar in the stem and the height of the plant is the result of selection or dictated by physiological or genetic constraints. In this regard, the research conducted at the University of Florida Suwannee Valley Agricultural Extension Center (SVAEC) near Live Oak, Florida, USA [14] highlighted the role of shading on the accumulation of sugar in the hybrids of sweet sorghum with contrasting heights, and the results achieved showed that the increased concentration of sugar in the sweet sorghum is not conditionated on the fact that the plants are high, which makes it feasible to develop short lines with high sugar content.

Table 1. Morphological and quality qualities of the sorghum hybrids cultivated under the conditions of the psamosoils

No var.	Hybrid	Plants height cm	Diameter of the stem (mm)	% Sugar in the milk phase-wax of the grain	Leaf area index (LAI)	Dry substa nce (%)
1	SHAŞM 1	297	22.08	15.8	9.62	26.08
2	SHAŞM 2	299	21	16	9.85	24.13
3	ES ATHENA	274	18.4	16.9	9.66	23.35
4	BMR GOLD	227	19.83	14.8	8.55	22.59
5	SUPERSILE	286	19.75	15	8.28	23.37
Average		276.6	20.21	15.54	9.192	23.9
Minimum		227	18.4	14.8	8.28	22.59
Maximum		299	22.08	16.7	9.85	26.08

Analyzing the diurnal variation of the photosynthesis rate, determined during the flowering period of the plant (Figure 1), a maximum value was found at lunchtime, which, at the hybrids *Es Athena, BMR Gold* and *Super Sile*, decreased slightly in intensity at 15 o'clock. At the Hybrids *Shaşm 1* and *Shaşm 2* the values of photosynthesis recorded at 15 o'clock were below the level of values at 9 o'clock, because in these hybrids the values of the stomata conductance were very small, the closure movements of the stomatas influencing the exchange of gases at the leaf level. The average daily values were between 22.07-31.69 μmol $CO_2/m^2/s$, the best results being recorded at the *Super Sile* hybrid.

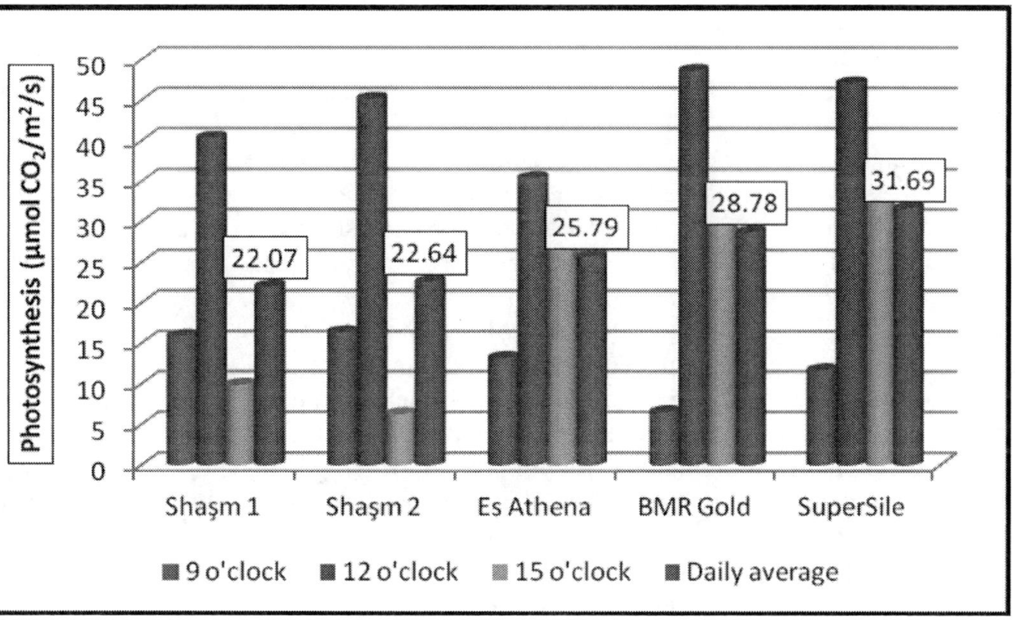

Fig. 1. The diurnal variation of photosynthesis in hybrids of sorghum cultivated in the conditions of psamosoils

The foliar sweating process presented a dynamic similar to the photosynthesis process, with the maximum diurnal being recorded at 12 o'clock, with values ranging from 2.69 mmol $H_2O/m^2/s$ to the *Shaşm 1* hybrid and 5.55 mmol H_2O/m2/s to the *SuperSile* hybrid. The average daily values ranged from 1.91 to 3.64 mmol $H_2O/m^2/s$ (Figure 2).

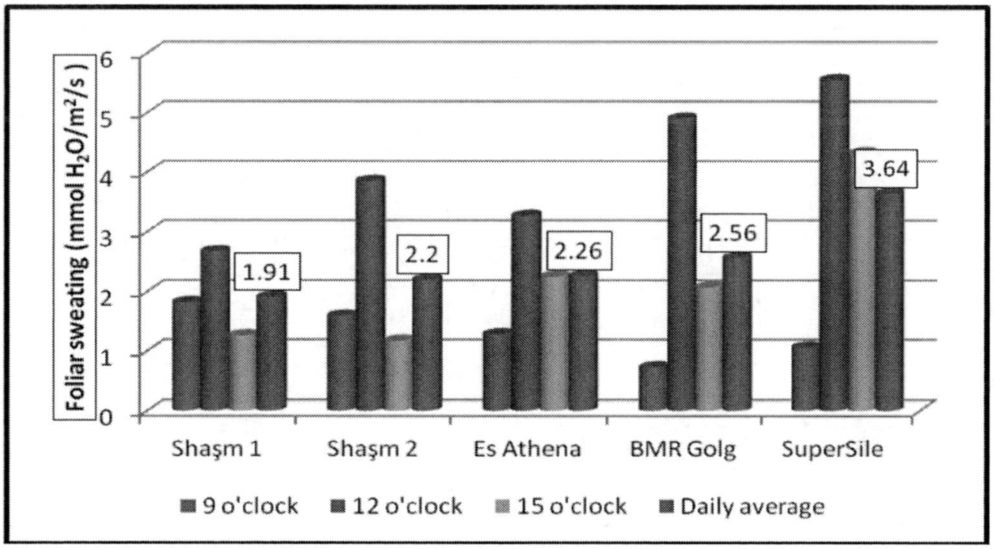

Fig. 2. The diurnal variation of foliar sweating of sweet sorghum hybrids

In this phenophyase of vegetation, the intensity of the two physiological processes (the plant photosynthesis and the foliar sweating) were positively correlated, the plants with intense photosynthetic activity also presenting higher levels of sweating. The hybrid of sweet sorghum *Shaşm 1* presented the best adaptability to the conditions of the psamosoils in southern Oltenia, recording the lowest rate of sweating, with a daily average of 1.91 mmol $H_2O/m^2/s$ correlated with a large percentage of dry substance accumulated in the leaves (26.08%). Sorghum is one of the cultivated plants that requires the smallest quantity of water. One of its peculiarities is the fact that it has a CO_2 absorption mechanism "in C_4", which gives it better photosynthesis performance, therefore better efficiency in drought or high temperatures. In addition, thanks to its very dense root system (which can descend to more than 2 meters in deep and well-prepared soil), sorghum can absorb and use more efficiently the water and nutrients in the soil. Its total water requirements (soil reserves + rain + irrigation) is around 400 – 500 mm [7]. Water maintains the cells in the state of turgescence and ensures the acid-base balance of the plants. In the photosynthesis process, chlorophyll cells activate normally when their protoplastics are at a hydration degree of 80-85%. In other species, for example barley, this process stops at 80% and in beans at 82%, while in sorghum and millet, it also works at 35% . The efficiency of transpiration (ET), an indispensable phenomenon associated with drought tolerance of plants is the amount of biomass produced (kg dry matter) per unit of transpired water (m^3) [6, 15]. Improving the efficiency of transpiration, especially in dry environments, can have a major impact on improving productive yield, because high transpiration efficiency will allow plants to either postpone the symptoms of the stress or produce more biomass from the same amount of soil moisture available or a combination of both [16]. Improving genetic training as well as crop management practices can lead to greater transpiration efficiency. The values of the efficiency of transpiration at the physiological maturity of the sorghum were between 3.21 and 7.55 kg/m^3 [11, 15]. However, little is known about the efficiency of foliar sweating in sorghum and how, or if, it changes with the stage of growth. The production results on biomass obtained at harvest, in the milk wax phase, highlighted a level of yields in the range 61400-90150 kg/ha, with a maximum of the *Shaşm 2* hybrid, which made a production difference of 12976 kg/ha, compared to the average of hybrids. *BMR Gold* and *SuperSile* hybrids produced below the witness level with significantly negative differences (Table 2).

Table 2. Production of biomass obtained hybrids of sweet sorghum cultivated under the conditions of the psamosols

No var.	Hybrid	Production		Diference of energy (Mcal/ha)	Net energy growth (Mcal/ha)	Energy consumption per kg (%)
		Biomass kg/ha	Energy Mcal/ha			
1	SHAŞM 1	87200	86415.2	9947.66	9748.707	88.69
2	SHAŞM 2	90150*	89338.65	12871.11	12613.69	85.84
3	ES ATHENA	81870	81133.17	4665.63	4572.317	94.36
4	BMR GOLD	65250^0	64662.75	-11804.8	-11568.7	117.83
5	SUPERSILE	61400^0	60787.94	-15620.1	-15307.7	125.03
Average (Mt.)		77174	76467.54	Mt	Mt	100
LSD 5% = 1107 0 kg/ha						
LSD 1% = 16100 kg /ha						
LSD 0.1% = 24150 kg/ha						

The energy equivalent used to calculate the energy obtained at the sweet sorghum is 0.991 Mcal/kg biomass [9]. The calculation of the energy consumption was also taken into account in the amount of 0.02% in the collection of the crops increase. Compared to the average of hybrids in which the energy value of 76467.54 recorded Mcal / ha, three sweet sorghum hybrids was recorded increases in energy. The lowest energy consumption per kg of product was recorded in the *Shaşm 2* hybrid (85.84%)

The intra-group competition between plants is carried out during the development of the foliar system and the root system, and the results show that higher increases in energy biomass are achieved as the plant is cultivated in a complex similar to that of it's origin, [12]. Bioenergy is, in principle, energy obtained from biomass, because, while growing, plants capture solar energy and carbon dioxide with which they build the organic matter in which energy is stored. Studies show that the use of bioenergy in the EU could double or triple without damaging the environment and without reducing the production of food, feedingstuffs or raw materials. In the notion of bioenergy are identified the following types: biomass, bioethanol, biodiesel, biogas. There are many potential sources of bioenergy that are currently under-exploited, such as biomass of certain plants, which incorporated into the soil can contribute to reducing the inputs in agriculture. The increase in the bioenergy efficiency of plants, by choosing the appropriate assortment of a particular area, could make a substantial contribution to reducing energy security problems generated by global climate changes [2].

The production results recorded at the sweet sorghum were positively correlated, distinct significantly with the leaf area index determined in the flowering phase of the plant (Figure 3).

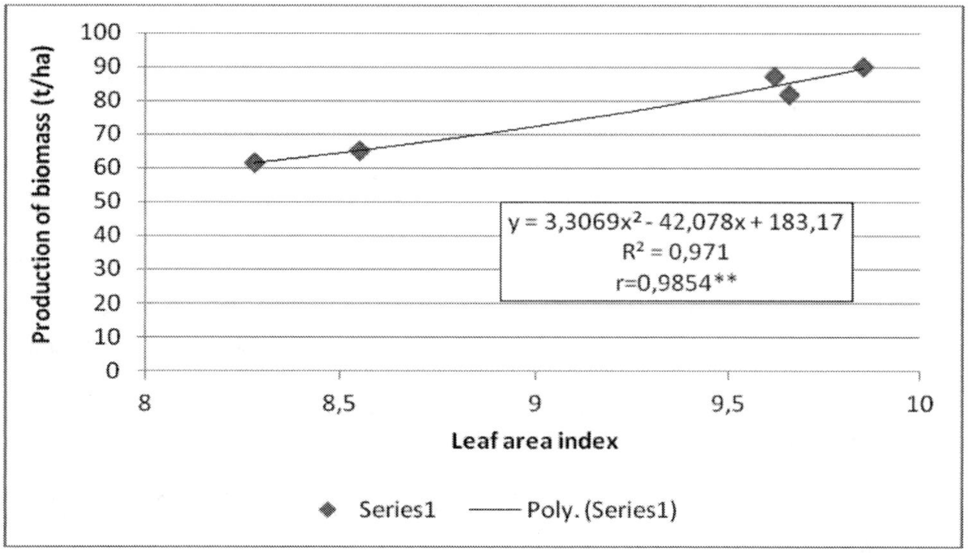

Fig. 3. Correlation between the foliar surface index and the production of biomass obtained by the hybrids of sweet sorghum cultivated under the conditions of the psamosoils

4. Conclusions

1. The measurements of sugar in the strains carried out in the milk wax phase of the grain revealed values of 14.8-16.7%, with a maximum at the hybrid *Es Athena*.
2. The daily average values of photosynthesis recorded in the 5 sweet sorghum hybrids were contained between 22.07-31.69 µmol $CO_2/m^2/s$, with the best results at the *Super Sile* hybrid.

3. The *Shaşm 1* hybrid presented the best adaptability to the conditions of the psamosoils in southern Oltenia, recording the lowest rate of transpiration, with a daily average of 1.91 mmol $H_2O/m^2/s$ and a large percentage of the dry substance accumulated in the leaves (26.08%).
4. The production results of biomass obtained at harvest, in the milk wax phase, highlight a level of eficiency in the range 61400-90150 kg/ha, with a maximum of the *Shaşm 2* hybrid, which made a difference in production compared to the average of hybrids and energy consumption per kg within the limits 85.84-125.03%
5. The production results recorded at the sweet sorghum were positively correlated, distinct significantly with the leaf area index determined in the flowering phase of the plant (r = 0.9854**).

Acknowledgement

This work was supported by a grant of the Romanian Ministery of Research and Innovation CCDI - UEFISCDI, "COMPLEX SYSTEM OF INTEGRAL CAPITALIZATION OF AGRICULTURAL SPECIES WITH ENERGY AND FOOD POTENTIAL", project number PN-III-P1-1.2-PCCDI-2017-0566, Contract no. 9PCCDI / 2018, within PNCDI III.

References

1. A. Almodares, M. R. Hadi, H. Ahmadpour, *Sorghum stem yield and soluble carbohydrates under different salinity levels.* African Journal of Biotechnology, vol. **7**, pp. 4051– 4055 (2008).
2. A. Karp, I. Shield, *Bioenergy from plants and the sustainable yield challenge*, DOI: 10.1111/j.1469-8137.2008.02432.x., Journal New Phytologist, vol. **179,** Issue 1, pp.15–32, (2008).
3. B.V.S. Reddy, S. Ramesh, S. P. Reddy, B. Ramaiah, P.M. Salimath, R. Kachapur, *Sweet sorghum - a potential alternative raw material for bioethanol and bio-energy*. Int Sorghum Millets Newsl, vol. **46,** pp. 79–86, (2005).
4. C. Babeanu, Gh. Matei, A. M. Dodocioiu, *Evaluation of sugar content in stem juice from sweet sorghum hybrids grown in south west of Romania.* Annals of the University of Craiova - Agriculture, Montanology, Cadastre Series,vol. **47**, pp 25-29, (2017).
5. Gh. Matei, *Study on yield features of sweet sorghum hybrids grown in south west of Romania.* 16th International Multidisciplinary Scientific GeoConference SGEM 2016, www.sgem.org, SGEM2016 Conference Proceedings, ISBN 978-619-7105-68-1 / ISSN 1314-2704, Book 6, vol. **1**, DOI: 10.5593/SGEM2016/B61/S25.104, pp: 783-790, (2016).
6. H. Kun-Jun, W. D. Pitman, K. Misook, F. D. Donal, W. A. Montgomery, E. Michael, Mc Cormick, G. Aita, 2012, *Ethanol production potential of sweet sorghum assessedusing forage fiber analysis procedures*, GCB Bioenergy, vol. **5**, doi: 10.1111/j.1757-1707.2012.01203.x, pp. 358–366, (2013).
7. https://www.sorghum-id.com/ro/conservarea-resurselorin-apa-sorgulo-solutie-sigura, Sorghum ID - *The Future is on the sorghum*-the First European Congress of Sores – a founding act: "*Sorghum in Europe*: A real potential", Bucharest (2016).
8. I. Antohe, *Achievements in the improvement of sorghum*, Annals. N.A.R.D.I. Fundulea, Vol. **75**, 2007, JUBILEE VOLUME, pp. 137-157, (2007).
9. I. Tesu, V. Baghinschi, *Energy and agriculture*, Ed. Ceres, Bucharest, (1984).
10. K. S. Vinutha, L. Rayaprolu, K. Yadagiri, A. V. Umakanth, J. V. Patil, P. Srinivasa Rao, *Sweet Sorghum Research and Development in India: Status and*

Prospects. Sugar Tech, DOI 10.1007/s12355-014-0302-9, ISSN: 0972-1525, Vol. **16 (2)**, pp:133–143 (2014).

11. M. Balota, W.A. Payne, W. Rooney, D. Rosenow, *Gas exchange and transpiration ratio in sorghum.* Crop Science, vol. **48**, pp. 2361–2371, (2008).
12. S. N. Olson, K. Ritter, W. Rooney, A. Kemanian, B. A. McCarl, Y. Zhang, S. Hall, D. Packer, J. Mullet, *High biomass yield energy sorghum: developing a genetic model for C4 grass bioenergy crops*. Biofuels, Bioproducts and Biorefining, vol. **6**, issue 6, pp. 640–655, (2012).
13. S. R. Khalil1, H. A. Hussien, S. R. Abazied, *New Approach for Using Different Forms of Sweet Sorghum Syrup in Cake Making.* World Journal of Food Science and Technology, vol. **2**, No. 2, doi: 10.11648/j.wjfst.20180202.11, www.sciencepublishinggroup.com, pp. 25-32, (2018).
14. S. Shukla, T. J. Felderhoff, A. Saballos, W. Vermerris, *The relationship between plant height and sugar accumulation in the stems of sweet sorghum (*Sorghum bicolor *(L.) Moench),* Field Crops Research, vol. **203**, 1, pp. 181-191, (2017).
15. Z. U. Xin, R. Aiken, J. Burke, *Genetic diversity of transpiration efficiency in sorghum*. Field Crops Research, vol. **111**, pp. 74–80, (2009).
16. Z. Xin, C. Franks, P. Payton, J.J. Burke, *A simple method to determine transpiration efficiency in sorghum*. Field Crops Research, vol. **107**, pp. 180–183, (2008).

Cost structure – an important indicator of innovation activity in agro-industrial complex (by the example of crop production)

Ivan Voiku[1,*], and *George* Varlamov[1]

[1]Pskov State University, 180000 Lenin Square 2, Pskov, Russia

Abstract. In agro-industrial complex, cost structure is of strategic importance, because it allows to determine the level of innovation activity. The study revealed diversified influence of the innovation factor (acceleration of scientific and technological progress) on the cost structure and, first of all, on the change in the ratio of share of direct labor and materialized labor in favor of the latter. Implementation of the achievements of scientific and technological progress in enterprise activities leads to a reduction in the cost of direct labor and, at the same time, to an increase in its value due to the growth of qualification requirements. Furthermore, the innovative development of enterprises assumes the implementation of intensive technologies and the acquisition of more productive equipment. In the research, all the main agricultural crops in the Pskov region (Russia) are divided into groups according to the share of direct labor and materialized labor. The effectiveness of innovative development is determined by:

- ratio of the savings from reduction of direct labor costs to the effect from implementation of the innovations, which have affected this reduction;
- ratio of the savings from reduction of material costs and energy costs to the effect from implementation of the expensive technological innovations, which have affected this reduction.

1 Introduction

Organization and development of any kind of activity is associated with the certain costs of current (operational) and capital (investment) nature.

It is commonly believed that capital expenditures are associated with the solution of strategic tasks, and current expenditures – with the solution of tactical problems of business activity of an enterprise, namely, with the acquisition of raw materials and supplies, maintenance of personnel, maintenance of material and technical base, etc.

However, the current costs that form the cost of production should be considered not only as the most important economic category, but also as a qualitative indicator of the level of use of all resources (of current and capital nature) at the disposal of an enterprise [1].

* Corresponding author: voiku-ivan@yandex.ru

It is customary to distinguish between individual and industry average cost. In contrast to the individual cost, which is due to the specific conditions of operation of an enterprise, the industry average cost is defined as weighted average unit cost in the industry.

Cost structure, which means its composition by elements or items and their share in the total cost, is characterized by the following relations [2]:

- between fixed costs and variable costs;
- between basic expenses and overhead expenses;
- between manufacturing expenses and business (nonmanufacturing) expenses;
- between direct expenses and indirect expenses;
- between human labor and materialized labor, etc.

In the field of crop production, due to biological, edaphoclimatic, agricultural features, production has sharply seasonal, time-stretched nature. For individual crops, costs for production outlay in the reporting year, but the products are harvested in the following reporting year (for example, in a winter grain crop production, etc.), some technological processes are carried out before cultivation, not to mention the output of agricultural crops. The production time of individual crops is more than one year, i.e. these crops produce a yield for 2 - 3 years. All these features of production technology require specific costs, a significant portion of which in each reporting period is a work in progress [3].

Cost structure is in motion and is influenced by [4]:

- specifics (features) of an enterprise;
- level of concentration, specialization, cooperation, combination and diversification of production;
- acceleration of scientific and technological progress;
- some other factors.

On the basis of the first factor enterprises are classified as: labor-intensive enterprises (a large share of labor costs in the cost structure), material intensive (a large share of material costs), capital-intensive (a large share of amortization charges); energy-intensive (a large share of fuel and energy in the cost structure) [5].

2 Methods

The innovation factor (acceleration of scientific and technological progress) affects the cost structure in many ways [6]. However, the main impact is to change the ratio of the share of human labor and materialized labor in the cost structure in favor of the latter.

The implementation of the achievements of scientific and technological progress in enterprise activities leads to a reduction in the cost of human labor and, at the same time, to an increase in its value due to the growth of qualification requirements. The effectiveness of innovative development is determined by a ratio of the savings from reduction of human labor costs to the effect from implementation of the innovations, which have affected this reduction.

Furthermore, the innovative development of enterprises assumes the implementation of intensive technologies and the acquisition of more productive equipment. The effectiveness of innovative development is determined by a ratio of the savings from reduction of material costs and energy costs to the effect from implementation of the expensive technological innovations, which have affected this reduction [7].

3 Results and Discussion

In crop production, the main share of the costs for production is costs for maintenance of capital assets, labor costs, costs for seeds, fertilizers and plant protection products [8].

The share of materialized labor (material costs and amortization charges) dominates in the cost structure of crop production in the Pskov region. In 2012 this share was 67.77%, in 2017 it already was 77.83% (Fig. 1) [9].

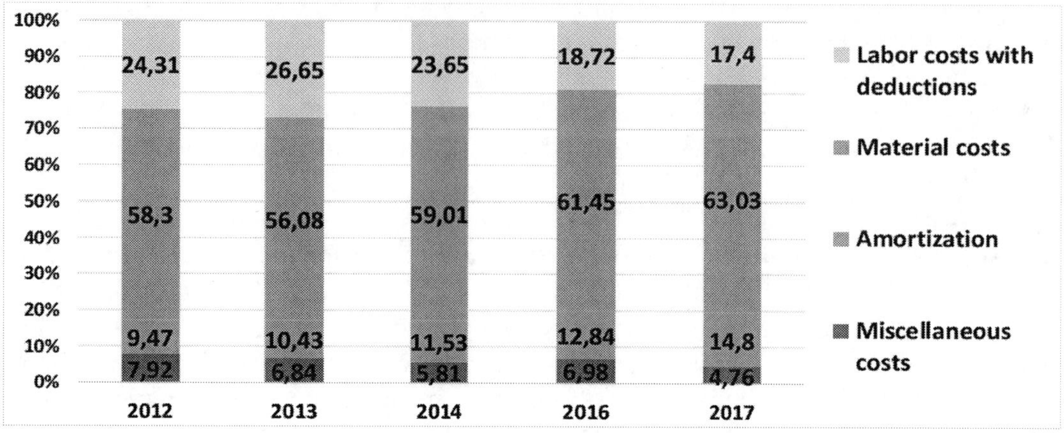

Fig. 1. Changes in the cost structure of crop production in the Pskov region, %.

In contrast, the share of human labor decreased by 6.91% and amounted to 17.4% in 2017.

During the research period the material intensity of crop production also changed its structure (Fig. 2).

The greatest growth in the structure of material costs is shown by such items as chemical crop protection products, purchase of seeds and planting material (including elite seeds), as well as payments for services and works performed by third parties. The share of costs for equipment maintenance and purchase of mineral fertilizers is growing.

The share of fuel and electricity costs, cargo transportation and spare parts and materials for maintenance is reducing.

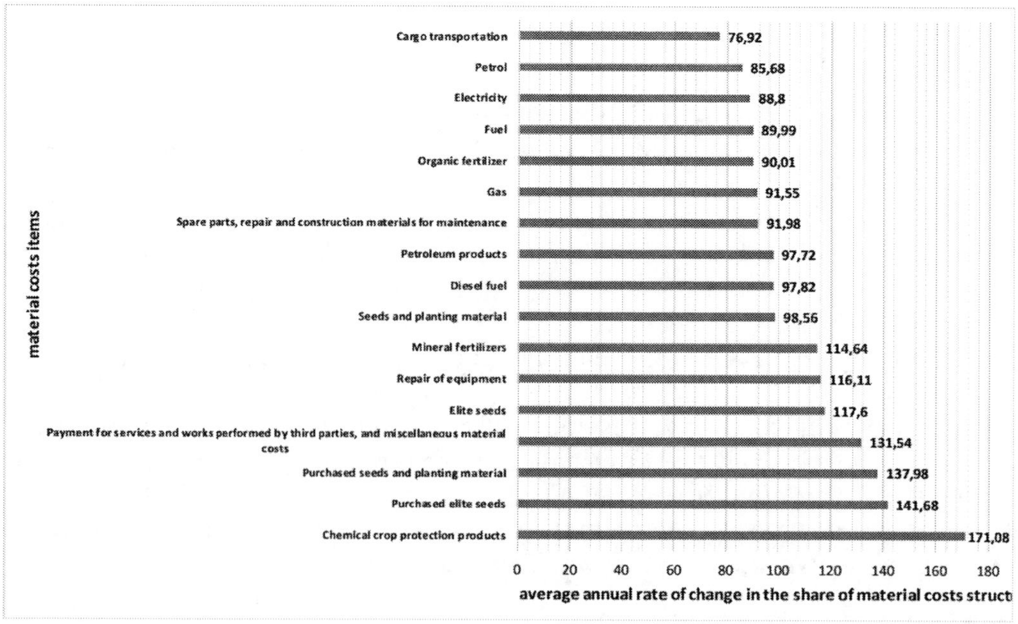

Fig. 2. Average annual rate of change in the share of costs in the material costs structure in crop production of the Pskov region.

It is worth noting that the cost structure of the basic agricultural crops in the Pskov region is heterogeneous (Fig. 3).

A relatively high share of labor costs in the production cost, indicating the insufficient level of mechanization of the operations in crop production, presents in the production of such crops as fruits, potatoes, open-ground vegetables and protected-ground vegetables [10].

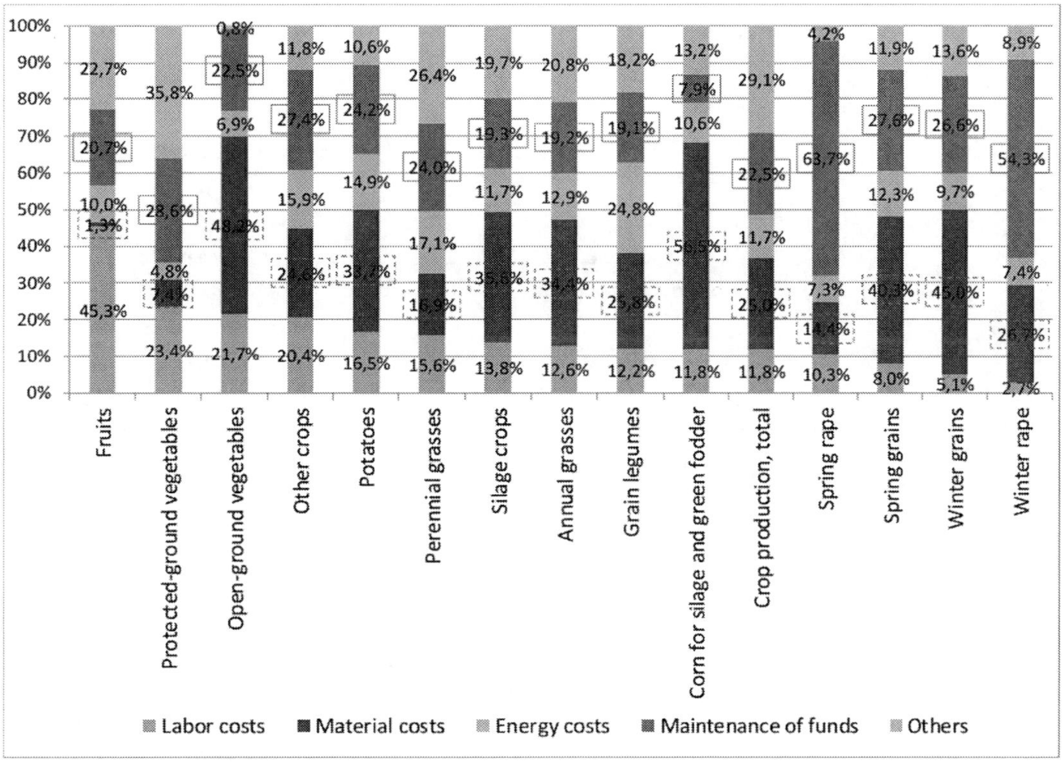

Fig. 3. Structure of the costs for cultivation of the basic crops in the Pskov region, %.

Under a high level of mechanization of grain and rape production, the share of labor costs in the cost structure is much lower, but the share of costs for the maintenance of capital assets (amortization, fuel and lubricants) is higher [11].

Depending on the share of human and materialized labor, all basic agricultural crops in the region can be divided into four groups (Fig. 4):

1. Group A: crops with a low share of human labor (up to 25%) and a low share of materialized labor (up to 50%) in the production cost;
2. Group B: crops with a high share of human labor (more than 25%) and a low share of materialized labor (up to 50%) in the production cost;
3. Group C: crops with a low share of human labor (up to 25%) and a high share of materialized labor (more than 50%) in the production cost;
4. Group D: crops with a high share of human labor (more than 25%) and a high share of materialized labor (more than 50%) in the production cost;

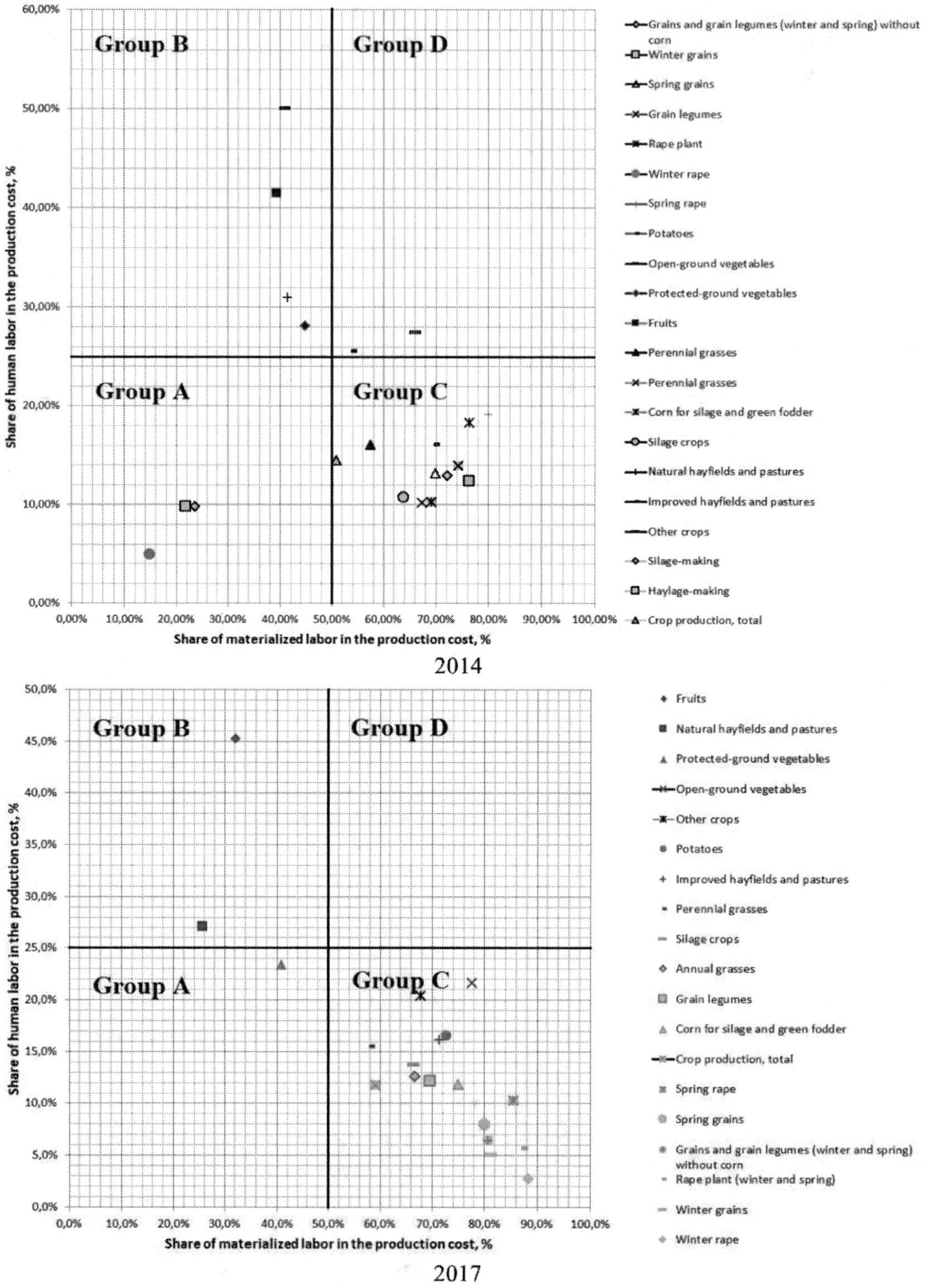

Fig. 4. Grouping of the agricultural crops of the Pskov region according to the share of human and materialized labor in the production cost in 2014 and 2017.

For the period from 2014 to 2017 the following changes have taken place in the distribution of the basic crops of the region:

1. Group A was freed from haylage-making, silage-making and winter rape and, at the same time, had been joined by protected-ground vegetables;
2. Group B was freed from protected-ground vegetables and other crops, preserving fruit crops, natural hayfields and pastures;

3. Group D was freed from open-ground vegetables, improved hayfields and pastures;
4. Group C accumulated the biggest variety of crops in crop production of the region.

Besides the heterogeneity of the cost structure of the basic crops in the Pskov region, a ratio of the share of human and materialized labor of municipal units of the region also varies (Fig. 5).

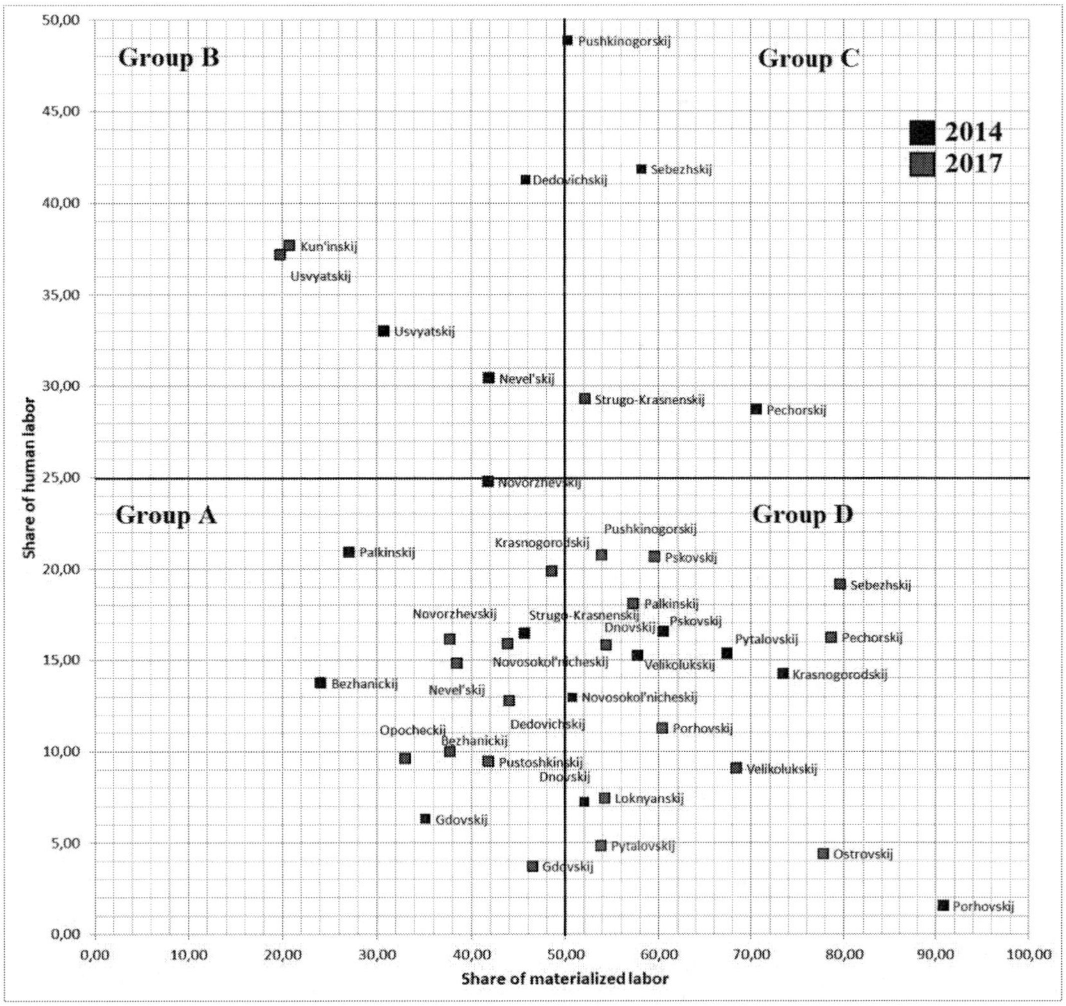

Fig. 5. Grouping of the municipal units of the Pskov region according to the share of human and materialized labor in the crop production cost in 2014 and 2017.

In 2014, there is a fairly strong scatter of municipal units among all groups, including groups with an increased value of the share of human labor.

In 2017, besides the distribution of the basic agricultural crops of the region, there were also such changes as an increase in the concentration of municipal units in the groups with a decreased value of the share of human labor and an increased value of materialized labor (groups A and C).

Undoubtedly, the reduction of energy intensity and the growth of capital intensity against the background of the advanced growth of material intensity of crop production, the high concentration of agricultural crops and municipal units in Groups A and C indicate a certain level of innovative development of the enterprises in the industry.

4 Conclusions

As the important indicator of innovation activity in the regional agro-industrial complex, the cost structure also allows to identify the main reserves for cost reduction. It is a reduction of the production cost that is one of the main conditions for accelerating innovative development. It allows enterprises of the industry to form savings (a source of expanded reproduction of the fixed production assets).

The study was performed with the financial support from Russian Fundamental Research Fund and the Administration of the Pskov region in relation to the scientific project № 18-410-600003 "Development of a methodological approach to evaluation of the effectiveness of subsidization as a mechanism to increase the innovation activity of crop production in the Pskov region".

References

1. E.G. Lezina, Engineering technologies and systems **17(1)**, 211-213 (2007)
2. S.A. Adajkina, T.A. Salimova, Engineering technologies and systems **3-4**, 20-24 (2000)
3. Z.A. Kostina, Modern problems of science and education **5**, (2009) URL: http://www.science-education.ru/ru/article/view?id=1259 (access date: 01.04.2019)
4. E.G. Saranceva, Engineering technologies and systems **3-4**, 51-55 (1999)
5. A.A. Kislov, A.F. Kislov, Far Eastern Agricultural Bulletin **2(38)**, 87-91 (2016) DOI:10.24411/1999-6837-2016-00031
6. G.G. Maslov, A.B. Hejfec, Polythematic Network Electronic Scientific Journal of Kuban State Agrarian University **115**, 1089-1110 (2016)
7. I.I. Piunovskij, V.R. Petrovec, N.I. Dudko, Bulletin of the Belarusian State Agricultural Academy **1**, 98-101 (2016)
8. N.M. CHernysheva, E.S. Altynbaeva, Technics and technology of food production **43(4)**, 204-211 (2016)
9. *Unified interdepartmental information and statistical system* http://www.fedstat.ru/indicator/data.do
10. V.P. Zotov, O.A. Kon'shina, E.V. Neustroeva, S.G. CHernichenko, Technics and technology of food production **40(1)**, 120-125 (2016)
11. YA.D. Van Mansvel't, S.K. Temirbekova, Agricultural biology **52(3)**, 478-486 (2017)

E3S Web of Conferences **112**, 03016 (2019) https://doi.org/10.1051/e3sconf/201911203016
TE-RE-RD 2019

Kinematic parameters determination of the drilling equipment designed to work in slope degraded terrains

Alexandra Liana Vişan [1*], *Gabriel Constantin* Bogdanof[1], *Dumitru* Milea[1], *Radu Ciupercă*[1] and *Gabriel* Gheorghe[1].

[1]INMA Bucharest, Research Development Innovation Department, 6 Ion Ionescu de la Brad, Romania

Abstract. This paper aims to present the cinematic parameters determination of a drilling equipment design to work in slope degraded terrains, based on field experimental data acquired in normal exploitation condition and also in heavy environmental conditions. This equipment is developed in order conduct forestation and rehabilitation works of degraded lands and to be incorporated on innovative planting technology (degraded soil processing, levelling and terrace surface modelling, planting settlement, etc). This technology answers to the need to implement measures of environmental rehabilitation and enhancement of slope-degraded forestlands, by development several performing equipments that answer to technological needs of this works (land stability, terrace dimensions, type of power source and seedlings planting variety) in order to rise to the performances and to respect the labour safety standards under difficult working conditions. Also, here in presented equipment has a great impact on forestry and pomiculture works, and also on environmental technical works meant to improve soil quality, lowering the surface and depth erosion, rainwater capitalization, lowering the dust pollution and natural calamity risk, and at last to rehabilitate the inland flora.

1 Introduction

In Romania, lately are made considerable efforts and a large campaign, by civil society and to stop the illegal deforestation and to protect protected areas, but also the virgin forest and natural parks, Fig. 1. The national territory, covered with forest has decreased at 28.95 % (approx. 7 million ha), figure that is lower than the UE average that is 43 %. A significant part of it, approx. 415,000 ha, belong to small owners (areas smaller than 100 ha) which are poorly administered and exposed of illegal cuts. If we refer to virgin and quasi-virgin forests, areas in which the nature has its one rules of development and without human intervention, is estimated at 3% from the total forest lands, compared to 12 % form 1974, and many of them represent secular oak forests were included in UNESCO Mondial Patrimony. [1]

* Corresponding author: alexandrion1982@gmail.com

The yellow areas from Fig.1, represent low risk level for framing sector and involves form authorities' point of view, close monitorization, the red areas represents significant risk level for framing and involves control plans foundation regarding the implementation EUTR (European Union Regulation on Wood Trade), respectively to: evaluate periodic the risks in those area; to identify them and to establish the appropriate measures to mitigate them.[2]

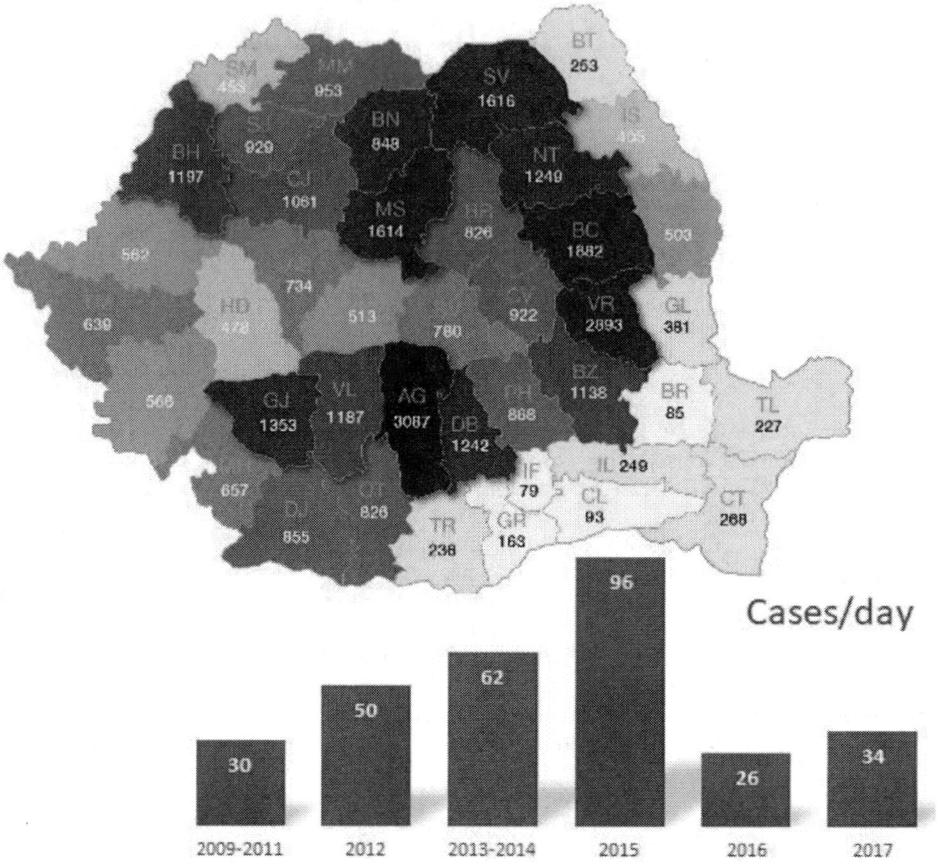

Fig.1 Romanian reginal map of illegal deforestation in 2015 and its evolution until 2017 [3,4]

2 Methods to mitigate the environmental and ecologic risks

Making the correlation between those two official statistics, it can be concluded that, the regions with highest illegal cutting cases/day are the regions with highest risk (respectively Gorj- GJ., Vâlcea - VL., Arges - AG., Vrancea - VR., Bacau-BC., Neamt - NT., Suceava - SV., Bistrita-Nasaud - BN. and Mures - MS.) which are mainly situated in hilly and mountain landscapes, fact which implies that the risk assessment depends on soil erosion and landslides. Analysing, also the maps regarding those evaluations [5, 6], was noticed that those regions are also with high risk and for this reason the mitigation measures that can be implemented in deforestation terrains can be adequate.

Apart from the legislative measures that are supported by WWF (World Wide Fund for Nature), another only way to mitigate the forest loss and the environmental risks, is to restore the forests on degraded lands by, two technologic processes, respectively afforestation and reforestation. The difference of those terms is the afforestation represents conversion of other land-uses into forest, while reforestation is re-establishment of forest formations. [7]

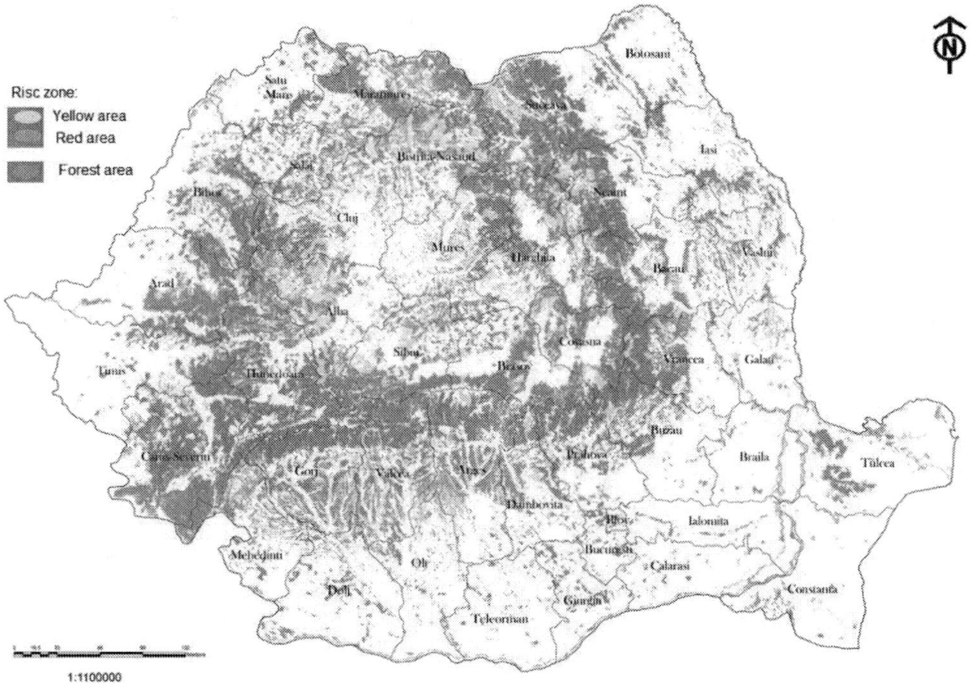

Fig.2. Risk map regarding the illegal cuttings [4]

Taking in to consideration the fact that every area has a specific fauna that makes part of an ecologic group, INMA Bucharest has develop an innovative afforestation technology that can be adapted to technical-forest measures to regenerate or to plant autotune trees and shrubs in order to increase the productivity and to reduce the human labour. [8, 9, 10] More than that, the mechanized operations developed had taken in to consideration the soil quality, because, first of all the ground must ensure the aggregate equipment stability during operation, secondly the operator safety and thirdly the optimal condition to develop the seedlings. In order to re-establish the soil stability (in slope degraded terrains – mountain and hilly areas) it is necessary to be accomplish specific works that have a high efficiency in diminishing superficial leakage and retention water. The most efficient slopes proved to be terraces with platform in the contraption, especially during the establishment period (spring and autumn) and for this reason the drilling equipment comply specific requirements and must incorporate special systems in order to adapt to terrace angle and seedling type (with nude rooted or with earth bale) as: Auger position adjustment system used to make platting pits on sloping land [11], system for rapid mounting and damping the axial shocks of the drill for making the planting holes [12]; deviations compensation device during the drilling process using a guided auger by parallelogram mechanism [13].

In order to establish the drilling aggregate performance were performed several experimental tests in relevant conditions, namely in ground land and in cultivated land. The firs condition (ground land) is representative because in some cases the land owner do not consider to make land improvement works (scarifying, milling and levelling) and the other one when the soil it requires them.

3. Experimental procedure and equipment's used

The aggregate is made from small gauge power source, 318 BERTOLINI motocultivator type of 19 HP, equipped with an innovative drilling equipment - BM [7,11] designed to work in difficult soil conditions in order to generate seedling dwellings. In Fig.3, are presented three imagines regarding the tested equipment: a – drilling equipment, b- working aggregate, c – specific adjustments for auger inclination.

a) b) c)

Fig.3. The drilling equipment working in aggregate with the power source

The BM equipment weights max. 112 kg and is provided with two earth augers of Ø 300 and Ø 400 mm, that can work at 300 mm and 400 mm depth. The augers are guided by the adjusting position system (designed for max. 30^0 left and right inclination), in order to achieve vertical planting holes regardless of the terrace counter-slope. The auger working speed is controlled by the human operator, using the motocultivator acceleration handle and its powered only when the auger is positioned on the soil surface, the maximum rotation speed is 135 rot. min^{-1}. This equipment is required to be handled by two workers who have been previously trained and familiarised with this equipment. One to serve the motocultivator - controlling the auger speed, and the other the BM equipment - controlling the auger working parameters (the hole depth and the earth drilling speed).

The data from the experimental field activity was carryout the measuring instruments from Table 1.

Table 1. Measuring instruments used on experimental field.

Measuring instruments	**Measuring range**	**Precision**
Weighing machine	0 ÷150 [kg]	0.01 [g]
Digital timer	0 ÷ 30 [min]	0.2 [s]
Soil Compact Meter SC 900 - Spectrum	0÷45 [cm], 0÷7000 [kPa]	±1.25 [cm], ±103 [kPa]
Hygrometer HH-2	0 ÷ 50 %	± 3 %
Roulette	0 ÷ 8 [m]	0.5 [mm]
Dynamometer	0 ÷ 300 [daN]	±5 [daN]
Level indicator	1.5 [m]	
Linear	2 [m]	
Graduated glass	500 [ml]	

The first step in the experimental procedure was to measure the physic-mechanical soil properties as: compaction and humidity measurements at different depths. The measurements were made for two land strips: on cultivated (CL) and grounded land (GL), and the results are presented in Fig.4.

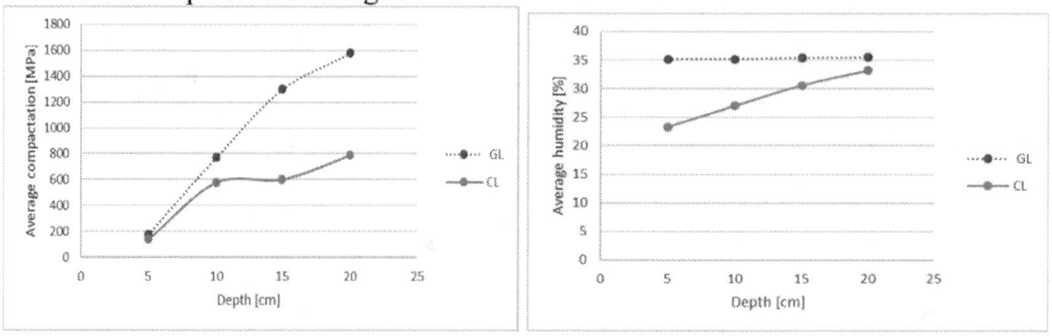

Fig.4. The soil proprieties for grounded land - GL and cultivated land – CL.

 https://doi.org/10.1051/e3sconf/201911203016

After that, the aggregate is moved into the experimental field, where had been created terraces with various slope, respectively 0, 15 and 20 ^{0}, and for each case are made three determinations. In Fig.5.a, is presented an image from the experimental field for a terrace of 20 ^{0} slope, on which was previously cultivated (milled and levelled).

Another case is the one on which the aggregate is used on slope lands were have not been made soil works and for this reason is covered with vegetation. This case study is important to test the technical limitations and working limits. In Fig.5.b, is presented an image from the testing ground that had a slope of 20 ^{0}.

a) b)

Fig.5 Images from the experimental test fields

The data collected from this activity had been centralised and analysed according to scientific literature methodologies.[14]

Experimental results and discussions

The most relevant experimental data are presented in the graphs from Fig. 6.

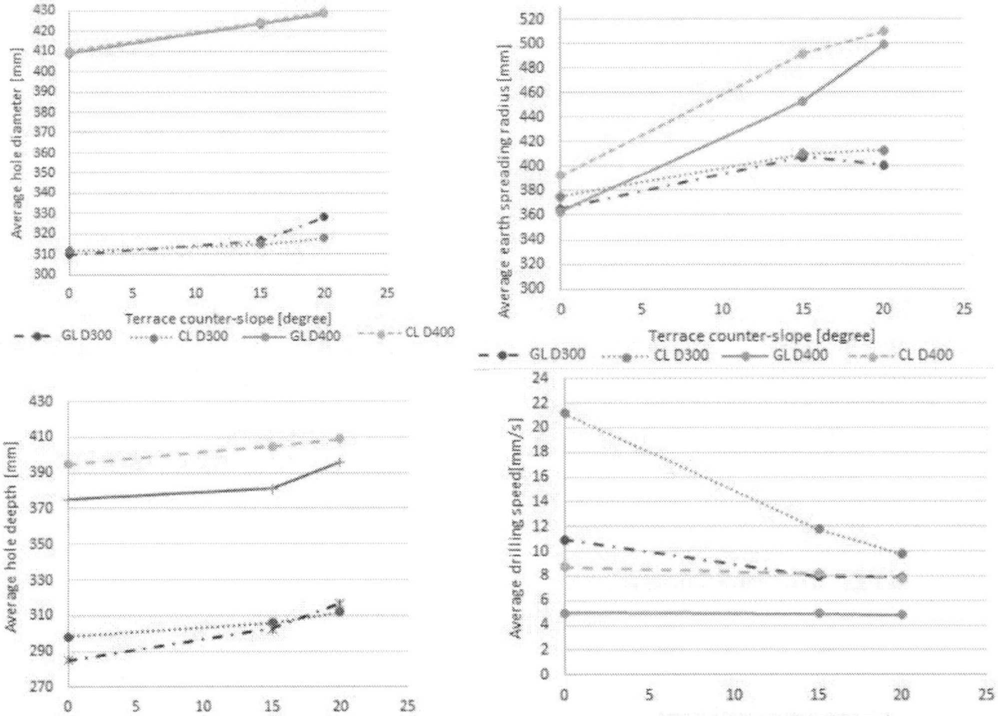

Fig.6 Most relevant working parameters determined in the field research with terrace slope.

Appling the mathematical algorithms for drilling equipment's on the above results, were determine the qualitative exploitation indices, Table 2. The power consumed during the working process is influenced by auger rotation moment (max. gear motor speed of output shaft) and by the transmission mechanical efficiency (that in our case it is evaluated at 0.8). The global specific resistance coefficient for drilling is 2 and 3 kgf/cm^2, for CL and respectively GL, and the gear motor speed is 135 rot/min.

Table 2. Estimation of BM qualitative exploitation indices.

Qualitative indices	Soil type	Terrace counter-slope					
		0^0	15^0	20^0	0^0	15^0	20^0
Auger diameter, mm		**300**			**400**		
Average depth, cm	CL	29.8	30.6	31.2	39.5	40.5	40.9
Drilling period, s		15	26	32	45	49.3	52.3
Auger advancement, cm/rot		0.88	0.52	0.43	0.39	0.37	0.35
Auger rotation moment, kgf/cm^2		1247.6	739.1	612.3	980	917.2	873.1
Average depth, cm	GL	28.5	30.26	31.66	37.5	38.1	39.6
Drilling period, s		28.8	38	40	75.4	76.8	81.6
Auger advancement, cm		0.44	0.35	0.35	0.22	0.22	0.22
Auger rotation moment, kgf/cm^2		1087.55	875.15	869.86	971.71	969.26	948.16

From the above results it can be concluded that, the auger rotation moment depends by soil physical-mechanical properties: the global specific resistance coefficient for drilling is 2 kgf./cm^2 (CL) and of 3.5 kgf./cm^2 (GL).

Knowing that, the rotational resistant moment M_v of the radial augers with two radial blades mounted on two constant helices and a conical point with two helical drilling channels, can be determined the vertical resistant force R_v that is composed from R_r- earth splitter resistance force and R_d – earth splitter deformation and displacement.

The radial forces that act on the same level plane, on auger blades, at central point are annulated, so remain only the vertical force R_v and its moment M_v, that have the mathematical expressions from equation (1) and (2).

$$R_c=2[dR'_t*\sin\tau - dR'_d*\cos(\alpha+\varphi)],\ daN \quad (1)$$

$$M_v=2r*[dR'_t*\cos\gamma_v*\cos\tau+dR'_d*\sin(\alpha+\varphi)*\cos\gamma_v],\ daN/cm \quad (2)$$

Where: α - the inclination cutting blade; γ_v - the angle of upper cutting blade surface in accordance with radial direction; τ - the angle generated by dRt with the suiting earth surface, and the elementary forces dR'_t and dR'_d which can be calculated with equation (3) and (4), where the terms have the next signification: k'_t - the specific resistance force of earth splitter, daN/cm; k'_d - the deformation force of the earth splitter, daN/cm; ds –the conic cutting surface length, cm; a_s – the earth splitter section, cm^2.

$$dR'_t= k'_t* ds \quad (3)$$

$$dR'_d= k'_d* a_s \quad (4)$$

Replacing the elementary mechanical measures are elaborated the equation (5) and (11).

$$M_v=0.125\ (D_b^2-D_v^2)*i_c*[k_t*\sin\tau-k_d*(S/i_c)*\cos(\delta_{oc}+\varphi_c)],\ daN/cm \quad (5)$$

Which contain the terms: D_b – auger diameter, cm; D_v – head diameter, cm; τ - the earth detachment splinter angle with auger horizontal plane, ^{0}; φ_c – the friction cuts angle with earth, ^{0}; δ_{oc} – the upper surface cut angle with horizontal plane, ^{0}; S- the auger advances at one rotation, cm/rot; i_c – the number of auger blades. Cons derating the fact that, it is known only the vertical auger moment can be determined only one parameter at one time of earth splitter specific resistance force k_t, equation (6), which it is dependent of 3 variables.

$$k_t = f\,(M_v,\ k_d,\ S),\ daN/cm^2(daN/sq\ cm) \qquad (6)$$

By replacing the constructive parameters for thouse two augers in the mathematical expression (6) are obtained: equation (7) for 300 mm auger and equation (8) for 400 mm.

$$k_t^{'} = (M_v/892.16)*0.65+444.29*k_d*S,\ daN/cm^2\ (daN/sq\ cm) \qquad (7)$$

$$k_t^{''} = (M_v/1592.15)*0.65+792.89*k_d*S,\ daN/cm^2(daN/sq\ cm) \qquad (8)$$

In order to draw these functions which, have to vectors (M_v and S) and one undefined variable k_d, it was made a mathematical assumption, k_d = 1, so to this variable to be a contestant and unitary value, in order to not influence the function.

As it can be noticed in Fig.7, the k_d evolution variates very much in cultivated land case and for large auger diameters decreases its tendency. In the case of grounded land, the variation is smaller and for big diameters is almost constant on all terrace slope, is not influenced by land inclinations.

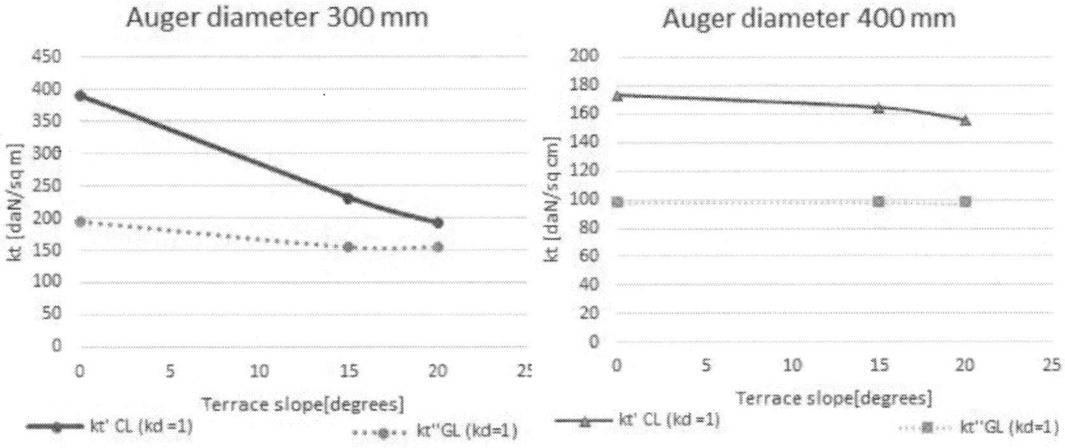

Fig.7 Dependency of k_t function with terrace slope, when k_d = 1 for different auger dimensions.

In order to draw also the dependency of earth splitter deformation force k_d, it must make the same procedure, to establish the mathematical equations from two auger type (9) and (10) and to draw the charts so the k_d to have a minimum influence (k_d = 1).

$$k_d^{'} = (892.16*0.65*k_t\text{- } M_v)/444.29*S,\ daN/cm^2(daN/sq\ cm) \qquad (9)$$

$$k_d^{''} = (1592.15*0.65*k_t–M_v)/792.89*S,\ daN/cm^2(daN/sq\ cm) \qquad (10)$$

From Fig. 8, it can be concluded that this factor has a strange evolution and for this reason must be made many other research activities for those augers and in the similar testing conditions. The correlation between k_t and k_d is difficult to make, because are different order of magnitude and for this reason have no intersection points.

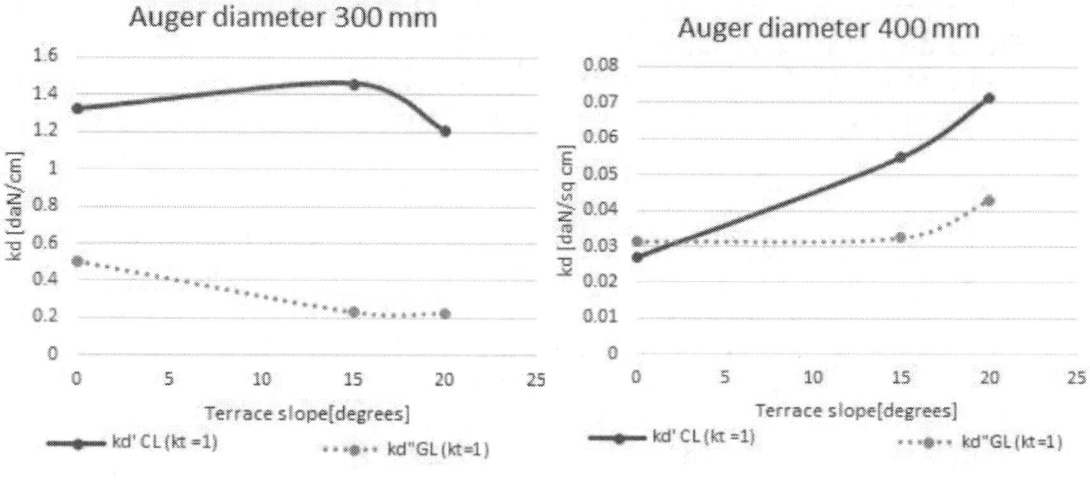

Fig.8 Dependency of k_d function with terrace slope, when $k_t = 1$ for different auger dimensions.

In order to determine the characteristics of auger vertical resistance force R_c, was used the mathematical expression (11) that also is influenced by k_t and k_d, variables that are dependent between them and hard to determine in our case, because we do not have enough data, so they were replace with M_v and after that where replaced same of the auger geometric construction parameters, and in this way was obtain the equation (12) and graphs from Fig. 9.

$$R_c = 0.5*(D_b - D_v)*i_c*[k_t*\sin\tau - k_d*(S/i_c)*\cos(\delta_{oc}+\varphi_c)],\ daN \quad (11)$$

$$R_c = [0.5\ (D_b - D_v)*i_c*M_c\ ]/0.125*(D_b^2 - D_v^2),\ daN \quad (12)$$

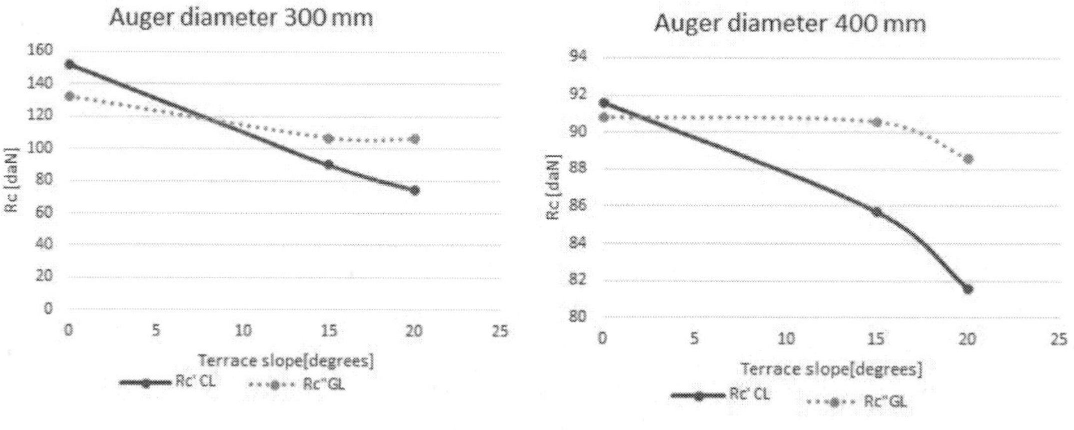

Fig.9 Dependency of R_c with terrace slope and auger dimensions.

The R_c has a linear tendency when the auger is working in GL condition especially for larger diameters, for CL conditions the variation is greater.

The main conclusion is that, all the values presented in this paper are influenced by the human factor, respectively by the operator physical force that is identified as R_V, and the experience of the operator in those working conditions, respectively the soil composition that mainly is the same during the exploitation. This factor also influences the power consumption and the aggregate productivity.

Acknowledgements

This work was supported by a grant of the Romanian Research and Innovation Ministry, through Programme 1 – Development of the national research-development system, subprogramme 1.2 – Institutional performance – Projects for financing excellence in RDI, contract no. 16PFE and research was conducted on project "Integrated management of works on agricultural, wine and fruit farms", financed by NUCLEU Program - PN 19 10 01 05.

References

1. https://www.greenpeace.org/romania/ro/campanii/paduri/Activitati/raport-taieri-ilegale-2015/Date-i-cifre-despre-situaia-tierilor-ilegale-de-arbori-in-2015/;
2. http://www.wwf.ro/ce_facem/paduri/combaterea_taierilor_ilegale/?gclid=Cj0KEQjwgODIBRCEqfv60eq65ogBEiQA0ZC5-b8b2hWfNErfUQ3uYdzEP2sXZfe1TsFH8MBX57rLB-AaAiEl8P8HAQ;
3. https://www.rri.ro/ro_ro/raport_privind_taierile_ilegale_de_paduri_in_romania-2586909;
4. http://www.wwf.ro/_core/general.cfc?method=getOriginalImage&uImgID=%26%2AB8%2A%20%2EW4%0A;
5. https://www.scribd.com/doc/56047299/alunecari-de-teren;
6. https://stoianconstantin.wordpress.com/category/energie-eoliana/;
7. C.G. Bogdanof, A.L. Visan, D. Milea, R. Ciuperca, A.I. Petrisor, SIMI 2018 proceedings book, 127, (2018);
8. A.L. Vişan, G.C. Bogdanof, D. Milea, G.V. Ciobanu, AGIR Buletin no.4, 62, (2016);
9. A.L. Vişan, G.C. Bogdanof, D. Milea, G.V. Ciobanu, C. Mircea, HERVEX Proceeding, 383, (2016);
10. C.G. Bogdanof, V. Moise, A.L. Visan, G.V. Ciobanu, Engineering for rural development, Jergava, 653, (2017);
11. C.G. Bogdanof, A.L.Vişan, A. Paun, D. Milea, OSIM patent application A0075,(2016)
12. C.G. Bogdanof, A.L. Visan, G.V. Ciobanu, A., D. Milea OSIM patent application A01001/12.12.2016
13. C.G. Bogdanof, A.L. Vişan, A. Păun, Milea D. OSIM patent application A0056, (2017)
14. C. Ciocirdea, Masini de spat gropi, Creres Publisher,51, (1973);

Researches on drying the clover through aeration to obtain hay

Ana Zaica[1*], *Ancuța* Nedelcu[1], *Radu* Ciupercă[1], *Alexandru* Zaica[1] and *Alexandra-Liana* Vișan[1]

[1] INMA Bucharest, Romania

Abstract. Harvesting, preparation and conservation of hay is a frequent method for harvesting fodder from grasslands and forage crops, especially from the hill and mountain areas. The greatest losses of nutrients are obtained by traditional drying on the stubble of the hay. These losses are in the case of unfavourable weather even at 50 ... 60%, the drying time increasing to over 6 ... 8 days. This paper presents the results of the experimental researches achieved within a new technology for shortening the time from harvesting the clover to obtaining the hay at storage humidity. A pilot ventilation installation with a monitoring and control system for working processes was used to carry out the research. During the experiments, the parameters of the aeration agent were measured: temperature and humidity, as well as clover mass characteristics: humidity, temperature, drying uniformity.

1 Introduction

The directions for development and profitability of the agro-food production, the increase of the volume and the quality of the agricultural production by diminishing the losses in the zoo-technical sector can only be achieved by upgrading, modernizing the machinery and the technical equipment used for carrying out the basic operations within these technologies. These are possible only by making new types of specialized machinery and equipment with high reliability and low operating costs that ensure the application of technologies for the harvesting and preservation of fodder plants with superior technical and economic parameters.

Of particular importance is the contribution of meadows, consisting of pastures and hayfields, to the production of fodders necessary for the development of the zoo-technical field. Because a feed with a floral composition and complete nutritional value, in turn determines a balanced composition in proteins, vitamins, salts and other substances are the basic feed of the animals.

Fodder plants represent an important part of plant production and are subordinated to general principles and methods of agro-technical. However, they have their own biological, ecological and technical particularities, which distinguish them from other agricultural

*PhD. Stud. Eng. *Ana* Zaica: zaica_ana@yahoo.com

plants involved in human nutrition, although some of them are used as feed. We can speak of a direct relationship agricultural plants - fodder plants - animal nutrition, [1-3].

Perennial legumes (such as alfalfa, clover, sparse, orchard) in the meadows have the most important role in producing a high-quality feed rich in protein, plus the extraordinary attribute of atmospheric nitrogen, living in symbiosis on the roots with a bacterium called Rhizobium sp. This miraculous process benefits organic farming, [4].

Leguminous plants represent an important group of plants that participate in the planting of the vegetation of the meadows, but in a smaller proportion than the grasses. Legumes are typically 5 to 10 percent of grassland vegetation and have a high economic value due to valuable chemical composition, high yields, and high digestibility. The degree of consumption is very high in most legumes, both green and hay, [5-7].

Clover due to the high content of nutrients is highly appreciated in the daily fodder of animals, especially dairy cows, and under optimum harvesting technology; high quality hay is produced. The nutritional value of the green clover is maximal in the early flowering phase, which corresponds to the optimal harvesting period. Clover thistle is also rich in protein and calcium and contains large amounts of carotene when it is preserved and preserved under some conditions. In animal feed it is used similarly to lucerne. [8].

The optimal harvesting period for clover and lucerne corresponds to the beginning of the flowering phase when the protein content is maximum and corresponds to a plant humidity of about 75 … 80% and the production per hectare is satisfying.

Obtaining valuable feeds, as well as composition, requires special attention to the exploitation of meadows through the use of harvesting and conservation methods and technologies with minimal losses and reduced costs. Climatic and landscape conditions also allow grass and leguminous fodder plants to be grown to conserve fodder throughout the year.

Harvesting and preparing, then preserving grass and fodder feed in the form of hay are the basic work to get a final product with a nutritional value closest to the original feed of the green feed. In order to complete the drying of the clover fodder in the warehouses, special ventilation systems with cold air or hot air are required, suitably sized according to the farm's feed requirements, [9].

2 Methodology

Preparing the clover hay using the ventilation method with cold air or hot air in the process of aeration drying by means of the ventilation system consists in removing water from the feed in a shorter time by using the air streams introduced by the fan. In this way, losses of nutrients and vitamins, as well as mechanical losses, are much lower and the harvested hay is green, pleasant aroma and a high content of vitamins.

The pilot plant (IVF) for finishing the drying of the hay is designed to reduce the humidity of fodder stored from humidity of 35 ... 40% at humidity of approx. 17% by ventilation with cold or heated air for optimal long-term conservation and obtaining of quality feeds in the final phase of feed drying technology, Figure 1. In order to heat the ventilation air, the installation uses solar radiation, which is heat converted with the aid of solar panels and transported via aluminium tubes [10].

Fig. 1. Phases of feed drying technology by air ventilation.

For that, determinations and measurements were carried out following the variation of the main parameters:

- atmospheric air temperature and humidity, used as an aeration agent during the drying process of the feed;
- air temperature and humidity in the solar panel, part of the installation;
- the temperature and humidity in the feed mass (clover) deposited on the plant platform for drying for optimal storage;
- the duration of the drying process by ventilation.

In order to measure the humidity of the feed during the aeration process, samples from 4 different areas of the platform Figure 2 were collected to trace if the clover hay was uniformly dried.

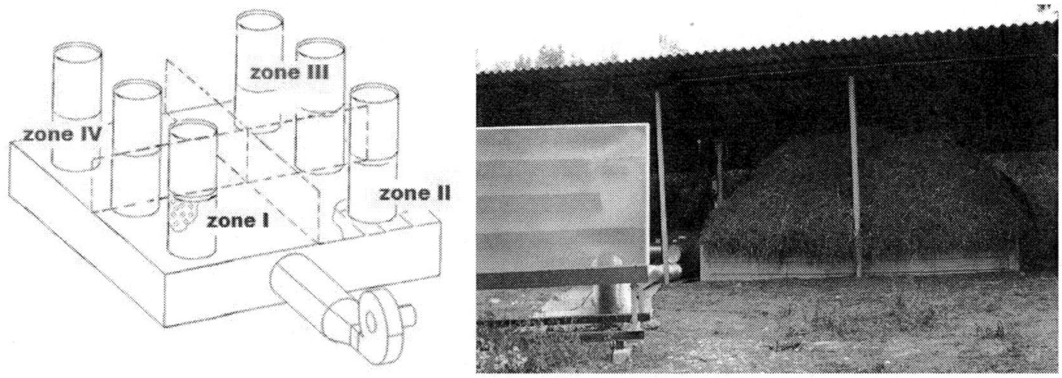

Fig. 2. Schematic diagram of the platform, component part of the installation, with division by areas and installation loaded during experimental research.

The test period is characterized by high summer atmospheric temperatures (July) and low air humidity, the material undergoing the aeration process is composed of clover fodder plants harvested from an experimental lot.

The clover drying operation deposited on the platform begins when the ventilation system is switched on and the fan suction air (cold or heated) is introduced through the intake channel into the uniformization chamber from where it is diffused into the mass of the layer of fodder stored. A layer of material with a height of at least 2 m is thick enough to prevent rapid flow of air through the plant mass.

The ventilation program adapts to weather conditions and storage parameters on the experimental plant platform.

In the technology of drying the heated air feed, it is necessary during the aeration process to continuously control the temperature in the feed mass. It must not rise above 30°C. When hot air was used for ventilation drying, the drying process ends with cold air to ensure that the temperature of the feed mass is balanced with atmospheric air temperature, thus avoiding any condensation.

During experimental research on experimental research, to achieve the objectives are establish and determines the functional characteristics and working indices presented in Table 1.

Table 1. Determination of functional characteristics and working indices

No. crt.	Specification		UM	The determined value
1	Fodder type		-	White clover
2	The drying period		-	July
3	Humidity of fodder when loaded on the drier		%	45,38
4	The height of the fodder layer		m	2
5	Active dimensions of the drying platform	Total length	m	6
		Width	m	6
		Height	m	0,65
6	The total surface of the drying platform		m^2	36
7	The overall height of the hay		m	4,6
8	The initial fodder loaded on the platform (at loading humidity)		kg	3000
9	Number and brand of fans used		buc.	1 buc., tip HM 90 T4 5,5
10	Main Technical Features Fan HM 90 T4 5.5	The power	kW	5,5
		Supply voltage	V	380
		Maximum air flow	m^3/h	45000
		The pressure produced (according to the fan diagram)	mm H_2O	35 mm H2O at the flow rate of 25000 m^3/h 60 mm H2O at the flow rate of 10000 m^3/h
		Fan tube diameter on the platform	mm	900
11	Number of routing tubes for air in the hay		buc.	6
12	Dimensions of routing tubes for air	Diameter	mm	250
		The length	mm	2000
13	Type of the temperature probe in the feed mass		-	Thermal resistance PT100 with temperature control and control panel
14	Technical characteristics for thermo resistance	The maximum temperature	°C	100
		Length of the stem	mm	950
Data relating to the air heating system required for ventilation				
15	Type of installation and main technical features	Type	-	Solar panel
		Type of collectors	-	Solar planers mounted on the ground
		Mounting mode	-	Southern exhibition, mounted on the ground on a special stand beside the hay platform
		Horizontal angle	grade	45
		Total capture area	m^2	540
16	Type of connection to the air heating system and fan		-	Aluminium tube
17	Fan suction tube diameter		mm	400

Therefore, the quantitative and qualitative loss in hay harvesting and preservation is as small as possible, it is necessary to meet the following requirements to control the action of factors that influence the ventilation process proportionally:

- harvesting the plants in optimal phenophase;
- not to exceed the optimal harvesting age;
- shortening as much as possible the time elapsed from the beginning of the harvest to the hay at the optimal humidity (below 17 ... 18%);
- high - performance execution of agricultural works within the technology;
- execution of high quality works by observing imposed agro technical requirements (cutting height, reduced quantitative and qualitative losses for each operation, etc.);
- the use of equipment, machines and aggregates specific to working on sloping land and compliance with the conditions imposed in this case;
- introducing in hay harvesting and preservation technologies the mowing condition of crushing, breaking or defibration, works that shorten the drying time;
- reducing the shelf-life of mowed fodder by using modern dry-drying facilities.

3 Results

The results from the application of the ventilation experiment experimental methodology for experimentation will be presented in order to determine some of the parameters that influence the drying process.

During the experimental research period, the variation of air parameters, humidity and temperature, is presented as data in Table 2 and Table 3.

Table 2. Results of atmospheric air humidity measurements

Day	U [%] air atm.					The average humidity, %
	hour **7.30**	hour **9.30**	hour **11.30**	hour **13.30**	hour **15.30**	
1	88	83	65	52	61	69.8
2	92	86	68	71	59	75.2
3	75	71	52	58	55	62.2
4	82	64	51	37	33	53.4
5	85	58	47	32	38	52.0
6	78	74	60	44	71	65.4
7	64	64	45	37	34	48.8
8	85	75	69	69	59	71.4

During the studied period, the influence of the atmospheric conditions on the drying process is observed in the case of the material placed on the drying platform being the clover.

Knowing from the literature that clover is a leguminous plant, which contains more nutritive substance but has a rather high moisture content. During rainy periods, the drying process will be carried out only by ventilation with cold air for 30 minutes. at 2h. Ventilation is short-lasting so the feed does not absorb excess humidity in the air.

The data in Table 3 highlights high atmospheric temperatures throughout the day, characterized by a warm weather but with a rather high humidity during the morning.

Table 3. Results of atmospheric air temperature measurements

Day	**T [°C] air atm.**					**The average temperature, °C**
	hour **7.30**	hour **9.30**	hour **11.30**	hour **13.30**	hour **15.30**	
1	21	19	29	31	31	26.2
2	19	21	23	21	30	22.8
3	19	23	24	20	21	21.4
4	17	22	28	31	32	26.0
5	18	27	26	26	33	26.0
6	20	21	26	32	22	24.2
7	21	21	27	30	30	25.8
8	21	24	25	30	28	25.6

The results of the humidity and temperature measurements in the clover feed are shown in Table 4, and the data interpretation was performed by plotting the graphs in Figure 3 and Figure 4.

Table 4. The results of the humidity and temperature measurements of the clover fodder

Day	**Mass moisture of clover-fodder samples [%]**				**Average humidity**	**Duration of the aeration process** [h]	**Temperature in the mass of clover-fodder[° C]**				**Average temperature**
	zone I	zone II	zone III	zone IV			zone I	zone II	zone III	zone IV	
1	33.28	35.26	29.32	51.30	37	8	19	21	18	18	19
2	27.63	34.22	30.62	30.19	31	5	18	21	18	19	19
3	15.72	16.78	36.75	15.79	21	4	18	24	18	19	20
4	20.20	26.19	18.81	22.68	22	3	19	20	18	19	19
5	15.86	27.16	16.10	31.59	23	3	19	19	20	19	19
6	13.16	21.64	26.87	14.38	19	7	25	26	23	24	25
7	13.85	18.58	15.12	14.52	16	5	23	19	20	24	22
8	12.16	16.92	11.55	12.50	13	2	25	24	21	23	23

Graphical data mapping shows how the moisture content of the clover material stored for drying in the four areas set as a benchmark during tests varies, thereby determining the uniformity of dryness and shortening the duration of the process.

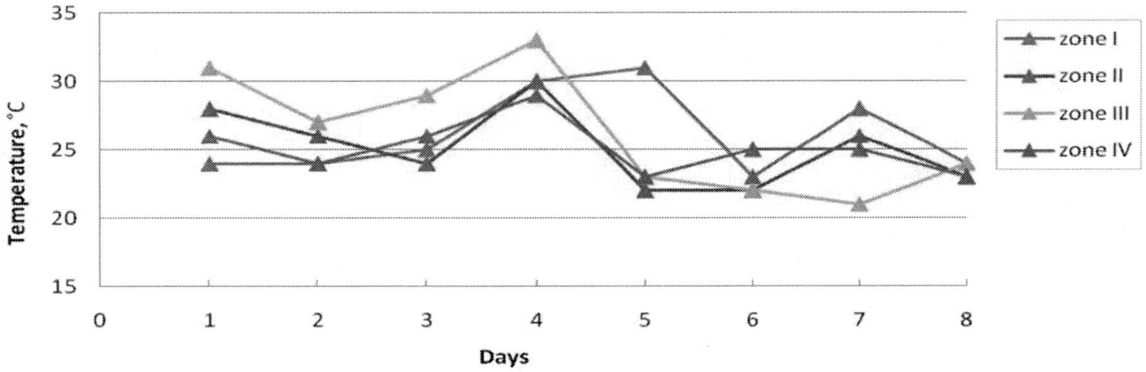

Fig. 3. The temperature variation in the mass of the clover-fodder during the aeration process in platform areas.

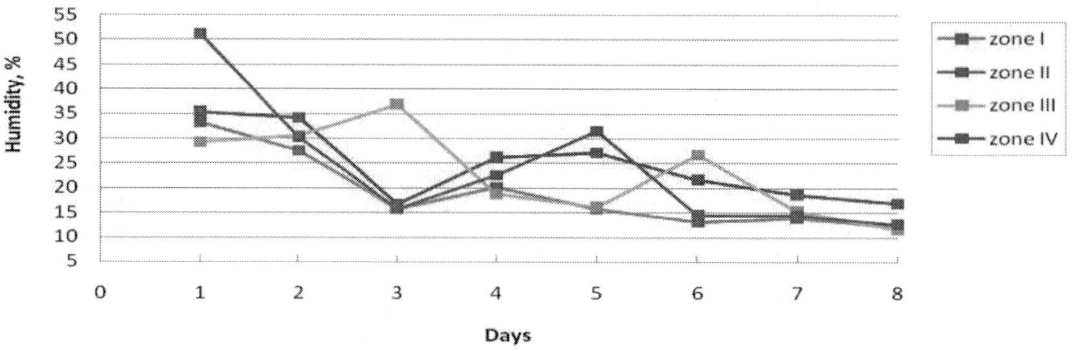

Fig. 4. Variation of the moisture content of the deposited material (clover) during the aeration process in different areas of the platform.

It can be noticed that the temperature variation recorded in the material reaches certain warning points on the material warming in zone 1 (Figure 3) on days 5 and 7, and in zone 3 during the first 4 days of ventilation, the temperature is kept at values that could affect the fodder quality. These high values, which exceed the average of 30°C in feed, are reflected in keeping the initial moisture of the deposited material in zone 3 during the first days of the aeration drying process, even with the slight increase tendency from 30.62 to 36.75%.

Zone 1 (Figure 4) best performs the drying process by loss of feed moisture, instead the most spectacular decrease in humidity is recorded in platform 4 (Figure 4), about 40% loss of moisture during the same period, 8 days.

The comparative analysis of the results of the aeration process over the ventilation drying trial periods under experimental research is expressed in the values and observations in Table 5.

Table 5. Comparative analysis of the obtained results

Characteristics	**UM**	**Fodder**	**Observations**
Type	-	white clover	
Humidity at harvest	%	74,07	
Date of harvesting		july	
Drying time on the ground		Aprox 2 days	Successive turns of the furrow
Drying time	-	09.07: ora 14^{00} 22.07: ora 12^{00}	
Drying time on the feed platform	days	8	
Drying time of starch forage	hours	45	
The initial feed load loaded on the platform (at loading humidity)	kg	3000	
Storage humidity on the platform	%	45,38	Zone variations: 45,38…51,30
Moisture after ventilation	%	12,50…16,92	
Electricity consumption	kWh	199	
The amount of hay obtained when drying is complete	kg	1973	

4 Conclusions

Green fodder at harvest contains a large amount of water, generally between 70% and 85%. For winter storage, it should be subjected to aeration drying so that its moisture content is reduced to below 18%, so storage will be done in optimum conditions without the occurrence of undesirable phenomena (eg. molds).

Experimental research and prior studies can provide a number of advantages for using the aeration drying process in the appropriate installations as follows: it allows to obtain a high quality hay (with a high nutritive value, with a pleasant smell, close to the grass of origin) due to shortening the shelf life of the feed after mowing and reducing dependence on weather conditions during harvesting.

Compared to cold air drying and heated air drying, it has the following advantages:

- reduces dependence on atmospheric conditions during harvesting;
- shortens the drying time of feed;
- the initial moisture content of the dried fodder can be 10 to 15% higher, which leads to an even better hay quality.

By supplementary drying of the hay with mechanical ventilation systems, the loss of nutrients up to 25 ... 30% or even 50 ... 70% in adverse weather conditions is reduced.

Conditions for drying fodders differ from one species to another, depending on their subsequent destination. These conditions refer to the temperature of the drying agent, the humidity and the duration of the drying process.

This work was supported by a grant of the Romanian Research and Innovation Ministry, through Programme 1 – Development of the national research-development system, sub-programme 1.2 – Institutional performance – Projects for financing excellence in RDI, contract no. 16 PFE.

References

1. N. Dumitrescu, *Pajişti degradate de eroziune şi ameliorarea lor*, (Ed. Ceres, Bucharest, 1979).
2. P. Mănişor, *Mecanizarea lucrărilor de producere a furajelor*, (Ed. Ceres, Bucharest, 1982).
3. http://www.scrigroup.com/tehnologie/merceologie/Factori-care-influenteaza-oper63115.php.
4. T. Maruşca, V. Blaj, M. Rusu, *Tehnologii de creştere a valorii pastorale pentru pajiştile montane*, Program ADER 2020, Project 1.3.3./2011, UMPP – ASAS, Bucureşti, (2012).
5. V. Mocanu, I. Hermenean, *Mecanizarea lucrărilor agricole pe pajişti - Tehnologii, maşini şi echipamente*, Ed. Univ. Transilvania, Braşov, (2013).
6. V. Neculăiasa, I. Dănilă, *Procese de lucru şi maşini agricole de recoltat*, (Ed. A92, Iaşi, 1995).
7. A. Muimba-Kankolongo, *Chapter 10 - Leguminous Crops, Food Crop Production by Smallholder Farmers in Southern Africa Challenges and Opportunities for Improvement*, 173-203, (2018).
8. V. Mocanu, *Tehnologie modernizată pentru culturile specifice zonei Făgăraş în vederea instalării stării de agroclimax*, (Ed. Capo-Lavoro, Braşov, 2011).
9. A. Nedelcu, G. Lazar, R. Dragan, V. Ciobanu, ACTA, fas. 4, (2012).
10. A. Nedelcu, R. Ciuperca, L. Popa, E. Voicu, A. Zaica, INMATEH - Agri. Eng. **43**, 81-86 (2014).

Aspects regarding FEM simulation of stress in hammer mill working tool

Georgiana Moiceanu[1,*], *Gigel* Paraschiv[1], *Edmond* Maican[1], *Mirela* Dincă[1], *Iulia* Găgeanu[2] and *Vlad Nicolae* Arsenoaia[3]

[1]University POLITEHNICA from Bucharest, Spl. Independentei no 313, Sector 6 Bucharest, Romania
[2]National institute of research - development for machines and installations designed to agriculture and food industry – INMA Ion Ionescu de la Brad Blv. No. 6, Sector 1 Bucharest, Romania
[3] University of Agricultural Sciences and Veterinary Medicine Iaşi, Mihail Sadoveanu Alley, No. 3, Iaşi, 700490, Romania

Abstract. Considering the need for development and improvement of biomass processing equipment, researchers bring contribution by making suggestions to design also through simulation and modelling of different stresses that occur on equipment's working tools. In this paper the stress in hammer mill working tools was analysed using FEM simulation with the help of SolidWorks 2016 Premium, projecting the geometric model and FEM simulation for a MC 22 hammer mill. The hammer mill has a four bolts rotor on which are the hammer disposed. The hammer mills rotor frequency is about 2940 rot/min. An important result in FEM simulation was the fact that the maximum reaction in the hammer joint was 11635 N. After FEM simulation experimental research were done using shredded miscanthus and energetic willow, the mill being equipped alternately with 4 types of hammers, with corners processed differently.

1 Introduction

According to data provided by Eurostat, renewable energy produced from biomass represents about 60%. Also, in 2016 the total amount of biomass used for bioenergy was 140Mtoe, 96% of which was obtained in the European Union and the other 4% was imported outside EU [1]. Considering this fact scientists are continuously researching for the optimum process to transform biomass in bioenergy. The main purpose of all process improvements is to reduce the energy consumption also on biomass harvesting process. [2].

Considering that the size reduction process is very high energy consumer for biomass processing, scientists tried to determine the optimal size reduction level. They used miscanthus and switchgrass as raw material and applied a combination of modelling and experimental tests. Their conclusions expressed the fact that cost would not increase if the grinded particle size is in the range of 4 to 6 mm. [3].

As specialized literature shows that the machine tool dynamic it is very important in analysing different elements of machining cutting operation such as simulating the cutting

[*] Corresponding author: moiceanugeorgiana@gmail.com

forces and making the model. In order to improve the equipment performance, the elements mentioned are very important, an input in chatter vibration modelling [4 – 7].

Also, scientists are researching through different modelling techniques the dynamics of the machine tool. In paper [8] researchers used a 3D Finite Element Modelling of holder tool assembly in order to obtain stability in milling process. Scientists presented the fact that they applied a 3D FE modelling technique instead of 1D beam modelling. This experiment presented the advantage of connecting element without requiring lumped stiffness. Their results expressed the fact that the number of experiments required were reduced, the iterative identification process was not necessary and also though experimental validation it could be said that proposed modelling technique achieves high accuracy in toolkit dynamics prediction [8].

Considering a hammer mill in paper [9] scientists analysed through numerical modelling the computational fluid dynamics. The grinding process is highly influenced by the flow characteristics in a hammer mill. A MB 7.5 hammer mill was used for the study which also had 4 hammers, a rotor distributed from 4 rows and a 4 mm sieve opening dimension. The fluid flow in the hammer mill was considered turbulent and unsteady because of the rotating of the hammer mill inside the grinding chamber that in known to be turbulent. After simulation the conclusions expressed a higher fluid velocity at the circumference of the grinding chamber and the total pressure inside had a normal distribution [9]. Also, scientists stated that the design of hammer mill structure can be improved by parameters with high reliability. This statement was possible after they used SolidWorks software for hammer mill modelling [10].

Continuing the analysis on hammer mills, from the point of view of modelling and simulation, in paper [11], researchers analysed the design of a hammer mill crusher and made an analysis of the rotor shaft assembly. Both the shaft and the rotor assembly of the hammer mill were modelled using ANSYS. The data obtained presented a deflection of shaft in Ansys analysis of 0.007 mm while in theoretical calculation this deflection was 0.009 mm. Also, the maximum stress identified by applying ANSYS modelling was 36.15 N/mm^2, a value much smaller that yield strength of the material of 45 N/mm^2, which led to the conclusion that the shaft will not fail under the conditions applied [11].

Also, except modelling and simulation scientists contributed to better understanding the biomass processing process, through mathematical model and analysis [12,13].

Because this processing process involves different stages an element analysed frequently by researchers consists in different aspects that can be used in obtaining the raw material for pelletizing, the influence of a various of recipes for obtaining the best biomass pellets and also different methods of lignocellulosic biomass pre-treatment used for biofuel production [14, 15, 16]. All these aspects are relevant in analysing the overall process of biomass processing process and the way it could be used for obtaining the optimum process to transform biomass in bioenergy.

Considering all presented in this paper the stress in hammer mill working tools was analysed using FEM simulation with the help of SolidWorks 2016 Premium, projecting the geometric model and FEM simulation for a MC 22 hammer mill. The simulation was accompanied experimental research done using shredded miscanthus and energetic willow.

2 Material and methods

In order to analyse the particle grinding process following hits from the hammers at the entry time of the grinding chamber, we consider the hypothesis that the hammer rotor spins with the ω=constant, and the centrifugal force maintains the hammers in a radial position. Starting from the conversation laws for impulse and momentum, and also from the hypothesis according to which the hammer angular speed before the shock $\omega 1 = 0$, and the direction of the shock is perpendicular on the hammer edge, the fact that during hammer hits with material particles, it was seen that the hammers get an oscillation movement inverse to the rotor spin, and are disposed following a α angle. So, in this case the collisions between the hammer and the material particles (biomass). Knowing these data as well as the equations for speed after the shock for both the particles and hammer, in this paper finite element simulation for hammer mill working organ stresses accompanied by experimental data, were realized.

2.1 Simulation process

For the equipment used for biomass mechanical processing to respond to economic and viability stresses, the models realized by designers and constructors need to go through a series of stages like modelling, simulation and computer aided analysis.
Design process is described by differential equations with partial derivates. Resolving these equations is realized through following three distinct directions: finite difference method, finite element method, frontier element method.
During the paper, finite element method (FEM) was applied for hammer mill working organs structural analysis. FEM simulation can be realized, in general, only after clarifying some aspects like: beneficiary requests, imposed costs, available materials and technology, product durability. So, for a certain product/part/assembly, we can take some restrictions into considerations: static load and/or dynamic, deformation maximum values, different values of safety coefficients, permitted tolerances, vibration frequencies, product life span, material weight, part shape and its inertia moments, rigidity at different stresses, static and/or dynamic stability, behaviour at different simultaneous loads, etc.
For hammer mill working process simulation, we used Solidworks 2016 Premium software, which permits both design as well as resolving difficult issues in a quick and efficient manner. In order to solve the paper's problem, we used a working station with the following hardware resources: processor Intel I7 64-bit, 3.6 GHz, Videoboard Nvidia 2 Gb, RAM Kingstone 12 GB, operating system Windows 10 Profesional SSD 572 GB.
The first phase for this modelling process was to generate the geometric model, followed by generating the meshing model. Working organ geometry was realized with the help of SolidWorks software and is presented in figure 1 a.
Working organs were created starting from known dimensions for discs, hammers and the rotation shaft.
Assembled model was initially subjected to cinematic and dynamic analysis, with the help of Motion module from SolidWorks. For analysis, the value for rotor speed was chosen at 3000 rpm (usual value in practice). A stabilization for the rotor was observed from the start until the maximum speed in just 1.2 s. Before applying graphical simulation for movement, assemble components between which it exists a 3D contact were mentioned.
Working precision was of 10^{-4}, and the maximum iteration number was of 50.

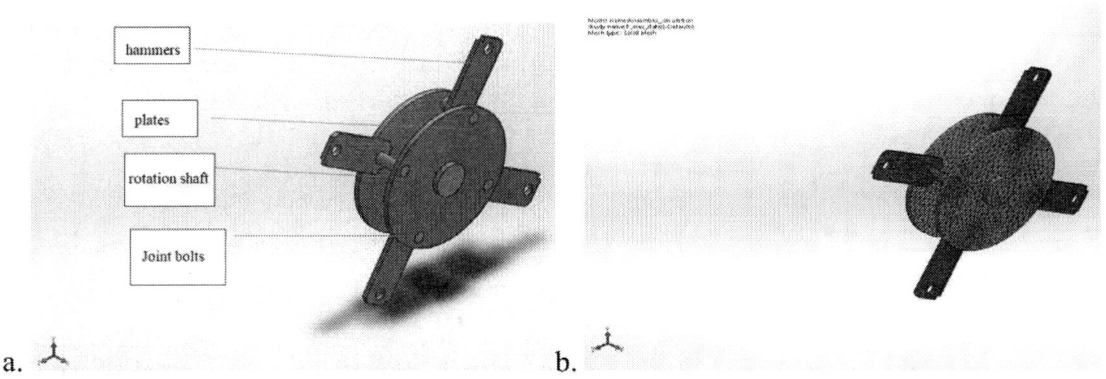

Fig. 1. a. Assembly geometric model. B. Finite element meshing of the model

2.2 Experimental research

Used material during tests was obtained from *Miscanthus x Giganteus*, harvested from the experimental field of the INMA Bucharest institute, as well as from the dried grinded mix of energetic willow (Salix Vinimalis). Both materials were kept in optimum conditions so that the physical-mechanical characteristics, especially the moisture content, remain mostly unchanged. For experimental tests, 5 kg of both miscanthus and energetic willow were used. Particle sizes were measured for the harvested material in a chopped state and average particle dimensions were determined.
So, miscanthus particles had average dimensions of 123-127 mm, while energetic willow had dimensions of 25-47 mm (over 85% of the material).
Equipment used for biomass grinding was a MC-22 hammer mill from the INMA Bucharest testing laboratory.
The hammer mill had a 0.33-0.50 $kg \cdot s^{-1}$productivity on grinding corn stalks, respectively 0.22-0.42 $kg \cdot s^{-1}$ on grinding packaged hay, using a sieve with Ø4 mm orifice size. MC-22 hammer mill is equipped with a 22-kW electrical engine, it has a 220 mm hammer rotor diameter and a rotor length of 500 mm. All 24 hammers are set up in a parallel distribution and have a length of 135 mm. The mills rotor frequency applied was 3000 rpm, 2850 rpm, 2700 rpm, 2550 rpm, 2400 rpm.. The hammer used for testing are presented in figure 2.

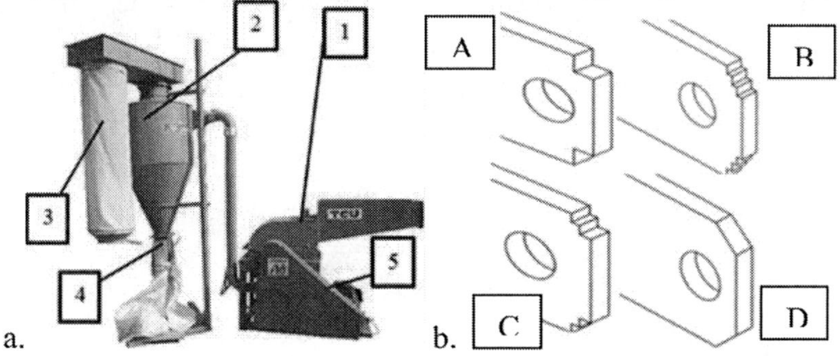

Fig. 2. a. Hammer mill MC 22 (hammer mill; 2. exhauster; 3. cyclone with support and dust collector bag; 4. grinding material evacuation vent; 5. electrical engine; 6. grinded material.), b. four types of hammer (A – one step hammer; B – two steps hammer; C- three steps hammer; D – triangle hammer)

3 Results and discussions

3.1 Simulation process

During the finite element simulation, the maximum reaction in the hammer joint was determined. From the resulting graph presented in figure 3 we can observe a stabilization for hammer oscillations after 2 seconds. The number of frames per second was set at 12000. Determined reaction force represents the force with which the joint bolt acts on the hammer, on the contact surface between the two.
If we select the point in which the reaction is maximum in the hammer articulation, we can observe that it was of 11635 N. Also, from analysing the realized move resulted the speed variation graph.
Starting from the value of reaction, a static analysis was realized, out of which the movements, tensions, as well as frequencies for hammer motion were resulted. In case of static analysis, the subjected material to analysis breaks in the spot where the calculated efforts overcome the material rupture limit.

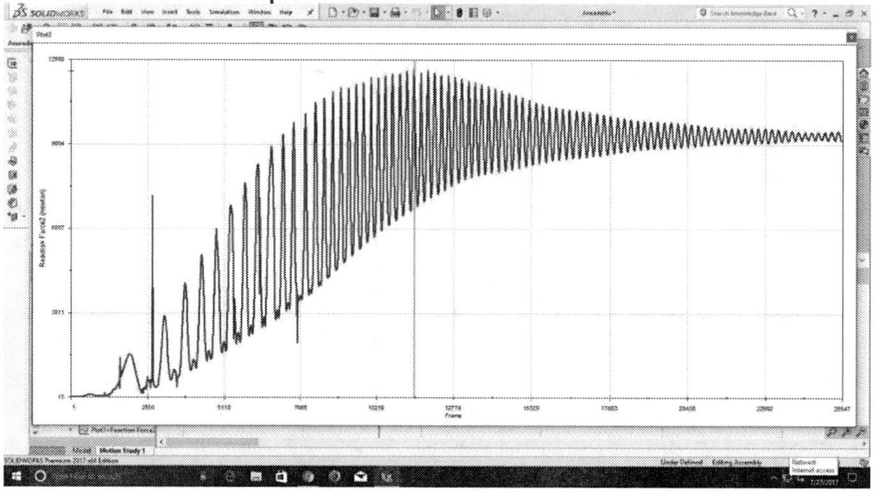

Fig. 3. Reaction force in the hammer joints

Frequency analysis was realized in order to identify if there are any inherent frequencies that coincide with the rotor functioning frequency or if they are lower. We could see that the first inherent frequency for the assembly is 5395.8 Hz, a much superior value to the rotor functioning frequency 314 rad/s.
The model following finite element meshing presented 24279 elements with 44120 nods. For finite element simulation, inside the programme, the model in which the parts were fixed together, the stress positioning and movement restrictions were defined.

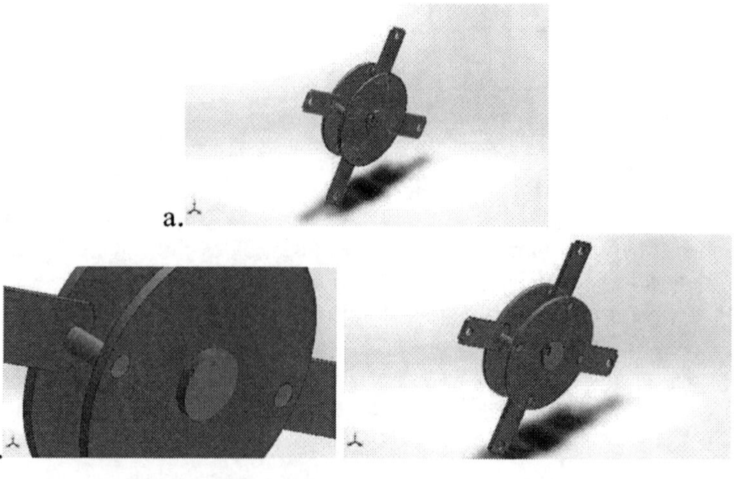

Fig.4. a. Positioning stresses and movement restrictions b. Lateral movement restriction; c. Radial movement restriction

After applying load and fixing conditions for the assembly, the simulation was realized. From the results obtained we could see a maximum deformation of 0.036 mm at the tip of the hammers, due to centrifugal force during the experiment (fig.5). Also, the stresses represent half of the elasticity limit, and the equivalent unitary effort is manifested on the hammers on the surface contact of each hammer and each shaft.

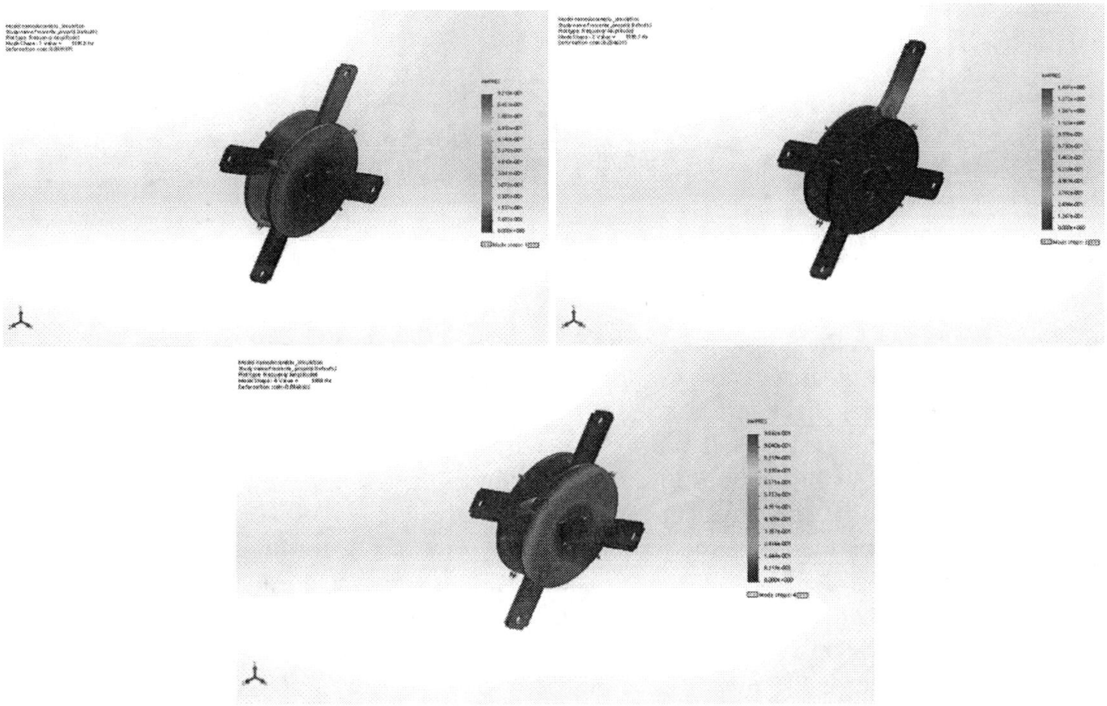

Fig.5. Assembly deformation through applying frequency analysis for different inherent frequencies

3.2 Experimental research

Experimental researches which follow the finite element analysis of stresses that appear in the hammer mill working organs present data and resulted curves for the two-step hammer (figure 2b). These point out, for both types of biomass, the way in which the consumed specific energy varied according to the diameter of sieve orifices and the hammer rotor speed, as well as to the average particle dimension variation in relation to the rotor speed.

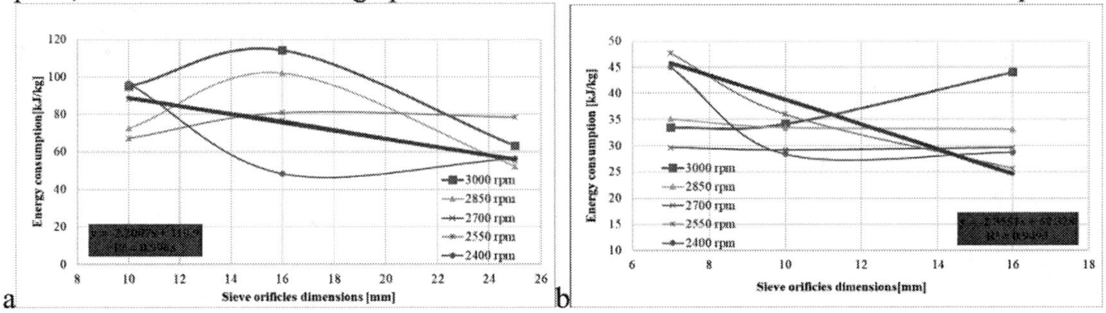

Fig.6. Consumed specific energy variation in relation to sieve orifice dimensions for the two-step hammer (fig 2b), a. Miscanthus X Giganteus, b. Salix Vinimalis

From figure 6, we can observe a normal energy variation in relation to the sieve orifices, also, speeds of under 2700 rpm, but if we disregard the obtained values for the sieve of

ϕ 16 mm, we can say that both small and large orifices for the sieves can be used, but in this situation, the determining factor is given by the destination of the grinded biomass (for pellets, briquettes or for usual burning).

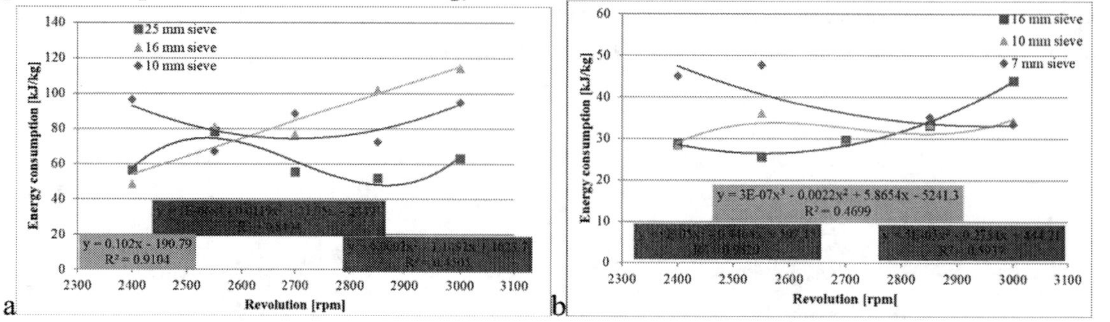

Fig.7. Consumed specific energy variation during grinding in relation to rotor speed for the two-step hammer type B, a. Miscanthus X Giganteus, b. Salix Vinimalis

In the case of miscanthus biomass, hammers with two steps showed non-uniform variation for the specific energy curves for grinding, according to the rotor speed, and observing that there are speed values that can't be used for optimizing the process, due to the fact that they do not fit in mathematical rules. With abstraction to some rotor speed values, the process could be optimized within certain limits, according to the used sieve and the destination of the grinded material if the physical properties remain unchanged. In this case, a relative normal variation for the consumed energy when grinding according to the sieve orifice dimensions was recorded for the ϕ16 mm, with values of 47 kJ/kg to 114 kJ/kg. Also, we mention the fact that for the ϕ10 mm sieve, the minimum specific energy consumption (approx. 75 kJ/kg for miscanthus biomass) was recorded for speed of 2700 rpm, because for smaller or larger speeds the energy consumption was higher.
Variation curves for the two-step knive and the mix of chopped batch of salix vinimalis presents a high energy consumption for the ϕ16 mm at 3000 rpm and a minimum point for energy consumption for the ϕ10 mm sieve and speeds of 2400 and 2700 rpm. We can thus conclude that optimizing the grinding process is possible for each sieve, but at different rotor speeds, if two-step hammers are used (figure 2b).

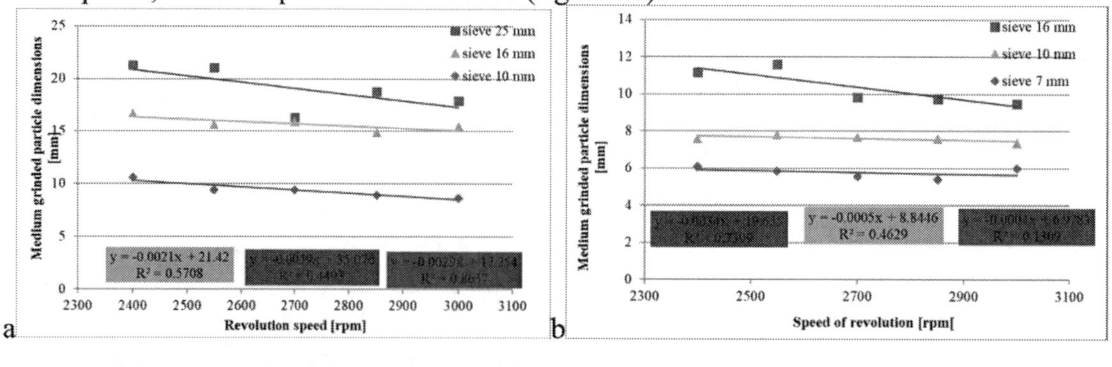

Fig.8. Particle average dimension variation in relation to the rotor speed for two step hammer, a. Miscanthus X Giganteus, b. Salix Vinimalis

From the presented graphs we can say that distribution of variation curves for grinded particle dimensions, in the case of type B hammer, miscanthus, for which the most values are under the dimension of sieve orifices, is normal. Still, these values are relatively close to the dimension of orifices which proves once again the important influence of the air current realized by the mill vent.

In the case of Salix Vinimalis biomass, curve tendency is also in a decreasing pattern if the 3000 rpm speed for the ϕ7 mm is disregarded. Also, in the case of the sieve with ϕ7 mm there also exist average speeds (2700 – 2850 rpm) to which the grinded average dimensions are smaller, which could recommend them for practice use.

4 Conclusions

Grinding vegetal biomass is one of the factors which influence the final product quality destined to consumers as pellets/briquettes. It is very important that the material which represents the biggest component from the agropellet/briquette creation process, is uniformly grinded and the dimension permits an optimum working flux during the following processes from the technological line. Still, we can express the following conclusions:

- Through finite element simulation of the hammer mill working process, a maximum deformation of 0.036 mm at the tip of the hammers due to centrifugal force during functioning was concluded;
- Tensions were half the limit of elasticity, and the equivalent unitary effort is manifested on the contact surface of the hammer with the shaft or with the joint.
- Maximum reaction for hammer joint was of 11635 N;
- For miscanthus biomass, the consumed energy variation in relation to the used sieve orifice diameter and the hammer mill speed, presents an overall decreasing tendency;
- The same general tendency was also observed for willow biomass, even if there are values of the rotor speed which present a random variation;

Dimensions of the grinded material and the energy consumption are influenced mainly by the future material destination, but also by its initial properties, as well as by the adopted working regime.

The work has been funded by National Grants GNaC 2018 ARUT, Internal Research Grant, through the Financial Agreement 06/15.10.2018.

References

1. Eurostat 2016
2. Igathinathane C., Womac A.R., Sokhansanj S. Effect of Angle Cut on Corn Stalks Mechanical Cutting Strength and Energy. An ASABE Meeting Presentation, Paper number 1008644, Presentation written for presentation at the 2010 ASABE Annual, (2010)
3. Shastri, Y.N., Miao, Z., Rodríguez, etc. Determining optimal size reduction and densification for biomass feedstock using the BioFeed optimization model. Biofuels, Bioproducts and Biorefining 8(3),DOI: 10.1002/bbb.1476, pp. 423-437 (2014)
4. Altintas, Y.: Manufacturing Automation: Metal Cutting Mechanics, Machine Tool Vibrations, and CNC Design. 2nd ed. USA: Cambridge University Press, (2012)
5. Euan, I.G., Ozturk E., Sims N.D.: Modeling Static and Dynamic Cutting Forces and Vibrations for Inserted Ceramic Milling Tools. Procedia CIRP.8:564-9, (2013)
6. Scippa A., Grossi N., Campatelli G.: Milled Surface Generation Model for Chip Thickness Detection in Peripheral Milling. Procedia CIRP. 8:450-5 (2013)
7. Grossi N., Sallese L., Scippa A., Campatelli G.: Chatter Stability Prediction in Milling Using Speed-varying Cutting Force Coefficients. Procedia CIRP. 14:170-5, (2014)
8. N. Grossi, F. Montevecchi, A Scippa, G. Campatelli., 3D Finiate Element Modeling of Holder – Tool Assembly for Stability Predintion in Milling, ELSEVIER - ScienceDirect, Procedia CIRP 31, pp 527 – 532 (2015)

9. I. Budacan, I. Deac, Numerical Modeling of CFD model Applied to a Hammer Mill, Bulletin USAMV, serie Agriculture 70 (1), pp 273 – 282, (2013)
10. Z. Y. Wu, S. Zhang, T. Gao, "Design of 3D Modelling for the New Hammer Mill Based on the SolidWorks", Applied Mechanics and Materials, Vol. 224, pp. 320-324, (2012)
11. E. V. Kumar, Design and Analysis of Rotor Shaft Assembly of Hammer Mill Crusher, International Journal of Engineering and Management Research, Vol. 3, Issue 2, pp. 22-30, (2013)
12. Voicea I., Voicu Gh., Cardei P., Vlăduţ V., Biriş S. - A Mathematical Model of Chopped Miscanthus Briquetting Process, The 8th International Conference INTER-ENG 2014 Interdisciplinarity in Engineering, 9 - 10 October 2014, "Petru Maior" University of Tîrgu Mureş, Procedia Technology (19) 2015, pages 750–757 ISSN: 2212-0173)
13. Voicea I., Vlăduţ V., Cârdei P., Matache M., Găgeanu I., Voicu Gh., Popescu C., Paraschiv G., Kabas O - Compacting process and mathematical analysis of miscanthus briquettes expansion, Proceedings of the 43 international symposium on agricultural engineering "Actual Tasks on Agricultural Engineering", pag. 667‚- 676, ISSN 1333-2651, Opatija – Croaţia (2015)
14. Găgeanu I., Voicu Gh., Vlăduţ V., Voicea I. - Experimental research on the influence of recipes used on the quality of biomass pellets, 16th International Scientific Conference "Engineering for rural development", pag. 785-791, ISSN 1691-5976, 24‚26.05, Jelgava, Latvia (2017)
15. Dincă M., Ferdeş M., Paraschiv G., Ungureanu N., Vlăduţ V., Moiceanu G., Zăbavă B. - Lignocellulosic biomass pretreatment for biofuel production, Anniversary Conference "Durable Agriculture – Agriculture of the future" Craiova, 9-10 November 2017, Annals of the UNIVERSITY OF CRAIOVA - AGRICULTURE, MONTANOLOGY, CADASTRE, Series, Vol. 47, No. 2 / 2017, Working Group No. 4 - MANAGEMENT, AGRICULTURE MECHANIZATION AND CADASTRE, pp. 310-316, ISSN 1841-8317, ISSN CD-ROM 2066-950X (2017);
16. Găgeanu I., Chiţoiu M., Găgeanu G., Vlăduţ V., Voicu Gh., Bunduchi G., Grigore I. - Aspects on obtaining the raw material for pelletizing, Scientific International Conferences - The 12th ANNUAL MEETING "DURABLE AGRICULTURE – AGRICULTURE OF THE FUTURE" Craiova, 17-18 November 2016, ANNALS OF THE UNIVERSITY OF CRAIOVA - AGRICULTURE, MONTANOLOGY, CADASTRE, Series, Vol. 46, No. 2 / 2016, Working Group No. 4 - Management, Agriculture Mechanization and Cadastre, pp. 379-382, ISSN 1841-8317, ISSN CD-ROM 2066-950X (2016)

E3S Web of Conferences **112**, 03019 (2019) https://doi.org/10.1051/e3sconf/201911203019
TE-RE-RD 2019

Theoretical research on kinematic and constructive parameters of roller presses for sweet sorghum

Mihai Olan[1*], *Gheorghe* Matei[2], *Livia* Apostol[3], *Vasile* Jinga[4], *Steluța* Lipianu [4,5], *Paul* Găgeanu[1], *Ana* Zaica[1], *Iulian*Voicea[1] and *Oana* Cristea[1]

[1]INMA Bucharest, Romania
[2]University of Craiova, Romania
[3]IBA Bucharest, Romania
[4]ICDPP Bucharest, Romania
[5]USAMV Bucharest, Romania

Abstract. The overall objective of the project is to develop an eco-innovative system for pressing sweet sorghum by the generalised object method specific to the technical creation applied in this field. (*Vitalie Belouasov*). The representative technical solutions of the companies recognised in this field are known and it is necessary to use a creativity method to analyze the existing solutions, to assess technical and economic characteristics by methods specific to value analysis but also to find new constructive variants and choose the optimal variant that will then be developed by the design-research activity for the realisation of the technical documentation for the execution of an industrial prototype. The generalized object of the technical creation is in the form of a cylindrical morphological matrix, each sector representing a solution (sector represented by a triple assembly BiCjDk, the total number of solutions being: N=i·j·k solutions, some of which are known, others unknown. The incompatible ones must be eliminated, while the others must be analyzed, as they can lead to particularly efficient constructions after their complete solving.

1. Introduction

Because of the energy crisis by lowering oil resources and of air pollution, research has been made to find safe substitutes. Of these, the bioethanol produced from sweet sorghum is of special interest in this field, [1]. There is a large amount of sugar in sweet sorghum stem juice, [2]. The sweet sorghum stem juice is rich in fermenting sugar and is a desirable material for alcoholic fermentation, [3].

Based on the analysis of the existing technologies and equipment for juice extraction from the sweet plant stems, the opportunity to use roller presses for squeezing the juice from sweet sorghum stems is argued and the basic parameters of the press, which would ensure the obtaining of large quantities of raw sweet sorghum juice used to make monatomic alcohols and other products, are presented, [4;5].

* PhD. *Mihai* Olan: vibrobloc@yahoo.com

The achievement of an optimal productivity related to the electric energy consumed in the case of pressing the sweet sorghum stems is determined by the kinematic study of the working parts' operating systems and can be obtained by analysing and optimizing the parameters that determine this technological process, [6].

By analysing the forces that act on the pressed material and determining the power consumed when pressing the material between the rollers, the energy consumed per unit time and the resulting productivity are obtained.

We will also analyse the variation of the working capacity depending on the thickness of the raw material layer the roller press is fed with and the rollers angular speed, but also the power variation of the driving electric motor according to the force required for driving the rollers, [7].

In the paper, the kinematic analysis of the roller press driving components is made, [8;9]

2 Materials and methods

2.1 Theoretical presentation of sorghum stems entrainment between two rollers

To obtain juice from sweet sorghum stems, we chose the press with three rollers having equal diameters and rotational speeds. In this case, the thickness of the raw material layer, denoted by h, is pressed at thickness s1 at the primary passage between the upper roller and the primary pressing and entraining roller and then reaches thickness 2 at the final pressing between the first roller and the last roller, thicknesses given by the distances between rollers.

According to the literature we will notice that the forces with which the rollers act on the layer of stems for each zone are directed perpendicularly to the contact surface and can be reduced to a resultant force P. Decomposing this force in two perpendicular directions, namely of material compression and of material movement, we obtain the components Py that compresses the material and Px that pushes the material from the passage. At the same time, friction forces, which are applied tangentially to the rollers surface, act on the material, forces that can be presented by resultant F. We will decompose the resultant force F and obtain the components Fx, which entrains the material between the rollers and Fy, which contributes to the pressing of the material. Fig.1, [3]

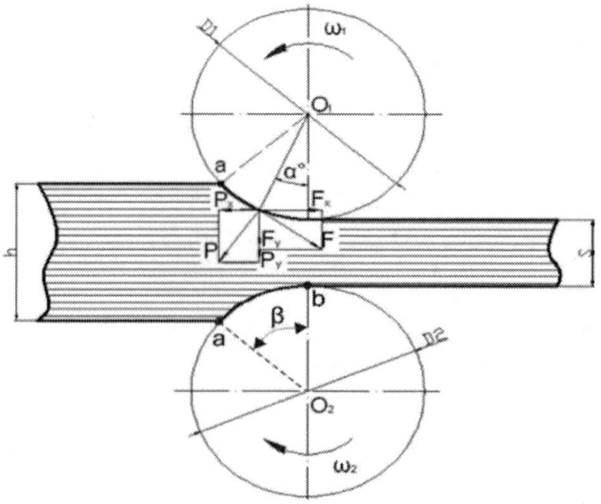

Fig. 1. Theoretical presentation of forces that arise in the pressing process, [2; 4].

2.2 Analysis of sorghum stems entrainment between three rollers

To extract juice from sweet plant stems, the most commonly used system is built in the three-roller version with equal diameters and rotational speeds and with a high degree of extraction.

Figure 2 shows the scheme of this pressing method. The raw material layer, with the initial thickness h=30...40 mm, is pressed in two technological phases up to a final thickness of s=5...7 mm, equal to the value of the passage between the rollers.

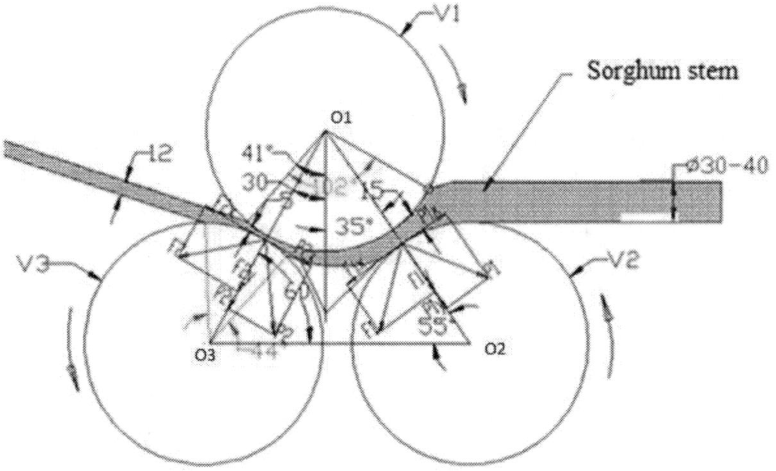

Fig.2. Scheme of forces application to the pressed material in the case of the three-roller press.

According to [1], the forces with which the rollers act on the layer of stems for each zone are directed perpendicularly to the contact surface and can be reduced to a resultant force P which will consist of the summing of the two forces **P1** and **P2** projected vertically in the contact area of each pair of drums. If we decompose these forces in two perpendicular directions, namely of material compression and of material movement, we obtain the components **P1y**; **P2y** (compressing the material) and **P1x**; **P2x** (pushing the material from the passage). At the same time, friction forces, which are applied tangentially to the rollers surface, act on the material, forces that can be presented by resultant **F1 and F2**.

By decomposing the resultant force **F1 and F2**, we obtain the components **F1x** respectively **F2x** which entrain the material between the rollers and **F1y** respectively **F2Y** which contribute to the pressing of the material.

By analysing the pressing process we will find that when the rollers get nearer simultaneously with the decrease of space s, the contact angle α increases (if thickness **h**=const.=30 mm) and the application point of forces **P1** and **P2** moves away from the centreline O_1O_3 and O_2O_3.

This leads to the increase of the components **P1x, P2x** from the value **P1y, P2y**. When forces **P1x** and **P2x** exceed the friction forces component **F1x** respectively**F2x**, the roller sliding begins and material movement stops. [1]

As it results from the above, the condition of entraining the material between the rollers is:

$$Fx \geq Px \text{ or } F\,cos\alpha \geq Psin\alpha, \text{ results } P\,tg\varphi cos\alpha \geq P\,sin\alpha, \quad (1)$$

For the application shown in Figure 2, respectively:

$$F\,cos30° \geq P\,sin30° \quad (2)$$

We'll have:

$$Fx0.866 \geq Px0.5, \text{ so } F \geq 1.73P \tag{3}$$

It can be seen that the thickness h and the compression degree Δ_h of the entrained layer, for pressing between two rollers, depend on the rollers diameter D and the friction coefficient ($f=tg\varphi$) of the material on the roller surface. Therefore, the increase in diameter D and the friction coefficient $tg\varphi$ increases the values of the layer thickness h and its thinning Δ_h, which improves the conditions of entraining the stem layer and increases the productivity of the pressing process Q and juice squeezing degree GE.

2.3. Productivity and liquid extraction degree when pressing the material between 3 rollers for the application shown in Figure 2

The volume of material passing, in a unit of time, through a passage between two rollers having equal peripheral speed, considering that sliding is missing, is determined from the relation:

$$V = b \cdot h \cdot v \tag{4}$$

where: b is the length of the roller (width of the pressing area) measured in mm;
h – thickness of the layer entrained between rollers measured in mm;
v – peripheral speed of the rollers, measured in m/s.[3]
The mass of the material processed in a unit of time (productivity) can be determined, knowing the bulk density (volumetric mass) φ of the material (kg/m^3):

$$Q_{mv} = V\varphi = b \cdot h \cdot \varphi \cdot \delta \tag{5}$$

We will notice that the material layer compresses from the initial thickness h (max.40 mm) to the final thickness s (6…7 mm, passage value) when the material passes through the rollers. Following the compression process, the release of the liquid from the material is obtained and the increase of the squeezed material volumetric mass.

If the final distance exceeds 15 mm, conditions are created for uneven pressing of the raw material layer: the larger diameter stems are pressed harder than those with a smaller diameter. In turn, the increase in force P reduces stem pressing unevenness, however, providing maximum values in the case of the distance within the limits of 6…7 mm.

It is observed that the pressure increase significantly rises the compression coefficient of the vegetal mass:

$$K = \gamma \cdot b/m \tag{6}$$

and is the ratio between the volumetric mass of bagasse and of raw material respectively.

2.4. Determining the power consumed when pressing the material between rollers

When pressing the material, it is necessary not only to achieve press operation with safe stem entraining, high productivity and high juice extraction (squeezing) degree, but also minimal energy consumption.

As mentioned before, it is necessary to act on the material with a force **P** normal to the contact surface to ensure that the stems are entrained and pressed by the rollers. In order to operate the rollers with speed according to [1], a moment of torsion is required:

$$M_t = 2P \cdot ¥ \cdot l \tag{7}$$

Where: **¥** is the boom coefficient;

α, β- angles of application of the force *P* and of the entraining-pressing sector, respectively;

$\mathbf{l_p}$, **l** – the lengths of boom applying force *P* and of the entraining-pressing sector, respectively.

According to the literature [3], the values of the boom coefficient are within the limit ¥ = 0.12÷0.67, then the moment of torsion will be:

$$M_t = p \cdot m \cdot b \cdot R \cdot \Delta h \tag{8}$$

In this case, the power consumed for carrying out the process of the green mass entraining and processing will be:

$$N = Mt \cdot \omega = p \cdot m \cdot b \cdot R \cdot \Delta h \cdot \omega \tag{9}$$

2.5 The kinematic scheme of the three-roller press for the production of sorghum juice will be shown in Figure 3

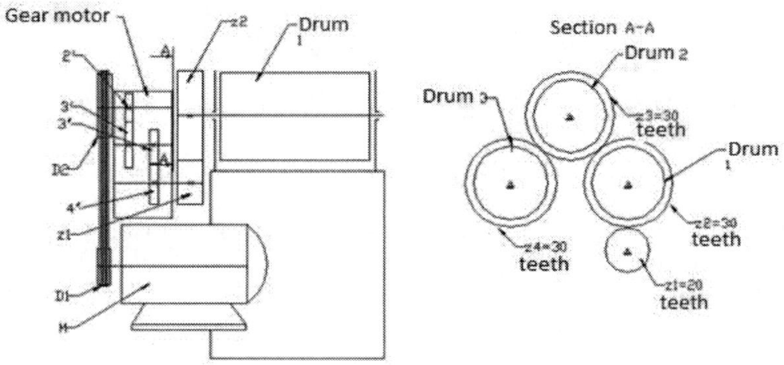

Fig.3. Kinematic scheme of the three-roller press.

Calculation of kinematic chain.

Gear ratio between belt pulleys:

$$i_{RC1,RC2} = \frac{\omega_{1R}}{\omega_{2R}} = \frac{D_2}{D_1}, \omega_{2R} = \omega_{1R}\frac{D_2}{D_1} \tag{10}$$

Gear ratio to the reducer:

$$i_{2'4'} = i_{2'3'} \cdot i_{3"4'} = \frac{\omega_{2'}}{\omega_{3'}} \cdot \frac{\omega_{3"}}{\omega_{4'}} = \frac{\omega_{2'}}{\omega_{4'}} = \left(-\frac{z_{3'}}{z_{2'}}\right) \cdot \left(-\frac{z_{4'}}{z_{3"}}\right) = \left(\frac{z_{3'}z_{4'}}{z_{2'}z_{3"}}\right) \tag{11}$$

$$i_{2'4'} = \frac{\omega_{2'}}{\omega_{4'}} = \frac{z_{3'}z_{4'}}{z_{2'}z_{3"}} \tag{12}$$

$$\omega_{4'} = \omega_{2'} = \frac{z_2 z_{3"}}{z_{3'}z_{4'}};\ (\omega_{2'} = \omega_{2R}) \tag{13}$$

Gear ratioto drum 1:

$$i_{12} = \frac{\omega_1}{\omega_2} = -\frac{z_2}{z_1}, \omega_2 = -\omega_1\frac{z_1}{z_2}, (\omega_1 = \omega_{4'}) \tag{14}$$

Gear ratio between drums 1 and 2:

$$i_{23} = \frac{\omega_2}{\omega_3} = -\frac{z_3}{z_2}, \omega_3 = -\omega_2 \frac{z_2}{z_3}, \text{or}(\omega_3 = -\omega_2), (z_2 = z_3) \quad (15)$$

Gear ratio between drums 2 and 3:

$$i_{34} = \frac{\omega_3}{\omega_4} = -\frac{z_4}{z_3}, \omega_4 = -\omega_3 \frac{z_3}{z_4}, \text{or}(\omega_4 = -\omega_3), (\omega_4 = \omega_2) \quad (16)$$

Notations:

ω_{1R} - motor angular velocity

$$\omega_{1R} = \frac{\pi \cdot n_{mot}}{30}; \quad (17)$$

ω_{1R} - belt pulley 2 angular velocity;

n_{mot} - motor speed;

Note: For drum 1 we have gear 2;
For drum 2 we have gear 3;
For drum 3 we have gear 4.

3 Results

3.1 Calculation of productivity and electric motor power to drive the three-roller press

Using the relation (4) and replacing the set values, the volume of the material to be pressed in the unit of time results:

$$V = 400\,mm \cdot 15\,mm \cdot 0.150 m/s = 1800\,mm^3/s, \text{ respectively } 3.34\ m^3/h \quad (18)$$

To determine the productivity value using the relation (5), it will result:

$$Q_{mv} = 3.24 m^3/h\ x\ 600\ kg/m^3 = 1944\ kg/h \quad (19)$$

Because of the uneven feeding that is made manually for this equipment, we will consider a 40% feed rate and will result in a pressing capacity of the sorghum stems of approx. 400...500 kg/hour.

In order to calculate the electric motor power to drive the roller press, we will replace the numerical values in relations (8 and 9) and we will get:

Moment of torsion:

$$Mt = 2 \cdot 15000\ N \cdot 0.12 = 3600\ Nm \quad (20)$$

Angular velocity:

$$\omega = 3.14 \cdot 12/30 = 1.256\ rad/s \quad (21)$$

Power consumed:

$$N = 3600 \cdot 1.256 = 4.520\ W \quad (22)$$

We will choose an electric motor with the power of 5.5 kW.

3.2. Calculation of kinematic chain

For the presented case, the kinematic chain is the following:

Pressing roller N rotations =1470 rpm x 160/450mm (D1/D2 belt transmission) x 1/18 (Gear ratio of parallel shaft reducer) x 20/30 (Z1/Z2-gear ratio with cylindrical gears) x 30 x 30 (Z3/Z4-gear ratio between each pair of drums) = 19.358 rpm.

3.3. Making diagrams for the variation of some working parameters when pressing sweet sorghum

Reference values:
- rollers width, m. B_0=0.4; $r_{0.12}$
-material layer height, m=0.15
-angular velocity, rad/s. ω_0=1.256
Mathematical relations used:
Flow rate of processed material:

$$P(Fs, r, n, \omega) := Mt(Fs, r, n) \cdot \omega \tag{23}$$

Pressing drum radius, m:

$$Q(b, h, w, r) := b \cdot h \cdot v(\omega, r) \tag{24}$$

Working capacity:

$$C(b, h, \omega, r, \rho) := Q(b, h, \omega, r) \cdot \rho \tag{25}$$

Moment of torsion, Nm:
Power consumed by the motor, W:

$$Mt(Fs, r, n) := n \cdot Fs \cdot r \tag{26}$$

After the application of the mathematical relations for the numerical values set in the design theme we obtained the following diagrams.

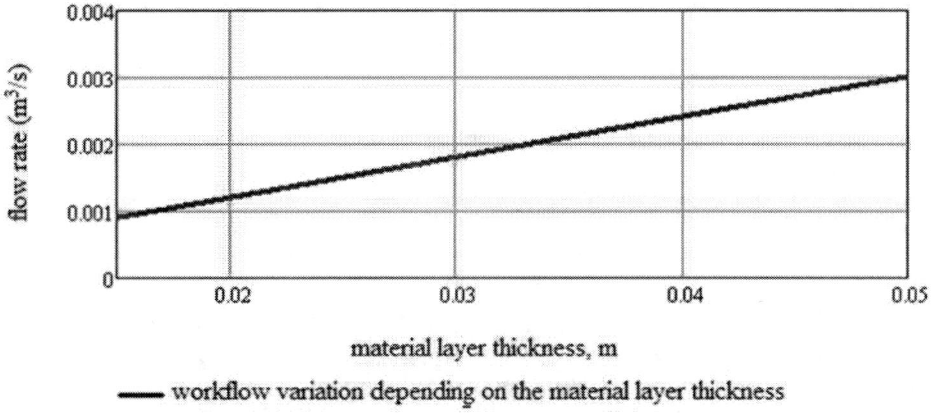

Fig. 4. Workflow variation depending on the material layer thickness, for the reference values of the model

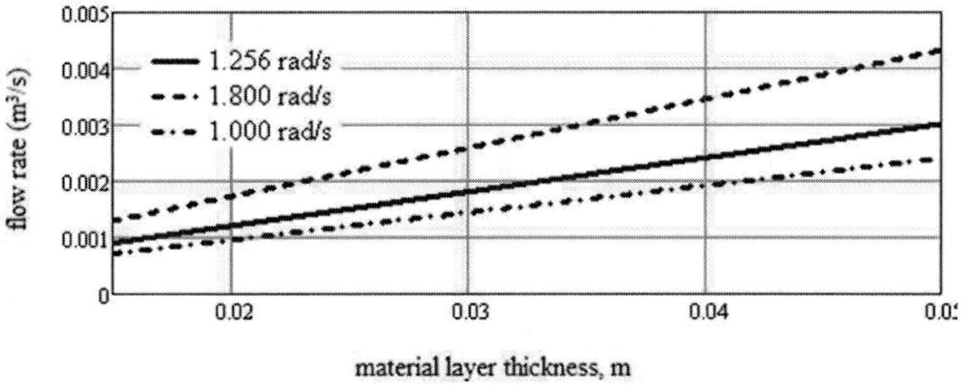

Fig. 5. Workflow variation depending on the material layer thickness, for three values of the drum's angular velocity5

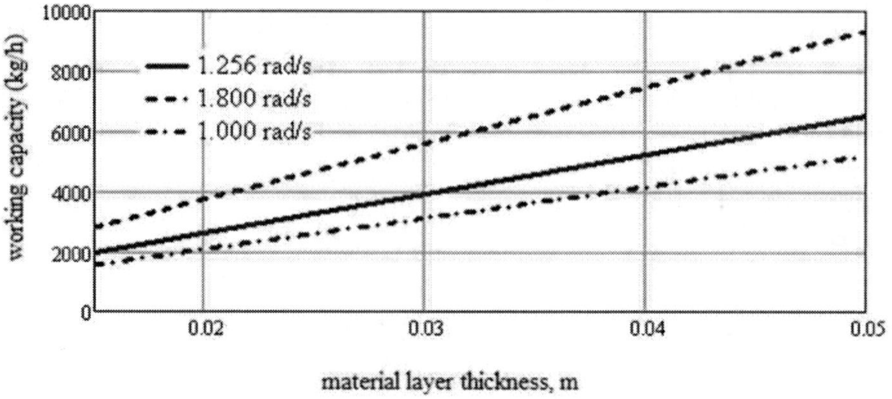

Fig. 6. Working capacity variation depending on the material layer thickness, for three values of the drum's angular velocity

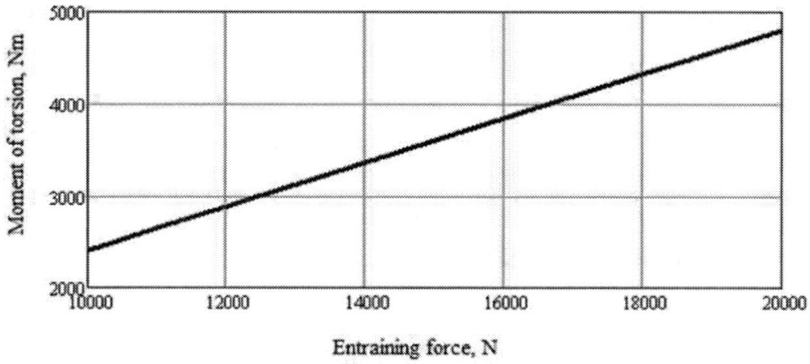

Fig. 7. Variation of the moment of torsion depending on the entraining force.

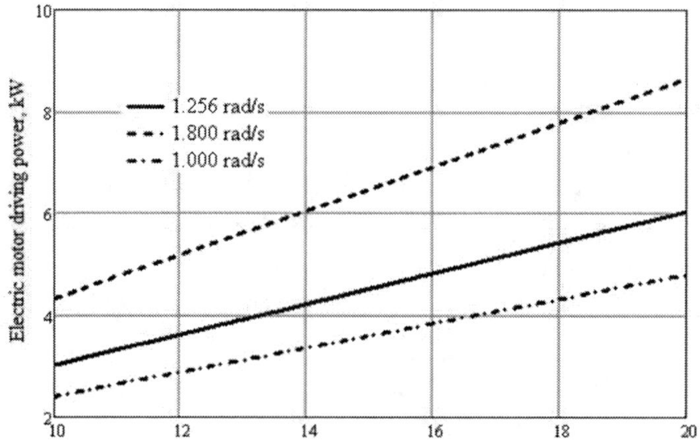

Fig. 8. Variation of electric motor power depending on the entraining force.

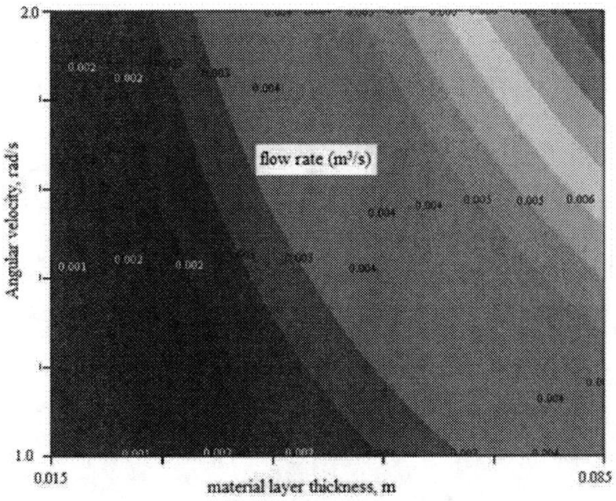

Fig. 9. Variation of the material flow rate depending on material layer thickness and angular velocity on the isocline map.

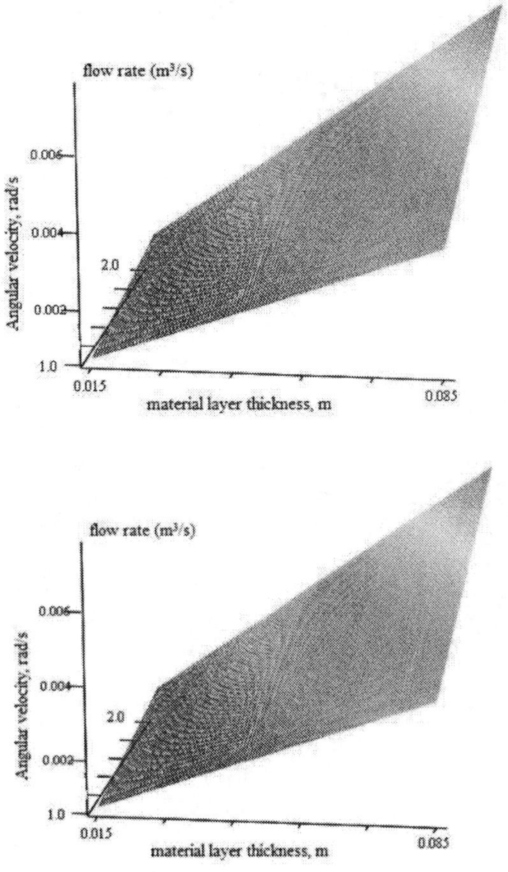

Fig. 10. Variation of the material flow rate depending on material layer thickness and angular velocityby surface representation.

The analysis of the study carried out establishes the following theme of constructive design for a functional prototype to be used in the research project:

Three-roller press, belt transmission and parallel shaft reducer drive, electric motor power 5.5 kW, drum diameter 240 mm, drum length 400 mm, pressing capacity 400-500 kg/hour, determined by gearmotor and roller length but also by the uniformity of manual feed. (Olan M. et al 2018).

3.4. Realization of the three-roller press construction project

The three-roller press (Figure 11) is intended for pressing sorghum stems to obtain the juice to be used for the production of bioethanol. [1, 10]

The installation for pressing sorghum stems is composed of a support (1) on which is mounted a pressing unit (2) with three rollers (3,4,5) which are driven together by two gears (6; 7 and 8; 9) and the rotation movement is received by means of a pinion (10) from a parallel shaft reducer (11), a belt pulley transmission and some trapezoidal belts which take up the rotation movement from an electric motor (12) and transmits it to that parallel shaft reducer (11).

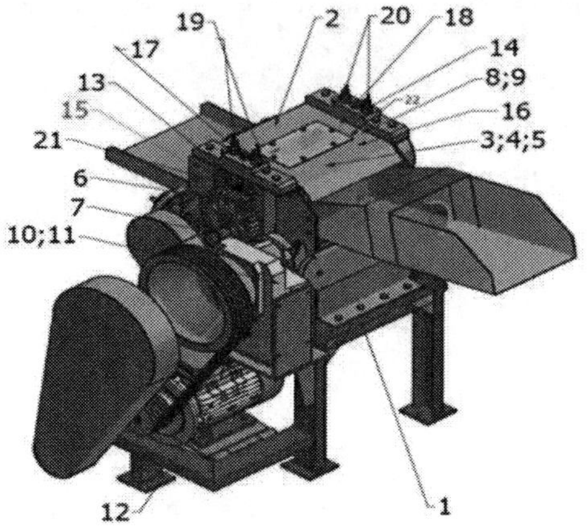

Fig. 11. Press for sorghum stems.

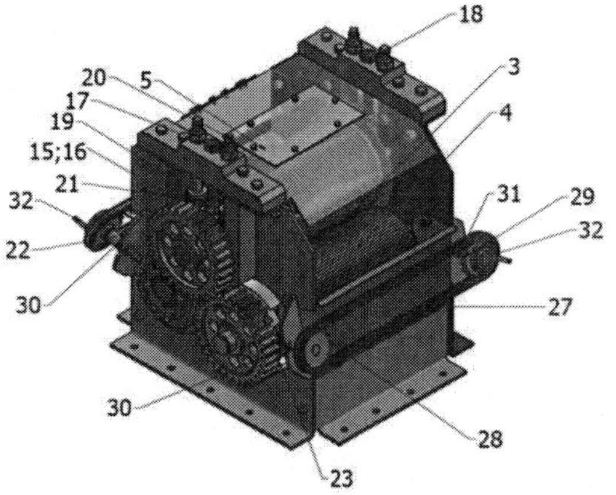

Fig.12. Three-roller pressing unit.

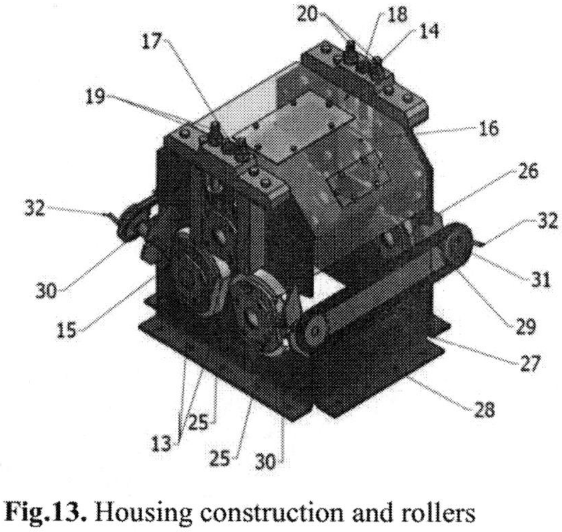

Fig.13. Housing construction and rollers adjusting mechanisms

The pressing unit has two vertical guides (13; 14) on which a pressing roller (3) slides having on the work surface some special profiles for entraining the sorghum stems and circular channels with a triangular section for pressing the sorghum. The two bearings (15; 16) supporting the pressing roller (3) are vertically adjusted with a hexagonal screw (17; 18) and two threaded rods (19; 20) on which some sets of disc springs are mounted (21; 22) which maintain an elastic pressure during the working process by directly pressing on some bearings supporting this pressing drum. [3; 5]

Under the pressing roller there are mounted two working rollers (4; 5) having special profiles on axial and circular (R and T) direction to the working surface, which aim to increase the entraining capacity of the sorghum stems and also the final pressing on two working directions which will increase the degree of juice squeezing from the sorghum stems. The two working rollers are moved by means of bearings (23; 24) and of mechanisms on two special guides (25; 26) to increase or decrease the distance between each of these rollers and the upper pressing roller (3).

For the identical movement of the two bearings supporting each of the two working rollers (4:5), the installation is provided with a Gall chain transmission (27) and two pinions (28; 29) mounted on the threaded rods (30; 31) which move the bearings (23; 24) on each end of the working rollers (4; 5) and which is actuated by one handwheel (32) for each transmission from the two working rollers. For fine adjustment, each of the two threaded rods (30, 31) has a graduated drum. [1]

The juice obtained after sorghum pressing operation is collected in a vessel placed at the bottom of the installation and then transported to a tank. [5]

To increase adherence to stems entrainment, a technical solution is the progressive increase by 1-2% (2.5-5mm) of the three rollers diameters in the following order: roller 2; roller 1 and roller 3 according to Figure 14.

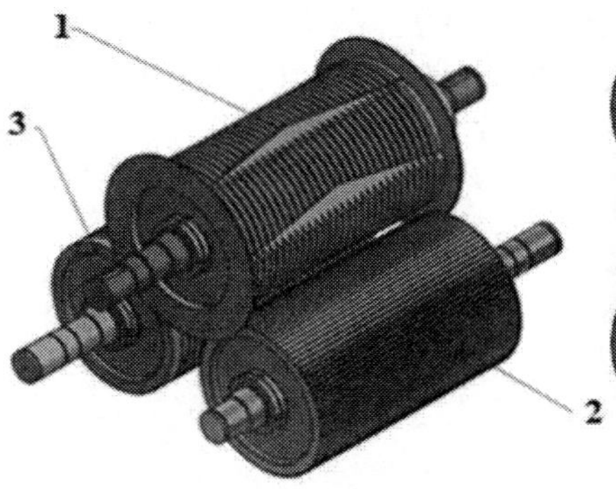

Fig.14. Constructive form of pressing rollers.

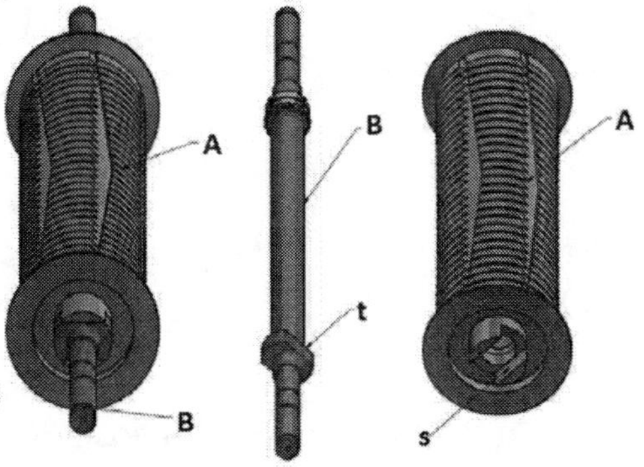

Fig.15. Dismountable roller construction.

Another technical solution for reducing friction and increasing the degree of stem entrainment, the three rollers will have different shapes according to Figure 6.

Thus:

- roller 1 will have channels with a triangular profile on a radial direction to the axis of rotation and for the purpose of entraining are made 4-6 special millings, of triangular shape, which will form stem entraining teeth; [3].
- roller 2 will have channels with a triangular profile on an axial direction to the axis of rotation but perpendicular to the direction of roller 1 channels, obtaining both the entrainment of stems and their primary pressing;
- roller 3 will have channels with a profile similar to roller 1 but displaced by half a step to allow tooth full on empty pressing between rollers 1 and 3 but also pressing in a direction perpendicular to the pressing between the rollers 1 and 2, thus obtaining a rectangular network for pressing the stems and extracting the juice, which will increase the extraction rate. [2]

To save materials and workmanship and for safety in operation in case of wear, the three rollers (1, 2 and 3) are executed in the version with a removable shaft. [1]

4 Conclusions

In this article, the main constructive data underlying the design of three-roller presses for the production of sweet sorghum juice necessary for bioethanol production have been presented and analyzed. The influence of the main parameters on the power consumed has also been analyzed. Having available the studies on the analyzed constructive solutions and the computational elements necessary for this type of installations, the constructive project of the prototype installation for sorghum pressing was prepared, the prototype variant Figures 11;12 and 13 with the following technical characteristics:

Three-roller press for sweet sorghum, belt transmission and parallel shaft reducer drive, electric motor power 5.5 kW, drum diameter 240 mm, drum length 400 mm, pressing capacity 400-500 kg/hour, determined by gearmotor and roller length but also by the uniformity of manual feed.

This work was supported by a grant of the Romanian Research and Innovation Ministry, through Programme 1 – Development of the national research-development system, sub-programme 1.2 – Institutional performance – Projects for financing excellence in RDI, contract no. 16 PFE.

References

1. V. Jinga, V. Vlăduţ, E. Marin, D. Manole, *Cultura de sorg o soluţie pentru agricultura dobrogeană*, ISB-INMA TEH' 2018, 951-954 (2018).
2. J. Fei, J. Chawhuaymak, R. Mark, Z. Werner, and K. L. Ogden, *Efficient extraction method to collect sugar from sweet sorghum,* Journal of Biological Engineering, **7**, (2013).
3. I. Caba, E. Laza,V. Vlăduţ, E. Marin, N.Ungureanu, S. Biriş, I. Găgeanu, I. Voicea, S. Isticioaia, M. Olan, Gh. Matei, *Sweet sorghum – an alternative for romanian farmers in the conditions of the current climate change*, ISB-INMA TEH' 2018, 961-970, (2018).
4. V. Cerempei, I. Habasescu, *Technical solutions for the production of liquid biofuels, Institute of agricultural technique* "MECAGRO", (INNOVATIVE TECHNOLOGIES, Republic of Moldova, 2014).
5. I. Dumitru, D.E. Iuga, M. Cristea, A.I. Grigore, L. Vladuţoiu, Gh Stroescu., M.R. Oprescu, V. Vlăduţ, I. Voicea, L-D. Popa, S. Boruz, F.D. Sîrbu, *Considerations on energetic crops potential*, 47th CONFERENCE OF ESNA, **48**, 257-264, (2018).
6. R. Oprescu, I. Voicea, V. Vlăduţ, I. Găgeanu, D. Cujbescu, C. Persu, D. Iuga, S.Şt. Biriş, I. Dumitru, Gh. Stroescu, L-D. Popa, S. Boruz, *Considerations on the extraction technology of sugar sorghum juice*, 47th CONFERENCE OF ESNA, **48**, 367-374, (2018).
7. A.I. Grigore, I. Grigore, L. Cristea, M. Cristea, L. Vlăduţoiu, I. Voicea, V. Vlăduţ, M. Epure, P. Andrei, C. Popescu, S. Boruz, Gh. Matei, *Considerations on the importance of cultivating sorghum to the biofuel industry for producing bioethanol*, 47th CONFERENCE OF ESNA, **48**, 283-288, (2018).
8. M. Olan, V. Vlăduţ, A. Păun, A. Vişan, P. Găgeanu, I.Voicea, I. Găgeanu, *Analysis of technique current stage in the field of sorghum pressing equipment designed to produce the raw juice necessary to obtain the bioethanol*, ISB-INMA TEH' 2018, 221-228, (2018),
9. V Vlăduţ., S. Bungescu, *The fondation of the technology for mechanizing the saccharate sorghum crop and specific et optimizing solution*, Vth International Symposium "YOUNG PEOPLE AND MULTIDISCIPLINARY RESEARCH", Sect B - Biotechnology, Agriculture, Veterinary Medicine, 369-381, (2003).
10. Gh. Matei, V. Vlăduţ, M. Vladu, R.L. Pânzaru, P. Săvescu, C. Sălceanu - *Study on the influence of mineral fertilization on the biomass yield of sweet sorghum cultivated in the Caracal plain*, ISB-INMA TEH' 2018, 1021-1026, (2018) .

E3S Web of Conferences **112**, 03020 (2019) https://doi.org/10.1051/e3sconf/201911203020
TE-RE-RD 2019

Researches on the accumulation and transfer of heavy metals in the soil in tomatoes - *Solanum lycopersicum*

Mihaela Nițu[1], *Augustina* Pruteanu[1,*], *Despina Maria* Bordean[1,2)], *Carmen* Popescu[4,5)], *Gyorghy* Deak[6)], *Mădălina* Boboc[6)] and *Gabriel* Mustățea[3)]

[1] INMA National Institute of Research - Development for Machines and Installations designed to Agriculture and Food Industry, Bucharest
[2] USAMVBT Banat's University of Agricultural Sciences and Veterinary Medicine "King Mihai I of Romania", Timisoara
[3] IBA National Institute of Research and Development for Food Bioresources, Bucharest
[4] S.C. HOFIGAL Export Import S.A.
[5] Faculty of Medicine, Pharmacy and Medical Dentistry, "Vasile Goldiș" Western University, Arad
[6] INCDPM National Institute for Research and Development in Environmental Protection, Bucharest

Abstract. Heavy metals in contaminated soils have benefited from a considerable attention due to the possible risks for the human body.
The current study has investigated the accumulation and transfer coefficient for three heavy metals (Cu, Pb, Zn) found in the contaminated soil with three concentrations (c1=1.5%, c2=3.0%, c3=4.5%, c4=6.0%), obtained by mixing the three metals, in the tomato fruit. The highest accumulation in the tomato fruits was recorded for zinc, then copper and the smallest for lead, for all four concentrations used. The transfer coefficient decreases as the concentration of heavy metals increases, so that for high heavy metals concentrations, the values of the transfer coefficient are very low, and for small heavy metals concentrations in the soil, the values for the transfer coefficient are higher.
The assessment of accumulation and transfer of heavy metals in the fruits of tomatoes grown in the contaminated soil has concluded that all concentrations of the copper, lead and zinc mix have shown a low risk for human consumption.

Key words: Heavy metals, tomato, contamination, pollution

1 Introduction

Environmental pollution has become one of the most debated issues of contemporary times. One of the current problems of mankind is the accumulation of heavy metals in agricultural soils. The harmful metals in the soil can come from the base rock, from solid or liquid waste landfills, agricultural inputs and industrial and urban emissions. This accumulation of heavy metals determines the contamination of agricultural soils, having consequences on food quality and safety [1, 2].

* Corresponding author: pruteanu_augustina@yahoo.com

Fruits and vegetables are an important component of food, with input of carbohydrates, proteins, vitamins, minerals and fibers indispensable to human health [3].

Also, being are the most exposed to environmental pollution due to airborne load, they represent the major source of toxic metal intake by human beings. They accumulate heavy metals in their edible and inedible parts and can cause various dysfunctions to humans [4].

Good functioning of the human body is also due to the presence of heavy metals in a small amount, but a large amount of heavy metals present in the body can lead to the appearance of certain diseases. Some metals present in fruits and vegetables are biochemically and psychologically important for health but, without exceeding the recommended concentrations, when they can cause metabolic disturbances [5, 6].

In this respect, information on heavy metal concentrations in fruits and vegetables, as well as their nutritional input are essential for assessing the health risk for beings [7].

In paper [8] was shown that almost half of the average intake of Pb, Cd and Hg through ingestion of food is owed to plant category (vegetables, fruit and cereals). Therefore, vegetarians, through the high consumption of fruits and vegetables, which can be contaminated with metals, are most exposed to exceeding the tolerable daily intake.

Also, in paper [9], was analyzed and evaluated both concentrations of heavy metals in the soil and the fruit of tomato. Plants were grown in places with soils contaminated with metals, and the results showed that there was a large transfer of soil metals into the plants. All sites studied had Fe, Cr in small amount and Pb, Zn, Cu in larger quantity.

In paper [10], the authors analysed the toxic heavy metals from irrigation water, soil and vegetative parts of tomato plants. In irrigation water and in soil were found elements such as Cu, Pb, Cr, Fe, Mn, Na, K, Ca and Mg, and Zn. The results of analysis have indicated that toxic metals Fe, Pb, Co, Cr, Mn, Zn and Cu are accumulated in the roots of tomato plants. The most toxic metal, Pb, was found in the roots and in the stalks of tomatoes. Fe was accumulate in all vegetative parts of the tomato plant and in tomato fruit, were found of Zn, Mn, Cr, Pb and Cu. Also, tomato (Solanum lycopersicum) seeds were planted and grown in pots in greenhouse under optimum conditions.

Paper [11] analyses the ability of cereals and vegetables to absorb potentially toxic heavy metals from the sludge treatment – soil system. It was found that cereals and vegetables accumulate less Cd in the leaves compared to the strains, and tomatoes, accumulate more Mn, even than barley or beans. Also, a high iron content was observed in the soil, the other elements being within normal limits. The highest transfer factor values were determined for iron, copper and lead having low values.

In paper [6], an investigation was conducted to determine the toxic concentration of heavy metals in wastewater, irrigated soil, their cumulation in vegetables and the potential risks for human illness. The researches showed that the concentration of Cr in residual waters, Cd, Zn in soil and plants was found well above the tolerable doses. The Health Risk Index (HRI) was found to be > 1 for Pb in okra (Abelmoschus esculentus), while the remainder of the heavy metals (Cr, Cd, Ni and Zn) were marked with less than 1 indicating the relative absence of ingestion risk for the vegetables. The study suggested that even low concentrations of heavy metals in wastewater pose risks to human health through accumulation over a longer period of time.

In Romania, the authors of paper [12] studied the concentration of six heavy metals (Zn, Cu, Pb, Ni, Cd and Cr) in a few vegetables (potato, onion, garlic, carrot, parsley, and beans) grown in the gardens of inhabitants of the Ferneziu neighbourhood in the Baia Mare mining area (NW Romania) to characterize the transfer of heavy metals from the soil into vegetables.

Concentrations of heavy metals varied among the investigated vegetables (potato, onion, garlic, carrot, parsley, salad, tomato and beans) grown in the gardens of the inhabitants due to their different absorption capacity. The total fraction of heavy metals in the soil followed the

Zn> Pb> Cu> Ni> Cr> Cd sequence, the results showing that Cd, Zn and Ni had the highest transfer factor.

The transfer of heavy metals from contaminated soil to plants is the main route of human exposure to metal contamination. This work investigated the accumulation and the transfer coefficient of three heavy metals (Cu, Pb, Zn) with four concentrations (c1 = 1.5%, c2 = 3.0%, c3 = 4.5%, c4 = 6.0%) obtained by mixing the three metals, from the contaminated soil in the tomato fruit.

2 Material and method

In the paper, the tomato fruit was studied to identify the levels of heavy metals absorbed during plant growth after planting them in agricultural soils contaminated with mixtures of heavy metals (copper, zinc and lead) at different concentrations.

Six kg of soil was mixed with 60 ml of copper, zinc and lead solution in equal proportions for each concentration: 1.5%, 3.0%, 4.5% and 6.0%. To obtain the mixtures of solutions with Cu, Pb, Zn for each of the concentrations of 1.5, 3.0, 4.5 and 6.0% individually prepared, equal parts were taken from each solution, element and concentration, and mixed until homogeneous, thus resulting the mixture. For each concentration of the mixture of Cu, Zn, Pb, the contaminated soil was divided into three pots (3 repetitions) in which a seedling of tomato was planted. Heavy metal loading was performed only by initial loading without supplementing until harvest.

The reference samples consisted in planting three tomato seedlings in three different pots, adding uncontaminated soil (fertile soil). During experimental research, the plants were watered weekly to maintain soil moisture and implicitly plant growth.

2.1 Soil samples

Basic properties of the farmland soil under study were: pH 6.0; total nitrogen 1.9 %, total phosphorus 0.5%, total potassium 0.9%, electrical conductibility 1.2, particle elements of over 20 mm maximum 5%, moisture 14.7% [13].

2.2 Plant material

For this study, tomato (*Solanum lycopersicum*) was chosen as research material (Figure 1) because it is a nutrient-rich fruit.

Tomato, scientifically called *Solanum lycopersicum*, is a plant of the Solanaceae family, genetically close to the line of pepper, potato, eggplant and physalis alkekengi. It is a perennial plant, cultivated at almost all latitudes, often grown in temperate climates as an annual plant, often reaching 1 to 3 m in height, with a grassy stem, growing like any creeping plant, on the stems of other plants, on sticks or fences. The leaves of the plant are 10-25 cm long, pen-composite in shape, both the stem and the small branches, respectively the leaves being covered with animal hair-like excrescences. Flowers are small, about 1-2 cm. Tomato has an acidic pH of 5.5. From a botanical point of view, tomato is a fruit [14].

Fig. 1. Tomato (Solanum lycopersicum).

2.3 Determination of heavy metals in the contaminated soil

The farmland soil is contaminated with mixture of heavy metals (Pb, Cu, Zn) in this study, with four concentrations.

The method by which the heavy metals (Cu, Zn, Pb) from soil samples (uncontaminated, contaminated with mixtures solutions of 1.5 %, 3.0 %, 4.5 % and 6.0 % concentration) was represented by x-ray fluorescence analysis.

An x-ray fluorescence (XRF) spectrometer was used, measuring the peak line emission was obtained a value, which represented the concentration of the analyte. The concentration of the sample to be analysed was correlated with the net measured value [13].

2.4 Determination of heavy metals in tomatoes (fruit)

The plant samples were dried (60° C, 72 h), grinded and then were digested with nitric acid (65 %) in a microwave digestion system as follow:

-weigh 300 mg of the sample into the digestion vessel and add 7.5 ml of nitric acid.

-heat in the microwave oven with the following program.

• First step: T 150° C, power (%) – 50, time - 5 min.

• Second step: T 190° C, power (%) – 70, time 5 min.

The metal content was measured using a flame atomic absorption spectrometry (FAAS, GBC 932AA or GFAAS, GBC Savant AAZ) [15].

2.5 Transfer Coefficients of system Soil-Plant

The transfer factor is an important parameter for soil and plant. Transfer coefficient values show both the biological availability of plants to absorb more or less heavy metal and the amount of metal retained by the soil. Low transfer coefficients indicate that the metals are sown in soil colloids, and high transfer coefficients show the presence of large metals in a particular plant species and indicate that the plant can be successfully used in soil remediation [9].

Due to the fact that plants have the capacity to accumulate metals in the soil, the ratio between metal concentration in the plant and in the soil was measured. Using equation (1) [16], the transfer coefficient (tf) was determined.

$$tf = \frac{heavy \quad metal \quad concentrati \qquad plant}{heavy \quad metal \quad concentration \quad in \quad sol} \tag{1}$$

The value of a transfer factor of 0.1 indicates that the plant excludes metal from its tissues, while the value of a transfer coefficient of 0.2 indicates the pollution plant through human activities [17], and therefore environmental monitoring is necessary [18].

3 Results and discussions

In fig. 2, 3 and 4 are shown the average values of the results of the three repetitions on the content of metals (Cu, Pb, Zn) from soil samples (mg / kg) contaminated with mixed solutions of different concentrations (1.5%, 3.0%, 4.5%, 6.0%).

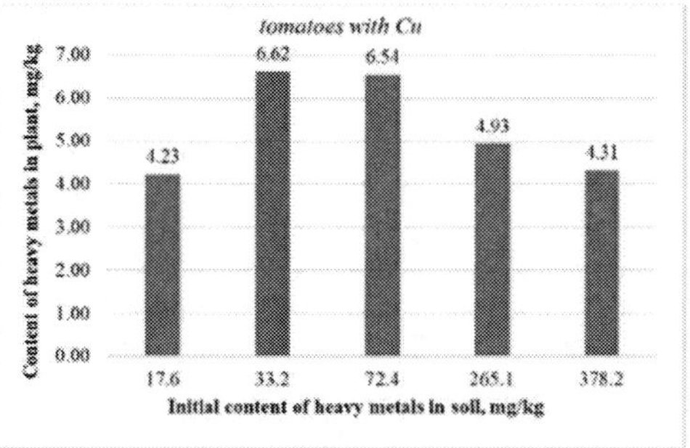

Fig. 2. Concentrations of heavy metals (Cu) in soil and in plant.

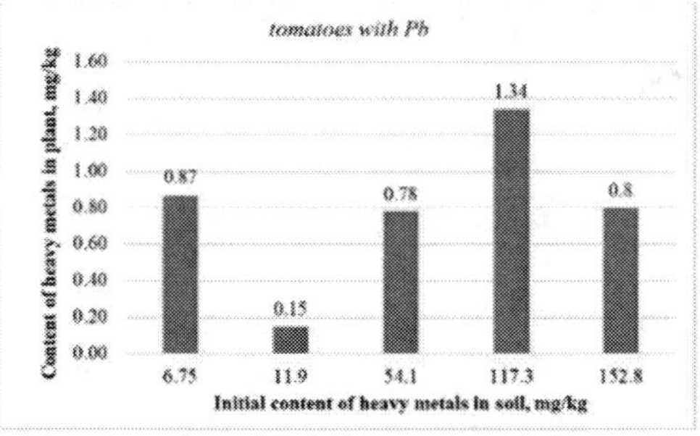

Fig. 3. Concentrations of heavy metals (Pb) in soil and in plant.

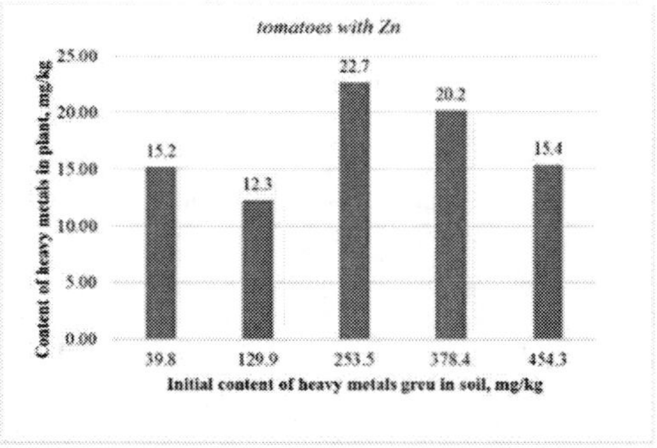

Fig. 4. Concentrations of heavy metals (Zn) in soil and in plant.

Analysing figures 2, 3 and 4 with the content of heavy metals in the soil and their transfer into plants (fruit), the following are observed:

- in case of Cu, the content drops in the fruit as the soil content increases;
- in the case of Pb, the content is progressively increased in fruit for the first three concentrations of 1.5% (11.9 mg / kg), 3% (54.1 mg / kg) and 4.5% (117.3 mg / kg). At the maximum concentration of 6% (152.8 mg / kg) in soil, the Pb content decreases in fruit;
- for Zn, the 3% concentration in soil (253.5 mg / kg) is maximal in fruit (22.7 mg / kg). For the other Zn concentrations in the soil, the quantity found in the fruit is lower;
- for uncontaminated soil (control sample), it is found that there are quantities present in the fruit for each of the three metals. It was therefore found that at the highest concentration (6%), for each of the three metals, their content in the tomato fruit tends to have values approximately equal to those in the fruit grown in the uncontaminated soil.

Using the numerical data from the experiments and represented graphically in figures 2, 3, 4, the interpolation of the metal mixture data into the tomato fruit was performed.

$$Cp = \left(C_{Cup}, C_{Pbp}, C_{Znp}\right) \quad (2)$$

$$Cp = \left(C_{Cup}(C_{Cus}, C_{Pbs}, C_{Zns}), C_{Pbp}(C_{Cus}, C_{Pbs}, C_{Zns}), C_{Znp}(C_{Cus}, C_{Pbs}, C_{Zns})\right) \quad (3)$$

where, C_p - Cu, Zn, Pb mixture content on the plant; C_{Cup} - Cu content in the plant; C_{Pbp} - Pb content in the plant; C_{Znp} - Zn content in the plant; C_{Cus} - Cu in the soil; C_{Pbs} - Pb content in the soil; C_{Zns} - Zn in the soil.

The linear regression corresponding to this data set will take the form:

$$C_{Cup} = \alpha_{Cup} + \beta_{Cup}C_{Cus} + \gamma_{Cup}C_{Pbs} + \theta_{Cup}C_{Zns} \quad (4)$$
$$C_{Pbp} = \alpha_{Pbp} + \beta_{Pbp}C_{Cus} + \gamma_{Pbp}C_{Pbs} + \theta_{Pbp}C_{Zns} \quad (5)$$
$$C_{Znp} = \alpha_{Znp} + \beta_{Znp}C_{Cus} + \gamma_{Znp}C_{Pbs} + \theta_{Znp}C_{Zns} \quad (6)$$

Table 1 - Coefficients of the interpolation data for tomato fruits

Function Coefficients					
α_{Cup}	3.57	α_{Pbp}	1.097	α_{Znp}	16.094
β_{Cup}	$3.632 \cdot 10^{-3}$	β_{Pbp}	-0.012	β_{Znp}	-0.166
γ_{Cup}	-0.098	γ_{Pbp}	0.058	γ_{Znp}	0.514
θ_{Cup}	0.032	θ_{Pbp}	$-9.766 \cdot 10^{-3}$	θ_{Znp}	-0.035

where, α, β, γ, θ - function coefficients for Cu, Pb and Zn in the plant.

Due to the fact that the data in figures 2, 3 and 4 does not have linear distributions, many of the values taken from outside the interpolation points have negative values.

The results regarding the transfer factor (tf) of the metals in the soil in the tomato fruit are graphically represented in figures 5, 6 and 7.

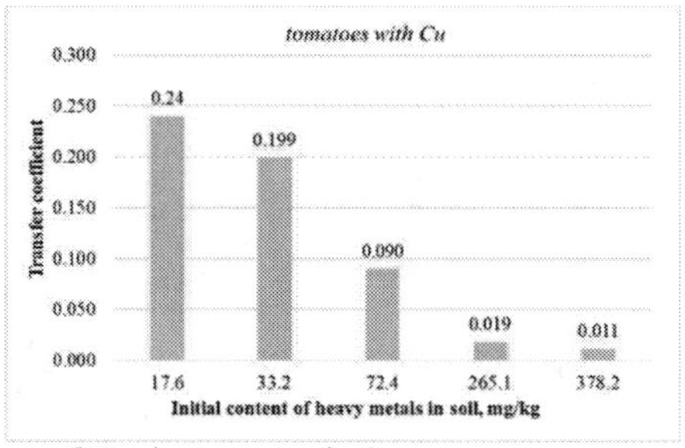

Fig. 5. The transfer coefficient of Cu in tomato fruit.

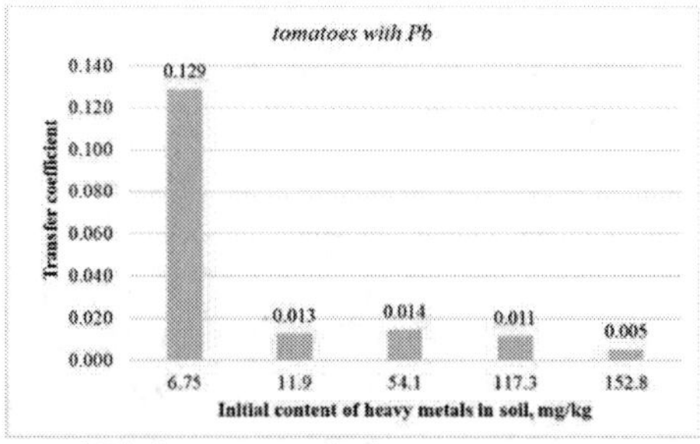

Fig. 6. The transfer coefficient of Pb in tomato fruit.

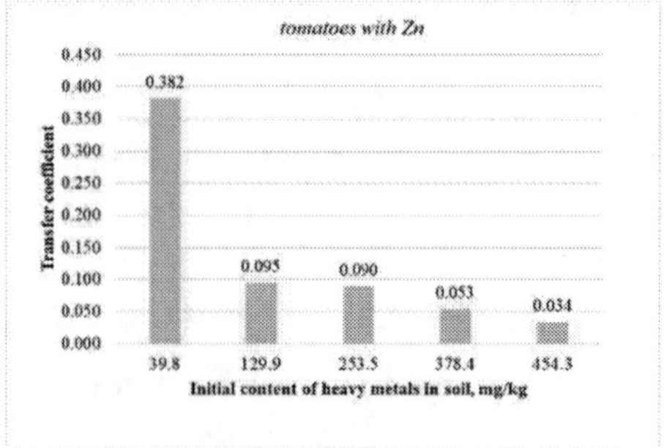

Fig. 7. The transfer coefficient of Zn in tomato fruit.

From figures 5, 6 and 7 it is observed that heavy metals found in tomato fruits grown in contaminated soil at different concentrations show a decrease in the heavy metal transfer coefficient from the soil to the plant along with an increase of the initial heavy metal concentrations in the soil.

Therefore, for tomatoes, the bioaccumulation is even weaker as the initial heavy metal concentration in the soil is higher.

4 Conclusions

As a result of the experimental results obtained after soil contamination with heavy metals and their absorption by the plants, the following conclusions were drawn:

- the most absorbed metal was zinc, regardless the concentration, then copper and the least lead;

- the transfer factor is biggest for Cu, followed by Zn and has a fairly low value for Pb;

- a tendency of increase of the final heavy metal content was noticed in the tomato fruit and a decrease of the transfer coefficient of heavy metal in the fruit at the end of the vegetation period;

- because vegetables are known to intake and accumulate traces of heavy metals from the contaminated soil, the detection in tomato leaves was not surprizing.

Results obtained for *Solanum lycopersicum* show that the experiments and statistical models can provide the basis for the construction of dynamic mathematical models that can simulate the life of the plant.

Acknowledgement

This paper was financed by support of Executive Agency for Higher Education, Research, Development and Innovation Funding, Exploratory Research Programme, PN-III-P4-ID-PCE-2016-0860, contr. 174/ 08.08.2017, Research on the development of some mathematical models to evaluate the impact of soil contamination on fruits and vegetables – CONTAMOD.

References

1. Y. N. Jolly, A. Islam, S. Akbar, Transfer of metals from soil to vegetables and possible health risk assessment, SpringerPlus, **2**, 385–391(2013)
2. Z. Huang, X.-D. Pan, P.-G. Wu, J.-L. Han, Q. Chen, Heavy metals in vegetables and the health risk to population in Zhejiang, China, Food Control **36**, 248-252, (2014)
3. A. Cherfi, S. Abdoun, O. Gaci, Food survey: levels and potential healthrisksofchromium, lead, zinc and copper content in fruits and vegetables consumed in Algeria, Food Chem Toxicol 7**0**:48–53, (2014)
4. L. Jarup, Hazards of heavy metal contamination, Br Med Bull.; **68**:167–82, (2003)
5. A. Maleki, M.A. Zarasvand Heavy metals in selected edible vegetables and estimation of their daily intake in Sanandaj, Iran, Southeast Asian j Trop Med Public Health.;**39(2)**:335–340, (2008)
6. A. Fawad, U. Hidayat, K. Ikhtiar, Heavy metals accumulation in vegetables irrigated with industrial influents and possible impact of such vegetables on human health, Sarhad Journal of Agriculture, Volume **33**, Issue 3, 489-500, (2017)

7. C. Roba, C. Roşu, I. Piştea, Al. Ozunu, C. Baciu, Heavy metal content in vegetables and fruits cultivated in Baia Mare mining area (Romania) and health risk assessment, Environ Sci Pollut **23**, pp. 6062–6073, (2016)
8. E. Islam, X. Yang, Z. He, Q. Mahmood, Assessing potential dietary toxicity of heavy metals in selected vegetables and food crops, Journal of Zhejiang University Science B, **8(1)**: 1-13), (2007)
9. F.I. Bassey, C.M.A. Iwegbue, G.E. Obi-Iyeke, G.O. Tesi, A.R. Rotu, O.A. Gobe, A.I. Tsafe, Heavy metals in soils and tomatoes grown in urban fringe environment in Asaba, Delta State, Nigeria, Nigerian Journal of Basic and Applied Science, **22**(1&2): 27-31), (2014)
10. N. M. Salem, L. S. Albanna, A. M. Awwad, Toxic heavy metals accumulation in tomato plant (Solanum lycopersicum), ARPN Journal of Agricultural and Biological Science, vol. **11**, no. 10, 399-404, (2016)
11. A. Laţo, I. Radulov, A. Berbecea, K. Laţo, F. Crista, The transfer factorof metalsin soil-plant system, Research Journal of Agricultural Science, **44** (3), 67-72, (2012)
12. C. Roba, C. Rosu, I. Pistea, A. Ozunu, C. Baciu, Heavy metal content in vegetables and fruits cultivated in Baia Mare mining area (Romania) and health risk assessment, Environ Sci Pollut Res, **23**:6062–6073, (2016)
13. A. Pruteanu, D. M. Bordean, V. Vlăduţ, M. Niţu, I. Găgeanu, Cercetări privind contaminarea solului cu metale grele şi remanenţa acestora în legume, Analele Universităţii din Craiova, seria Agricultură – Montanologie – Cadastru (Annals of the University of Craiova - Agriculture, Montanology, Cadastre Series) Vol. **48**, (2018)
14. A. Ardelean, Gh. Mohan, Medicinal Flora of Romania, All Publishing, Bucharest, (2008)
15. I. Gergen, Chemical and physicochemical methods in the control of vegetable agrifood products quality, Orizonturi Universitare Publishing House, Timişoara, (2003)
16. M. Nedelescu, D. Baconi, A. Neagoe, V. Iordache, P. Constantinescu, A. M. Ciobanu, I. A. Vardavas, M. Stan, M. Vinceti, M. A. Tsatsakis, Environmental metal contamination and health impact assessment in two industrial regions of Romania, Science of the Total Environment **580**, pp 984–995, (2017)
17. S. Khan, R. Farooq, S. Shahbaz, M.A. Khan, M. Sadique, *Health risk assessment of heavy metals for population via consumption of vegetables*, World Applied Sciences Journal, **6**(12): 1602-1606, (2009)
18. N. Sponza, Karaoglu, *Environmental geochemistry and pollution studies of Aliaga metal industry district*, Environment International, **27**: 541-553, (2002)

Structural and kinematic analysis of a shutter from monocoque trailer

Lucreția Popa[1,*], *Ștefan-Iulian* Moise[2], *Victor* Moise[3], *Vasilica* Ștefan[1] and *Iulian* Tabără[3]

[1]INMA Bucharest, 6 Ion Ionescu de la Brad, Bucharest, Romania
[2] Etudes et Productions Schlumberger, Paris, France
[3] POLITEHNICA University of Bucharest, Romania

Abstract. The bodies of trucks or trailers are provided with automatic folding shutters to ease the discharge of the transported materials. Typically, these shutters are powered by mechanisms that have active motor groups such as RRTaR or RR-RR-RTaR (i.e. motor dice or triads) and passive motor groups such as RRR, RR-RR-RR etc. If the technological forces are high and the relative speeds between the elements forming the steering coupler are small, the motor groups are formed by rigid bars and hydraulic cylinders. If relatively high gear ratios are required, then pneumatic cylinders may be used. If required relatively high speeds in the driving couplings, then you can use pneumatic cylinders. This paper presents structural and kinematic analysis of a shutter from a trailer used to transport bulk agricultural products, discharge towards the back. For the kinematic analysis of the folding shutter mechanism, computational procedures written in the Matlab syntax are used. The results of the calculations are presented in tabular and diagrams. After the cinematic analysis of the mechanism, it is possible to perform its cinetostatic analysis, i.e. the determination of the reactions from the kinematic couplers, as well as the hydraulic force in the hydraulic cylinder. Results can be used for the dimensional verification of the chosen elements.

1 Introduction

Trailers for the transport of bulk agricultural products may have different forms of construction, in terms of the way of unloading the cargo.

They can download the products tipping on the side, right, left or on both sides, but there are constructive variants where the unloading is done backwards.

The trailers, which are built in the monocoque construction, perform only the rear discharge and are fitted with a shutter by manual opening, but the newer trailers are equipped with automatic opening shutters.

The automatic shutter of the trailer provides operator comfort, who is no longer required to descend from the tractor to open the shutter, return to the tractor to command the

* Corresponding author: Lucreția Popa, lucretia_popa@yahoo.com

https://doi.org/10.1051/e3sconf/201911203021

unloading and then lower down the shutter after unloading the cargo, after returning the body trailer to the travel position.

Fig. 1. Two sides tipping trailer.

Fig. 2. Monocoque trailer by automatique shutter.

Making a mechanical model that includes structural and kinematic analysis of a shutter fitted to the monocoque trailer is important in terms of construction and functioning.

The kinematic analysis of the mobile shutter drive mechanism gave useful information on the positions, speeds and the accelerations of the mechanism elements, without taking into account the forces acting on them.

In this paper, structural and kinematic analysis of the mechanism that perform the opening of the mobile shutter will be presented.

2 Material and method

In order to determine the kinematic parameters of the elements of the mechanism, it is necessary to know the structural groups in its composition, the dimensions of the elements, the initial position of the mechanism, as well as the kinematic parameters in the motor coupling. Figure 3 shows the schematic kinematic mechanism for tilting the shutter of trailer SR8, design to INMA Bucharest. The mechanism is surprised in any position.

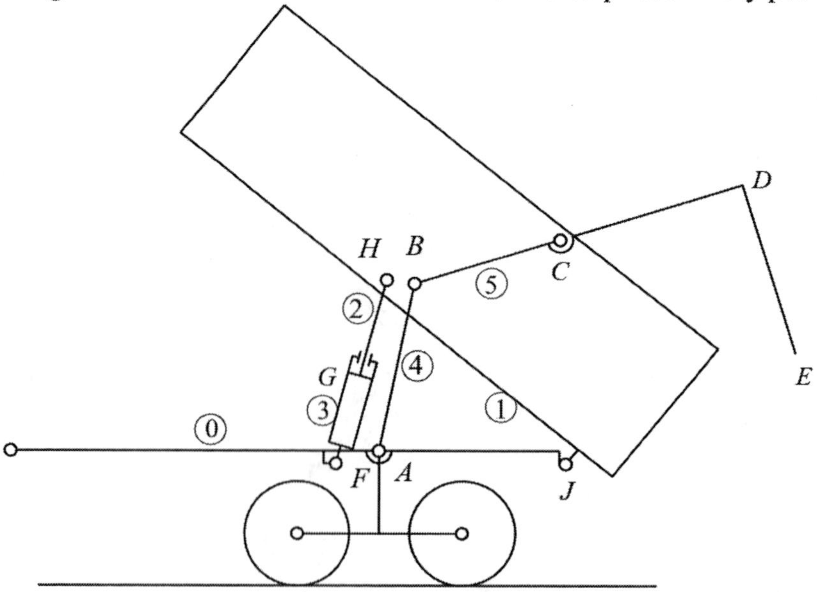

Fig. 3. The kinematic scheme of the mechanism for acting the trailer shutter.

2.1 Structural analysis of the mechanism

The structural analysis of the mechanism consists in establishing kinematic couplers, determining the mobile elements, establishing the degree of mobility, drawing up the structural scheme, establishing the structural structures of the mechanism, constructing the multipolar scheme and establishing the structural relationship.

Observing the relative movements between the elements of the mechanism, it is found that it has the following inferior couplings: $A(0R4)$, $B(4R5)$, $C(1R5)$, $F(3R0)$, $G(2T3)$, $H(1R2)$ and $J(1R0)$.

The number of superior couplings is zero.

The mobile elements of the movement are: $1(C,H,J)$, $2(G,H)$, $3(F,G)$, $4(A,B)$, and $5(B,C)$.

Taking into account the number of moving elements and the number of kinematic couplings, the degree of mobility of the mechanism results [1-6], namely M = 1, meaning only one independently parameter is required so that the elements of the mechanism having unique determined movements.

Figure 4 shows the structural diagram of the climbing mechanism of the shutter, and Figure 5 presents the multipolar scheme and its structural relationship.

Figure 5 shows that the mechanism consists of *Z0* base, motor dyad $RRTaR(1,2,3)$, and passive dyad $RRR(4,5)$.

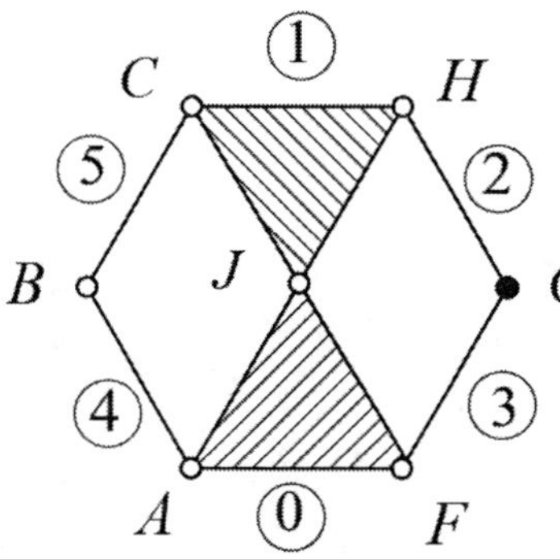

Fig. 4. Structural scheme of the mechanism.

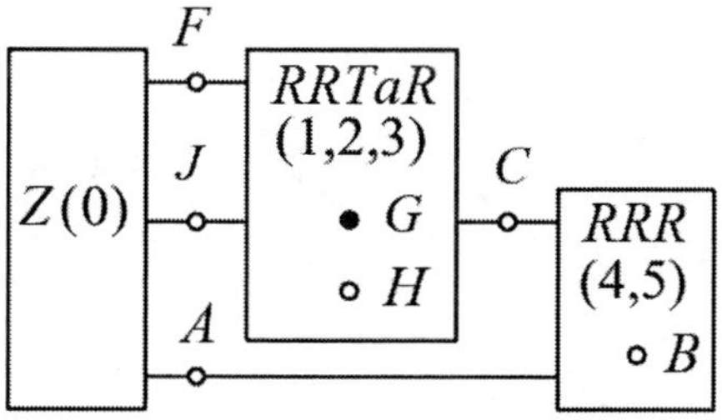

$Z(0) + RRTaR\ (1,2,3) + RRR\ (4,5)$

Fig. 5. Multipolar scheme of the mechanism and structural relationship.

2.2 Kinematic analysis of the mechanism

Figure 6 shows the kinematic diagram of the mechanism, with the position parameters being highlighted.

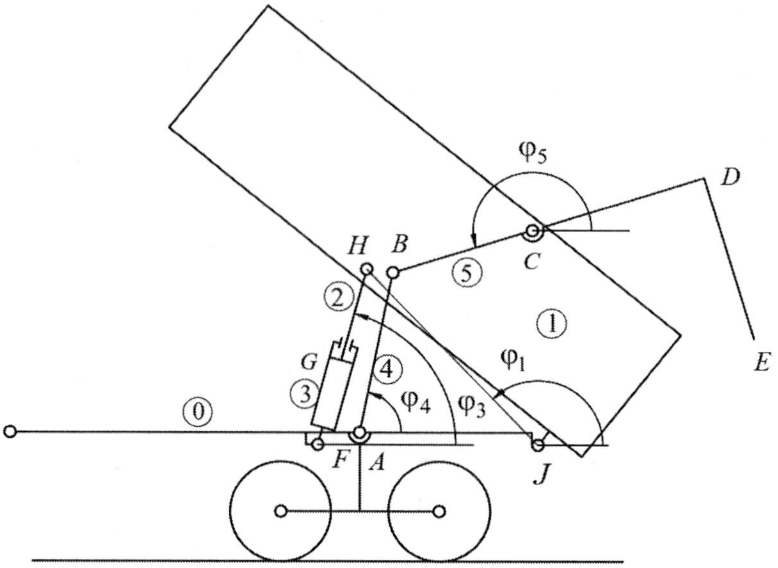

Fig. 6. The kinematic scheme of the mechanism, with highlighting position of parameters.

For the kinematic analysis of the mechanism are used calculation procedures written by the authors in the Matlab syntax.

The theoretical study was applied using the constructive parameters of a trailer equipped with automatic shutter, designed to INMA Bucharest, namely SR-8, showed in Figure 7 [7].

Fig. 7. Trailer with payload capacity of 8 tons on transport position.

Fig. 8. Trailer with payload capacity of 8 tons on tipping position.

For the kinematic analysis of the mechanism, the following numerical data were used (fig.6): $AB = 1.270$ m, $BC = 1.000$ m, $CJ = 1.580$ m, $FH_{initial} = 0.216$ m, $HJ = 1.520$ m, $Piston\ stroke = 1.150$ m, $XA = 0.000$ m, $YA = 0.000$ m, $XF = -0.200$ m, $YF = -0.185$ m, $XJ = 1.340$ m, $YJ = -0.050$ m, $\angle(\overline{JC}, \overline{JB}) = 0.769838$ rad, $Lifting\,time = 20$ sec.

Transmission function in the hydraulic cylinder: lrsin (sinusoidal acceleration - zero acceleration - sinusoidal acceleration).

The numeric values of the angles done by vectors $\overline{JH}$, $\overline{FH}$, $\overline{AB}$ and $\overline{CB}$ with the positive sens of axis *OX* are presented In Table 1.

Table 1. The numeric values of the angles done by vectors $\overline{JH}$, $\overline{FH}$, $\overline{AB}$ and $\overline{CB}$ with the positive sens of axis *OX*.

Position	fi 1	fi 2	fi 4	fi 5
0	3.0884	1.4685	2.1850	3.2122
1	3.0882	1.4686	2.1848	3.2123
2	3.0871	1.4689	2.1836	3.2125
3	3.0843	1.4698	2.1804	3.2132
4	3.0792	1.4712	2.1745	3.2145
5	3.0712	1.4729	2.1655	3.2165
6	3.0602	1.4743	2.1530	3.2192
7	3.0465	1.4749	2.1374	3.2226
8	3.0304	1.4743	2.1190	3.2266
9	3.0125	1.4724	2.0985	3.2310
10	2.9938	1.4692	2.0769	3.2358
....				
40	2.4084	1.2285	1.3526	3.4322
41	2.3880	1.2188	1.3241	3.4428
42	2.3686	1.2096	1.2965	3.4535
42	2.3510	1.2012	1.2710	3.4637
44	2.3359	1.1940	1.2489	3.4728
45	2.3239	1.1883	1.2309	3.4805
46	2.3151	1.1841	1.2177	3.4862
47	2.3093	1.1813	1.2090	3.4901
48	2.3062	1.1798	1.2042	3.4922
49	2.3051	1.1793	1.2024	3.4930
50	2.3049	1.1792	1.2022	3.4931

Figure 9 shows the variation diagrams of the angles done by vectors $\overline{JH}$, $\overline{FH}$, $\overline{AB}$ and $\overline{CB}$ with the positive sens of axis *OX*.

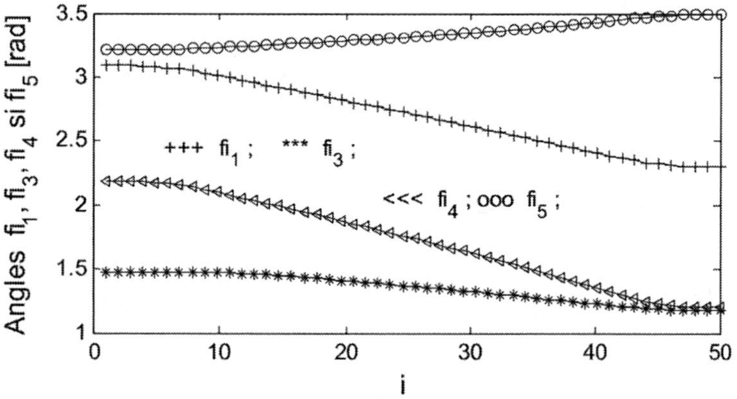

Fig. 9. Variation diagrams of the angles done by vectors $\overline{JH}$, $\overline{FH}$, $\overline{AB}$ and $\overline{CB}$ with the positive sens of axis *OX*.

Figures 10 and 11 show variation diagrams of velocity and angular acceleration of the moving parts.

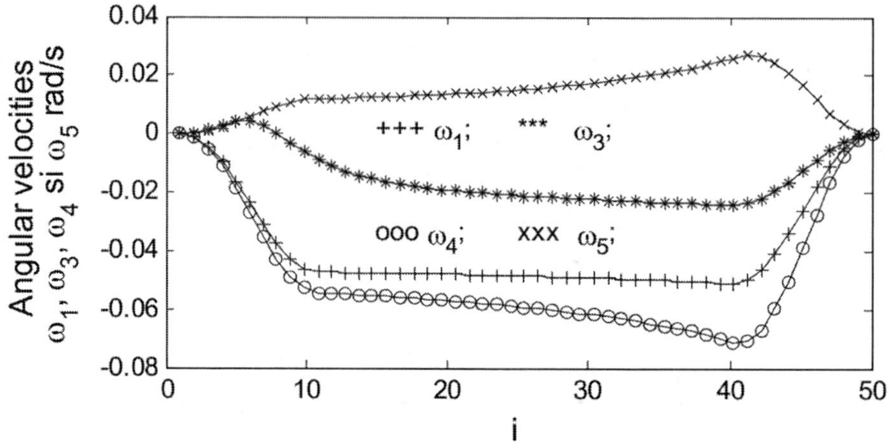

Fig. 10. Angular velocity variation diagrams of mobile elements.

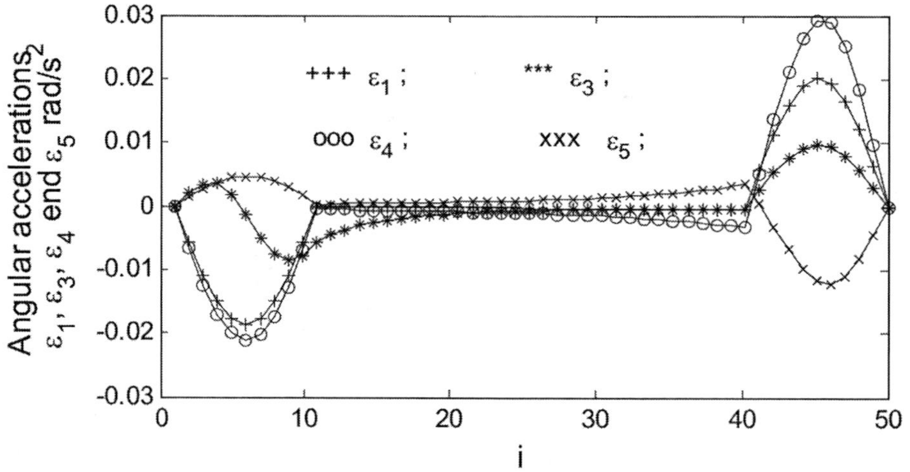

Fig. 11. Angular acceleration variation diagrams of mobile elements.

3 Conclusions

The kinematic analysis of the mechanism provides useful information regarding the tipping speed of the trailer body and, implicitly, the cargo. A high-speed causes damage to the automatic shutter, and a small one causes a long discharging time.

The kinematic analysis made by the simulation enables the proper dimensioning of the elements of the mechanism, so that in the key positions the shutter is at the level that provide the discharge section.

The kinematic analysis realized in this paper is useful for the dynamic analyse, which will be perform in another research.

Acknowledgement

This work was supported by a grant of the Romanian Research and Innovation Ministry, through Programme 1 – Development of the National Research-Development System, subprogramme 1.2 – Institutional performance – Projects for financing excellence in RDI, contract no. 16PFE.

References

1. I.I. Artobolevski, *Théorie des Mécanismes et des Machines (Theory of Mechanisms and Machines)*. (Ed. Mir, 1977)
2. C. Duca, Fl. Buium et al. *Mechanisms*, (Gh. Asachi Publishing House, 2003).
3. V. Moise, E. Maican, Şt. I. Moise, *Numerical methods in engineering* (Bren Publishing House, 2003)
4. V. Moise, I. Simionescu, M. Ene et al. *Analysis of Applied Mechanisms*, (Printech Publishing House, 2008)
5. Chr. Pelecudi. *The precision of the mechanisms*, (Publishing House of the Academy of RSR, 1975)
6. I. Simionescu, V. Moise, *Mechanisms*, (Technical Publishing House, 1999)
7. https://www.inma.ro/Prez_CDI/SR_8.pdf

 https://doi.org/10.1051/e3sconf/201911203022

Modelling by Finite Element Method of the twin-screw of a press for obtaining grape seed oil

Carmen Vasilachi[1*] *and* *Sorin-Stefan* Biriş[2]

[1]National Institute of Research - Development for Machines and Installations Designed to Agriculture and Food Industry – INMA, Bucharest, Romania
[2]University Politehnica Bucharest, Faculty of Biotechnical Systems Engineering, Romania

Abstract. This paper is about getting grape seed oil using a twin-screw press. Grape seed oil is a by-product of the wine industry due to the need to reduce the amount of waste, but at the same time it has been found that this oil is an excellent ingredient that can be used in many industries. It is assumed that the use of twin-screw is more efficient than single-screw extraction due to the larger pressures that are created inside the press chamber. This study focuses on the finite element method analysis of how the press process is carried out. Using the finite element method, we can get an overview of how the pressure acts on the sides of the twin-screw, how it influences the temperature but also other process factors. When we have all these, we can begin optimizing the twin-screw press process.

1 Introduction

Due to the imminent exhaustion of the fossil resources which are at present being excessively used in the petrochemical industry a huge amount of effort is being made to discover a viable new alternative, a renewable solution for raw materials to be used instead of these resources[1]. In order for this alternative to be considered it has to include usage of renewable resources, it has to have an accessible price, it needs to be easily accessible by an industry, it needs to be produced widely across the globe and it should be considered practically unlimited. Another further benefit which is desired from this research is that the derived products of these renewable raw materials (E.g.: polyurethanes, lubricants, fuels) are to be biodegradable and also to have a proper integration with the natural cycle,[2,3,4].

Taking into consideration the situation from the petrochemical industry where more environmentally friendly alternatives are being looked for, the extraction of oils from vegetable materials is starting to become an opportunity worth exploiting. Grape seed oil can be used in industrial applications as well as a rich source of bioactive compounds (E.g.: essential fatty acids, tocopherols, tocotrienols, polyphenols, flavonoids, tannins and phytosterols), [5].

* Corresponding author: carmen.vasilachi@gmail.com

These bioactive compounds in oils can have a various composition depending on the grape variety, on the maturity level on the conditions and locations in which the grapes have been grown as well as the type of extraction technique used. The extracted grape seed oil benefits human health greatly due to its high unsaturated fatty acids content alongside natural antioxidants compounds. Similar to other cold pressed oils the grape seed oil is used in food, cosmetics and pharmaceutical industries mostly due to its composition and properties, [6].

As a sub-product, grape seed oil is obtained through harvesting the grape marc which in turn is part of the wine industry. Throughout the year in the wine industry a little over 20% of the grapes become waste material out of which between 38-52% are dried calories seeds from the dry matter. Between 8 and 20% of oil can be obtained from this quantity of grape seeds and it's also worth mentioning that this would be the most economic and efficient way to deal with the unused leftover material, [7]. As an example we can use the numbers reported by Castilla-La-Manca which is a region located in the center of Spain that runs one of the largest wine producing industry in the world. Their annual waste sums up about 100000 tons of grape seeds over one year, [8]. In order to extract grape seed oil from the seeds a mechanical process can be used and obtain the oil through screw extrusion or a solvent process can be used and obtain the oil in this way. The most preferred extraction method which coincidently is also the efficient one because it ensures the superior quality of the extracted oil is the mechanical pressing. Through mechanical pressing we will benefit from low process temperatures without the solvent but unfortunately we will also obtain low levels of oil, usually between 55 and 95% depending on the processed material, [9]. The mechanical screw extraction has been a subject of study for a wide variety of oil producing seeds (such as linseed, canola and grape seed). These studies have shown that the factors which contribute to increase the temperature and pressure in the screw press chamber benefit the oil yield greatly. By changing the nozzle's diameter through which the solid parts are being discharged and also by adjusting the rotation speed of screw inside the press we can adjust the factors to benefit the production of grape seed oil. It was noticed throughout testing that the lower the screw speed and the smaller the nozzle opening is the more oil is yielded, [10]. There are other ways of increasing the oil yield which consist of seed pre-treatments such as degreasing, cooking, moisturizing conditioning. Moisturizing conditioning has an essential role to play in the extraction of oil since the optimal moisture content which determines the oil yield can be discovered based on the seed variety. Cooking the seed before extraction could potentially cause oil degradation even though it increases the oil yield. Because of the many pre-treatments downsides cold pressing is usually to go-to method preferred because of the retention of healthier components such as antioxidants which increase the nutritional value of the pressed oil. It is however possible to increase the composition and quality of the oil by making sure that during the pressing process the temperatures reached would stay constantly between 60 and 68 ° C, [11,12].

Just as mentioned above, mechanical single screw extractions will have a lower yield result. Taking this matter into consideration has made researchers try to implement new ways of increasing the yield of pressed oil by using a method called twin-screw extrusion.

The first twin-screw extruder was developed by Coignet in 1869 for the concrete industry. The first oilseed oil twin-screw extractors were developed in the 1990s, [13].

The twin-screw extrusion method is an innovative piece of technology which has continuously growing applications in industries such as the grain industry, the polymers industry, the pet food industry and the paper industry. Even though at present this technology is not yet applied in the seed oil extraction industry it presents itself as a very promising technique to be used for a sustainable extraction of vegetable oil and also as a way of ruling out some of the problems found in single-screw extraction methods such as the relatively small amount of oil extracted, the excessive heat generated during the extraction because of friction, the high energy consumed by the process and the deterioration present in the oil and clod quality, [14].

It is considered very likely that the twin-screw press could be a solution to the above mentioned problems since it provides a much higher transport force and a much better shredding and mixing interface between the two worm screws. After performing a set of lab tests to see how the twin-screw press and the single-screw press handle sunflower seeds when trying to obtain sunflower oil through this method the tests showed that the twin-screw technology has a much greater efficiency when extracting oil from oilseeds, it yields a high-quality oil and meal even if the materials are not the type that could be easily machined as in the case of peeled sunflower seeds pressed by a single-screw press. Another advantage of the twin-screw press is that it makes it possible to discard the pre-treatments such as cooking and peeling the seeds and also the roasting and dripping pre-treatments as well. While performing the tests it was noticed that the energy consumption during the process was much lower than the one for single-screw presses. After multiple experiments it was noted that the twin-screw press has a better performance at milling when the seeds are small just like sesame and rape seeds are, but it is still unknown whether the press can perform just as good when string low-oil seeds are used like soybean seeds.

To get a more intense mechanical machining of the seeds a design with counter-rotating and intermittent screws was thought of, a design which generates longer retention times and high shearing pressure at the upper interconnection of the screws, [14]. This is an important machining advancement in this industry since the ability to configure the screw with a wide variety of elements that have a varying thread profile and also making it possible to have different working sections into the same extruding machine is an important and beneficial change. As such an advancement of this sort is presented in literature as having been developed for sunflower seeds presses for the extraction of sunflower oil where it was noticed that the sunflower seeds have undergone thermomechanical pressures during the extraction process, [15]. The below Figure 1 presents the structure of an extruder and that of the twin-screw profile.

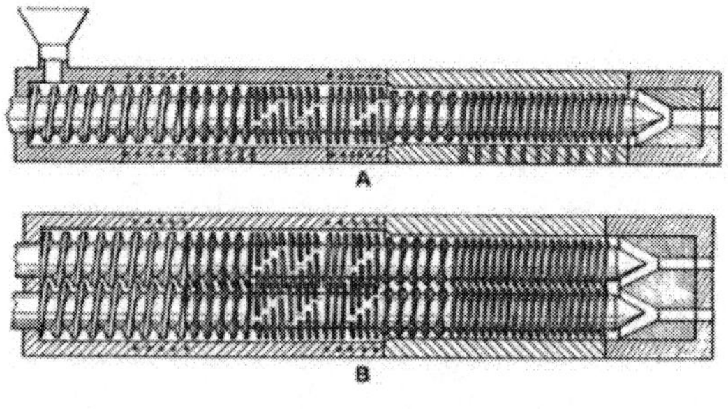

Fig. 1. The twin-screw extruder design comprising two different pressing sections and developed; A. Schematic view through a vertical plane running through one of the screw axes; B. Schematic view through a transversal plane showing both screw axes [13].

Because of the possibility of having different configurations for the twin-screws the press becomes more versatile. The first element present in the configuration of the right screw is a section with a certain pitch which is special for its ability to ensure constant material transportation and a high material pumping capacity and which is also used in the extruder feed or in the compression section. On the left screw the pitch configuration is referred to as a reversed screw member. This element exerts a transportation force on the material present in the press which is opposite to the direction flow which in turn causes intense shearing of the material. As a conclusion successively positioning different screw combinations in the press results in a more effective material compression, the creation of a dynamic plug and a good separation which is caused by the efficient pressure build-up, [16].

The objective of this study was to show how the compression of the double screw during the pressing process works and to show that an extruder with two co-rotating screws can be used to more effectively extract grape seed oil by mechanical pressing. The study's subject also proposes the evaluation of the effects that the screw configuration, the screw pressure and the screw rotation speed have on oil extraction efficiency.

2 Materials and methods

In this study, grape seeds were used as raw material. Initially grape seeds have a moisture content between 40-50% and must be reduced to about 5-8% in the drying chamber. In the literature, the following texture parameters for grape seeds were obtained (Table 1): seed break force (N, as F_s), seed break energy (mJ, as W_s), seed N (N.mm^{-1}, as E_s) and seed deformation index (%, as DI_s).

In this study, we chose the mechanical properties of the material, grape seeds, to be part of the D density class where we encounter the greatest breaking force.

The main objective of this paper was to find a way to highlight how forces and deformations propagate into the double screw during the compression process in order to optimize the energy consumption required for processing. Therefore, it is now possible to try using the finite element method. This method is the most advanced engineering tool for calculating the numerical and mathematical modelling of complex phenomena involving the propagation of stress and strain fields in continuous environments. In this paper is presented a two-dimensional FEM model for the analysis of double screws subjected to compression through the axial and lateral directions.

Tabel 1 Seed mechanical properties of sorted Cabernet-Sauvignon grapes harvested at different density class [17]

Density class	F_s [N]	W_s [mJ]	E_s [N/mm]	DI_s [%]
A	44.5±7.8	11.9±3.2	69.7±15.4	31.4±5.7
B	42.2±9.1	11.8±4	70.1±13.1	29.7±5.9
C	45.7±7.9	12.6±3.5	77.2±13.9	30.2±4.9
D	47.5±9.4	12.6±3.4	83.2±13.5	28.2±4.2
E	44.6±9.2	12.3±5.3	81±14	28.9±7.6

All data are expressed as average value ± standard deviation. Fs = berry seed break force, Ws = berry seed break energy, Es = berry seed Young´s modulus of elasticity, DIs = berry seed deformation index (distance of seed break point/seed height x 100). A = 1081 kg.m^{-3}, B = 1088 kg.m^{-3}, C = 1094 kg.m^{-3}, D = 1100 kg.m^{-3}, and E = 1107 kg.m^{-3}

To obtain a double-screw FEM model, we used ANSYS Polyflow software that provides advanced fluid dynamics technology to solve different tasks in different industries. ANSYS Polyflow software is generally used to minimize physical prototypes in product manufacturing or to reduce thickness variation to improve product quality to substantially reduce costs and save time. Product quality is improved by running trial and error processes with ANSYS Polyflow rather than by testing changes on the production line.

A twin-screw model was created in the Catia program, after which it was imported into the ANSYS Polyflow software. For this study a piece of double screw was chosen to determine the flow rate generation. Dimensions are given in Table 2.

Table 2. Dimensions of screws and barrel.

Element	**Dimensions [mm]**
Screw pitch	40
Centre distance of screws	33.9
Barrel bore diameter	42
Flight depth	7.3
Clearance	1

The material from which the two screws were made was selected from the Ansys library. Figure 2 shows the properties of AISI 1045 Steel.

Material	
Material Name	AISI 1045 Steel, cold drawn, 19-32 mm (0.75-1.25 in) round
Fluid	**False**
Density	**7850000 g/m^3**
Ultimate Strength (Pa)	**625000000 Pa**
Elastic Modulus (Pa)	**2,0500E+11 Pa**
Shear Modulus (Pa)	**8,0000E+10 Pa**
Poisson's Ratio	**0,29**
Thermal Conductivity (W/m-K)	**49,8 W/m-K**

Fig. 2. Properties of AISI 1045 Steel

Figure 3 shows how the mesh superposition technique (MST) is used to simplify the mesh generation of the flow domain. Moreover, in order to avoid pressure peaks in regions were a large number of geometrical constraints exist, we modify the mass conservation equation so that the fluid becomes slightly compressible.

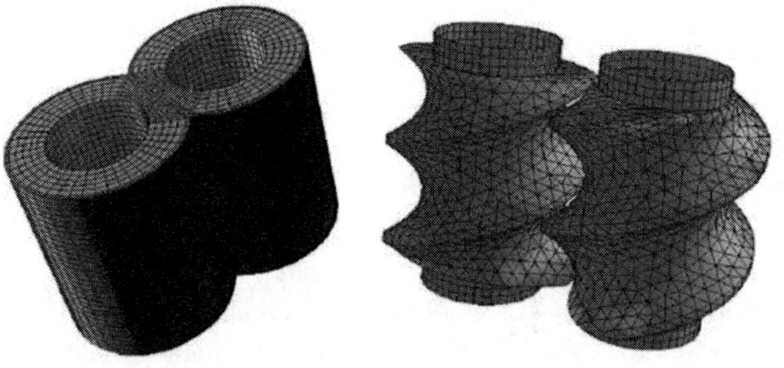

Fig. 3. Mesh of the flow domain and the two screws.

It is essential to avoid incompatibilities between the motion of the moving parts and the boundary conditions imposed on the boundaries of the flow domain. In this example, we take into account the clearance between the screws and between the barrel wall and the screws. In order to obtain a good calculation of the pressure, it is important to mesh correctly the clearance area.

3 Results and discussion

Using the ANSYS Polyflow, there were obtained following results: the velocity field in a cutting plane parallel to rotation axes (fig. 4), pressure distribution on each screw (fig.5) and along the barrel wall (fig. 6)and also we observe unphysical pressure peaks in the clearance between the screws (fig.7) which it indicates that the mesh is not fine enough in this area. These results help us to better understand the phenomena that happen during the double-screw press process.

Figure 4 illustrates a cutting plane parallel to the rotation axis, which shows dispensing the material along the two screws. In areas with red and orange, the space between the two screws is the smallest and clutter of material is created.

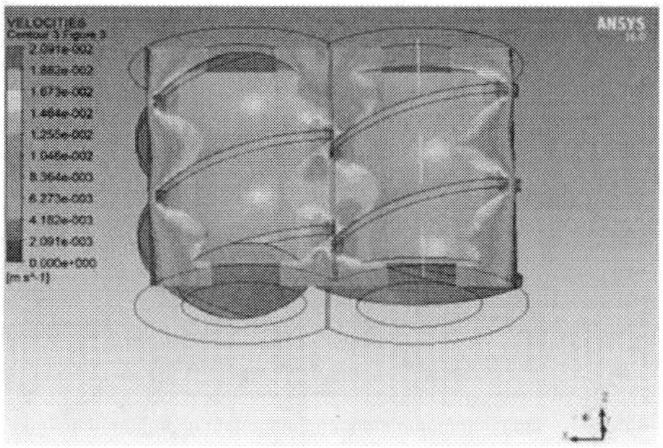

Fig. 4. Velocity field in plane.

Also, in the plane parallel to the axis of rotation, the pressure distribution on each screw is observed. In Figure 5 the area marked with red shows where the pressure reaches the highest value. These areas coincide with the crushing area of grape seeds.

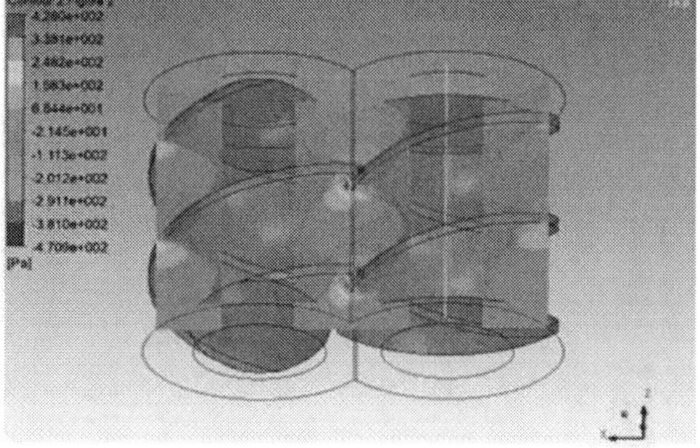

Fig. 5. Pressure field in plane.

 https://doi.org/10.1051/e3sconf/201911203022

In contact with the pressed material is the wall of the pressing chamber. Figure 6 shows how the pressure acts during the pressing process in the feed direction of the material.

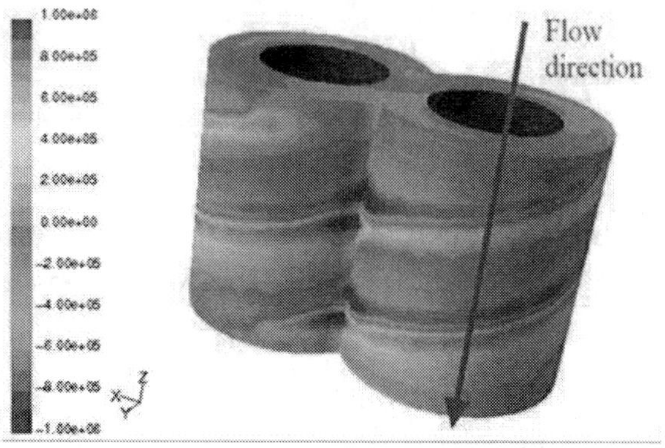

Fig. 6. Pressure field along barrel wall.

During the pressing process, the grape seeds advance into the pressing chamber according to the speed of rotation of the screws, figure 7. At a higher rotational speed, it will increase the working capacity of the press but will decrease the oil yield.

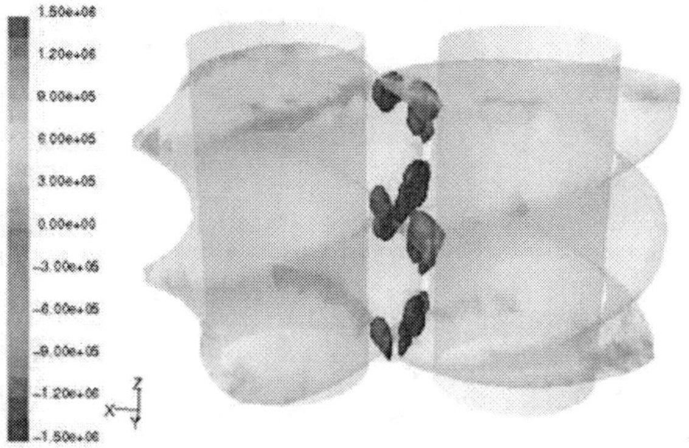

Fig. 7. Pressure peaks between screws.

In the direction of material advancement, the press screws suffer deformations over time, differently depending on the area in which we want to look. Larger deformations are in the cake escape area due to the exhaust nozzle that helps create more pressure, and smaller deformations are in the feed area. In figure 8, we can also see deformations of the twin-screw contour.

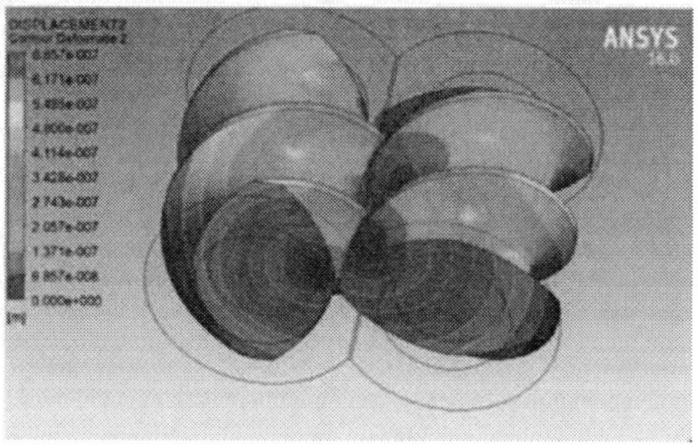

Fig. 8 Deformations of the twin-screw contour.

4 Conclusions

In the pressing process, the most important parameter is the pressure that is influenced by both the characteristics of the press and the characteristics of the material.

As expected, the pressure is high on the side of the screw pushing matter, while it is lower on the other side. In Figure 5, we notice red-marked pressure peaks in the space between the screws, indicating that there is a crushing of grape seeds. During the pressing process, the temperature increases due to the friction, which is a parameter that can lead to deformation of the screws and at the same time affects the characteristics of the oil.

Pressing grape seeds with a twin-screw press is a better option because the seeds have a higher amount of cellulose and a smaller quantity of oil and by using these presses, the yield of the oil is higher due to higher pressures used.

Acknowledgement

This work was supported by a grant of the Romanian Ministery of Research and Innovation CCDI - UEFISCDI, Project for financing excellence in RDI, contract no. 16PFE, within PNCDI III.

References

1. Achawangkul, Y., Maruyama, N., Hirota, M., Chaichana, C., Sedpho, S., Sutabutr, T. Evaluation on environmental impact from the utilization of fossil fuel, electricity and biomass producer gas in the double-chambered crematories. *Journal of Cleaner Production.*(2015)
2. Dutta, S., Karak, N., Saikia, J.P., Konwar, B.K. Biocompatible epoxy modified bio-based polyurethane nanocomposites: Mechanical property, cytotoxicity and biodegradation. Bioresource Technology 100, 6391-6397.(2009).
3. Wu, X., Zhang, X., Yang, S., Chen, H., Wang, D., 2000. Study of epoxidized rapeseed oil used as a potential biodegradable lubricant. JAOCS, Journal of the American Oil Chemists' Society 77,561-563.
4. Owsianiak, M., Chrzanowski, Ł., Szulc, A., Staniewski, J., Olszanowski, A., Olejnik-Schmidt, A.K., Heipieper, H.J., 2009. Biodegradation of diesel/biodiesel blends by a consortium of hydrocarbon degraders: Effect of the type of blend and the addition of biosurfactants. Bioresource Technology 100, 1497-1500.

5. Rombaut, N., Savoire, R., Thomasset, B., Castello, J., Hecke, E. Van, & Lanoisellé, J. 2015. Optimization of oil yield and oil total phenolic content during grape seed cold screw pressing. *Industrial Crops & Products*, *63*, 26–33.
6. Akin, Göü., Karuk elmas, Şü.Nihan., Arslan, F.N., Yilmaz, İ., Kenar, A. Chemometric classification and quantification of cold pressed grape seed oil in blends with refined soybean oils using attenuated total reflectance–mid infrared (ATR–MIR) spectroscopy, LWT – Food Science and Technology(2018)
7. Rombaut, N., et al., Optimization of oil yield and oil total phenolic content during grape seed cold screw pressing. Ind. Crops Prod. (2014).
8. Juan C. de Haro, Irene Izarra, Juan F. Rodríguez, Ángel Pérez, Manuel Carmona,2016. Modelling the epoxidation reaction of grape seed oil by peracetic acid. Journal of Cleaner Production 7197, 10.1016/j.jclepro.2016.05.015.
9. Singh, J., Bargale, P.C., 2000. Development of a small capacity double stagecompression screw press for oil expression. J. Food Eng. 43, 75–82.
10. Savoire, R., Lanoisellé, J.-L., Vorobiev, E., 2013. Mechanical continuous oilexpression from oilseeds: a review. Food Bioprocess Technol. 6, 1–16.
11. Lutterodt, H., Slavin, M., Whent, M., Turner, E., Yu, L., 2011. Fatty acid composition, oxidative stability, antioxidant and antiproliferative properties of selected cold-pressed grape seed oils and flours. Food Chem. 128, 391–399.
12. Maier, T., Schieber, A., Kammerer, D.R., Carle, R., 2009. Residues of grape (Vitis viniferaL.) seed oil production as a valuable source of phenolic antioxidants. Food Chem.112, 551–559,
13. Bouvier, J.M., Guyomard, P., Method and Installation for Continuous Extraction of a Liquid Contained in a Raw Material. EP 0914243 B1, 14 May (1996).
14. S. Isobe, F. Zuber, K. Uemura, A. Noguchi, „A New Twin-Screw Press Design for Oil Extraction of Dehulled Sunflower Seeds", Journal of the American Oil Chemists' Society, vol. 69, nr. 9, 884-889, 1992.
15. Kartika, I.A., Pontalier, P.Y., Rigal, L., 2010, Twin-screw extruder for oil processing of sunflower seeds: thermo-mechanical pressing and solvent extraction in a single step. Ind. Crops Prod. 32, 297e304.
16. Dufaure, C., Leyris, J., Rigal, L., Mouloungui, Z., 1999a. A twin-screw extruder for oil extraction: I. Direct expression of oleic sunflower seeds. J. Am. Oil Chem. Soc. 76, 1073e1079
17. L. Rolle, F. Torchio, B. Lorrain, S. Giacosa, S. Río Segade, E. Cagnasso,V. Gerbi, P. Teissedre,2012, Rapid methods for the evaluation of total phenol content and extractability in intact grape seeds of cabernet-sauvignon: instrumental mechanical properties and ft-nir spectrum, J. Int. Sci. Vigne Vin, 46, n°1, 29-40, France

Researches on the higher valuation of sugar beet pulp from sugar production to obtain premixes for animal feed

Elena Mihaela Nagy *[1], *Constantin* Coța[1], *Nicolae* Cioica[1], *Zoltan* Gyorgy[1] and *Teodora Deac*[2]

[1]INMA Bucharest, Branch Cluj-Napoca, 400436, 59 Al. Vaida Voievod Str., Cluj Napoca , Romania
[2]Technical University of Cluj-Napoca, Faculty of Mechanical Engineering, 400641, 103-105 Muncii Bdul, Romania

Abstract. The sugar beet pulp from the sugar production is, thanks to the easily assimilable carbohydrate content, an important source of feed, especially for polygastric animals. Its efficiency in terms of assimilation and biological effect can be substantially improved by the addition of macroelements (Ca, Mg, P) and microelements (Zn, Co, Fe, Mn, Cu). These elements react with the sugar beet pulp compounds to form specific chelates. In order to preserve and maintain the nutritional value of the premix, the realization process involves a drying operation. In view of this, experiments have been carried out to determine the maximum moisture content that ensures preservation and optimum drying conditions. This paper presents how the percentage of micro- and macro-elements used and the drying temperature influence the drying speed of the premix

1 Introduction

The pulp resulting from the technological process of sugar production by processing sugar beets represents a valuable feed with a high energy potential and increased efficiency in the nutrition of highly productive milk cows, bulls and rams on intensive fattening.

From the energy point of view, the nutritional value of the sugar beet pulp is approaching that of a good quality hay with the remark that it is assimilated much easier and with a superior efficiency: 90-95%.

From a chemical point of view, the sugar beet pulp resulting from the sugar production process is a macromolecular complex of the beet cell membrane consisting of protopectin and represents 5% of the processed beet. The protopectin under the action of pectolytic enzymes in the digestive tract is hydrolysed, both in acidic and basic media, passing into pectinic acid and low molecular weight polypectins, soluble in water, which are absorbed easily and at high velocity [1-3].The nature and chemical structure of sugar beet pulp carbohydrates determines their easy transformation into compounds that enter into milk composition, with a positive effect on both the level of production and the fat content that increases by 8-11%.

* Corresponding author: nagy_m2002@yahooco.uk

Also the presence of sugar beet pulp in the feed ration of bulls and rams subjected to intensive fattening leads to the elevation of the qualitative level of proteins [4-11]. High-efficiency transformation of sugar beet pulp carbohydrates directly into useful compounds contributes to the sensitive reduction of greenhouse gas emissions (carbon dioxide and methane) produced by ruminants during digestion with positive effect on the environment [12] . Furthermore, the practice has shown that partial substitution of grains starch (corn, barley, etc.) with sugar beet pulp is beneficial as it increases efficiency of using the fibrous feed [12].

Although the addition of beet pulp in animals feed has important advantages, the practice has shown that there are still restrictions on the quantities administered since, long-term, both milk production and the metabolic health of animals can be affected [4].

The efficiency of beet pulp in the food ration of polygastric animals can be greatly improved by the addition of macroelements (N, P, Ca, Mg) and microelements, especially Zn^{2+}, which has a positive effect on fertility and Co^{2+} that facilitates the synthesis of vitamins from complex B in the final intestinal tract. The presence of macroelements and microelements in fodder ration is beneficial [13] especially because by reacting with polyglucuronic acids, organo-mineral complexes are formed which facilitate the absorption of Ca^{2+}, Mg^{2+}, PO_4^{3+}.

The seasonal functioning of sugar factories that processe sugar beets, correlated with the large quantities resulting as a by-product (5% of the quantity of beet processed) requires pulp processing by drying for preservation over longer periods of time [13,14, 15].

In this paper we present an experimental study showing the influence of temperature and additions of macroelements (Ca^{2+}, Mg^{2+}, PO_4^{3+}) and microelements (Fe^{3+}, Zn^{2+}, Co^{2+}, Cu , Mn^{3+}) on drying speed of the pulp resulting from sugar beet processing in sugar factories.

2 Materials and methods

The experimental study was carried out with sugar beet pulp resulting from TEREOS Romania Ludus sugar factory. As a source of Ca we used industrial lime, for Mg we used dolomite lime resulting from calcination of Vaslobeni dolomite in laboratory at 850 °C. As phosphorus source (P) we used phosphoric acid 85% quality p.a. from NORDIC CHEMICALS Cluj Napoca. Microelements (Co, Zn, Mn, Fe) were added as sulfates quality p.a. purchased from NORDIC CHEMICALS Cluj Napoca.

Fig. 1. Moister analyzer AXIS 100

The drying study was carried out with a moisture analyzer AXIS 100, using thermo-gravimetrical method, The analyzer has a maximum capacity of 100 g, precision weighing of 0,001 g, precision humidity of 0,01%, maximum drying temperature 160 °C and heating by two halogen lamps (Fig. 1.). Samples of 5 g of four different compositions were subjected to drying at four different temperatures (60, 70, 85 and 95 °C): P1-pressed beet pulp, P2-pressed beet pulp with lime ($Ca(OH)_2$), P3-pressed beet pulp with lime and phosphoric acid (H_3PO_4), respectively P4 pressed beet pulp with lime, phosphoric acid, dolomite lime (CaOMgO) and microelements (Co, Fe, Mn, Cu, Zn).

Compositions, in organic parts, CaO, P_2O_5, MgO and microelements, of the four samples subjected to drying, are shown in Table 1.

Table 1. Sample composition relative to dry matter

Sample	Organic parts [%]	CaO [%]	P_2O_5 [%]	MgO [%]	Microelements
P1	18.7	-	-	-	-
P2	70	30	-	-	-
P3	65.63	28.5	5.85	-	-
P4	61.5	23.2	5.5	9.8	Ca=1.2ppm Zn=2.1ppm Cu=0.75ppm Fe=1.5ppm Mn=0.25ppm

The AXIS STAT software, with which the moisture analyzer is provided, allow to automatically measure the sample mass every two minutes as well as to collect and store the measured values on the PC connected to the moisture analyzer.

3 Results and discussions

The technological process of drying the sugar beet pulp resulting from sugar production is complex. It consists of several elementary processes: mass transformation (chemical reactions between carbohydrates and CaO, P_2O_5 and MgO), heat transformation (water evaporation, heat consumption on the evaporation surface), mass transfer (water vapor diffusion through crust of dry material to the external surface of the beet pulp particles, diffusion of water vapor from the external surface of the particle in the volume of the gaseous phase) and heat transfer (heat transfer from the heating agent to the external surface of the particle, heat transfer through the crust of dry material to the external surface of the still wet core).

Depending on the working conditions the drying process can be carried out after any of the elementary processes listed. Moreover, it is possible that the process takes place according to a combined model, under the conditions that two elementary processes have speed of the same order of magnitude [16, 17]. Choosing the type of dryer, designing, automating, conducting with the computer and optimizing the drying process imply the knowledge of the law after which the drying process takes place. In order to elucidate these goals, experimental researches are needed to highlight how the main parameters influence the process. In this context, in the paper, the influence of temperature and additions of macro and microelements on drying was studied.

The diagrams in Fig. 2, 3, 4 and 5 show, in comparison, the drying curves for each sample P1, P2, P and P4 at different drying temperatures..

From these graphs we can observe both the moisture variation in time relative to the dry matter for four samples, at four temperatures, and the high influence that the drying temperature has on the time necessary for the complete elimination of humidity, time whose values are shown in Table 2.

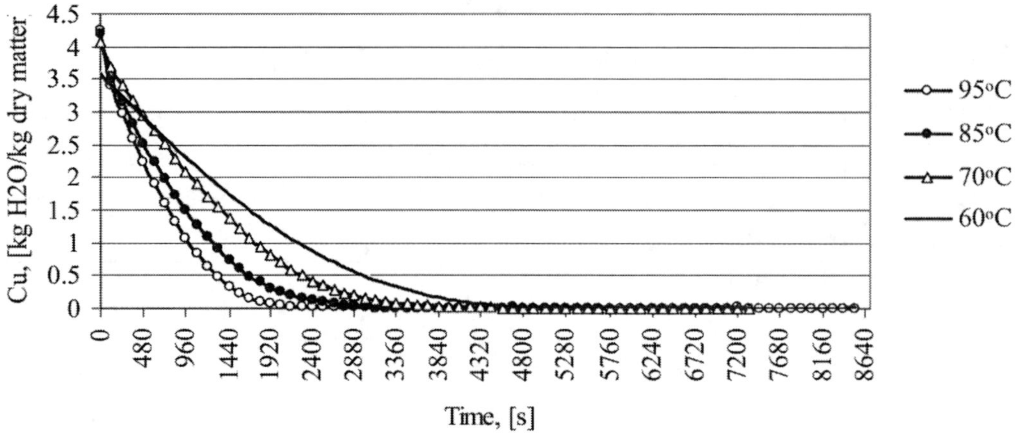

Fig.2. Drying curves for sample P1

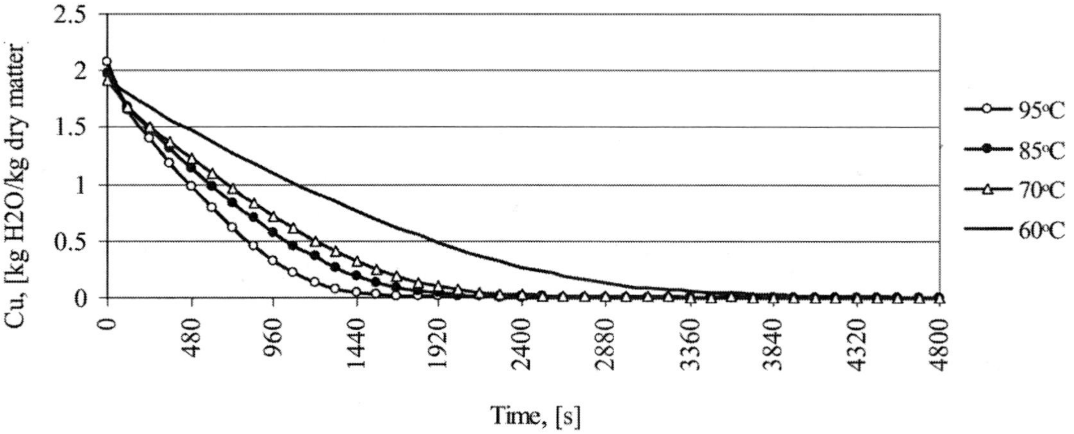

Fig.3. Drying curves for sample P2

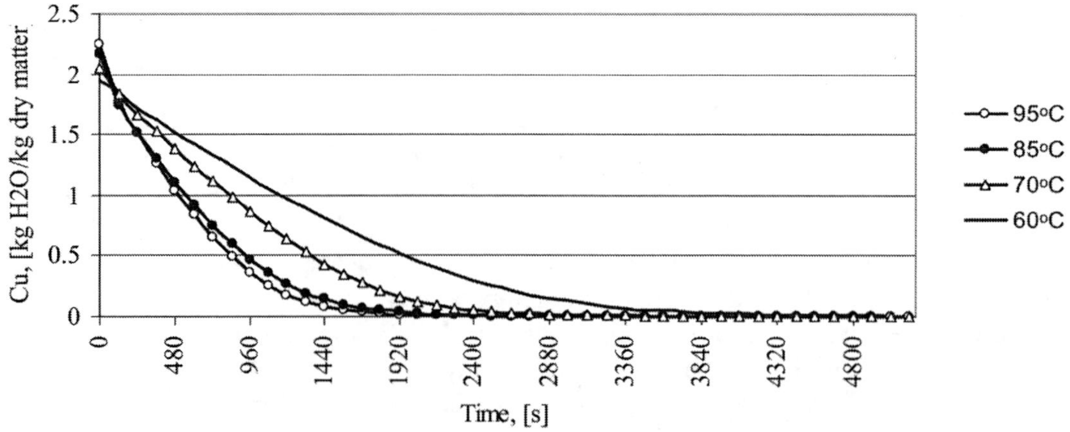

Fig.4. Drying curves for sample P3

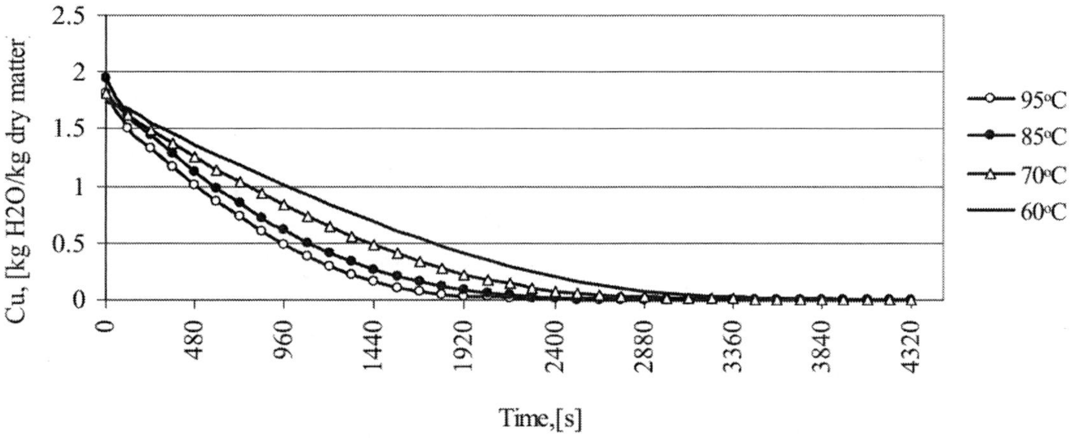

Fig.5. Drying curves for sample P4

The moisture content of samples at any point during drying decreases as the drying temperature increases.

Table 2. Influence of temperature and additions on the time required to reach steady state

Drying time, [s]	τ			
Temp, [°C]	60	70	85	95
P1	8400	4920	3720	3480
P2	4680	4320	3600	3360
P3	4560	4200	3360	3240
P4	4200	3720	2760	2400

Table 2 shows that in case of sample P1 (pressed beet pulp), the increase in the drying temperature from 60°C to 70°C leads to a 40% reduction of the time required to completely eliminate the moisture. At temperatures above 70 °C this time decreases further but to a lesser extent.

Regarding the influence of the addition of macro and microelements on pressed beet pulp, we found that this is manifested especially at the drying temperature of 60°C for the sample P2 where the addition of lime leads to a 42% decrease of time required to reach the steady state. Addition of dolomite lime in P4 sample composition leads, at the same temperature, at a drying time necessary to reach the steady state by 50% lower than in the case of P1 sample. In the case of the other three temperatures the addition of lime and dolomite lime leads to 24-30% reductions of the time required to completely eliminate the moisture. The addition of phosphoric acid and microelements alters the duration of reaching the steady state of the drying process, but their influence is less pronounced.

It is noted that for all four compositions at 60°C the time required to reach the steady state is high and the value of this time at 95°C does not differ significantly from that at 85 °C.

4 Conclusions

Addition of macro-elements (lime, dolomite lime and / or phosphoric acid) and microelements (Co, Fe, Mn, Cu, Zn) in the feed premixes formula based on beet pulp resulting as a byproduct from the process of manufacturing sugar, positively influences its drying for preservation.

The positive influence these additions have on the drying process can be explained by the fact that they modify the porous structure of the material, thus enhancing the internal water vapor diffusion processes.

The greater contribution of dolomite lime to reducing the time required to reach steady state, especially at low temperatures, is due to an exothermic reaction of dolomite lime with the water from the organic part of the premix, which speeds up its drying.

The optimal temperature at which the drying process of the premix must be carried out falls within the range of 70-85°C. Temperatures in this area ensure both short drying times and therefore low energy consumption and preservation of the nutritional value of the premix. In this way, good efficiency can be obtained both at the drying of the sugar beet pulp premix for preservation and for the efficient use in feed rations of animals, especially polygastric animals.

This work was supported by a grant of the Romanian Research and Innovation Ministry, through Programme 1 – Development of the national research-development system, sub-program 1.2 – Institutional performance – Projects for financing excellence in RDI, contract no. 16PFE and by Contract 5N/07.02.2019.

References

1. F.N. Domsa, L. Iliescu, *Tehnologia zaharului*, (Ed.Tehnica, Bucuresti, 1973)
2. Z. Bubnik, P. Kadlec, D. Urban, *Sugar Technologists Manual: Chemical and Physical Data for Sugar Manufacturers and Users*, (Ed Bartens, Berlín, 1995)
3. A. Lenart, J.M. Flink, Int. J. of Food Sci. & Tech., **19** (1), 45, (2007)
4. M. Münnich, F. Klevenhusen, Q. Zebeli, Animal Feed Sci. and Tech., **241**, 75, (2018)
5. A.P.A. Monteiro, J.K. Bernard, J.R. Guo, X.S. Weng, S. Emanuele, R. Davis, G.E. Dahl, S.Tao, J. of Dairy Sci., **100** (2), 1063, (2017)
6. R.E. Rauch, P.H. Robinson, L.J., Erasmus, Animal Feed Sci. and Tech., **177** (3-4), 180, (2012)
7. A. Jokić, Z. Zavargo, N. Lukić, B. Ikonić, J. Marković, J. Dodić, J. Grahovac, J. on Proces. and Energy in Agri., **17** (1), 24, (2013)
8. Ch.V. Boucque, B.G. Cottyn, F.X. Buysse, Livestock Prod. Sci., **5** (2), 171, (1978)
9. M. Münnich, R. Khiaosa-ard, F. Klevenhusen, A. Hilpold, Q. Zebel, Animal Feed Sci. and Tech., **224**, 78, (2017)
10. P. Kelly, Animal Feed Sci. and Tech., **8** (1), 1, (1983)
11. P. Ertl, Q. Zebeli, W. Zollitsch, W. Knaus , J Dairy Sci. **99** (2), 1228, (2016)
12. I. Sanchis, L. Luján, C. Ferreira, F. Martínez, Animal Prod. Sci., 55(1), 56 (2015)
13. M. A. Salgado, A. Lebert, H.S. Garcia, J. J. Bimbenet, Drying Tech. An Int. J., **12** (4), 955, (1994)
14. O. Levenspiel, *Tehnica reactiilor in ingineria chimica*, (Ed.Tehnica, Bucuresti, 1967)
15. L. Yang, Z.Hu, L.Yang, S. Xie, M. Yang, INMATEH-Agricultural Engineering, **55** (2),53, (2018)
16. C. Calistru, C. Leonte, *Tehnologia substantelor anorganice*, (Ed. Didactica si Pedagogica, Bucuresti, 1972)
17. Z. Dancea, *Nutriția animală și elemente de nutrițiea omului*, (Ed TODESCO, Cluj-Napoca, 2005)

A review on phytoremediation as an ecological method for in situ clean up of heavy metals contaminated soils

Nicolae Cioica *[1], *Cătălina*Tudora [2], *Dorin* Iuga [2], *György* Deak[4], *Monica* Matei [4], *Elena Mihaela* Nagy [1], and *Zoltan* Gyorgy [1]

[1] INMA Bucharest, Branch Cluj-Napoca, 59 Al. Vaida Voievod Str., Cluj Napoca, Romania
[2] INMA Bucharest, 6 Ion Ionescu de la Brad Bd., Bucharest 1, Romania
[3] IBA Bucharest, 5 Baneasa Ancuta, Bucharest 2, Romania
[4] INCDPM Bucharest, 294 Splaiul Independentei, Bucharest 6, Romania

Abstract: Heavy metals are among the most common types of contaminants in agricultural soils, especially those bordering the cities, due to the uncontrolled use of sewage sludge, compost, mining waste and chemical fertilizers. Excessive accumulation of heavy metals, which do not degrade over time, adversely affects crop yields by decreasing microbial activity and fertility of contaminated soils. Also, excess of heavy metals in the soil poses a serious threat to plant and animal health and, through their entry into the food chain, to human health. For this reason, the decontamination of soils contaminated with heavy metals has become a necessity. This review presents the current state of phytoremediation research as the most cost-effective method of in-situ environmental decontamination of soils contaminated with heavy metals.

1 Introduction

The phenomenon of urbanization, rapid industrialization and intensive farming in recent decades has led to the accumulation of a significant amount of heavy metals both in agricultural land bordering the cities and in the countryside because of uncontrolled disposal of household waste, treatment sludge, mining waste and the excessive use of chemical fertilizers. The accumulation of heavy metals, which do not degrade over time, some of them (Cd, Pb, Hg) being toxic in any quantity, however small and others (Zn, Cu, Mn, Mo, Ni, Co) being beyond acceptable levels as essential micronutrients adversely affecting crop yields by decreasing microbial activity and fertility of contaminated soils. Also, excess of heavy metals in the soil poses a substantial danger to plant and animal health and, through their entry into the food chain, to human health [1-7]. For this reason, the decontamination of soils contaminated with heavy metals has become a necessity.

Decontamination can be done by conventional methods (chemical or physical) or by biological methods (plants, algae, microorganisms, etc.).

*Corresponding author: ncioica@yahoo.com

Conventional technologies produce different pollutants and are unprofitable [1].

Of the various biological technologies existing so far, *phytoremediation*, as a technology based on solar energy, is considered the most environmentally friendly and very cost-effective. Phytoremediation, also called botanical bioremediation, is an ongoing research area and consists in reducing or eliminating the toxic effects of heavy metals in contaminated soils by their assimilation or immobilization by some specially selected green plants, with the help of associated microorganisms in the soil [*8*].

Depending on the response to the presence of heavy metals in the soils where they grow or are grown, the plants can be divided into three categories [9]:

- exclusion plants, which limit the access of heavy metals from soil only to roots and restricts the transport and assimilation into their aerial parts. Such plants can be used as phytostabilisers;
- bioindicators, in the case of which the absorption and transport of heavy metals to their aerial parts is regulated so that the concentration of heavy metals in the plant reflects the concentration of heavy metals in the soil. They have a limited extraction potential and are used to biologically monitor soil contamination;
- accumulators which allow the access, transport and entry of heavy metals from soil into their aerial parts to a level which does not affect their health and normal development. Of these, there are hyperaccumulating plants that can concentrate large quantities of heavy metals in their aerial parts, being successfully used for the remediation, by phytoextraction, of soils contaminated with heavy metals.

2 Phytoremediation techniques applied to soils contaminated with heavy metals

Depending on the way of fixing, absorbing, accumulating or eliminating the heavy metals by the plants used, there are three phytoremediation techniques applicable for the remediation of soils contaminated with heavy metals: *phytoextraction, phytostabilisation* and *phytovolatilisation*, the mechanism of which is presented in Figure 1.

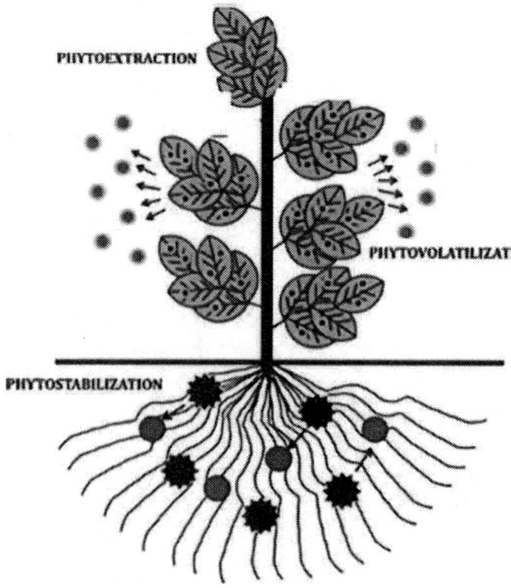

Fig.1 Phytoremediation strategies for soils

2.1 Phytoextraction

Soil remediation by phytoextraction implies the absorption of the contaminant through plant roots, its transport and subsequent accumulation in its aerial parts, followed, at the appropriate time, by harvesting and clearing the ground of plant biomass. Plants are sown or transplanted into the soil contaminated with heavy metals and then maintained by appropriate agricultural practices. The roots of these plants absorb, from the soil, the soluble compounds resulting from the processing by the soil microorganisms of the metals and translocate them to the aerial shoots where they accumulate. After sufficient development, the aerial parts of the plants are harvested and cleared, resulting in permanent removal of heavy metals from contaminated soils.

A large number of researches indicate that there are many hyperaccumulating metallophyte plant species that have a tremendous genetic potential for removing toxic metals from the soil so that they can play a major role in the phytoextraction process [10].

These include approximately 20 hyperaccumulators for As, 10 for Cd, 14 for Pb, 30 for Co, 34 for Cu, 320 for Ni, 30 for Se, 11 for Zn and 10 for Mn [11, 12].

Table 1 presents a part of the hyperaccumulating metallophyte plants that have been studied and presented in the literature, some with a high phytoextraction yield both in terms of the number and amount of heavy metals assimilated.

Table1. Hyperaccumulating plants used to study the phytoremediation of soils contaminated with heavy metals

Plant	Heavy metal	References
Arabidopsis halleri	Zn	[13]
Thalaspi caerulescene	Zn	[14]
Brassica juncea L.	Cu, Zn, Pb	[15, 16, 17, 18, 19]
Brassica napus L.	Cu, Zn, Pb	[16, 20]
Brassica oleracea, Raphanus sativus	Zn, Cd, Ni, Cu	[1, 21]
Cardaminopsis halleri	Zn, Pb, Cu, Cd	[22].
Cicer aeritinum L.	Cd, Pb, Cr, Cu	[23, 24,25]
Cucumis sativus L.	Pb	[15]
Eichhornia crassipes L.	Cr, Zn	[26]
Euphorbia cheiradenia	Pb, Zn, Cu, Ni	[27]
Helianthus annuus (Sunflower)	Pb, Cd	[28]
Jatropha curcas L.	Cu, Mn, Cr, As, Zn, Hg	[29]
Lantana camara L.	Pb	[30]
Lavandula vera L.	Pb	[31]
Lens culinaris Medic.	Pb	[32]
Lepidium sativum L.	As, Cd, Pb,	[33]
Lactuca sativa L.	Cu, Mn, Zn, Ni, Cd, Pb, Co	[34, 35]
Miscanthus x giganteus	Cu, Ni, Pb, Zn	[36]
Oryza sativa L.	Cu, Cd	[37]
Minuartia verna, Agrostis tenius	Pb	[14]
Pelargonium	Pb	[38, 21]
Pisum sativum L.	Pb, Cu, Zn, Ni, As, Cr	[39, 40, 41]
Potentilla griffithii	Zn	[42]
Rapanus sativus L.	Cd, Fe, Pb, Cu	[33]
Salvia sclarea L.	Pb, Cd, Zn	[31]
Spinacia oleracea L.	Cu, Ni, Pb, Zn, Cr	[43]
Sorghum bicolor L.	Cd, Cu, Zn	[44, 45, 46]
Trifolium alexandrinum	Zn, Pb, Cu, Cd	[47]
Viola principis	Pb	[48]
Zea mays L.	Cd, Pb, Zn, Cu	[43, 49, 50]

Compared to common plants, hyperaccumulating metallophyte species are able to accumulate heavy metals, in their aerial parts, in quantities of 100-500 times larger than normal plants, without effect on their development performance [51]. There are no general standards to establish the level of heavy metals from which metallophytes are considered to be hyperaccumulators [52]. Contaminant levels that different types of hyperaccumulating metallophytes are able to accumulate differ for different metals. The authors of a recent review on the subject consider that the species of plant which, when grown on soils contaminated with heavy metals, accumulate in their aerial parts, in relation to the dry matter, >10,000 mg / kg Mn and Zn, >1000 mg / kg Cu, Ni, As and Pb or >100 mg / kg Cd and Se, are part of the hyperaccumulating metallophytes category [53].

The potential of hyperaccumulating metallophyte species to absorb and accumulate heavy metals is of great ecological importance. Mechanisms of heavy metal hyperaccumulation comprise biochemical and biophysical transformations. Unfortunately, these mechanisms and the processes involved in the accumulation of metals are still little studied in depth and the mechanism of physiological transformations are not entirely known [52]. Some recent research on physiological hyperaccumulation processes has partly explained the mechanisms involved in metal hyperaccumulation.

There are two parameters that indicate the efficiency of phytoextraction. The first is the bioconcentration factor that indicates the efficacy of a plant species to accumulate a heavy metal from the contaminated soil in its tissues and is expressed by the ratio of the concentration of the metal accumulated in the harvested plant biomass to the concentration of the same metal in the soil. The second factor is the translocation factor that indicates plant efficiency in translocating accumulated metal from roots to shoots and is expressed as the ratio of the metal concentration in plant shoots to the metal concentration in plant roots [54]

A successful phytoextraction is conditioned by increased values of these two parameters, combined with increased metal xylem loading and detoxification of metals in plants [9]. Physiological studies showed that heavy metal xylem loading and its transfer to aerial parts of plants is mediated by a series of transport proteins that are in most cases found in plasma or intracellular membranes [52]. Also, the tolerance for heavy metals absorption of hyperaccumulating metallophytes is given by the type and amount of chelators, such as the phytochelatins and metallothioneins, they contain and which contribute decisively to detoxification by vacuole sequestration of heavy metals [55, 56].

Once the mechanisms and factors that govern the absorption and transport capacity of metallophyte plants are known, one can proceed to the mathematical modeling of heavy metal bioaccumulation [57, 58]. Mathematical modeling of heavy metal bioaccumulation seeks to know the mechanisms of this process and to identify ways to limit or optimize the bioaccumulation and phytoremediation processes [59]. The model can predict the optimal harvesting as well as the subsequent use of these metallophytes depending on the plant type and absorption of heavy metals [59].

Plants used in phytoextraction must have some special features including: high growth rate, large volume of aerial biomass, broadly distributed and highly branched root system [47], supraunitary, as high as possible, bioconcentration factor and translocation factor of target heavy metals from soil, high tolerance to the toxic effects of target heavy metals, good adaptation to environmental and climatic conditions in the respective area, increased resistance to pests and pathogens, easy cultivation and harvesting, aversion for herbivores in order to avoid contamination of the food chain [47, 55]

Although phytoextraction leads to a permanent removal of metal from contaminated soils, phytoextraction is especially recommended for those soils that are contaminated with heavy metals at moderate levels, as most plant species are not capable to decontaminate heavily polluted soils [60]. For instance, phytoextraction can be applied for Pb decontamination by

Brassica juncea only for soils having a Pb contamination degree less than 1500 mg / kg [52, 61]

Like metallophyte plants, there are also metallophyte algae that also have a large capacity to absorb heavy metals and so there is appreciable potential to use them to remove heavy metals from polluted soils.

Table 2 presents some algae species that researchers used to remove heavy metals from contaminated sites.

Table 2. Soil algae used in the study on phytoremediation of soils contaminated with heavy metals

Alga	Heavy metal	References
Ascophyllum nodosum	Ni, Pb	[62]
Cladophora fascicularis	Pb (II)	[63]
Cladophora glomerata	Zn, Cu	[64]
Cladophora glomerata, Oedogonium rivulare	Cu, Pb, Cd, Co	[65]
Cymodocea nodosa	Cu, Zn	[66]
Fucus vesiculosus, Laminaria japonica	Zn	[67]
Oscillatoria quadripunctulata, Oscillatoria tenius	Cu, Pb	[68]
Sargassum filipendula	Cu	[69]
Sargassum natans	Pb	[62]
Spirogyra hyalina	Cd, Hg, Pb, As	[70]

Possible mechanisms for heavy metal tolerance, adsorption, transport and storage vary depending on various factors such as algal anatomy and growth medium [1]. In the case of algae, metal ions are adsorbed over the cellular surface by physical adsorption and then they are slowly transported into the cytoplasm by chemisorption [62].

2.2 Phytostabilisation

Soil depollution through phytostabilisation involves the use of certain plant species to immobilize contaminants in the soil and groundwater by absorption and accumulation in plant tissues, root adsorption or precipitation in the root area, which prevents their migration to the soil as well as their displacement through erosion and deflation [25, 71, 72]. Thus, unlike phytoextraction, phytostabilisation does not have the role of decontaminating the soil but is intended to reduce environmental contamination.

For phytostabilisation, it is especially recommended to use exclusion plants that allow heavy metals from soil to enter only in their roots, protecting their aerial parts against contamination with these metals. Such metallophytes are the *Agrostis tenius* and *Festuca rubra* plant species used for the phytostabilisation of soils contaminated with Pb, Zn and Cu [73]. It was also found that in the rhizosphere zone of the *Silene vulgaris* and *Agrostis capilaris* species Cd forms compounds with sulphides and Pb is converted into insoluble product as phosphate, which recommends them as phytostabilisers on the land contaminated with these heavy metals [14, 52].

To enhance the immobilization of heavy metals in the soil, phytostabilisation can be applied together with other decontamination methods, such as the application of organic amendments that increase the number of roots and multiplication of microorganisms in the contaminated soil [1, 52].

Studies on medicinal and aromatic plants showed that some of them such as *Ocimum basilicum L.*, *Melissa Officinalis L.*, *Valeriana officinalis L.*, *Calendula officinalis L.* and *Matricaria chamomilla L.* are exclusion plants for Pb, Zn and Cd, which recommends them to be used for the phytostabilisation of soils contaminated with these metals [34, 74].

Alimurgic species, which are edible wild plants that grow spontaneously in natural pastures, can also be used as phytostabilisers: *Cichorium intybus L.*, for soils contaminated with Zn, and *Taraxacum officinale* for soils contaminated with Cd [75].

2.3 Phytovolatilisation

Phytovolatilisation is a phytoremediation technique in which plants absorb volatile metallic contaminants from the soil and metabolically processed them into less toxic compounds, which then, through the process of plant transpiration, are released as vapours into the atmosphere [22]. This technique is applicable among heavy metals for Hg and in the case of metalloids for Se and As, but there are few studies that demonstrated the performance of applying this technique to remedy contaminated soils [8].

Gaseous transformation of heavy metals from contaminated soil is based on specific mechanisms that are coordinated within plants by certain specific enzymes or genes.

Brassica juncea and *Arabidopsis thaliana* are capable of making the conversion of inorganic Se into organic selenoaminoacids (selenomethionine and selenocysteine) followed by biomethylation to form dimethyl selenide or dimethyl diselenide which are volatile and can be released into the atmosphere [52, 76].

However, there are few metallophyte plants that are naturally capable of transforming metals into volatile form. Therefore, in order to increase the ability to volatilize heavy metals, a number of genetically modified plants have been made [14].

3 Optimizing phytoremediation of soils contaminated with heavy metals

From the information presented so far, it results that phytoremediation techniques applicable to soils polluted with heavy metals have a depollution potential that can be used for practical purposes. However, there is a great need to improve these techniques, which can be achieved by *optimizing agricultural practices, chemical assistance of phytoremediation and the association of metallophyte plants with microbial strains* [77].

Optimizing agricultural practices refers to both land preparation and plant cultivation as well as to the maintenance and harvesting of the plant aerial part in the case of phytoextraction. Soil preparation consists of performing operations to ensure optimal density and aeration, as well as water drainage. While some crops are sown directly in the field, there are others where seedlings must be obtained that can be planted in the field in spring when there is no longer a frost hazard. Given the unfavourable conditions for plant germination in soils polluted with heavy metals, it is recommended that the seed density be doubled compared to the one used in standard agriculture [78]. Crop maintenance during plant growing consists of weed and pest control, fertilization and irrigation depending on the weather conditions and the water needs of the crops. As one of the objectives of the phytoextraction is to ensure an aerial part mass as large as possible, the optimal harvesting periods for phytoremediation crops may differ from the harvesting periods considered suitable for conventional agricultural purposes. Crop rotation and the practice of associated crops are also important ways to increase the phytoremediation yield of soils contaminated with heavy metals [79].

Chemical assistance of phytoremediation consists of the use of synthetic chelating agents such as ethylenediaminetetraacetic acid (EDTA), N-(2-Hydroxyethyl)ethylenediaminetriacetic acid (HEDTA), diethylenetriaminepentaacetic acid (DTPA) and ethylenediamine-N,N′-bis(2-hydroxyphenylacetic acid) EDDHA which increase the amount of heavy metal assimilable by plants from the soil and thereby improve phytoextraction yield [11, 77]. It has to be noted that the excessive use of this type of synthetic chelates can lead to an increase in the concentration of heavy metals in levigations collected in the area under the plant roots and it is therefore recommended to use other chelating agents such as nitrilotriacetic acid (NTA) or citric acid [80].

The association of metallophyte plants with microbial strains improve the precipitation, reduction, oxidation and absorption of heavy metals in the soil and help plants developing suitable even under conditions of heavy metal stress [14, 52].

In the literature, these strains are called metallophyte growth-promoting rhizobacteria. It has been found that heavy metal-resistant microorganisms protect plants from the harmful effects of heavy metal and even increase the absorption of metals by hyperaccumulating plants.

Many studies provided evidence that the association of microorganisms with rhizosphere leads to a yield of heavy metal accumulation in hyperaccumulating metallophytes up to 4 times higher compared to yields obtained in sterilized soil [81].

Table 3 shows some species of microbial strains that researchers used for microbial assisted phytoremediation.

Table 3. Microbial strains used in the study of microbial assisted phytoremediation

Microorganism	Heavy metal	References
Aspergillus niger, Aspergillus awamori, Aspergillus ussamii, Rhizopus delemar, Penicillium brevicompactum, Saccharomycopsis lipolytica, Candida blankii, Saccharomyces cerevisiae, Hansenula schneggie, Debaromyces senii	Co, Cd, Cu, Zn	[82]
Pseudomonas veronii	Cd, Zn, Cu	[83]
Kocuria flava	Cu	[83]
Yarrowia lipolytica	Cd, Ni, Co, Zn	[84]
Geobacillus toebii subsp. Decanicus	Cd, Cu, Co, Mn	[85]
Bacillus cereus Bacillus subtilis	Mn, Ni, Co, Cu, Pb	[86]

The two main techniques for the microbial inoculation of the rhizosphere are soil drainage and direct inoculation with microorganisms of seeds or plant roots before sowing or planting [87].

Bacteria introduced into the contaminated soil increase the mobility and bioavailability of heavy metals in it by lowering the pH, increasing soil redox potential, producing plant growth stimulants and chelating compounds of heavy metals such as biosurfactants, organic acids, siderophores. [52, 88].

4 Conclusions

Phytoremediation is an ecological and very cost-effective technology, based on solar energy and aims at reducing or eliminating the toxic effects of heavy metals in contaminated soils by their assimilation or immobilization by some green plants or algae.

In the case of heavy metals, phytoremediation can be achieved by phytoextraction, phytostabilisation or phytovolatilisation, techniques that differ in the way the plant reacts to the contaminant.

Studies in the field showed that phytoremediation efficiency depends on many soil, contaminant and plant characteristics such as soil physicochemical properties, the bioavailability of heavy metals in the soil, the type and amount of exudates released by microorganisms and plant roots as well as the capacity of plants to take over, translocate, accumulate and detoxify these metals.

To improve phytoremediation efficiency, use of genetically modified metallophyte plants, optimization of agricultural practices, chemical assistance of phytoremediation and association of metallophyte plants with microbial strains are sought.

This work was supported by Ministry of Research and Innovation, Projects code: PN-III-P4-ID-PCE-2016-0860, PN-III-1.2-PCCDI/Contract 11/31.03.2018 and Core Program/Contract 5N/07.02.2019.

References

1. K. K. Yadav, N. Gupta, V. Kumar, J.K. Singh, Indian J. Environmental Protection, **37**, 65 (2017)
2. A. Pruteanu, M. Hărmănescu, V. Vlăduţ, A. Muscalu, E. Marin, Annals of the University of Craiova - Agriculture, Montanology, Cadastre. **45(2)** (2015)
3. A. Pruteanu., D. M. Bordean., I. Voicea, V. Vlăduţ, I. Gageanu, Proceedings of the 18th International Multidisciplinary Scientific GeoConference SGEM, **18** (2018)
4. A. Pruteanu, D. M. Bordean, N. Vrînceanu, V. Vlăduţ, M. Niţu, I. Găgeanu, I. Caba, Annals of the University of Craiova - Agriculture, Montanology, Cadastre. **48(2)** (2018)
5. A. Pruteanu, D.M Bordean, V. Vlăduţ, Proceedings of the 47 International Symposium Actual Tasks on Agricultural Engineering, 135 (2019)
6. P. Ekta, N. R. Modi, J Pharmacogn Phytochem, **7(4)**, 1485 (2018)
7. M.J.A. Ali, O. Al-Tamimi, Open Science Journal **3(1)**, 1 (2018)
8. A. Sumiahadi, R. Acar, *IOP Conf. Series: Earth and Environmental Science,* **142**, 1 (2018)
9. A. J. M. Baker, P. L. Walke, J Plant Nutr. **3**, 643 (1989)
10. T. Jakovljević, I. R. Redovniković, A. Laslo, Zastita Materijala, **57 (3)**, 496 (2016)
11. M. Ghosh, S.P. Singh, Appl. Ecol. Environ. Res. **3(1)**, 214 (2005)
12. N. K. Niazi, B. Murtaza, I. Bibi, M. Shahid, J. C. White, M.F. Nawaz, S. Bashir, G. Murtaza, *Environmental Materials and Waste: Resource Recovery and Pollution Prevention* (Academic Press, Cambridge, 2016)
13. F. J. Zhao, E. Lombi, T. Breedon, S.P. McGrath, Plant Cell and Environment, **23**, 507 (2000)
14. A. Vasavi, R. Usha, P.M. Swamy, Jr. of Industrial Pollution Control **26 (1)**, 83 (2010)
15. R. Takeda, Y. Sato, R. Yoshimura, S. Komemushi, A. Sawabe, *Proceedings of the Annual International Conference on Soils, Sediments, Water and Energy, 11, 293* (The Barkeley Electronic Press, 2006)
16. M. Turan, A. Esringu, Plant Soil Environ., **53(1)**, 7 (2007)
17. A. Singh, M. H. Fulekar, T*he Plant Family Brassicaceae: Contribution Towards Phytoremediation,* (Springer Science+Business Media, Amsterdam, 153, 2012)
18. G. N. Koptsik, Eurasian Soil Sci., **47**, 923 (2014)
19. H. Sharma, Phytoremediation of lead using Brasica juncea and Vetiveria zizanoides, Int. J. Life Sci. Res., **4(1)**, 91 (2016)
20. J. Park, J-Y. Kim, K-W. Kim, Geosystem Engineering, **15(1)**, 10 (2012)
21. A. J. M. Baker, R. D. Reeves, S. P. McGrath, *In situ Bioremediation: Applications and investigations for hydrocarbon and contaminated sites remediation* (Butterworth-Heinemann, London 1991)
22. S. Sharma, S. Tiwari, A. Hasan, V. Saxena, L. M. Pandey, 3 Biotech., **8(4)**, 1 (2018)
23. P. A. Wani, M. S. Khan, A. Zaidi, Australian J. Exp. Agric., **47**, 712 (2007)
24. P. S. Satvat, A. B. Mahinrakar, IJTES, **2(3)**, 338 (2011)
25. S. Dasgupta, P. S. Satvat, A. B. Mahinrakar, IJTES **2(3)**, 238 (2011)
26. V. K. Mishra, B. D. Tripathi, J. Hazard Mater, **164**, 1059 (2009)
27. A. Chehregani, B.E. Malayeri, Int. J. Agric. Bio. **9 (3)**, 462 (2007)

28. K. A.Alaboudi, B. G..Ahmed, https://www.sciencedirect.com/science/article/pii/S0570178318300174 - !Annals of Agricultural Sciences, **63(1)**, 123 (2018)
29. S. K. Yadav, A. S. Juwarkar, P. Kumar, P. R. Thawale, S. K. Singh, T. Chakrabarti, Biosource Tech., **100**, 4616 (2009)
30. F. O. Alaribe, P. Agamuthu, Ecological Engineering, **83**, 513 (2015)
31. V. R. Angelova, *Proceedings of the XV Balkan Mineral Processing Congress,* Sozopol, Bulgaria, 1045 (2013)
32. P. A. Wani, M. S. Khan, Bacteriology J., **2(4)**, 66 (2012)
33. S. Gunduz, F. N. Uygur., I. Kahramanoglu, Herald J. Agric. Food Sci. Res., **1(1)**, 1 (2012)
34. A. K. K. Achakzai, Z. A. Bazai, S. A. Kayani, J. Bot. **43(6)**, 53 (2011)
35. A. K. Quainoo, A. Konadu, M. Kumi, J. Bioremed. Biodeg. **6(1)**, 1 (2015)
36. A. Kocoń, B. Jurga, Environ Sci Pollut Res. **24**, 4990 (2017)
37. P. Li, X. Wang, T. Zhang, D. Zhou, Y. He, J. Environ. Sci., **20**, 449 (2008)
38. M. Arshad, J. Silvestre, E. Pinelli, J. Kallerhoff, M. Kaemmerer, A. Tarigo, M. Shahid, M. Guiresse, P. Pradere, C. Dumat, Chemosphere **71**, 2187 (2008)
39. A. Malecka, A. Piechalak, I. Morkunas, Acta Physiol Plant, **30**, 629 (2008)
40. A. Hegedusova, S. Jakabova, A. Vargova, O. Hegeus, T. J. Pernyeszi, Nova Biotechnologica, **9(2)**, 125 (2009)
41. S. Sharma, P. Sharma, P. Mehrotra, JAmSci. **6(6)**, 43 (2010)
42. P. J. Hu, R. L. Qiu, P. Senthilkumar, D. Jiang, Z. W. Chen, Y. T. Tang, F. J. Liu, Environ. Exp. Bot., **66**, 317 (2009)
43. M. R. Abhilash., S. Srikantaswamy, D. Shiva Kumar, K. Jagadish, L. Shruthi, Int. J. Applied Sci. **4(1)**, 129 (2016)
44. A. P. Pinto, A. M. Mota, A. de Varennes, F. C. Pinto, Sci Total Environ. **326**, 239 (2004)
45. V. R. Angelova, R. V. Ivanova, V. A. Delibaltova, K. I. Ivanov, J. Agric. Sci. Technol. JAST-A, **1(5)** 693 (2011)
46.S. Muthusaravanan, N. Sivarajasekar, J. S. Vivek, T. Paramasivan, Mu. Naushad, J. Prakashmaran, V. Gayathri, Omar K. Al Duaij, Environ Chem Lett, **16(4),** 1339 (2018)
47. H. Ali, E. Khan, M. A. Sajad, Chemosphere, **91**, 869 (2013)
48. X. Wan, M. Lei, J. Yang, Catena, **148(1)**, 67 (2016)
49. T. Tiecher, C. A. Ceretta, P. A. A. Ferreira, C. A. Lourenzi, T. Tiecher, E. Girotto, F. T. Nicoloso, H. H. Soriani, L. De Conti, T. Mimmo, S. Cesco, G. Brunetto, Geoderma, **262**, 52 (2016)
50. A. Mojiri, J. Biol. Environ. Sci. **5(13)**, 17 (2011)
51. V. Sheoran, A. S. Sheoran, P. Poonia, Pedosphere, **26(2)**, 148 (2016)
52. S. Khalid, M. Shahid, N. K. Niazi, B. Murtaza, I. Bibi, C. Dumat, *Journal of* Geochemical Exploration, **182**, 247 (2017)
53. A. Mahar, P. Wang, A. Ali, M. Kumar, A. Hussain, Q. Wang, R. Li, Z. Zhang, Ecotoxicol. Environ. Saf., **126**, 111 (2016)
54. M. Rezvani, F. Zaefarian, AJAE, **2(4)**, 114 (2011)
55. H. Ali, M. Naseer, M. A. Sajad, Int. J. Env. Sci., **2 (3)**, 1459 (2012)
56. M. Shahid, C. Dumat, S. Khalid, N. K. Niazi, P. M. C. Antunes, Rev. Environ. Contam. Toxicol, **241**, 73 (2016)
57. R. Sfiru, P. Cardei, V. Vladut, M. Matache, *International Symposium ISB-INMA-THEH Procedings*, Bucharest, Romania, 435 (2017)
58. D. M. Bordean, V. Vlăduţ, I. Caba, I. Gogoasa Alda L., C. Moldovan, D. Moigradean, L, Pirvulescu. Animal Science and Biotechnologies, **50**, 162 (2017)
59. P. Cardei, C. Tudora, *7th International Scientific Conference Engineering for Rural Development*, Jelgava, Lithuania, 893 (2018)

60. M. Sabir, E. A. Waraich, K. R. Hakeem, M. Öztürk, H.R. Ahmad, M. Shahid (Soil Remediation and Plants, Elsevier Inc. 2015)
61. M. J. Blaylock, J. W. Huang, *Phytoremediation of Toxic Metals: Using Plants to Clean up the Environment* (John Wiley & Sons Inc, New York, NY 2000)
62. Z. R. Holan, B. Volesky, Biotech. Bioeng., **43 (11)**, 1001 (1994)
63. L. Deng, Y. Su, H. Sua, X. Wang, X. Zhu, J. Hazard. Mater. **143 (1-2)**, 220 (2007)
64. J. Vymazal, Uptake of lead, chromium, cadmium and cobalt by Cladophora glomerata, Bull. Env. Contam. Toxicology, **44 (3)**, 468 (1990)
65. S., *J.* Dwivedi, Advanced Laboratory Res. Bio., **3 (3)**, 195 (2012)
66. A. Sánchez, A. Ballester, M. L. Blázquez, F. González, J. Muñoz, A. Hammaini, FEMS Microbiol Rev. **23(5)**, 527 (1999)
67. E. Fourest, B. Volesky, Appl. Biochemistry Biotech., **67 (3)**, 215 (1997)
68. K.V. Ajavan, M. Selvaraju, K. Thirugnanamoorthy, Pakistan J. Bio. Sci., **14 (16)**, 805 (2011)
69. T.A. Davis, B. Volesky, R. H. S. F. Vieira, Water Res., **34 (17)**, 4270 (2000
70. J. I. Kumar, C. Oommen, J. Env. Bio., **33 (1)**, 27 (2012)
71. B. V. Tangahu, S. R. S. Abdullah, H. Basri, M. Idris, N. Anuar, M. Mukhlisin, International Journal of Chemical Engineering, **2011 (31)**, 1 (2011)
72. H. Singh, A. Verma, M. Kumar, R. Sharma, R. Gupta, M. Kaur, M. Negi, S.K. Sharma, Austin Biochem, **2(2)**, 01 (2017)
73. M. A. Galende, J. M. Becerril, O. Barrutia, U. Artetxe, C. Garbisu, A. Hernández, J. Geochem. Explor. **145**, 181 (2014)
74. E. Masarovicova, K. Kralova, M. Kummerova, Acta Physiol Plant, **32**, 823 (2010)
75. M. Bandiera, C. Dal Cortivo, G. Barion, G. Mosca, T. Vamerali, Sustainability, **8(4)**, 357, (2016)
76. G. S. Bañuelos, H. F. Mayland, Ecotoxicol. Environ. Saf. **46**, 322 (2000)
77. A. Vassilev, J. P. Schwitzguébel, T. Thewys, D. van der Lelie, J. Vangronsveld, ScientificWorldJournal **4**, 9 (2004)
78. T. Vamerali, L. Marchiol, M. Bandiera, G. Fellet, N.M. Dickinson, P. Lucchini, G. Mosca, G. Zerbi, Italian J Agron, **7**, 323 (2012)
79. P. Kidd, M. Mench, V. Álvarez-López, V. Bert, I. Dimitriou, W. Friesl-Hanl, R. Herzig, J. O. Janssen, A. Kolbas, I. Müller, S. Neu, G. Renella, A. Ruttens, J. Vangronsveld, M. Puschenreiter, Int J Phytoremediation, **17**, 1005 (2015)
80. J. Vangronsveld, R. Herzig, N. Weyens, J. Boulet, K. Adriaensen, A. Ruttens, T. Thewys, A. Vassilev, E. Meers, E. Nehnevajova, D. van der Lelie, M. Mench, Environ Sci Pollut Res **16(7)**, 765 (2009)
81. L. Jennifer, J. L. Wood, W. Liu, C. Caixian Tang, A. E. Franks, AIMS Bioengineering, **3(2)**, 211 (2016)
82. K. Tsekova, A. Kaimaktchiev, A. Tzekova, Bioaccumulation of heavy metals by microorganisms, Biotechnol Biotechnol Equip., **12(2)**, 94 (2014
83. O. B. Ojuederie , O. O. Babalola, Int J Environ Res Public Health., **14(12)**, 1504 (2017*)*
84. M. Strouhal, R. Kizek, J. Vacek, L. Trnková, M. Němec, Bioelectrochemistry, **60**, 29 (2003)
85. S. Özdemir, E. Kılınç, A. Poli, B. Nicolaus, Biorem J, **17(2)**, 86 (2013)
86. S. Banerjee, R. Gothalwal, P. K. Sahu, S. Sao, India. Adv Biosci Biotechnol. **06(02)**, 131 (2015)
87. S. Zhang, M.S. Reddy, J.W. Kloepper, Plant Soil, **262**, 277 (2004)
88. A. Ullah, S. Heng, M.F.H. Munis, S. Fahad, X. Yang, Environ. Exp. Bot., **117**, 28 (2015)

Improving the composting of waste materials from agricultural farms, a step towards sustainable agriculture

Ioan Caba.[1]*, *Despina* Bordean [2], *Gheorghe* Matei [3], *Monica* Matei [4] *and Ion* Pavel [5]

[1]INMA Bucharest, Branch Timisoara, Romania
[2]USAMVB Timisoara, Romania
[3]University of Craiova, Romania
[4]INCDPM Bucharest, Romania
[5]INOE 2000- IHP Bucharest, Romania

Abstract. The European Union is working hard to find effective ways for agriculture to remain clean and sustainable. To achieve this objective, the capacity and capability of the research and development system in the agricultural field is increased. Promoting and supporting the development of innovative technologies and their application on agricultural farms of any size or type for sustainable development. The use of innovative technologies for the production of natural fertilizers, besides the complete elimination of the use of chemical fertilizers, is a desideratum to be pursued. Natural fertilizers bring added value in terms of crop yields. Several technical variants of composting from the vegetal waste produced at an agricultural farm are known. Since the cost of obtaining natural fertilizers differs greatly depending on the method chosen. We want to find through this work simple solutions to make the production of natural fertilizers more efficient, solutions that are available to farmers.

1. Introduction

It is known that in modern agriculture high yields per hectare can only be obtained with the use of chemical fertilizers that are widely used either in liquid or granular form. Once they are administered, reach the soil and by absorption into the root system of the plants, ensuring their more pronounced development. Unfortunately, many of these chemicals also come into groundwater or even in flowing water if they are not being administered on the ground properly. This leads to infestation of nitrite and nitrate waters. Infested drinking water can lead to major health problems for population. An alternative to farmers reach exists, the use of compost obtained on their own farm. Made in its own farms, with raw materials obtained by cultivating technical plants, cereals, and so on, to which organic matter is added as a manure from animals raised on these farms, mixing with

* Coresponding author: *inmatm_caba@yahoo.com*

different types of organic waste also produced in their own households or collected from neighborhood. You can also add to the compost and chopped wood remnants (branches, leaves) or mowing grass. All of these components are well chopped, mixed and left on the ground in the form of a string. This is where microorganisms, bacteria, and different types of fungi work to transform the mixture called compost into an organic fertilizer [1].

Fig.1. Placement of the mixture of organic materials in the field for composting

Under temperate climates, composting is severely hampered by heavy winters, so composting over time can last between six and nine months. Through this paper, we want to propose simple techniques for speeding up production, so that the batch production period is reduced to about four months, using techniques that are at the fingertips of farmers, avoiding unbearable expenses.

2. Methodology

2.1. Formation and composition of the organic soil

Disintegration and alteration processes have transformed compact rocks into water and air stores materials, containing simple chemical elements such as salts, oxides and hydroxides, and clay compounds. Together with the water they created conditions for installing the first forms of vegetation. Vegetation transforms the mineral substance into an organic substance by the phenotype of photosynthesis. Thus, a new "organic matter" component [1] appears on the surface and the loose layer of the lithosphere.

Parallel to the photosynthesis process, the process of decomposition of organic matter takes place in nature, especially under the action of microorganisms. The decomposition may be partial when the vegetal debris is transformed into humus or total, when it is called mineralization and the vegetal debris is decomposed to the final products (CO2, H2O, NH3, NO3-a.s.o)[1].

These processes of synthesis of the organic matter in the mineral matter and of the decomposition of the dead vegetal organisms into the mineral matter, in the nature, realize the so-called "biological circuit of the matter." The biological circuit of the matter consists in the fact that it ensures the permanent resumption of the life process, giving the soil an essential property fertility [1]. Higher plant formations, which provide the synthesis of organic matter together with microorganisms, which cause the decomposition of organic matter, make up biocenoses. Biocenoses are specific to the different environmental conditions [1]. Microorganisms, by way of purchasing the food, are divided into autotrophies and heterotrophs. Autotrophic micro-organisms get their food directly from the mineral matter, using as a caloric energy source, resulting from the chemical processes that take place in the soil. The heterotrophic microorganisms get their food by decomposing organic matter up to simple elements.

2.2 The role of organisms in soil formation

The main groups of microorganisms that activate in the synthesis and decomposition of organic matter are: bacteria, fungi and actinomycetes.

Along with these microorganisms, algae, lichens and animal protozoa microorganisms also contribute to the transformation of organic matter, but to a lesser extent. Vegetable macro organisms, represented by grassy vegetation and woody vegetation, are the ones that synthesize most of the organic matter subject to humification. Animal organisms contribute to the humification process in a much smaller proportion compared to vegetal organisms [2].

2.3 The origin of organic matter in the soil

The organic matter in the soil, mostly vegetal, is mainly formed from the debris of the upper plates such as roots, branches, leaves, strains, fruit, seeds, etc. Along with these, but to a much lesser extent, there are animal remains and microorganisms [1].

The amount of organic matter undergoing decomposition processes depends on the type of vegetation and the climatic zone in which it develops.

Under the conditions of our country, the grassy vegetation can leave 10-20 t / ha of organic residues, mostly from the roots, which are spread within the soil in a thickness of about 100 cm per year. The annual cultivated plants leave 3-4 t / ha organic soil, consisting of roots and stalks of stalks known as stubble. A larger amount of organic remains remains in the soil from alfalfa and clover, which each year bring 3-12 t / ha of roots.

Forest vegetation can leave a quantity of organic mass estimated annually at 2-4 t / ha, mostly on the surface of the soil in the form of a litter. The litter is made up of leaves, branches and bark fragments, having a thickness of 1 to 3 cm under the woods and 3 to 6 cm under the hardwood forests [1].

Microorganisms (bacteria, fungi, actinomycetes, etc.) and animals in the soil also bring a significant amount of organic matter, which amounts to 1-4 t / ha per year.

All of these data confirm the statement that we have enough organic raw materials available at any farm for the purpose of composting.

2.4 Transformation, decomposition of organic residues and composting

The processes of decomposition of organic residues by microorganisms depend on the nature of debris, environmental conditions and the nature of microorganisms. Organic mattery transformation takes place in three main stages: hydrolysis, oxidation-reduction reactions and total mineralization.

Hydrolysis causes the decomposition of complex organic substances into simpler but organic products. The products resulting from hydrolysis are further subjected to oxidation-reduction processes and transformed into either simple organic substances or mineral compounds. Oxido - reduction continues to subject hydrolysis products to intense decomposition processes, resulting in simple organic or even mineral compounds.

Oxide - reducing the hydrolysis products of the protein substances form organic acids, aliphatic acids, alcohols, ammonia, carbon dioxide, water, methane, hydrogen sulphide, a.s.o.

The products resulting from the hydrolysis of the carbonates by oxidation reduction processes are transformed into oxides, volatile organic acids, aldehydes, alcohols, carbon dioxide, water, methane, a.s.o.

The products resulting from the hydrolysis of lignins and oxidation - reducing substances form phenols, quinones, carbon dioxide, water, a.s.o.

Hydrolysis products of lipids and resins subjected to oxidation reduction processes are converted into unsaturated acids, oxides, volatile organic acids, hydrocarbons, carbon dioxide, water, a.s.o.

Mineralization represents the final phase of decomposition of organic residues and results in the formation of mineral compounds. In the aerobic medium, different acids (HNO_2, HNO_3, H_2SO_4, H_3PO_4 etc.) are combined with the bases in the soil (Ca, Mg, Na, K etc) and the corresponding salts result, whereas in the anaerobic environment are obtained as final products: CH_4, H_3PO_4, H_2, a.s.o. There are also end products that form under both aerobic and anaerobic mineralization conditions like Ca, H_2O, CO_2, NH_3.

The decomposition of organic residues is greatly influenced by their composition, taking place rapidly or very slowly, with the formation of intermediate or final products, as will be shown below.

Water-soluble carbonates are easily decomposed into final products (CO_2, H_2O).

Water-insoluble carbonates (cellulose and hemicellulose), by hydrolysis, are converted into aminosaccharides and uronic acids, and by oxidation reduction, aldehydes, alkali, carbon dioxide, water, methane, a.s.o. are obtained.

Lignin and tanning substances are very resistant to decomposition, with the tendency of residual accumulation in the soil. By hydrolysis, compounds of the polyphenol type are obtained, and by the oxidation-reduction reactions the polyphenols are passed into phenols and quinones. These intermediate compounds have an important role in the formation of humus.

Protein substances in soil generally break down easily. By hydrolysis they pass into peptides and amino acids, and they are converted to organic acids, fatty acids, alcohols, ammonia, carbon dioxide, water, methane, hydrogen sulphide, a.s.o. by oxidation reduction. Peptides and amino acids, together with phenols and chimes, make up basic compounds that participate in the formation of soil humus.

Fatty substances, waxes and resins slowly break down and hydrolyze pass into glycerin and fatty acids, and oxidation-reducing reactions give rise to unsaturated acids, oxides, volatile organic acids, hydrocarbons, carbon dioxide, water, a.s.o.

3. General aspects of the properties and composition of the compost

The compost contains in a balanced form a lot of useful components:

- Mineral elements in a well-assimilated species.
- Enzymes that ensure the transformation of organic residues into nutrients.
- Substances that prevent the spread of pathogens.
- Phytohormones, which improve plant growth and resistance to stress.

This type of organic fertilizer contains 2-4 times more humus than cow dung. Its advantages include good moisture capacity, friability, compatibility with other organic fertilizer types, no significant input energy in production and use. The ability to sell surplus products will allow you to recover costs and get a certain income [3].

3.1 Compost production

In the conventional composting world, the unwritten rule is that one ton of inputs results in 0.75 m3 of compost, the weight of which varies depending on the moisture content, but is usually about 0.5 t. In other words, 50% of the mass is lost, part of the nitrogen is lost as ammonia, but if the process is well managed, the loss of nitrogen is minimal. This varies depending on the nature of the inputs and the system used. The higher the ratio of carbon-rich materials to those rich in nitrogen, the higher the final product weight as a percentage of the input weight [4].

Of course, the final weight and volume of the product varies depending on the origin of the food, the loom used, etc., but the above-mentioned intuitive rule is a handy method to quickly calculate production [5].

In composting, mixtures of nitrogen and carbon-rich materials are made at first and nothing is added afterwards. We can also use the raw material and shredded paper or cardboard, compost can be converted with a small (in volume) of 5% poultry manure [4].

The result of this process is a weight of the finished product closer to 50% of the initial input weight.

To produce compost we need some living conditions:

- hospitable living environment;
- source of food;
- adequate humidity (water content greater than 50% by weight);
- appropriate aeration;
- extreme temperature protection.

Some materials are suitable to create a good environment for composting, while others lack one or more of the above features and must be used in different combinations. Table 1 provides some data on the absorption of each material, the leaching potential and the carbon-nitrogen ratios (C: A) that can be obtained from composting.

Table 1. Materials used for composting

No. crt.	Material	Absorption	Loosening Potential	C/A Report
1.	Animal manure	medium-good	good	22-56
2.	Domestic waste	good	medium	58
3.	Silage corn	medium-good	medium	38-43
4.	Hay	low	medium	15-32
5.	Straw	low	medium-good	48-150
6.	Paper from the municipal waste stream	medium-good	medium-good	127-178
7.	Bark hard essence	low	good	116-436
8.	Bark soft essence	low	good	131-1280
9.	Corrugated cardboard	good	medium	563
10.	Waste paper fibers	medium-good	medium	45 -250
12.	Sawdust	low-medium	low-medium	142-750
13.	Strong wood chips	low	low	451-819
14.	Soft wood chips	low	low	212-1313
15.	Leaves (dry, free)	low-medium	low-medium	40-80
16.	Corn strains	low	good	60-123

If available, crushed paper or paperboard is an excellent component, especially when combined with typical organic resources in the farm, such as straw and hay. However, organic producers should be careful and ensure that such materials are not restricted in accordance with their organic certification standards [4].

Other material in this category is the paper-making residue with a high absorption capacity and small particle size that perfectly complements the high ratios and also has good aggregation properties of straw, hay, shrubs, or shavings [6,7]. In general, it should be noted that the selection of compost materials and their mixing in the necessary quantities is the key to success.

Recipient blends used to create compost are essential in achieving a bio-quality fertilizer. They offer protection in the event of extreme temperatures, the necessary level and consistency of humidity and an adequate supply of oxygen. Fortunately, given their critical importance in the process, good mixes are generally easy to find on farms. The most difficult criterion to satisfy adequately is usually absorption, because most straw and even hay are not good at retaining moisture. This can be easily addressed by mixing sweepings from cattle, horses or sheep with straw.

Fig. 2. Crushing and mixing the raw materials used to make the cores

The process of creating compost mixtures should not be a burden, it can be done manually with a fork (small operations), a bucket tractor (larger operations), a technological trailer for mixing the feed or, if necessary, the shredders of large plant waste.

The latter equipment would be suitable only for large commercial composting operations where a high level of product efficiency and quality is required [6,7].

There are regions where seaweed, human biowastes from purification stations of large localities, etc. are used as raw material for the composting.

3.2 Humidity

The mixture of material used in composting must be capable of retaining sufficient moisture as the microorganisms and bacteria have a favorable environment in which they can live and multiply. A moisture content of less than 50% is dangerous. Except for extreme heat or extreme cold, the lack of adequate humidity may stop or increase the duration of composting.

Fig. 3. Wetting the compost during the aeration operation.

The range of ideal moisture content for materials in conventional composting systems is between 45-60%, depending on the geographical area and the multi-annual climate [4, 5, 8].

3.3 Aeration of compost

Aeration (passive or forced) has many different functions during the composting process:

- supply of oxygen to keep microorganisms alive
- removal of carbon dioxide
- reducing water to allow material to dry
- temperature control to prevent values higher than 70 °C - 80 °C

The need for oxygen depends on the activity of the microorganisms. Very important for the supply of oxygen is the mixture of materials introduced into the heap, especially the quantity, but also the size and distribution of the interstices. During the process of composting, the oxygen concentration in the pile decreases. To supply more oxygen, ventilation or material returns are used. Ventilation systems either force the air to enter the heap so that the pressure in the interior is higher than that in the outside environment, or it reduces the pressure in the heap.

Fig. 4. Tube for ventilation of compost.

Fig. 5. Ventilator

Both systems can increase the concentration of oxygen in the heap by mechanically pumping the air. The return of the material is also a possibility for increasing the oxygen concentration. The turntable can be turned manually or using mechanized devices. The combination of natural aeration and the return of compost piles is a very good method. During winter it is recommended to use the aeration duct for the advantages of mechanical devices:

• there is no need to return the snow covered area;

• warm-air compost produced by solar panels can be heated without cost, keeping microorganisms active, reducing the time required for composting;

• the fan can operate with current obtained from photovoltaic panels, green current, without current costs, without using the national distribution network;

• usability on any part of the farm.

Fig. 6. Panels for heating the air mass circulated during aeration of compost.

Fig. 7. Photovoltaic panels

3.4 Digestion of the compost

The optimum digestion for the corresponding composting activity is the total porosity of 48-60%, of which capillary porosity of 30-36% and of non-capillary (aeration) of 18-24%. At this level, the compost contains 1/3 air and 2/3 water. The apparent density of the compost under these conditions is between 1.0 and 1.4 g / cm3 of the "seated" soil.

A properly worked soil, 50-150 kg / ha of nitric nitrogen accumulates through the activity of nitrifying microorganisms and 23-55 kg / ha of nitrogen through the activity of nitrogen fixation bacteria. In a too loose soil, the process of aeration and intense mineralization of organic matter and soil humus increases, an unwanted phenomenon.

Fig.8. Towing machine for loosening and wetting the compost.

3.5 Temperature control

Active organisms are producing thermal energy. This energy is measurable in turn. By setting the temperature inside the heap, we can conclude as much time as possible until the composting process is complete. The temperature of the external environment also influences the decomposition process. The rate of decomposition is determined by the internal temperature of the compost, as the activity of the microorganisms increases with the temperature increase, which should never pass 80 °C. Temperature is an appropriate value for determining the degradation phase and the degree of decomposition of compost. The curve characteristic of temperature variation is presented according to time passage, (fig. 9), which does not exceed 6 months. Through additional storing, aeration and wetting activities, this range can be reduced to almost 4 months.

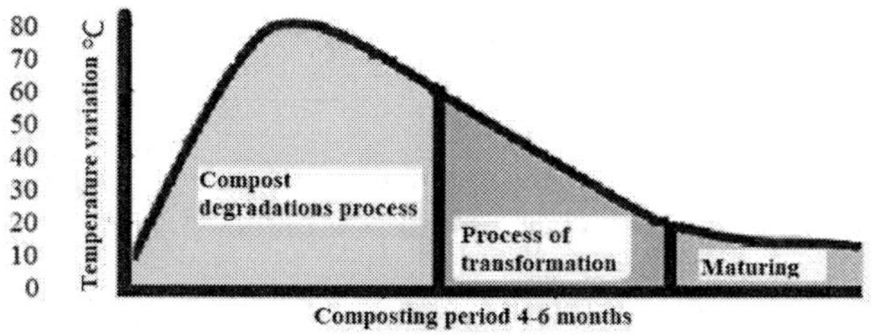

Fig. 9. Variation of compost temperature over time

3.6 pH level

The activity of microorganisms is closely related to the pH level of the substrate introduced. Good for biological activity are pH values ranging from 7 to 11.

If the passes below 7 lead to a reduction in speed in the first stages of the process. If the pH is below 5, it may notice a strong inhibition in the initial phase (up to rapid temperature rise). Therefore, the period between collection, storage and commencement of treatment at the station should be short. Lack of adequate oxygen during collection and storage leads to an uncontrolled natural anaerobic digestion, which causes low pH values [9,11].

3.7 Carbon / nitrogen ratio

The ratio of carbon to nitrogen (not specific analyzes carried out in a specialized laboratory) compost material influences the rate of decomposition process. The optimal carbon ratio (C): nitrogen (N) should be between 1 and 35: 1. If the proportion is below 10: 1, growth is inhibited by the lack of carbon, and if it exceeds 40: 1, too little nitrogen is available. Outside these limits 1:10 and up to 40: 1, the microorganism cannot grow and grow. In fact, the activity of these microorganisms is unchanged, but without an increase in population the time required for the composting process expands, an unwanted phenomenon [10,11].

4. Conclusions

Based on the data and materials submitted, composting is an environmentally beneficial process that has great potential as a component of organic and sustainable agriculture. Agricultural specialists claim that the fertilizer thus obtained is one of the best fertilizers because it contains concentrations of beneficial bacteria and other microorganisms, biologically active stimulants for plants [10]. The product contains the necessary set of macro and micro nutrients, enzymes, soil antibiotics, vitamins, growth hormones and humic substances. The compost contains nitrogen, phosphorus, potassium, calcium and a host of beneficial microorganisms and bacteria than those we usually find in the upper layers of the soil. It is used as a natural fertilizer for soil being mixed with earth. In contrast to chemical fertilizers that feed the plant directly, but damages the soil, matured compost enriches the fertility of the soil as a whole [12].

Using compost generates multiple benefits for farmers:
- most of the waste and vegetal remains and animal manure present on an agricultural farm are used without costly investment in raw materials;
- significant production increases are achieved because plants easily assimilate nutrients;
- water retention increases in soil (the amount of water required for irrigation decreases by about 30%);
- it is not toxic, does not burn plants, has no restrictions on use, can be used in any crop, greenhouse or field, with excellent results;

- the compost obtained can be used directly on the farm by distributing it on the surface of the plant, reducing or even eliminating the investment in the purchase of traditional fertilizers (fungicides, insecticides, etc.).
- rebuild soils affected by long-term use of chemicals. By repeated application, the soil will be completely repopulated with the microorganisms beneficial to the plants destroyed by chemicals used over the years.
- the application of compost improves the soil structure, aerates the soil and makes it easy to process, which reduces the cost of fueling the agricultural machinery, indirectly reduces soil compaction through multiple passes.
- the administration of the compost does not require investments in new machinery, it is possible to use the existing ones on the agricultural farms: solid chemical fertilizer machines, wood scraping machines, forage mixers, etc.

Acknowledgement

This work was supported by a grant of the Romanian Research and Innovation Ministry, through Programme 1 – Development of the national research-development system, subprogramme 1.2 – Institutional performance – Projects for financing excellence in RDI, contract no. 16PFE.

References

1. N. Bernier, J.F. Ponge, *Dynamic formation of humus during silvicogenetic cycle in a mountain spruce forest*, Soil Biology and Biochemistry J., (1994)
2. R. Hempfling, H. R. Schulten, R. Horn, *Relevance of humus composition to physical / mechanical stability of agricultural soils: a mass spectrometry and direct pyrolysis study*, Analytical and Applied Pyrolysis J., (1990)
3. P. Bogdanov, *Commercial vermiculture*: how to build a prosperous business with red rum, VermiCoPress J., Oregon, U.S.A. (1996)
4. GEORG, *The feasibility of organic and transitional organic farming development for the processing of organic waste and municipal waste using large-scale vermicompostation*, Organic Earth Resource Group, Halifax, Nova Scotia, Canada, (2004)
5. R. Rink., *Farm composting guide,* Ed. Natural Resources Service, Agriculture and Engineering, NRAES-**54**, (1992)
6. C. Elvira, M. Goioechea, et al., *Bioconversion of sludge solids from cellulose-paper through earthworms*, Bioresources Technology J., UK, (1996)
7. C. Elvira, L. Sampedro, *Vermicomposing of sewage sludge in the paper industry through nitrogen-rich materials*, Biochim. Bio. Sol. J., (1997)
8. J.Dominguez, C. A Edwards, *Effects of stocks and moisture content on growth and maturation in Eisenia andrei (Oliogochete) in pork waste*, Biochim. Bio. Ground J., (1997)
9. W. Bidlingmaier, *Biologische Verwertungstechnologien*, C. S., D. for W. M. - Bauhaus – Weimar U., G., Deutschland, (2009)
10. Canellas, et al., *Humic acids isolated from earthworm compost amplifies root elongation, side roots and H + -ATPase activity of the plasma membrane in maize roots*, Plant Physiology J., (2002)
11. C. Bachert, W. Bidlingmaier, S. Wattanachira, *Handbook on the production of compost in graded uncovered strings*, E. C. N., Weimar G., ISBN 3-935974-23, (2008)
12. I. Voicea, V. Vladuţ, P. Cardei, M. Matache, I. Gageanu, Gh.Voicu, C. Popescu, G. Paraschiv, O. Kabas, *Compacting process and mathematical analysis of miscanthus briquettes expansion,* Proceedings of The **43** International Symposium on Agricultural Engineering "Actual Tasks on Agricultural Engineering", Opatija – Croatia, pag. 667÷676, (2015).

Greenhouse heating by using an installation of biomass gasification

Evelin Laza[1]*, *Liliana* Dumitrescu[2], *Madalina* Boboc[3] and *Georgiana* Moiceanu[4]

[1]INMA Bucharest –Timisoara Branch, Romania
[2]INOE 2000-IHB Bucharest, Romania
[3]INCDPM Bucharest, Romania
[4]UP Bucharest, Romania

Abstract. The use of renewable energies has seen a significant increase in energy demand in agriculture, in competition with solid, liquid or gaseous fossil fuels. Wood and other forms of biomass including energy crops, agricultural and forest biomass are transformed into energy through thermal, biological or physical processes. Thermo-chemical conversion, biomass gasification, is the most attractive technology that offers a high conversion efficiency compared to direct burning or rapid pyrolysis.

1. Introduction

Generator gas, called gas, can be used directly for thermal applications: cooking machines and heating installations, where changes have been made to ensure an optimal combustion air excess, and after proper filtration at the generation of electricity by feeding internal combustion engines, Stirling engines, or in the perspective of combustion cells[1,2,3,4].

The gasification of the biomass produces a generator gas with an average calorific value of 5 MJ / Nm^3, with a power conversion of the wood mass in the 70-85% range, is at least 3 times more efficient than direct combustion. Conversion efficiency depends on: the design gas-generator solution, the composition, dimensions and humidity of the fuel, as well as the generator operating load [1,4,5]. Air conditioning of greenhouses, heating and ventilation is an essential condition for achieving economically efficient production but is requiring large-scale installations at the same cost, which has made it more and more abandoned. At present, traditional systems with trunking networks that occupy a lot of space, make it hard for operators and little mechanization, contain a great deal of metal and cost a lot. Obviously, this system has advantages in terms of high heat inertia, heating and radiation, high operating safety and lower maintenance personnel [1].

The main issue with regard to the fuel used for heating greenhouses is that of the price on which production costs depend. The second is the level of pollution. Combustion fossil fuels increase the concentration of CO_2 in the earth's atmosphere, spill into the environment particles of smoke, NO_x and uncharged hydrocarbons that all contribute to increased

* Corresponding author: eveline_anda@yahoo.com

environmental pollution. The smallest pollution is obtained when gaseous fuels are consumed and combustion is controlled to reduce as much as possible the production of pollutants [1].

Most small and medium greenhouses cannot benefit from methane gas, LPG or diesel, but more often CLU and fuel oil, are much cheaper. Combustion of CLU or naphtha is done with the release of pollutants and requires hot or steam heating systems that are expensive as investment and maintenance [1, 2].

As an alternative to these energy sources in the last decade, it is found that the energy produced from the gasification of wood or agricultural biomass is increasingly used in rural thermal applications where there is a sufficient amount of gas-based biomass at a cost of use to reduce [1, 2, 3].

Taking into account the seasonality of greenhouse heating, the heat source needs a very small initial investment or outside of the greenhouse season to be used in other applications to lower the share of the investment's depreciation [1, 3, 4].

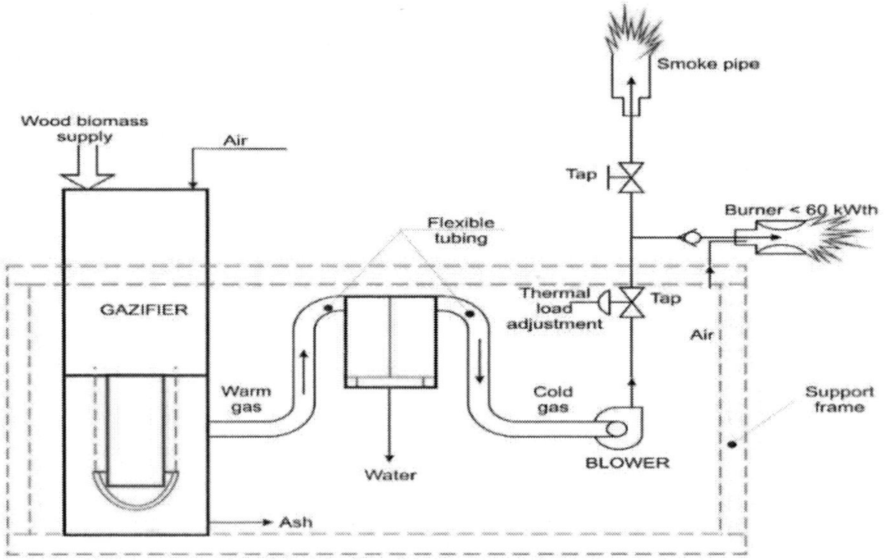

Fig. 1. Sketch of the heating system with producer gas.

2. Methodology

In the cold mixing chamber, a mixture of a recirculated air flow from the inside and an outside air flow is made. It is located at the bottom, at ground level, and consists of external air intake valves mounted on the outside wall and recirculated air intake valves mounted inside the greenhouse. The recirculated air / outside air ratio is controlled by the simultaneous actuation and the actuator that receives a command signal from the automatic steering system.

The gas-fired burner, specially designed for combustion of low calorific gases, is supplied with a gaseous flow rate at a pressure of 6 to 20 mbar adjusted with the regulating valve and a pre-heating combustion air flow rate that is regulated with a flap.

The heating agent is guided into a polyethylene tube stretched across the greenhouse and hanging in the ceiling. The tube has clamp-shaped cutouts that distribute hot air in the form of jets oriented so as to wash the exterior walls and achieve good mixing with the greenhouse air to smooth the temperature inside. The side walls of the greenhouse, are provided with the exhaust flow, equal to the outside air flow introduced into the greenhouse. The self-closing flaps are adjusted to maintain an overpressure in the greenhouse in the 20 ... 60 mbar range, depending on the construction of the greenhouse and the current crop.

The biomass was transported to the location of use, where it was stored in a covered enclosure to dry naturally to 10-20% relative humidity. For gasification, the biomass was minced to the size of 1-5 cm and the hopper loaded and unloaded it to the top of the gas generator until it reached the upper level measured with a capacitive level transducer that outputs a signal. The supply was done sequentially, with constant operating periods, to increase the durability of the conveyor [6-13].

Fig. 2. Overview of the producer gas heating system

Fig. 3. Biomass material used for testing

Fig. 4. Front view of the producer gas heating system

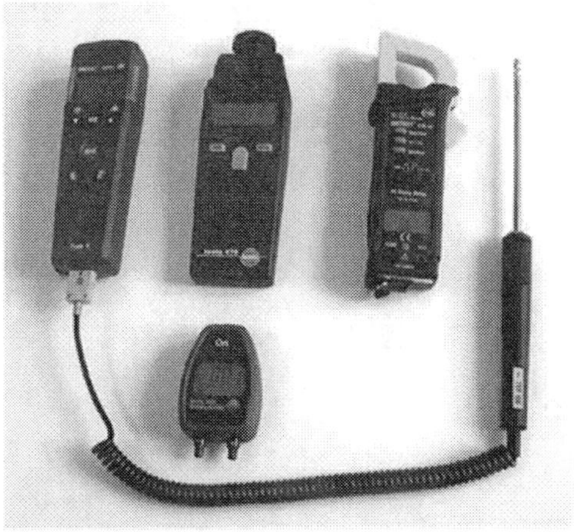

Fig. 5. Apparatus used for testing

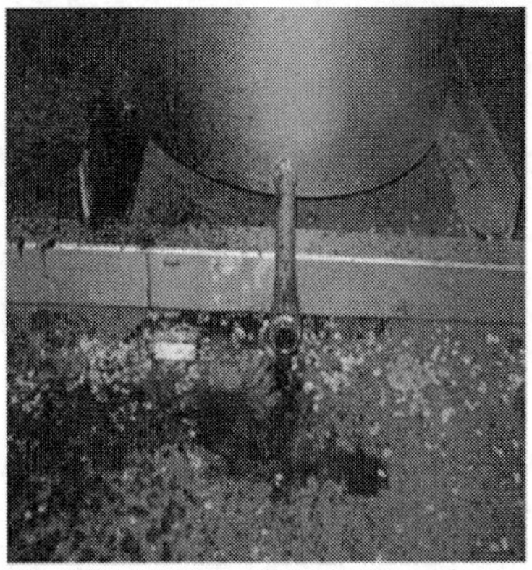

Fig. 6. Evaporation of condensate from the producer gas generator

To start the fire in the gasifier, the blower was used to create an air-sucking airflow through the biomass from the fuel tank. When the wood gas leaves the gasification unit, all the oxygen pressed down with the air through the fire tube was chemically modified and contained carbon monoxide (CO) and water (H_2O). Wood gas can not burn unless combined with a correct amount of oxygen. The tightness of the gas burner unit is absolutely essential.

Ideally, while wood gas enters the burner, it must be mixed with air in a 1: 1 or 1.1: 1 ratio (air to gas) in volume.

The fire tube is filled with coke up to 100 mm above the grid. The fuel tank is filled with dry woody biomass, the lid is closed and fully open the air valve.

For the determination of the kinematic, technological and electric parameters (power voltage, current absorbed by electric motors) the following tools were used:

Table 1. Instrument or apparatus used

No.crt.	**The name of the instrument or apparatus**	**Range of measurement**
1	Metal roulette	0 ÷ 2 m
2	Callipers 200 × 0,1	0 ÷ 200 mm
3	Callipers with deepth 200 × 0,1	0 ÷ 200 mm
4	Gradient scale	0 ÷ 500 mm
5	Chronometer	- minutes - seconds
6	Laboratory balance	-
7	BENQ digital camera	-
8	Sound Level Meter Type 2237 B vibrometer	30 ÷ 140 dB
9	Technical thermometer with solid capillary	- 20 ÷ +30 °C
10	Differential electronic manometer type TESTO 506	0 ÷ 1 mbar 1 ÷ 10 mbar 10 ÷ 200 mbar
11	MICRODUST PRO digital dust detector	0,001÷2.500 mg/m^3
12	Digital TESTOTERM 4000 Air Flow Meter, Temperature and Flow Meter (Testovent)	0,4 ÷ 40 m/s - 120 ÷+1.300 °C 10÷500.000 m^3/h
13	Digital apparatus for measuring electrical parameters type UNITEST CHB 140	0 ÷ 400 A 0 ÷ 600 Vc.a. 0 ÷ 400 Ω
14	Digital thermometer with laser type UNITEST 94009,	- 50 °C ÷ 0 °C 0 °C ÷ +200 °C +200 ^{C}÷+1.333 °C
15	Gas analyzer Code 99F20006 Tempest 50	O_2/CO/NO

The general behavior of the gasifier heating system was observed and the determination of the production capacity and the quality of the combustion operation in order to determine the optimal technological parameters for the design of a 50 kWth producer gas heating system.

For this purpose, within the limit of the quantity of product available for testing, weighing 150 kg biomass waste of oak, european beech (fagus) and fir, with a moisture content of less than 20% of the pieces with a length between 20 ÷ 50 mm were weighed for the preparation of ten batch of gasifier and one gaseous filter batch [11-13].

The products corresponding to a batch were manually loaded into the fuel tank, the fire tube, the gasifier grid and the gas filter, the blower started, and then passed the primary coke production cycle into the. Next, after 10 ÷ 15 minutes, during the normal gas production phase (attempted by igniting the exhaust gas to the tube), the burner of the installation is switched on.

The flow of gas produced by the gasifier has been determined based on the measurements after the filter and gas-producing are recorded in the following table:

Table 2. The flow of gas produced by the gasifier

No. crt.	The measured feature	M.U.	The measured value		
			I	II	III
1.	The gas velocity	m/s	0,5	0,9	2,09
2.	Diameter of pipeline	mm	90	90	90
3.	The rate flow of produced gas	Nm^3/h	9,1	18,3	48

The burned gas flow rate produced by the gasification system was determined on the basis of the measurements after the burner are recorded in the following table:

Table 3. The burned gas flow rate produced by the gasification system

No. crt.	Measured or calculated feature	M.U	The measured value		
			I	II	III
1.	Speed of gas burned	m/s	0,5	0,8	2,09
2.	Diameter of pipeline	mm	180	180	180
3.	The gas flow burned	m^3/h	36	72	188

The lower calorific value of the producer gas was determined on the basis of the chemical composition of the gas at the different biomass components are recorded in the following table:

Table 4. The lower calorific value of the producer gas

Chemical composition of gas in %							Inferior heat power. kcal/ m^3
Wood Biomass	CO	H	CH_4	C_nH_{2n}	CO_2	N_2	
Dry oak	17,8	15,9	3,2	0,4	16,1	-	1296
Dry beech	18,4	16,5	3,1	0,5	15,1	48,3	1321
Dry fir	16,1	18,1	2,2	0,8	15,7	56,1	969
Wood charcoal	32,2	7,3	2,8	-	2,6	57,1	1363
Charcoal brown briquettes	24	16	1	-	11	50	1223

Biomass mass flow rate (kg/h), determined by weighing, was between 20 (Beech) and 22 (Fir) kg / h so on average 21 kg / h. It results in a specific biomass consumption of 21 kg / h: 60 kW_{ht} / h) = 0.35 (kg / kW_{ht}).

The duration of the normal operating cycle was timed and was between 10 and 15 minutes.

The pressure drop through the gasifier, Δ_{p1} [mm H_2O] were determinated by measuring the depressions before and after the gasifier being in the limits of 30 ... 50 mm H_2O so an average of 40 mm H_2O.

The pressure drop through the filter, Δ_{p2} [mmH_2O] were determined by measuring the depressions before and after the gasifier being in the limits of 15 ... 25 mm H_2O, an average of 20 mm H_2O.

The temperature of the producer gas at the outlet of the gasifier was determined with the gas velocity in the specially designated locations using the TESTOTERM 4000 (Testovent) digital speedometer, temperature and air flow meter and was around 80°C.

The producer gas temperature at the filter outlet was determined with the same apparatus at the specially designated locations of the filter and was around 35°C. The producer gas temperature at the burner inlet was determined under similar conditions and was around 28°C.

3. CONCLUSIONS

There is required legislative and organizational measures to reduce pollution and emission of greenhouse gases and dependence on crude oil imports more expensive and uncertain about oil being a source of exhaustion.

A solution is the use of biomass resources for the production of solid fuel, liquid fuel and gas. The use of biomass for the production of heat by gasification has the advantage of using a renewable energy resource that does not increase the concentration of CO_2 in the earth's atmosphere. The use of biomass for heating greenhouses has the advantage of reducing production costs.

Applying low energy efficient technologies, allowing long-term industrial plants to operate, will help reduce the cost of heating greenhouses all year round.

As a result of those presented above it can be seen that the heating of greenhouses with relatively cheap energy produced from the thermo - chemical gasification of wood or agricultural biomass is an economic solution for small and medium - sized greenhouses located in areas with high potential for gas - biomass collection.

Acknowledgement

This work was supported by a grant of the Romanian Research and Innovation Ministry, through Programme 1 – Development of the national research-development system, subprogramme 1.2 – Institutional performance – Projects for financing excellence in RDI, contract no. 16PFE.

References

1. A. Romanek, *Local Greenhouse Biomass Heating System*, AGRAL Program, INMA Bucharest, (2004)
2. Europeean Comission Directorat-General for Energy (DG XVII), *Combustion and Gasification of Agricultural Biomass – Technologies and Applications*, Thermie Program Action BM **40**, CEEETA-PRATEX, Lisabona, (1995)
3. H.E.M. Stassen, *Knoef, Small Scale Gasification Systems,* Biomass Technology Group BV, Olanda, (1997)
4. Ministry of Energy and Environment, *Wood for Energy Production, Technology-Environment-Economy*, Center for Biomass Technology, Denmark, Report, pp.54-59, (1999),
5. R. Sims, J. Gigler, The Brilliance of Bioenergy *Small project using biomass*. Renewable Energy World, Vol. **5**, Nr. 1, pag. 57-63, (2002).
6. I. Voicea, Gh. Voicu, P. Cardei, V. Vlăduţ, S. Biriş, *A Mathematical Model of Chopped Miscanthus Briquetting Process,* The 8th International Conference INTER-ENG 2014 Interdisciplinarity in Engineering, 9 - 10 October 2014, "Petru Maior" University of Tîrgu Mureş Romania, Procedia Technology (**19**), ISSN: 2212-0173, pages 750–757, (2015)
7. I. Voicea, V. Vlăduţ, P. Cârdei, M. Matache, I. Găgeanu, Gh. Voicu, C. Popescu, G. Paraschiv, O. Kabas, *Compacting process and mathematical analysis of miscanthus briquettes expansion,,* Proceedings of the **43**th International Symposium on Agricultural

Engineering *"Actual Tasks on Agricultural Engineering",* Opatija – Croatia, ISSN 1333-2651, pag. 667÷676, (2015)

8. M. Chiţoiu, Gh. Voicu, G. Paraschiv, G. Moiceanu, V. Vlăduţ, M. Matache, E. Marin, G. Bunduchi, A. Danciu, I. Voicea, I. Găgeanu, *Energy consumption of a hammer mill when chopping miscanthus stalks,* Proceedings of the **44**th International Symposium on Agricultural Engineering *"Actual Tasks on Agricultural Engineering",* Opatija – Croatia, pag. 215÷224, (2016)
9. M. Chiţoiu, Gh. Voicu, G. Paraschiv, G. Moiceanu, V. Vlăduţ, M. Matache, I. Grigore, G. Bunduchi, I. Voicea, I. Găgeanu, *Comparison of energy consumption of a hammer mill during chopping miscanthus stalks and energetic willow,* Proceedings of the **45**th International Symposium on Agricultural Engineering "Actual Tasks on Agricultural Engineering", Opatija – Croatia, pag. 259÷268, (2017)
10. I. Găgeanu, Gh. Voicu, V. Vlăduţ, I. Voicea, *Experimental research on the influence of recipes used on the quality of biomass pellets,***16**th International Scientific Conference "ENGINEERING FOR RURAL DEVELOPMENT", pag. 785-791, (2017)
11. G. Moiceanu, M. Chiţoiu, Gh. Voicu, Gh. Paraschiv, V. Vlăduţ, I. Găgeanu, M. Dincă, *Comparison between miscanthus and willow energy consumption during grinding,* Proceedings of the **46**th International Symposium on Agricultural Engineering "Actual Tasks on Agricultural Engineering", Opatija – Croaţia, pag. 369÷378, (2018)
12. N. Ungureanu, V. Vlădut, Gh. Voicu, M. Dincă, B. Zăbavă, *Influence of biomass moisture content on pellet properties – review,*17th International Scientific Conference "Engineering for Rural Development", Jelgava, Latvia, pag. 1876-1883, (2018)
13. G. Moiceanu, Gh. Voicu, Gh. Paraschiv, V. Vlăduţ, P. Cârdei, M. Dincă, *Relationships analysis between the grinding parameters of Miscanthus giganteus stalks using a hammer mill,* Proceedings of the **47** th International Symposium on Agricultural Engineering "Actual Tasks on Agricultural Engineering", Opatija – Croatia, ISSN 1333-2651, pag. 399-408, (2019).

Soil compaction under the wheel of a sprayer

Nicoleta Ungureanu[1], *Valentin* Vlăduț[2*] and *Dan* Cujbescu[2]

[1]Politehnica University of Bucharest, Department of Biotechnical Systems, Splaiul Independentei nr. 313, sector 6, Romania
[2]National Institute of Research – Development for Machines and Installations Designed to Agriculture and Food Industry – INMA Bucharest, Ion Ionescu de la Brad nr. 6, sector 1, Romania

Abstract. Soil degradation by artificial compaction is recognized by the European Union as a major environmental and agricultural problem. Artificial compaction has worsened with the intensification of mechanized agriculture where heavier machinery is used, often moving on soils with high moisture. Experimental research was designed to determine the influence of five wheel loads and tire inflation pressures, on the contact area, the shape of footprint and the contact pressure, under the wheel of a machine for high accuracy application of phytosanitary treatments in orchards. It was found that the only situation when compaction does not occur is when the tank is empty (2.45 kN wheel load), at lowest tire inflation pressure of 100 kPa. Subsoil compaction (at 0.3-0.4 m) occurs when the sprayer machine's tank is filled with different amounts of liquid. With empty tank, the sprayer only causes topsoil compaction.

1 Introduction

Artificial compaction is a large scale, negative phenomenon that raises serious concerns over the sustainability of soils. Compaction develops as a consequence of using larger agricultural machines, increased number of soil works, short rotations and intensive pasturing of livestock [1]. Each mechanized agricultural work, from the preparation of the soil for crop establishment until its harvest, is involved in increasing the degree of artificial compaction of the agricultural soil [2].

Consequences of the artificial compaction on soil properties and especially on soil fertility are strongly negative. Compaction reduces the biota activity, soil porosity and permeability. At the same time, soil strength increases and its structure is broken. In terms of environment, there are increased risks of erosion, surface landslides, pesticides and nutrients flow into groundwater. Some agronomic consequences of soil compaction are: reduced plant capacity to assimilate the nutrients and water, restricted root penetration into the subsoil, decline in crop yields, and last but not least, high resistance to soil processing and high fuel consumption [2-5].

It has been reported that there are 68 mil. hectares of compacted soil globally only because of the traffic of agricultural machinery [6]. In Europe, over 33 mil. hectares of soil are compacted. Eastern Europe has 20 mil. hectares of compacted soil (37.5% of its

* Corresponding author: *valentin_vladut@yahoo.com*

agricultural soil) [7]. In 2017, Romania had reported a total of 6.5 mil. hectares of arable soils artificially compacted and 2 mil. hectares affected by natural compaction [8].

The artificial compaction of agricultural soil is due to a variety of factors that are related to both the characteristics of the agricultural machinery and work (wheel load; tire inflation pressure; footprint area which in turn has variable size and shape; the pressure on the soil or contact pressure; machinery speed; the number of passes; tire size; tire slippage) and the soil properties (soil moisture, soil type, texture, structure, and moisture) [9, 10].

Tire inflation pressure has a tremendous influence on the contact area and also on the contact pressure, the latter being also influenced by the wheel load [11, 12, 13]. Tire inflation pressure mostly influences the stress distribution up to 100 mm deep, and has a small influence on the stress distribution in the subsoil (below 300 mm) [14].

Wheel load is dynamic, as it can act over the soil from a few seconds (for example, at soil-plowing tool interaction or load applied to the soil at the passage of the tire) and up to a few minutes (often found in forestry) [15]. Over the past three decades, larger agricultural machinery have been built, but this has inevitably increased wheel load without reducing the contact pressure to prevent or minimize the artificial compaction [16].

The intense use of heavy agricultural machinery leads to the distribution of stresses at greater depths in the soil. 10-15 tons/axle harvesters produce compaction of the soil under the arable layer (at 25 cm deep) [17]. To avoid compaction below the arable layer (0.2-0.3 m deep), wheel load should be under 4-6 tons on one axle, for wet mineral soil at tire inflation pressures of 50 kPa. For tandem tires, the limit is 8-10 tons on wet soils. On dry soils, wheel load can be higher without causing subsoil compaction [18]. Repeated traffic of light agricultural vehicles on the same tracks with loads of less than 3 tons per axle produces subsoil compaction [19].

The contact pressure should be less than 100 kPa during field works and 300–350 kPa during transportation on field roads [20]. Contact pressure higher than 80 kPa leads to soil compaction and and will prevent root development and plant growth [21]. One tractor pass with 0.15 MPa contact pressure will have effects at the depth of 0.3–0.4 m, and multiple passes have effects to 0.6 m depth [22].

Loads applied by the agricultural machinery are distributed to different soil depths, through the footprint between the soil and tire (Figure 1), resulting in artificial topsoil and /or subsoil compaction. Contact area is the part of the tire that is in contact with the soil. It is computed as ratio of the wheel load and the tire inflation pressure. If the agricultural machinery is stationary, the tire is still loaded but only statically, and in this case, the static contact area is the contact area between the tire and the rigid or deformable surface [23].

The larger the footprint, the contact pressure is smaller and the stress is distributed at lower depths in the soil, minimizing rut depth and the compaction intensity.

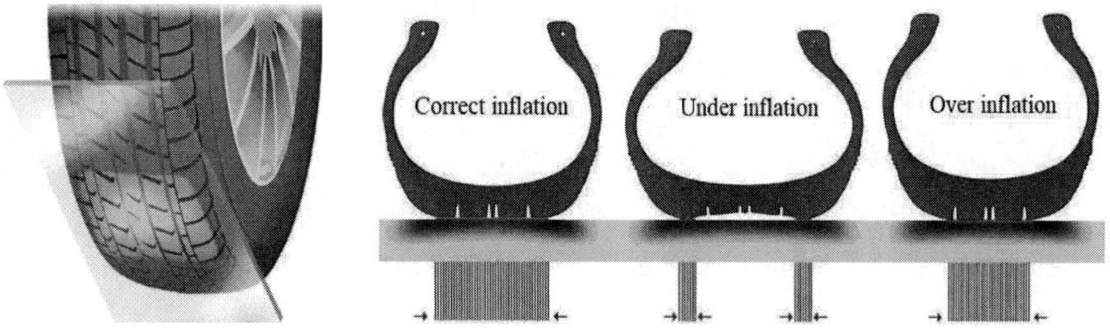

Fig. 1. Footprint detail and aspects at different tire inflation pressures [24, 25]

Tires are flexible, so the footprint can vary when the machinery is in motion than when it is static. In many studies, observations on the footprint were made in static experiments, in which footprint size and shape, and pressure distribution are mainly influenced by tire inflation pressure and by wheel load.

2 Material and method

Testing was carried out on an experimental field at INMA Bucharest. The aim was to determine the contact area under the influence of different wheel loads and tire air pressures.

It was tested the contact area and shape of footprint under the left wheel of the MSL sprayer machine for the high-precision application of phytosanitary treatments in fruit plantations, in aggregate with the Universal 445 tractor. Both the sprayer wheels and the front wheel of the tractor are equipped with Danubiana Superfront Tractor tires, size 6.00-16, profile F-2. The total weight of the spraying machine with the empty liquid tank is 4.9 kN (2.4517 kN on each wheel).

Fig. 2. U445 tractor in aggregate with MSL sprayer

Five sets of tests were carried out. In the first set, the tank was empty (wheel load 2.45 kN), then it was progressively filled with 250 liters of water (wheel load 4.9 kN), 500 liters (wheel load 7.36 kN), 750 liters (wheel load 9.81 kN) and up to a maximum capacity of 1000 liters (wheel load 12.26 kN).

Wheel load was determined using the RW-10PRF weighing platform with electronic indicator. In each test, a compressor and a manometer were used to obtain five tire inflation pressures, from 100 to 300 kPa, in increments of 50 kPa.

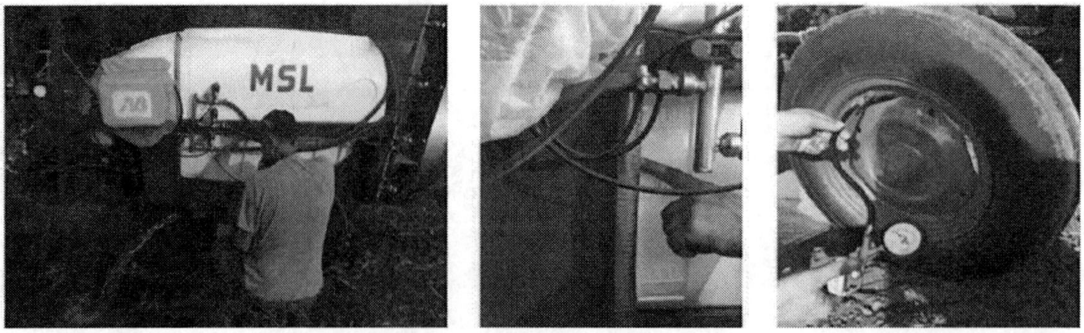

Fig. 3. Setting of wheel load and measuring tire inflation pressure during testing

Contact pressure between the wheel and the soil, as well as the contact area, were measured by the Tekscan pressure sensor (Figures 4 and 5) with dimensions 600 x 500 mm. For data acquisition, the VersaTek Handle electronic system and a laptop were used.

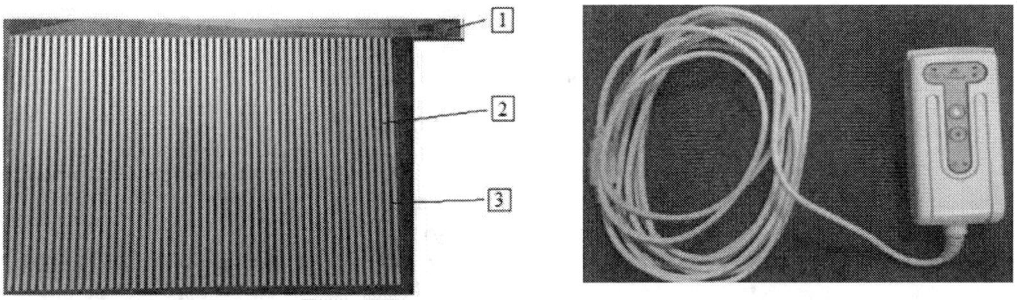

Fig. 4. Mesh-type Tekscan pressure sensor and VersaTek Handle data acquisition system
1 – connection to data acquisition system; 2 – sensels; 3 – connecting threads between sensels

Fig. 5. Field preparation for testing and placement of the pressure sensor before wheel passing

During the tests, the tractor moved backwards (so its wheels did not cross the pressure sensor) and stopped when the wheel of the MSL spraying machine reached the surface of the sensor, allowing data saving in the I-Scan software.

Figure 6 shows how the sprayer's tire behaved with the tank loaded with 250 liters of water (wheel load 4.9 kN) at the tire inflation pressure of 100 kPa.

Fig. 6. Testing with wheel load of 4.9 kN and tire inflation pressure of 100 kPa

3 Results and discussion

Results were analysed with the I-Scan software, which records the pressure distribution in the footprint, the individual values and the graphical variation over time of the contact area, the contact pressure, maximum pressure recorded, force on the soil, etc.

Data for both analysed input and output parameters are presented in Table 1.

Table 1. Data recorded with the pressure sensor under the wheel of the MSL sprayer.

Tire inflation pressure (kPa)	Wheel load (kN)	Contact area (m^2)	Contact pressure (kPa)
100	2.45	0.0315	78
150		0.0243	101
200		0.0210	117
250		0.0203	121
300		0.0184	133
100	4.90	0.0288	170
150		0.0229	214
200		0.0218	225
250		0.0188	261
300		0.0177	277
100	7.36	0.0422	175
150		0.0417	177
200		0.0377	195
250		0.0358	206
300		0.0349	211
100	9.81	0.0619	159
150		0.0546	180
200		0.0539	182
250		0.0513	191
300		0.0481	204
100	12.26	0.0602	204
150		0.0541	227
200		0.0525	234
250		0.0509	241
300		0.0463	265

The variations of the contact area were plotted, under the influence of the three factors: tire inflation pressure, wheel load and the contact pressure.

Figure 7 shows the variation of the contact area at variable wheel loads and constant tire inflation pressures. The contact area increases approximately linearly, proportional to the increase in wheel load, disregarding tire inflation pressure.

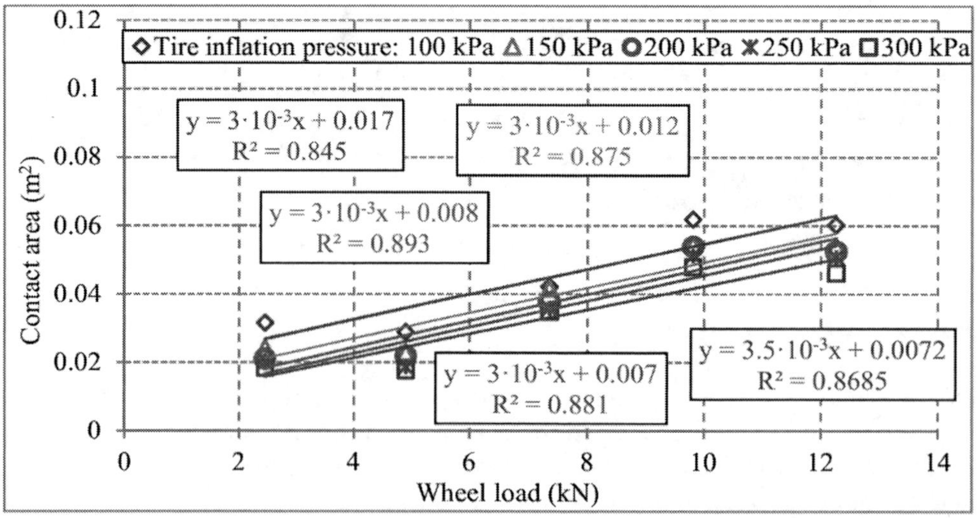

Fig. 7. Changes in MSL sprayer-soil contact area, for variable wheel loads and constant tire inflation pressures

There is a large correlation of experimental data with the linear variation law (over 0.845 in all analysed cases), showing that the contact area is directly proportional to the wheel load increase. At the lowest wheel load (2.45 kN) the contact area is about 0.0184-0.0315 m^2, and these values increase to 0.0436-0.0602 m^2 for 12 kN load and for 100-300 kPa tire inflation pressures. The regression lines are not perfectly parallel, so they do not have the same slope, but this lack of parallelism can be due to the specific testing conditions (the appearance and geometry of the tire, tire stiffness or elasticity, and also by the intersection of the sensitive lines of the sensor with the ribs of tire). If the variation of the contact area is proportionally increasing with the wheel load, it is proportionally decreasing with the tire inflation pressure.

The variation of the contact area in relation to variable tire inflation pressure, for constant wheel load, is presented in Figure 8. If the tire inflation pressure increases, the contact area decreases proportionally with it, but the decrease is insignificant. The values follow exponential and power distribution laws, with high correlation coefficients (R^2 = 0.942-0.977). The regression curves follow a certain parallelism.

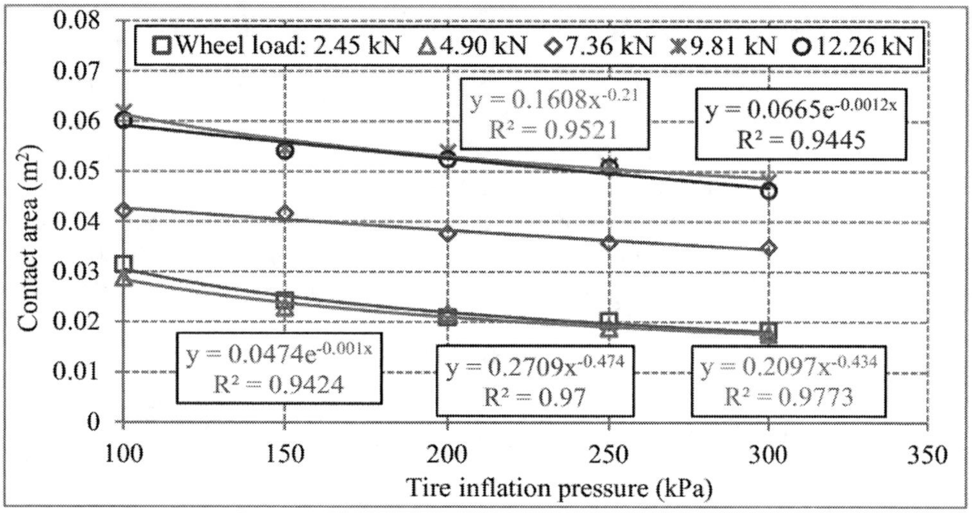

Fig. 8. Changes in MSL sprayer-soil contact area, for variable tire inflation pressures and constant wheel loads

At the variation of contact area with variable tire inflation pressure at constant wheel loads, the range of contact area values is wider. At 100 kPa tire inflation pressure, the contact area increases almost twice, from 0.0315 m^2 at 2.45 kN load to about 0.0602 m^2 for 12.26 kN wheel load.

Figure 9 shows the changes in tire-soil contact area depending on the contact pressure, for five wheel load values. It was obtained a perfect correlation of experimental data with the power variation law (R^2 = 0.999-1). The variation curves are distinctive and in a relatively wide range of values. For 2.45 kN load, the regression line is found in the left bottom corner (due to low values of contact area and contact pressure). The range of contact areas for 12.26 kN load are placed in the upper right corner of the graph with a wide range of contact areas and contact pressures (both of them have higher values). Even though the contact pressure domains overlap partially for the filled tank, it can be noticed that the contact area values are relatively distinct. The contact pressure varies between 78-133 kPa for 2.45 kN wheel load, being related to contact areas of 0.0184-0.0315 m^2, and these domains become wider for higher wheel loads.

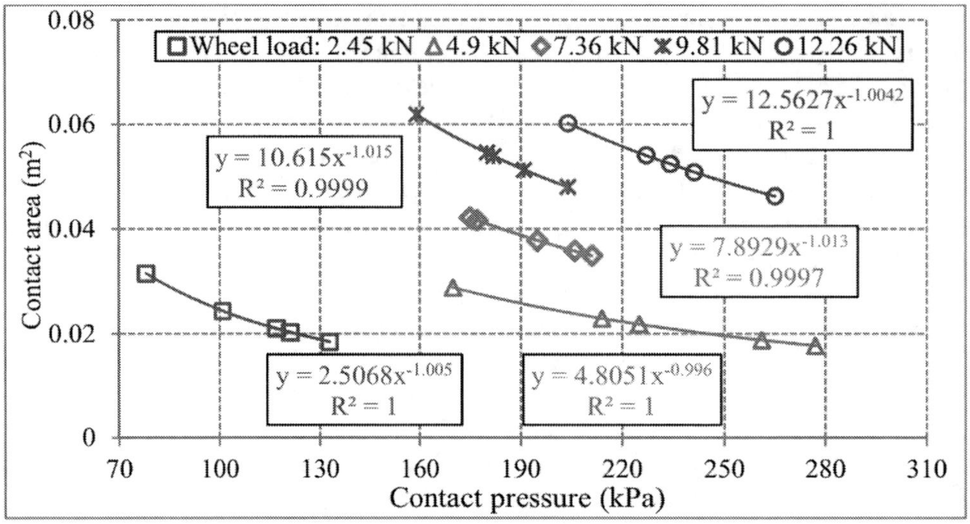

Fig. 9. Contact area variation with MSL tire-soil contact pressure and wheel load

Thus, at 12.26 kN load, for contact pressures of 204-265 kPa, contact areas are between 0.0463-0.0602 m^2, with smaller values at higher contact pressure. All regression curves have a slightly downward slope, the values of this slope being similar from one wheel load to another (the regression curves are approximate symmetrical).

Figure 10 shows the variation of the contact pressure with tire inflation pressure for the tested wheel loads. The regression curves have an increasing slope and are slightly parallel, except for the curve corresponding to the wheel load of 4.9 kN. The best correlation of the experimental data with the force type variation law was obtained for wheel load of 4.9 kN, where the correlation coefficient had a high value ($R^2 = 0.9775$).

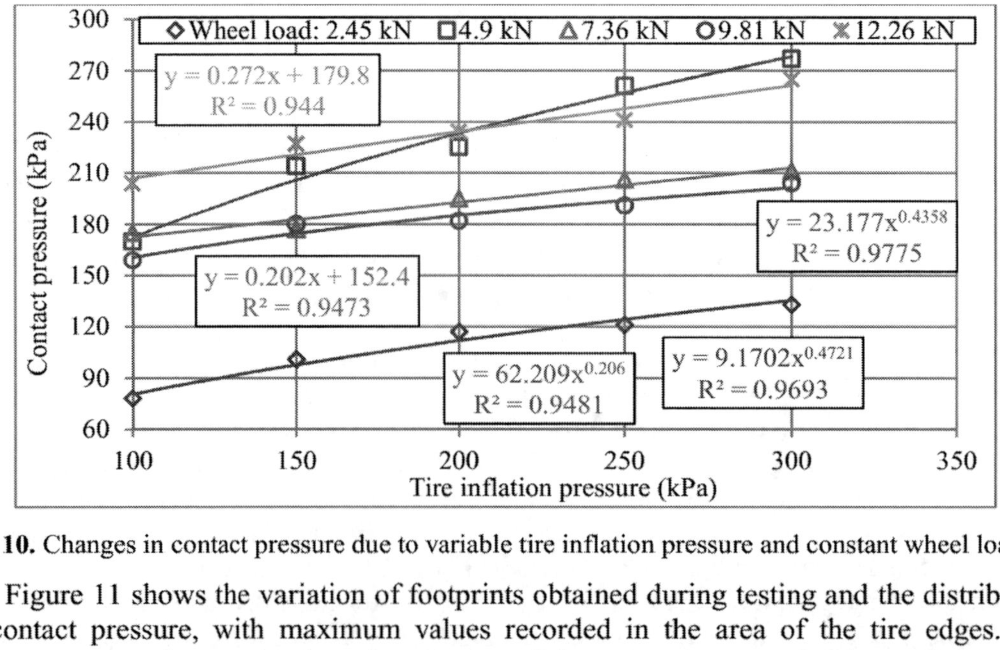

Fig. 10. Changes in contact pressure due to variable tire inflation pressure and constant wheel load

Figure 11 shows the variation of footprints obtained during testing and the distribution of contact pressure, with maximum values recorded in the area of the tire edges. The contact pressure is computed as the average of the pressures recorded by each sensitive element of the Tekscan Industrial Sensing pressure sensor which was in contact with the tire during testing. Studying the footprints, the tendency is obvious: the increase in tire inflation pressure results in decreased contact area. As wheel load is distributed in a smaller surface, an increase in the contact pressure is obtained.

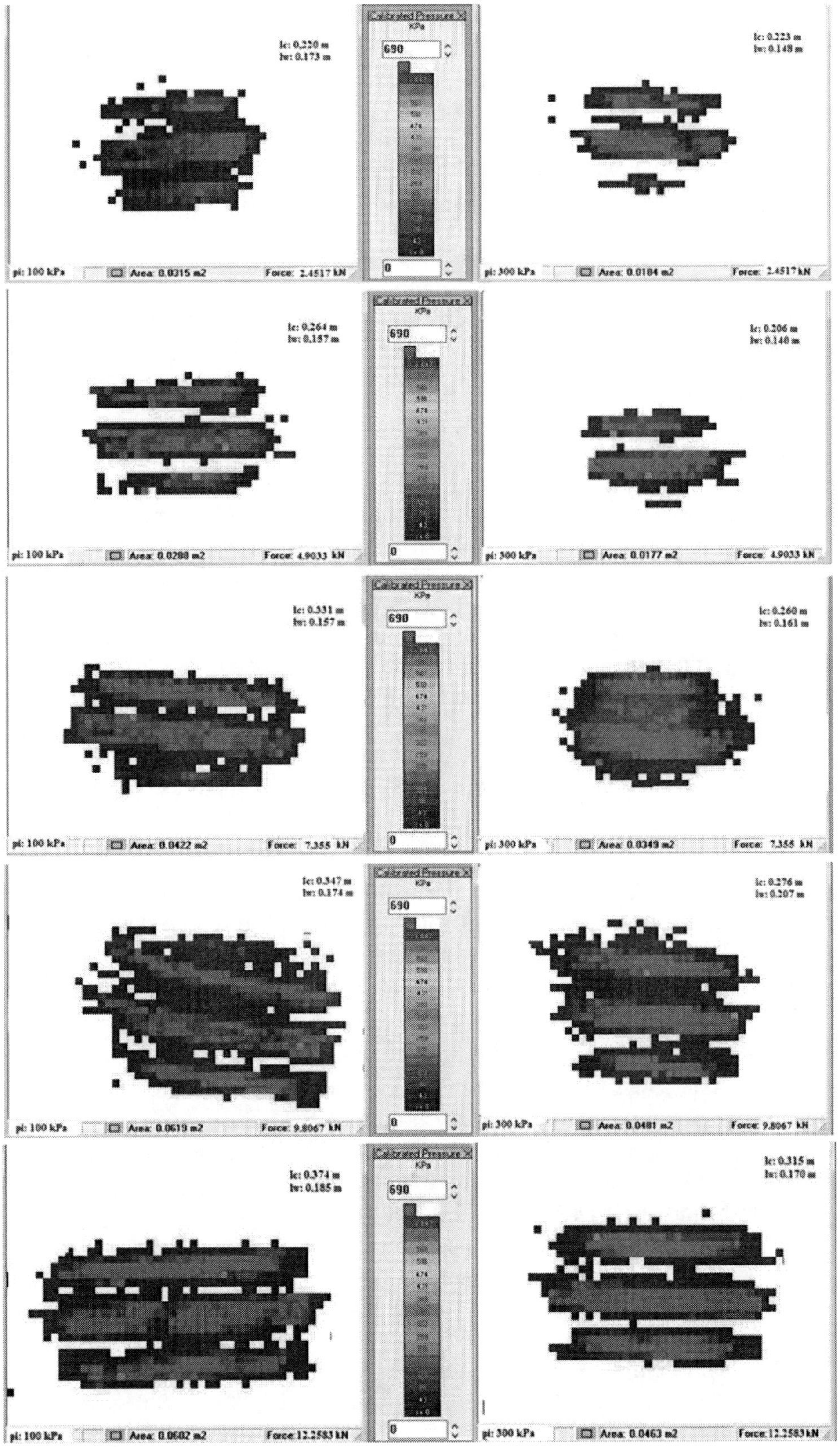

Fig. 11. Changes in footprint shape during testing with increasing wheel loads (tire inflation pressure: left side – 100 kPa, right side – 300 kPa)

Also, as wheel load increases, under constant tire inflation pressures, a significant increase in contact area was achieved. At 100 kPa tire inflation pressures, the contact area increased almost three times, from about 0.184 m^2 for a load of 2.45 kN to about 0.0463 m^2 for wheel load of 12.26 kN. Footprint shape also changed visible during the tests. If at the minimum wheel load (2.45 kN) and tire inflation pressure (100 kPa) the footprint is somewhat elliptically-shaped, at the same tire inflation pressure but at maximum wheel load (12.26 kN), the shape of footprint tends to become rectangular. If wheel load is kept constant, and the tire is inflated with minimum tire pressure, the footprint is rectangular; increasing the tire inflation pressure leads to the decrease of contact area, and also in a change of footprint shape. At low wheel load the shape of the footprint is elliptical, and at high wheel load the footprint becomes rectangular shaped with slightly spherical corners.

4 Conclusions

When using higher tire inflation pressures during soil works, farmers will get smaller contact areas and as a result, soil deformation will increase and stresses will be distributed to greater depths. The ideal situation is to use low tire inflation pressures, because thus the tires will deform more, contact area will be larger, contact pressure will be lower, soil deformation will be lower and the stresses will will be distributed at lower depths.

Unless the case of wheel load of 2.45 kN and the tire inflation pressure of 100 kPa, in all other situations, contact pressures higher than 80 kPa have been obtained, which means that in all these cases the agricultural soil will be compacted. In the analyzed cases, subsoil compaction (0.3–0.4 m depth) occurs only when the spraying machine tank is filled with liquid. With empty tank, the sprayer would only cause topsoil compaction. However, proper management practices, such as choosing the lightest sprayer, maintaining proper tire inflation pressure, monitoring soil moisture, using the same tracks and limiting unnecessary passes over a field can reduce compaction risks when applying phitosanitary treatments.

Acknowledgement

This work was supported by a grant of the Romanian Research and Innovation Ministry, through Programme 1 – Development of the national research-development system, subprogramme 1.2 – Institutional performance – Projects for financing excellence in RDI, contract no. 16PFE.

References

1. European Academies Science Advisory Council (EASAC). Policy report 36. Opportunities for soil sustainability in Europe. ISBN: 978-3-8047-3898-0 (2018)
2. K.Ç. Selvi, M. Sağlam, O. Dengiz, Geostatistical approach to determine the effects of different soil tillage methods on penetration resistance in a clayey soil. INMATEH – Agricultural Engineering **52** (2): 19–26 (2017)
3. M.F. Nawaz, G. Bourrié, F. Trolard, Soil compaction impact and modelling. A Review, Agron. Sustain. Dev. Springerlink (2012)
4. Y. Tekin, B. Kul, R. Okursoy, Sensing and 3D mapping of soil compaction, Sensors **8**: 3447- 3459 (2008)
5. S.Şt. Biriş, V. Vlăduţ, Use of Finite Element Method to determine the influence of land vehicles traffic on artificial soil compaction, Book: Water Stress **1**: 179–198, 1st Edition, Editors: Ismail Md. Mofizur Rahman and Hiroshi Hasegawa. InTech Publishing. Rijeka, Croaţia (2012)
6. M.A. Hamza, W.K. Anderson, Soil compaction in cropping systems. A review of the nature, causes and possible solutions. Soil & Tillage Research **82**: 121–145 (2005)

7. A. Alaoui, E. Diserrens, Mapping soil compaction – a review, Current Opinion in Environmental Science & Health **5**: 60–66 (2018)
8. Annual report on the state of the environment in Romania, 2017. National Agency for Environmental Protection, Bucharest (2018)
9. S.Şt. Biriş, N. Ungureanu, D. Cujbescu, Modelling of soil compaction under heavy-duty tractors. INMATEH – Agricultural Engineering **57** (1): 261–270 (2019)
10. S.Şt. Biriş, N. Ungureanu, E. Maican, G. Paraschiv, Gh. Voicu, M. Manea, FEM model for the study of interaction between the driving wheel and the rolling track for agricultural land vehicles. Proceedings of the 39th International Symposium „Actual Tasks on Agricultural Engineering", 95-106 (2011)
11. T. Batey, Soil compaction and soil management – a review, Soil Use and Management **25**: 335–345 (2009)
12. K. Xia, Finite element modeling of tire/terrain interaction: Application to predicting soil compaction and tire mobility, Journal of Terramechanics **48**: 113–123 (2011)
13. S.Şt. Biriş, L. Savin, N. Ungureanu, E. Maican, V. Vlăduţ, I. Ganea, I.L. Caba, Researches on the use of Finite Element Method to optimize the wheel tyres exploitation for agricultural vehicles. INMATEH - Agricultural Engineering **39** (1): 5-12 (2013)
14. J. Arvidsson, T. Keller. Soil stress as affected by wheel load and tyre inflation pressure. Soil Till. Res. **96**: 284–291 (2007)
15. T. Keller, M. Lamandé, S. Peth, M. Berli, J.Y. Delenne, W. Baumgarten, W. Rabbel, F. Radjaï, J. Rajchenbach, A.P.S. Selvadurai, D. Or, An interdisciplinary approach towards improved understanding of soil deformation during compaction. Soil & Tillage Research **128**: 61–80 (2013)
16. J.J.H. Van den Akker, T. Hoogland, Comparison of risk assessment methods to determine the subsoil compaction risk of agricultural soils in The Netherlands. Soil & Tillage Research **114**: 146–154 (2011)
17. Fountas S., Paraforos D., Cavalaris C., Karamoutis C., Gemtos T., Abu-Khalaf N., Tagarakis A., A five-point penetrometer with GPS for measuring soil compaction variability. Computers and Electronics in Agriculture **96**: 109–116 (2013)
18. N. Lozano, M.M. Rolim, V.S. Oliveira, U.E. Tavares, E.M.R. Pedrosa, Evaluation of soil compaction by modelling field vehicle traffic with SoilFlex during sugarcane harvest. Soil & Tillage Research **129**: 61–68 (2013)
19. G.F. Botta, A. Tolon Becerra, F. Bellora Tourn, Effect of the number of tractor passes on soil rut depth and compaction in two tillage regimes. Soil & Tillage Research **103**: 381–386 (2009)
20. V. Vlăduţ, M. Matache, S. Bungescu, Ş.Şt. Biriş., The determination of soil deformations as a result of the applied pressure by the tractors rolling parts and the auto propelled agricultural machines. Scientific Papers, Faculty of Agriculture, USAMVB Timişoara **40** (2): 505–512 (2008)
21. Y. Tekin, R. Okursoy, Development of a hydraulic-driven soil penetrometer for measuring soil compaction in field conditions. Journal of Applied Sciences **7**: 918–921 (2007)
22. B. Badalikova, Influence of Soil Tillage on Soil Compaction. Chapter in book In: Soil Engineering, Soil Biology, Amity University Uttar Pradesh, Noida, UP, India, **20** (230): 19–30 (2010)
23. D. Wulfsohn, Soil-tire contact area. Advances in Soil Dynamics. ASABE **3**: 59–84 (2009)
24. https://www.discounttire.com/store-locator (accessed 15.04.2019)
25. Contact patch - the risks due to under and over inflation, https://www.tyreleader.co.uk/tyres-advices/tyre-contact-patch (accessed 15.04.2019)

Experimental and numerical simulation research of sedimentation process in stationary column of aqueous suspension of solids

George Ipate[1], *Gabriel* Musuroi[1], *Gabriel-Alexandru* Constantin[1], *Elena Madalina* Ştefan[1], *Bianca* Zabava[1] and *Marina* Pihurov[1]

[1] University POLITEHNICA of Bucharest, Biotechnical Systems Department, 313 Spl. Independentei, Romania

Abstract. Our study proposes a new mathematical model for predicting the gravitational settling velocity of sediment particles, a measure to evaluate water clarity, to provide information to environmental engineers, first responders dealing with environmental emergencies or farmers, for the most rapid and efficient management responses. New and simplified simulation program are developed to the study sedimentation process. The mathematical model proposes a first-order differential equation who is solved by a numerical method algorithm. The proposed formula has the highest degree of predictive accuracy when compared with experimental data.

1 Introduction

The static sedimentation method is very often applied in the mechanical liquid-solid separation stage to domestic and industrial effluent treatment, drinking water treatment. Also, sedimentation of small particles into a fluid is a long-standing issue and with many practical applications [1]. The study of sedimentation of solid particles in aqueous suspension is performed for both stationary and dynamic systems (streams with different flow directions) [2]. Among the parameters that influence particle behavior in suspension in the sedimentation process are density of solid particles, the size and shape of the solid particles, density and viscosity of the fluid. Likewise, the structure of the flock is a parameter that greatly influences the rate of sedimentation of the mineral particles. Sedimentation process and factors that affect it have been reviewed extensively. The influences of the particle size and shape on the particle terminal velocity have been investigated in detail for some regular geometries such as spheres, disks, cylinders, and isometric particles [3].

The behavior of solid particles in sedimentation is strong influenced by the concentration of the suspension and the more or less pronounced tendency of the particles to influence each other during sedimentation. The static sedimentation process can be addressed by considering four sedimentation regimes: a) decanting of the isolated particles; b) decanting of the coagulated particles; c) global sedimentation (decantation); d) sedimentation by sludge compression. The phenomenon that is taking place differs greatly from one regimen to another. Therefore, their separate study is a simplification of the treatment of the static settling process.

Richardson and Zaki [4] investigated the sedimentation of spherical particles in liquid-solid suspensions under the influence of gravity; they determined the reliance of the sedimentation velocity on the void proportion and have developed an expression to predicted the rate of settling of suspensions in form $v_e = v_i * \varepsilon^n$, where v_e is the sedimentation or fluidisation velocity, ε is the porosity, and n, is an exponent. Argument n, shown in Tab. 1, was described as a dependence of the flow regime, expressed by the number Reynolds Re_t, and the ratio of the particle diameter to the sedimentation cylinder d/D.

Table 1. Values of the exponent n recommended for calculating sedimentation velocity.

$Re_t<0.2$	n=4.65+19.5d/D
0.2< Ret <1	n=(4.35+17.5d/D) Ret $^{\pm 0.03}$
1< Ret<200	n=(4.45+18d/D) Ret$^{\pm 0.1}$
200< Ret<500	n=4.45 Ret$^{\pm 0.1}$
Ret> 500	n=2.39

Di Felice and Kehlenbeck in their research [5] demonstrated that the Richardson-Zaki formula is an excellent instrument in describing the profile of sedimenting undiluted suspensions. As the solid grain, and some fluid appended to them, move to the bottom, some fluid must move at the top, rising in this way the drag force and subsequently

reducing the balancing settling velocity [2,4,5]. White and Verdone [6] present a numerical model used for the design of settling reservoir and calculations of limiting flow whit the equations in a dimensionless layout. They extended the example for a cylindrical separator to the case of a conical decanter geometry and demonstrated that this type can handle a increased range of solids charging.
The purpose of this paper is to achieve a solid particle sedimentation behavior simulation program using the Python language to study the influence of changes on the physical parameters involved and the effect these changes make on the corresponding graphs. Program codes are very easy to use, with direct application to solve real engineering problems, and increase students' motivation to recognize, process and apply the information they receive.

2 Material and methods

If a graduated suspension pours a concentration suspension sufficient but not too high, the following phenomena can be observed: Large particles are deposited rapidly and deposited at the bottom of the graduated cylinder (Fig. 1). Just below the surface of the liquid immediately appears a separation surface (solid-liquid interface) which separates the clear liquid or almost clear of a lower area formed by the suspension. This interface moves from top to bottom with constant speed (the straight portion of the sedimentation curve). Then the velocity decreases, the solids are collected (assembled) on the bottom of the tube and the sedimentation rate decreases, the sedimentation curve developing asymptotically parallel to the time axis.

2.1 Experimental setup

In this investigation, settling analysis were carried out in a 100 cm^3 transparent scaled glass cylindrical tank. Size and height of the cylinder were 3.0 and 21.5 cm, respectively. It was loaded with a specific mass of the particle and water. For tests has been used an aqueous suspension of the dish cleaning powder as follows: in 100 mL of tap water was added an amount of 5 g of powder.

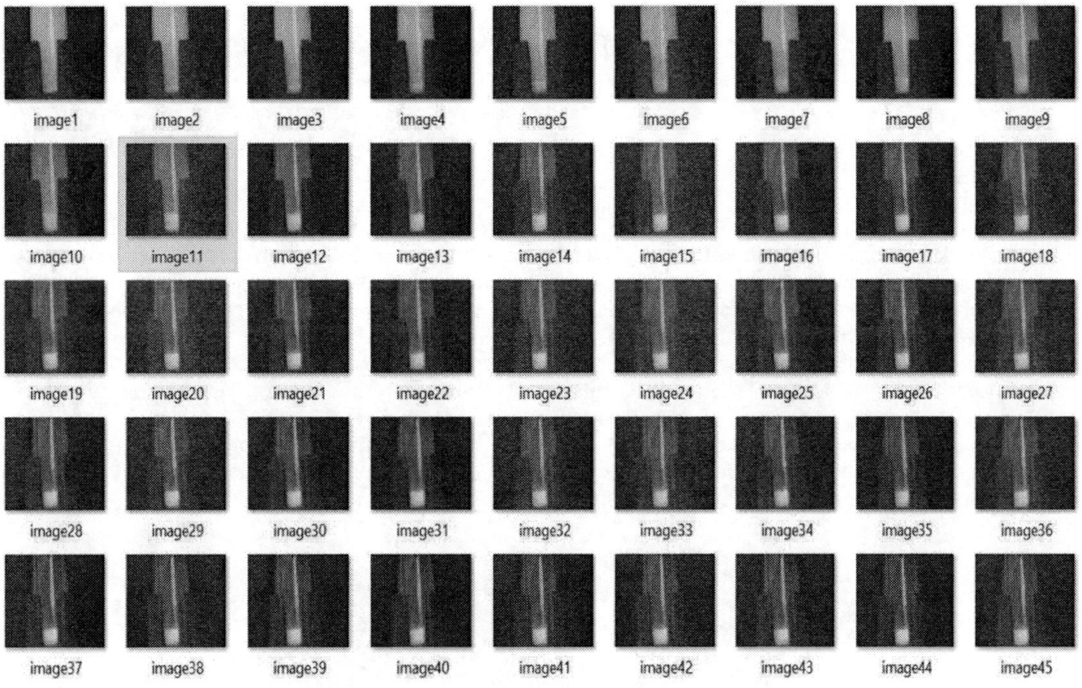

Fig. 1. Image capture of sedimentation process in stationary column.

By measuring the time evolution of the interface height, the sedimentation curve can be drawn. At first it is possible to observe the sedimentation of coarse particles at speed big fall, superior to other smaller particles. After one sometime begins to notice in the upper part of the cylinder an area of clear liquid, and at the bottom of the suspension area and area compression of the sediment at the bottom of the cylinder. It is noted over time the downward trend of the interface line position between the volume of clarified liquid and the volume of suspension still unsettleable. Progressively, clarified interface clear water-suspension descends, while suspension-concentrated sediment fixed interface is lifted and at a particular time these interfaces attain the similar height and overlap. In the present instance, in column maintain only two regions, namely, clarified water and concentrated sludge settled zone [2]. The entire sedimentation process can be easily observed from recorded capture shown in Figure 1.

2.2 Programming language

The study of the dynamic behavior of physical systems is done with mathematical models in the form of ordinary differential equations. The program for study solid-liquid separation by gravitational sedimentation was developed using the Pyhton programming language (vers 3.4). This program has a very user-friendly interface, as users only need to enter only relevant information to calculate complete system solution differential equations. The model describing the settling velocity and interface height is a first-order linear system that are solved in Python with the Scipy.integrate package using function ODEINT. This program can be an extremely effective tool academic community of professors and students, and others.

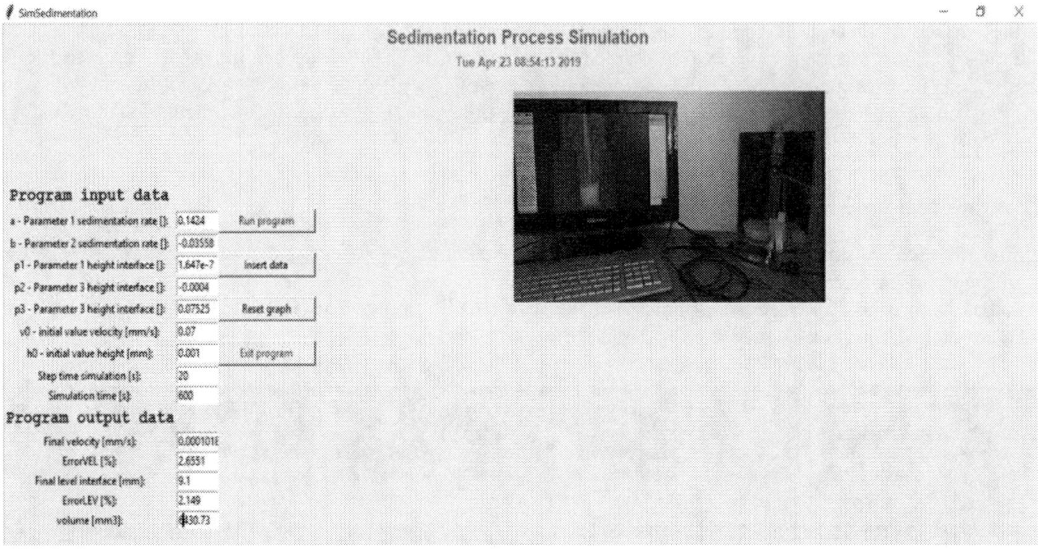

Fig. 2. GUI Python program.

Figure 2 illustrates the friendly user interface program of the settling velocity and height interface simulation model. The Tkinter package that is available to Python programmers provides a robust and independent windowing toolkit to create Graphical User Interface. By modifying the input data in mathematical models, (for initial time t_0=0 to final time t_{sim}) settling rate values and solid-liquid interface height can be predicted. Firstly, users can select initial condition (v_0 and h_0) to solve ODE's by press "Input data" button. The initial values v_0 and h_0 of the dependent variables v(t) and h(t) are specified in the columns B, rows 6 and 7 for the differential equation. Once the "Run program" button is clicked, the simulation run, and the numerical values will be calculated automatically based on the integration time step. Solve button is associated with Python code that is written to compute the solution of the ODE's using Adams/BDF method with an automatic stiffness detection and switching and after that, to graphically plot these values. Also, the average error between numerical and experimental value is calculated and printed in rows 11 and 13 of column B. By pressing the "Reset graph" button, the results obtained by running the simulation program are deleted.

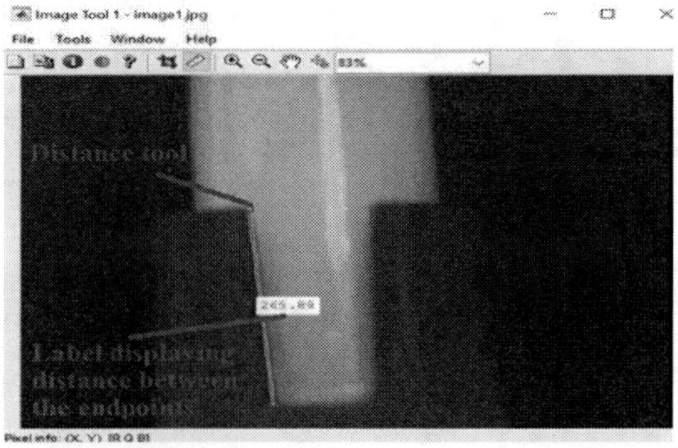

Fig. 3. Distance measurement by image processing method.

The method of digital image processing using the Image Processing Toolbox in MATLAB was used for the experimental data (measurement) of the liquid-solid interface. The distance between two pixels in a digital image is a quantitative amount. The advantage of defining Euclidean Distance between two coordinate points is that it is intuitive, and the main disadvantage is the presence of a cost great computing due to the use of radical but also from the cause of the resulting incomplete value and the necessary interpolation. For the solid-liquid interface level measurement, the "Distance Tool" was used, representing a traceable, resizable, overlapping axle that measures the distance between the two endpoints of the line. The Distance tool displays the measured distance in a superimposed over-line text tag. The tools specify the distance in the data units determined by the XData and YData properties, which are default pixels. Figure 3 shows an example of Measurement of Distance on an Axis. Capturing the images at a time interval of 20 seconds was done with a Raspberry Pi camcorder equipped with an 1/4-inch OmniVision OV5647 5-megapixel sensor. This camera can capture high-quality images as well as unbelievable 720p HD frames at 60 frames per second (fps) or Incorporating a range of advanced technologies and useful features in a rugged and compact body at 1080p HD at 30fps.

2.3 Numerical model

Supposing uniform particle density and size, the solids fall down at the same rate, traveling a distance ***dh*** during the settling time period ***dt***. The grain settling velocity is given by [4]:

$$v\,(t) = dh(t)/dt \tag{1}$$

The cylindrical column tank is occupied by the solid particle volume given by:

$$V(t) = h(t)\cdot A \tag{2}$$

where A is the cross segment zone of the settling column.

The settling velocity of solid particles in the sedimentation process described in the scheme in Figure 1 can be also modelled, according to the law of motion, by the following first-order differential equations:

$$dv/dt = a\cdot b\cdot \exp(b\cdot t) \tag{3}$$

where a, b are parameters that depend on the particle size and shape, suspension concentration and flow regime. Relationship (3) is a derived form of the equation proposed by Richardson-Zaki. The model predicts that the distribution of particle settling velocities evolves toward a steady state.

The solid-liquid height interface can be modelled, by the following first-order differential equations:

$$dh/dt = 3\cdot p_1\cdot pow(t,2)+p_2\cdot t+p_3 \tag{4}$$

where p_1, p_2 and p_3 are parameters determined by regression of experimental data.

3 Result and discussion

The terminal settling velocity of the dish cleaning powder, quantified by the settling column, is 1.14 cm/h (Figure 4). This amount is much littler than the numerical simulation value. According to Zhaowei and Belovich, this little variance is most likely induced by particle inertial effects, and this are hard to estimate. It is excluded to prevent the convective movement of the particles in the water-solid mixture. Particle inertia affect the deposition velocity of both small and macro level. Convection can also be reduced by slowly pouring the suspension into the cylinder [7].

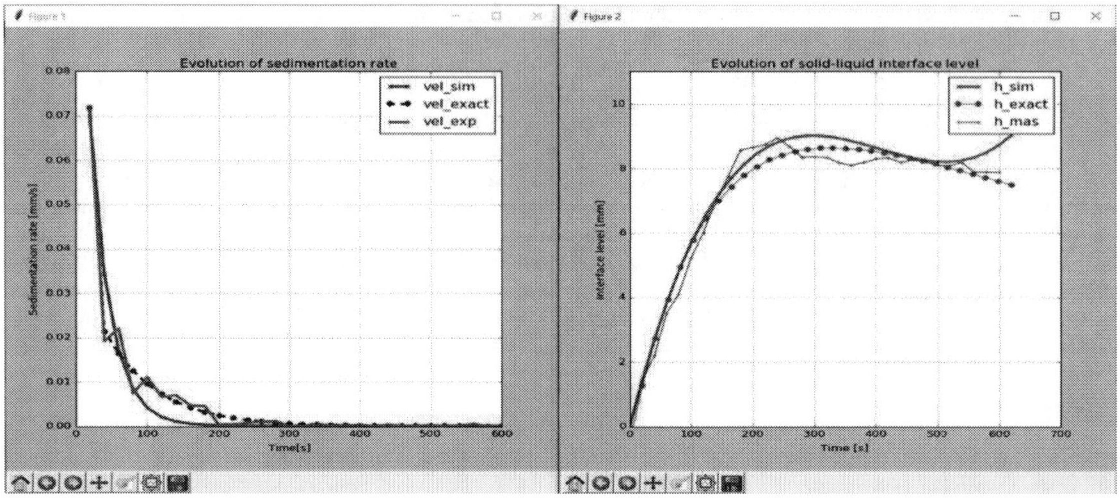

Fig. 4. Batch sedimentation curves for polydisperse suspensions.

The full solution is shown in Figure 4 (left). The magenta continuous line represents the experimentally determined settling curve, while the continuous blue line represents the settling curve curve obtained by numerical simulation. A second chart of the similar data, this time applying the analytical calculated data of settling velocity, is shown in Fig. 4 (left) whit discontinues black pointed line.

The sedimentation curve has two distinct parts, respectively two lines straight lines connected to each other. Extending the two straight segments of the curve, representing the tangents traced to the curve at the appropriate points the time t = 0 and the final time form an angle. The intersection of the bisector the angle of the tangents with the sedimentation curve represents the point defines the critical height and critical time values to be computed sedimentation rate [2]. In short, from the upper side of the cylinder to the floor, solids concentration, gently grows and at the floor it reaches a maximum [8]. Figure 4 (right) shows the curves of the liquid-solid interface according to time; the continuous red line shows the values obtained by the numerical solution, the continuous line with green dots represents the analytical solution of the model, while the blue line is the curve determined experimentally by the image processing method. The fair accordance between the data and the predicted values evidently suggest that Eq. (4) and (5) provides a faithful method to establish suitable values for both settling velocity and height solid-liquid interface.

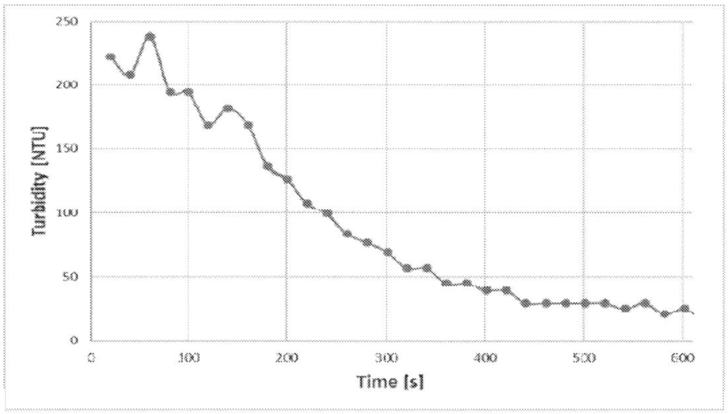

Fig. 5. Time evolution of water turbulence at a distance of 55 cm from the bottom of the cylinder.

The water clarification curve corresponding to the height of 55 cm from the bottom of the cylinder in which the turbidity of water was measured was plotted in Figure 5. Also, the diminished settling velocities allow significant effect for sediment conveyance modelling near to, and among, sheet flow covering and in the swash region [11]. Future research will extend the application of the model to estimate the sedimentation rate of suspensions consisting of small and light particles, or heavier and larger particles. It is also necessary to carry out experiments with various suspensions, more complex, for further testing of different models.

Conclusions

To study the sedimentation process of solid particles in a suspension for this work, a simulation program was developed in the Python environment. This gave the opportunity to obtain indirect information about the points in the decanting curves obtained for different concentrations of solid matter. The settling tank mentioned here supply an affordable, rapid, and truthful method for calculating particles settling velocities and such can differentiate the settling velocities. The numerical solution have been certify using dust particles with recognized physical characteristics and triggered in less than 5% error matched with the experimental values obtained for settling velocity of particles determined when wise settling test with various feed solids concentration in suspensions unless flocculant. Therefore, it can be concluded that the dynamic simulation model can be used to estimate the suspension behavior in a decanter and the simulation results are reliable.

Acknoledgement

The work has been funded by Ministry of National Education and Research through the Executive Agency for Higher Education, Research, Development and Innovation Funding, within the project entitled „Ensure the sustainable development of the base of agro-mechanical practical applications in gardens, solariums, vineyards and orchards (PRAGRODEZ)", CNFIS-FDI-2019-0288.

References

1. E.M. Tory, R.A. Ford, Advances in Fluid Mechanics III, WIT Press Ed., Stochastic simulation of sedimentation, 663-672, (2000).
2. V. V. Safta, M. Dilea, G. A. Constantin, J. Mech. Eng. Aut., Plotting the clarifying curves and determination of the specific amount of settled material during the initial period to sedimentation in stationary column of aqueous suspensions of solids, **3**, 203-208, (2013).
3. M. H. Nasab, JME, A new approach for obtaining settling velocity in a thickener using statistical regression: A case study, **7(1)**, 47-56, (2016).
4. J.F. Richardson, W.N. Zaki, Trans. Inst. Chem. Eng., Sedimentation and fluidisation - Part I, **32**, 35-53, (1954).
5. R. Di Felice, R. Kehlenbeck, Chem. Eng. Technol, Sedimentation Velocity of Solids in Finite Size Vessels, **23 (12)**, 1123-1126, (2000).
6. D.A. White, N. Verdone, Chem. Eng. Sci., Numerical modeling of sedimentation process, **55(12)**, 2213-2222, (2000).
7. M. Zhaowei, J. M. Belovich, Biotechnol. Progr., A simple apparatus for measuring cell settling velocity. **26(5)**, 1361-1366, (2010).
8. G.A. Parsapour, M. Hosseini-Nasab, M. Yahyaei, S. Banisi, Sep. Purif. Technol., Effect of settling test procedure on sizing thickeners, **122,** 87-95, (2014)
9. H. Z Ha, S. Liu, Can. J. Chem. Eng., Settling Velocities of Polydisperse Concentrated Suspensions, **80(5)**, 783 – 790, (2008).
10. J. Martin, N. Rakotomalala, D. Salin, Phys. Fluids, **7 (10)**, Accurate determinaion of the sedimentation flux of concentrated suspensions, 2510-2512, (1995).
11. T.E. Baldock, M.R. Tomkins, P. Nielsen, M.G. Hughes, Coastal Engineering, Settling velocity of sediments at high concentrations, **51**, 91 – 100, (2004).

Determination of extensibility for certain types of pretzels dough

Mariana Gabriela Munteanu[1*], *Gheorghe* Voicu[1], *Elena Mădălina* Ştefan[1] and *Gabriel Alexandru* Constantin[1]

[1]Politehnica University of Bucharest, Faculty of Biotechnical Systems Engineering, Romania

Abstract. Determination of pretzels dough is essential to analyse the technological process, the dough handling behaviour and properties of the finished product. In various processing steps such as kneading, dividing, shaping and baking the dough undergoes different mechanical deformation, including the extent, compression or shear. In the literature there are several methods for analysing these properties. These include empirical rheological methods, such as recording dough evolution during kneading, extensibility measurement or fundamental rheological methods. The study in this paper propose to investigate the behaviour of different types of test dough extensibility and make estimates of the values of the moment (time) and maximum force that wick dough is deformed to appearance of fracture. The maximum elongation of the dough wick is determined for the same values of its initial geometry as well as the other test parameters for two types of dough, one with a composition of 45% water, 10% oil and 45% flour, and another with a composition of 55% water, 10% oil and 35% flour. Preliminary results have found that a dough with more water has a greater tensile strength than the dough with the same amount of oil and less water.

1 Introduction

The dough extensibility is a rheological property which begins to form even during the dough kneading. If during kneading the elastic properties are not formed, then the dough will not have the proper rheological behaviour in fermentation, modelling or baking processes, [1]. In more general terms, the extensibility of the dough is its property of stretching, without breaking, [2,3]. This property does not depend only on the quality and amount of gluten in wheat flour, depends also on the amount of water added in the mixture, [4,5].

The main method for determining the dough extensibility is standardized with AACC Method no. 54-10.01/ ISO 5530-2/ ICC-Standard no. 114/1 and involves stretching a dough wick and drawing extensographic curves with the help of Brabender Extender, [6].

Results on dough stretching help determine its rheological properties, the amounts of ingredients to obtain it, including the possibility of processing and baking, [7]. A very extensible dough is used in the bread making process because it is considered easier to

* Corresponding author: munteanumaya@yahooo.com

process, and the less extensible dough is used for other bakery products such as pretzels, sticks, grissins, [6].
The tensile strength of the dough is recorded in the Brabender Extender system data recorder. The recording system has the role of plotting extensographic curves which allow the determination of certain characteristics:

- resistance to stretching due to maximum resistance, Rm and resistance to constant deformation corresponding to the time of stretching at 50 mm, R50. Each of the average values is expressed by approximately 5 UE;
- the extensibility is due to the displacement of the dough from the time of the curve recording until the moment the sample breaks and is indicated on the extensogram by a smooth fall to zero or by a sudden fall; each of the mean values is expressed by approximately 1 mm;
- energy is defined as the area under the curve recorded and describes the effort required to stretch the dough wick; where: m_a represents the dough mass [mm].

$$W = \frac{Aria}{m_a} 10^3 \tag{1}$$

- the R/E ratio is noted with δ and is the ratio of maximum resistance Rm or resistance to constant deformation, R50 and extensibility, [8]

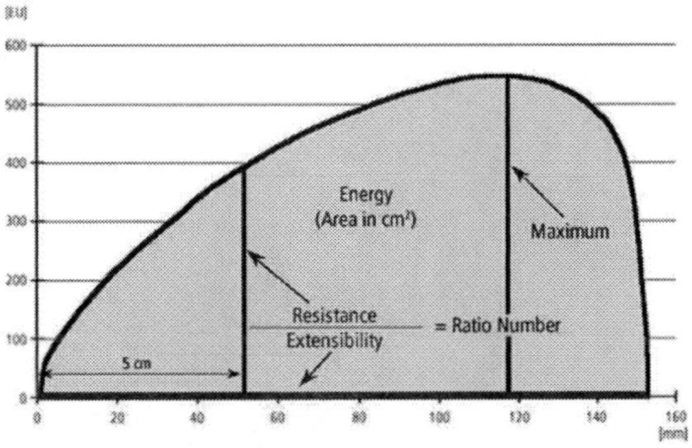

Fig. 1 Extensograph evaluation [9,10]

2 Methodology

To quantify the extensibility of the dough, the Brabender extensograph is generally used or other comparable test devices, [6,9,11]. In all situations, the force used to stretch the dough wicks up to breakage, when it is reached at the maximum dough extensibility level, [12,13]. The dough used to determine extensibility with the Brabender extensograph is obtained with the help of the Brabender farinograph, by the method AACC 54-21 (2000) and must have the characteristics of a normal consistency dough [5,14]. The dough is prepared from 300 g flour, water and 1,2 % salt, and a 150 g sample is divided from the total dough which is shaped like a wick (long form). The wick of dough is placed on racks dough and let rest in thermostat for 45, 90, 135 and 180 minutes. After the thermostat period, the dough wick is subject to stretching until it breaks, [6,8].
Determinations result in an extension curve whose characteristics indicate the rheological properties of the dough (figure 1). The rheological properties of the dough depend on the next steps in the established technological flow, [8].
Two types of dough were prepared for the proposed study:

- type I - dough with composition of 45% water, 10% oil from the total amount of flour;
- type II - dough with composition of 55% water and 10% oil from the total amount of flour.

Besides the amount of water and oil, to obtain each type of dough was also added the flour purchased from the trade (1 kg), kitchen salt (0.012 kg), yeast (0.005 kg), sugar (0.010 kg). Flour had a moisture of 9,56 % and a hydration capacity of 65,7 % which was determined with the help of the Brabender farinograph. For blending the ingredients and kneading the dough was used spiral arm and planetary motion kneader, DitoSama. Mixing the ingredients was done for 2 minutes in the first velocity stage of the kneading arm, and subsequently switched over for 10 minutes in the second stage of the velocity to carry out the kneading process of the dough. After completing the kneading, the dough was divided into pieces whose mass was 100 g.

To perform the extensibility test, pieces of dough were moulded in the form of wicks with a length of 400 mm and a diameter of about 16 mm and placed on the fixed crosspiece of the HOUNSFIELD H1KS Mechanical Testing Machine. The apparatus used during the experiments was equipped with a hook, a force cell of 10 N and a clamping system for the end of wicks on the fixed crosspiece of the device. To record the maximum force that has been applied for stretching to break the dough wick was used a computer on which the Qmat software was installed.

After attaching wick on the fixed crosspiece and hanging the hook on the movable traverse, the wick was lifted vertically at a speed of 50 mm / min. The movement of the hook was made until the dough wick broke into one of the tear areas a, b, c, d, e from figure 2. At the time of breaking the dough wick in the Qmat software acquisition system the maximum force at which the breakage occurred and the hook movement vertically was recorded for each wick sample (figure 3).

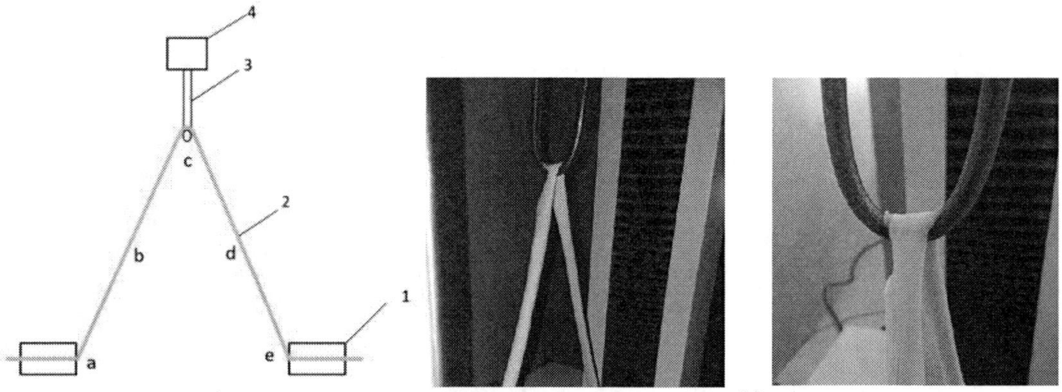

a) b)

Fig. 2. a) Traction test who showing dough extensibility and breakage zones (a, b, c, d, e); 1 attaching the wick holder; 2. wick; 3. hook; 4. the force cell (10 N);. b) an example

After recording data for all samples of the two types of dough, the information has been input into Excel and processed so as to identify the moment of dough breakage and displacement from the start of the lifting stroke to the maximum breaking force.

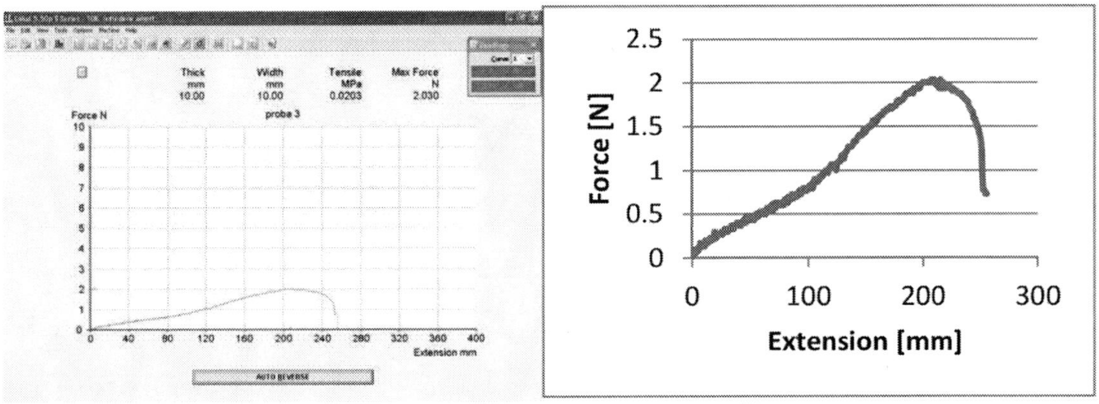

a) b)

Fig. 3 Example of image extracted during experiments
a) Image of the Qmat software; b) Image from the Excel program

To determine the modulus of elasticity (E) the linear regression analysis of the experimental points recorded since the "zero" moment to the point where the stress-strain curve shows a sharp curve before the maximum point. This is clearly visible on the graph of experimental points traced in Excel, because it shows a roughly linear variation up to the maximum force point.
The angle of the horizontal regression line represents the modulus of elasticity (see relationship 2).

$$E = \frac{L \cdot tg\alpha}{A_0} \qquad [N/m2] \qquad (2)$$

where: L- length of the wick after stretching; A_0 – the initial area of the wick section.

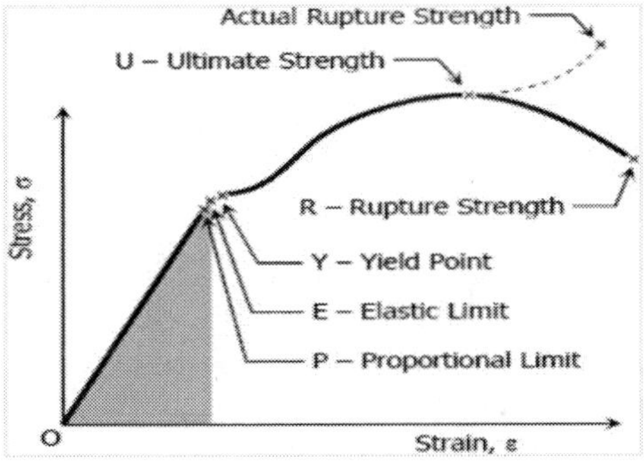

Fig. 4 Diagram "stress-strain" [15]

To determine the length of the dough corresponding to the moment of reaching the maximum force the value of the angle β formed by it at the point "c" in Figure 1 was first found, applying the calculation scheme in Figure 5 and it was taken into account to drop 100 mm of the length of the wick, since the ends of the wicks have been inserted into the holder of the dough wicks.

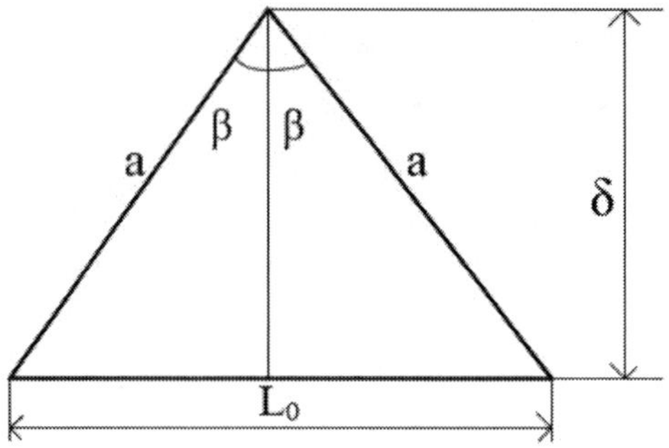

Fig. 5 Calculation scheme for determining the relative length of the dough

$$tg\beta = \frac{L_0}{2\delta} \quad (2)$$

where: L_0 represents the length of the wick after the catch [mm]; δ represents the movement of the wick vertically [mm].
After finding the angle β the value of the wick extension was calculated (L), with relationship (4) and (5).

$$a = \frac{L_0}{2si} \quad [\text{mm}] \quad (4)$$

namely, $L = 2a$ [mm] (5)
Knowing the initial length (L_0) and length at stretching (L) of the wick, could be calculated deformation (ε), and dough resistance (σ) to stretch using relationships (6) and (7).

$$\varepsilon = \frac{2L - L_0}{L_0} \quad [\text{mm}] \quad (6)$$

$$\sigma = \frac{F}{A} = \frac{F \cdot L}{A_0 \cdot L_0} \quad [\text{N/m}^2] \quad (7)$$

3 Results and discussions

Experimental results recorded for maximum strength, respectively the displacement of the hook corresponding to the maximum force is shown in Table 1 and in the graphs of Figures 6 and 7. Graphs have been processed using Excel.

Table 1 The experimental results recorded following determinations

	Dough type I				Dough type II			
	Sample				Sample			
	1	**2**	**3**	**4**	**1**	**2**	**3**	**4**
Extension [mm]	200	205	215.2	234	272.8	190	217	188.9
Maximum force [N]	1.53	1.9	2.03	2.17	2.67	2.67	2.67	2.67

Fig. 6 Variation of force and maximum hook displacement for dough type I

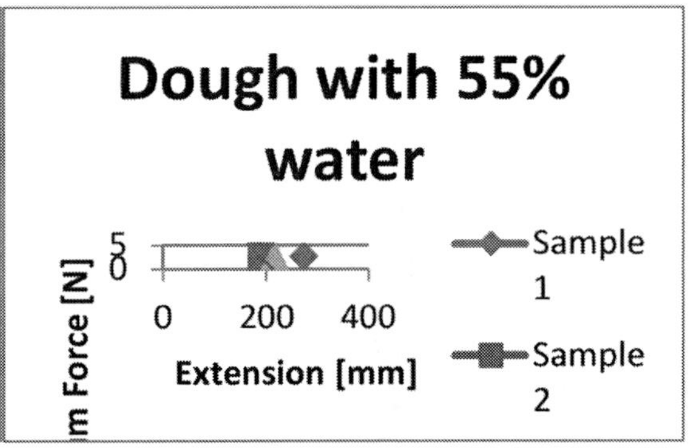

Fig. 7 Variation of force and maximum hook displacement for dough type II

The charts in Figures 6 and 7 shows the variation of the force values and the maximum hook displacement for dough types I and II and from what can be seen from these graphs for type I dough, the maximum force values for the four samples fall within the range 1.53 - 2.17 N, and for type II dough, the strength values remained unchanged for all four dough samples (2.67 N). Even if the force values remained unchanged for type II dough, it can be seen that samples of this type of dough required a higher breaking strength compared to dough type I. In the case of hook displacement values for type I dough, they fall within a much narrower range, i.e. 200 - 234 mm, compared to the hook displacement for type II dough samples, i.e. 188.9 - 272.8 mm. By this analysis it can be pointed out that the samples of type II dough had a much greater extensibility until the moment of flowing and reaching the breaking point compared to the Type I dough samples.
Elastic modulus values (E) determined by linear regression, the length of the wick after stretching (L), the deformation values (ε) and the resistance of the dough to the stretch (σ) are shown in Table 2.

Table 2 Physical characteristics and experimental data obtained for pretzels dough wicks

Dough type	Sample	L0·10-3 [m]	L·10-3 [m]	ε [mm]	E [MPa]	σ [MPa]
Dough with 45% water	1	300	500	0.66	0.069	0.05
	2	300	508.5	0.69	0.91	0.064
	3	300	526.3	0.75	0.094	0.07
	4	300	566	0.88	0.1	0.081
Dough with 55% water	1	300	638.28	1.27	0.072	0.11
	2	300	500	0.66	0.123	0.088
	3	300	535	0.78	0.079	0.094
	4	300	482.33	0.6	0.098	0.085

To highlight the increase in dough extensibility along with increasing the amount of water in the dough the mean of the modulus of the elastic modulus was made and the length of the dough to the stretch and were shown in Figures 8.

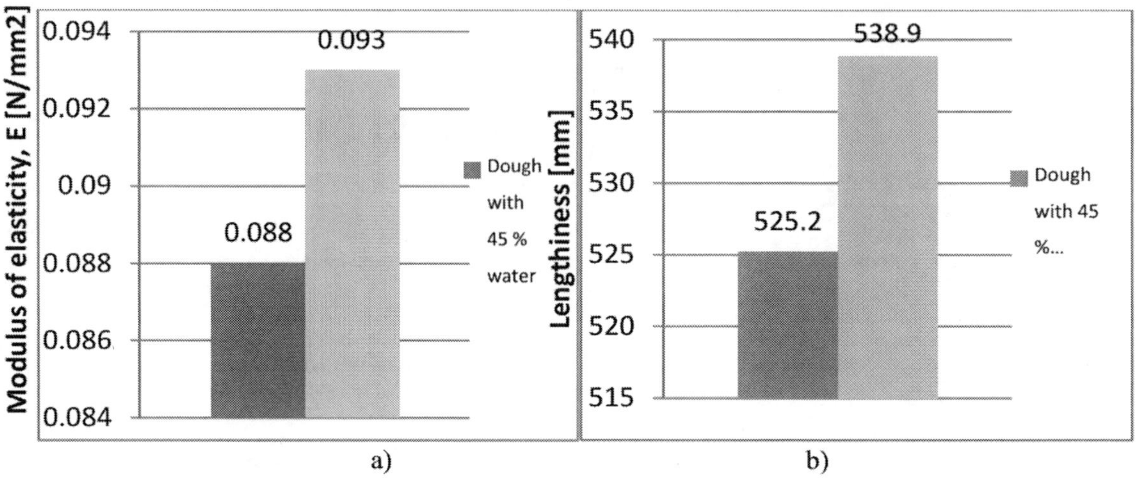

Fig. 8 a) Evidence of the average elastic modulus difference for the two types of dough; b) Evidence of the difference in the average length of the wick dough to the stretch for the two types of dough

4 Conclusions

The experimental data show a deformation of the dough within the limits of 0.66 - 0.88 until it reaches the breaking strength for dough with 45% water and within 0.6 - 1.7 for a dough with 55% water.

Modulus of elasticity of dough with 45% water and 10% oil has much lower values compared to 55% water and 10% oil.

It can be concluded that a greater amount of water in the dough leads to an increase in dough extensibility, respectively, to easier handling of the dough wicks without breaking or thinning during moulding.

A dough with less water will break faster than one with a higher water content.

References

1. E. Chiang, https://bakerpedia.com/test-dough-extensibility-why-you-need-to/, (2017)
2. ***https://www.foodtechcorp.com/tms-extensibility-fixture , accessed in April 2019
3. Gh. Constantin, *Resarch on Rheological Behavior of Wheat Flour Dough for Bakery Applications* – Phd. Thesis, Bucharest, (2011)
4. Wheat Marketing Center, *Wheat and Flour Testing Methods: A Guide to Understanding Wheat and Flour Quality*, Kansas State University, (2008)
5. M.J. Patel, S. Chakrabarti-Bell, J Food Eng, **115**, 371- 383, (2013)
6. S. Chakrabarti-Bell, M.J. Patel, Int J Food Eng, **12**, (2016)
7. S. Chakrabarti-Bell, M.J. Patel a, J. Ng, W.E. Hawkins, J Food Eng, **209**, 36-44, (2017)
8. G.G. Codina, *The rheological properties of the wheat flour dough*, (Ed. AGIR, Series "Food Engineering", Bucharest, (2010)
9. ***https://www.brabender.com/en/food/products/rheometers/inspect-rheological-dough-properties-extensograph-e/, accessed in April 2019
10. ***http://www.wafermatic.com/wheattesting.html, accessed in April 2019

11. B. Dunnewind, E.L. Sliwinski, K. Grolle, T. Van Vliet, J Texture Stud, **34**, 537-560, (2004)

12. Ş. Tietze, M. Jekle, T. Becker, Trends Food Sci Technol, **57**, 1-10, (2016)

13. D. Nash, S. P. Lanning. P. Fox, J. M. Martin, N. K. Blake, E. Souza, R. A Graybosch, Cereal Chem., **83**, 255 - 258 (2006)

14. S. Md. Zaidul Islam, Y. Hiroaki, K. Sun-Ju, M. Chie, T. Shigenobu, H. Naoto, N. Takahiro, Food Sci. Technol. Res., 14 (2), 211 – 216 (2008)

15. ***http://www.forohistarmar.com.ar/foro/showthread.php?tid=670, accessed in April 2019

Heavy metals contamination of the soil – water – vegetables chain in the Ilfov region

Gabriel Mustățea[1*], *Nastasia* Belc[1], *Elena Loredana* Ungureanu[1], *Radu* Lăcătușu[2], *Jana* Petre[3], *Augustina* Pruteanu[4]

[1] IBA - National Research and Development Institute for Food Bioresources, Bucharest
[2] ICPA - National Research and Development Institute for Soil Science, Agrochemistry and Environment
[3] ECOIND - National Research and Development Institute for Industrial Ecology
[4] INMA - National Institute of Research - Development for Machines and Installations designed to Agriculture and Food Industry, Bucharest

Abstract. Dietary exposure to several heavy metals such as Cd, Cr, Pb, As represents a risk to human health through the consumption of vegetables. Due to the possible risks for the human body, contamination of soils as well as drinking and irrigation water has been the subject of many researches.
In this study, the analysis of heavy metals content from soil, water and vegetable samples from households in Ilfov County it was carried out.
As a result, no contamination of soil and water with heavy metals was observed.
Cadmium content is above the maximum allowable limit for all analyzed vegetables, excepting cabbage and green onion. Two lettuce samples have an alarmingly high content of lead, the values being even 10 times higher than the maximum admissible limit of 0.3 mg/kg. The high values obtained for chromium content in lettuce and radish samples raise serious questions about possible contamination. Arsenic content values ranged between 0.87 and 7.69 mg/kg, which also represents high values.
Based on the transfer factor (TF) calculated the strongest accumulation of the metals was in lettuce.

Key words: Heavy metals, contamination, soil, water, vegetables

1 Introduction

Ilfov County is the smallest county in Romania with a total area of 158.328 hectares and is located in the Arges, Ialomita and Mostistea river basins. This county surrounds Bucharest as a ring. On the territory of this county there is a significant industrial and economic activity with a very good evolution. Representative industries of this county are: the food industry, tobacco processing, leather and footwear, paper and cardboard, rubber and plastics processing. The food, beverage and tobacco industries account for 72.5% of the total value of the county's total production.

* Corresponding author: gabi.mustatea@bioresurse.ro

In the last 20 years there has been a socio-economic development of the county in an accelerated pace. This development, besides the positive aspects, also presents the disadvantage caused by the increase of sources of anthropogenic pollution of the environment.

There are many studies regarding the accumulation and origin of heavy metals in soil and the potential ecological hazards associated with this process [1-3]. Researches had attracted attention to the effects of heavy metals on human health. Among many toxic elements, lead, arsenic, and cadmium are considered to be potential carcinogens and are associated with the development of several diseases, especially cardiovascular, kidney, nervous system, blood, and bone diseases [4-8].

Heavy metals enter soil through a variety of sources, including include urban, industrial aerosols created by combustion of fuels, metal smelting, and other industrial activities. Other sources of soil contamination of heavy metals are excessive dose of pesticides, micronutrient fertilizers, and manures such as sewage sludge [9, 10].

Heavy metals can become a health risk via consumption of contaminated vegetables, fruits and drinking water. Vegetables are a necessary component of the human diet, providing a source of essential nutrients, antioxidants, dietary fibers and metabolites.

Absorption and accumulation of heavy metals in vegetables are influenced by many factors, including: concentration of heavy metals in soil, composition and intensity of atmospheric deposition, including precipitations, phase of plant vegetation, irrigation with wastewater, using organic and mineral fertilizers with the load of heavy metals, or application of pesticides [11].

The aim of this paper is to study the contamination with potential carcinogenic heavy metals (As, Pb, Cd and Cr) in soil – water – vegetables chain in the Ilfov region.

2 Materials and methods

This study aim was to investigate the degree of contamination by carcinogenic and toxic substances (heavy metals) of the water used for watering crops, of the soil of agricultural use as well as of the grown vegetables in some households from Ilfov County (Copaceni, Jilava, Berceni, Glina).

2.1 Water samples

The water samples collected were representative samples of the water quality at the time of sampling and the sampling site. Six samples of water from the wells of some households from the above mentioned places were analyzed.

Determination of heavy metals in water was done by induction coupled atomic emission spectroscopy (ICP-OES) according to SR EN ISO 11885:2009 [12]. Water samples were mineralized in acidic medium to eliminate interference and then were analyzed by ICP-OES.

2.2 Soil samples

Regarding soil analysis, 11 soil samples from the upper horizon (0 - 20 cm) of agricultural soils (vegetables) from the places above mentioned above were collected (Fig. 1).

Fig. 1. Places where soil and vegetables samples were collected

Thus, was analyzed the total content of chromium (Cr), cadmium (Cd), lead (Pb) and arsenic (As) in the hydrochloric acid solution obtained after solubilizing the residue resulting from the melting of soil samples with a mixture of perchloric acid ($HClO_4$) and nitric acid (HNO_3) (Milestone Digestor Method; SR ISO 11047: 1999 [13]).

2.3 Vegetable samples

11 samples of vegetables (lettuce – 4 samples, green onion – 2 samples, cabbage – 2 samples, spinach – 1 sample, radish – 2 samples) corresponding to the 11 soil samples listed above were harvested and analyzed (Fig. 2).

Fig. 2. Analyzed vegetables samples

2.4 Heavy metals analysis of vegetable samples

Analysis of the heavy metal content involves two phases: preparation of the samples and proper dosing by atomic absorption spectrometry. Sample preparation was carried out by microwave mineralization according to SR EN 14084: 2003 [14] using a Berghoff MWS-2 microwave oven. The mineralization process was carried out according to the program described below:

Step 1 – 145 ^{0}C – 10 minutes – 80 % power;
Step 2 – 160 ^{0}C – 10 minutes – 80 % power;
Step 1 – 190 ^{0}C – 20 minutes – 80 % power;

The content of lead (Pb), cadmium (Cd), total chromium (Cr) and arsenic (As) was determined using a graphite furnace atomic absorption spectrometer (GF-AAS) AAnalyst 600 - Perkin Elmer, provided with Zeeman background noise correction.

3 Results and discussions

In Table 1 are shown the values of the heavy metals content (Pb, Cd, Cr, As) from the water samples. All water samples were taken from wells in different households, having depths between 7 and 15 m.

Table 1. Heavy metals analysis of water samples

Code	Place	Type	Appearance	Heavy metals content (μg/L)			
				Pb	Cd	Cr	As
W1	Copaceni	Groundwater (well) – 8 m depth	Clear, without suspended solids	< 1	< 0.12	1.2	< 0.2
W2	Copaceni	Groundwater (well) – 7 m depth	Clear, without suspended solids	< 1	< 0.12	< 1	< 0.2
W3	Jilava	Groundwater (well) – 7 m depth	Clear, without suspended solids	< 1	< 0.12	< 1	< 0.2
W4	Berceni	Groundwater (well) – 7 m depth	Clear, without suspended solids	< 1	< 0.12	2.3	< 0.2
W5	Berceni	Groundwater (well) – 8 m depth	Clear, without suspended solids	< 1	< 0.12	2.6	< 0.2
W6	Glina	Groundwater (well) – 15 m depth	Clear, without suspended solids	< 1	< 0.12	< 1	< 0.2

The maximum admissible concentrations (Pb – max 10 μg/L; Cd – max 5 μg/L; Cr – max 50 μg/L; As – max 10 μg/L), according to Law no. 458/2002 [15] on the quality of drinking water, updated version, for none of the 4 metals analyzed were exceeded. All the values obtained were extremely low, showing no contamination.

In Table 2 are shown the values of the heavy metals content (Pb, Cd, Cr, As) from the soil samples.

Table 2. Heavy metals analysis of soil samples

Code	Place	Type	Relief unit	Heavy metals content (mg/kg)			
				Pb	Cd	Cr	As
S1	Copaceni	Aluviosoil	Arges-Sabar Meadow	37	0.532	26	2.46
S2	Copaceni	Aluviosoil	Arges-Sabar Meadow	12	0.423	36	3.62
S3	Copaceni	Aluviosoil	Arges-Sabar Meadow	9	0.426	34	3.39
S4	Copaceni	Aluviosoil	Arges-Sabar Meadow	17	0.383	36	3.84
S5	Jilava	Preluvosoil	Bucharest Plain	25	0.420	27	2.63
S6	Jilava	Aluviosoil	Bucharest Plain	14	0.472	28	3.05
S7	Berceni	Preluvosoil	Bucharest Plain	20	0.318	43	3.80
S8	Berceni	Preluvosoil	Bucharest Plain	22	0.300	32	3.21
S9	Berceni	Preluvosoil	Bucharest Plain	26	0.324	33	3.97
S10	Berceni	Faeoziom	Bucharest Plain	19	0.420	33	3.99
S11	Glina	Cernoziom	Bucharest Plain	18	0.351	30	3.39

Concerning the heavy metal content, normal values were obtained for As, Cd, Cr and Pb. The lowest values, ranging between 0.300 and 0.532 mg/kg, were obtained for cadmium, while the highest values, ranging between 26 and 43 mg/kg were obtained for total chromium. The analyzed soils are not polluted with heavy metals.

11 samples of vegetables: 4 samples of lettuce, 2 samples of green onion, 2 samples of cabbage, 1 sample of spinach and 2 samples of radishes were analyzed (Figure 2). The results obtained for the heavy metals content (Pb, Cd, Cr and As) are presented in Table 3.

Table 3. Heavy metals analysis of vegetable samples

Code	Place	Vegetable	Heavy metals content (mg/kg)					
			Pb	ML_{Pb}	Cd	ML_{Cd}	Cr*	As*
S1	Copaceni	Lettuce	**3.65**	0.3	**0.86**	0.2	**11.17**	3.47
S2	Copaceni	Cabbage	< 0.002	0.3	0.20	0.2	1.11	4.68
S3	Copaceni	Lettuce	< 0.002	0.3	**1.92**	0.2	< 0.0013	3.22
S4	Copaceni	Green onion	< 0.002	0.1	< 0.0002	0.2	< 0.0013	3.59
S5	Jilava	Lettuce	< 0.002	0.3	**1.350**	0.2	**9.75**	7.69
S6	Jilava	Cabbage	< 0.002	0.3	< 0.0002	0.2	< 0.0013	0.87
S7	Berceni	Spinach	< 0.002	0.3	**1.66**	0.2	**2.27**	1.70
S8	Berceni	Radishes	< 0.002	0.1	**0.26**	0.05	**7.70**	1.32
S9	Berceni	Green onion	< 0.002	0.1	< 0.0002	0.2	0.74	2.64
S10	Berceni	Lettuce	**10.05**	0.3	**0.65**	0.2	**60.79**	5.39
S11	Glina	Radishes	< 0.002	0.1	**0.11**	0.05	**18.74**	0.88

ML_{Pb} – Maximum allowed limit for lead (Pb), stated by Regulation (EU) no. 1881/2006;
ML_{Cd} – Maximum allowed limit for cadmium (Cd), stated by Regulation (EU) no. 1881/2006;
* - for chromium (Cr) and arsenic (As) there are no maximum allowed limits;

The lead content in green onion, radishes, cabbage and spinach samples are extremely low, well below the maximum permitted level laid down in Regulation (EU) no. 1881/2006 [16]. Two lettuce samples (S1 and S10) showed extremely high lead content, exceeding the maximum permitted level laid down in Regulation (EU) no. 1881/2006.

Regarding cadmium content, excepting green onion and cabbage samples, all other recorded values exceed the maximum permitted level laid down in Regulation (EU) no. 1881/2006.

Despite the fact that there is no maximum limit for chromium, the extremely high values recorded for lettuce and radish samples raise serious questions about possible contamination.

Arsenic content values ranged between 0.87 mg/kg and 7.69 mg/kg. For this element there are no maximum limits.

The plants ability to take up the metals from soil was measured by the ratio between the concentration of element in plant and in soil, representing the biological absorption coefficient (BAC) or the transfer factor (TF) [17, 18]. TFs for each sample and each element are presented in Table 4.

Table 4. Transfer factors (TFs) for each heavy metal

Type	Transfer factor (TF)			
	TF_{Pb}	TF_{Cd}	TF_{Cr}	TF_{As}
Lettuce	0.1	**1.62**	**0.43**	**1.41**
Lettuce	0	**4.51**	0	**0.95**
Lettuce	0	**3.21**	**0.36**	**2.92**
Lettuce	**0.53**	**1.55**	**1.84**	**1.35**
Cabbage	0	**0.47**	0.03	**1.29**
Cabbage	0	0	0	0.29
Green onion	0	0	0	**0.93**
Green onion	0	0	0.02	**0.66**
Spinach	0	**5.22**	0.05	**0.45**
Radishes	0	**0.87**	0.24	**0.41**
Radishes	0	0.31	**0.62**	0.26

The higher transfer factor of heavy metal indicates the stronger accumulation of the respective metal by that vegetable. A transfer factor of 0.1 indicates that plant is excluding the element from its tissues [19]. The greater the transfer factor value than 0.50, the greater the chances of vegetables for metal contamination by anthropogenic activities will be and so the need for environmental monitoring of the area will be required [20]. Analyzing results it can be observed that the strongest accumulation of the metals is in lettuce.

4 Conclusions

The study shows the possible contamination with heavy metals of water, soil and vegetables samples collected from different places in Ilfov County.

Regarding water samples analyzed, no exceedances of the maximum admissible limits for any of the 4 metals analyzed were recorded.

For the soil samples, normal values were obtained for As, Cd, Cr and Pb, showing no contamination.

Excepting cabbage and green onion samples, cadmium content is above the maximum allowable limit for all other vegetables. Two lettuce samples have an alarmingly high content of lead, the values being even 10 times higher than the maximum admissible limit. Also, the strongest accumulation of the metals is observed in lettuce. Despite the fact that there is no maximum limit for chromium, the extremely high values recorded for lettuce and radish samples raise serious questions about possible contamination. Arsenic content values ranged between 0.87 mg/kg and 7.69 mg/kg. For this element there are no maximum admissible limits.

Acknowledgement

This work was supported by Ministry of Agriculture and Rural Development through project ADER 8.1.4.

References

1. H. Ha, J.R. Olson, L. Bian, et al., Analysis of heavy metal sources in soil using kriging interpolation on principal components. Environ Sci & Technol 48, 4999–5007 (2014).
2. F. Li, J.H. Huang, G.N. Zeng, et al., Spatial risk assessment and sources identification of heavy metals in surface sediments from the Dongting Lake, Middle China. J Geochem Explor 132, 75–83 (2013).
3. J. H. Huang, F. Li, G.M. Zeng, et al., Integrating hierarchical bioavailability and population distribution into potential ecorisk assessment of heavy metals in road dust: A case study in Xiandao District, Changsha city, China. Sci Total Environ 541, 969–976 (2016).
4. A.K. Krishna, K.R. Mohan, Distribution, correlation, ecological and health risk assessment of heavy metal contamination in surface soils around an industrial area, Hyderabad, India. Environ Earth Sci 75 (2016).
5. F. Noli, P. Tsamos, Concentration of heavy metals and trace elements in soils, waters and vegetables and assessment of health risk in the vicinity of a lignite fired power plant. Sci Total Environ 563, 377–385 (2016).
6. S. Izhar, A. Goel, A. Chakraborty, et al., Annual trends in occurrence of submicron particles in ambient air and health risk posed by particle bound metals. Chemosphere 146, 582–590 (2016).
7. S.Z. Cao, X.L. Duan, X.G. Zhao, et al., Health risks of children's cumulative and aggregative exposure to metals and metalloids in a typical urban environment in China. Chemosphere 147, 404–411 (2016).

8. L. Jarup, Hazards of heavy metal contamination. Brit Med Bull 68, 167 (2003).
9. S. Singh, M. Kumar, Heavy Metal Load of Soil, Water and Vegetables in Peri-Urban Delhi. Environ. Monitor. Assess. 120(1–3), 79-91 (2006).
10. B. Arti, S. ShivDhar, K. Amit, Heavy Metal Contamination of Soil, Irrigation Water and Vegetables in Peri-Urban Agricultural Areas and Markets of Delhi, Water Environment Research 87(11), 2027-2034 (2015).
11. R. Lacatusu, A. R. Lacatusu, Vegetable and fruits quality within heavy metals polluted areas in Romania, Carpth. J. of Earth and Environmental Sciences 3(2), 115-129 (2008).
12. SR EN ISO 11885:2009. Water quality - Determination of selected elements by inductively coupled plasma optical emission spectrometry (ICP-OES).
13. SR ISO 11047:1999. Soil quality - Determination of cadmium, chromium, cobalt, copper, lead, manganese, nickel and zinc - Flame and electrothermal atomic absorption spectrometric methods.
14. SR EN 14084:2003. Foodstuffs. Determination of trace elements. Determination of lead, cadmium, zinc, copper and iron by atomic absorption spectrometry (AAS) after microwave digestion.
15. Law no. 458/2002 on the quality of drinking water.
16. Commission Regulation (EC) No 1881/2006 of 19 December 2006 setting maximum levels for certain contaminants in foodstuffs.
17. K. Khan, Y. Lu, H. Khan, M. Ishtiaq, S. Khan, M. Waqas, et al., Heavy metals in agricultural soils and crops and their health risks in Swat District, northern Pakistan.Food Chem. Toxicol. 58, 449–458 (2013).
18. Y.N Jolly, A. Islam, S. Akbar, Transfer of Metals from Soil to Vegetables and Possible Health Risk Assessment 2, 385 (2013).
19. M. Thornton, E. Farago, Geochemistry of Arsenic.In: C. O. Abernathy, R. L. Calderon and W. R. Chappell, Editors, Arsenic, Exposure and Health Effects, Chapman & Hall, London, pp: 27 (1997).
20. D. Sponza, N. Karaoglu, Environmental geochemistry and pollution studies of Aliaga metal industry district. Environ. Int. 27, 541-553 (2002).

 https://doi.org/10.1051/e3sconf/201911203031

Cereal supply chain waste in the context of circular economy

Nastasia Belc[1,*], *Gabriel* Mustatea[1], *Livia* Apostol[1], *Sorin* Iorga[1], Valentin-Nicolae Vlăduţ[2] and *Claudia* Mosoiu[1]

[1] National R&D Institute for Food Bioresources, IBA Bucharest

[2] INMA National Institute of Research - Development for Machines and Installations designed to Agriculture and Food Industry, Bucharest

Abstract. Taking into consideration the circular economy context, from cereal production and processing result by-products and residues which can be transformed into the new raw materials. Straw, husks, brans, flours, bread waste, confectionary waste, so on, can be re-used using different processes such as: extraction, fermentation, microorganism cultivation, for obtaining added value products. These new products obtained can be: biofuels, enzymes, biodegradable material food contact, single cell protein, bio-adsorbent, nanoparticles, bio alcohol, bioactive compounds like fibres, phytochemicals, minerals, so on. This paper is a short review regarding sharing knowledge and good practices in implementing circular economy within food systems, specifically, cereal supply chain.

Key words: cereal, circular economy, waste, loss, bio-technology

1 Introduction

Cereal production and processing are one of the most important sectors of agri-food industry knowing that the cereal food products cover over 20% of daily diet [30]. More, the cereal products are the basis of all Food Pyramid that were developed and proposed in different studies. Cereal products are also vectoring to increase the consumption of dietary fibre in the daily diet. By changing the eating habits, replacing white bread, pasta, biscuits so on, with wholemeal products, an improvement of food cereal-based health benefits may be achieved. In many studies, the role of dietary fibres in nutrition is pointed out, especially in prevention against chronic non-communicable diseases [25].

The increasing demand for food as a consequence of a growing global population is direct connected with an increasing of quantity of food waste. UN Sustainable Development Goals include the reducing of food waste which is one of the most serious environmental, economic and social issues [23].

One of the aims of waste management in agri-food industry is to improve resource efficiency while protecting the environment. For this purpose, management strategies and measures for the proper reduce waste or transform it into new raw materials are needed. These management strategies are included within the circular economy system which is an industrial system that is regenerative or restorative by intention and design (producing

[*] Corresponding author: nastasia.belc@bioresurse.ro

no waste and pollution). A circular economy is characterized by several keys characteristics and enabling factors related to efficiency of using resources, reducing of emissions, minimizing losses and residues, reusing and recycling of products [4].

Different studies for the management of food food supply chain offer opportunities to biofuels, biodegradable plastics, alcohols, antioxidants, In this respect, food waste is a source of very valuable compounds such as: macronutrients - proteins, carbohydrates and lipids, and and micronutrients or bioactive compounds which can be further transformed into new added value products . Different studies for the management of food waste along the supply chain offer opportunities to biofuels, biodegradable plastics, alcohols, enzymes and nanoparticles production, so on, among many others [24].

2 Cereal supply chain waste as a new raw material resource

Huge quantities of waste are generated along the food chain during harvesting, transport, processing, storage, distribution of retailing. Baiano (2014), estimated that about 12.9% of food waste are generated from the cereal processing and manufacturing.

In medium, as well as in high income countries, wheat is the main crop for milling and baking industries [8]. Approx. 35% of the total production account cereal loss and waste in North America, Europe, and industrialized Asia. Along the cereal food chain approx. 10–12% of the total production is lost in North America and Europe, while, in industrialized Asia total loss and waste amount reaches up to 18% [8].

The cereal supply chain waste estimates show that, for example, wheat straw is produced in SUA in amount of 57,000 tonnes/year [34], rice husk in amount of 120,000 tonnes/year at the global level [14] and cereal waste, in general, in amount of 40,000-45,000 tonnes/year in Europe [8].

Each kg of pasta produces approx. 1.98 kg of loss and waste throughout its entire lifecycle is mentioned in a 2019 study of Ludovica Principato et al. This waste is consisting in 1.65 kg by inedible parts, that it means about 83.4% of the total loss and waste. To determine re-usable waste and inevitable losses and to identify the corrective actions whenever required it is needed to analyse entire process of pasta production related to the composition of waste across the stages of food chain [23].

Within the cereal chain some by-products and waste can be recovered, such as the following: sesame husk and rice bran for obtaining dietary fibre [17], oat waste for extraction of antioxidants [27] wheat bran for fructans [36], and Brewers' spent grain for ferulic acid production [16].

3 The new Circular Economy Policy Package

Published in December 2015 by the European Commission, *Closing the loop* — An EU action plan for the circular economy is a strategy with the main aims to support the transformation of linear economy in a circular one.. Earlier, in September, in the same year, 2015, the UN through, the 2030 Sustainable Development Goals, adopted a food waste target per capita at the retail and consumer levels and reducing food losses along the food chain [39]. The new Circular Economy Package consists in several documents: an EU Action Plan for the Circular Economy, a timetable for achieving the actions (in Annex to the Action Plan) and several legislative proposals: Directives on Waste including Packaging Waste or on Landfill.

In this respect, at the level of EU, calculation methods for recycling rates and definitions have been harmonized and simplified. EC is encouraging to have targets for re-use and

recycling for municipal waste by weight, to 60% by 2025 and, 65% by 2030. EC is also encouraging to increase the economic incentives for the products with better design, through provisions on extended producer responsibility schemes. At the same time, for the packaging material waste, specific targets were established for different materials – plastic, wood, glass, paper and cardboard but also general reuse and recycling targets waste to 65% by 2025 and 75% by 2030.
The ensuring the separate collection of bio-waste is needed, including biodegradable gardens waste, food and kitchen waste from caterers, restaurants, retail premises and households where it is economically and environmentally appropriate and practicable from the technically point of view [39].

4 Opportunities to meet Circular Economy

Food waste can be transformed into a raw material to be further used for different purposes.

The bio-refinery concept means to obtain from the agri-food side products, in cascade, commercial added value products and energy. This concept is seen as a sustainable option by the scientific community. The products like bio-fuels, bio-fertilizers, biomass or chemical compounds are obtained by bio-refinery process, in fact, through biotechnological transformation of agri-food waste via anaerobic digestion, fermentation and composting technologies [26].

Table 1 Examples of transforming cereal producing and processing waste into new products [38]

Food waste	Product	Country where research was developed
Bread waste	Sourdough	The Netherland
Oat husks from milling operations	Bio-diesel	Finland
Waste from the confectionery site's manufacturing processes		
Agricultural waste		
By-products from bakery and confectionary factories		
Bakery and confectionary factories	Bio-ethanol	Finland
Agricultural residues	Food and pharma ingredients	Finland
	Soil improvement and organic fertilizers	
	Microbial feed protein	Most of EU countries

For implementing the circular economy both, at the European and national levels it is needed to have an integrated approach based on the triangle: consumer-company-natural environment.

Circular economy means changing people habits, mentalities, especially of the policy decision makers at both public administration and company levels. The most important issue is that the national policy related to Circular Economy to be operational across governmental, entrepreneurial and consumers levels.

Further, some examples of products are given, that can be produced using cereal production and processing waste.

4.1 Bio-fuels production

Many favourable outcomes, coming from the recent literature showed that researches on different cereal waste, such as straw and bran, rich in simple sugars and complex carbohydrates have been demonstrated generation capacity of bio-fuel producing of such substrates. Fuel ethanol was produced from plant biomass since almost one hundred years ago. Transformed through enzymatic hydrolysis of lignocellulose, the plant biomass releases fermentable sugars which are further converted by fermentation process, into bio-ethanol.

The 'biofuel' term comprises a wide variety of products such as biodiesel, biogas, bioethanol, biobutanol, bio ether, biohydrogen,and syngas [35].

Separate studies showed the biohydrogen has been produced using wheat straw or sweet sorghum. In all studies, hydrogen was produced through dark fermentation using Enterobacter, *Bacillus*, and *Clostridium* as the most used microorganisms [29]. Genetic improvements of the fermentative capacity of microorganisms resulted in better biohydrogen yields [14, 3, 12].

4.2 Bio-ethanol production

Bio-ethanol has been derived from the microbial supporting fermentation of agriculturally based carbohydrates waste. Because the similar properties and characteristics of petroleum, including high-octane number, ethanol is used as an alternative liquid fuel. Its blend can achieve the same anti-knock effect as that of petroleum.

Ethanol can also be used in the production of plastics, especially polyethylene. First-generation feedstock, derived from agricultural cereal (wheat, rice, sweet sorghum etc.) has limited applications for ethanol production mainly because of issues related to land use and food security.

By the action of commercial enzymes like amylases (α-, β-, glucoamylase etc,), carbohydrates (starch or lignocellulosic substrates like rice, potato, and sugarcane) are converted to simple sugars through "saccharification process" and further, by fermentation, involving yeast like *S. cerevisiae* and its enzymes – invertase and zymase - to ethanol production. [31, 11]. For bio-ethanol production, *S. cerevisiae* utilises only hexose sugars (Balat, 2011).

Fungi and bacteria have been also used for fermentation.

Table 2. Performance data of ethanol production from cereal processing waste.

Cereal Waste	Yeast	Glucose yield (g/100g FW)	Ethanol yield(g/g FW)	References
Rice husk	*S. cerevisiae*	49	0.98	Saha and Cotta, 2008
Rice straw	*S. cerevisiae*	60.0 g/L	12.34	Sukumaran et al. (2009)
Wheat straw	*S. cerevisiae*	54.96 g/L	25.14	Han et al. (2009)

4.3 Bioactive compounds

One of by-products milling industry is rice bran. Rice bran contains proteins,fibre, minerals,and vitamins, as well as phytochemicals such as polyphenols and tocopherols. The rice bran consumption has been reported to have health benefits by lowering cholesterol with effects on the health of cardiovascular system and antitumor effects. Fivefold increased the antioxidant activity of bread was obtained by addition of 30% rice bran to wheat flour (Irakli et al.). By this addition, the increasing of phenolic content leads to an overall acceptable bread, although the vitamin E content was reduced.

4.4 Nanoparticles

A new research area is to use food processing residues such as, wheat husk and rice bran, to obtain nanomaterials.. As renewable resources, starch, cellulose, xylan or chitosan have been widely used to produce stable nanoparticles [15]. The same cereal based resource, rice husk is a very good raw material to obtain nanoparticles on the basis of silica found in its composition.

In this sense, there are several new methodologies for using rice husk in order to produce nanoparticles.. Silica was extracted from rice husk and was used for anchorage of Pt and Ni nanoparticles, *in situ*. By using a non-ionic surfactant (Span 40) and a cationic surfactant (CTAB), rice husk silica (RHS) texture was tuned. The non-ionic surfactant and rice husk silica (Span 40 RHS) had been immobilised Ni particles on its surface and revealed high dehydrogenation activity and performance in acetaldehyde production [9].

4.5 Biodegradable Plastics

A perfect replacement for petroleum-derived plastics are polyhydroxyalkanoates (PHAs). The main barrier in the commercialisation of PHAs is the high operational cost of its production [22]. For the production of PHAs and poly-3-hydroxybutyrate (PHB), agri-food waste is used as substrate having the advantage of abundance of lignocellulosic materials and very low value.

The agri-food waste coming from wheat, corn, rice, barley or other cereal food chain is rich in carbohydrates which are enzymatic transformed into fermentescible sugars and further fermented by a series of bacteria and fungi such as *Lactobacillus*, *Streptococcus*, *Leuconostoc*, and *Enterococcus* or *Monilina*, *Mucor*, and *Rhizopus* to lactic acid producing [13,5].

Some studies showed that PHB is produced by *Burkholderia sacchari*, through sugars (e.g. glucose, xylose and arabinose) metabolization. It was also tested the efficacy of wheat straw hydrolysate used as a raw material for PHB production. Some experiments demonstrated that it was a cells accumulation of 60% g PHB/g cell dry weight with a yield of 0.19 g/g when *B. sacchari* was grown on wheat straw hydrolysate as the only carbon source [2].

The microenvironment had the greatest influence on PHA production it was the conclusion of Venkata et al., 2013, after an optimisation study on PHA production when it was used mixed aerobic and anaerobic cultures.

Other studies showed the possibility to obtain PHB from rice straw hydrolysate, wheat bran and potato waste or wheat straw hydrolysate using *Bacillus firmus, Halomonas boliviensis and,* respectively, *Burkholderia sacchari* (28, 32, 2).

4.6 Bio-adsorbents development

For producing adsorbent for wastewater by-products and residues from agriculture and forest have been used, having the advantage of price and composition and being considered as a sustainable alternative for the synthesis of quality activated carbons [6, 19, 21]. Among the residues from agriculture and forest it can be mentioned: rice and wheat bran, rice and wheat husk, corn cob, maize, etc. which have been demonstrated their suitability as substrates for use as adsorbents for water treatment.

Table 3. Waste based activated carbon for wastewater treatment. Adsorption capacities. (adapted after A. Nayak, Brij Bhushan [18]).

Adsorbent	Adsorbate	Adsorption capacity mg/g	Optimum conditions, pH, time, °C	References
Rice husk (treated with H_3PO_4)	Cd(II)	102	6	Zafar et al., 2007
Rice husk (sulfuric acid)	Se(IV)	41,15	na	El-Shafey, 2007
	Cd(II)	40,92	na	
	Zn(II)	19,38	na	
	Hg(II)	384,62	-	
Rice husk	Cd(II)	73,96	6.5	Ye et al., 2010
Rice husk (alkali treated)	Cd(II)	125,94	6.5	
Rice husk ash	Pb(II)	39,74	6	Akhtar et al., 2010
	Cd(II)	39,87	6	
	Zn(II	39,17	6	
	Cu(II)	40,82	6	
Wheat bran (chem mod)	Pb(II)	62	5, 20 min.	Farajzadeh and Monji, 2004
	Cr(III)	93	5, 20min.	
	Ni(II)	12	5, 20min.	
Wheat bran	Pb(II)	87	4-7, 60 min, 60°C	Bulut and Baysal, 2006
Wheat straw (chem mod)	Cr(VI)	322,58	55°C	Chen et al., 2010
Rice husk (H_3PO_4 treated)	Cd(II)	102	pH-6	Zafar et al., 2007
Rice husk	Cd(II)	73.96	pH-6.5	Ye et al., 2010
Rice husk (alkali treated)	Cd(II)	125.94	pH-6.5	
Rice husk ash	Pb(II)	39.74	pH-6	Akhtar et al., 2010
	Cd(II)	39.87	pH-6	
	Zn(II	39.17	pH-6	
	Cu(II)	40.82	pH-6	
Rice husk (sulfuric acid)	Se(IV)	41.15	Na	El-Shafey, 2007
	Cd(II)	40.92	na	
	Zn(II)	19.38	na	
	Hg(II)	384.62	-	
Wheat straw (chem mod)	Cr(VI)	322.58	temp-55°C	Chen et al., 2010
Wheat bran	Pb(II)	87	pH=4-7, 60 min, 60°C	Bulut and Baysal, 2006
Wheat bran (chem mod)	Pb(II)	62	pH-5, 20min.	Farajzadeh and Monji, 2004
	Cr(III)	93	pH-5, 20min.	
	Ni(II)	12	pH-5, 20min.	

5 Conclusions

Several ideas for re-using of cereal supply chain waste were presented in this study. More studies are necessary in order to optimize the additional technologies for processing waste. It is important to evaluate the entire process of biorefinery, the main process and additional ones, including all technological flows, in order to be "clean", without any solid or liquid residues or by-products. Also, it is important to obtain the maximum added value products while optimizing other resources consumption such as water and energy.

References

1 Baiano, A., 2014. Recovery of bio-molecules from food wastes — a review. Molecules 19, 14821–14842
2 Cesário, M.T. et al. (2014) Enhanced bioproduction of poly-3- hydroxybutyrate from wheat straw lignocellulosic hydrolysates. N. Biotechnol. 31, 104–113
3 Chong, P.S. et al. (2013) Enhancement of batch biohydrogen production from prehydrolysate of acid treated oil palm empty fruit bunch. Int. J. Hydrog. Energ. 38, 9592–9599
4 Danu C. M., Nedeff V., Circular Economy in Romania within European context, Studies and Scientific Researches. Economics Edition, No 21, 2015, doi: http://dx.doi.org/10.29358/sceco.v0i21.321
5 Gholami, A., Mohkam, M., Rasoul-Amini, S., Ghasemi, Y., 2016. Industrial production of polyhydroxyalkanoates by bacteria: opportunities and challenges. Minerva Biotechnol. 28, 59–74.
6 Gupta, V.K., Nayak, A., Agarwal, S., 2015a. Bioadsorbents for remediation of heavy metals: current status and their future prospects. Environ. Eng. Res. 20, 1–18.
7 Gupta, V.K., Nayak, A., Bhushan, B., Agarwal, S., 2015b. A critical analysis on the efficiency of activated carbons from low cost precursors for heavy metals remediation. Crit. Rev. Environ. Sci. Technol. 45, 613–668.
8 J. Gustavsson, C. Cederberg, U. Sonesson, Global Food Losses and Food Waste Swedish Institute for Food and Biotechnology (SIK), Gothenburg (2011)
9 Hassan, S.A. et al. (2015) Various characteristics of multi-modi-fied rice husk silica-anchored Ni or Pt nanoparticles as swift catalytic systems in some petrochemical processes. J. Taiwan Inst. Chem. Eng., http://dx. doi.org/10.1016/j.jtice.2015.08.001
10 Irakli, M. et al. (2015) Evaluation of quality attributes, nutraceutical components and antioxidant potential of wheat bread substi-tuted with rice bran. J. Cereal Sci. 65, 74–80
11 Kim, S., Dale, B.E., 2004. Global potential bioethanol production from wasted crops and crop residues. Biomass Bioenergy 26, 361–375.
12 Kotay, S.M. and Das, D. (2008) Biohydrogen as a renewable energy resource – prospects and potentials. Int. J. Hydrog. Energ. 33, 258–263
13 Li, X., Chen, Y., Zhao, S., Chen, H., Zheng, X., Luo, J., Liu, Y., 2015. Efficient production of optically pure l-lactic acid from food waste at ambient temperature by regulating key enzyme activity. Water Res. 70, 148–157.
14 Liu, C.H. et al. (2013) Biohydrogen production by a novel inte-gration of dark fermentation and mixotrophic microalgae cultivation. Int. J. Hydrog. Energ. 38, 15807–15814
15 Luo, Y. et al. (2015) Green synthesis of silver nanoparticles in xylan solution via Tollens reaction and their detection for Hg2+. Nanoscale 7, 690–700
16 Mussatto, S.I. et al. (2006) Brewers' spent grain: generation, characteristics and potential applications. J. Cereal Sci. 43, 1–14

17 Nandi, I. and Ghosh, M. (2015) Studies on functional and antioxi-dant property of dietary fibre extracted from defatted sesame husk, rice bran and flaxseed. Bioactive Carbohydrates Dietary Fibre 5, 129–136
18 A. Nayak, Brij Bhushanb, An overview of the recent trends on the waste valorization techniques for food wastes, Journal of Environmental Management 233 (2019) 352–370
19 Nayak, A., Bhushan, B., Rodriguez-Turienzo, L., 2018b. Recovery of polyphenols onto porous carbons developed from exhausted grape pomace: a sustainable approach for the treatment of wine wastewaters. Water Res. 145, 741–756.
20 Nayak, A., Bhushan, B., Rosales, A., Rodriguez Turienzo, L., Cortina, J.L., 2018a. Valorisation potential of Cabernet grape pomace for the recovery of polyphenols: process intensification, optimization and study of kinetics. Food Bioprod. Process. 109, 74–85.
21 Nayak, A., Bhushan, B., Gupta, V., Sharma, P., 2017. Chemically activated carbon from lignocellulosic wastes for heavy metal wastewater remediation: effect of activation conditions. J. Colloid Interface Sci. 493, 228–240.
22 Obruca, S. et al. (2015) Use of lignocellulosic materials for PHA production. Chem. Biochem. Eng. Q 29, 135–144
23 Ludovica Principato, Luca Ruini, Matteo Guidi, Luca Secondi, Adopting the circular economy approach on food loss and waste: The case of Italian pasta production, https://doi.org/10.1016/j.resconrec.2019.01.025
24 Rajeev Ravindran and Amit K. Jaiswal, Review Exploitation of Food Industry Waste for High-Value Products, Trends in Biotechnology, January 2016, Vol. 34, No. 1
25 M. Sajdakowska, J. Gebski, S. Zakowska-Biemans, M. Jezewska-Zychowicz, Willingness to eat bread with health benefits: habits, taste and health in bread choice, p u b l i c health 1 6 7 (2 0 1 9) 7 8e8 7
26 Schieber, A., Stintzing, F.C., Carle, R., 2001. By-products of plant food processing as a source of functional compounds-recent developments. Trends Food Sci. Technol. 12, 401–413.
27 Serea, C.P. and Barna, O. (2011) Phenolic content and antioxi-dant activity in milling fractions of oat. Cancer 7, 8
28 Sindhu, R. et al. (2013) Pentose-rich hydrolysate from acid pre-treated rice straw as a carbon source for the production of poly- 3-hydroxybutyrate. Biochem. Eng. J. 78, 67–72
29 Tan, T. et al. (2010) Current development of biorefinery in China. Biotechnol. Adv. 28, 543–555
30 Mohammed Al-Thani, Al-Anoud Al-Thani, Nasser Al-Mahdi, Hefzi Al-Kareem, Darine Barakat, Walaa Al-Chetachi, Afaf Tawfik, and Hammad Akram, An Overview of Food Patterns and Diet Quality in Qatar: Findings from the National Household Income Expenditure Survey, Cureus 2017 May; 9(5): e1249, doi: 10.7759/cureus.1249
31 Thomsen, A.B., Medina, C., Ahring, B.K., 2003. Biotechnology in ethanol production. In: Larsen, H., Kossman, J., Petersen, L.S. (Eds.), New and Emerging Bioenergy Technologies, vol. 2. Riso National Laboratory, Denmark, pp. 40–44.
32 Van-Thuoc, D. et al. (2008) Utilization of agricultural residues for poly (3-hydroxybutyrate) production by Halomonas boliviensis LC1. J. Appl. Microbiol. 104, 420–428
33 Venkata Mohan, S. and Venkateswar Reddy, M. (2013) Optimi-zation of critical factors to enhance polyhydroxyalkanoates (PHA) synthesis by mixed culture using Taguchi design of experimental methodology. Bioresour. Technol. 128, 409–416

34 Zheng, Y. et al. (2012) Feasibility of filamentous fungi for biofuel production using hydrolysate from dilute sulfuric acid pretreat-ment of wheat straw. Biotechnol. Biofuels 5, 50
35 Yang, X. et al. (2015) Current states and prospects of organic waste utilization for biorefineries. Renew. Sust. Energ. Rev. 49, 335–349
36 Verspreet, J. et al. (2015) Purification of wheat grain fructans from wheat bran. J. Cereal Sci. 65, 57–59
37 Yu, J. and Stahl, H. (2008) Microbial utilization and biopolyester synthesis of bagasse hydrolysates. Bioresour. Technol. 99, 8042–8048
38 Ingredients for a Circular Economy, FoodDrinkEurope, e-News, 2016
39 Circular Economy Package, Dec. 21, 2015, http://researchbriefings.parliament.uk/ResearchBriefing/Summary/

Researches on obtaining vegetable oils as a source of alternative energy

Paul Găgeanu[1], *Iuliana* Găgeanu[*1], *George* Bunduchi[1] and *Alexandru* Zaica[1]

[1]INMA Bucharest, 013811, 6 Ion Ionescu de la Brad, Bucharest, Romania

Abstract. Biodiesel type fuel is widely used on a large scale in most of the European Union. Carbon dioxide is eliminated by combusting biodiesel, but not toxic emissions. Due to the more drastic European ecological demands to reduce the quantity of toxic emissions from automobiles, there is an increased interest to use biodiesel, because it is a bio-degradable and non-polluting fuel. Even by mixing diesel fuel with biodiesel in a percentage of 20%, the emission of toxic gases will be greatly reduced. Vegetable oils used as biofuels are obtained from oleaginous plant crops, without any prejudice to food security. The use of biodiesel can also bring benefits by prolonging the lifetime of diesel engines, ensuring a better lubrication. The paper presents a series of researches on obtaining vegetable oils from various oleaginous seeds (rapeseed, camelina, hemp) using a special installation for producing oils through cold pressing.

1 Introduction

According to European standards, Romania, together with all European countries, must take measures to replace fossil fuels used in transport with biofuels, which would represent 20% by 2020. EU Directive 2003/30 / EC provides the promotion of the use of biofuels and other renewable fuels, particularly in the transport sector, as part of EU policy to reduce energy import dependency and reduce greenhouse gas emissions [1].

Given that fuel consumption in transport is steadily increasing, generating significant amounts of greenhouse gas emissions, widespread use of biofuels is an important solution. However, it faces several obstacles: in Europe, their cost remains much higher than that of fossil fuels, and their impact on the environment remains limited as they are currently used in limited quantities. In addition, producing enough quantities to meet the consumption targets imposed in Europe is a real challenge. Research efforts should therefore be made to use a considerable amount of vegetal fuels [2].

Seed processing has seen significant progress in recent years in terms of endowment with high performance machinery, equipment and plant. This has been virtually dictated by qualitative development through continuous improvement of process technologies.

The knowledge of the physical characteristics of vegetable oils in general and rapeseed and camelina oils in particular is useful to specialists and workers in the field for the

[*] Corresponding author: iulia.gageanu@gmail.com

evaluation of the technological and functional parameters of the machinery that ensures their obtaining and for the establishment of technological parameters of the process of purification [3-6].

The possibility of using them in blending without the need for a dedicated distribution network is an important advantage over other alternative fuels, such as liquefied petroleum gas (LPG) or natural gas for vehicles (NGV), obtained through a distribution network specific and adaptation of vehicles.

Vegetable oils and fats are found in nature in plant tissue, being concentrated in seeds, pulp, or fruit kernels, tubers, or germs. For our country, the main raw material for obtaining vegetable oils is represented by oleaginous plants (sunflower, rapeseed, camelina).

Attempts to use vegetable oils as fuels are older than internal combustion engines. "Biodiesel" can cut CO_2 emissions by 70% compared to petroleum-based fuel, thus reducing greenhouse gases, which contribute to a great extent to global warming [11-14].

Biofuels compatible with Diesel engines are obtained from vegetable oils, especially rapeseed oil, sunflower, soybean, hemp, camelina etc. Flax or castor oil can also be used, but costs are higher.

The paper addresses the importance, the production and use of vegetable oils as biofuels. Romania has a sufficient land fund to cultivate farmland with oilseeds specific for the production of biofuels, especially rapeseed, hemp and newer cameline (for biokerosene).

1.1 Material and method

In the process of obtaining vegetable oils, oilseeds are subjected to technological treatments that provide them with the optimal quality to obtain oil with maximum yields and minimum expenses.

Vegetable oils can be obtained by cold or hot pressing the oilseeds and extracting the liquid component from them. The seed pressing process is carried out in 2 phases: in the first stage, the seeds are dried and crushed, and in the second phase the seeds are pressed cold or passed through a heating phase and then pressed.

The technological scheme for obtaining vegetable oil is shown in Figure 1.

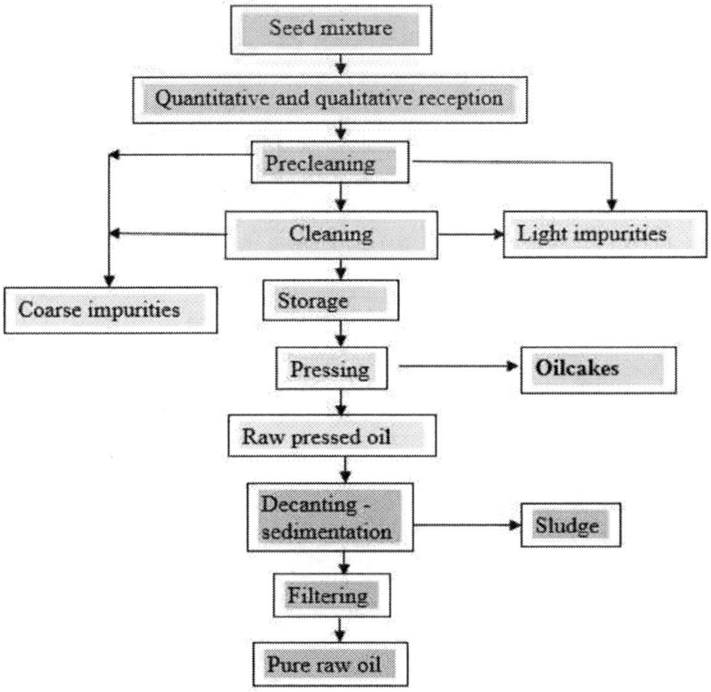

Fig. 1. Technological diagram for obtaining the oil [8]

Separation of oil from seed of oleaginous plants takes place under the influence of external forces, resulting in raw press and broken oil (cakes, pellets) [15-17].

Increasing the pressure on the grinding should be gradual because otherwise the fine particles of milling stop the capillaries and block the oil drain. Pressing can be considered as capillary filtration, a phenomenon that is expressed using the relation:

$$V = \frac{\pi P d \tau}{128 \eta l} [m^3] \qquad (1)$$

where: V - volume of separated oil, [m³];
P – applied pressure [daN/cm²];
l - the length of the capillaries [m];
τ - duration of pressure application [s];
d – the distance between the oil outlet plates [m];
η - yield [%].

The oil separation process is positively influenced by increasing the P, d, and τ values and decreasing η and l.

The calculation of the power required to drive the auger is done with the relation:

$$P = \frac{M_t * n}{95.500} \qquad (2)$$

where: M_t - is the twisting moment applied to the spindle shaft [daNm];
n - is the revolution speed [rot/min];
P - is power required to drive [kW].

The moment of twisting is evaluated considering that the resultant F of the pressure forces exerted by the oleaginous material on the spirals of the auger is actuated on the spindle, which is calculated with the relation:

$$F = \frac{\pi}{4}(D^2 - d^2)P \qquad (3)$$

where: P - is the extraction pressure that takes values in the range 25-28 MPa or 40-200 MPa, depending on the press type - with one or two sections;
D - is the external diameter of the spool of the auger in the pressing zone;
d - is the inside diameter of the pressing area.

The force F is normal to the angle of the auger. β_m .

The force required to push the material along the spindle of the auger is marked with H. As the material is moved by the force H, its pressing occurs, developing the axial resultant pressure forces so that in the case of lack of friction.

$$H = H_0 = F \tan \alpha_m \qquad (4)$$

The time to perform the pressing process according to the constructive and functional characteristics of the press is between 40 and 200 seconds. The pressing process is influenced by the rotation frequency of the auger, the physico-chemical properties of the seed from which the oil is extracted and the way the product is exhausted (cakes or pellets) [9,10].

For cold pressing, the rotation of the auger is between 8 and 36 revolutions per minute.

Taking into account the above mentioned, technological fluxes were designed for 12 lines of oil extraction from seed oil plants consisting of 1, 2 or 3 presses whose pressing capacity is 130-150 kg / h of pressed seeds, depending on the species.

These can be successfully used in a small or medium-sized farm. Depending on where you want the installation to be installed, it can be used in horizontal or vertical flow. A vertical flow installation is more expensive to buy, but exploitation is cheaper with the advantage of gravity.

Starting from the operation diagram with 3 horizontal presses in the technological flow, a seed processing plant with a capacity of 450 kg seed / h consisting of 3 (three) modules (Figure 2) was used:

Fig. 2. Installation for extracting the oil

Fig. 3. Seed preparation module

Fig. 4. Oil extraction module

Fig. 5. Oil purification module

2 Results

The results obtained after pressing the camelina seeds are presented in table 1. The values shown in the table represent the average from three tests.

Table 1. Results for the extraction of oil from camelina seeds.

No.	Parameter	MU	Press I*	Press II*	Press III*	Total
			Average	Average	Average	
1	Sample weight	kg	5	5	5	15
2	Processing time	s	135.67	136.33	134	136.33
3	Productivity	kg/h	132.68	132.04	134.32	133.05
4	Oil quantity	kg	1.71	1.71	1.70	1.706
5	Extraction degree	%	34.13	34.13	34.07	34.11
6	Current intensity	A	13.17	13.17	13.13	13.16
7	Current tension	V	380	380	380	380
8	Cos φ	-	0.78	0.78	0.78	0.78
9	Power	kW	6.76	6.76	6.75	6.76
10	Specific energy consumption	kWh/t	51.07	51.19	59.25	50.80

The products obtained after processing Camelina seeds are shown in Fig. 6 (main product) and 7 (by-product).

https://doi.org/10.1051/e3sconf/201911203032

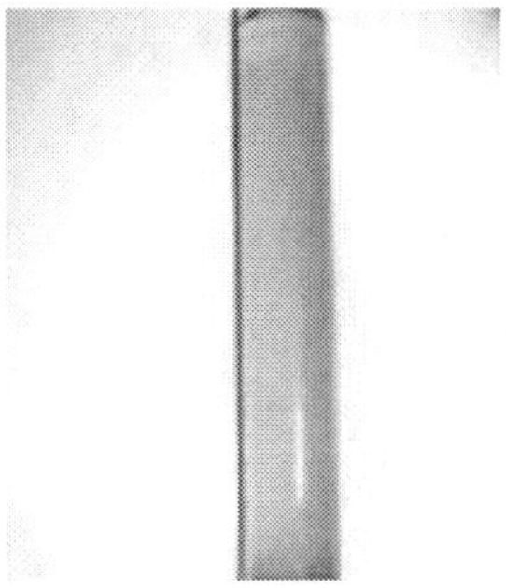

Fig. 6 Decanted camelina oil

Fig. 7. Camelina pellets

The unfiltered and unheated oil obtained through the process was subjected to a test for determining the dynamic viscosity. The results are presented in table 2. The same test was conducted for the heated oil, the results being shown in figure 8.

Table 2. Dynamic viscosity of the unfiltered and unheated oil

No.	Viscosity η [mPa.s]	Temperature t [°C]
1.	40.9	32.5
2.	48.2	28.3
3.	50.3	27.3
4.	53.0	26.0
5.	54.7	25.2
6.	56.2	24.6
7.	57.9	23.9
8.	59.4	23.2

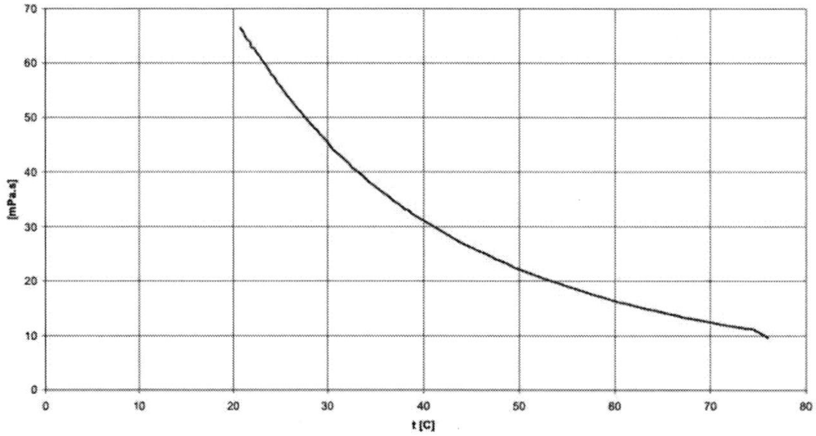

Fig. 8. Dynamic viscosity variation with the temperature for the heated oil

The dynamic viscosity has a tendency to decrease when the temperature is higher, both for the unheated and the heated unfiltered oil obtained.

Another important parameter verified for the unfiltered oil was density (fig. 9).

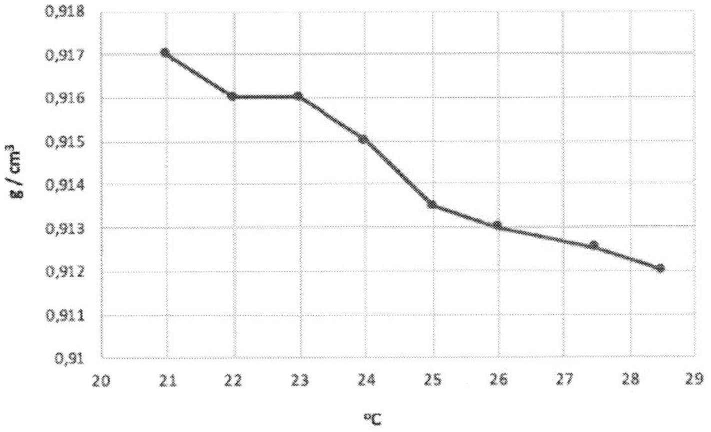

Fig. 9. Density variation with the temperature for the unfiltered and unheated oil

Density also had the tendency to slowly decrease (from 0.917 to 911 g / cm^3) when the temperature increased from 21 to 28.5 °C.

3 Conclusions

The production and use of biodiesel from vegetable oils is done within a complex system, integrated with strong interconnections between all levels of the system.

Within this system, besides the main levels, the secondary ones, as well as the use of the by-products (pellets, cakes), which also have a significant share in establishing the economic efficiency, stand out. It is also noted that oleaginous plants are also melliferous plants.

The quantity of oil obtained depends largely on the nature and quality of the pressed oleaginous material. The dynamic viscosity and the density of the oils obtained decreased when increasing the temperature, a fact that translates in a need to keep a low temperature during the process.

This work was supported by a grant of the Romanian Research and Innovation Ministry, through Programme 1 – Development of the national research-development system, subprogramme 1.2 – Institutional performance – Projects for financing excellence in RDI, contract no. 16PFE.

References

1. Singh J., Bargale P. C. - Development of a small capacity double stage compression screw press for oil expression, în Journal of Food Engineering, no.43, 2000, p.75-82.
2. Ciubota-Rosie C., Ruiz J.R., Ramos M.J. - Biodiesel from Camelina sativa: A comprehensive characterisation, Fuel 105 (2013) 572–577;
3. Fröhlich A., Rice B. - Evaluation of Camelina sativa oil as a feedstock for biodiesel production, Industrial Crops and Products 21 (2005) 25–31.
4. Copeland R., Belcher W.M. (AG Processing Inc). Improved method for refining vegetable oil. PCT Patent Application WO 00/31219 (2000).
5. Kartika A., Pontalier P. Y., Rigal L., „Extraction of sunflower oil by twin screw extruder: Screw configuration and operating condition effects", Bioresource Technology, no.97, 2006, pp.2302-2310.
6. Soriano N., Narani A. - Evaluation of Biodiesel Derived from Camelina sativa Oil, J Am Oil Chem Soc (2012) 89:917–923.

7. Constantin I. and others, „Elaboration of execution documentation for vegetable oil extraction installation (functional model)", Research Report, UP Bucharest, Mechanical Department, 2008, p.11-14;
8. Gageanu P., Paun A., Vladut V., Danciu A. - Vegetable oils - as pure power source of environment protection of gas emissions resulting following their utilization in agricultural farms, , bulletin of university of agricultural sciences and veterinary medicine Cluj-Napoca, nr. 66 (1 – 2) / 2009, – pp. 332;
9. Gageanu P., Vladut V., Paun A., Chih I., Biriş S. Pure plant a source of alternative energy, Actual tasks on Agricultural Engineering, Opatija – Croatia, 22- 25 februarie 2011, pp. 141 – 152;
10. Gageanu P., Vladut V., Gageanu G., „Vegetable oils, present and future biofuel", Terra Nostra Publisher, Iasi, 2012, pp. 83-104;
11. Biriş S., Manea M., Paraschiv G., Vlăduţ V., Bungescu S. - *Power calculus elements for oil presses,* Proceedings of the 37 International Symposium On Agricultural Engineering *"Actual Tasks on Agricultural Engineering",* pp. 475÷484, 2009, ISSN 1333-2651, Opatija - Croatia;
12. Ionescu M., Voicu Gh., Biriş S., Ungureanu N., Vladut V., Voicea I., Persu C. - *About the oil expression process using screw presses withstrainers oil outlet,* Proceedings Of The 43 International Symposium On Agricultural Engineering *"Actual Tasks on Agricultural Engineering",* pp. 513÷524, 2015, ISSN 1333-2651, Opatija – Croatia;
13. Ionescu M., Voicu Gh., Biris S.Şt., Vladut V., Ungureanu N., Stefan E.M., Dinca M., Matache M. - *Oil extraction using a vertical press with piston,* Proceedings Of The 44 International Symposium on Agricultural Engineering *"Actual Tasks on Agricultural Engineering",* pp. 249÷258, 2016, ISSN 1333-2651, Opatija – Croaţia;
14. Ionescu M., Vladut V., Ungureanu N., Dinca M., Zabava B.St., Stefan M. - *Methods for oil obtaining from oleaginous materials,* Scientific International Conferences - *The 12th Annual Meeting "Durable Agriculture – Agriculture of the Future"* Craiova, 17-18 November 2016, Annals Of The University Of Craiova - Agriculture, Montanology, Cadastre, Series, Vol. 46, No. 2 / 2016, Working Group No. 4 - Management, Agriculture Mechanization And Cadastre, pp. 411-417, ISSN 1841-8317, ISSN CD-ROM 2066-950X;
15. Ganea-Christu I., Bunduchi G., Gageanu G., Marin E., Gageanu I., Matei Gh., Vladut V. - *Considerations on obtaining camelina oil by cold pressing,* Scientific Symposium with International Participation *„Sustainable Development in Agriculture and Horticulture- Second edition"* Craiova, 12-13 November 2015, Annals of the University of Craiova - Agriculture, Montanology, Cadastre, Series, Vol. 45, No. 2 / 2015, ISSN 1841-8317, ISSN CD-ROM 2066-950X, pp. 67-70;
16. Vladut V., Pirna I., Postelnicu E., Manea D., Bungescu S. - *The methodic concerning determining of oil content from vegetables products,* Annals of Craiova University - Agriculture, Montanology, Cadastre, vol. XXXVII/B 2007, pp. 373-382, ISSN 1841-8317, Craiova – Romania;
17. Vladut V., Gangu V., Bajenaru S. Biris S., Paraschiv G. - *Researches regarding the determination of oil content from vegetables products,* Annals of Craiova University - Agriculture, Montanology, Cadastre, vol. XXXVII/B 2007, pp. 363-372, ISSN 1841-8317, Craiova – Romania;
18. Biris S., Covaliu C., Paraschiv G., Vladut V., Atanasov At. - *Research on valuable recovery of oil from grape seeds,* 3nd International Conference on Thermal Equipment, Renewable Energy and Rural Development, TE-RE-RD 2014, pp.

187-192, Mamaia, Constanţa – Romania, June 12 - 14, 2014, ISSN 2359-7941, ISSN-L 2359-7941.

Researches on the freezing of blueberries and green bean pods using an experimental model of a fast freezing equipment

Cristian Sorica[1,*], *Marian* Vintilă[2], *Ion* Pirnă[3], *Ion* Grigore[1], *Mario* Cristea[1], *Laurenţiu* Vladuţoiu[1], *Elena* Sorica[1], *Irina* Neacşu Bălănescu[4,5], *Cristina* Grosu[6], *Marius* Raischi[7] and *Marga* Grădilă[4]

[1]INMA Bucharest, Romania

[2]ICDIMPH-Horting Bucharest, Romania

[3]ASAS, Romania

[4]ICDPP Bucharest, Romania

[5]USAMV Bucharest, Romania

[6]INCDTP Bucharest, Romania

[7]INCDPM Bucharest, Romania

Abstract. In order to conserve perishable food and extend the permissible storage and marketing period, one of the best methods that uses artificial frost is freezing. The action of low temperatures leads to a slowing down of the modifying processes, both nutritionally and organoleptically. The researches presented within the paper aims to determine some specific parameters for the freezing process using an experimental model of a fast freezing equipment with liquid nitrogen. Experimental researches use two species of horticultural products, namely blueberries and green bean pods. The results indicate that the values of total freezing time, average linear freezing rate and liquid nitrogen consumption for quick freezing of blueberries were higher than those recorded for green bean pods. Also, with the lowering of the set temperature inside the freezing chamber, the freezing time from 0 °C to the temperature of -15 °C decreases, resulting in lower freezing times, higher average linear freezing rates and liquid nitrogen consumptions with increasing tendency.

1 Introduction

In order to extend the permissible period of storage and marketing of perishable foodstuffs, a number of conservation methods and techniques can be applied [1-6]: preservation by using low temperatures (refrigeration, freezing, modified atmosphere); preservation by using high temperatures (pasteurization, sterilization); conservation by reducing the water content of the product (dehydration, concentration); preservation by the use of inhibitors of

*Cristian Sorica: cri_sor2002@yahoo.com

depreciation phenomena (natural antimicrobials, bacteriocins, sugar addition, acidification, chemical preservatives etc.); preservation by unconventional methods (ionizing radiation, high pressure, high voltage electrical impulses, non-ionizing radiation, ozone, ultrasound, photodynamics, etc.). Of all these processes, the use of artificial cold (refrigeration, freezing) leads to the best results when it is desired to keep the attributes and quality as close as possible to those of the fresh product [7-9].

For long-term storage, in order to use the products in the off-season, freezing is frequently used. By freezing the admissible storage time of the products is increased by over 5 ... 50 times the preservation by refrigeration [10].

Generally, freezing occurs at temperatures between -18 °C and -40 °C. Depending on the temperature at which it occurs, freezing can be:

- slow, carried out at temperatures of -18 °C ... 20 °C for about 80 hours;
- semi-rapid, at temperatures between -20 °C and 40 °C (in cold rooms or freezing tunnels) and lasting about 60 hours;
- fast, conducted at temperatures of -30 °C to -35 °C and lasting up to 24 hours;
- ultrafast, carried out at -35 °C to -40 °C and lasting about 3 hours.

The freezing rate determines the formation of ice crystals as follows: slow freezing ("classic") leads to the formation of a small number of ice crystals having large dimensions, while rapid freezing with cryogenic liquids generates a large number of small size crystals. Large crystals resulting from slow freezing cause deformation, dislocation, and perforation of cell walls, and when defrosting these products, juice losses are high. In the case of rapid freezing with cryogenic fluids, tissue changes due to the formation of ice crystals are less damaging, and defrosting juice losses are lower because much of the water remains in the cells [11, 12].

The quality of frozen fruits and vegetables depends to a certain extent on the characteristics of the raw material used. Compatibility with freezing is a function of the species, climate and soil characteristics when the fruit and vegetables were harvested, crop technologies (fertilization, irrigation etc.). Some species such as lettuce, tomatoes, radishes, table grapes are not suitable for freezing, because during the freezing process there are important losses of turgidity, discoloration and loss of flavor. The conditions that vegetables and fruits have to fulfill to be preserved by freezing are determined by: compatibility with freezing preservation, widespread acceptance by consumers, high technological yields, virus resistance, mold, physiological defects and insects, uniformity of maturation, good handling resistance, pronounced color and flavor, firm texture with as little as possible deformation and softening tendency to thaw, as little juice as possible in defrosting. One of the modern freezing techniques is that of cryogenic freezing using liquid nitrogen (N_2) as a cooling substance [11, 13, 14]. Compared to conventional freezing techniques, the cryogenic method has the following advantages: rapid freezing, inhibition of bacterial growth, limited dehydration, considerable reduction of quantitative losses, optimal nutritional preservation, maintenance of the appearance and taste of the food.

Among the fruit species with the widest use for freezing can be remembered berries, strawberries, cherries, peaches and apricots, and among vegetables: green peas, green beans, spinach and carrots.

2 Materials and methods

Two species of horticultural products, namely blueberries and green bean pods, were used in the experimental research. (Fig. 1).

Fig. 1. Species of horticultural products used in the experimental research

For horticultural products fast freezing it has been used an experimental model of fast freezing equipment with liquid nitrogen, ECR-0, intended for cryogenic freezing of different food products (Fig. 2).

Fig. 2. Fast freezing equipment – ECR [15]

The equipment is provided with temperature sensors that allow monitoring and permanent control of process parameters, namely:

- Type K Thermocouple (chromel–alumel), probe diameter φ1.5 mm, measuring range -100 °C...+30 °C, allows temperature measurement in the centre of the product;
- Type K Thermocouple, with temperature transfer element, measuring range -100 °C...+30 °C, allows temperature measurement on the external surface of the product;
- Thermal resistance, type TTR Pt 100, measuring range -200 °C...+30 °C, allows temperature measurement inside the freezing chamber.

The chamber is provided with an access door, thermal insulated and provided with two sealing gaskets and de-icing resistances to prevent ice formation and to facilitate access to the chamber at the end of the freezing cycle. The locking system is provided with two clamps and a handle to obtain a uniform and firm closing. In order to achieve a forced convection of the atmosphere in the chamber in order to improve the thermal transfer from the product to the thermal agent, a fan was provided inside the chamber. The equipment allows operation in two working modes, namely:

- automatic operation mode, with setting the chamber temperature limit at which ventilation starts and liquid nitrogen purging stops;
- manual operation with continuous liquid nitrogen purging.

All measured parameters during the working cycle were recorded on the SD memory card of the Programmable Logic Controller (PLC) within the automation installation of the equipment.

The weight of each sample was set at 750 g, 250 g on each of the 3 trays. For each sample, the mass and dimensional characteristics were determined as follows:

- for blueberries: the average bean mass, average height and the maximum equatorial diameter were determined as the average of five random measurements from the product;
- for green bean: the pods were fragmented up to lengths between 35 and 65 mm. For these, the average mass, average length, average width and average thickness of the pods fragments were measured as the average of five random measurements from the product.

For products with regular geometric shapes, the thermal center (TC) coincides with their geometric center. The smallest distance between the TC thermal center and the outer surface of the product is noted with δ_0 and is an important parameter for determining the freezing rate (Fig. 3). For blueberries, δ_0 was approximated by *h/2* and for green bean pods with *g/2*.

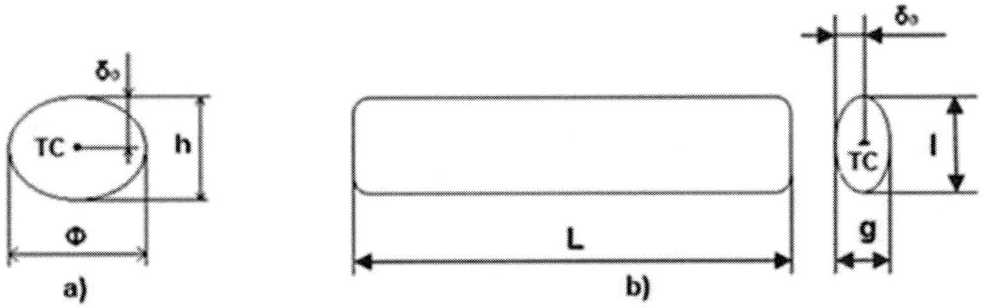

Fig. 3. Dimensional characteristics: a) – blueberries; b) – green bean pods

The freezing process is considered complete when the temperature in the product thermal centre reaches -15°C.

The parameters followed during the freezing process were as follows:

- Mass of the liquid nitrogen container used for freezing;
- Freezing time from 0 °C to -15 °C;
- Total freezing time.

Following the processing of experimental data, the following energy indexes were determined:

- Average linear freezing rate;
- Liquid nitrogen consumption for a freezing cycle.

To characterize a freezing process in terms of cooling intensity, it is considered as criterion, the average linear freezing rate given by the relation:

$$w_m = \frac{\delta_0}{\tau_0}, [cm/h] \quad (1)$$

where: $\delta_0$is the smallest distance between the thermal centre and the outer surface of the product, [*cm*];

τ_0 - the freezing time from a uniform initial temperature of 0°C to the temperature to be achieved in the thermal centre, [*h*].

Depending on the average freezing rate, w_m, the International Institute of Refrigeration recommends the following classification of freezing methods [16]:

- slow freezing $w_m<0.5$ *cm/h*;
- fast freezing $w_m = 0.5 \ldots 3$ *cm/h*;
- very fast freezing $w_m= 3 \ldots 10$ *cm/h*;
- ultrafast freezing $w_m = 10 \ldots 100$ *cm/h*.

Liquid nitrogen consumption for a freezing cycle was determined by the difference between the mass of the liquid nitrogen container before and after the freezing process was completed.

3 Results and discussion

Following the processing and interpretation of the experimental data, the following results were obtained regarding the characterization of the two species of horticultural products used:

Table 1. Characteristics of the analysed products

No.	Characteristic	UM	Value of parameters determined in the tests	
			Blueberries	Green bean pods
1.	Sample mass	*g*	250x3=750	200x3=600
2.	Average mass of a product	*g*	1.72	1.29
3.	Maximum equatorial diameter (for blueberries), Φ	*mm*	15.30	-
4.	Bean average height (for blueberries), h	*mm*	12.07	-
5.	Average length of the pod fragment (for green bean pods), L	*mm*	-	48.77
6.	Average width of the pod fragment (for green bean pods), l	*mm*	-	8.79
7.	Average thickness of the pod fragments (for green bean pods), g	*mm*	-	6.75
8.	The smallest distance between the thermal centre and the outer surface of the product, δ_0	*mm*	6.04	3.38

Process parameters, during the freezing process, can be determined by direct reading on the touch screen of the command and control panel or by post-processing the data recorded on the SD memory card. Freezing times were determined based on the data recorded on the SD memory card (Fig. 4), as follows:

- the freezing time from 0°C to -15°C was determined by the difference between the total timer recorded at the end of the freezing process (-15°C) and timer recorded when the temperature reached 0°C in the product's thermal centre;

- the total freezing time was determined by direct reading of the total timer recorded at the end of the freezing process.

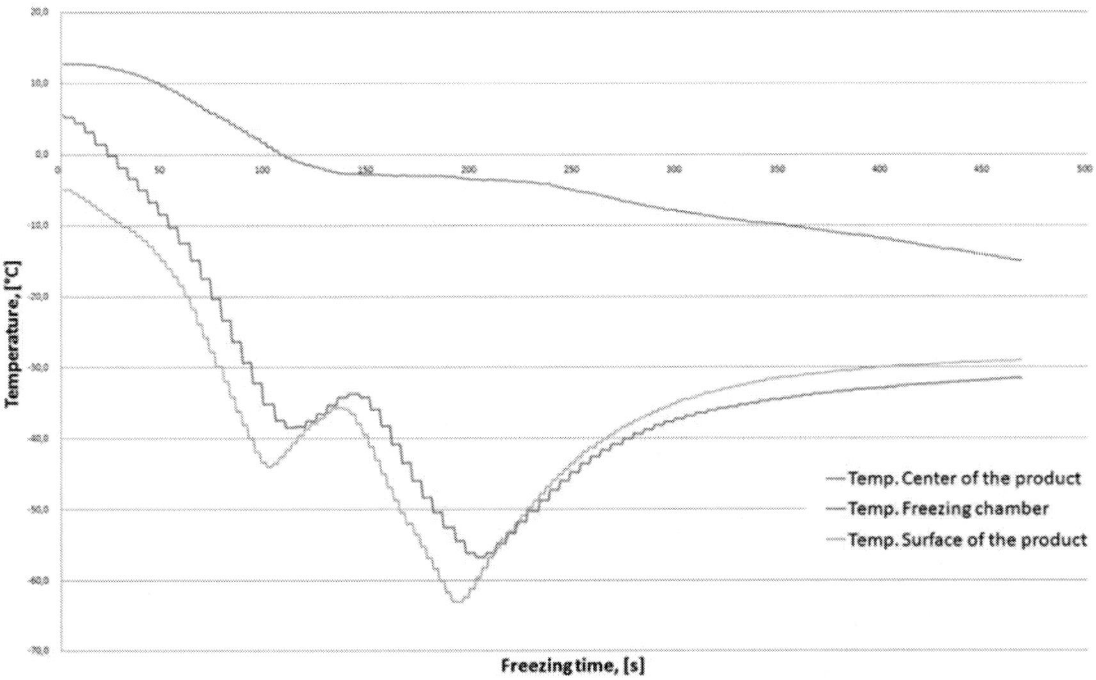

Fig. 4. Determination of the process temperatures based on data recorded by sensors

The parameters followed during the freezing process and the energetic indexes of fast freezing equipment are shown in Table 2.

Table 2. The parameters followed during the freezing process

No.	**Characteristic**	**U.M.**	**Value of parameters determined in the tests**					
			Blueberries			**Green bean pods**		
			The chamber temperature limit set to [°C]					
			-30	**-35**	**-40**	**-30**	**-35**	**-40**
1.	Freezing time from 0°C to -15°C	*s*	355	250	234	383	239	187
2.	Total freezing time	*s*	467	364	343	433	303	255
3.	Average linear freezing rate for an operating cycle	*cm/h*	6.13	8.70	9.29	3.18	5.09	6.51
4.	Liquid nitrogen consumption for an operating cycle	*kg*	2.30	3.06	2.16	0.90	1.60	2.50

Graphical reprezentation of the results is shown in Fig. 5.

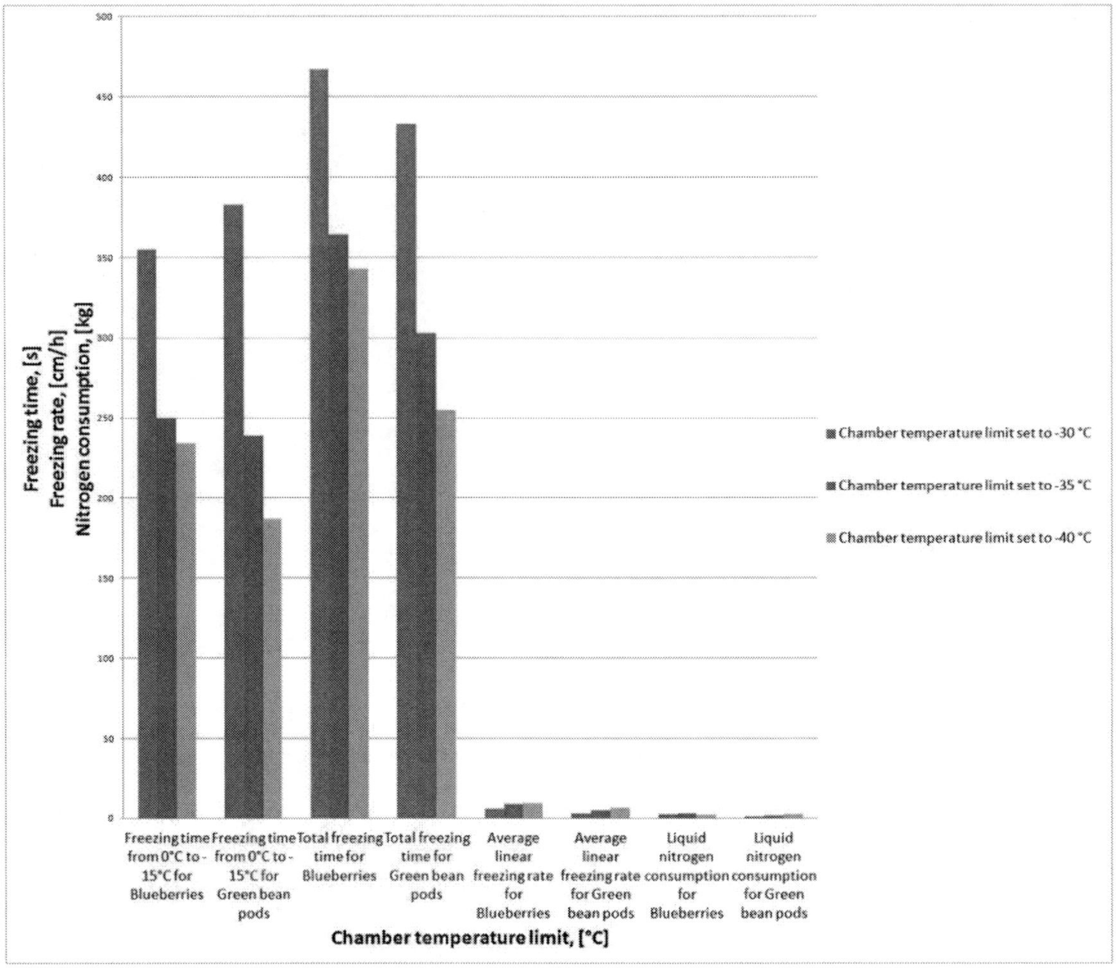

Fig. 5. Graphical reprezentation of the results

The total freezing times and the average linear freezing rates depending on the selected temperature limits within the chamber, are presented, both for blueberries and green bean pods, in Fig. 6 and 7.

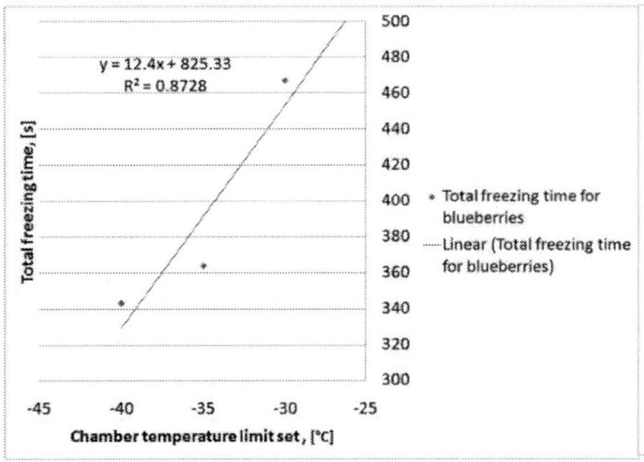

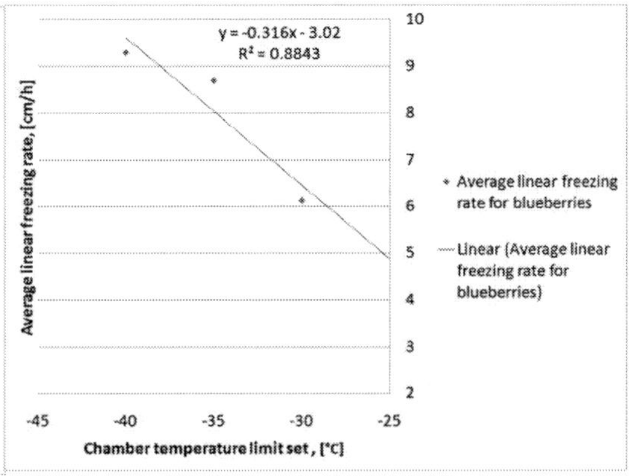

Fig. 6. The total freezing times and the average linear freezing rates depending on the selected temperature limits within the chamber, for blueberries

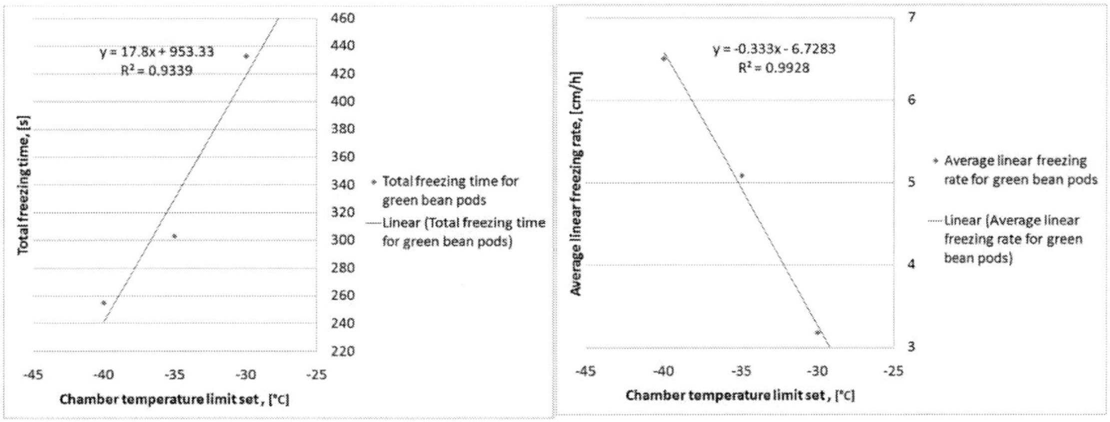

Fig. 7. The total freezing times and the average linear freezing rates depending on the selected temperature limits within the chamber, for green bean pods

Linear regressions performed using Excel, allowed the identification of some linear functions, which estimate the variation of the total freezing time and the average linear freezing rate depending on the selected temperature limit within the chamber, with coefficients of determination varying from 0.87 to 0.99.

Samples aspect before and after freezing is shown in Table 3:

Table 3. Samples aspect during the experimentation

	The chamber temperature limit set to		
	-30 °C	**-35 °C**	**-40 °C**
Blueberries			
Fresh product			
Quick frozen product		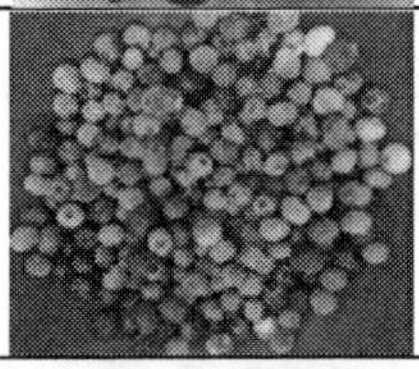	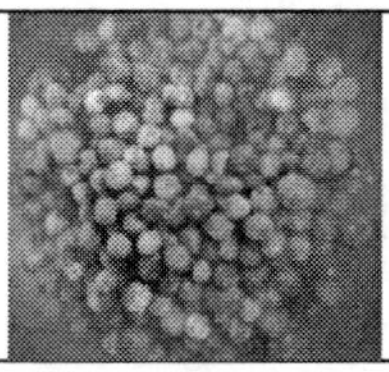
Green bean pods			
Fresh product			
Quick frozen product	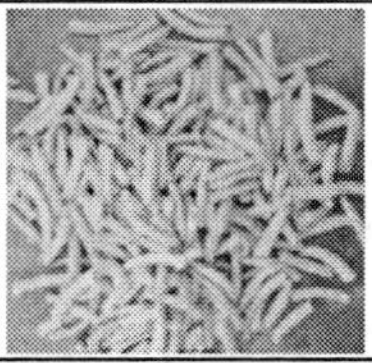	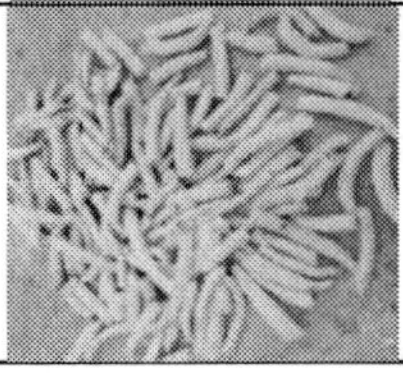	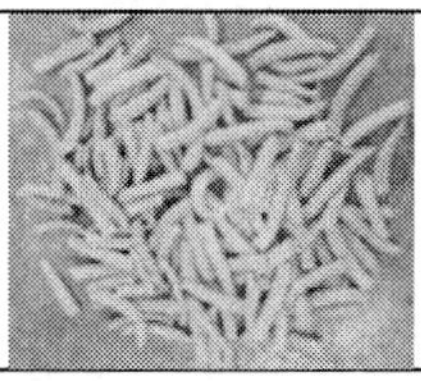

4 Conclusions

Following the determinations made during the experimentation and data processing, the following results were obtained:

- with the lowering of the set temperature inside the freezing chamber, the freezing time from 0 °C to the temperature of -15 °C decreases, resulting in lower freezing times;
- with the lowering of the set temperature inside the freezing chamber, higher average linear freezing rates are obtained;
- with the lowering of the set temperature inside the freezing chamber, the liquid nitrogen consumptions record an increasing tendency;
- the total freezing time for blueberries was higher than the freezing time for green bean pods, regardless of the set temperature inside the freezing chamber;
- the average linear freezing rate for blueberries was higher than the freezing rate for green bean pods, regardless of the set temperature inside the freezing chamber;
- the liquid nitrogen consumption for freezing blueberries was, overall, higher than the nitrogen consumption for green bean pods.

5 REFERENCES

1. A. Alhamdan, B. Hassan, H. Alkahtani, D. Abdelkarim, M. Younis, *Cryogenic freezing of fresh date fruits for quality preservation during frozen storage*, Journal of the Saudi Society of Agricultural Sciences, Volume **17**, Issue 1, Pages 9-16, (2018);
2. A. Alhamdan, B. Hassan, H. Alkahtani, D. Abdelkarim, M. Younis, *Freezing of fresh Barhi dates for quality preservation during frozen storage*, Saudi Journal of Biological Sciences, Volume **25**, Issue 8, Pages 1552-1561, (2018);
3. A. Bigliaa, L. Combaa, E. Fabriziob, P. Gaya, D. R. Aimoninoa, *Case studies in food freezing at very low temperature*, Energy Procedia 101, Pages 305–312, (2016);
4. C. Bilbao-Sainz, A. Sinrod, M. J. Powell-Palm, L. Dao, G. Takeoka, T. Williams, D. Wood, G. Ukpai, J. Aruda, D. F. Bridges, V. C.H. Wu, B. Rubinsky, T. McHugh, *Preservation of sweet cherry by isochoric (constant volume) freezing*, Innovative Food Science & Emerging Technologies, Volume **52**, Pages 108-115, (2019);
5. M. J.Sousa-Gallagher, A. Tank, R. Sousa, *The Stability and Shelf Life of Food (Second Edition)*, Woodhead Publishing Series in Food Science, Technology and Nutrition, Pages 399-430, (2016);
6. M. Ingeaua, T. Prisecaru, I. Pirna, C. Sorică, *Analysis of a convective drying process of plums*, INMATEH Agricultural Engineering, Volume **46**, no. 2 / 2015, e-ISSN 2068-2239, p-ISSN 2068-4215, Pages 115 – 124;
7. M. S. Aghdam, S. Bodbodak, *Food Bioprocess Technology*, Volume **7**, Issue 1, Pages 37- 53, Springer US, p-ISSN 1935-5130, e-ISSN 1935-5149, (2014);
8. X. Cao, F. Zhang, D. Zhao, D. Zhu, J. Li, *Effects of freezing conditions on quality changes in blueberries*, Journal of the Science of Food and Agriculture, Volume **98**, Issue 12, ISSN 0022-5142, e-ISSN 1097-0010, Pages 4673-4679, (2018);
9. P. K. Jha, E. Xanthakis, S. Chevallier, V. Jury, A. Le-Bail, *Assessment of freeze damage in fruits and vegetables*, Food Research International, Available online, (2018);
10. P. Niculita, *Refrigeration technique and technology in agri-food fields*, Didactic and Pedagogical Publishing House R.A, Bucharest, ISBN 973-30-5719-3, (1998);
11. P. J. Fellows, *Food Processing Technology (Fourth Edition): Principles and Practice*, Woodhead Publishing Series in Food Science, Technology and Nutrition, Pages 885-928, (2017);
12. C. Kennedy, *Food Preservation Techniques*, Woodhead Publishing Series in Food Science, Technology and Nutrition, Pages 228-240, (2003);

13. V. Damian, C. Iosifescu, G. Coman, *Environmentally friendly method for freezing raspberries and blueberries using liquid nitrogen*, Environmental Engineering and Management Journal, Volume **9**, Issue 10, ISSN 1582-9596, Pages 1319-1322, (2010);
14. V. Damian, C. Iosifescu, G. Coman, M. Dragan, E. O. Constantin, *Theoretical and experimental study on cryogenic freezing of berries*, Advances in Communications, Computers, Systems, Circuits and Devices, European Conference of Systems-Proceedings, ISBN 978-960-474-250-9, ISSN 1792-6637, Pages 171-174, (2010);
15. Sectoral Research Program of the Ministry of Agriculture and Rural Development, ADER 16.2.1: Researches on thermal, physical properties, heat and mass transfer coefficients of horticultural products for the optimization of freezing technologies applied on cold chains, 2015 – 2018;
16. L. Bøgh-Sørensen, Recommendations for the processing and handling of frozen foods, IIR, (2006).

E3S Web of Conferences **112**, 03034 (2019) https://doi.org/10.1051/e3sconf/201911203034
TE-RE-RD 2019

Structural analysis of deep soil loosening machine MAS-65

Mihai Gabriel Matache[1*], *Remus Marius* Oprescu[1], *Dragos Nicolae* Dumitru[1] and *Gabriel Valentin* Gheorghe[2], *Dan* Cujbescu[1], *Cătălin* Persu[1]

[1]INMA Bucharest, Testing Department, 013813 Bucharest, Romania

[2]INMA Bucharest, Soil Laboratory, 013813 Bucharest, Romania

Abstract. Deep soil loosening machine MAS 65 is destined to work soil at depths exceeding 45 cm, thus the machine's frame is subjected to loads which could affect its structural integrity. Within this paper a static structural analysis was performed on the machine's 3D model using finite element method and strain and stress distribution maps were created. Using the Von Mises stress map there were identified several critical points which could fail during normal exploitation conditions and which should be monitored by strain gages during field tests in order to prevent structural failure.

1 Introduction

Agricultural machinery are self-propelled working machines or driven from alternative power sources, designed for performing a series of works within agricultural processes, according to agricultural-biological and technical-economical requirements imposed.

Generally, conception of an agricultural machine starts from its physical operations which have to be executed during operation. Conception could be new, innovative which takes into consideration the state of the art of the working technologies (e.g. technologies for realization of no tillage soil works [1]), but also could be one inspired after a similar operation performed by an existent machine that we want to improve its performances [2]. The resistance structure, frame or chassis, is designed and dimensioned according to a strength reserve which assures the good development of the activity, choosing for it an adequate material [3] (so to be able to withstand maximum strain which does not affect the work quality and maximum loads in the structure which cannot cause damage). This primary structure is subjected to a first process of structural analysis for an improved or even optimal design [4, 5, 6]. This structural analysis process includes the following stages: structural modelling, static analysis, frequency analysis, dynamic simulation, fatigue analysis. After all, or some of these stages the final theoretical dimensioning of the structure could be finalized.

Because of the different loads frequency and amplitude characteristics and their number of cycles sustained by the mechanical resistance structure of agricultural machinery, these

* Corresponding author: gabimatache@yahoo.com

could fail functionally or structurally. There are numerous examples in current practice which confirm the apparition possibility of this phenomenon. It is a known fact that the failing process of diverse structures, manufactured with diverse materials, is very complex and that there are fundamental differences between failing due to static loads and failing because dynamic loads. Thus, there were elaborated principles, methods of experimental research and calculus specific to both types of solicitations.

Failing or breaking because static solicitations is produced on a structure when it reaches the yield tension of a material or it is exceeded the limit breaking tension of the respective material.

Failing or breaking because dynamic solicitations is produced in function of the shock solicitation when it's reached the yield point of a material or it's exceeded the limit breaking tension of the respective material or by material fatigue phenomenon. Also, of great importance are the material binding procedures (with screws and nuts, with rivets or through welding) [7].

Structures fail when under-dimensioned, if there are exceeded the specified loading characteristics in their user manual or if they were designed to fail as a precautious measure. In the design stage is very important to dimension the structure for operating in complete elasticity conditions.

2 Material and method

2.1. Structural model of MAS-65 deep soil loosening machine

Structural model of MAS-65 deep soil loosening machine was constructed in view of preliminary identification of critical points within its resistance frame. The model was developed using finite three-dimensional elements. The software used for modelling was SolidWorks [8], starting from the already developed experimental model of the machine. Thus, have been identified the principal components of the resistance frame: central beam, longerons, three-point linkage system to the tractor, support of the working body and the working body. Also, were identified the metallic profiles from which the experimental model was constructed through welding and we measured their dimensions, allowing to develop the structural model presented in figure2.

Fig. 1 MAS-65 experimental model in aggregate with New Holland TD80 tractor.

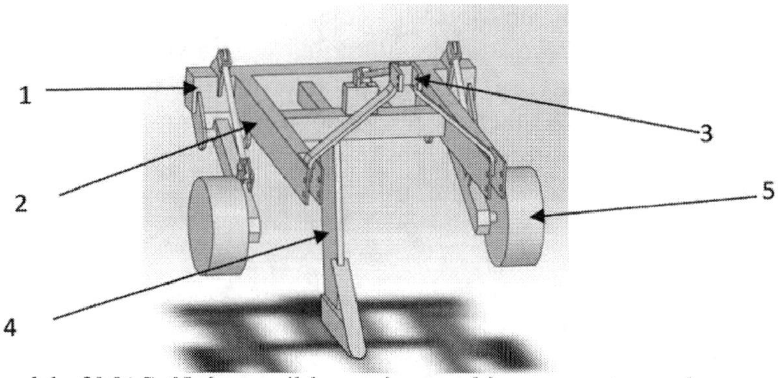

Fig. 2 Structural model of MAS-65 deep soil loosening machine: geometry and component elements (1 – central beam, 2 - longerons, 3 – three-point linkage system, 4 – working body organ support and working body, 5 – wheels for establishing working depths).

The material from which was constructed the experimental model chosen also for the structural model was carbon steel [9], with the following characteristics:
- yield strength $\sigma_c = 2.20594 \cdot 10^8$ N/m^2;
- ultimate tensile strength $\sigma_r = 3.99826 \cdot 10^8$ N/m^2;
- Young modulus $E = 2.1 \cdot 10^{11}$ N/m^2;
- Poisson coefficient $\nu = 0.28$;
- material density $\rho = 7800$ kg/m^3;
- transversal elasticity modulus $G = 7.9 \cdot 10^{10}$ N/m^2

2.2. Study properties in static regime

We performed a structural analysis in static regime, using the solid type discretization. The discretized structure had a total number of 113081 nodes, with 59248 standard elements. Elements dimension was 54.5167 mm with a tolerance of 2.72584 mm. Finite elements analysis was performed in Solidworks. In figure 3 is presented the discretized structure:

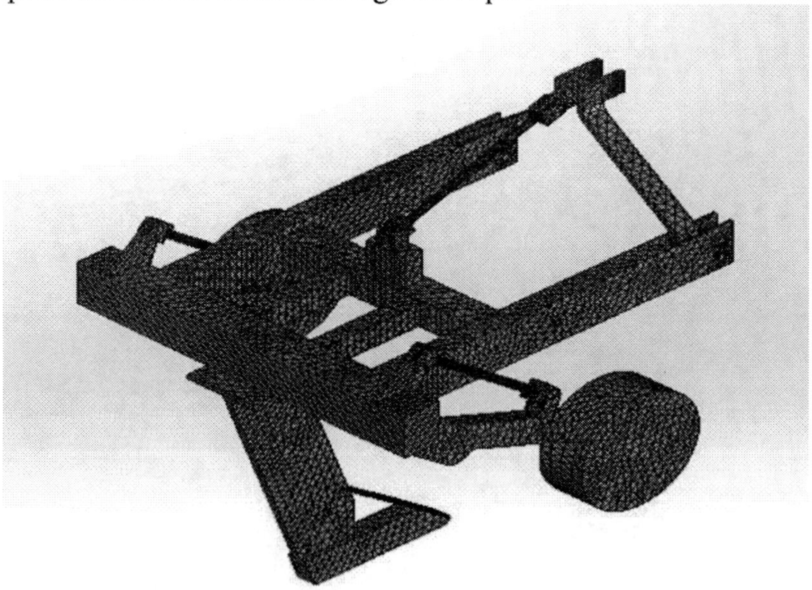

Fig. 3 Discretization of the MAS-65 structural model

Loads and boundary conditions: the three fixtures in which were cancelled all the three translations (on orthogonal axes Ox, Oy și Oz) are corresponding to the three-point coupling system to the tractor existent on the physical model and were applied according to the following figure:

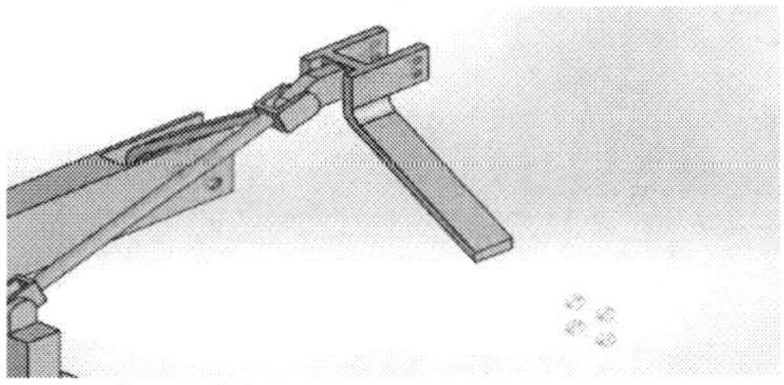

Fig. 4 Fixtures of MAS-65 structural model.

The loading forces were applied on the working body, which consisted of 2 subassemblies (figure 5), on longitudinal direction in the plain corresponding with the forward movement of the machine. For the loading force on the working body chisel was selected a value of 9126 N while for the loading force on the edge of the working body was selected a value of de 4447 N, after some estimation calculus according to [10]. Thus, the total reaction force in the fixture points achieved the value of 13573 N.

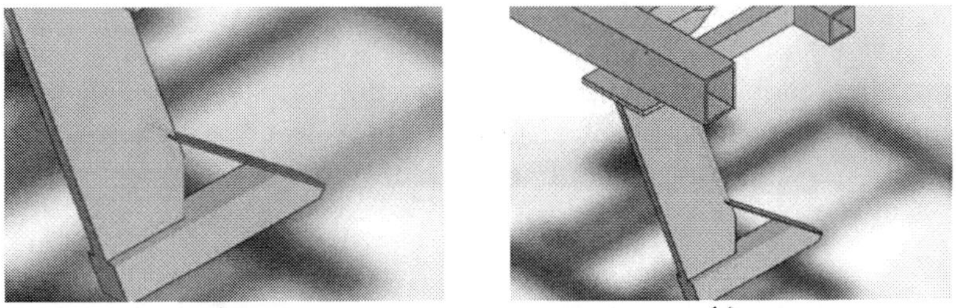

a) b)

Fig. 5 Loading forces on the working body: a) chisel of the working body; b) edge of the working body.

3 Results of the resistance study in static regime

The minimum and maximum equivalent tensions (Von Mises) in the structural model were determined subsequent to the static analysis according to the fixtures and loads anteriorly presented. The maximum value of the equivalent tension was of 1.77389e+08 N/m^2 and was located in the central beam at the junction point with the working body organ, node 22155. The minimum value recorded was of 9793.72 N/m^2 and was located in node 40892 of discretized structure.

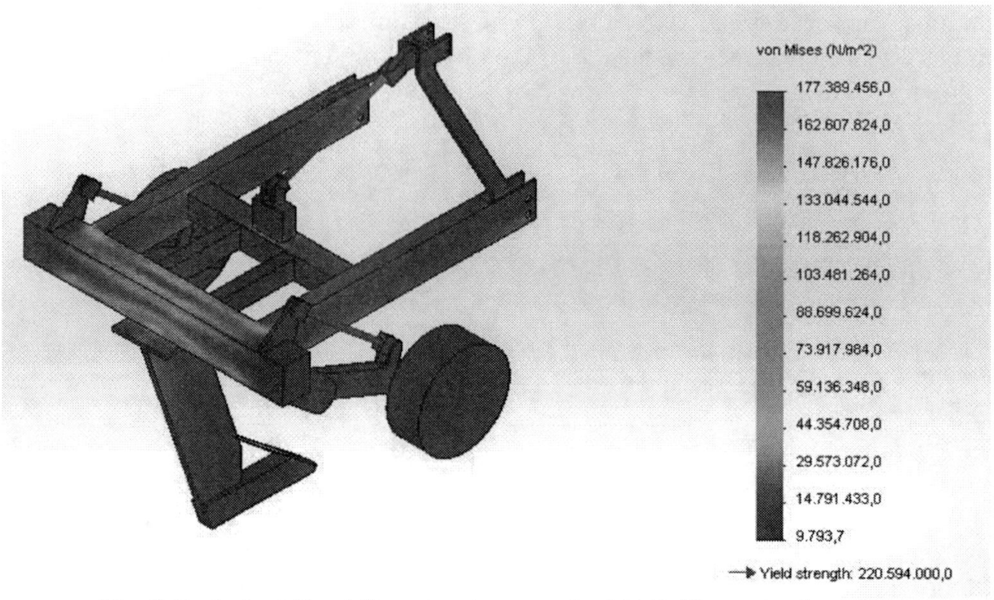

Fig. 6 Equivalent Von Misses tension map for MAS-65 structural model.

In figure 7 is graphically displayed the distribution of relative displacement values URES resulted in the structural model. Representation was realized on the deformed body of the structure. It could be observed the maximum values were located in the working body extremum, and the minimal ones in the fixture points to the tractor. The maximum value of the relative displacement resultant was of 5.79 mm, located in node 21748 of the discretized structure.

Fig. 7 Relative displacements map for MAS-65 structural model.

In the following figure is presented the map of equivalent strain ESTRN resulted after the static study. A minimum value of 5.65876e-008 m/m was recorded in element 37669 and a maximum value of 0.000444062 m/m in element 35453.

Fig. 8 Specific equivalent strain map for MAS-65 structural model (m/m).

After analysing the obtained results there were identified 14 critical points which could be subjected to critical failure in real exploitation conditions, for which the equivalent Von Misses tension was maximum. In figure 9 and 10 there are presented the 14 points located on the MAS-65 structure. The points were named M_i, with $i=1\div14$, the dimensions being given in millimetres. These points could be monitored using strain gages in real working conditions, using data acquisition.

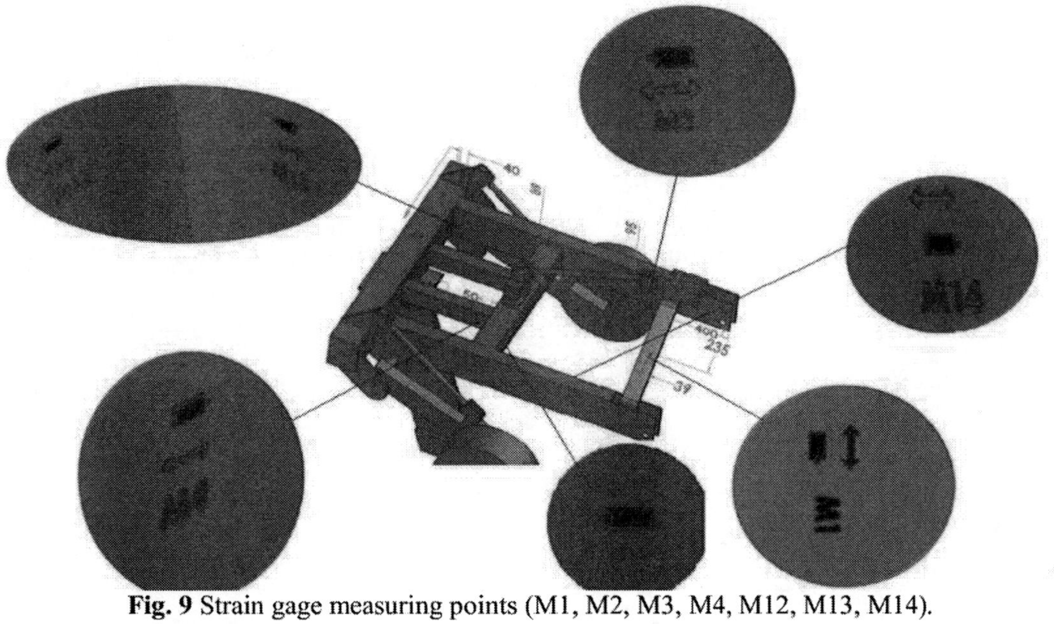

Fig. 9 Strain gage measuring points (M1, M2, M3, M4, M12, M13, M14).

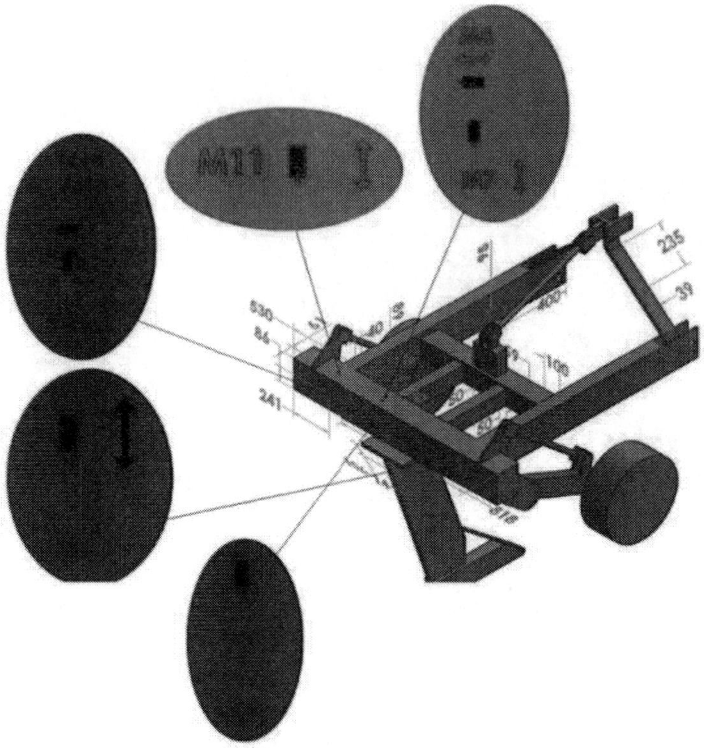

Fig. 10 Strain gage measuring points (M5, M6, M7, M8, M9, M10, M11).

In the following table there are presented the values obtained for the strain on axes Ox, Oy and Oz for the 14 critical points identified after finite element analysis of the structural model of MAS-65 deep soil loosening machine in static regime. There are presented all the three reference directions because for strain measuring could be used unidimensional strain gages or rosettes strain gage applied in the identified points. There were reported the values corelated with the direction of the strain gages so that to make possible their experimental value comparison after experiments will be performed.

Table1. Values of strain determined after FEM simulation

Critical point identification	Ox strain– ε_x (μm/m)	Oy strain – ε_y (μm /m)	Oz strain – ε_z (μm /m)
M1	64.35	147.31	45.3
M2	-	-	-99.34
M3	-98.99	-	-
M4	-	-	-47.35
M5	-	-271.14	-
M6	-	-28.64	-
M7	-	-	22.52
M8	97.81	-	-
M9	-	19.22	-
M10	36.33	-	-
M11	-345.25	-	-
M12	44.73	-	-
M13	-	-	-39.24
M14	-	-	-135.63

Conclusions

After the performed analysis there were identified critical areas of the resistance structure of the MAS-65 deep soil loosening machine. It was observed that the beam from the back of the equipment, which linked the two longerons of the resistance frame and on which was mounted the working body support, it was the most solicitated. Also, the chisel of the working body is heavily stressed in exploitation conditions.

After finite element analysis performed on the structural model of the equipment there were identified 14 critical points which could be chosen for applying of strain gages in view of future experimental analysis. Also, based on such recorded signals the finite analysis model could be validated.

Acknowledgement: This work was supported by a grant of the Romanian Research and Innovation Ministry, through Programme 1 – Development of the national research-development system, subprogram 1.2 – Institutional performance – Projects for financing excellence in RDI, contract no. 16PFE.

References

1. I.D. Vlăduţ, S. Şt. Biriş, et al. - *State of the art regarding conservative tillage technologies*, Annals of the University of Craiova - Agriculture, Montanology, Cadastre, **Vol. 44**, (2014)
2. L. Vlăduţoiu, V. Vlăduţ, I. Voiculescu, et al. - *The increase of active bodies of agricultural machines in work by hardening*, Proceedings of the 43 International Symposium on Agricultural Engineering "Actual Tasks on Agricultural Engineering", pp 153÷164, (2015)
3. K. Martensen - *Progress în typical materials for agricultural machinery*, Maschinenfabrik Bernard Krone, Spelle, Germany, http://www.clubofbologna.org/ew/documents/Martensen-paper.pdf.
4. R. Sfîru, V. Vlădut, P. Cârdei, R. Ciupercă , M. Matache, V. Stefan, N. Ungureanu, - *Technical aspects regarding axles calculation of road transport means*, Lambert Academic Publishing, (2014)
5. V. Vlăduţ, S. Şt. Biriş, et al. - *The verification of stress by FEM analysis/ mechanical testing of a traction bar*, Proceedings of the 43 International Symposium on Agricultural Engineering "Actual Tasks on Agricultural Engineering", pp 141÷152, , (2015)
6. E. Voicu, I. Cojocaru, G. Ciurel, S. Popescu - *Experimental researches regarding the dynamic and energetic optimisation of the fodder trailed harvester CTF in aggregate with U-650 tractor*, pp 73-79, INMATEH, **Vol. 20**, No. 2, (2007)
7. C. Th. Ploscariu– phd thesis: *Researches and contributions regarding utilization and heat treatment of low alloy steels used for welded structures of Romanian helicopters*, (2010)
8. SolidWorks Software Documentation.
9. C. Dumitrescu – *Treaty of science and material engineering*, **Vol. 1, 2, 3**, (2007)
10. V. Scripnic, P. Babiciu – *Agricultural machinery*, (1979)

E3S Web of Conferences **112**, 04001 (2019) https://doi.org/10.1051/e3sconf/201911204001
TE-RE-RD 2019

Analysis and Optimization of a Piezoelectric Energy Harvester

Claudia Borzea[1,*], and *Daniel* Comeagă[2]

[1]Romanian Research and Development Institute for Gas Turbines COMOTI, Automation and Electrical Engineering Department, 061126 Bucharest, Romania
[2]University Politehnica of Bucharest, Mechatronics and Precision Mechanics Department, 060042 Bucharest, Romania

Abstract. The paper aims to assess and improve the performances of a multilayer piezoelectric MEMS device for vibrations harnessing. Two operating modes are possible: at resonance and outside resonance. In some applications it is not possible to operate at resonance, functioning being mostly at low frequencies in a quasi-static regime. An Euler-Bernoulli classic beam theory mathematical model was studied for estimating the behaviour of multilayer piezoelectric generators, in terms of deflection and voltage, at functioning under resonance frequency. The analytical results were compared with the finite element method simulation in COMSOL Multiphysics. The main goal of this study is to obtain an accurate model for engineering design purposes, with simple analytical equations and ease of use, but with predictable errors. The study proved the usefulness of the derived model but also its limitations. It also proves the need to improve the model using plate theory, for sensors with high width/height ratio.

1 Introduction

Different transduction mechanisms (piezoelectric [1-4], electromagnetic [4-6], electrostatic [7,8]) have been studied for converting vibrations into electricity, but piezoelectric energy harvesting has received the most attention due to the high power density and ease of application [1]. Hybrid harvesters using combinations of piezoelectricity with either of the other two methods have also been studied [9,10]. Piezoelectric materials have the property that, when subjected to mechanical strain, they produce an electrical charge proportional to the stress applied, unlike electromagnetic devices, which exhibit hysteretic responses.

Piezoelectric harvesters use the direct piezoelectric effect for converting mechanical vibrations into electrical energy, and tap this energy by connecting the harvester electrodes with an electric circuit [3]. For obtaining maximum electrical response, it is preferable to excite a given harvester at its fundamental frequency (first bending mode). At resonance, a small excitation can produce large vibration response, a sharp voltage peak occurring at that frequency. Operating at resonance enables energy harvesters to be placed on equipment producing high vibrations (engines, compressors, turbines, etc.).

* Corresponding author: borzeaclaudia@gmail.com; claudia.borzea@comoti.ro

Usually, a cantilever structure is preferred due to high electrical energy to transducer mass and volume ratio. To improve the output, an inertial mass is placed on the free end. Designing such a transducer involves deriving a lumped parameters model.

In other applications, the mechanical energy results from exerting a force on a point on the transducer. Such systems are currently integrated in suspensions and vibration isolation elements, under stairs, in highways, etc. The transducer does not support the external mass and force, this being the role of a flexible mechanical suspension, but is actuated during suspension movement. In this case the regime is quasi-static, the operating frequency being very low comparing with the resonance frequency.

2 Mathematical model

Simple, accurate and easily developable mathematical models are necessary for analysing piezoelectric structures. The best choice is performing a finite element method (FEM) simulation of a 3D model. This option gives very accurate results, however the main drawback is that electronic circuitry integration is not simple. Often it is more convenient to elaborate a concentrated parameters model of the assembly integrating the piezoelectric devices [2]. Previous studies have helped to derive simple and precise analytical models for a piezoelectric actuator and have also shown the modelling errors [11].

From engineering design point of view, approximate formulas are necessary, for ease of use during design and optimisation stages. These models can be validated by FEM simulations and experimental studies, as we did in the present paper.

The computations did not consider the materials yield strengths when applying the variable force. The purpose of the work herein was only of assessing the general behaviour of the harvester. The maximum voltages and deflections that can be obtained are limited by material yield strengths.

The simplest models are based on theory of elasticity hypotheses, which are applied to homogenous multilayered cantilever structures (Fig. 1 a)), with stress and strain exerted only on beam direction [12]:

- Bending takes place on a normal plane to the maximum dimension of the beam;
- The points on median line displace only on normal direction to this line.

Euler-Bernoulli Beam Theory assumptions are also considered [12]:

- A straight line or a cross-section, which is initially perpendicular on the median line of a layer, also remains perpendicular after deformation;
- The perpendiculars on the median lines have the same rotation angles for all layers, resulting in a complete transmission of the specific deformations to the separation surfaces between layers.

The usual equations are presented hereinafter.

A) Layers' constitutive piezoelectric equations

The system of equations for mechanical stresses and electrical displacements is:

$$\begin{cases} T^{(k)}(x,z) = Y^{(k)} \cdot S^{(k)}(x,z) - d_T^{(k)} \cdot Y^{(k)} \cdot E^{(k)}(x,z) \\ D^{(k)}(x,z) = d_T^{(k)} \cdot Y^{(k)} \cdot S^{(k)}(x,z) + \epsilon_S^{(k)} \cdot E^{(k)}(x,z) \end{cases} \tag{1}$$

where: $H_{(k-1)} \leq z \leq H_k$; T – Mechanical stress; Y – Young's modulus; S – Strain; s – Elastic constant; E – Electric field intensity; D – Electrical displacement; d – Piezoelectric charge/strain constant; ϵ – Permittivity; k – Layer number; x, z – Coordinates on x, z axes.

a) 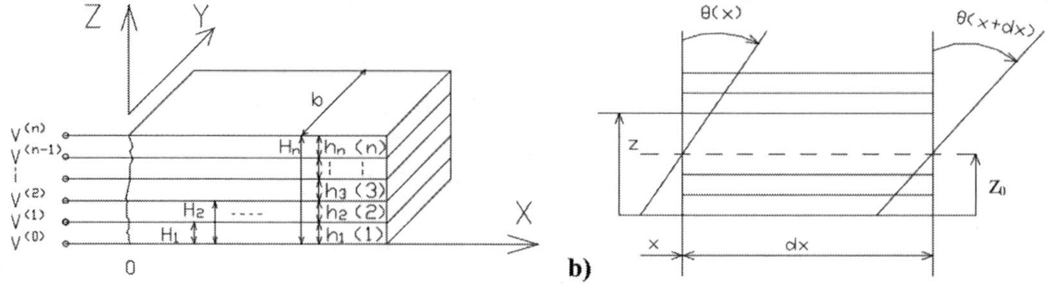

b)

Fig. 1. a) Multilayer element [12]; **b)** Deformation of infinitesimal element [12]

Since only x direction components were taken into consideration for vectors and tensors, it is assumed that $s^E = \frac{1}{YP}$.

B) Specific deformations and section displacements

For a beam with length dx, its cross-surface deformation occurs as in the schematic representation in Fig. 1 b). At z_0 coordinate, there is a line whose length dx stays constant after deformation (neutral fibre), the sectioning surfaces having the relative rotations $\frac{d\theta}{dx}\,dx$. We note $H_i = \sum_{j=1}^{i} h_j$. The strain and stress in the fibre at z coordinate are:

$$S(x,z) = (z - z_0)\frac{d\theta(x)}{dx} \qquad T(x,z) = Y(z - z_0)\frac{d\theta(x)}{dx} - d_T \cdot Y \cdot E(x) \tag{2}$$

C) Cross-surface equilibrium condition

The forces equilibrium condition on axial direction and the momentums equilibrium condition in report to the neutral axis, within the section at x coordinate, are:

$$N(x) = b\frac{d\theta(x)}{dx}\left[\sum_{i=1}^{n} Y^{(i)} h_i \left(\frac{H_i + H_{i-1}}{2}\right) - z_0 \sum_{i=1}^{n} Y^{(i)} h_i\right] - b\sum_{i=1}^{n} d_T^{(i)} Y^{(i)} \Delta V^{(i)}(x) \tag{3}$$

$$M^{(i)}(x,z) = Y^{(i)}(z - z_0)^2 \frac{d\theta}{dx} - d_T^{(i)} \cdot Y^{(i)} \cdot E^{(i)}(x,z) \cdot (z - z_0) \tag{4}$$

where: N – External reduced forces in cross-section; M – External reduced bending momentum in cross-section; $\Delta V^{(i)}$ – Electric potential of layer (i); $V^{(i)}$ – Electrical potential on layers separation surface (i).

Knowing the external axial force and bending momentum, the expressions for z_0 and $\frac{d\theta}{dx}$ can be inferred from equations (3) and (4).

In this case, equation (4) becomes:

$$\frac{d\theta(x)}{dx} = \frac{M(x)}{K_M} + \frac{M_{piezo}(x)}{K_M} = \frac{M(x)}{K_M} + + \frac{b}{K_M} \cdot \sum_{i=1}^{n}\left[d_T^{(i)} \cdot Y^{(i)} \cdot H_{medium}^{(i)} \cdot \Delta V^{(i)}(x)\right] \tag{5}$$

The momentum-(inverse of bending radius) constant K_M and the piezoelectric momentum M_{piezo}, calculated in relation to the neutral axis, are given by:

$$K_M = b \cdot \left[\sum_{i=1}^{n} Y^{(i)} \cdot h_i \cdot \left(z_0^2 - z_0(H_i + H_{i-1}) + \frac{H_i^2 + H_{i-1}^2 + H_i H_{i-1}}{3}\right)\right] \tag{6}$$

$$M_{piezo} = b \cdot \left[\sum_{i=1}^{n} d_T^{(i)} \cdot Y^{(i)} \cdot (H_{medium}^{(i)} - z_0) \cdot \Delta V^{(i)}\right] \tag{7}$$

Using the expressions for z_0 and $\frac{d\theta}{dx}$, the electric displacement D and the electric charge Q accumulated on the electrodes can be calculated.

The charge accumulated on an electrode, for constant beam width b, and the electric displacement are:

$$Q = \int_0^1 D^*(x)dA = b\int_0^1 D^*(x)dx \quad ; \quad D^*(x) = \frac{d}{s^E}\cdot S^*(x) + \epsilon^S E^* \tag{8}$$

where: Q – electrical charge; D – electrical displacement; b – beam width; d – charge/strain constant; s – elastic constant; E – Electrical field intensity; S – Strain; ϵ – Permittivity. Superscript * has the significance of the value measured on the electrode.

2.1 Trimorph using direct piezoelectric effect

Commonly, three-layered elements are used, with an intermediate passive metallic layer and two piezoelectric layers bonded on the faces of the central support layer, made of the same material, with opposite or identical polarization directions (Fig. 2 a) and 2 b)).

Two configurations are usual:

a) two piezoceramic active layers polarized in opposite directions (Fig. 2a). For this configuration, $d^{(1)} = d_{31}$, $d^{(2)} = 0$, $d^{(3)} = -d_{31}$. The linear model condition is met for $V = 0$ and $V^{(0)} = -V$, $V^{(3)} = V$ (common command mode). The electric potentials are applied as schematised in Fig. 2a).

b) two piezoceramic active layers polarized in the same direction (Fig. 2b). For this configuration, $d^{(1)} = d_{31}$, $d^{(2)} = 0$, $d^{(3)} = d_{31}$. The linear model condition is met for $V^{(1)} = V^{(2)} = -V$ and $V^{(0)} = V$, $V^{(3)} = V$ (common command mode).

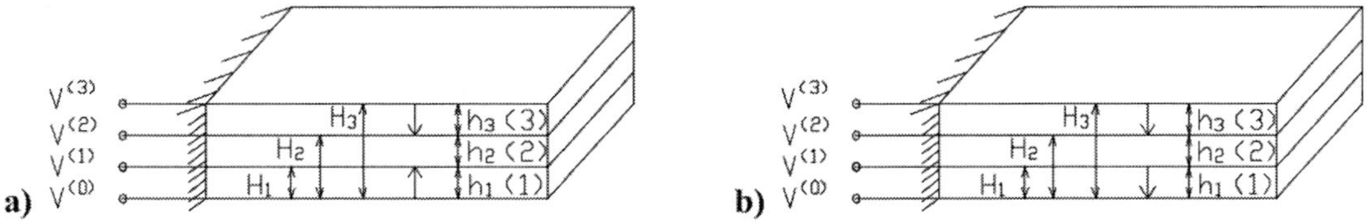

Fig. 2. Three layered structure polarised in: **a)** opposite directions and **b)** identical direction [12]

The stiffness constants are:

$$K_M = b\cdot\left\{\sum_{i=1}^{3}\left[Y^{(i)}\frac{(z-z_0)^3}{3}\Big|_{H_{i-1}}^{H_i}\right]\right\} = b\cdot\frac{Y^S h_S^3 + 2Y^P h_p(4h_p{}^2 + 6h_p h_s + 3h_s{}^2)}{12} \tag{9}$$

where: $h_1 = h_3 = h_p$, $h_2 = h_s$ and $Y^{(1)} = Y^{(3)} = Y^P$, $Y^{(2)} = Y^S$; $H_0 = 0$, $H_1 = h_p$, $H_2 = h_p + h_S$, $H_3 = 2h_p + h_S$.

For trimorph, the electric induction in the vicinity of the superior electrode of layer 3 is:

$$D_{sup} = \frac{d}{s^p}\cdot S^{(p)}\Big|_{\substack{superior\\ electrode}} - \epsilon^p\frac{\Delta V}{h_p} + \frac{d}{s^p}\cdot S^{(p)}\left(h_p + \frac{h_S}{2}\right)\frac{\partial\theta(x)}{\partial x} \tag{10}$$

where: h_p – height of piezoelectric layer; h_s – height of central support layer; θ – rotation angle; ϵ^p – piezoelectric permittivity.

The electric induction in the vicinity of the inferior electrode of layer 3 is:

$$D_{inf} = \frac{d}{s^p}\cdot S^{(p)}\Big|_{\substack{inferior\\ electrode}} - \epsilon^p\frac{\Delta V}{h_p} + \frac{d}{s^p}\cdot S^{(p)}\frac{h_S}{2}\frac{\partial\theta(x)}{\partial x} \tag{11}$$

The overall electrical charge generated in superior layer 3 is:

$$Q = \frac{Q_1 + |Q_2|}{2} = -\frac{\epsilon^p bl}{h_p}\cdot\Delta V + \frac{db}{s^p}\left(h_p + h_s\right)\frac{1}{2}[\theta(l) - \theta(0)] \tag{12}$$

For trimorph with fixed constraint at one end and free at the other end, we obtain:

$$Q = -\left[\frac{\epsilon^p bl}{h_p} - \frac{1}{K}\left(dY_p b(h_s + h_p)\right)^2 \cdot l\right] \cdot \Delta V + \frac{db}{s^p}\left(h_s + h_p\right) \cdot \frac{1}{2K}\left[M \cdot l + \frac{Fl^2}{2} + \frac{pl^3}{6}\right] \quad (13)$$

where: l – beam length; b – beam width; d – charge constant; Y – Young's Modulus; p – uniformly distributed pressure; F – Concentrated force at free end; M – concentrated bending momentum at free end; K – equivalent beam stiffness, given by equation (6).

The total charge measured on the terminals depends on how the layers are connected, being $Q_{total} = 2Q$ for parallel connection (Fig. 2a) and $Q_{total} = Q$ for series connection (Fig. 2b). For usual case $p = 0$, $M = 0$.

3 Finite element method simulation

In order to evaluate the accuracy of the models and to estimate the modelling errors, FEM simulations were performed in COMSOL Multiphysics, for a three-layered cantilever structure, in the configuration from Fig. 2 b). The FEM 3D model designed consists of a beam with two PZT-5H (Lead Zirconate Titanate) piezoelectric layers of 0.5 mm thickness, and a 1 mm thick central metallic layer, of a high-strength alloy steel.

Two sets of simulations were performed, static and dynamic. In both cases a cantilever structure was considered, with fixed constraint at one end and free in rest. In the case of static simulations, a variable force was applied, between 0 ÷ 30 N, uniformly distributed on the surface of the free end (values of 0 ÷ 200 N/m), perpendicular on the surface of the composite structure. We evaluated the displacement, the total voltage in the two piezoelectric layers and the maximum mechanical stress (von Mises stress and the component of the stress tensor on *x* direction), calculated as functions of force.

Numerical simulations using finite element method were conducted, assuming:

- complete anisotropic 3D model;
- plane stress: thin plate (less likely, assumes width much smaller than height);
- plane strain: thick plate (more realistic, assuming width much larger than height).

Several options were also considered for declaring how the bar was clamped, both at the fixed and free end. The possible influence of motion and deformation blocking are relevant both for a correct design of structure coupling with neighbouring elements in the vibrating equipment, and also for simplifying the simulation model. The complete declaration of all components leads to a very high number of finite elements, and therefore to higher computational resources (in terms of processor, drive speed and very high memory requirement), leading to a high computing time.

The simulations were performed considering two intertwined physics (Fig. 3), namely:

- ***Solid Mechanics***, where layers materials are assigned, the material of layers 1 and 3 being declared piezoelectric. As well, fixed constraints and loads (force) are declared.
- ***Electrostatics***, where boundary conditions and initial conditions are declared and a conservation condition of the electric charge is introduced for layers 1 and 3. The two physics are coupled by activating the Multiphysics – Piezoelectric effect option, which allows simultaneous numerical solving of equations from the theory of elasticity, electrostatics and constitutive equations of the piezoelectric material.

In the case of 2D simulation in plane stress and plane strain hypotheses, similar models were used but the mathematical model of Solid Mechanics physics is simplified assuming a plane state of mechanical stresses (specific to a very thin structure) or a plane state of specific deformations (specific to a structure of infinite width).

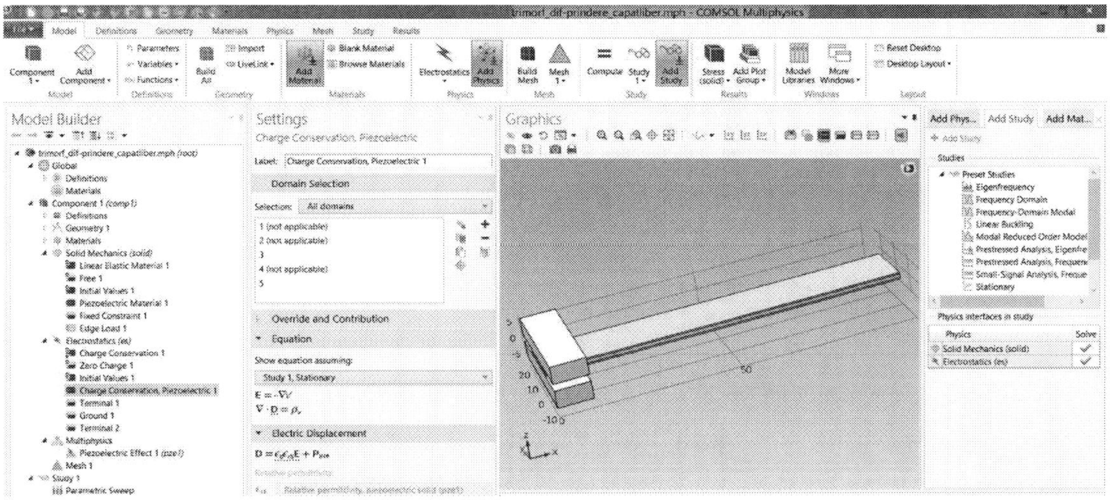

Fig. 3. Declaring physical phenomena and coupling mode in COMSOL Multiphysics 3D simulation

4 The Influence of Modelling Hypotheses

The results of FEM simulations and the values calculated with the theoretical relations in the previous section are presented in the subsequent graphs. The theoretical relationship for output voltage results from equation (13) in the hypothesis of null electric charge.

Hereinafter, considering a trimorph with ratio $b/(2hp+hs)=10$, the influence of clamping declaration was analysed for the fixed constrained end. Fastening between clamping plates (with condition of null displacements on exterior superior and inferior faces) was compared with the case when declaring null displacements for the cross-sectional surface of the three layered beam. The two possibilities were analysed, combined with the cases of cantilever with free or blocked tip.

We evaluated, for all the considered values of the force applied at the free end on a direction perpendicular to the surface (*z*-direction): the electric field in the piezoelectric layers, the voltage differences between central layer declared as ground and the upper and lower layer declared as terminals (based on which the total output voltage was computed), the deformations (based on which we computed the average deflection of the upper free end line on the superior edge), the distribution of stress tensor component on *x* direction evaluated in the piezoceramic layers, and the maximum overall stress in the piezoceramic layers, computed with von Mises relations (in the following figures, a force of 100 N/m).

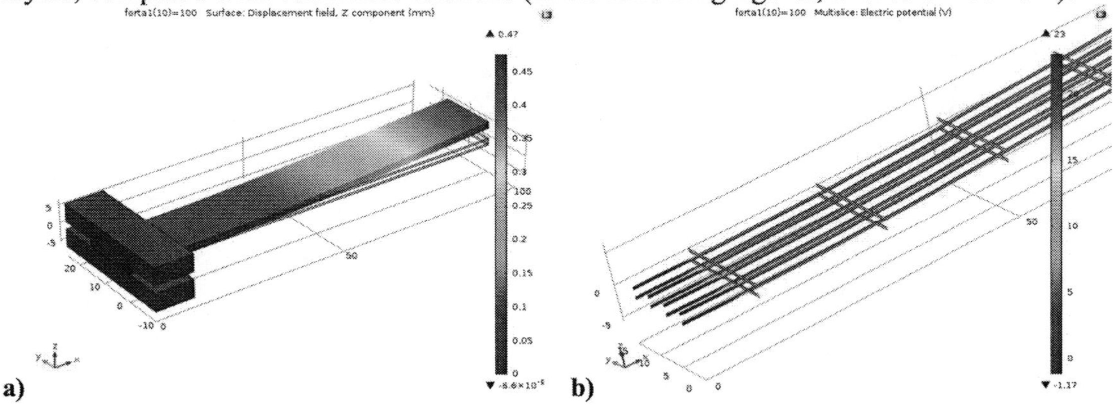

Fig. 4. Evaluating the parameters of interest: **a)** Deflection; **b)** Electrical potential

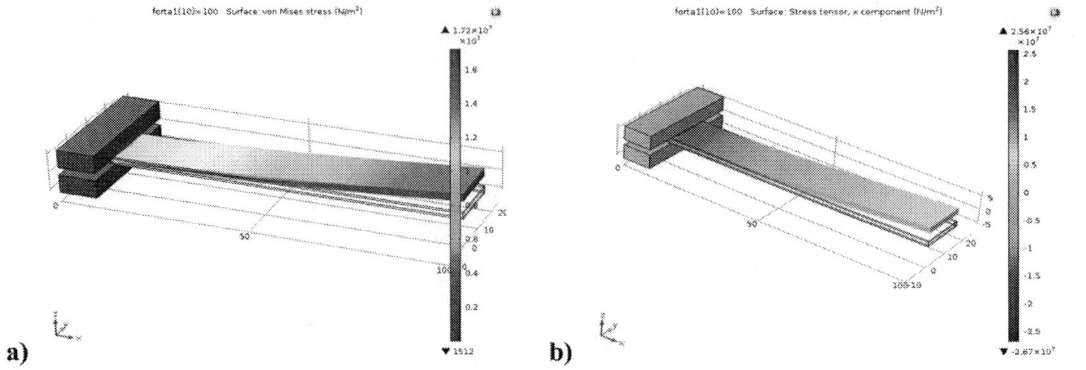

Fig. 5. a) Total mechanical von Mises stress; **b)** Stress tensor component on *x* direction

The computed values for displacement and output voltage, as well as the maximum values of the stress tensor component on *x* direction and the von Mises stress, both evaluated in the piezoelectric layers, are presented in the subsequent graphs.

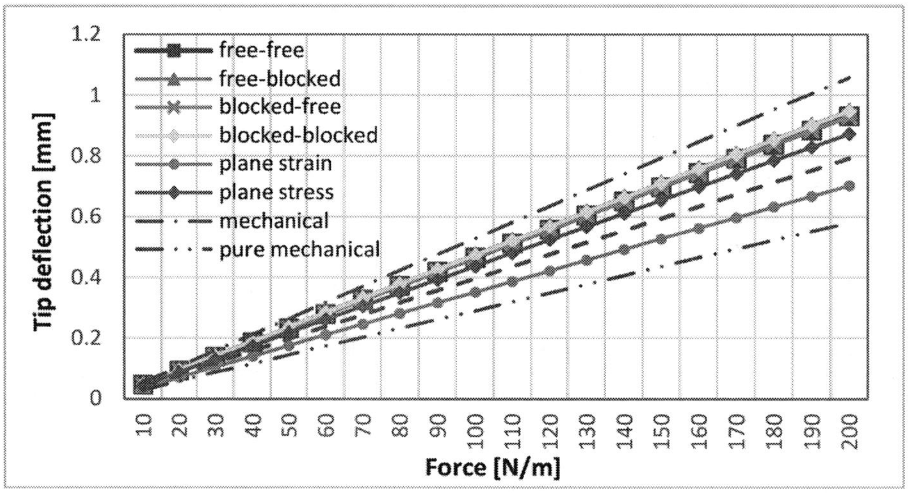

Fig. 6. Free end displacement depending on the external force applied

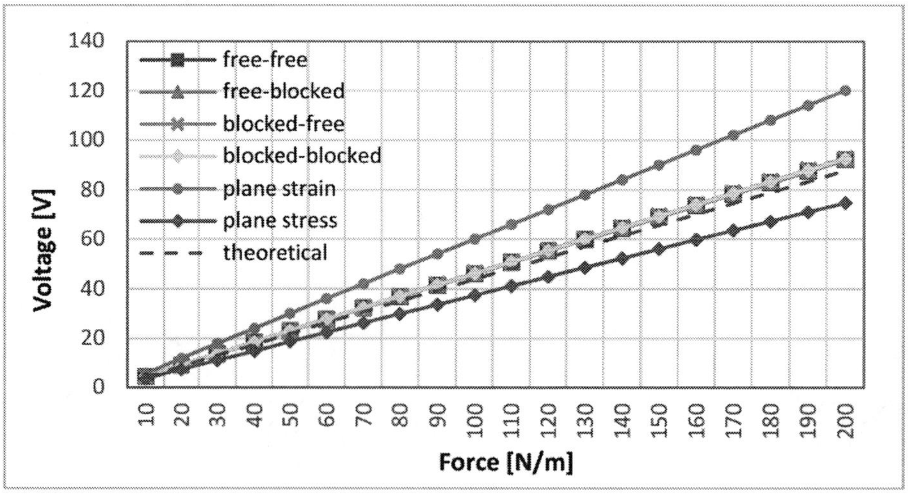

Fig. 7. Voltage depending on the external force applied

In the case of displacements (figure 6), the plane stress model underrates by about 5% and the plane strain model by about 20%. The theoretical displacements are situated between those of 2D models. If for voltages (figure 7) there are no relevant differences between the analytic model and the 3D models (under 3%), higher differences appear when using 2D models (the plane strain model overrates by about 30% and the plane stress model underrates by about 10%). In the case of von Mises stresses (figure 8), the plane stress model overrates negligibly (below 2%) and the plane strain underrates by about 15%.

For x-direction component of the mechanical stress tensor (figure 9), the 2D models (plane stress and plane strain) give similar results and very close to the theoretical model, but they are all significantly underrated comparing with the 3D models. These are grouped into models with a mobile tip subjected to deformations, with blocked curvature and respectively free. The last category leads to mechanical stress having very close value to the mechanical anisotropic model, about 20% smaller than the first group (which coincides with the mechanical isotropic model), and followed by about 20% smaller with 2D models.

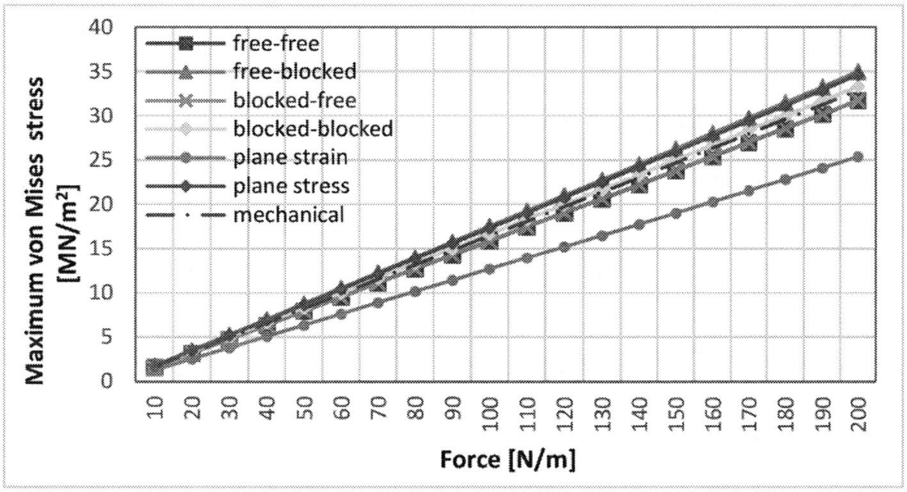

Fig. 8. Von Mises stress depending on the external force applied

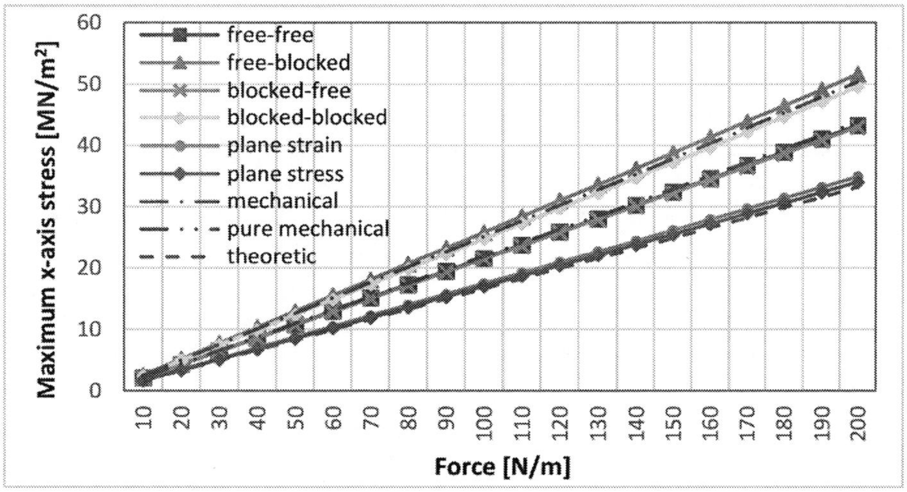

Fig. 9. Mechanical stress tensor component on x direction depending on the external force applied

These results lead to the idea of an underrating of the mechanical stresses in isotropic or theoretical models and in the 2D models, which are consistent with previous studies on modelling of piezoelectric actuators [11].

Differences can occur due to blocking free end deformation, but also due to not taking into account the plate assumptions. The current use of increasingly large piezoelectric structures, possible to manufacture as a result of technological advances, may amplify the errors due to neglecting the plate type models. Consequently, we analysed the influence of harvester width on the dimensions evaluated above, presented in the subsequent section.

5 Validity of beam models for large width structures

The first question is the independency of displacement in relation to beam width, assumed in the theoretical model. In the first case, the cantilever beam was considered fastened between two rigid parts in the clamping section, with free bending of the rest of the structure, on the in-plane directions y and x. The second case supposes constraining the bending of the free end, by means of other two bars.

Mechanical load is defined as the constant per width (total force will increase with increasing width, but it should lead, according to the theoretical model, to constant values of tip deflection, voltage and mechanical stress).

The ratio between width b and the overall thickness *(2hp+hs)* was varied. A variation is obtained for the output voltage (figure 10 a), the deflection at the free end (figure 10 b), the maximum von Mises stress in the piezoelectric layers (figure 10 c) and the maximum x-axis component of the stress tensor in the piezoelectric layers (figure 10 d). Theoretically, according to the linear model, all these parameters should be independent of beam width b.

Analysing the output voltage variation, we observed that the theoretical model is a good approximation for smaller width beams. For high width beams the voltages increase, with a higher value for blocked curvature of free end. The values of output voltage tend to stabilize as the width grows. The maximum voltages and deflections are limited by the lowest material yield strength.

The deflections of the free end on perpendicular direction on the trimorph were evaluated as average deflection of the superior edge. The deflections for constricted end were evaluated on the superior edge of the mobile end.

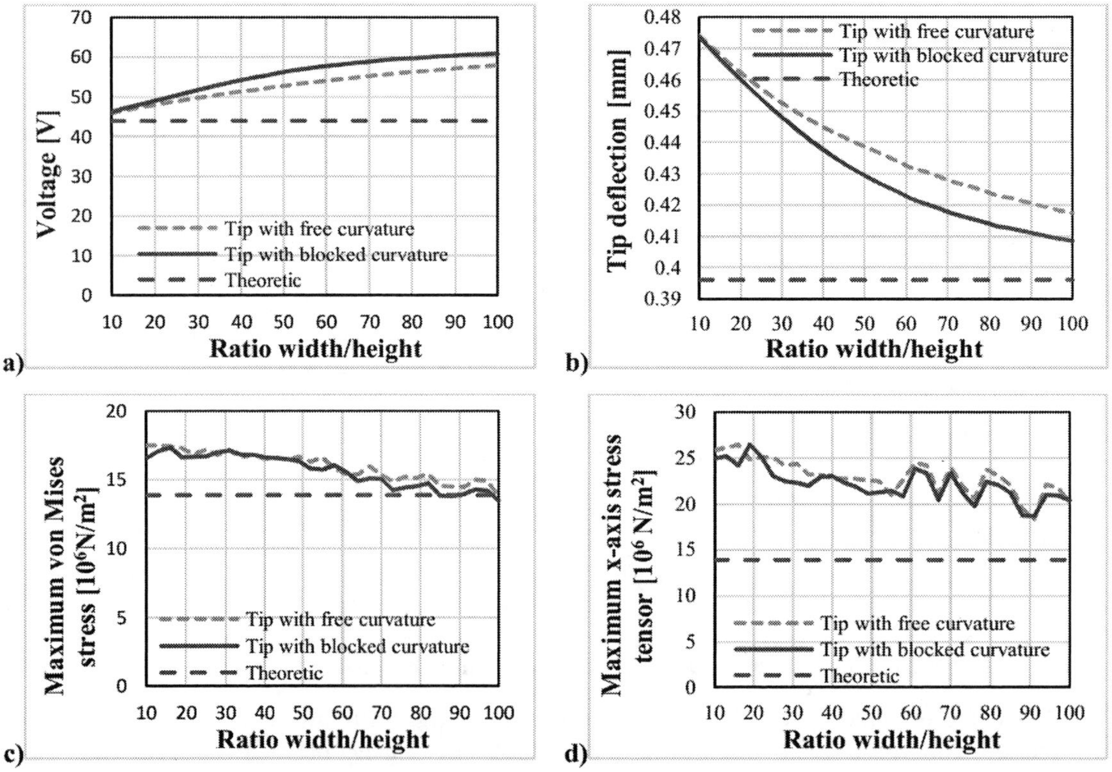

Fig. 10. a) Total voltage; **b)** Tip displacement; **c)** Maximum von Mises stress and **b)** Maximum x-axis stress tensor component, vs. width/height ratio, computed for constant external force

We notice that for high *b/(2hp+hs)* ratio, in case of free curvature tip, the deflection value for the superior edge is higher than for blocked curvature tip. For small *b/(2hp+hs)*, both are higher than the theoretical values, but tend towards these for a very wide beam.

High von Mises stresses occur mainly not due to the piezoelectric bending in *xOz* plane, but due to the clamping that constrains the bending in *yOz* plane [11].

There are significant differences between FEM simulations for small *b/(2hp+hs)* ratios. At high ratios, differences do exist, but very small, negligible, for von Mises stress, and smallest but not negligible for *x*-axis stress component. Nevertheless, both FEM simulations show differences more than 50-60% higher than the theoretical values calculated analytically with Euler-Bernoulli beam theory model.

6 Conclusions

The voltage values for 3D modelling are not dependent on the mechanical clamping mode and are comparable to the Euler-Bernoulli values for small width/height ratios, but however increase for large ratios, tending to stabilize as the ratio increases. Displacements for 3D modelling are higher than those for the Euler-Bernoulli model.

The mechanical von Mises maximum stresses in the piezoelectric layers for 3D simulations are comparable to the theoretical ones, but the component in the *x*-direction of the mechanical stress tensor depends on the free end clamping mode and is higher for 3D modelling than for the analytical model. Simplified 2D models lead to overrated stress (for plane strain) or significant stress (for plane stress). 2D models can be used for mechanical computations (displacement and von Mises stress), mainly for plane stress model.

At large beam widths, the behaviour of the harvester changes significantly: voltage increases and tends to stabilize, displacement decreases and tends to stabilize and the mechanical stress decreases with width. The theoretical model offers a good approximation of voltages at small widths but not for displacements and mechanical stresses at large widths. When not properly considering the sizes, the theoretical model underrates these parameters. An improved model, considering the plate theory, should be analysed.

The computations did not consider the yield strength limitations when applying the variable force, the purpose only being of assessing the general behaviour of the trimorph.

References

1. T. Silva, M. Clementino, A. Erturk, C. De Marqui, Mechatronics **54**, 133-143 (2018).
2. D. Gibus, P. Gasnier, A. Morel, S. Boisseau, A. Badel, *PowerMEMS 2018 Conference.*
3. A.M. Matos, J.M. Guedes, K.P. Jayachandran, H.C. Rodrigues, Comput. & Struct. **192**, 144-156 (Elsevier, 2017).
4. L. Dal Bo, P. Gardonio, J. of Sound and Vibr. **433**, 385-424 (2018).
5. P.V. Malaji and S.F. Ali, Appl. Phys. Letters **111**, no. 8, 083901 (2017).
6. K. Yamaguchi, T. Fujita, K. Kanda, K. Maenaka, J. of Phys.: Conf. Series **660**, 012116 (2015).
7. Z. Saddi, H. Takhedmit, A. Karami, P. Basset, L. Cirio, J. of Phys.: Conf. Series **773**, 012111 (2016).
8. Y. Li, M. Misra, S. Gregori, *2016 IEEE International Conference on Electronics, Circuits and Systems ICECS* (2016).
9. R. Toyabur, J. Kim, J. Park, *J. of Phys.: Conf. Series* **1052**, 012051 (2018).
10. H. Madinei, H. Khodaparast, S. Adhikari, M. Friswell, Shock & Vibr., Aircr./Aerosp., Ener. Harv., Acous. & Opt. **9**, 189-195 (2016).
11. C.I. Borzea, C.D. Comeagă, *The 11th International Symposium on Advanced Topics in Electrical Engineering ATEE* (Bucharest, 2019).
12. C.D. Comeagă, *Materiale şi structuri active piezoelectrice* (*Piezoelectric Active Materials and Structures*), 21-40 (Bucureşti, Printech, 1999).

E3S Web of Conferences **112**, 04002 (2019) https://doi.org/10.1051/e3sconf/201911204002
TE-RE-RD 2019

The concept of sustainability applied to an industrial sludge deposit

Mariana-Florentina Ştefănescu[1], *Alexandru* Dobrovicescu[1], *Iuliana-Marlena* Prodea[1], and *Nicoleta* Sporea[1*]

[1]University POLITEHNICA of Bucharest, Faculty of Mechanical Engineering and Mechatronics, 313 Spl. Independenţei, 060042 Romania

Abstract. The paper presents an environmental impact study in a real case: a landfill containing hazardous substances, held by an economic operator in an area near a city. The sludge storage was monitored taking into account three main impact factors: soil, groundwater and leachate. The actual environmental status of the storage area is assessed by calculating these impact factors before and during ten years of storage and the global pollution index.

1 Introduction

Industrial waste deposits are encountered in all industrial processes. Almost every industry either generates hazardous waste or uses products from industries that produce hazardous waste. All sludge from industrial wastewater treatment contains hazardous substances which, depending on their content and percentage, endanger the health of population and the environment. The presence of heavy metals is a real problem for the environment.

The way these wastes are landfilled must be pursued through a monitoring activity of the environmental factors affected by the presence of waste. The values of the impact factors showing that the environment is not affected give to the industrial activities, responsible for generating and managing these wastes, the status of sustainable development. In the literature [1-6] there have been studies on the impact of deposits from various industrial sources.

The monitoring of environmental factors for the exploitation phase is based on the provisions of the current legislation elaborated on the basis of Directive 1999/31/EC on the landfill of waste, as subsequently amended and supplemented and Directive 2008/98/EC on waste disposal of certain directives [7-9].

The paper presents the environmental impact of a sludge deposit resulting from the purification of the wastewater obtained in the galvanizing process of the drawn steel wire. The sludge is generated from two sources of wastewater: the galvanizing thermal process and the neutralization process of acidic waters by the precipitation of heavy metals as hardly soluble hydroxides. The sludge thus separated has high water content (about 85-95%). Therefore, before being evacuated, the sludge cake is subjected to a dehydration process, so that its final humidity is max. 60%.

* Corresponding author: nsporea@yahoo.com

The sludge storage cell is made in accordance with legal requirements with a waterproofing barrier with clay, bentonite and high density geomembrane.

The impact of the deposit on the environment during the exploitation period is determined by comparing the values of the environmental factors in the reference samples before the site is made and during the storage period. Thus, water and soil samples were taken from the storage area and the adjacent area before the industrial sludge was stored and during ten years of storage. The monitored environmental factors were: soil, groundwater and leachate. The presence of heavy metals (Cu, Fe, Zn) that exceeded the alert limits at the beginning of storage was found. The explanation was given by the presence of a second non-compliant deposit of substances from a steel plant that ceased activity without being monitored and treated. The quality of environmental factors has been assessed by the individual *Pollution Indices*, which determined the global effect of the deposit on the environment through the *Global Pollution Index* (IpG) chart. The conclusion of the study is that the environment is subjected to human activity within acceptable limits.

2 Investigation methods

The environmental impact produced by sludge during the exploitation period was established using the *Illustrative Method of Assessing the Environmental Quality Status* [10]. This method assesses the impact of industrial activity on each environmental factor as well as the associated impact of environmental factors. The quality of each environmental factor is framed by a *creditworthiness scale*, by providing notes that show proximity or distance from the ideal state. The *creditworthiness scale* is expressed by numbers from 1 to 10, in which 1 represents the irreversible and particularly serious deterioration of the analysed environmental factor, and 10 the natural state, unaffected by the industrial activity. The credit scores for each environmental factor are based on the norms in force.

The quality of environmental factors is estimated by the *Pollution Index*, Ip, which is calculated by dividing the maximum concentration of the measured pollutant, C_{max}, to the maximum allowable concentration according to the legislation in force, C_a:

$$Ip = \frac{C_{\max}}{C_a} \quad (1)$$

The values of the Pollution Index, the corresponding creditworthiness notes and the correlation with the effect on the environment are presented in Table 1.

Table 1. Pollution Index, the corresponding creditworthiness notes and the correlation with the effect on the environment [10].

Pollution Index (Ip)	Creditworthiness note	The effect on the environment
0	10	Environment unaffected by human activity, natural environment.
(0.0 – 0.2]	9	The environment affected by human activity, without quantifiable effects.
(0.2 – 0.7]	8	The environment is affected within the permissible: level 1. Alert limit: with potential effects.
(0.7 – 1.1]	7	The environment is affected within acceptable: level 2 limit. Intervention threshold: significant effects.
(1.1 – 2.0]	6	The environment is affected within acceptable limits: level 1. Alert limit: with potential effects.
(2.0 – 4.0]	5	The environment is affected within acceptable limits: level 2. Intervention level: significant effects

Pollution Index (Ip)	Creditworthiness note	The effect on the environment
(4.0 – 8.0]	4	The environment is affected within the admissible level: 3 limits. Harmful effects are highlighted.
(8.0 – 12.0]	3	The environment is degraded to level 1. The effects are lethal at average exposure times.
(12.0 – 20.0]	2	The environment is degraded level 2. The effects are lethal at short exposure times.
> 20.0	1	The environment is inappropriate for life forms.

One of the most important steps is to identify the full impact effect of pollution factors analysed. The Creditworthiness notes for each environmental factor in the analyzed area serve to graphically plot a chart as a method of simulating synergic effects.

The ideal state is represented graphically by a regular geometric shape - area S_0 - with the sides having the value of 10 Creditworthiness notes for the analyzed environmental factors. The values of the corresponding Creditworthiness notes of each measured environmental factor are represented in the ideal state graph. An irregular geometrical figure having a surface (S_r) smaller than ideal one (S_0) results. The global pollution status of the environment is evaluated by the Global Pollution Index (Ip_G) calculated with the relationship:

$$Ip_G = \frac{S_0}{S_r} \tag{2}$$

When there are changes in environmental factors, Ip_G will have higher supraunit values as the surface geometry figure reduces. For the Global Pollution Index, a scale of 1 to 6 was admitted (Table 2).

Table 2. The Global Pollution Index (Ip_G) of an ecosystem and its effect on the environment [10].

Global Pollution Index (Ip_G)	The effect on the environment
1	Environment unaffected by human activity, natural environment.
(1 – 2)	Environment subjected by human activity within acceptable limits.
[2 – 3)	Environment subjected by human activity that causes discomfort to life forms.
[3 – 4)	Environment affected by human activity, causing disturbances to life forms.
[4 – 6)	Environment seriously affected by human activity, dangerous to life forms.
≥ 6	Degraded environment, improperly for life forms.

The method has the advantage that it offers the possibility of establishing a direct link between the quality of the environment and the health of the population.

3 Experiments

Objectives pursued are: monitoring the technological waste characteristics; soil monitoring in the area of influence of the deposit - by sampling 4 points (on each side of the deposit) at two different depths (5 cm and 30 cm); ground water monitoring - through control drilling at least three points in the flow direction (one point upstream and two downstream from the warehouse); leachate monitoring.

Heavy metals are defined as factors of concern/interest that represent the hazardous part of the waste. And these are: Fe, Cu and Zn that are found in the waste of the technological process studied.

3.1 The characterization of the sludge produced by the treatment of wastewater from the technological process

The monitoring period of the deposit began at the opening and took place over a period of ten years. The galvanizing thermal process has used the same raw materials and technologies throughout all the monitoring period. This resulted in the same average daily composition of the sludge containing dangerous substances – heavy metals (Table 3).

Table 3. The chemical composition of the sludge at the beginning of the storage operation and after ten years.

Measured indicator	Permissible limit values (mg/kg s.u.) [7]	Measured values (mg/kg s.u.)	
		at the opening	after ten years
As	25	<0.05	<0.05
Ba	30	0.03	0.07
Cd	5	0.01	<0.01
Cr total	70	0.03	0.015
Cu	100	<0.04	0.1
Hg	2	<0.05	<0.05
Mo	30	<0.01	0.16
Ni	40	0.88	0.66
Pb	50	<0.05	<0.05
Sb	5	<0.05	<0.05
Se	7	<0.1	0.12
Zn	200	2.47	0.56

By analysing the composition of the sludge (Table 3), much lower concentrations are observed compared to the values set in the legislation, which makes the sludge produced by treating the wastewater, from the technological process, to be stored as it is, being conform with Best Available Techniques for the Waste Treatments Industries [8, 9].

3.2 Land quality monitoring before storage

The deposit is located in the immediate vicinity of the factory which produces the sludge. There are no natural waters or rivers or lakes on the territory related to the studied objective. The samples were taken from each side of the deposit at 2 different depths (5 and 30 cm) and contain heavy metals. The depths were chosen so that any contamination in the depth could be estimated. Samples P1, P3, P5 and P7 were taken from a depth of 5 cm and samples P2, P4, P6 and P8 from a depth of 30 cm. Following the soil samples analysed taken prior to the commissioning of the deposit, copper, lead, zinc, manganese are present under the normal limits established by the regulations, as well as ferric oxide, which is not regulated by normative acts. The composition of the soil is shown in Table 4.

Table 4. Land quality before storage: measured values vs. limits.

Sample	Side of the deposit	Pollutant determined	Limits [7-9] (mg/kg s.u.)			Measured values (mg/kg s.u.)	
						depth (cm)	
			normal values	alert thres-hold	inter-venetion thres-hold	5	30
P1, P2	East	Cu	20	100	200	19.4	11.2
		Pb	20	50	100	8.3	1.1

		Zn	100	300	600	37.9	19.4
		Mn	900	1500	2500	534	261
		Fe_2O_3 (%)	-	-	-	5.56	2.6
P3, P4	North	Cu	20	100	200	17.2	9.6
		Pb	20	50	100	5.8	0.75
		Zn	100	300	600	46.5	23.4
		Mn	900	1500	2500	484	213
		Fe_2O_3 (%)	-	-	-	5.27	2.1
P5, P6	South	Cu	20	100	200	15,7	16.3
		Pb	20	50	100	3,9	4.2
		Zn	100	300	600	69,5	21.9
		Mn	900	1500	2500	311	302
		Fe_2O_3 (%)	-	-	-	3.05	2.98
P7, P8	West	Cu	20	100	200	13.9	14.1
		Pb	20	50	100	2.4	2.5
		Zn	100	300	600	80.3	25
		Mn	900	1500	2500	324	325
		Fe_2O_3 (%)	-	-	-	3.19	3.12

3.3 Land quality monitoring after ten years of storage and environmental impact

Values measured after 10 years of storage (Table 5) are below the normal values shown in Table 3, except for the Cu indicator. The concentration of this indicator at both the 5 cm and the 30 cm depths, for all samples, does not exceed the alert and intervention thresholds as is shown in Table 4. In Table 5 are also calculated the values of the Pollution Index.

Table 5. Land quality monitoring after ten years of storage

Sample	Side of the deposit	Pollutant determined	Measured values (mg/kg s.u.) depth (cm)		Pollution Index (Ip) depth (cm)	
			5	30	5	30
P1, P2	East	Cu	**37.8**	**38.8**	**1.89**	**1.94**
		Pb	13.4	12.5	0.67	0.625
		Zn	58.9	61.7	0.589	0.617
		Mn	575	584	0.639	0.649
		Fe_2O_3 (%)	128	123	-	-
P3, P4	North	Cu	**35.2**	**37.4**	**1.76**	**1.87**
		Pb	9.4	9.8	0.47	0.49
		Zn	46.7	48.3	0.467	0.483
		Mn	556	567	0.618	0.63
		Fe_2O_3 (%)	132	137	-	-
P5, P6	South	Cu	**31.5**	**33.7**	**1.575**	**1.685**
		Pb	11.6	11.8	0.58	0.59
		Zn	53.2	56.3	0.532	0.563
		Mn	537	541	0.597	0.601
		Fe_2O_3 (%)	125	130	-	-
P7, P8	West	Cu	**27.8**	**28.2**	**1.35**	**1.41**
		Pb	8.4	9.6	0.42	0.48
		Zn	71.3	78.4	***0.713***	***0.784***
		Mn	588	571	0.653	0.634
		Fe_2O_3 (%)	119	121	-	-

The highest level of pollution of the land, calculated with normal values (Table 4) and relation (1), is recorded in the eastern part of the deposit (P1, P2) for Cu. These values correspond to Creditworthiness note 6: *The environment is affected within acceptable limits: Level 1. Alert limit: potential effects.* For the elements Pb and Mn the level of pollution correspond to Creditworthiness note 8: *The environment is affected within the permissible: level 1. Alert limit: with potential effects.* Zinc pollution level has the creditworthiness note 6, with the exception of the western area where, for both depths, the creditworthiness note is 7: *The environment is affected within acceptable: level 2 limit. Intervention threshold: significant effects.*

3.4 Groundwater monitoring

Water samples were taken by carrying out three drillings in the direction of flow of groundwater (a point located upstream and two downstream from the deposit) in accordance with environmental regulations before commencement of storage (reference values) and every six months of operation. In Table 6 the results are presented only for drilling 2, as the synthesis of the results is presented in full in Table 7. The following indicators were measured: pH, total ionic Fe, chlorides, CCO-Cr, CBO_5, ammoniacal nitrogen, nitrates, sulphates, copper, zinc.

Table 6. Groundwater monitoring: measurement values vs. reference values.

Date of sam-pling (mon ths)	pH	CCO-Cr mg/l	CBO_5 mg/l	A-mm-onia-cal nitro-gen mg/l	Nitra-tes mg/l	Chlo-rides mg/l	Sul-phates mg/l	Fe mg/l	Cu mg/l	Zn mg/l
Point drilling 2										
Reference values										
before storage	7.27	24	7.56	0.42	0.92	220.6	515.1	0.47	0.04	0.01
Experimental measurements										
after 6	6.71	**33**	**8.85**	**0.54**	**0.98**	**236.4**	**529.7**	**0.58**	**0.05**	**0.02**
after 12	**7.38**	16	7.41	**0.47**	**0.98**	**232.5**	**543.4**	**0.48**	**0.05**	**0.02**
after 18	**7.39**	18	**7.82**	**0.48**	0.82	211.6	**521.8**	0.44	**0.05**	**0.03**
after 24	**7.42**	21	**9.2**	0.4	0.72	204.3	505.8	0.41	**0.07**	**0.04**
after 30	**7.4**	20.4	**9.04**	**0.44**	0.77	211.5	510.7	0.4	**0.06**	**0.05**
after 36	**7.33**	19.7	**8.14**	**0.81**	0.56	192.3	527.6	0.38	0.03	0.01
after 42	**7.31**	21.3	**8.93**	0.36	0.34	66.5	58.3	0.13	0.03	0.01

Measurements have shown an abnormal excess from reference values in the early years since the start of the warehouse operation, although the technological process has remained unchanged. The most significant differences, in comparison with baseline values, were recorded in the first 42 months at all three drilling points for almost all indicators (bold values in Table 6). Table 7 shows the values of the Pollution Index for the highest overcomes of the reference values and the effects of their presence on the environment.

The environmental impact for all monitored parameters is in Creditworthiness note 7: *The environment is affected within acceptable: level 2 limit. Intervention threshold: significant effects* and Creditworthiness note 6: *The environment is affected within acceptable limits: Level 1. Alert limit: potential effects*, except Zn. This indicator has, in drilling point 2, Creditworthiness note 4: *The environment is affected within the admissible level: 3 limits. Harmful effects are highlighted.*

Table 7. The Pollution Index for the highest overcome of reference values and the corresponding Creditworthiness note.

pH	**CCO-Cr mg/l**	**CBO_5 mg/l**	**A-mm-onia-cal nitro-gen mg/l**	**Nitra-tes mg/l**	**Chlori-des mg/l**	**Sul-phates mg/l**	**Fe mg/l**	**Cu mg/l**	**Zn mg/l**
Point drilling 1									
Pollution Index (Ip)									
1.048	1.104	1.087	1.156	1.090	1.243	1.071	1.231	1.75	2
Creditworthiness note									
7	*6*	*7*	*6*	*7*	*6*	*7*	*6*	*6*	*6*
Point drilling 2									
Pollution Index (Ip)									
1.021	1.375	1.217	1.928	1.065	1.072	1.055	1.234	1.75	5
Creditworthiness note									
7	*6*	*6*	*6*	*7*	*7*	*7*	*6*	*6*	***4***
Point drilling 3									
Pollution Index (Ip)									
1.031	0.915	1.137	1.159	1.145	1.113	1.055	1.267	1.25	1
Creditworthiness note									
7	*7*	*6*	*6*	*6*	*6*	*7*	*6*	*6*	*7*

In order to analyse the impact on the territory related to the studied objective, it must be said that there is also a non-permeable deposit nearby, from another industrial operator, operable at the time of construction of our deposit, which had stockade steel slag.

This proximity with steel slag from the non-permeable deposit and operating in the first 4 years at the time of measurements is the cause of abnormal values for Fe, Cu and Zn factors in the first 7 measurements performed during the first 42 months of monitoring. The leachates caused by rain have increased the percentage of Fe in groundwater. The reference value for Fe (0.47) was also influenced by the leachate in the non-impermealized deposit.

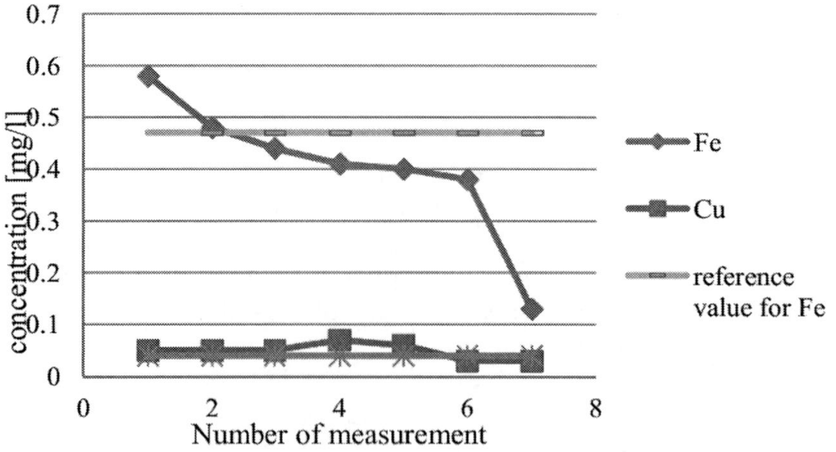

Fig. 1. Concentration in Fe and Cu in groundwater during the first 42 months of monitoring (point drilling 2)

Figure 1 and 2 show the highest Fe input from the non-permeable deposit (obtained at the second drilling) in groundwater, compared to Zn and Cu participations. These parameters begin to decrease after closing the steel slag deposit, exhibiting variations below the reference values (Table 6).

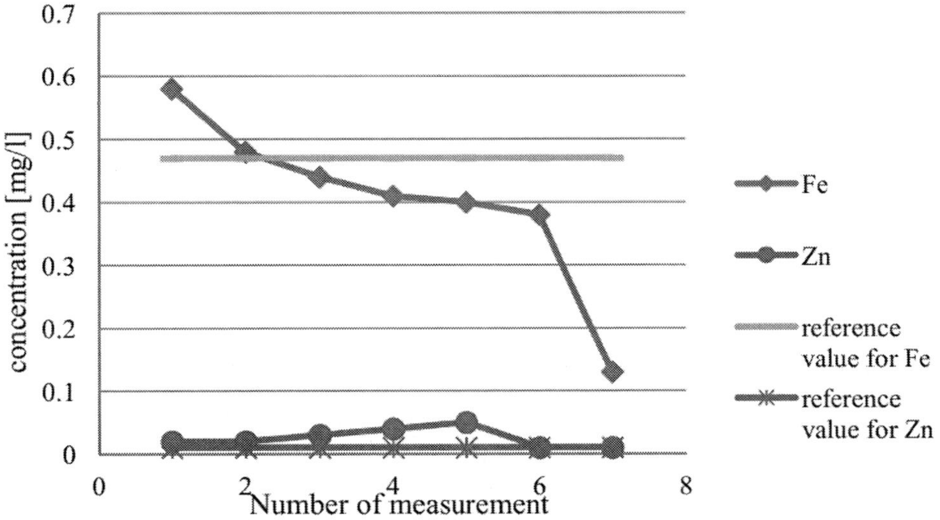

Fig. 2. Concentration in Fe and Zn in groundwater during the first 42 months of monitoring (point drilling 2)

3.5 Leachate monitoring

It is also necessary to monitor the leachate captured in the collector basin. The parameters to be monitored and the frequency of their determination is the following: total ion iron, copper, zinc - once a quarter every year beginning with the fourth year of depositing. The results are shown in Figures 3 and 4.

Table 8 presents the reference values for concentrations of heavy metals in the leachate, the Pollution Index calculated for the maximum measured values (Fig. 3 and 4) and the creditworthiness notes.

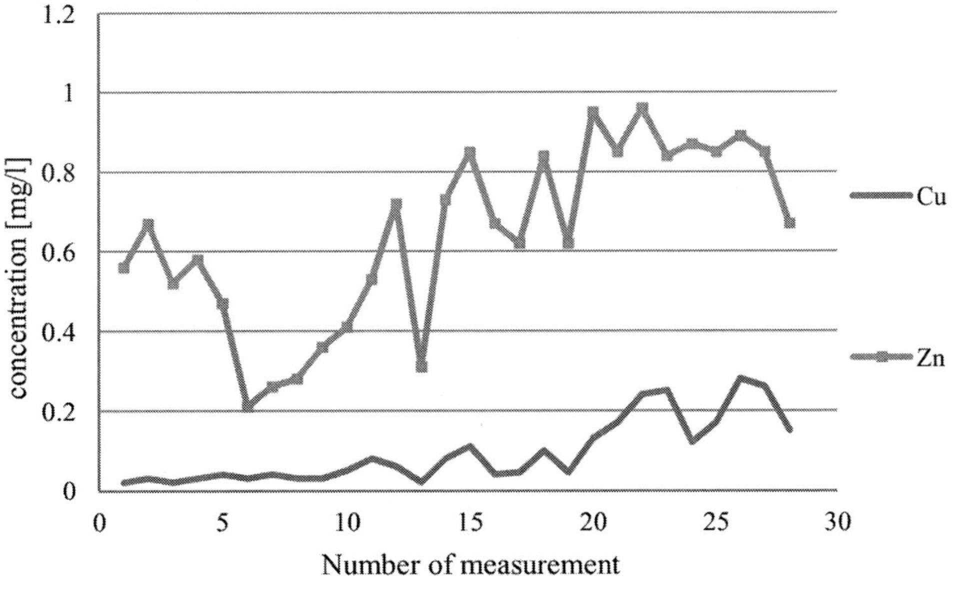

Fig. 3. Results of leachate monitoring for Cu and Zn

https://doi.org/10.1051/e3sconf/201911204002

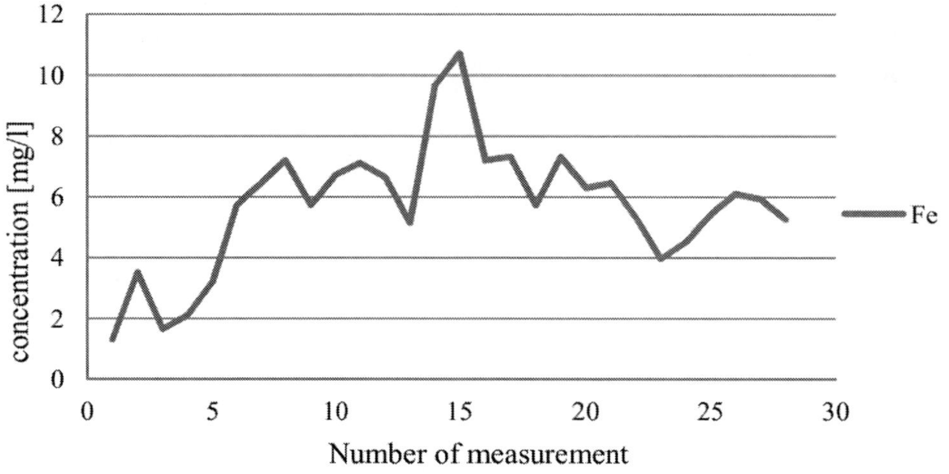

Fig. 4. Results of leachate monitoring for Fe

Table 8. The values for concentrations of heavy metals in the leachate, Pollution Index and the creditworthiness notes.

	Cu	Zn	Total ion Fe
Reference value from legislation in force [mg/l]	10	10	Not specified
Maxim value of measured concentration [mg/l]	0.28	0.96	10.72
Pollution Index (Ip)	0.028	0.096	-
Creditworthiness note	9	9	-

The effect of the Cu and Zn content from the leachate produced by sludge deposition results from the analysed process of technology is included in the Creditworthiness notes 9 (Table 1): *The environment affected by human activity, without quantifiable effects*.

3.6 The cumulative effect of factors

In order to determine the global impact of the measured pollutants on the environment, the Global Pollution Index (Ip_G) chart is built. The geometric figure is a triangle with data for three quality factors: soil, groundwater and leachate (Fig. 5).

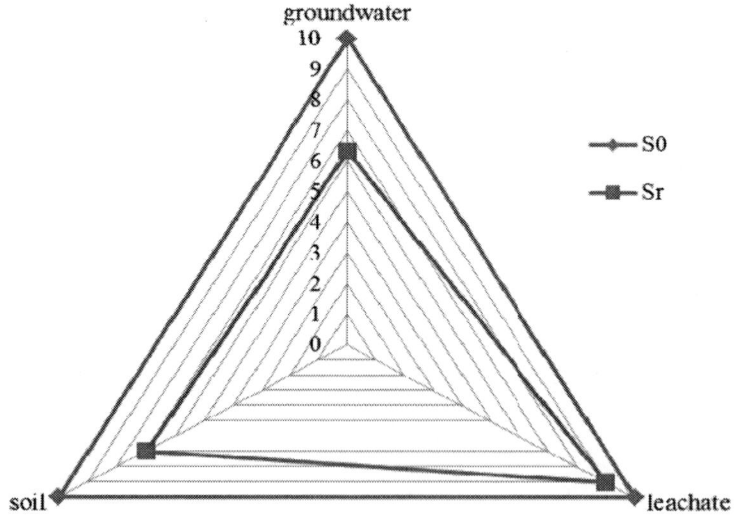

Fig. 5. Global Pollution Index (Ip_G) chart

These values were design on Creditworthiness notes chart (Fig. 5) resulting an asymmetrical triangle. The value of the Global Pollution Index (Ip_G) calculated with Eq. (2) is 1.438. This value means that the effect of the heavy metals from the sludge deposit on the environment, for the period analysed, is in acceptable limits (Table 2).

4. Conclusions

Sustainable development in the industrial sector is assessed, among other things, by how produced waste affects environmental factors.

To assess the impact of sludge on the environment, four factors are taken into account: the sludge composition, soil, groundwater and leachate produced in the landfill. The results obtained by the measurements, before sludge storage and during ten years of storage, showed the presence of dangerous substances in the sludge, the most significant effect being in the form of heavy metals (Cu, Zn and Fe), substances that were transferred to the environment: soil, groundwater and leachate. With the working method supported by the literature, method of the Creditworthiness notes, it has been interpreted the effect of each element on the environment by comparing the ideal / reference value and the real value at a certain time.

The study's results show that in the first 42 months soil and groundwater have undergone changes by aggregating the effects of the sludge deposit and the influence of a non-permeable deposit situated nearby, steel slag deposit that was closed after 42 months of opening the study. Over the next six years, the level of heavy metal contamination decreases to values close to the reference levels (Fig. 1 to 4). The global effect of heavy metals on the physical environment (ground and groundwater) was established using the Global Pollution Index (Fig. 5) and the result is: *Environment subjected by human activity within acceptable limits* (Table 2).

References

1. U.M. Uwem, D.M. Sunday, A.F. Wosilat, U.U Jonah, Science Journal of Analytical Chemistry, **7**, 1 (2019)
2. M.F.L. Barbosa, A.B.S. Pironcelli, C.A. Silva et al., Asian J. Civ. Eng. (2019)
3. U.M. Uwem, D.M. Sunday, Chemical Science International Journal, **23**, 4 (2018)
4. H. Bauman-Kaszubska, M. Sikorski, Journal of Ecological Engineering, **19**, 6 (2018)
5. L. Chiriac, Buletin AGIR, 4 (2016)
6. C. Preda, Dissertation, (UPB, 2016)
7. ***NTPA 001/2002 Normative on the setting of pollutant loading limits for industrial and urban waste water when discharged into natural receptors
8. ***Reference Document on Best Available Techniques Reference on Emission from Storage (2006)
9. ***Reference Document on Best Available Techniques for the Waste Treatments Industries (2006); Final Draft (October 2017)
10. V. Rojanschi, F. Grigore, V. Cimos, *Valuer's Guide and Environmental Auditor's Guide* (2008)

E3S Web of Conferences 112, 04003 (2019) https://doi.org/10.1051/e3sconf/201911204003
TE-RE-RD 2019

Theoretical analysis of the dynamics of a nacelle drive system in order to optimize the system energy

Liliana Dumitrescu[1], *Polifron Alexandru* Chiriță[1], *Corneliu* Cristescu[1], *Ioan* Lepădatu[1] and *Evelin* Laza[2]

[1]Hydraulics and Pneumatics Research Institute INOE 2000-IHP, Bucharest, Romania
[2]National Research - Development Institute for Machines and Installations Designed to Agriculture and Food Industry

Abstract: The paper presents the hydraulic system for the operation of a lifting platform in two constructive variants. One of the drive schemes is with energy recovery. A hydraulic, energy and economic analysis is performed for the two proposed drive schemes using the numerical simulation technique. The conclusions are useful in design and exploitation. The platform can be used for various applications in agricultural farms or construction.

1 Introduction

Lifting of loads in industry and agriculture is possible with different machinery, some of which are commonly called nacelles or lifting platforms. In agriculture, these fixed or mobile platforms are designed for specific works such as dressing trees cutting vineyards, harvesting fruit on orchards, etc. or handling large loads (bales of fodder, bags of cereals or fertilizers etc.) view of them storage or distribution. They can also be used by people with reduced mobility, for easy access to different institutions.

Fig. 1. Mobile platform used in viticulture for harvesting grapes [1]

Fig. 2. Mobile platform used to cleaning and cutting the fruit trees in the spring [2]

The lifting mechanism may be a telescopic arm or a metal frame in the form of a parallelogram or scissors (X-shaped) which is acted by single or double action hydraulic cylinder. Parallelogram actuation mechanism allows for large strokes depending on the number of parallelograms used, and for larger loads, two or more actuating cylinders are used.

Fig. 3. Variants of lifting platforms with scissor lifts produced by Haulotte

For mobile platforms, it is recommended that the primary energy source be made up of a set of electric batteries at the expense of the polluting thermal engine; In this case, reducing consumption leads to extending the operating time between 2 battery charging and reducing the power consumption of the batteries.

2 Function

In the case of *the classic drive scheme* (Figure 4a), the hydraulic pump (1) driven by the electric motor (2) supplies hydraulic oil to the lifting cylinder via the check valve (4) and the proportional distributor (5). The rod of the cylinder (7) extends and drives the drive mechanism (scissor type) to raise the load. At the lowering phase, the pump no longer supplies flow, and due to the weight of the platform, the oil in the lifting cylinder piston chamber passes to the tank through the proportional distributor (5) and through the throttle (8) to adjust the speed to the descent. The potential energy resulting from the lifting of the platform and the load (or only the platform, if there is no load) is not used in the descent phase in any way, with no apparatus or extra system to recover it. The diagram further includes a valve (6) which provides lowering of the load by accidental manual control and a safety valve (3). The sense valve (4) allows the platform to remain locked in the event of damage (hose break, accidental power cut).

The energy recovery system (Figure 4b) comprises, in addition to the components 1÷7 of the classic system, a pneumohydraulic accumulator (8) whose role is to store energy during the descent of the platform and to provide it at the lifting, a safety valve (3.2) which allows the hydraulic oil to pass to the reservoir if the pressure limit in the hydraulic accumulator is exceeded, and a pressure multiplication cylinder (9), which is intended to raise the level of pressure entering the accumulator, (4.1) and (4.2), distributors (5.1), (5.2), (5.3).

When the platform rise, the pump supplies oil to the lifting cylinder through the check valve, and at lowering the flow rate out of the piston chamber of the cylinder lifting passing through the distributor (5.1), the other paths are closed (valve (6), the distributor (5.2), check valve (4.1)), enters the chamber of the piston of the multi-cylinder (9) and determines its displacement. At the platform descent, the flow from the multiplier cylinder rod chamber (9) now having the increased pressure passes through the valve (4.2) and enters the accumulator (8) which stores the potential energy resulted. For a new lifting of the platform, only the energy stored in the accumulator can be used; for this, the distributor (5.2) is engaged and the flow from the accumulator reaches the lifting cylinder piston chamber (7). The flow from accumulator does not return to the multiplier cylinder due to the sensing valve (4.2). The oil from the piston chamber and rod chamber of multiplier cylinder flows through the distributor (5.1) or (5.3) to the tank to allow the multiplier cylinder to raise the fluid pressure to a new descent of the platform.

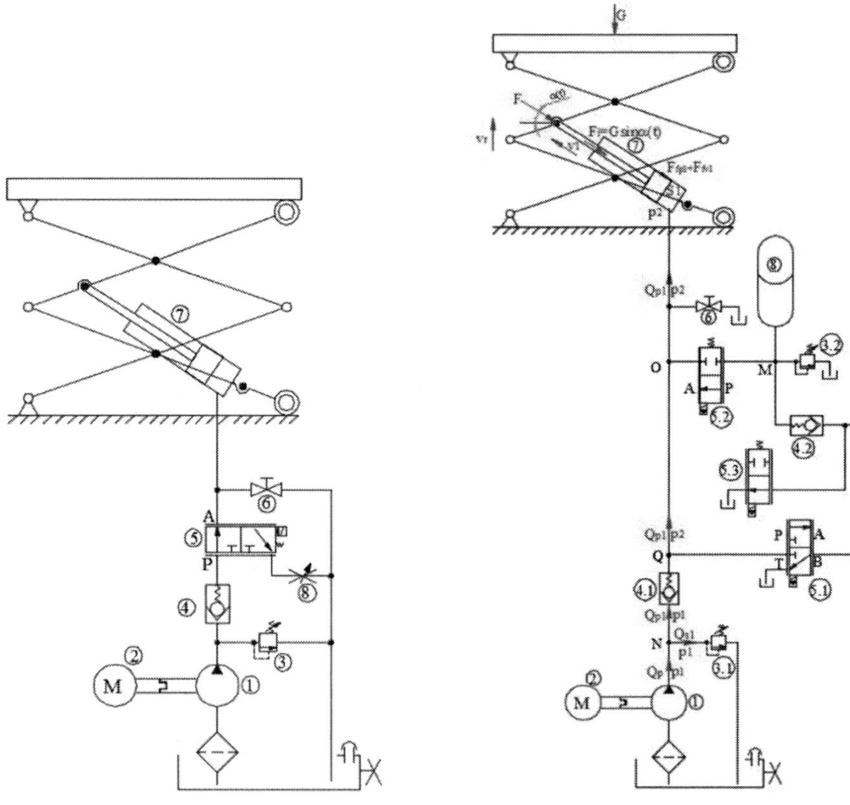

Fig. 4. Diagram of the lift platform drive system
a) No energy recovery b) With energy recovery – the lifting phase

3 Numerical simulation of the system with energy recovery and studying its dynamic behavior [3-6]

For the hydraulic energy recovery system that includes a pneumatic-hydraulic accumulator and distribution devices in the drive schematic diagram, a numerical simulation using AMESim was performed to estimate the maximum amount of energy, in percentage form, that could be recovered and which technical features the system could develop if only recovered energy is used, taking into account that any element introduced in the schematic diagram for achieving energy recovery and reuse also produces a local loss of energy.

Knowing the maximum value of energy that can be recovered in a lifting / lowering cycle, an economic analysis can be made to approximate the amount of energy recovered for a certain number of hours of operation. In other words, an estimation of the energy efficiency of the system that justifies or not the necessity for certain expenses in order to procure the equipment required for energy recovery, namely accumulator, multiplication cylinder, directional control valves, regular valves, etc., has been achieved by simulation.

3.1 Stage lifting without accumulator (fig. 4b)

The pump supplies in the system a Q_p flow at the pressure p_1. The sensing valve (4.1) located between the pump and the lifting cylinder produces a pressure drop equal to p_2 - p_1 in the circuit. The Q_{p1} flow through the sense valve is the flow that acts on the lift cylinder and moves the piston causing the platform to lift; at the entrance to the lifting cylinder, the oil pressure is p_2. Part of the flow, Q_{s1}, passes through the safety valve (3.1), being a laminar flow rate to the tank if the pressure exceeds the set value at the system safety valve.

3.2 Phase of platform descending and accumulator charge (fig. 5)

Platform descent is carried out under the action of weight, the entire volume of Q_{c1} oil in the piston chamber passes through the distributor (5.1) to the piston chamber in the pressure multiplier cylinder; at the output of the multiplier cylinder, the reduced volume of oil, Q_{c2}, has a higher pressure, p_5. The oil then passes through the check valve (4.2) and enters the accumulator at a pressure p_6, loading it. If the value of the pressure p_6 exceeds the value at which the safety valve (3.2) is set, part of the discharge flow from the multiplier cylinder will pass through the safety valve (3.2) and be discharged to the tank, Q_{s2}.
Simulation of this phase of operation has led to the determination of the theoretical maximum value of the amount of energy recovered from the energy used in the first phase.

3.3 Phase of platform lifting with reuse of energy stored in accumulator (fig. 6)

In the phase of platform lifting with reuse of the recovered energy from the accumulator, the flow of pressurized oil that supplies the hydraulic cylinder of the platform comes from both the accumulator (8), by switching the directional control valve (5.2) to position 2, and the pump (1) through the check valve (4.1), and it has the value $Q_{c1} + Q_{p1}$.
But the theoretical analysis of this phase, carried out by the authors through simulation, focused on the system's ability to function (lifting of the platform) using only the energy stored in accumulator (8), without the energy input of the pump (1). Thus, data has been obtained on the degree of achieving the working height when the cylinder is supplied with fluid only from the accumulator and taking into account the local loss in this system that does not exist in the classical system.

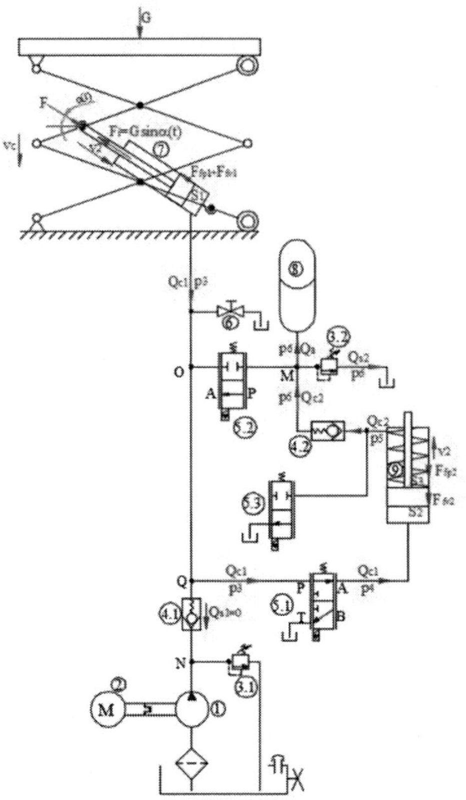

Fig. 5. Hydraulic drive system to lowering platform

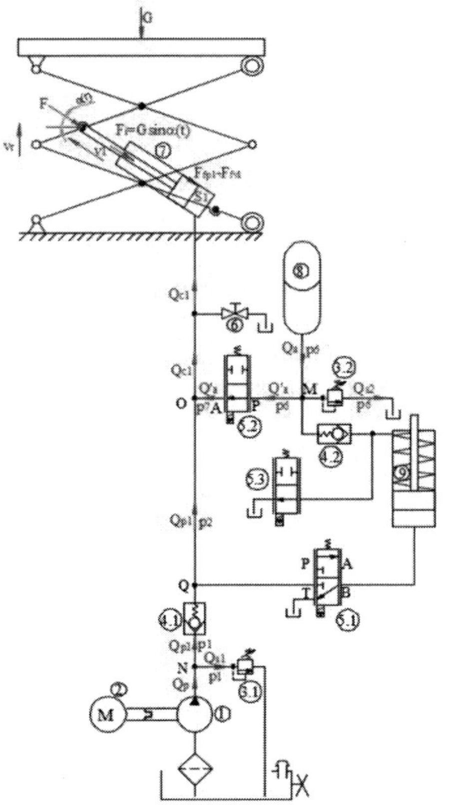

Fig. 6. Hydraulic drive system for lifting the platform with energy recovery

The system parameters for which the modeling exercise was performed are: the load: 2000 kg; the platform weight: 150 kg; the weight of an articulated bar: 30 kg; hydraulic platform cylinder stroke: 2500 mm (ϕ120 /ϕ110); accumulator volume: 20 l; min /max accumulator pressure: 112 bar /400 bar; multiplier cylinder stroke: 2500 mm (ϕ120 /ϕ90); pump capacity: 100 cm^3 /rot; motor speed: 1000 rpm; maximum safety valve pressure: 250 bar; gravitational acceleration: 9.81 m /s^2; hydraulic oil absolute viscosity: 51 cP; hydraulic oil temperature: 40^0 C.

4. Results

The simulation process was carried out over a 60 second interval, which was divided into 3 sections corresponding to the lifting /lowering /lifting with recovered energy. On the first part (Section I with the duration of 0-16 seconds), the platform is lifted only with the pump, the second part (Section II with the duration between 16 - 42 seconds) the platform is lowered and the accumulator is charged, and on the third portion (Section III with a duration of 42 to 60 seconds), the platform is lifted using only energy from the accumulator.
Figure 7 shows the time variation of platform lifting height (curve a), and time variation of rod cylinder speed (curve b). From the graph it is noted that on Section I the platform rises to a height of 10 m at a speed of 1 m /s ... 0.6

m /s, then falls below its own weight, on Section II, up to a height of approx. 1 m with an almost constant descent rate of 0.3 m /s, and then, in Section III, the platform rises again to a height of approx. 6 m, but only with accumulator power. It is noticed that the speed on Section II (lower) is lower because the platform falls below its own weight, time in which the oil flow passes the circuit up to the accumulator by charging it, but at a pressure obtained through the multiplier cylinder.

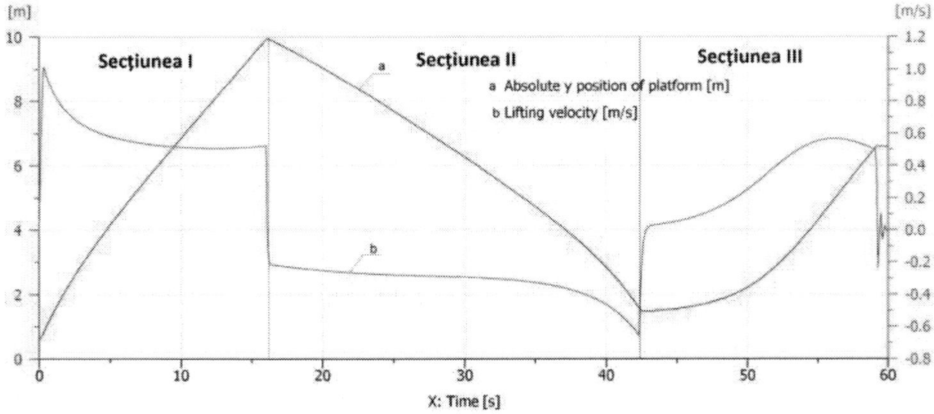

Fig. 7. Variation of platform lifting height and rod cylinder speed according to time

Figure 8 shows pressure and flow variation versus time for the hydraulic lifting cylinder. On Section I, cylinder pressure (curve a) has a maximum value at the beginning of the platform lifting, especially in the first 2-3 seconds, and it corresponds to the overcoming of a maximum force consisting of useful force, friction forces, inertia forces, etc., then the pressure drops to the nominal value and again increases towards the end of the first section, when the platform reaches the maximum height, that is 10 m. On the descent (Section II), the pressure has an almost constant slope at a value around 60 bar because charging of the accumulator takes place, so we have pressure in the circuit different from the tank pressure. On Section III, the pressure rises above the value in the previous section but without reaching the maximum value. As for the hydraulic cylinder flow (curve b), it is provided exclusively by the pump and has a constant value of 100 l/min, on Section I, then it drops to zero, on Section II, in order to allow lowering of the platform (neither the pump nor the accumulator delivers flow), and on Section III, there is only one flow, that is, only the flow provided by the accumulator; the pump does not work and therefore it does not deliver flow into the system.

Figure 9 shows the evolution of the force developed by the cylinder rod (curve a), speed of the cylinder rod (curve b) and piston stroke (curve c) for the 3 phases (lifting of the platform, lowering and lifting with energy recovery).

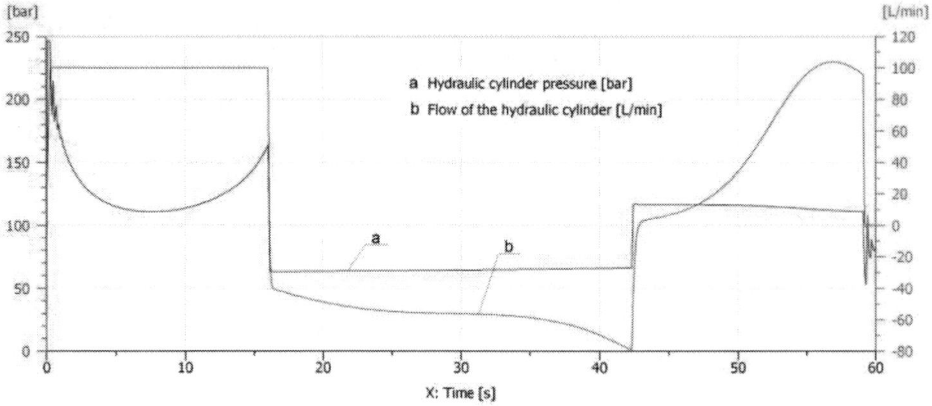

Fig. 8. Pressure and flow variation versus time for the hydraulic lifting cylinder

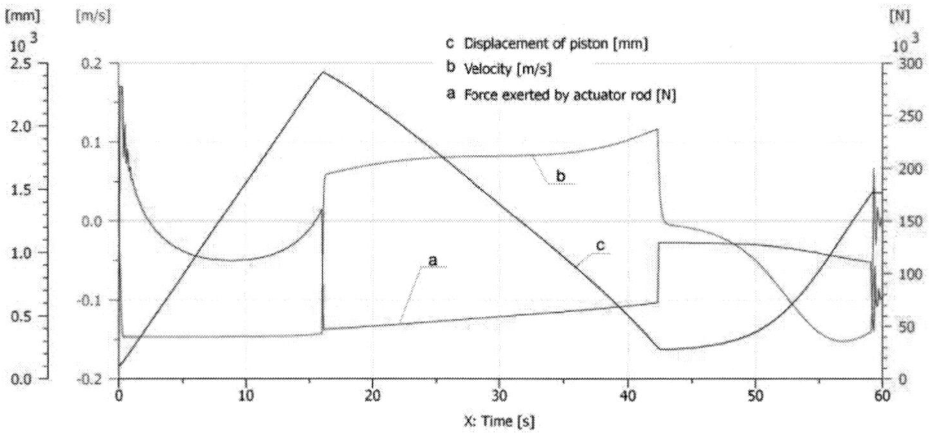

Fig. 9. Variation of the force developed by the cylinder rod, of the cylinder rod speed and piston stroke versus time

Synthesis of the theoretical results obtained by simulation for the drive system of a lifting platform with and without energy recovery is given in Figure 10 and Figure 11 below:

Figure 10 shows the graph of power provided by the pump to the system (curve a), the power received and returned to the system by the accumulator (curve b), the power of the lifting cylinder (curve c) and the graph of platform positioning versus time, in dotted line (curve d).

From this figure one can note that lifting of the platform (Section I) at a height of 10 m is carried out by the hydraulic lifting cylinder which achieves maximum force with the flow and pressure provided by the pump. On Section II the pump does not work, the platform descends under its own weight, and the power of the lifting cylinder on descent is transferred to the accumulator thus achieving the recovery of potential energy, which can be seen from the similarity of *curves b* and *c* on Section II. On Section III lifting of the platform is performed exclusively with the energy recovered in Section II which is supplied from the accumulator and raises the platform up to approx. 7 m, but starting from an initial height different from zero (approx. 1 m).

 https://doi.org/10.1051/e3sconf/201911204003

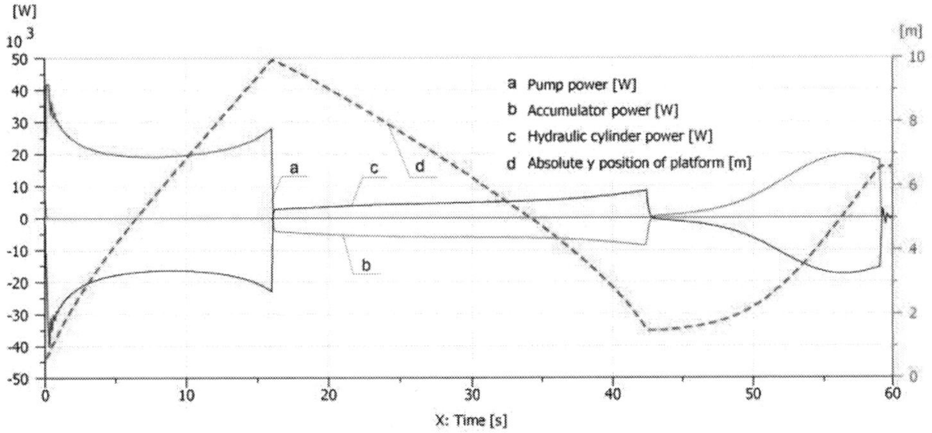

Fig. 10. Variation of power rates and height along the 3 phases

In figure 11 a can note that the energy supplied by the pump to the system on Section I (the area outlined in red), is partially used (87%) by the hydraulic lifting cylinder to accomplish the mechanical work corresponding to the lifting of the platform, so the total power consumption is of 87% (the blue shaded area).

In the case of optimizing the hydraulic schematic diagram by introducing a pressure multiplier cylinder and an accumulator, connected within the system via directional control valves and normal valves, there is achieved an energy efficiency that consists of recovering the potential descent energy, amplifying, storing and reusing it. But from the percentage of power consumed of 87% only a part can be recovered and stored in the accumulator. According to Figure 11 b, the graph of power reused in Section III, which is actually the power consumed by the cylinder at a new lifting of the platform, overlapped the graph representing the power consumed by the hydraulic lifting cylinder in Section I, and this resulted in a percentage of energy recovery of 42%.

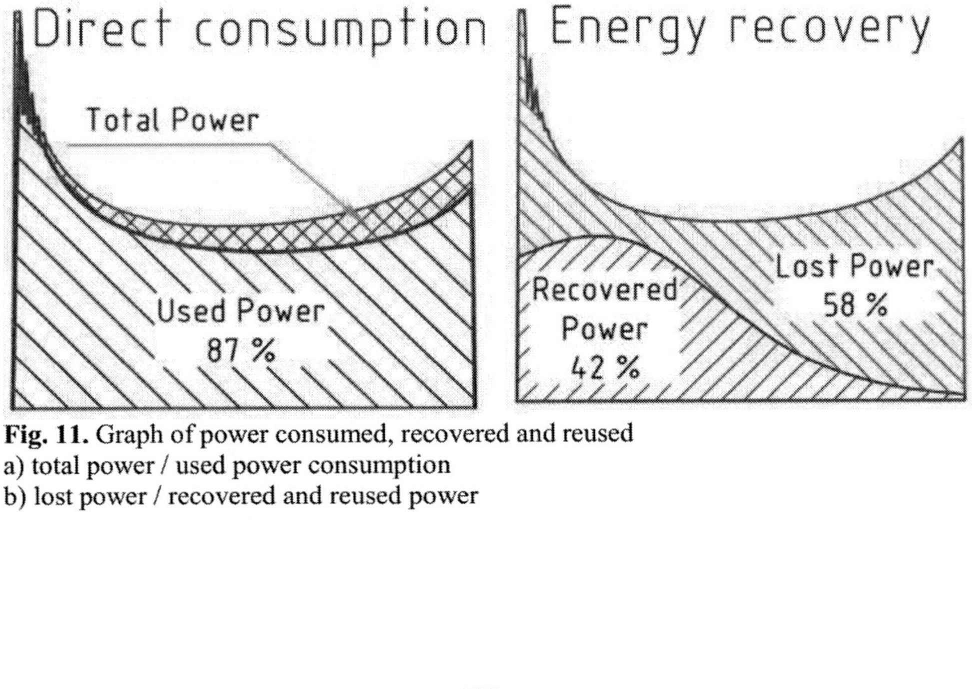

Fig. 11. Graph of power consumed, recovered and reused
a) total power / used power consumption
b) lost power / recovered and reused power

5 CONCLUSIONS

1. Simulation of the hydraulic lifting / lowering system of the platform shows that it is possible to increase the energy efficiency of the hydraulic scheme by using an accumulator storage battery and a pressure multiplier cylinder in the scheme.
2. The percentage obtained from the theoretical research (42%) represents a maximum value, achievable in a particular situation, namely: the descent of the platform (Section II) is not fully realized, up to the minimum quota, because at a new elevation due to the inclined position of cylinder, the overcoming of the force (the useful force needed to lift the platform, the friction and inertia forces, etc.) can not be achieved only with the force from the recovery of the energy. If, however, the recovered energy were large enough to overcome the force, what would remain of it after the overthrow of the forces would lead to a very small races.
3. On Section III, unimportant of the energy recovery mode, it is always necessary to use the pump to achieve the full lifting height (max. 10 m) for proper operation of the drive system.
4. The energy efficiency of the hydraulic system for lifting the platform is maximal if an energy supplier type management program (pump /accumulator) is used on Section III as follows: at the start of platform lifting, only the pump is used up to a quota that depends on each system (1 m in this case), then the pump stops and the platform continues to be lifted with the energy recovered and stored in the accumulator (up to 7 m in the case of the study), after which the pump is coupled again to achieve the required / maximum height.
5. The optimization of the hydraulic scheme of the platform's drive system by using recovered energy can be achieved for other purposes than its re-use when lifting the platform, such as moving the platform to and between working points, to illuminate working area when using the platform for farm work in the field (in this case the platform is powered by a thermal engine) etc.

Bibliography

[1] http://www.eqinto.eu/product/remorci-inox-recoltat-struguri/
[2] http://www.eqinto.eu/product-category/vii-si-livezi/nacele-autopurtate/
https://www.1234.ro/ro/platforma-fixa-de-ridicare-hw500l.html
[3] N. Vasiliu, D. Vasiliu, *Acționări hidraulice și pneumatice*, **I**, (2004)
[4] A. Bureček, L. Hružík, M. Vašina, EPJ, **114**, 6, (2016)
[5] T. C. Popescu, D. D. Ion Guță, C. Călinoiu, Sci. Bull., **73**, 12, (2011)
[6] I. A. Hazem, B. B. Samsul, S. M. Bashi, H. M. Mohammad, Aust. J. Basic & Appl. Sci., **3(4)**, 9, (2009)

Transformer lifetime management by analyzing the content of furan and gas dissolved in oil

Irina Alina Chera Anghel[1*], *Elena* Gatman[2]

[1]Junior Researcher, Energy Research and Modernization Institute – ICEMENERG, Bucharest, Romania
[2] Engineer, Energy Research and Modernization Institute – ICEMENERG, Bucharest, Romania

Abstract. In order to see the transformer insulation status, the temperature profile, the degree of polymerization (DP), the furan compound content, the carbon dioxide, and carbon monoxide concentration are monitored. By evaluating the degree of polymerization (DP) of the cellulose paper insulation and furans in the determination of the insulating oil, it can estimate the lifetime of the transformer. Analysis of cellulose insulation paper to see the degree of polymerization (DP) can be performed safely only when the transformer is interrupted. However, being a non-invasive test, determination of furane content in an oil-insulating oil is often used as an alternative method for measuring the DP average of insulation paper. In order to analyze the technical condition of a transformer, furan content in oil should be used and interpreted as a primary chemical indicator in conjunction with the results of carbon oxide from the dissolved gas analysis, the acidity and moisture results from the chemical tests, the methanol content from the chromatographic tests and other monitoring techniques.

1. Introduction

Power transformers are among the main cost elements of the power industry, with transmission and power generation functions. They are very reliable devices and with the passage of time they can be defective, and so we can denote this mode of failure, the end of life of the transformation [1]. Organic materials used in the transformer are subject to aging processes which lead to the gradual degradation of their physical, chemical and electrical properties. The most important of these organic materials is the composite insulation system, ie cellulosic materials (mainly paper and cardboard) impregnated with insulating liquid (mainly mineral). The aging of cellulose/oil insulation systems is based on three main processes.

The first process that causes aging is pyrolysis, which is initiated by temperature as the main driving force behind the chemical reactions [2,3]. Following the other hydrolysis processes, initiated by the presence of water and oxidation. Thus cellulose lifetime can be maximized by careful monitoring and maintenance. Proper monitoring of the state of the aging parameters of the insulation system could be achieved by improving the replacement planning, thus avoiding the consequences of a rapid and unforeseen transformer failure.

The lifetime of cellulose insulation is determined by mechanical strength and the best method for assessing this property is the degree of polymerization (DP) [4]. From the literature, it follows that when the DP is below 250, the end of life is reached and this DP value indicates the minimum power considered necessary to resist a short circuit.

Since it is not possible to monitor the DP values of transformers in operation, laboratory models simulating aging of paper/oil insulation under accelerated conditions allow a feasible way to analyze the effect of different oils and key aging parameters: temperature, oxygen, and water on cellulose aggressiveness [5]. Due to the increase in the furans content in the insulating oil and the moisture content in the solid insulation, we decided to perform the DP test to find out the state of solid insulation degradation.

* irina.chera@icemenerg.ro; irina.alina.chera@gmail.com.

There are a lot of power transformers from electrical utilities around the world that are approaching the end of their design life. The state of degradation of the insulation materials in transformers is a major concern for these old transformers, degrading them at higher operating temperatures in the presence of oxygen and humidity. Currently, a number of modern diagnostic techniques are used to evaluate the isolation status of older transformers. Frequency analysis is used for transformer oil, dissolved gas analysis (DGA), polymerization measurement and furan analysis by high-performance liquid chromatography (HPLC) [6,7].

Case studies

In various articles from the literature, there are presented case studies of the diagnostic status of the transformer where the presence of furan is a primary indicator of chemical degradation of insulating paper. Interpretation of the results of this analysis must be carried out only in conjunction with other analyses and determinations, with the operating regime (in particular by monitoring the oil temperature of the transformer) and based on consistent periodic analyses [8].

Due to the complexity of cellulose degradation mechanisms and the multitude of factors that can influence the process, the dependence between the concentration of furan compounds in the insulating oil and the lifetime of the transformer is not fully elucidated. Here are some of the most interesting aspects of bibliographic research done.

Factors that have a significant influence on the formation of 2-furaldehyde (which is the most stable and soluble furan) in oil it has:

• the initial concentration of 2-furaldehyde in the insulating oil is determined by the ratio of the formation/degradation rate to a certain temperature;

• the presence of moisture significantly influences the distribution of furan compounds between paper and oil (favors 2-FAL formation);

• temperature influences the formation/degradation of 2-furaldehyde. Operating temperature variations due to different load conditions (cyclic or constant) significantly influence the formation of furans;

• the type of oil is also very important. Additive oil inhibitors have a certain protective effect on paper thermosetting and lead to slowing down the formation of furan compounds, compared to non-additive oils;

• heat treated paper produces a lower amount of furan compounds than normal Kraft paper;

• transformer type (paper/oil ratio) and cooling conditions are very important. Increasing the amount of paper involves the formation of furans (different temperatures of exposure);

• changing the oil causes a lower furnace content over a period of time. But, as an equilibrium reaction, the initial concentration will be reached again.

Internationally, there are more studies on the model used to obtain a correlation between oil content furans and the degree of polymerization of paper. There is still no single correlation, supported by all groups of researchers who have dealt with these studies [9].

However, there is a common element, all the accepted relationships correlate the degree of polymerization of the paper with the furfural or 2-furaldehyde (2-FAL) content of the oil [7]. This is due to the fact that all the other compounds of furan resulting from the degradation of the paper: 5-hydroxymethylfurfural (5-HMF), 5-methyl furfural (5-MEF) and 2-acetylfuran (2-ACF) are less stable than 2-FAL under the specific conditions of the transformer, and tend to decompose in a relatively short time in 2-furaldehyde (2-FALs) [10].

Chendong s.a. [11], found the following linear relationship between the furfural concentration in logarithmic scale and the degree of polymerization:

$$\log [2\text{-FAL}] = 1{,}51 - 0{,}0035 \text{x} DP \quad \textbf{(1)}$$

where: 2-FAL – the concentration of 2-furaldehyde, mg/l
DP – degree of polymerization

Experiments by De Pablo [11,4] showed that not all paper degrades at the same speed. They found the following relationship between the furans concentration and the degree of polymerization:

$$DP = 7100 / (8{,}88 + 2\text{-FAL}) \quad \textbf{(2)}$$

Pahlavanpour [11], found the following relationship between furan concentration and degree of polymerization:

$$DP = 800 / [(0{,}186 \times 2\text{-}FAL) + 1] \quad \textbf{(3)}$$

Studies of Mulej in Slovenia [12] established equations (4) and (5) correlating the concentration of furfural degree of polymerization of paper, depending on the embodiment of the experiments.

Equation (4) was obtained based on the experimental results of artificial aging performed in several laboratories

$$\log 2\text{-}FAL = - 0{,}0023 \times DP + 1{,}237 \quad (4)$$

At the same time, the degree of polymerization and the content of furfural were determined for samples of paper and oil taken from 25 partially open transformers, which led to equation (5):

$$\log 2\text{-}FAL = - 0{,}0026 \times DP + 0{,}727 \quad (5)$$

The first research found that the concentration of 2-furaldehyde in oil obtained by artificially aging paper is higher than that determined in the oil in the working transformer, whose paper has the same degradation [12, 13].

The results of the study are presented in Table 1. Also, Table 1 includes the values for the degree of polymerization of the paper based upon the calculation performed by specialists ICEMENERG using the above relations.

Table 1. Correlation of degree of polymerization with furans content

Nr. crt.	Equipment	Content of dissolved CO, ppm	2-FAL content, ppm		The degree of polymerization, calculated according to the respective IEC 450 equations				
				IEC 450	ec. 1 Chendong	ec. 2 Pablo	ec.3 Pahlavanpour	ec.4 Mulej	ec. 5 Mulej
1	Trafo 132/66/11 Kv	4573	9,0	237	158,8	397,1	299,2	122,9	-
2	Trafo 66/11 kV	5495	12,0	108	124,1	341,3	247,5	68,6	-
3	Trafo 66/11 kV	5887	7,5	298	182,4	435,6	310,0	157,4	-
4	Trafo 132/66 kV	1541	6,6	345	197,3	461,0	359,1	181,5	-
5	Trafo 220/132 kV	417	1,0	780	431,4	724,5	774,5	537,8	279,6

The analysis of the resulting data presented in Table 1:

• the values obtained for the degree of polymerization calculated with the above formulas show a variation over a very wide range from the values determined according to IEC 450 [14];

• the values of the degree of polymerization calculated by the Pahlavanpour formula are the closest to the values determined according to IEC 450, then come to the values calculated with the formulations of De Pablo, Chendong, Mulej (formula 4) and Mulej (formula 5);

• the content of 2-furaldehyde (2-FAL) determined in the oil is lower, the degree of polymerization calculated with the formula is any more real than the value determined by IEC 450;

• it is noted that the best correlation between the oil furans content and the degree of polymerization of the coating was obtained using the formula of Pahlavanpour;

• it is recommended to periodically determine the furans content during the operation of the transformer for the cumulative monitoring of the insulating cellulose state;

• the furans content should be used and interpreted as a primary chemical indicator in conjunction with the results of oil-gas analysis and other monitoring techniques;

• for values of 2-furaldehyde content higher than 1 ppm, Mulej's formula 5 can not be applied.

2. Results and discussions

To establish the preliminary limit of furans in the insulating oil was sampled and analyzed by multiple transformer oils. In Table 2, are given the transformers from which oil samples analyzed were taken and the type of oil (without additives/additives) with which they are equipped.

Table 2. Types of equipment

Nr. crt.	Equipment	Type oil
1	T 1 – st. 400 kV Pelican – ST Bucharest	Tr 25A
2	Trafo – CHE Tismana – SH Tg. Jiu	Tr 30.01
3	T 2 – st. 400 kV Brasov Nord – ST Sibiu	Tr 25A
4	AT 200 – st. 400 kV Iernut – ST Sibiu	Tr 25A
5	Trafo 1, CET Grozavesti	Tr 30.01
6	Trafo 400 MVA, Complex Energetic Rovinari	Tr 30.01

Oil samples were analyzed in two locations:

- in the Oil Laboratory of ICEMENERG, where physicochemical and electrical analyzes were performed;
- in the chemical laboratory, where the determined content of furan compounds.

For some equipment, there was introduced and dissolved gas analysis (the content of CO_2 and CO) in conjunction with oil which subsequently results in the determination of the content of furans.

The analysis of the characteristics determined and presented in Table 3 results in the following:

- both additive electro-insulating oils (Tr 25A) and unadapted (Tr 30) were analyzed;
- transformer life ranges in a wide range, between 4 ÷ 32 years;
- the analyzed oils showed different stages of degradation, from very low degradation - oil from the AT 200 Iernut to an advanced degradation - the oil from the trafo 1 CHE Tismana and the trafo 400 Rovinari;
- the total determined furans content is between 0.01 ppm ÷ 6,079 ppm;
- furans in the highest concentrations are 2-furaldehyde (0 ÷ 2,548 ppm) and 5-hydroxymethylfurfural (0 ÷ 5,862 ppm);
- based on the analyses performed, the dependence of the furans content on the degree of degradation of the oil was not revealed.
- using the above-mentioned formulas, the degree of polymerization of the solid insulation in the transformers for which the furan content of the oil was determined was calculated.

The results obtained are summarized in Table 3.

Table 3. The calculation of the content based on the degree of polymerization of 2-FAL

Nr. crt.	**Equipment**	**2-FAL, ppm**	**Degree of polymerization calculated with the formulas:**		
			ec. 1 Chendong	**ec. 2 Pablo**	**ec. 4 Mulej**
1	T 1, st. 400 kV Pelicanu	0,00	-	806,8	-
2	AT 200, st. 400 kV Iernut	0,00	-	806,8	-
3	Trafo 1, CHE Tismana	0,22	619,3	787,1	823,7
4	T2, st. 400 kV Braşov N.	0,05	803,2	802,3	1103,5
5	Trafo 1, CET Grozavesti	0,349	562,1	776,0	736,6
6	Trafo 400 MVA, Complex Energetic Rovinari	0,146	670,2	793,7	901,2

The analysis of the data contained in Table 3 shows that for the same content from 2-furaldehyde (2-FAL) different values are obtained for the degree of polymerization, according to the calculation formula applied.

It is noted that until the time of writing, the degree of polymerization of the insulating paper according to IEC 450 was not determined because the transformers from which the paper samples had to be taken were not yet out of operation.

3. Conclusions

The presence and concentration of furan compounds in the oil are dependent upon:

• operating temperature;
• type transformer;
• type of oil;
• the type of paper.

The following furan compounds result from the degradation of the paper: 2-furaldehyde (2-FAL), 5-hydroxymethylfurfural (5-HMF), 5 methyl furfural (5-MEF) and acetylfuran (ACF). Of these, the most stable is 2-FAL, the other furans tend to decompose in a relatively short time furfural.

The factors that have an influence on the formation of 2-furfural in the insulating oil are:

• the actual concentration of furfural in the insulating oil is a balance between the rate of formation/degradation at a given temperature.

• moisture content has a very important influence, as they influence the distribution between the paper and the oil furan compounds;

There are many studies on the patterns used to get a correlation between the oil content of the furans and the degree of papermaking. However, there is no correlation accepted by all groups of researchers who have dealt with these specific studies.

It is recommended to periodically determine the furans content during transformer operation for the cumulative monitoring of the insulating cellulose state.

Furans content must be interpreted as a chemical used in conjunction with the primary analysis results of gas dissolved in oil and other monitoring techniques.

References

[1] R. Sans John, K. Muge Bilgin, J. Kelley Joseph, *Large Scale Survey Of Furan Compounds in Operating Transformersand Implication for estimating Service Life.* IEEE Proc., Virginia, USA, June 7-10, (1998).
[2] V.Shrinet, M.J.Patel, *M.Ramoorty, Role of furan and DP analysis for the refurbishment of power transformers. CIGRE,* (2000).
[3] *** CIGRE, *Guide for Life Management Techniques for Power Transformers*. CIGRE WG 2.18, (2003).
[4] A.DE Pablo, *Furanic Compounds Analysis as a Tool for Diagnostic and Maintenance of Oil-Paper Insulation Systems*, CIGRE Symposium, Berlin, Paper 110-09, (1993).
[5] J. J. Kelly, *Transformer Fault Diagnosis by Dissolved-gas Analysis*, IEEE Trans. Ind. Appl., Vol. 16, pp. 777-782, (1980).
[6] J. Unsworth, F. Mitchell, *Degradation of Electrical Insulating Paper Monitored with High-Performance Liquid Chromatography*, IEEE Trans. Electr. Insul., Vol. 25, pp. 737-746, (1990).
[7] A. M. Emsley, X. Xiao, R. Heywood, M. Ali, *Degradation of Cellulosic Insulation in Power Transformers. Part 3: Effects of Oxygen and Water on Ageing in Oil*, IEE Proc. Sci., Measure. Techn., Vol. 147, pp. 115-119, (2000).
[8]. I. Höhlein, A. J. Kachler, M. Stach, S. Tenbohlen, T. Leibfried, *Transformer life management German experience with condition assessment*, CIGRE SC 12/A2, Merida, Kolloquium, June 2 – 4, (2003).
[9]. T. V. Oommen, T. A. Pervost, *Cellulose insulation in oil-filled power transformers: part II-maintaining insulation integrity and life*, IEEE Electr. Insul. Magz., vol. 22, no. 2., pp. 5-14, (2006).
[10]. R. Sanghi, *Chemistry behind the life of a transformer*, Resonance, June (2003).
[11]. K. Saha, *Review of a modern diagnostic technique for assessing insulation condition in aged transformers*, IEEE Transactions on Dielectrics and electrical insulation, vol. 10, no.5, October (2003).
[12]. M. Mulej, A. Vaarl, M. Končan-Grandnik, *Up-to-date experience on furans for transformer diagnostics*, 12th International Symposium on High Voltage Engineerings, Netherlands, (2003).
[13]. H. Lűtke, I. Hőhlein, A. J. Kachler, *Transformer aging research on furan compounds dissolved in insulating oil*, CIGRE, (2002).
[14].*** IEC 450, Measurement of the Average Viscometric Degree of Polymerization of New and Aged Cellulosic Electrically Insulating Materials, Amendment 1, May, (2007).

E3S Web of Conferences **112**, 04005 (2019) https://doi.org/10.1051/e3sconf/201911204005
TE-RE-RD 2019

Modelling and simulation of a waste sorting mechatronic system for multi-floor residential buildings

Liviu-Marian Ungureanu[1,*], *Ileana* Dugăeşescu[1] and *Elisabeta* Niculae[1]

[1] University Politehnica of Bucharest, Theory of Mechanisms and Robots Department, RO-060042 Splaiul Independentei, No.313 room JC111, Romania

Abstract. The paper presents modelling and simulation of a mechatronic sorting system for residential buildings where the "wet" garbage can be used as fuel in an energy generation system. The proposed system can be used for the implementation of a selective storage model in the actual garbage chute system. The proposed system will have five different storage capacities that can be recycled: paper, glass, plastic packaging, metal boxes and organic household waste.

1 Introduction

Waste collection is not done at source; instead, the garbage is thrown through a garbage chute system, where reaches the garbage storage room on the ground floor of the building from where it is loaded manually into garbage collection machines. Also, storage is a real problem [1-8] and in this paper, we present a solution that enhance actual systems for a selective, sorting and storage system to obtain selective packages that can be recycled.

In literature, many papers or studies present similar systems for waste management systems implemented in different countries [9-11, 16] dealing with automated sorting systems [12-13]. These solutions have some important disadvantages regarding costs, maintenance and use. Therefore, we propose a low-cost solution for garbage chute system with four or six distinct routes for dumping the waste that can be used in multi-floor residential buildings. The system consists of a garbage chute with door access from each floor and four storage devices for waste placed in a ground room.

The proposed system addresses the issue of selective collection of the following types of waste:

- glass, glass containers made for beverages and food, jars;
- newspapers, magazines, paper-cardboard packaging, folded cardboard boxes;
- plastic packaging, plastic beverage containers (PET) and dairy products. Containers for cosmetics and detergents, plastic and polystyrene cassettes, plastic dishes, non-contaminated household waste plastic bags, plastic food containers;
- metal packaging, such as cans, beverage doses;

* Corresponding author: ungureanu.liviu.marian@gmail.com

- organic household waste.

According with the legislation, until 2025 [14], all the buildings must have a room on the ground floor, properly arranged, with four special containers for paper, glass, plastic and metal, plus the household. But it is hard to believe that the inhabitants on the upper floors will come down to throw the garbage in the especially designed places. Nevertheless, it will be preferred that the convenient and fast solution as outlined in this paper.

2 Study on rotation of the mobile waste sorting element

From the point of view of waste sorting, this study is essential to determine the position of the garbage container. Elements of kinematic modelling of mechanisms were used [15, 17]. The kinematic modelling refers to the fundamental principle of forming mechanisms, according to which a mechanism is made up of an active modular group and a passive modular group, the number and nature of which are dependent on the degree of freedom and the number of independent contours. The degree of mobility puts into evidence the number of independent variables that define the state of motion of the mechanism. The position function is

$$f_i\,(x_1, x_2 \ldots x_k, y_1, y_2 \ldots y_j) = 0 \tag{1}$$

where k=1...M (M=degree of mobility), j=1...2N (2N=number equation of independent loop), x_k are independent variables and y_j are dependent variables.

From Eq. 1 using derivation in function of time results the velocity.

$$\sum \frac{\partial f_i}{\partial x_k}\dot{x}_k + \sum \frac{\partial f_i}{\partial y_j}\dot{y}_j = 0 \tag{2}$$

We can rewrite Eq. 2 as

$$A \left\| \dot{y}_j \right\| = B \tag{3}$$

Where

$$A = \left\| \frac{\partial f_i}{\partial y_j} \right\| \qquad B = \left\| -\frac{\partial f_i}{\partial x_k}\dot{x}_k \right\| \tag{4}$$

From Eq. 2 using derivation in function of time results the acceleration.

$$\sum \frac{\partial f_i}{\partial x_k}\ddot{x}_k + 2\sum\sum \frac{\partial^2 f_i}{\partial x_k \partial y_j}\dot{x}_k\,\dot{y}_j + \sum \frac{\partial f_i}{\partial y_j}\ddot{y}_j + \sum\sum \frac{\partial^2 f_i}{\partial y_j \partial y_k}\dot{x}_k\,\dot{y}_j = 0 \tag{5}$$

We can rewrite Eq. 5 as

$$A \left\| \dot{y}_j \right\| = C; \qquad C = \left\| \sum \frac{\partial f_i}{\partial x_k}\ddot{x}_k + 2\sum\sum \frac{\partial^2 f_i}{\partial x_k \partial y_j}\dot{x}_k\,\dot{y}_j + \sum\sum \frac{\partial^2 f_i}{\partial y_j \partial y_k}\dot{x}_k\,\dot{y}_j \right\| \tag{6}$$

Using the above relationships, several trajectories of the mobile waste sorting element were determined using a numerical software.

The mechanism of waste sorting is presented in Fig.1 and the trajectory will be generated by the mobile sorting element at different rotation angles in Fig. 2.

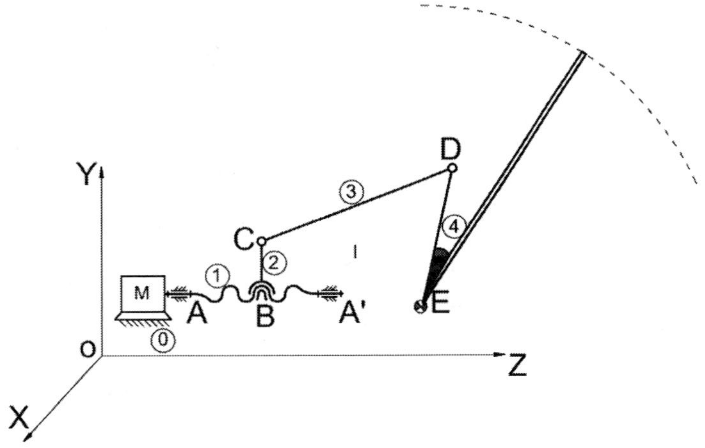

Fig. 1. Mechanism of waste sorting.

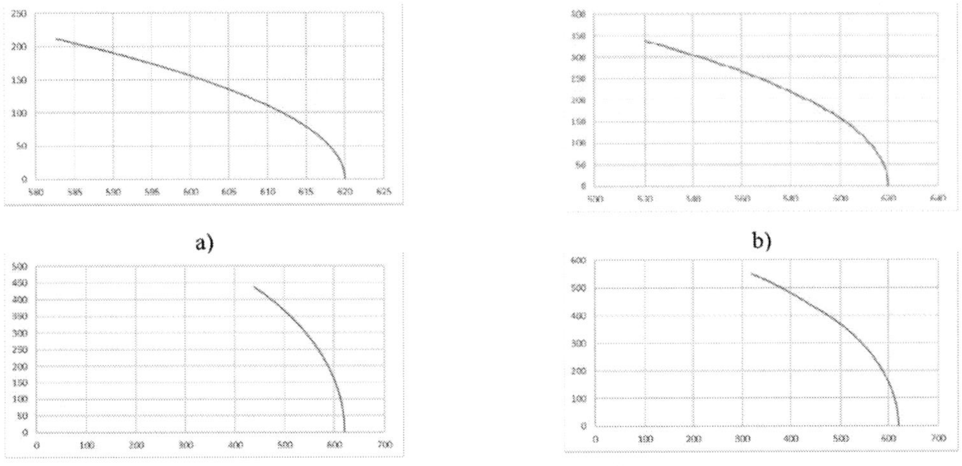

a) b)

c) d)

Fig. 2. Trajectory of the mobile sorting element at different rotation angles: a) 20^0 ; b) 30^0; c) 45^0 ; d) 60^0.

3 Modelling and simulation of a mechanical waste sorting system

The mechanical waste sorting system is made of five slots for: glass (green); metals (yellow); plastic (red); paper (blue) and household waste (the free fall section) as can be seen in Fig 3.

a) b) c) d)

Fig. 3. The prototype : a) CAD Model, b) realized small-scale model, c) top view, d) electronic part.

For kinematic modelling, it was used a 3D simulation software, before that it was realized each part and then combined into an assembly, resulting the desired waste sorting system.

In Fig. 4b we present how it was simulated the minimum and the maximum trajectory of garbage (interrupted line), to place the garbage container within these limits. The fixed place where the garbage container is located could not be established, as there are several variables regarding the waste: weight, density and form, but there are some tests in progress and the results will be included in a future study.

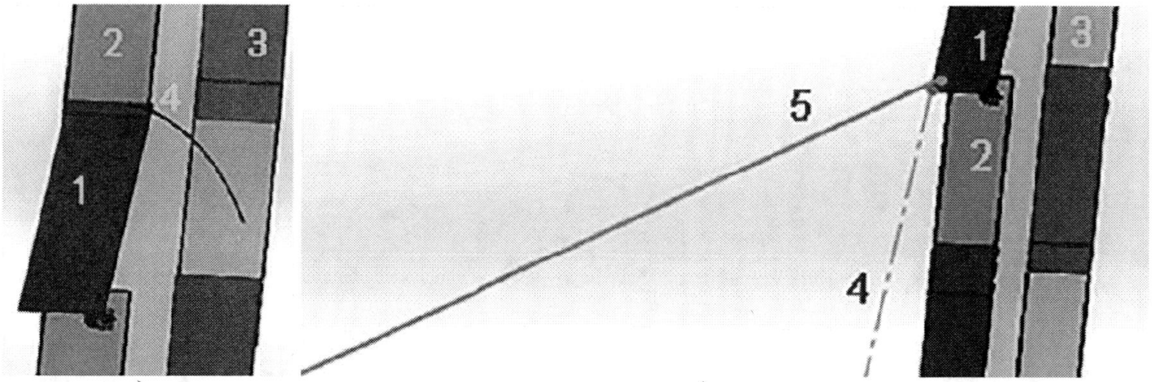

a) b

Fig. 4. Mobile sorting element: a) trajectory of the mobile sorting element (1-door; 2- front wall; 3- back door; 4 - trajectory), b) trajectory of the place garbage container (1-door; 2- front wall; 3- back door, 4 – minimum trajectory of garbage; 5 – maximum trajectory of garbage).

To see if the elements of the garbage selection system can take a specified load there was realized several FEM analyses.

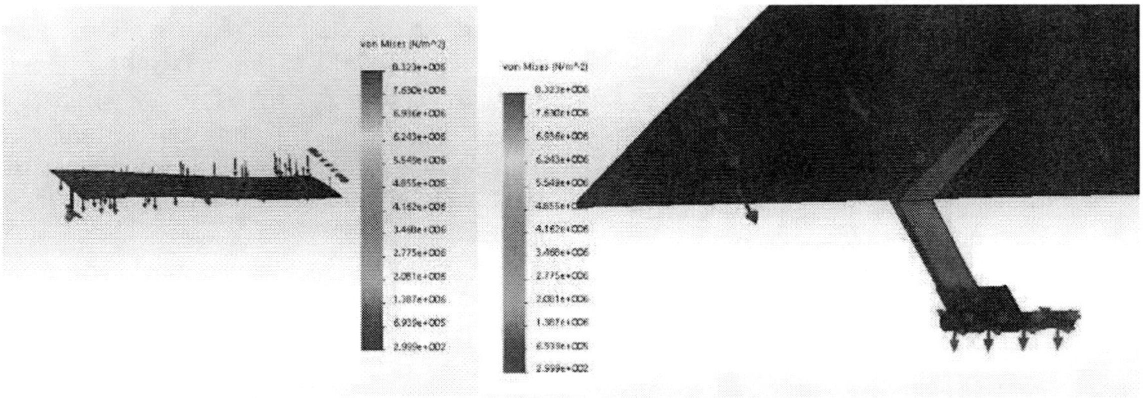

a) b)

Fig. 5. Mechanical stress obtained through the simulations: a) Mechanical load on the mobile sorting element; b) Mechanical load on the gripping element.

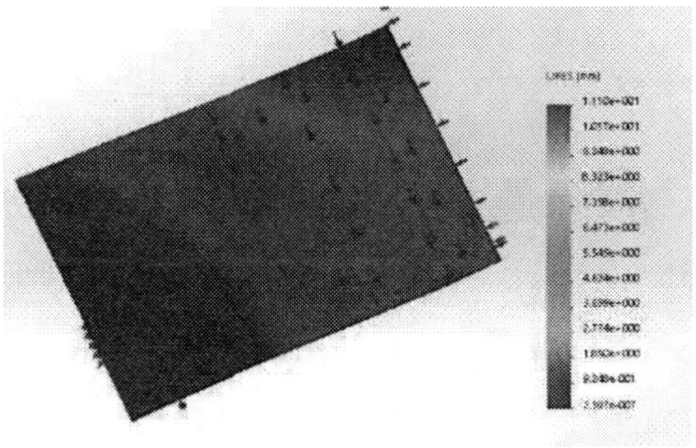

Fig. 6. Deformation of the mobile sorting element obtained using FEM.

These were focused on the actuation system and on the door. A 10N load is considered as constantly distributed over the sorting element, a sliding element at the side edge opposite the clamping member and a rotating element at the shaft of the engine (Fig, 5a and 5b). This is the most disadvantageous case where the entire load will fall on the engine spindle (which should not happen accidentally), for that mechanism of waste sorting has a screw nut elements. The deformation of the mobile sorting element is present in Fig.6. The second simulation considered a concentrate force of 10 N placed in the middle of the structure that is the most disadvantageous situation (Fig, 7a and 7b).

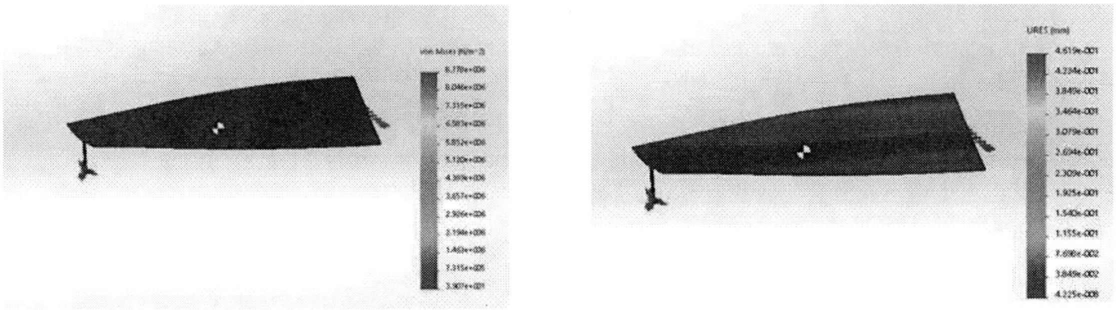

a) b)

Fig. 7. Results of the FEM simulation when a force of 10N is placed in middle of the clamp: a) Mechanical stress; b) Mechanical deformation.

As can be seen in Fig 7, the greatest deformation of the mobile sorting element occurs at the location where the element takes directly the weight of the trash. The maximum deformation of the mobile sorting element appears at the contact between the selector margin and the garbage chute. This is due to the unbalanced attachment to the gripping element. In order to be verify model we have developed a small-scale model of the mechanical sorting system because the solution needs a coherent approach to deliver a workable technical solution. This experimental model was imperative to implement the control and the electronic part of the mechanical sorting system.

The prototype (Fig. 2) is using four step motors type 42BYGHW609 with a power control done using a L298N driver. The automation of the system was made using the Arduino MEGA 2560 and four control buttons, one for each sort of waste.

Conclusions

In this article, was presented the 3D modelling and simulation of a mechanical waste sorting system. We verified the mathematical model using a specialized software, and it was realized the kinematic analyze. Also, a finite element analysis was performed to see the static mechanical behavior mobile of the sorting element .

To evaluate the theoretical results, a small-scale model was made which includes the automation software. In the near future, we want to test a real system in the existing system in the residential buildings.

References

1. L. Tralhao, J. Coutinho-Rodrigues, L. Alcada-Almeida, A multi-objective modeling approach to locate multi-compartment containers for urban-sorted waste, Waste Management, vol. **30**, pp.2418-2429, (2010)
2. W. S. Lu, H.P. Yuan, Exploring critical success factors for waste management in construction projects of China, Resources Conservation and Recycling, vol. **55**, pp. 201-208, (2010)
3. D. Mickael, D. Isabelle, M. Sebastien, A review of literature: Influence and communication applied to waste management, Pratiques Psychologiques, vol. **20**, pp. 211-230, (2014)
4. V. Yadava, A. K. Bhurjeed, S. Karmakara, A. K. Dikshit, A facility location model for municipal solid waste management system under uncertain environment, Science of the Total Environment, vol. **603-604**, pp. 760–771, (2017)
5. Y. L. Liu, P. X. Xing, J. G. Liu, Environmental performance evaluation of different municipal solid waste management scenarios in China, Resources Conservation and Recycling, vol. **125**, pp 98–106, (2017)
6. G. Aid, M. Eklund, S. Anderberg, L. Baas, L, Expanding roles for the Swedish waste management sector in interorganizational resource management, Resources Conservation and Recycling vol. **124**, pp. 85-97, (2017)
7. S. Moon, S. Han, P.R. Zekavat, L. E. Bernold, X. Wang, Process-waste reduction in the construction supply chain using proactive information network, Concurrent Engineering-Research and Applications, vol. **25**, pp. 123-135, (2017)
8. M. Kelly, D. Dowd, A review of construction waste management practices on selected case studies in Ireland, Proceedings of the Institution of Civil Engineers-Waste and Resource Management, vol. **170**, pp. 78-84, (2017)
9. J. Winkler, B. Bilitewski, Comparative evaluation of life cycle assessment models for solid waste management, Waste Management, vol. **27**, pp. 1021-1031, (2007)

10. A. Bernstad, J. L. Jansen, Review of comparative LCAs of food waste management systems - Current status and potential improvements, Waste Management, vol. **32**, pp. 2439-2455, (2012)
11. Z. Z. Wu, L. Y. Shen, A. T. W. Yu, XL. Zhang, A comparative analysis of waste management requirements between five green building rating systems for new residential buildings, Journal of Cleaner Production, vol. **112**, pp. 895-902, (2016)
12. S.P. Gundupalli, S. Hait, A. Thakur, A review on automated sorting of source-separated municipal solid waste for recycling, Waste Management, vol. **60**, pp. 56-74, (2017)
13. S. G. Paulraj, S. Hait, A. Thakur, Automated Municipal Solid Waste Sorting for Recycling Using A Mobile Manipulator, Proceedings of the ASME International Design Engineering Technical Conferences and Computers and Information in Engineering Conference, Vol **5a**, (2016)
14. Directive 2008/98/EC of the European Parliament and of the Council of 19 November 2008 on waste
15. A. Comănescu, D. Comănescu, I. Dugăeşescu, A. Boureci, *Bazele modelării mecanismelor*, Politehnica Press, pp 29-30, (2010)
16. T. Prisecaru, A. Semenescu, C.D. Comeaga, C. G. Alionte, "*Experimental integrated system used for combustion process and resulted flue gas analysis*", Metalurgia International, **14** (7), pp. 36-42, ISSN 1582-221, 2009
17. R. Mihai, L. A. Cartal, C. G. Alionte, T. C. Apostolescu, *The development of an electrical actuated planar parallel robot,* Romanian Review Precision Mechanics, Optics and Mechatronics, **44**, pp. 120-124

Demand-side Management Programs – A joint environmental protection action. Case study: The lighting system in the Campus of the University POLITEHNICA of Bucharest

Mircea Scripcariu[1] , *Ioan Sevastian* Bitir-Istrate[2], *Cristian* Gheorghiu[3*], *Ştefăniţă* Pluteanu[3] and *Aida Maria* Neniu[4]

[1]Associate Prof., Dept. of Energy Production and Use, University POLITEHNICA of Bucharest, Romania

[2]Associate Prof., Dept. of Energy Production and Use, University POLITEHNICA of Bucharest, Romania

[3]PhD. Student, Dept. of Energy Production and Use, University POLITEHNICA of Bucharest, Romania

[3]PhD. Student, Dept. of Energy Production and Use, University POLITEHNICA of Bucharest, Romania

[4]PhD. Student, Dept. of Energy Production and Use, University POLITEHNICA of Bucharest, Romania

Abstract. The current development of Smart Grids, combined with the pressure enforced by national legislation as a direct effect of the 2012/27/EU and the 2018/844/EU Directives and the ever-growing energy demand lead to a new set of challenges for both the end-users and the utility companies, under the form of optimizing the EPIs (Energy Performance Indicators), reducing the Environmental Impact and flattening the Power Demand Curve. In this paper, the economical viability and the potential environmental impact reduction obtained by implementing a DSM (Demand–Side Management) program in the Campus of the University POLITEHNICA of Bucharest were analysed. The energy boundary consisted of all the 26 Student Dorms and the main Significant Energy Use) consisted of the lighting system. Four different scenarios were conceived, taking into account the initial investment and the energy savings sharing between the end-user and the Utility Company. Based on these scenarios, a technical-economic model is presented. Using the data gathered on-site and the DSM program mode, relevant results were obtained and a development solution for such projects was proposed. In the final part of the paper, the predicted Environmental Impact Reduction was quantified and analysed, under the form of the Carbon Footprint generated, respectively avoided by implementing the most economically efficient DSM program development solution.

* Corresponding author: gheorghiu.cristian.k@gmail.com

1 Introduction

DSM consists in the proper planning, implementation and monitoring of energy use activities which are destined to encourage the end-users to modify and optimize the energy-consumption levels and behaviour.
In the past, the main purpose of implementing DSMs was to ensure an efficient energy supply and economically efficient solutions in order to delay the necessity of installing additional Power Generation sources [1], as a result of the ever-growing energy demand.
However, due to the changes that occur in the industrial sector, the emergence of new consumers (e.g. Office Buildings, Apartment Buildings etc.), DSM became a tool that can be used by the Utility Companies in order to improve the quality of their services.
In essence, DSMs are programs and activities that modify the power-demand curve by offering modern load-management solutions, in the form energy performance indicators improvement solutions.
As a cumulative result of the liberalization of the energy market, the energy efficiency targets and the targets set by the European Directives with regard to the reduction of the Carbon Footprint, DSMs are becoming more and more attractive. The main objective of this paper is to prove the necessity, the benefits and the economic efficiency of such Programs.

2 Romanian national legislation and strategy regarding energy efficiency

Improving the EPIs is the main objective of Romania's energy strategy, considering the major contribution that they have in assuring the end-user power supply, the competitive character of the aforementioned services and the reduction of the environmental impact.
The main EPI used by all the Member States is the EI (Energy Intensity) [2-3]. Thus, by following EU's Directives 2012/27/EU and 2018/844/EU, the necessity of improving the existing national policies is obvious, as it can be seen in Table 1.

Table 1. The potential energy consumption reduction for different end-use sectors [1].

Sector	Share in total energy consumption [%]	Energy Use Reduction Potential [%]
Industry	31	13
Buildings	36	41.5
Transport	22	31.5
Services	11	14

The case study is based on the Buildings Sector. The available EPIAs for the Buildings Sector differ depending on the building type and destination but cover the entire energy cycle – primary resources, generation, transport, distribution, supply and end-use.
The main energy use for buildings is associated to the HVAC (Heating, Ventilation and Air Conditioning) and the lighting systems.
Based on [1], 87.2% of the energy used in Buildings is used for heating, hot water and cooking, the difference of 12.8% being used lighting and powering up house appliances. Also, as [4] has shown, in 2015, 82.7% of final energy consumption in buildings was supplied by fossil fuels.

The estimated energy saving potential generated by the lighting systems is evaluated [4] at approximatively 160 EJ or 3,821.54 Mt.o.e. in the 2018 – 2040 time-frame, worldwide.
For the 2020-2030 time-frame, the European Union set up the energy efficiency target of 30% by 40% Carbon Footprint Reduction/27% Renewable Energy Share in Energy Consumption/30% Energy Efficiency Improvement set by the 2016 EU Commission package of measures, to which an environmental impact reduction of 40% is associated.
Romania's Energy Strategy is centred on the needs and interests of all end-users (household, commercial and public administration). It focuses on the fact that these needs and interests are diverse and perpetually changing and evolving [5]. In order to ensure the implementation of major EPIAs a stable and transparent methodology is required, based on a realist approach with regard to the set targets.

3 Theoretical aspects of DSM Programs

The implementation method of a DSM Program mainly depends on the objective that the said Program sets. As previously stated, the usual main objectives of DSM Programs is the flattening of the Power Demand Curve and the reduction of the energy consumption, but is also dependent on several other factors, such as:

- Financial restrictions of the end-user or the energy supplier;
- The existence / absence of financial incentives;
- The potential of achieving economical or social objective targets by implementing the Program;
- The wish of the energy supplier (or national Government) to persuade end-users to implement DSM Programs.

The fundamental characteristic of any DSM Program is the fact that different sectors of society perceive different groups of costs and benefits associated to them.
In order to economically evaluate the viability of a DSM Program a number of tests were developed and applied by various Energy Suppliers worldwide. The main that are briefly presented in Table 2 are all based on the classic Discounted Total Cost criterion.

Table 2. Tests used for the economic analysis of DSM programs [5].

No.	Test Name	Formula	Unit
1	Participants Benefit (PB)	$PB = \Delta B + S - C_{participants}$	EUR
2	Energy Supplier Benefit (ESB)	$ESB = AC - C_{ES} - S$	EUR
3	Total Resource Cost (TRC)	$TRC = AC - CES - C_{participants}$	EUR

where:
ΔB – the energy bill reduction;
S – stimulants paid by the energy supplier to the DSM Program participants;
$C_{participants}$ – the end-user share of the DSM Program Costs;
C_{ES} – the Energy Supplier share of the DSM Program Costs;
AC – Energy Supplier Avoided Costs as a result of implementing the DSM Program.

3.1 Participants Benefit (PB) Test

This first test includes the quantifiable costs and benefits generated by a DSM Program, from the end-user Point of View. It evaluates the economic attractiveness of the Program for the end-users, by determining the present value of the project with (1):

$$PB = \Delta B + S - C_{participants} \; [\,EUR\,] \tag{1}$$

The Program is considered economically efficient for the end-user if PB ≥ 0.

3.2 Energy Supplier Benefit (ESB) Test

This test is used to measure the total cost and benefits variation for the energy supplier as a result of implementing the DSM Program. The DSM Program effect on the energy supply company is a complex one, enabling multiple opportunities to cut the costs, to improve the annual cash flow and to maintain the current activity. The test is defined by the following formula (2) and it essentially measures the energy bill variation of all end-users:

$$ESB = AC - C_{ES} - S \; [\,EUR\,] \tag{2}$$

The Program is considered economically efficient for the energy supplier if ESB ≥ 0.

3.3 Total Resource Cost (TRC) Test

This test evaluates the total cost and benefit variation for the national economy, thus quantifying the impact that a DSM Program has on both the energy supply company and the end-users, by using the formula (3). This test offers a global image on the economic efficiency of the analysed DSM Program.

$$TRC = AC - C_{ES} - C_{participants} = ESB + PB - \Delta B \; [\,EUR\,] \tag{3}$$

The Program is considered economically efficient for the national economy if TRC$\geq$0.

4 The considered scenarios

Based on the aforementioned theoretical considerations, four analysis scenarios were considered, based on each side's share of the investment cost and on the way that the energy savings, quantified in the reduction of the monthly energy bill, is shared between the involved entities:

- Scenario A:
  - The investment cost is supported integrally by the end-user ($C_{participans}$ = I, C_{ES} = 0);
  - ΔB is constant for the entire analysis period;
- Scenario B:
  - The investment cost is supported integrally by the energy supplier ($C_{participants}$ = 0, C_{ES} = I);
  - The energy supplier does not directly recover the investment but by the indirect benefits generated by the DSM Program;
  - The end-user obtains the ΔB energy bill reduction for the entire analysis period.

- Scenario C:
  - The investment cost is supported integrally by the energy supplier ($C_{participants} = 0$, $C_{ES} = I$);
  - The share quota "y" is determined by optimizing the function in order to ensure the economic viability for both sides involved in the Program;
  - The energy savings are shared throughout the analysis period between the end-user and the energy supplier based on the (4) formula:

$$ESV_{ES} = y \cdot \Delta B \cdot D_F \text{ [EUR]} \tag{4}$$

where:
ESV_{ES} – energy supply company quota of the energy savings value;
y – the share of the end-user energy savings that are paid back to the energy supply company;
D_F – Discount factor.

- Scenario D:
  - The investment cost is supported by an "x" share by the end-user ($C_{participants} = x \cdot I$, $C_{ES} = (1-x) \cdot I$);
  - The "x" share is determined by the results of the test, in order to assure the economic viability for both sides involved in the Program.

When comparing the proposed scenarios with the theoretical DSM Program, one can observe that only the second scenario satisfies the definition of a Demand – Side Management Program. The other scenarios expand the classical definition of the DSM and facilitate the integration of the aforementioned concept with the ESCO (Energy Services Company) ever-growing market, thus making the Energy Supply Company able to compete against ESCOs or even form a sort of partnership with one of them.

5 Case Study: University POLITEHNICA Bucharest Student Dorms

In the case study the implementation of a DSM Program for the UPB (University POLITEHNICA Bucharest) Student Dorms lighting systems was analysed. In essence, the technical-economic viability of replacing the existing lighting systems – Incandescent bulbs – with energy efficient LED (Light Emitting Diode) lamps was analysed.
Firstly, the twenty six dorms were classified based on the dorm-room capacity, the results being presented in Table 3.

Table 3. Student Dorms in UPB Campus.

Dorm	Room Type	Dorm Type
P13, P14, P17, P18	4 beds + sink / shared bathrooms	1
P15, P16, P21, P22, P23, P24, P25, P26, P27	2 beds + sink / shared bathrooms	2
P1, P2, P3, P4, P5, P6, P7, P8, P9, P10, P11	4 beds + sink + bathroom	3

In order to properly analyse the viability of a DSM Program, the energy boundary must be well known and the main energy uses must be thoroughly identified.
In Fig. 1 the annual (2018) electricity consumption for all the twenty six dorms is presented.

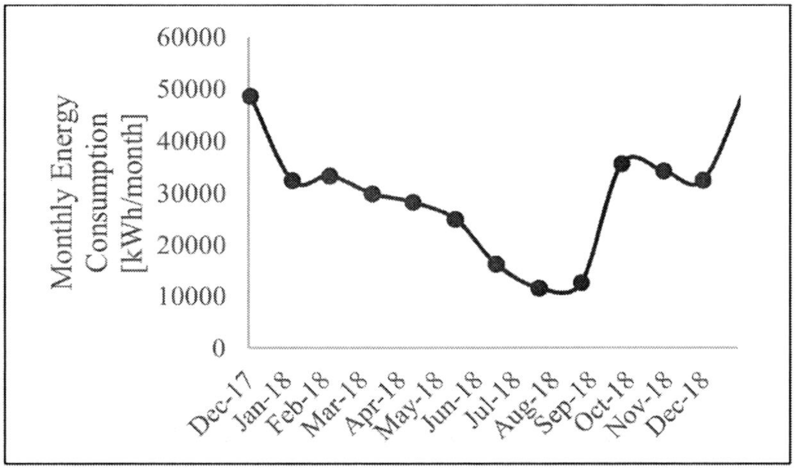

Fig. 1. 2018 electricity consumption of one of the 26 UPB Student Dorms

It can be observed from Fig. 1 that the electricity consumption registers a decreasing trend during the spring – summer period, with a minimum value in August, correlated to the minimum number of students still living in the dorms.

The analysed DSM Program targets the reduction of the lighting system electricity consumption, which amounts to a 12% percentage of the total electricity consumption. To address the bigger-share end-uses (such as household appliances), another DSM Program would have to be developed. The development would only be possible if UPB would own the aforementioned household appliances.

5.1 Technical-economic analysis

In order to determine the technical-economic viability of implementing the Lighting System DSM Program, the technical solutions were firstly determined. The technical and economic characteristics of the existing and the proposed lighting system is presented in Table 4.

Table 4. Existing and proposed technical solution [6-7-8]

Dorm	Dorm Type	Lighting Type	Installed Power [W/pcs.]
P13, P14, P17, P18	1	Existent : Incandescent	180
		Proposed : LED	47
P15, P16, P21, P22, P23, P24, P25, P26, P27	2	Existent : Incandescent	96
		Proposed : LED	47
P1, P2, P3, P4, P5, P6, P7, P8, P9, P10, P11	3	Existent : Incandescent	84
		Proposed : LED	41
HALLWAYS + BATHROOMS	-	Existent : CFL	36
		Proposed : LED	16

The economic analysis was based on the data inputs briefly presented in Table 5.

Table 5. Economic analysis input data.

Input type	Value	M.U.
Discount rate [9]	4,5	%/year
Analysis Period[10]	10	Years
Active Energy Price*	88.38	EUR/MWh
Reactive Energy Price*	12.84	EUR/MVArh
Rated Average Lifetime for Incandescent Bulbs	1,000	Hours
Rated Average Lifetime for LED lights	25,000	Hours

* without VAT (Value Added Tax)

By replacing the existent Incandescent Lights with Energy Efficient LED ones, depending on the type of dorm, an energy economy of approximatively 5,200 kWh/year/dorm can be achieved, as it can be seen in Fig. 2.

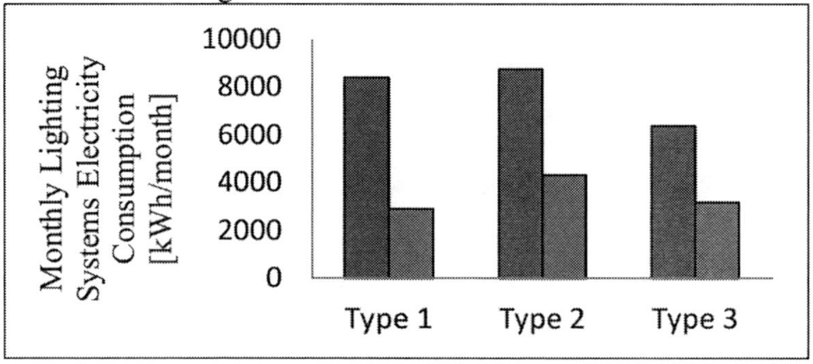

Fig. 2. Energy saving potential of replacing the existent Incandescent Bulbs (Red) with Energy Efficient, LED lights (Green)

In Fig. 3 the energy saving generated by implementing the LED light System is shown with regard to the total electricity consumption of all the twenty six dorms.

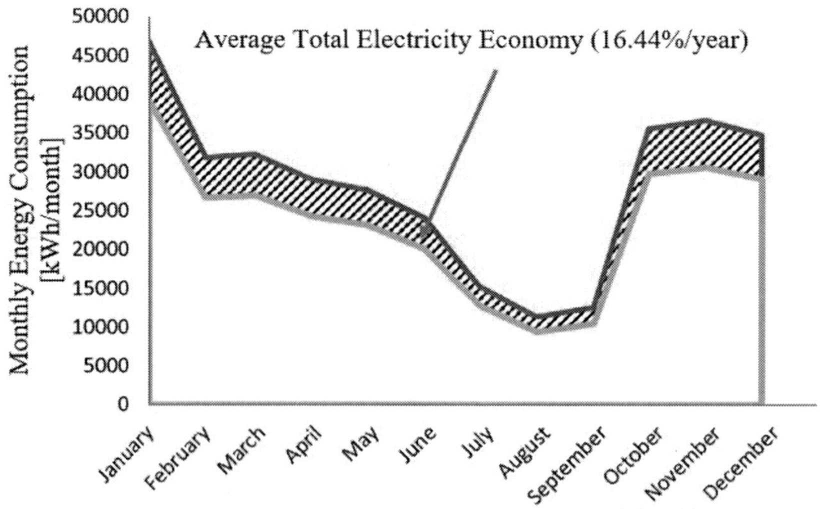

Fig. 3. Electricity consumption in the actual case (red) and in the DMS Program scenario (green) for one Dorm

By applying the DTC (Discounted Total Cost) criterion, the replacement of the existent lighting system with a LED based one would generate an economy of **ΔW_a = 1,184,147 kWh/year** (an average total electricity consumption of 16.44%/year, respectively an associated 55.55% lighting system economy) , respectively a **ΔDTC = 1,245,933 EUR** with an associated **SPP** (Simple Payback Period) of **2.42 years**.

5.2 Testing the DSM Program

In order to classify and test the economic viability [11-12-13] of the analysed DSM Program, the four scenarios which were presented in Chapter 3 were applied to the three-test methodology presented in Chapter 2.
The results are presented in Table 6.

Table 6. DSM Program Tests results.

Scenario	PB [EUR]	ESB [EUR]	TRC [EUR]
A	620,035	225,423	-122,751
B	968,210	-122,751	-121,751
C	484,105	361,354	-122,751
D	794,123	51,336	-122,751

From the end-user's economic perspective, the most attractive scenario is the 2nd one (B), in which the investment cost is fully supported by the energy supply company.
From the energy supply company economic perspective, the most attractive scenario is, as expected, the 3rd one (C), in which the investment cost is fully supported by the end-user.
It can be observed that the TRC test led to a constant value in all four scenarios, as it is used to compare the costs and benefits variation at the society-level (which, for a single program is constant, regardless of how the investment cost or the energy savings are shared).
The tests differentiate from one another by the costs supported by the end-users and by the energy supply companies. In Fig. 4 is shown the test value variation with regard to the share of the investment costs each side supports.

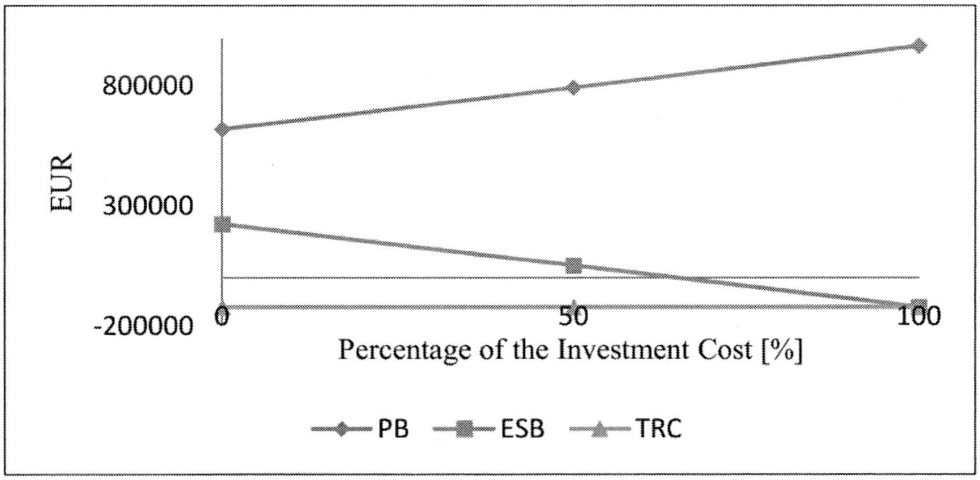

Fig. 4. The DSM Tests variations.

5.3 Environmental Impact Analysis

The Paris Agreement of 2015 (COP21) and the European Union climate change policies boost the development of low CO_2 emission technologies.
In order to quantify the Environmental Impact Reduction generated by implementing the analysed DSM Program, the conversion rate of 299,02 tons CO_2 / MWh [14] was considered. The results are presented in Table 7.

Table 7. Environmental Impact Analysis Results

Scenario	Annual Electricity Consumption [MWh/year]	Annual Carbon Footprint [tCO_2/year]	Annual Environmental Impact Reduction [tCO_2/year]
Incandescent Lights	7,478.4	2,236.2	-
LED Lights	6,294.3	1,882.1	354.1

The 354.1 tons of CO2/year carbon footprint reduction can mitigate the TRC negative result, as it amounts to a total reduction of 3,541 tons of CO_2 for the analysis period, or the equivalent of one typical passenger vehicle worth of carbon dioxide emissions for a year's worth of driving (18.500 km/year) [14-15].

6 Conclusions

Lighting has an important effect on the intellectual performance. Thus, by improving the quality of the light one can greatly improve the memorization, the logical thinking, the concentration and the overall intelligence level.
Both DSM Programs and the Romanian Energy Strategy are centred on the end-user. From other countries experience, Romania can adapt the existing models for promoting the energy efficient energy performance improvement actions by using the concept of demand side management.
The first step Romania needs to take in order to facilitate the implementation of DSM Programs consists in creating and developing a legislative framework which can stimulate the energy efficiency improvement and regulate the way in which the energy supply companies can invest in their end-users energy boundaries, from a fiscal point of view.
In the actual economic context of Romania, DSM Programs should be included in the National Energy Policies, oriented towards the efficient use of energy, supported by the main effects that it can generate:

- The decrease of primary energy sources imports (as a direct effect of decreasing the Power Demand and diminishing the Power and Energy Losses);
- The energy performance indicators improvement for the end-users (by utilizing energy efficient equipment and technologies);
- The sensible decrease of the Environmental Impact.

Based on the technical-economic analysis, the LED replacement DSM Program would be able to generate an energy saving of approximatively 1,184 MWh/year and a decrease of the Carbon Footprint of approximatively 354.1 tCO_2/year.
The reduction of the Peak Power Demand from an initial value of 433.84 kW to 197.03 kW will also lead to an (unquantifiable) benefit of delaying the development and redesign of urban distribution networks. This will also lead to a decrease in transited load (I, respectively RI^2) throughout the entire transport – distribution networks.

References

1. The Romanian Government, *National Action Plan in Energy Efficiency*, Official Gazette No. 273 bis. (2019)
2. The Romanian Parliament, *Energy efficiency Law no 121/2014* (04.08.2014)
3. The Romanian Parliament, *Electricity and Natural Gas no 123/2012* (2013)
4. United Nations environment – *Global Status Report 2017 – towards a zero-emission, efficient and resilient buildings and construction sector* (2018)
5. D. Scripcariu, PhD Thesis – *Improvements in the energy supplier – end-user relationship* (1997)
6. Ministry of Regional Development and Public Administration, *NP 061-02 Normative for the design and execution of artificial lighting systems in buildings* (2002)
7. *** Society of Light and Lightning CIBSE, *Code for Lighting*
8. *** Illuminating Engineering Society IES, *Lighting handbook: the standard lighting guide*
9. Public Procurement National Agency, *Order no. 2717/318/2018 for the revision of the discount rate* (2019)
10. *** http://www.e-ressource.eu - etalux® Lichtsysteme
11. N. Golovanov, I. Iordănescu, P. Postolache, C. Toader, R. Porumb, C. Popescu and L. Lipan, *Electric systems and industrial audit elements*, N'EGRO Publishing House (2008)
12. D. Scripcariu and M. Scripcariu, *Cost effectiveness analysis of Demand-Side Management programmes*, UPB Scientific Bulletin, Series C, **60**, pp. 291-301 (1998)
13. D. Scripcariu and M. Scripcariu, *A modern approach to the energy supplier – end-user relationship and the participation of other organizations in demand-side management programmes*, Power Engineering Magazine, **44**, pp. 125-129 (1996)
14. *** U.S. Energy Information Administration, https://www.eia.gov/
15. *** The World Bank, http://www.worldbank.org

The Environmental Impact Reduction obtained by implementing an Energy Management System. The advantages of using Energy Management and Energy Savings Standards when performing Industrial Energy Audits

Mircea Scripcariu[1], *Ioan Sevastian* Bitir-Istrate[2], *Cristian* Gheorghiu[3*] and *Aida Maria* Neniu[4]

[1]Associate Prof., Dept. of Energy Production and Use, University POLITEHNICA of Bucharest, Romania

[2]Associate Prof., Dept. of Energy Production and Use, University POLITEHNICA of Bucharest, Romania

[3]PhD. Student, Dept. of Energy Production and Use, University POLITEHNICA of Bucharest, Romania

[4]PhD. Student, Dept. of Energy Production and Use, University POLITEHNICA of Bucharest, Romania

Abstract. Energy audits are used world-wide for developing energy efficiency projects. Industrial consumers have complex energy supply, generation and distribution networks and a variety of energy use installations. Romanian industrial companies became more interested in the last years in implementing Energy Management Systems in accordance with ISO 50001 standard. This paper presents a comparison of using the current way of developing energy audits and the concepts provided by the ISO group of standards, by quantifying the environmental impact reduction generated by each methodology. Authors pointed out that current legislation does not fully match the rigors of the ISO 50001 group of standards when evaluating the Energy Baseline (EB), the Energy Performance Indicators (EPI) or the Energy Performance Improvement Actions (EPIA), thus leading to a lower global energy efficiency improvement in the hypothesis of implementing all the recommended EPIAs [1]. Identifying and developing energy efficiency measures following the recommendations of the energy management and energy savings group of standards may be more consistent and less risky for the industrial company, which in turn can lead to an overall improvement of the Carbon Footprint [2].

1 Introduction

[*] Corresponding author: gheorghiu.cristian.k@gmail.com

The improvement of energy efficiency is a strategic objective of the national energy policy [3]. The main deficiency of the energy performance improvement program that every industrial end-user that records an equivalent energy consumption of over 1.000 t.o.e. (tons of oil equivalent) has to submit to A.N.R.E. (Romanian Energy Regulatory Agency) consists of [4]: a poor correlation between the declared plan and the actual and the predicted energy consumption, including irrelevant actions, with respect to the total energy consumption and the type of energy that has the biggest share and the lack of relevant benchmarking data and the poorly reported energy management data.

Even though the regulatory authority constantly tries to improve the Energy Audit Guide that provides rules that every auditing licensed entity must follow, the lack of metering energy inputs/outputs, the impossibility to correlate energy flows with the final product and the lack of an energy performance financial support scheme that can benefit from the aforementioned energy performance improvement, leads to an unsystematic implementation of the identified EPIAs.

In this paper the potential of improving the quality of an Industrial Energy Audit by implementing an EnMS (Energy Management System) in accordance to the ISO 50001 group of standards was analysed. The main issues that derived from the general requirements of the ISO 50001 group of standards were the normalization of the EPIs, respectively the EnB.

2 Classic Energy Audit vs. ISO 50001 based Energy Audit

In order to point out the benefits derived from implementing an EnMS in accordance to the ISO 50001 group of standards, the differences between the two methodologies were compared.

For the case study the latest Energy Audit [5] that the industrial end-user reported towards A.N.R.E., for the reference period 2015 – 2017. The share of each energy form that make up the total energy consumption shown in Fig.1, based on the aforementioned Energy Audit. The share of each energy form was considered equal to the reference period monthly average.

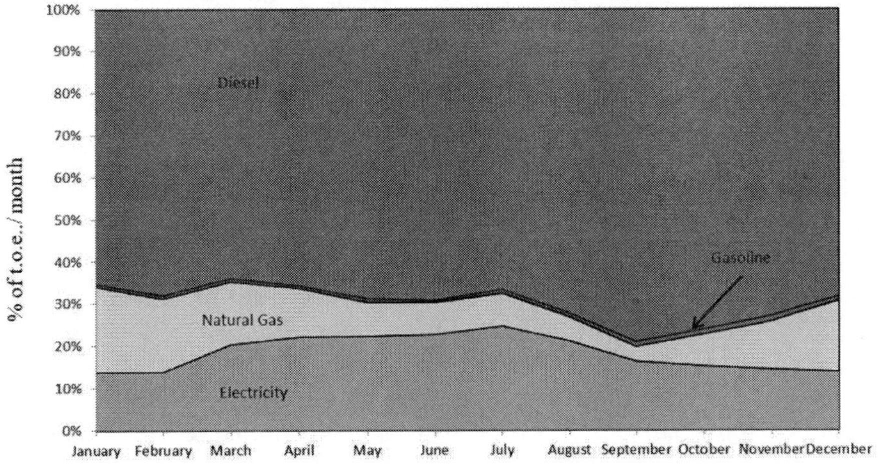

Fig. 1. The share of each energy form in the total, multiannual average energy consumption

As it can be easily seen in Fig. 1, Diesel holds the most significant share in the total energy consumption, averaging at about 72% of the total equivalent energy consumption. We will further propose and analyse the technical and the economic effect of an EPIA directed towards Diesel consumption.

Firstly, the *reference period* for the Diesel consumption - ***EPI*** (as seen in Table 1), respectively the ***EnB*** for the average monthly fuel consumption (as seen in Fig. 2) were determined, considering a conversion factor of 1.015 t.o.e./ton of Diesel [6]. The reference period for the case-study is 2015-2017 and the reporting period is 2018-2027.

Table 1. Reference period Diesel consumption, in tons and in t.o.e.

YEAR	M.U	Jan	Feb	Mar	Apr	May	Jun	Jul	Aug	Sept	Oct	Nov	Dec	TOTAL
2015	tons	0.60	0.28	1.11	0.44	1.41	1.31	1.81	0.47	0.86	0.57	0.34	0.77	9.97
2016		0.22	1.87	1.36	1.48	2.59	0.86	3.49	2.03	2.34	1.92	2.92	1.10	22.18
2017		1.40	1.04	1.56	1.82	2.76	1.96	2.07	2.92	1.80	1.62	0.75	0.86	20.56
YEAR	M.U	Jan	Feb	Mar	Apr	May	Jun	Jul	Aug	Sept	Oct	Nov	Dec	TOTAL
2015	t.o.e	0.63	0.29	1.17	0.46	1.48	1.38	1.90	0.49	0.90	0.60	0.36	0.81	10.47
2016		0.23	1.96	1.43	1.55	2.72	0.90	3.66	2.13	2.46	2.02	3.07	1.16	23.29
2017		1.47	1.09	1.64	1.91	2.90	2.06	2.17	3.07	1.89	1.70	0.79	0.90	21.59

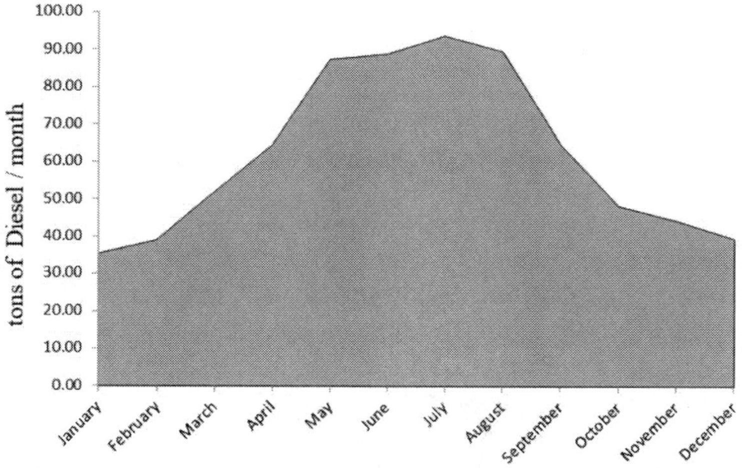

Fig. 2. The un-normalized EnB for the Diesel Consumption – EPI during the reference period (2015-2017)

The next step is proposing an appropriate ***EPIA*** to be evaluated in the two aforementioned scenarios – a Classic Energy Audit versus an ISO 50001 group of standards based Energy Audit. While keeping in mind that the Diesel consumption is generated by the cars fleet, which has the structure detailed in Table 2, the impact of replacing 44 of the 87 existing vans (3.5 to 7 ton) with 88 lighter, more efficient, 3.5 ton utility vehicles with an average consumption of 9 l/100 km was analysed. This ***EPIA*** is also supported by the seasonality of the ice-cream production, presented in Fig. 3, which can be correlated to an inefficient use of the carrying capacity of the actual cars fleet.

Table 2. The structure of the analysed industrial consumer utility vehicle fleet

Utility Vehicle Type	Quantity [Pcs.]	Fuel Type	Average Fuel Consumption [l/100 km]
Utility vehicle – under 3 tons	33	Diesel	8
Refrigerated van – under 3.5 tons	95	Diesel	12
Refrigerated truck – 3.5 – 7 tons	87	Diesel	20
Refrigerated truck – 7 – 12 tons	1	Diesel	21
Utility Truck Head	1	Diesel	35
Semitrailer	1	Diesel	3.5 l/h
Special shop utility vehicle	2	Diesel	8
Semi tanker	1	-	-

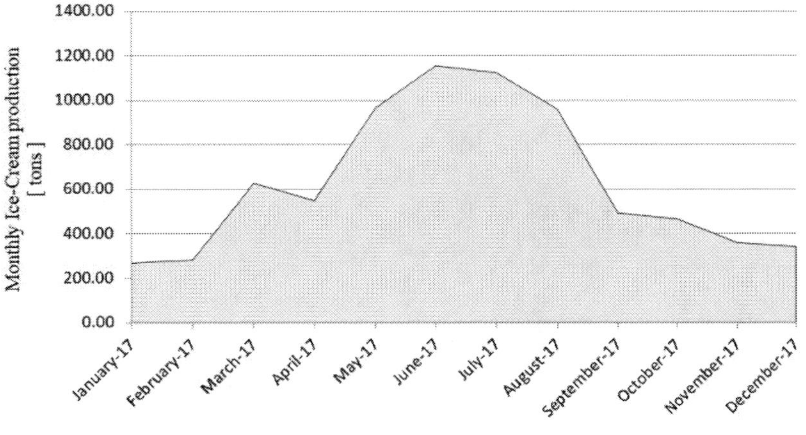

Fig. 3. The monthly production in 2017

For the technical and economic analysis, a series of hypothesis regarding the technical characteristics of the proposed replacement-vehicle, supported by an offer submitted by a utility vehicles producer were made, see Table 3.

Table 3. The technical characteristics of the proposed vehicles

Characteristic	Value	M.U.
Engine Displacement	2.998	[cm3]
Engine Power	156	[BHP]
Fuel Type	Diesel	[-]
Average Fuel Consumption	9	[l/100 km]
EU Emission Standard	Euro 6	[-]
Maximum Weight	3.5	[tons]
Refrigerator Box Temperature	0 – 2	[°C]
Normal Use Time[7]	6	[years]
Producer Warranty	2	[years]

Furthermore, for the economic analysis of the proposed ***EPIA*** the following input data was considered:

- The number of refrigerated trucks replaced each year: 4 pcs;
- The number of refrigerated vans bought each year: 8 pcs;
- The price of a refrigerated van: 24,000 EUR, without VAT;
- The price of a re-sold refrigerated truck: 10,000 EURO, without VAT;
- The yearly maintenance cost per van while in the warranty period: 300 EUR/van;
- The yearly maintenance cost per van while out of the warranty period: 500 EUR/van;
- The discount rate: r=9%/annum [8];
- The analysis period:2017-2027 – based on the fact that the investment in the ***EPIA*** began in August, 2017;
- The reporting period: 2018-2027 – this period is used to quantify and monetize the energy savings generated by the implementation of the ***EPIA***.

The 10 years reporting period will generate a number of risks such as: Diesel price evolution legislation, social and political, climate etc. and the high value of the investment in the proposed ***EPIA*** makes it necessary to split this payment over a longer period of time in order to maintain a reasonable degree of economic efficiency and be accepted by decision-makers.

2.1 Classic Energy Audit approach

In the Classic Energy Audit case, due to the industrial consumer inability to forecast the ***EPI***'s evolution, a hypothesis is made: the production and Diesel consumption are similar to the las year of the reference period for entire analysis period. Using the NPV (Net Present Value) criterion by applying the 1st formula, respectively the IRR (Internal Rate of Return) [9], and the C/B (Cost Benefit Analysis) described in equations 2 and 3, for a generally accepted interest rate r=9%/annum, the EPIA appears to be unattractive. The results are summarized in Table 4.

$$NPV = I_{annual} \cdot T_{tst} - I - O\&M \cdot T_{tst} \ [€] \quad (1)$$

where I_{annual} = annual economy generated by the *EPIA*, T_{tst} = discount factor, I = initial investment in the *EPIA*, O&M = operating and maintenance costs.

$$IRR = r_a + \frac{NVP_a}{NVP_a - NVP_b} \cdot (r_b - r_a) \ [\%] \quad (2)$$

where r_a = lower discount rate, r_b = higher discount rate, NPV_a = the NPV for r_a, NPV_b = the NPV for r_b.

$$B/C = \frac{undiscounted\ annual\ benefit}{initial\ investment\ cost} \ [-] \quad (3)$$

Table 4. The results of the technical-economic analysis – classic Energy Audit approach

Unit	Value	M.U.
NPV	- 383,789	[EUR]
IRR	-	[%/year]

B/C	- 2	[-]
Carbon Footprint Reduction	4,736.74	[tons CO_2/year]

The total Carbon Footprint reduction generated by the analysed EPIA in this scenario totalled a mere 4,736.74 tonnes of CO_2 for the entire *reporting period* for a conversion factor of 3.183 grams of CO_2 / kg of Diesel [10].

In Fig. 4, the predicted Diesel consumption as a result of implementing the analysed EPIA can be noticed to have an almost linear variation, due to the fact that neither the predicted production, nor the predicted Diesel consumption, were normalized.

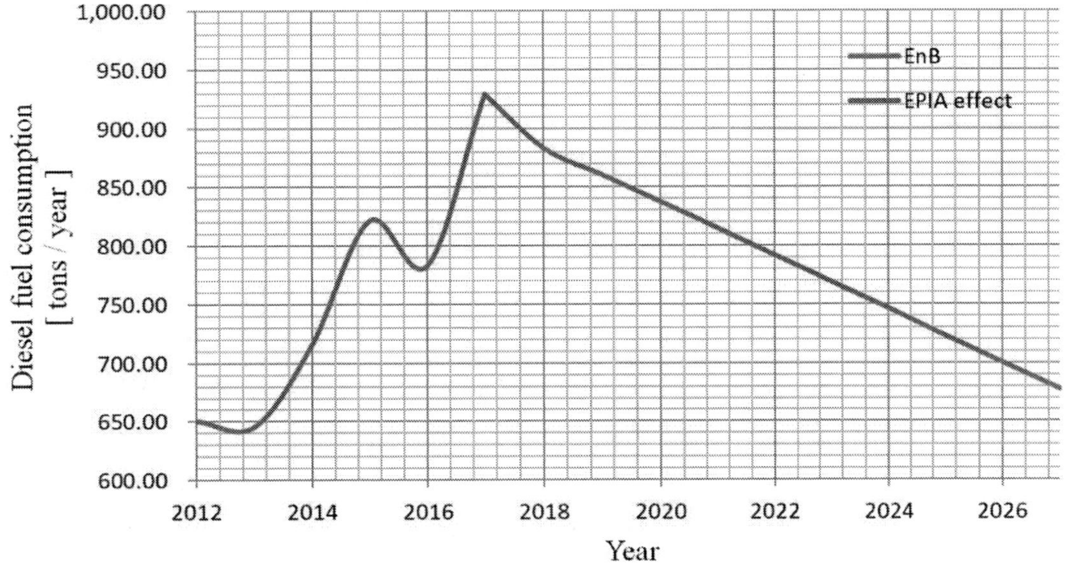

Fig. 4. The EPIA effect to the non-normalized Diesel Consumption

2.2 ISO 50001 standard based Energy Audit approach [11]

Because the linear regression method for normalizing the analysed EPIs lead, in some cases, to an unacceptable standard deviation, greater than 45%, we used *Python*'s scipy.interpolate.InterpolatedUnivariateSpline algorithm to determine the normalization function for both the main EPI – Diesel Consumption and the variable factors – Average outdoor temperature and average monthly food and non-alcoholic beverages expenditures per family – that are influencing it. In order to do so, a list of *static* and *variable factors*, presented in Table 5, was firstly set up.

Table 5. Static and variable factors [11-12] used for the normalization of the analysed EPI

FACTOR	TYPE	REFERENCE PERIOD			M.U.
		2015	**2016**	**2017**	
No. of production lines	Static	2	2	2	[pcs.]
Work shifts no.	Static	1	1	1	[shifts/day]
Average no. of employees	Static	542	586	623	[pcs.]
No. of deposits outside the audited energy boundary	Static	1	1	1	[pcs.]
Average outdoor temperature	Variable	See table 6			[°C]
Average monthly food and non-alcoholic beverages expenditures per family	Variable	665.16	653.22	699.67	[RON/family]

Table 6. Average outdoor temperature in the geographic area of the audited energy boundary

Month	2015	2016	2017	M.U.
January	2	0	-3	[°C]
February	4	9	3	[°C]
March	9	11	13	[°C]
April	16	20	15	[°C]
May	24	21	22	[°C]
June	27	29	28	[°C]
July	32	31	29	[°C]
August	31	30	31	[°C]
September	26	25	23	[°C]
October	15	13	15	[°C]
November	12	7	9	[°C]
December	7	2	6	[°C]

In order to properly forecast the predicted energy savings generated from implementing the analysed EPIA, a set of predicted values for the key *variable factors* have been considered – see Fig. 5 and Table 7.

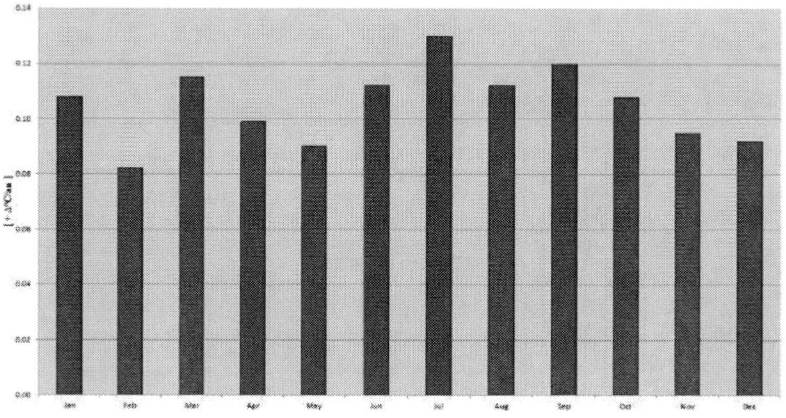

Fig. 5. The average temperature variation for the reporting period [13]

Table 7. The forecast of the average monthly food and non-alcoholic expenditures per family

Indicator	M.U.	2020	2025	2030	2035	2042
Average monthly food and non-alcoholic beverages expenditures per family	[RON/month]	711.45	823.91	1,134.6	1,567.65	2,256.13

Based on the second set of *variable factors*, the normalization function for the production during reporting period is presented in Fig. 6.

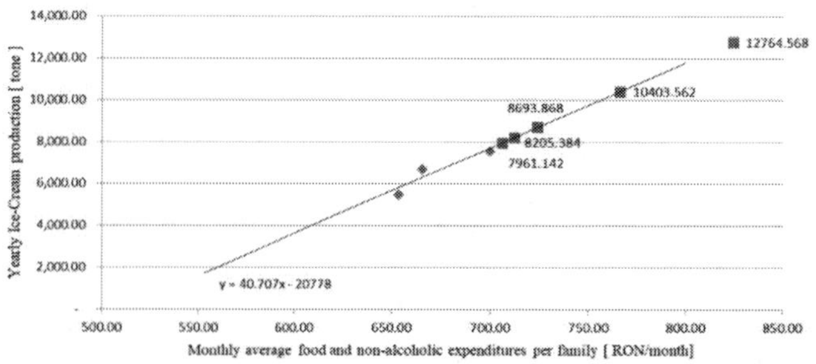

Fig. 6. The normalization of the predicted yearly Production with respect to the predicted increase of the monthly average food and non-alcoholic expenditures per family – linear regression

For determining the variation function of the production with respect to the monthly average exterior temperature (Production= f (temperature) ant the monthly average food and non-alcoholic expenditures per family (Production = f (expenditures)), the linear regression method was used. The determined linear functions were overlapped and the intersection of them was considered the normalized function of Production = f (temperature, expenditures).

By applying the double-normalized predicted yearly production to the aforementioned *Python* function, which fits a spline y=spl(x) of degree k to the provided x,y data via a Lagrange interpolation technique (4), the predicted Diesel consumption, during the reporting period was obtained.

$$P(x) = \sum_{j=1}^{n} P_j(x) = \sum_{j=1}^{n} \delta_{jk} \cdot y_k = y_j \quad (4)$$

where x = double normalized predicted yearly production, y = predicted Diesel consumption and δ = the *k* degree constant.

By applying the Lagrange interpolation technique, the normalized Diesel consumption – P(x) – was determined as being a 36th degree polynomial function.

The predicted production is limited to 10,000 tons/ year due to the no. of production lines, considered to be a *static factor* (see Table 5). Based on this newly obtained predicted Diesel consumption, the NPV / IRR/ B/C indicators were re-evaluated. The results are presented in Table 8.

Table 8. The results of the technical-economic analysis – ISO 50001 group of standards approach

Unit	Value	M.U.
NPV	9,070	[EUR]
IRR	10	[%/year]
B/C	0.05	[-]
Carbon Footprint Reduction	7,200.22	[tons CO_2/year]

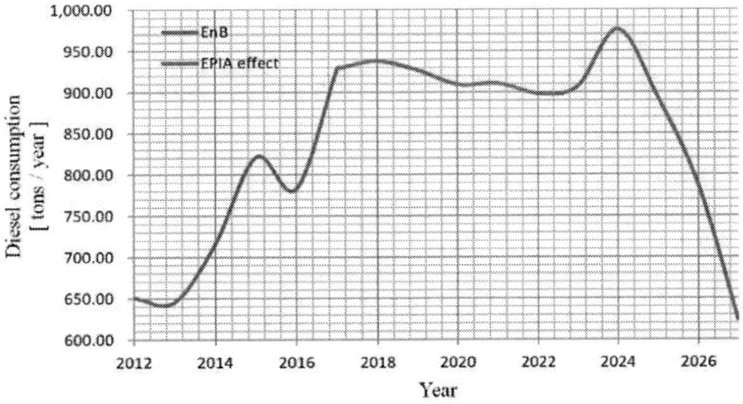

Fig. 7. The EPIA effect on the double-normalized Diesel consumptions, using ISO 50001 approach

The total carbon footprint generated by the analysed ***EPIA*** in this scenario totalled 7,300.22 tonnes of CO_2 for the entire *reporting period*.

In Fig. 7, the predicted Diesel consumption as a result of implementing the analysed ***EPIA*** can be seen to have a non-linear variation, due to the fact that both the predicted production and the predicted Diesel consumption were normalized. The Diesel consumption begins to have a linear variation starting from 2024 until the end of the reporting period, in 2027, because of the technological limitation that prohibits the production variable to further increase above the 10,000 tons/year threshold.

3 Conclusions

Normalizing the production with the average daily outdoor temperature and the monthly average expenditures per family on food and non-alcoholic beverages helps the decision-making for energy efficiency projects, both *variable factors* having an obvious influence on the quantity of items produced, distributed, sold and consumed.

By using the normalized *production* to analyse the main *EPI* – Diesel fuel consumption, the results of the NPV method were drastically affected in a positive manner. This can also be observed in Fig. 8, which depicts the predicted Diesel consumption during the reporting period in both analysed cases.

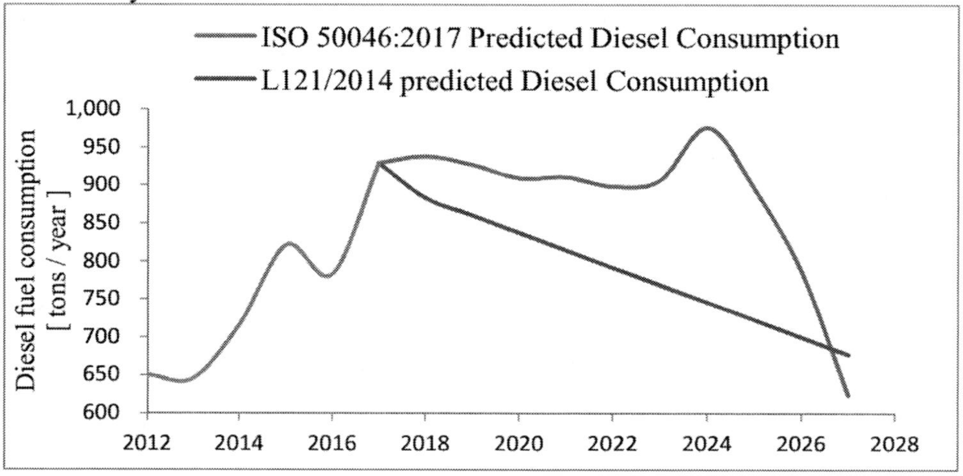

Fig. 8. The EPIA effect on Diesel consumptions during the reporting period without taking the static and variable factors into account (red), respectively with respect to the static and variable factors (green)

The main issues that authors encountered while applying the ISO 50001 standard proposed methodology were the inability to determine all the relevant *variable factors*, and the difficulty of determining the relevant *variable factors* values both from measurements and from official forecasts.

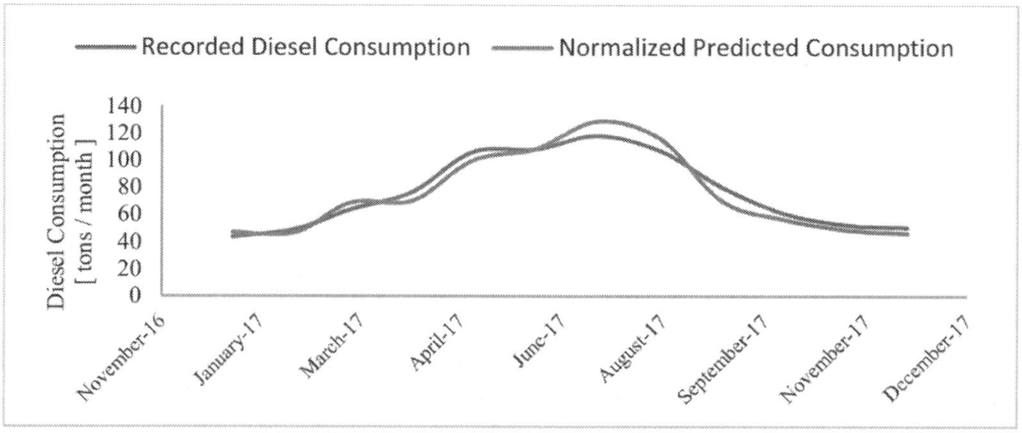

Fig. 9. The EPI – Diesel Consumption in 2017, based on its recorded values (blue) and the forecast, normalized values (red), based on the 2012 – 2015 production and outdoor temperature

As it can be observed in Fig. 9, the algorithm used to evaluate the effect to the EPIA leads in this particular case to an under 10% standard deviation, that lead to the conclusion that using a higher rank for the interpolating function when developing an EPI normalizing method is worth further researching. The opportunity of developing a specific methodology for evaluating the energy efficiency of each industry type is obvious, as it is clear that both the *static factors* and the *variable* ones are also specific to each type of consumer/industry.

Furthermore, it is strongly recommended that a better method of normalizing the different types of EPIs should be developed. The development of a Neural Network that can analyse all the data that an advanced monitoring and metering system can feed in and determine a daily / monthly / yearly set of normalization functions for the analysed EPIs could further improve the benefit of using the ISO 50001 standard group methodology both for implementing a proper and efficient EnMS and for the periodic Energy Auditing process.

References

1. I. Dzene, I. Polikarpova, L. Zogla, and M. Rosa, *Application of ISO 50001 for Implementation of Sustainable Energy Action Plans,* ENERGY PROCEDIA, **72**, pp. 111-118 (2015)
2. F. Marimon and M. Casadesus, *Reason to adopt ISO 50001 Energy Management System*, SUSTAINABILITY, **9** (2017)
3. The Romanian Parliament, *Energy efficiency Law no 121/2014* (04.08.2014)
4. A.N.R.E., *Decision 8/DEE/12.05.2015 – Energy Efficiency Improvement Program Model for industrial consumers* (2015)
5. S.C. ELSACO ESCO S.R.L., *Complex Energy Audit of the analysed industrial consumer* (2016)
6. ***, https://ec.europa.eu/eurostat/statistics-explained/index.php/Glossary:Tonnes_of_oil_equivalent_(toe)
7. The Romanian Government, *Government Decision no. 2139/2004 – Classification and Normal Use Time of fixed assets* (2004)
8. Public Procurement National Agency, *Order no. 2717/318/2018 for the revision of the discount rate* (2019)
9. Y. K. Shestopaloff, A.Yu. Shestopaloff – *Solving the Puzzle of IRR Equation. Choosing the Right Solution to Measure Investment Success*, AKVY PRESS (2011)
10. ***, (2017), https://www.eia.gov/tools/faqs/faq.php?id=307&t=11
11. International Standardisation Organization, *ISO 50001:2018 Energy Management Systems – Requirements with quidance for use* (2018)
12. International Standardisation Organization, *ISO 50047:2016 Energy savings – Determination of energy savings in Organizations* (2016)
13. R. Bojariu, M.V. Bîrsan, R. Cică, L. Velea, S. Burcea, A. Dumitrescu, S. I. Dascălu, M. Gothard, A. Dobrinescu, F. Cărbunaru, and L. Marin, *Climate Change – from the physics basics to risk and adaptation (in Romanian language)*, Printech publishing house (2015)

E3S Web of Conferences **112**, 04008 (2019) https://doi.org/10.1051/e3sconf/201911204008
TE-RE-RD 2019

Energy - Efficient Reconstruction of Public Lighting in Pravets Municipality from Bulgaria

Veselka Kamburova[1]†, *Violeta* Rasheva[2,*] and *Michail* Velikanov[3]

[1]Department of Heat, Hydraulics and Environmental Engineering, University of Ruse, Bulgaria
[2]Department of Industrial Thermal Engineering, University of Food Technologies, Bulgaria
[3]Encon Services International, USA

Abstract. Improving energy efficiency and reducing greenhouse gases is a central issue of the Europe 2020 strategy. Today enormous electric energy is consumed by the street lighting. Much of it could be saved through suitable energy-conservation measures (ECMs). The energy audit of the street lighting system in the Pravets municipality was carried out in compliance with the legislation in force in the EU and Bulgaria. Data about the present state of street lighting and electricity consumption for a period of 3 years were gathered and analysed. The baseline electricity consumption was defined. Classification of streets was done and the light class of the different categories streets was determined. The potential for energy savings was identified and a package of ECMs was developed. The annual electricity savings are 680 MWh (or 57,441 €/y) and the annual CO_2 emissions savings are 557,640 t. The investment for implementation of the envisaged ECMs is 414,840 €. The expected electricity savings compared to the baseline are 75%.

1 Introduction

Global energy policy aims at limiting climate change, energy security and competitiveness. The issue of energy consumption in buildings, industry and transport stands in the heart of the EU policy on energy efficiency. A number of studies have been carried out in the EU Member States to increase energy efficiency, environmental protection and the use of renewable energy sources in buildings and industrial plants [1-5]. Some authors have explored the possibilities of saving energy from lighting in different types of buildings [6-8]. But street lighting is also a great energy consumer. Scientific and technical literature does not sufficiently monitor the possibilities of electrical energy savings in street lighting. Today as electrical savings in street lighting as well as CO_2 emission reduction are very important. So there is a need for a detailed analysis of the savings in the existing street lighting systems and quality planning for new ones. Savings in the field of street lighting

† Deceased on 19.05.2019.

* Corresponding author: v_rasheva@abv.bg

could be better invested in other budget items in municipalities [9]. There are some attempts, in which the energy wastes of the street lights are reduced. A sensor light, which is controlled by the brightness sensor and the motion sensor, is sometimes used [10]. In recent years, Bulgaria has also been working to improve the state of street lighting systems and to reduce their electricity consumption. In this connection Bulgaria has harmonized European standard for street lighting [11] and in 2016 has published an Ordinance [12] according to which systems for outdoor artificial lighting located in settlements with over 20 000 inhabitants are subject to obligatory energy audits.

One of the basic requirements of European standard for street lighting is that the average brightness of the roadway does not fall below a specified operating brightness, because the task of outdoor lighting systems (street, park, hiking, advertising, architectural, decorative lighting, etc.) is to ensure the safe movement of motor vehicles and pedestrians. Street lighting improves visibility on the road and is a prerequisite for reducing fatal crashes. Also good street lighting leads to improved security of inhabitants and to decline in criminal activities. Thanks to it, a unique night view of towns and villages is created.

According to [13], municipalities are the owners of the street lighting systems. They invest in its reconstruction, build new sections, and pay the cost for electricity consumption. Costs for street lighting represent about 3% of the total costs of the municipalities.

The aim of this study is to evaluate the state of the street lighting systems in the municipality of Pravets and to identify the potential for reducing electricity consumption through appropriate ECMs, while at the same time provide the necessary normative parameters for illumination of the streets. This in turn would reduce the cost of municipal electricity consumption and CO_2 emissions into the atmosphere as well as would increase road safety.

2 Methodology

Pravets Municipality is situated in the western part of Bulgaria in the Sofia region. In 2017, the municipality has a territory of 324 km^2 and population of 7,801 inhabitants. The town of Pravets is an administrative centre of municipality and has a territory of 46 km^2 and population of 4,407 inhabitants. The municipality has 10 more villages besides the town of Pravets.

The energy audit of the existing street lighting system in municipality of Pravets covers the street lighting of the town of Pravets and the rest villages in the municipality and was carried out in 2017. The investigation involves the following steps and activities:

- Gathering basic information about the municipality, electricity consumption, the number, type and power of the street and park lighting, the number and condition of the transformer substations, the state of the street lighting supply network and the available electricity meters, the geometric parameters of the streets and the location of the street lighting poles;
- Assessment of the current state of street lighting system;
- Analysis of electricity bills, operating and maintenance (O&M) costs and operating time of street lighting system for a period of 3 years;
- Determination of baseline energy consumption by street lighting;
- Development of phototechnical classification of the street network according to [11] and selection of normative indicators;
- Determining the potential for electricity savings;
- Development of appropriate ECMs;
- Assessment of required investment costs;
- Assessment of the savings of electricity and the O&M cost after implementation of the ECMs;
- Techno-economic and ecological assessments of the proposed ECMs.

Assessments were performed on the basis of preliminary studies, analytical calculations and measurements were carried out using specialized measuring devices according to [14].

The next normative documents [11-12, 15-19] were used to assess the state of the street lighting in the municipality and to determine the ECMs.

3 Technical condition of street lighting in municipality of Pravets

The main supplier of electricity for the municipality is "CEZ Bulgaria" EAD, as deliveries are made through transformers 20/0.4 kV. In the street lighting panels there is no protection against atmospheric overvoltage and increased voltage due to zeroing. The existing street lighting system of the town of Pravets and settlements in the municipality is morally and physically obsolete, depreciated and energy-intensive and requires investment for reconstruction and replacement. The distribution network for street lighting is built consistently with the expansion of urban areas and in certain sections, updated according to the incidents on them and depending on the funds available for investment. The power supply is single phase, the network is airy, located on insulators and steel poles, and is mainly made with non-insulated conductor and partially with AL/R twisted conductors. Power supply via underground cables is applied only to a small number of lighting poles and park lighting within the town of Pravets.

The existing lighting is realizeded with the following technologies of lamps: mercury vapor (MV) with power of 125W; high pressure sodium (HPS) with power of 100W, 70W and 50W; compact fluorescent lamps (CFL) with power of 55W, 36W and 25W; decorative compact fluorescent lamps with power of 125W and lamps type "ball" with power of 85W.

The long-term operation of most lighting fixtures has led to their amortization - pollution, corrosion of the luminaires and their reflectors. In luminaires with mercury vapor lamps, most of the light (about 60%) is lost as it is emitted in the upper hemisphere, and causes considerable light pollution. These are the main reasons most of the luminaires to function with impaired working parameters of light output and thus with low efficiency. In some areas regulated norms of the lighting parameters are not met. In general, the existing lighting system uses obsolete technologies and is in poor technical and operational condition. The categorization, the photometric requirements for the different types of zones and the design of the street lighting network are in line with the old Bulgarian Standard [20], which is now replaced by the new European standard [11]. Under the requirements of the new standard, streets and parks are classified into separate categories depending on the required lighting of the area.

Most switchboards of street lighting are located in transformer substations. In accordance with the [13], these substations are owned by the electricity supplier. Distribution boards for street lighting must be removed outside of the supplier's property.

The control of street lighting is done mainly by automatic relays produced in Bulgaria. The control system operates on a predetermined time schedule, which is a relatively simple approach, but the main disadvantage is the lack of a centralized control system. The starting and switching devices for street lighting are located in the street lighting switchboards. Switching and protection equipment for street lighting are installed in low voltage panel inside a power substation. Most of the street lighting switchboards are not in good technical condition and need major overhaul, and some of them from replacement.

From the analysis of the current state of street lighting in the municipality of Pravets, it was found that the number of unlit streets is high and in certain sections of some illuminated streets there is a large irregularity of illumination. In this state, street lighting does not meet the requirements of the street lighting standard [11]. As a result of the street lighting survey the number, power and types of existing street lighting in Pravets and in the rest villages of the municipality respectively are presented in Tables 1 and 2.

Table 1. The number, power and types of existing street lighting in Pravets

№	Lighting type	Number of lighting	% of all	Lamp power + SRD, W	Total installed power, kW
1	HPS 50 W	90	5.38	64	5.760
2	HPS 70 W	510	30.50	87	44.370
3	HPS 100 W	69	4.13	120	8.280
4	CFL 25 W	426	25.48	25	10.650
5	CFL 36 W	54	3.23	45	2.430
6	CFL 55 W	116	6.94	65	7.540
7	HPS 150 W	132	7.89	175	23.100
8	type "ball" 85W	158	9.45	85	13.430
9	MV 125 W	75	4.49	140	10.500
10	LED 40 W	2	0.12	40	0.080
11	LED 60 W	40	2.39	60	2.400
	Total	**1 672**	**100**	-	**P_i = 128.540 kW**

Table 2. The number, power and types of existing street lighting in the villages of the municipality

№	Lighting type	Number of lighting	% of all	Lamp power + SRD, W	Total installed power, kW
1	HPS 50 W	447	46.18	64	28.608
2	CFL 15 W	18	1.86	15	0.270
3	CFL 25 W	5	0.52	25	0.125
4	CFL 36 W	498	51.44	45	22.410
	Total	**968**	**100**	-	**Pi = 51.413 kW**

4 Electricity consumption in the municipality

Data from the bills for electricity consumption in the municipality of Pravets for 3 years before the energy audit are presented in Table 3. The figures for the analyzed period show that the total electricity consumption for street lighting on the territory of the municipality is increasing. Electricity consumption in 2015 reaches 535,844 kWh per year, which is 18.8% more than in 2014. In 2016, electricity consumption continues to increase to 629,453 kWh per year or consumption increase is 39.6% compared to 2014. The increase in energy consumption concerns the town of Pravets and is mainly due to the expansion of the lighting system. Figures show that there is a drop in the energy consumption of some villages during the analysis period. This could be explained by the partial repairs of street lighting, where some luminaires have been replaced with new, energy-efficient ones. But in most cases, the drop in consumption is due to non-glowing lamps, disconnected power supply to lamps and other incidents.

Table 3. Electricity consumption in the municipality of Pravets for 3 years

Settlement	2014		2015		2016	
	kWh/y	BGN/y	kWh/y	BGN/y	kWh/y	BGN/y
Town of Pravets	254,295	42,088	382,436	64,841	509,323	84,292
Other villages	196,565	32,011	153,408	26,086	120,130	19,885
Total	**450,860**	**74,099**	**535,844**	**90,927**	**629,453**	**104,177**

The current technical and operational state of the street lighting system in the municipality of Pravets is the closest to those of 2016. Based on the above fact, 2016 was chosen to be representative.

During the audit, it was found that the operating hours of street lighting in the town of Pravets are 4,150 h/year and in the rest villages of municipality – 4,100 h/year. The total installed power in the town of Pravets is 128.54 kW. The annual electricity consumption for street lighting for year-round and full-night illumination at 100% working lamps is:

$$Eyear = 128.54*4,150 = 533,441 kWh/y \quad (1)$$

The annual electricity consumption according to the invoice data is 509,323 kWh/y. The difference between the calculated by equation (1) and the actual consumption gives information about the percentage of working lights in the town of Pravets. Consequently, the share of working street lighting in the town of Pravets is about 95.5% of total 1,672 number of luminaries. Applied per 1 inhabitant, the calculated specific electricity consumption is 121.04 kWh/year and the actual specific consumption is 115.57 kWh/year.

Similarly, the annual electricity consumption for street lighting for year-round and full-night illumination at 100% working lamps in the rest villages of the municipality is:

$$Eyear = 51.413*4,100 = 210,793 kWh/y \quad (2)$$

The annual electricity consumption according to the invoice data is 120,130 kWh/y. The difference between the calculated by equation (2) and the actual consumption gives information about the percentage of working lights. Consequently, the share of working street lighting in the villages of the municipality is about 57% of total 968 number of luminaries. Applied per 1 inhabitant, the calculated specific electricity consumption is 62.1 kWh/year and the actual specific electricity consumption is 35.39 kWh/year.

According to the National Statistical Institute [21], the average electricity consumption for street lighting per 1 inhabitant of the population is 29.12 kWh/year. Therefore the municipality of Pravets has great potential for electricity saving and appropriate ECMs should be offered.

5 Baseline of electricity consumption

It was established that in 2016 in the town of Pravets the street lighting system worked on an average of 11.37 hours per day or 4,150 hours/year, i.e. 100% of the theoretically determined 4,150 h and similarly in the villages of the municipality the street lighting system worked on an average of 11.25 hours per day or 4,106 hours/year, ie. 100% of the theoretically determined 4 100 h.

The theoretical period during which artificial lighting of external public spaces is necessary was determined on the basis of the Astronomical calendar of Bulgaria [22].

The baseline is the sum of the required installed capacity of all luminaires with an annual load of 4,150 h for the town of Pravets and 4,100 h for the rest villages of the municipality at 100% full-night and year-round illumination of street lighting. The implementation of these criteria leads to the satisfaction of normal requirements for brightness of different categories streets. It was established that the light-emitting indicators of the available luminaires constituting the municipal street lighting system do not meet the minimum requirements of the standard for any category of street. To achieve the standard brightness (luminance) criteria as well as the requirement for uniformity of illumination, it is necessary to install additional luminaires with the corresponding power. Table 4 shows the installed capacity of street lighting at baseline, both for the town of Pravets and for the rest villages in the municipality.

Table 4. Baseline electricity consumption in the municipality of Pravets

Settlement	Number of lighting		Installed power, kW		Base electricity consumption	
	Current value	Base value	Current value	Base value	kWh/y	kWh/y per 1 inhabitant
Town of Pravets	1,672	1,722	128.54	132.89	551,494	125.14
Other villages	968	1,429	51.413	86.819	355,958	104.87
Total	**2,640**	**3,151**	**179.953**	**219.709**	**907,452**	-

6 Classification of streets and selection of normative indicators

Street Lighting Standards [11, 19] place a major focus on efficient use of electricity and take into account the functional purpose of streets and roads and introduce light technical classification. The first step in the identification of the ECMs is the correct determination of the light class of the different categories streets in the settlement under [11, 19]. The road network in the town of Pravets consists of urban highways II class, class III arterial streets, a secondary street network including collection streets IV class and serving streets V class. The standard light-emitting indicators for the different categories of streets and pedestrian spaces in the town of Pravets are presented in Table 5.

Table 5. Lighting requirements for different categories streets and pedestrian spaces [15]

Light class	Lighting requirements for different categories streets				
	Lcp.	General unevenness	Longitudinal unevenness	Dazzle	Light environments
	cd/m2	V_0	V_L	TI (%)	SR
M3	≥ 1	$\geq 0,4$	$\geq 0,6$	≤ 15	$\geq 0,3$
M4	$\geq 0,75$	$\geq 0,4$	$\geq 0,6$	≤ 15	$\geq 0,3$
M5	$\geq 0,5$	$\geq 0,35$	$\geq 0,4$	≤ 15	$\geq 0,3$
M6	$\geq 0,3$	$\geq 0,35$	$\geq 0,4$	≤ 15	$\geq 0,3$
	Lighting requirements for pedestrian spaces				
Light class	**$E_{cp.}$ (lx)**			**E_{min} (lx)**	
S4	≥ 5			≥ 1	

7 Suggested energy conservation measures (ECMs)

The options for saving electricity ware identified on the basis of which a package of energy saving measures (ECMs) was developed as follows:

ECM 1: Replacement of all existing street luminaries (due to their low efficiency and strong amortization) with new LED luminaries - latest generation;

ECM 2: Installation of new LED luminaries in areas where there is no mounted lighting;

ECM 3: Installation of an appropriate control system of street lighting.

The development of LED technology is a prerequisite for high efficiency performance compared to conventional street lighting. For this reason, they are the most energy-efficient street lighting luminaire for the moment. The following types of LED street luminaries are selected for proposed ECMs in Pravets municipality: LED 55W; LED 40W; LED 27W; LED 14W; LED 6W and LED PARK 8W. The main results from the light technical calculations for different categories of streets in the municipality are presented in Table 6.

Table 6. Results from the light technical calculations for different categories of streets

Indicator	Operating luminance (cd/m^2)	General unevenness	Longitudinal unevenness	Dazzle	SR
Urban highways - LED 55W					
Normative value	1,0	0,40	0,60	15	0,30
Realized value	1,03	0,42	0,64	12	0,32
Arterial streets LED 40W					
Normative value	0,75	0,40	0,60	15	0,30
Realized value	0,78	0,44	0,68	12	0,33
Collection streets – 27W					
Normative value	0,50	0,35	0,40	15	0,30
Realized value	0,52	0,38	0,42	13	0,33
Serving streets – 14W					
Normative value	0,30	0,35	0,40	15	0,30
Realized value	0,32	0,38	0,42	13	0,33
Decorative lighting – 6W					
Normative value	0,20	0,35	0,40	15	0,30
Realized value	0,22	0,38	0,42	13	0,33
Park lighting – 8W					
Normative value	5 lux	minimum illumination – 1 lux			
Realized value	5,05 lux	maximum illumination – 1,1 lux			

Table 7. Distribution of LED street and park lighting by number and power

№	Street type	Number of lighting	Power, W	Total power, kW
Town of Pravets				
1	Urban highways	76+36=112	LED 55+ LED 27	4.18+0.972
2	Arterial streets	185	LED 40	7.4
3	Collection streets	94	LED 27	2.538
4	Serving streets	697	LED 14	9.758
5	Decorative lighting	426	LED 6	2.556
6	Park lighting	158	LED Park 8W	1.264
Total		**1672**	-	**Pi = 28.668**
Villages from the municipality				
1	Arterial streets	58+471=529	LED 40+ LED 27	2.32 +12.717
2	Collection streets	551	LED 14	7.714
3	Serving streets	377	LED 8	3.016

4	Park lighting	23+21=44	LED Park 8+14	0.184+0.294
Total		**1501**	-	**Pi = 26.245**

After the light technical calculations, the type and number of street lights for the municipality were determined and their distribution by category of streets is presented in Table 7. The total installed power of the lighting after the implementation of ECMs would be 54.913 kW and the annual electricity consumption would be 226,576.5 kWh.

A new automated street lighting control system will be built in Pravets. The distribution boards will be replaced by new ones located outside the electric substations. Some boards in the town of Pravets will be equipped with new meters, standard interface controllers and GPRS transmitters. The control system will provide information to a control station in real time. In case of incidents an alarm signal will be sent to the control station. These features will allow real-time monitoring of current electricity consumption and in case of incident, theft or other critical circumstances, the operator will be able to respond immediately. For other villages in the municipality, the automated street lighting control system will not have a significant effect on energy savings. This is explained by the low electricity consumption of street lighting. Due to the above facts, such type of automatic control system will not be implemented in other villages in the municipality.

8 Economic evaluations of energy conservation measures

The financial parameters of the project are calculated with specialized software Economy Software 6.26 - ENSI 2002-2004. The calculations were made at an average interest rate of 8% and inflation for 2016 - 0.1% according to [21].

The investment for implementation of the proposed package ECMs (auditing, designing and building of the new lighting system) is 414,840 €.

The average price of electricity without VAT for 2016 for two-rate metering (67% night rate and 33% daily rate) is 0.084 €/kWh.

The implementation of energy saving projects will reduce the amount of CO_2, emitted into the atmosphere. The savings from CO_2 were determined by the saved electricity and an emission factor of 819 g/kWh according to [23]. The annual electricity and CO_2 emissions savings after the implementation of new LED street lighting system compared to the baseline are presented in Table 8. The annual savings for street lighting are expected to be 57,441 €/y.

Table 8. Annual electricity and CO_2 savings after the implementation of ECMs

№	State	Number of lighting	Power	Annual electricity consumption	CO_2 emissions	Specific electricity consumption
			kW	kWh/y	tCO_2/y	kWh/y per 1 inhabitant
1	Baseline	3 151	219.709	907,451	743.20	116,33
2	After ECMs	3 173	54.913	226,577	185.56	29,04
3	Savings	-	164.796	680,875	557.64	87,29

After implementation of the ECMs, an effective street lighting maintenance plan will be developed, which will minimize the total annual O&M costs. The O&M costs for the street lighting in the municipality of Pravets according the baseline are 15,274 €/y. After the introduction of energy-efficient LED street lighting with a 5 year warranty, O&M costs are expected to be 4,021 €/y. Therefore the annual savings from O&M costs will be 11,253 €/y and the total annual savings will be 68,694 €/y. The payback period according to the achieved savings compared to baseline is 6.04 years. Net present value is 265,053 € and internal rate of return is 16%. After implementation of the ECMs, the specific energy consumption will decrease by 75.03% compared to the baseline.

9 Conclusion

The energy audit of the street lighting system in the Pravets municipality was carried out in compliance with the legislation in force in the EU and Bulgaria.

The proposed LED lights for street and park lighting are the latest generation and this ensures maximum reduction of electricity and CO_2 emissions and high operational reliability (minimum warranty of 5 years). The main part of the selected street and park lighting are with power of 8W, 14W and 27W and dimming during the night hours (from 23:00 to 5:00) is not provided for the following reasons:

- According to European and world practices, dimming for streets and pedestrian areas of categories (M6, M5 and S4) is not foreseen, as the value of brightness and illumination falls well below critical limit;
- The necessary initial investment to realize the nighttimes dimming function with 8W, 14W and 27W lighting is significant compared to the symbolic savings it would bring.

The overall effect of putting the new, energy-efficient LED street lighting into operation is as follows:

- decrease of the annual CO_2 emissions, compared to the baseline by 557,636 tons;
- decrease of the annual electricity costs, compared to the baseline by 57,441 €;
- annual specific electricity consumption for street lighting per 1 inhabitant is 29.04 kWh and is lower than the national average value – 29.12 kWh;
- reduction of the annual O&M costs of the street lighting by 11,253 €;
- total reduction of annual electricity consumption and O&M costs compared to baseline by 68,694 €;
- the new LED street lighting meets the requirements of EN 13201;
- the required investment amounts to 414,840 € and the payback period according to the annual electricity savings is 6.04 years;
- after implementation of the ECMs, the specific energy consumption in relation to the baseline will decrease by 75.03%.

References

1 N. Anisimova, The capability to reduce primary energy demand in EU housing, Energy and Buildings, **43 (10)**, pp. 2747–2751 (2011)

2 C.A. Balaras, A.G. Gaglia, E. Georgopoulou, S. Mirasgedis, Y. Sarafidis, D.P. Lalas, European residential buildings and empirical assessment of the Hellenic building stock, energy consumption, emissions and potential energy savings, Buildings and Environment, **42 (3)**, pp. 1298–1314 (2007)

3 S. Lechtenbohmer, A. Schuring, The potential for large-scale savings from insulating residential buildings in the EU, Energy Efficiency, **4(2)**, pp. 257–270 (2011)

4 G. Valtchev, Energy Efficiency and Protection of the Environment. *Scientific Conference, University in Town of Chanak Kale,* Turkey, (2010)

5 G. Lazaroiu, V. Berbece, L. Mihaescu, I. Pisa, G.-P. Negreanu, Biogas Production from Animal Protein Waste, an Environmental Protection Technology Coupled With Energy Production, Scientific Works of University of Food Technologies, Volume **64**, Issue 1, (2017)

6 C.K. Gana, A.F. Sapara, Y.C. Muna, K.E. Chongb, Techno-economic Analyses of LED Lighting: A Case Study in UTeM'sFaculty Building, Procedia Eng. **53**, pp. 208-216 (2013)

7 F. Frantozzi, F. Leccese, G. Salvadory, M. Rocca, M. Garofalo, LED Lighting for Indoor Sports Facilities: Can its Use Be Considered as Sustainable Solution from a Techno-economic Standpoint? Sustainability, **8**, 618-631 (2016)

8 G. Ciampi, A. Rosato, M. Scorpio, S. Sibilio, Retrofitting Solutions for Energy Saving in Historical BuildingLighting System, Energy Procedia **78**, pp. 2669-2674 (2015)

9 I. Ramljak, I. Ramljak, M. Sesar, Reconstruction of the Street Lighting Sustem as a Tool for Electrical Energy Savings, International Journal of Electrical and Computer Engineering Systems, Volume **7**, Number 2, pp. 71-81 (2016)

10 R. Velaga, A. Kumar, Techno-economic Evaluation of the Feasibility of a smart street sustem& A Case Study of Rural India, Procedia Social and Behavioral Sciences, **62**, pp. 1220-1224 (2012)

11 Bulgarian Standard BDS EN 13201Road lighting (2016) [in Bulgarian]

12 Ordinance № E-RD-04-05 for Determining the Indicators of Energy Consumption, Energy Characteristics of Enterprises, Industrial Systems and External Artificial Lighting Systems, and for Determining Terms and Conditions for Performing Energy Efficiency Activities and Preparing the Energy Savings Assessment of the Minister of Energy and the Minister of Economy (2016) [in Bulgarian]

13 Low of Energetics (2003) [in Bulgarian] available from: https://www.me.government.bg/en/library/zakon-za-energetikata-256-c25-m0-1.html

14 Ordinance № Е-РД-04-1 for circumstances subject to entry in the registers under the Energy Efficiency Act, its registration receiving information from these registers, the terms and conditions for obtaining a qualification from energy efficiency consultants of the Minister of Energy and the Minister of Regional Development and Public Works (2018) [in Bulgarian]

15 Bulgarian Standard BDS EN 60598-1 / 2 – Lighting (2015) [in Bulgarian]

16 Energy Efficiency Act (2015), Available from: https://www.seea.government.bg/documents/

17 Low of Spatial Planning (2001)[in Bulgarian]

18 Ordinance № 3 on the structure of electrical installations and power lines of the Minister of Energy (2004) [in Bulgarian]

19 CIE 115:2010 Lighting of Roads for Motor and Pedestrian Traffic (2010)

20 Bulgarian Standard BDS 5504-82 "Lighting of streets and pedestrian zones (1982) [in Bulgarian]

21 Statistical data (2016) [in Bulgarian] Available from: http://www.nsi.bg./bg/content/766

22 Astronomical Calendar (2016) [in Bulgarian] Available from:
http://nao-rozhen.org/astrocalendar/2016/Astronomical_calendar_2016.pdf

23 Ordinance № Е-РД-04-2 on the energy consumption and energy performance indicators for buildings of the Minister of Energy and the Minister of Regional Development and Public Works (2016) [in Bulgarian]

Experiments in use of bentonite for energy recovery of used oils

Mihai Dragne[1], *Elena* Pop *[1] *Cristina Ileana* Covaliu[2], *Ecaterina* Matei[3], *Ionel* Pîşă[1]

[1]University Politehnica of Bucharest Faculty of Mechanical Engineering and Mechatronics, Department of Thermodynamics, Engines, Thermal and Refrigeration Equipment, Romania;
[2]University Politehnica of Bucharest, Faculty of Biotechnical Systems Engineering, Department of Biotechnical Systems, Romania;
[3]University Politehnica of Bucharest, Faculty of Materials Science and Engineering, Department of Biotechnical Systems, Romania;

Abstract The paper proposes the development of advanced, integrated technologies for materials with properties of adsorbents (e.g. bentonite), and the use of these material in processes of used oil recovery with remediation effect and energy recovery. The main objective is to replace diesel fuel, which is an expensive fuel, with revitalized oils for energy production. For these, laboratory experiments are required in two important directions. The first tests are related to the determination the optimum concentration of bentonite for removal of metals from used oil. The revitalization of the used oil leads to the protection of the combustion plant, to the increase of its lifetime and polluting emissions reduction. The second set of experiments relates to the determination of energy characteristics of used oil and revitalized used oil for energy production. Elemental analysis, lower heating value, viscosity, density and ash analysis were determined. Experimental test of combustion process on a small scale boiler has been done and the results shows an improved combustion process of revitalized oil compares with used oil and the decrease of the pollutant emissions.

1 Introduction

Environmental pollution is a topical issue and trying to obtain non-polluting fuels or to recover waste materials for energy purpose concerns many researchers today. It is necessary to solve the problem of waste oil from road vehicles according to DIRECTIVE 2008/98/EC on waste and the national legislation in force: Decision no 235 of 7 March 2007. The legislation provides all operations by which waste oils transferred from the holder to the economic operators who can collect, recover or eliminate those oils or use them as fuel with adequate heat recovery. The present research studies the problem of waste oils from the automotive industry and the possibility of their regeneration with energy recovery. Motor oils are used to lubricate the components of the engine assembly and the

* Corresponding author: elena.pop@upb.ro

essential properties of it are: Greasing/lubrication to reduce friction between moving parts, cooling of the parts with which it comes into contact, especially the piston, cleaning of the parts on which residues of the combustion process are deposited (e.g. cylinders, segments, piston), protection against chemical corrosion of engine parts. The used motor oils are contaminated by contaminants and impurities resulted from undesirable oxidation processes: sediment, water, metallic particles and degraded additives [1]. There are several methods to remove these contaminants. The technology proposed in the present research is aimed for the process of revitalization of residual oils from motor engines by contacting them semi-dynamic, with natural/modifies nanostructured compounds [2]. In this regard we choose bentonite as nanostructured material. Bentonite powder plays an important role in the regeneration and clarifying of wastes [3]. Research regarding the wastewater treatment using bentonite clay due to chemical composition and crystal morphology, which are indicative of effective physicochemical adsorption [4] shows that the structural configuration (octahedral and tetrahedral) is effective in sorption by intercalation of the turbid materials (heavy metals) onto the sheets and porous surface of the bentonite clay. Bentonite has also been reported to have a high adsorption capacity for heavy metals due to its high specific surface area, small particle size, high porosity and high cation exchange capacity [5] The adsorption technique has been found to be superior to the other techniques for removal of heavy metals, in terms of cost, flexibility, simplicity of design, ease of operation, insensitivity to toxic pollutants and better removal efficiency [6].

2 Experimental determinations

2.1. Materials and methods

Samples analyzed are two types of used motor oils: one comes from an spark-ignition engine from a Renault vehicle brand, Symbol model, 1589 cm3 cylinder capacity, manufactured in 2008, and the other from a compression-ignition engine from a Volkswagen vehicle brand, Passat model, 1989 cm3 cylinder capacity, production year 2008.
As material for regenerating the used oil, activated Bentonite has been used. Activated bentonite is obtained by processing the crude bentonite $Al_2O_3{\cdot}4SiO_2{\cdot}4H_2O$ exploited from the calcium bentonite quarry from Orasu-Nou, activated by a specific process. Due to the many physical and chemical properties such as high water absorption capacity, high ion exchange capacity, high intercrystalline swelling in aqueous medium and high plasticity, bentonite can be appropriate for contaminants removal. The purpose of using bentonite in this experiment is related to its absorption capacity of the contaminants and regeneration of the waste motor oil Technical characteristics of bentonite are: Maximum granulation: rest on sieve Ø 0,063 mm - max.25% bulk density: 800-850 kg/m3, humidity: max. 10% pH: min.8, bentonite number: min. 0,85.

2.2 Physical and chemical characterization of used oil

The physico-chemical characteristics of motor oils change as a result of their exploitation [3]. The operating conditions lead to pollutants accumulation in the oil that change the quality of engine oil. Analyzes on density and viscosity of the different kind of motor oils (raw and used) were performed. Three samples have been considered: raw motor oil, waste oil from spark-ignition engine and waste oil from a compression-ignition engine
The carbon (C), nitrogen (N), hydrogen (H) and sulphur (S) contents of the waste motor oil samples were determined by a CHNS elemental analyzer EA 1110. Determinations were performed at ICEMENERG. The Lower Heating Power was determined according to ASTM D240-17;

Fig.1 Elemental analyzer EA 1110

2.3. Regeneration of used oil using bentonite

The regeneration was done on used crude MAC oil. Different amounts of bentonite (20g, 40g, 60g, 80g, 90g) were added in a volume of 400 mL of used oil. Each sample was homogenized for 2 weeks, 8 hours per day at room temperature and 300 rpm. Samples were taken from each set of experiments having different amounts of bentonite in order to quantify the amount of metal adsorbed by bentonite. The analysis of heavy metals was carried out by heat treatment of the samples at 800°C and the resulting ash was taken up for atomic absorption analysis.

2.4 Tests for combustion processes

The experimental researches were conducted in the Combustion laboratory of the Department of Thermodynamics, Engines, Thermal and Refrigeration Equipment, Faculty of Mechanical Engineering and Mechatronics of Politehnica University of Bucharest.

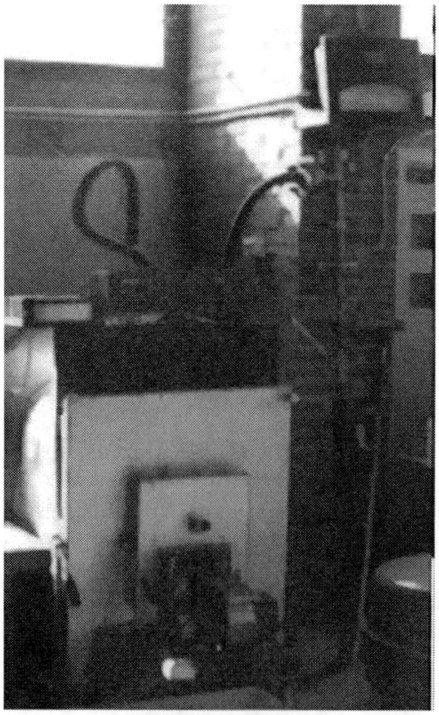

Fig 2 Experimental pilot boiler of 55 kW

Fig. 3 burnerAnyo-12

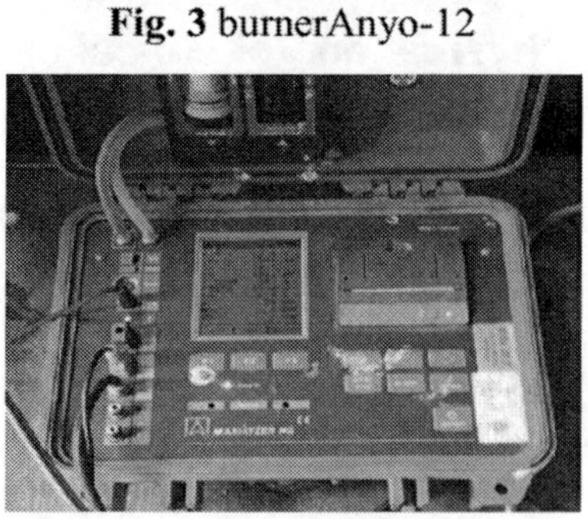

Fig. 4 Flue gas analyzer

MAXILYZER NG

For combustion tests a pilot boiler of 55 kW has been used. The boiler is equipped with a special burner for liquid which ensures the preheating of the fuel up to 80°C, temperature imposed by the high viscosity of the analyzed fuel.
For the tests, used diesel oil and regenerated used diesel oil after treatment with bentonite (P1 with 20g bentonite) has been used. For comparison of combustion process, we also used diesel fuel. During the experiments the pollutant emissions from the flue gases has been analyzed using a Maxilyser NG gas analyzer.

3. Results and discussion

High viscosity of used motor oil determined during the experiments (see table 1) imposes the necessity of used oil preheating for the combustion process.

Table 1. Viscosity and density of motor oil

	Type of sample /property	Temperature, t	Viscosity	Density, ρ
		°C	E^0	Kg/m^3
1.	Raw motor oil	75	2,4	840-850
2.	Used oil from spark-ignition engine	75	2,6	870-880
3.	Used oil from compression-ignition engine	75	2,7	840-850

The results of the elemental analysis are presented in the table below:

Table 2 Elemental analysis of the used oil

Type of sample	*C*	*H*	*N*	*S*	*LHV*
	%	%	%	%	MJ/kg
Used oil from spark-ignition engine	83,56	13,77	0,41	0	41,96
Used oil from compression-ignition engine	84,12	14,29	0,33	0	42,33

The composition of the used motor oil and the value of LHV is comparable with liquid fossil fuel as diesel fuel and this recommends it for energy recovery.
The analysis of ash content resulted after thermal treatment of used motor oil with different amount of bentonite in order to quantify the amount of metal adsorbed are presented below.

Table 3 Ash content analysis results - by atomic absorption analysis (AAS)

Amount of sample [g] / Metals content [mg/g]	P5 90 g	P4 80g	P3 60g	P2 40g	P1 20g
Cu	1.367	1.275	0.86	0.432	0.203
Zn	98.76	74.58	72.26	46.38	45.88
Fe	7.198	3.563	3.156	3.01	2.233
Cd	0	0	0	0	0
Pb	0	0	0	0	0
Ca	5.693	2.39	1.810	1.430	0.943

From the analysis of the results, we find that the sample P1 had the lowest content of metals so we chose this sample for the combustion experiments. Also regarding the possibility of removal of heavy metal content by adsorption on bentonite material, using the lowest amount of bentonite conducted to the highest adsorption of heavy metals from used oil.

The combustion process of the sample P1 was analyzed in terms of pollutants emissions and efficiency of the process and the results are presented below:

Table 4 Result of the combustion experimental tests

Type of fuel tested	**O_2***	**CO**	**CO_2**	**NO_X**	**T_{evac}**
	%	**ppm**	**%**	**ppm**	**°C**
Diesel Fuel	6	2869 - 3034	8,7	24 - 29	192
Raw Waste diesel oil	6	5032 - 5073	10	59 - 64	224
Regenerated waste diesel oil with bentonite	6	4146 - 4158	10	54 - 58	231

*All samples were corrected for standard oxygen concentration of 6%

The concentration of CO in flue gas is high because we used a burner design for CLU combustion coupled with a natural circulation boiler, strongly cooled and the burner's thermal power is lower than the furnace thermal power. It is mentioned that CO emitted in the flame does not burn below the temperature of 750 C, the temperature reached only in the final part of the furnace. Considering all this, it is recommended a boiler with chamotte insulation, uncooled which due to high combustion temperature will significantly reduce the final CO emissions. Analyzing the results of the combustion process, we note that the used oil regenerated with bentonite has better performance compared to raw used oil with a decrease of CO emissions of about 18%.

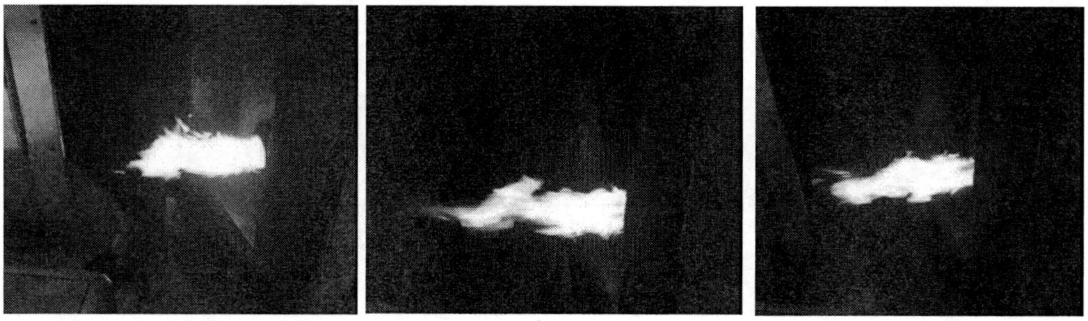

a b c

Fig. 5. Aspect of flame in the burning process
a. diesel oil; b. raw used oil; c. regenerated used oil

Due to the low volatile matter content of the used oil motor the flame presents pulses and a longer length than the reference flame (diesel oil). It is noted that the regenerated oil has attenuated pulses comparing with raw used oil and the ignitions closer to the spray nozzle.
The use of bentonite as a material to regenerate used engine oils proves to be an effective solution in lowering the metal content and improving the combustion process.
The reducing of metals concentration in used oil influence directly the burner lifetime by protecting the spray nozzle.
Improvement of the combustion process was achieved by higher flame stability and lower CO emissions. We expect that by better adaptation of the burner-furnace assembly to increase the efficiency of the combustion process
The results can be extrapolated to other types of used oils such as oils from the energy industry or waste oils from the food industry

References

1. Cornelia Stan, Cristian Andreescu, Marius Toma, *Some aspects of the regeneration of used motor oil*, Procedia Manufacturing, Vol. **22**, 2018, Pages 709-713
2. F. Dalla Giovanna, O. Khlebinskaia, A. Lodolo, S. Miertus, *Compendium of Used Oil Regeneration Technologies*, U.N.I.D.O., Trieste, (2003)
3. Shiva Salem, Amin Salem, Aylin Agha Babaei, *Preparation and characterization of nano porous bentonite for regeneration of semi-treated waste engine oil: Applied aspects for enhanced recovery*, Chemical Engineering Journal, Vol. **260**, 15 January 2015, Pages 368-376
4. Oupa I. Ntwampe and Kapil Moothi, *Removal of Heavy Metals Using Bentonite Clay and Inorganic Coagulants*, Heavy Metals, Chapter 3, Published: June 27th2018, page 33-51, http://dx.doi.org/10.5772/intechopen.76380
5. Kovo G.Akpomie, Folasegun A.Dawodu *Potential of a low-cost bentonite for heavy metal abstraction from binary component system*, Beni-Suef University Journal of Basic and Applied Sciences, Volume 4, Issue 1, March 2015, Pages 1-13
6. M.W. Amer, F.L. Khalili, A.M. Awwad *Adsorption of lead, zinc and cadmium ions on polyphosphate modified kaolinite clay*, J Environ Chem Ecotoxicol, 2 (1) (2010), pp. 01-08
7. Shabanzade, A. Salem, Sh. Salem, *Management of adsorbent content in waste motor oil regeneration by spectrophotometricalstudy and effective acidification in production of nano-porous clay*, Spectrochimica Acta Part A: Spectrochimica Acta Part A: Molecular and Biomolecular Spectroscopy, 202 (2018) 214-222
8. Tao Wua, Zhifen Wangb, Yanhua Tonga, Yongya Wanga, Luc R. Van Loonc, *Investigation of Re(VII) diffusion in bentonite by through-diffusion and modeling techniques*, Applied Clay Science 166 (2018) 223–229
9. Oprea I., Pîşă I., Mihăescu L., Prisecaru T., Lăzăroiu Gh., Negreanu G., *Research on the Combustion of Crude Vegetable Oils for Energetic Purposes,* Environmental Engineering and Management Journal, Vol.**8**, No.3, (May/June 2009), pg.475-482, indexat de ISI, Thomson Scientific Proceedings;
10. F. Danane, A. Ahmia, A. Bakiri, N. Lalaoui, *Experimental regeneration process of used oil* Revue des energies Renouvelables Vol. **17** N^0 2 (2014) 345-351
11. Temitayo E., Oladimeji, Jacob A. Sonibare, James A. Omoleye, Abiola A. Adegbola, Hilary I. Okagbue *Data on the treatment of used lubricating oil from two different sources using solvent extraction and adsorption,* Data in Brief 19 (2018) 2240-2252
12. Doaa I. Osman, Sayed K. Attia, Afaf R. Taman *Recycling of used engine oil by different solvent,* Egyptian Journal of Petroleum 27 (2018) 221-225

Halloysite nanotube applications for heavy metals removal from wastewater

Cristina Covaliu[1], *Gigel* Paraschiv[1], *Oana* Stoian[1],*Bogdan* Buzatu[1], *Costin* Mircea[2], *Camelia* Cristea[3]

[1]University Politehnica of Bucharest, Faculty of Biotechnical Systems Engineering, 313 Splaiul Independenței, 060042, sector 6, Bucharest, Romania
[2]National Institute of Research-Development for Machines and Installations Designed to Agriculture and Food Industry, 6 Ion Ionescu de la Brad, Bucharest, Sector 1, Romania
[3]University of Agronomic Sciences and Veterinary Medicine of Bucharest –Biotehnol.,59 Mărăşti Boulevard, District 1, 011464, Bucharest

Abstract. Halloysite is a clay mineral that it is found in nature being a non-toxic material. Halloysites are used in different industrial fields having an important potential. In this paper we reviewed the properties and structural characteristics of halloysites and their applications in wastewater treatment. Halloysites showed good results for heavy metals removal process from wastewater. Researches has been carried out on removing from wastewater heavy metals, such as: Cu(II), Fe(III), Cr(VI), Pd(II), Pb(II), Ag(I), Co(II), Zn(II).

1 Introduction

Halloysite nanotubes with negative charges on external surface represent a type of hollow tubular material measuring an outside diameter from 30 to 70 nm and an inner diameter of 20–30 nm [1, 2], which present excellent chemical and physical properties because of their abundant hydroxyl groups, large specific surface area, great biocompatibility and porous structure. Halloysite nanotubes are not toxic for humans [3, 4], so they are successfully applied for pharmaceutical and biomedical industries [5, 6] and have been applied in the adsorption process for dyes and heavy metals removal from wastewater [7–9].

Worldwide, in China, Australia, New Zealand, the United States, Brazil and Mexico have been found large deposits with halloysites [10]. Halloysite nanotubes have a length between the submicron scale and several microns, occasionally can exceed 30 μm [10], an outside diameter between approximately 30 and 190 nm, an inner diameter between almost 10 and 100 nm [11, 12].

2 General structure and principal properties of halloysite nanotubes

Halloysite nanotubes are aluminosilicate having the chemical formula $Al_2Si_2O_5(OH)_4{\bullet}nH_2O$, n having value 2 or 0 in case of hydrated or dehydrated halloysite. At temperature and humidity ambient conditions or by moderate heating the hydrated state can change into the dehydrated state irreversibly. Halloysite nanotubes can have different appearances such as thin or stubby and long or short. Halloysite nanotubes have layered and well crystallized structure composed of silicon dioxide tetrahedra present in the external layer and aluminum oxide octahedra present in the internal layer [13].

The external area of halloysite nanotubes has different electrical and chemical properties from the internal area due to the particular crystal structures. The internal area has a positive charge and a large number of hydroxyl groups (Al–OH), whereas the external area has a negative charge and a lower density of hydroxyl group (Si–OH). Actually, the external area of halloysite nanotubes is especially occupied by groups such as Si–O–Si and has a minority of Si–OH groups displayed on the possible defects or margins of the sheets. Thus, the oxygen atoms powerful electronegativity gives to the external area a negative charge. Meantime, the hydroxyl groups lower density on the external area together with the charged properties provide halloysite nanotubes well dispersion capacity in nonpolar polymers, giving a lot of advantages for the halloysite nanotubes / polymer nanocomposites preparation. In addition, halloysite nanotubes are considered as materials of reinforcing and flame retardancy in case of halloysite nanotubes / polymer nanocomposites due to their high appearance ratio between length and diameter and their borders against heat and mass transport [14].

3 Halloysite nanotubes-derived applications for heavy metal ions removal

Halloysites have been found as excellent alternatives for the conventional adsorbants used for the metal ions (Pb (II), Zn (II), Cu (II), Cd (II), Co (II) and Cr (II)) removal from wastewater. The removal of heavy metal ions from domestic and industrial wastewater or even drinking water has an environmental importance for reducing water pollution. Natural halloysite nanotubes without modifications can eliminate heavy metal ions from wastewater by physical and/or chemosorption mechanisms, metal speciation, site geometry, etc. Nevertheless, for the increasing of the heavy metal ions loading capacity, the nanotubes of halloysite are coated with different nanoparticles or functional groups. In Table 1 are showed the halloysite nanotubes applications used in the heavy metal ions removal from wastewater. Due to the layer structure, natural halloysite nanotubes were interpolated with active molecules in order to increase the reusability and loading capacity. For instance, intercalated halloysite nanotubes with sodium acetate were developed at different contact time and this development shows that the most intercalated trial was the most efficient for Cu (II) removal from wastewater having an adsorption capacity of 50 mg/g. [15]

Table 1. Halloysite nanotubes derived application used in heavy metal ions removal process from wastewater.

Adsorbant	Pollutant	Initial concentration (mg/L)	pH	Adsorption capacity (mg/g)	Ref.
Halloysite nanotubes / alginate	Cu(II)	50-100	-	74.13	16
Halloysite nanotubes / CH_3COONa	Cu(II)	10-200	6	52.3	17
Halloysite nanotubes / CH_3COO^-	Cu(II)	10-200	6	~50	15
Halloysite nanotubes / HBA	Fe(III)	10	4	45.54	18
Halloysite nanotubes / HDTMA	Cr(VI)	25-300	3-10	6.61	19
Halloysite nanotubes / KH-792	Cr(VI)	0-400	2-9	37.25	20

Table 1. Halloysite nanotubes derived application used in heavy metal ions removal process from wastewater (cont.)

Adsorbant	Pollutant	Initial concentration (mg/L)	pH	Adsorption capacity (mg/g)	Ref.
Halloysite nanotubes / murexide	Pd(II)	1	1-7	42.86	21
Halloysite nanotubes / PPy	Cr(VI)	1000	2-11	149.25	22
Halloysite nanotubes / PSA	Pb(II)	1	1-7	23.58	23
Natural halloysite nanotubes	Ag(I)	~50	-	3-6	24
Natural halloysite nanotubes	Co(II)	10	3-11	~6	25
Natural halloysite nanotubes	Zn(II)	10	2-9	9.87	26

The abbreviationsare: HBA - 2-hydroxybenzoic acid; HDTMA - hexadecyltrimethylammonium bromide; KH-792 - *N*-β-aminoethyl-γ-aminopropyl trimethoxysilane; PPy – polypyrrole; PSA - *N*-2-pyridylsuccinamic acid.

In order to remove Cr(VI) from wastewater was synthesized a polypyrrole-coated halloysite nanotubes nanocomposite through *in situ* polymerization process of pyrrole on halloysite nanotubes. The maximum removal capacity was 149.25 mg/g at pH 2 and 25 °C temperature. Nanocomposite was used for three cycles without loss its adsorption efficiency [22]. Besides removal capacity, the adsorption rate is also an important factor for the halloysite nanotubes adsorbants evaluation. In order to improve the adsorption rate for Cr(VI), was synthesized a new halloysite nanotubes - based adsorbant coated with the surfactant hexadecyltrimethylammonium bromide. Thus, the modified halloysite nanotubes showed an excellent adsorption for Cr(VI) reaching a maximum removal capacity of 90% in 5 minutes [19].In other articles, the adsorption properties of natural halloysite nanotubes were investigated for Ag [24], Co [25] and Zn [26] ions removal from wastewater.

Conclusions

Halloysite nanotubes - derived adsorbants presented excellent advantages such as reusability, easy operation, non-secondary pollution and lower cost in comparison with chemical methods for heavy metal ions removal.

Functionalized halloysite nanotubes proved a higher adsorption capacity (6.61–149.25 mg/g) for heavy metal ions than natural halloysite nanotubes (less than 10 mg/g), which is showed in Table 1.

Thereby, halloysite nanotubes are materials that have interesting properties being promising and cost-effective and it can be used in the composition of functional organic/inorganic nanocomposites. Halloysite nanotubes - derived properties can guarantee the wastewater treatment.

References

1. Y.M. Lvov, D.G. Shchukin, H. Möhwald, R.R. Price, ACS Nano, **2**, 814–820 (2008)
2. G. Cavallaro, A.A. Danilushkina, G.V. Evtugyn, G. Lazzara, S. Milioto, F. Parisi, V.E. Rozhina, F.R. Fakhrullin, Nanomaterials, **7**, 199 (2017)

3. X. Wang, J. Gong, R. Rong, Z. Gui, T. Hu, X. Xu, J. Agric. Food Chem., **66**, 2925–2933 (2018)
4. A.V. Stavitskaya, A.A. Novikov, M.S. Kotelev, D.S. Kopitsyn, E.V. Rozhina, I.R. Ishmukhametov, R.F. Fakhrullin, E.V. Ivanov, Y.M. Lvov, V.A. Vinokurov, Nanomaterials, **8**, 391 (2018)
5. G. Cavallaro, G. Lazzara, S. Milioto, F. Parisi, V. Evtugyn, E.; Rozhina, R. Fakhrullin, ACS Appl. Mater. Interfaces, **10**, 8265–8273 (2018)
6. G. Lazzara, G. Cavallaro, A. Panchal, R. Fakhrullin, A. Stavitskaya, V. Vinokurov, Y. Lvov, Curr. Opin. Colloid Interface Sci., **35**, 42–50 (2018)
7. Q. He, D. Yang, X. Deng, Q. Wu, R. Li, Y. Zhai, L. Zhang, Water Res., **47**, 3976–3983 (2013)
8. A. Moslehyani, M. Mobaraki, A.F. Ismail, T. Matsuura, S.A. Hashemifard, M.H.D. Othman, A. Mayahi, M. Rezaei DashtArzhandi, M. Soheilmoghaddam, E. Shamsaei, React. Funct. Polym., 95, 80–87 (2015)
9. V. Vergaro, E. Abdullayev, Y.M. Lvov, A. Zeitoun, R. Cingolani, R. Rinaldi, S. Leporatti, Biomacromolecules, 11, 820–826 (2010)
10. E. Joussein, S. Petit, J. Churchman, B. Theng, D. Righi, B. Delvaux, Clay Miner, **40 (4)**, 383-426 (2005)
11. P. Yuan, P.D. Southon, Z. Liu, M.E.R. Green, J.M., Hook, S.J., Antill, C.J. Kepert, J. Phys. Chem., **C 112 (40)**, 15742-15751 (2008)
12. P. Yuan, D. Tan, F. Annabi-Bergaya, W. Yan, D. Liu, Z. Liu, Appl. Clay Sci., **83-84**, 68-76 (2013)
13. E. Joussein, S. Petit, J. Churchman, B. Theng, D. Righi, B. Delvaux, Clay Miner., 40, 383–426 (2005)
14. M. Liu, Z. Jia, D. Jia, C. Zhou, Prog. Polym. Sci., **39**, 1498–1525 (2014)
15. S. Mellouk, A. Belhakem, K. Marouf-Khelifa, J. Schott, A. Khelifa, J. Colloid Interface Sci., **360**, 716–724 (2011)
16. Y. Wang, X. Zhang, Q. Wang, B. Zhang, J. Liu, Water Sci. Technol., **70**, 192–199 (2014)
17. S. Mellouk, S. Cherifi, M. Sassi, K. Marouf-Khelifa, A. Bengueddach, J. Schott, A. Khelifa, Appl. Clay Sci., 44, 230–236 (2009)
18. R. Li, Z. Hu, S. Zhang, Z. Li, X. Chang, Int. J. Environ. Anal. Chem., 93, 767–779 (2013)
19. J. Wang, X. Zhang, B. Zhang, Y. Zhao, R. Zhai, J. Liu, R. Chen, Desalination, 259, 22–28 (2010)
20. P. Luo, J.-S. Zhang, B. Zhang, J.-H. Wang, Y.-F. Zhao, J.-D. Liu, Ind. Eng. Chem. Res., 50, 10246–10252 (2011)
21. R. Li, Q. He, Z. Hu, S. Zhang, L. Zhang, X. Chang, Anal. Chim. Acta, 713, 136–144 (2012)
22. N. Ballav, H. J. Choi, S. B. Mishra, A. Maity, Appl. Clay Sci., 102, 60–70 (2014)
23. Q. He, D. Yang, X. Deng, Q. Wu, R. Li, Y. Zhai, L. Zhang, Water Res. 47, 3976–3983 (2013)
24. G. Kiani, Appl. Clay Sci., 90, 159–164 (2014)
25. J. Li, F. Wen, L. Pan, Z. Liu, Y. Dong, J. Radioanal. Nucl. Chem., 295, 431–438 (2013)
26. Y. Dong, Z. Liu, L. Chen, J. Radioanal. Nucl. Chem., 292, 435–443 (2012)

AUTHOR INDEX

AUTHOR INDEX

AUTHOR INDEX